MW01167179

ARSENIC

ARSENIC

Exposure Sources, Health Risks, and Mechanisms of Toxicity

Edited by

J. CHRISTOPHER STATES

University of Louisville
Louisville, KY, USA

Copyright © 2016 by John Wiley & Sons, Inc. All rights reserved.

Published by John Wiley & Sons, Inc., Hoboken, New Jersey
Published simultaneously in Canada

No part of this publication may be reproduced, stored in a retrieval system, or transmitted in any form or by any means, electronic, mechanical, photocopying, recording, scanning, or otherwise, except as permitted under Section 107 or 108 of the 1976 United States Copyright Act, without either the prior written permission of the Publisher, or authorization through payment of the appropriate per-copy fee to the Copyright Clearance Center, Inc., 222 Rosewood Drive, Danvers, MA 01923, (978) 750-8400, fax (978) 750-4470, or on the web at www.copyright.com. Requests to the Publisher for permission should be addressed to the Permissions Department, John Wiley & Sons, Inc., 111 River Street, Hoboken, NJ 07030, (201) 748-6011, fax (201) 748-6008, or online at http://www.wiley.com/go/permissions.

Limit of Liability/Disclaimer of Warranty: While the publisher and author have used their best efforts in preparing this book, they make no representations or warranties with respect to the accuracy or completeness of the contents of this book and specifically disclaim any implied warranties of merchantability or fitness for a particular purpose. No warranty may be created or extended by sales representatives or written sales materials. The advice and strategies contained herein may not be suitable for your situation. You should consult with a professional where appropriate. Neither the publisher nor author shall be liable for any loss of profit or any other commercial damages, including but not limited to special, incidental, consequential, or other damages.

For general information on our other products and services or for technical support, please contact our Customer Care Department within the United States at (800) 762-2974, outside the United States at (317) 572-3993 or fax (317) 572-4002.

Wiley also publishes its books in a variety of electronic formats. Some content that appears in print may not be available in electronic formats. For more information about Wiley products, visit our web site at www.wiley.com.

Library of Congress Cataloging-in-Publication Data:

Arsenic : exposure sources, health risks, and mechanisms of toxicity / edited by J. Christopher States.
pages cm
Includes index.
ISBN 978-1-118-51114-5 (hardback)
1. Arsenic–Health aspects. 2. Arsenic–Toxicology. I. States, J. Christopher.
RA1231.A7A7683 2015
615.9′25715–dc23

2015018789

Set in 10/12pt Times by SPi Global, Pondicherry, India

Printed in the United States of America

10 9 8 7 6 5 4 3 2 1

1 2016

CONTENTS

CONTRIBUTORS

Gavin E. Arteel, Department of Pharmacology and Toxicology, University of Louisville, Louisville, KY, USA

Susan R. Atlas, Physics and Astronomy Department, University of New Mexico, Albuquerque, NM, USA

Mayukh Banerjee, Department of Physiology, University of Alberta, Edmonton, AB, Canada

Aaron Barchowsky, Department of Environmental and Occupational Health, University of Pittsburgh, Pittsburgh, PA, USA

Pritha Bhattacharjee, Department of Environmental Science, University of Calcutta, Kolkata, India

Iain L. Cartwright, Molecular Genetics, Biochemistry & Microbiology, University of Cincinnati College of Medicine, Cincinnati, OH, USA

Yu Chen, Departments of Population Health and Environmental Medicine, New York University School of Medicine, New York, NY, USA

Harvey J. Clewell, III, Hamner Institutes for Health Sciences, Research Triangle Park, NC, USA

Karen L. Cooper, Pharmaceutical Sciences, University of New Mexico, Albuquerque, NM, USA

Arden D. Davis, Department of Geology and Geological Engineering, South Dakota School of Mines and Technology, Rapid City, SD, USA

Christelle Douillet, Department of Nutrition, University of North Carolina Chapel Hill, Chapel Hill, NC, USA

Ingrid L. Druwe, Department of Pharmacology and Toxicology, University of Arizona College of Pharmacy, Tucson, AZ, USA

Dominic B. Fee, Department of Neurology, Medical College of Wisconsin in Milwaukee, WI, USA

A. Jay Gandolfi, Department of Pharmacology and Toxicology, College of Pharmacy, University of Arizona, Tucson, AZ, USA

P. Robinan Gentry, Environ International Corporation, Monroe, LA, USA

Ashok K. Giri, Molecular and Human Genetics Division, Indian Institute of Chemical Biology, Kolkata, India

Laurie G. Hudson, Pharmaceutical Sciences, University of New Mexico, Albuquerque, NM, USA

Michael F. Hughes, Office of Research and Development, National Health and Environmental Effects Research Laboratory, US Environmental Protection Agency, Research Triangle Park, NC, USA

Elaina M. Kenyon, US EPA, NHEERL, Research Triangle Park, NC, USA

Brenee S. King, Human Nutrition, Kansas State University, Manhattan, KS, USA

R. Clark Lantz, Cellular and Molecular Medicine, University of Arizona, Tucson, AZ, USA

Michael Lawson, School of Earth, Atmospheric and Environmental Sciences and Williamson Research Centre for Molecular Environmental Science, The University of Manchester, Manchester, UK

Maryse Lemaire, Department of Oncology, Lady Davis Institute for Medical Research, McGill University, Montreal, QC, Canada

Ke Jian Liu, Pharmaceutical Sciences, University of New Mexico, Albuquerque, NM, USA

Koren K. Mann, Department of Oncology, Lady Davis Institute for Medical Research, McGill University, Montreal, QC, Canada

Matthew K. Medeiros, Department of Pharmacology and Toxicology, College of Pharmacy, University of Arizona, Tucson, AZ, USA

Patricia Ostrosky-Wegman, Instituto de Investigaciones Biomédicas, Universidad Nacional Autónoma de México (UNAM), Mexico City, Mexico

Somnath Paul, Molecular Biology and Human Genetics Division, Council for Scientific and Industrial Research – Indian Institute of Chemical Biology, Kolkata, India

David A. Polya, School of Earth, Atmospheric and Environmental Sciences and Williamson Research Centre for Molecular Environmental Science, The University of Manchester, Manchester, UK

Ana María Salazar, Instituto de Investigaciones Biomédicas, Universidad Nacional Autónoma de México (UNAM), Mexico City, Mexico

Nilendu Sarma, Department of Dermatology, NRS Medical College, Kolkata, India

Cara L. Sherwood, Arizona Respiratory Center, University of Arizona, Tucson, AZ, USA

J. Christopher States, Department of Pharmacology and Toxicology, University of Louisville, Louisville, KY, USA

Miroslav Stýblo, Department of Nutrition, University of North Carolina Chapel Hill, Chapel Hill, NC, USA

David J. Thomas, Integrated Systems Toxicology Division, USEPA, Research Triangle Park, NC, USA

Erik J. Tokar, Inorganic Toxicology Group, National Toxicology Program Laboratory, National Institute of Environmental Health Sciences, Research Triangle Park, NC, USA

Richard R. Vaillancourt, Department of Pharmacology and Toxicology, University of Arizona College of Pharmacy, Tucson, AZ, USA

Michael P. Waalkes, Inorganic Toxicology Group, National Toxicology Program Laboratory, National Institute of Environmental Health Sciences, Research Triangle Park, NC, USA

Cathleen J. Webb, Department of Chemistry, Western Kentucky University, Bowling Green, KY, USA

Fen Wu, Departments of Population Health and Environmental Medicine, New York University School of Medicine, New York, NY, USA

Yuanyuan Xu, Inorganic Toxicology Group, National Toxicology Program Laboratory, National Institute of Environmental Health Sciences, Research Triangle Park, NC, USA

Janice W. Yager, Department of Internal Medicine, University of New Mexico, Albuquerque, NM, USA

PREFACE

I became interested in arsenic toxicology in the late 1990s. I had engineered SV40-transformed human fibroblasts to express human cytochrome P450 IA1 for polycyclic aromatic hydrocarbon carcinogen activation studies. A biotechnology company was interested in licensing the cells but wanted me to test them with a variety of carcinogens to demonstrate specificity. Sodium arsenite was among the compounds. The SV40-transformed fibroblasts were sensitive to the arsenite but in a curious manner. The cells rounded up in what appeared to be mitotic arrest, and then the membranes "bubbled" like in apoptotic cells. At the time, the postulation of mitotic cells undergoing apoptosis was sheer heresy. Nonetheless, these odd observations set me off on a course investigating this enigmatic toxicant that never ceases to provide surprises and raise new questions.

Arsenic is the 20th most common element in the earth's crust, and its toxic potential has been known for millennia. Chronic exposure to arsenic, most commonly through natural contamination of drinking water, is a worldwide health problem. Arsenic has been number one on the ATSDR hazardous substances list for at least 15 years now. Over the past two decades, a vast amount of research has been performed on arsenic toxicity. Much of this research focused on carcinogenesis. However, more recently research has focused on the non-cancer disease endpoints of chronic arsenic exposure including cardiovascular disease (atherosclerosis and hypertension), and pulmonary, neurological, and ocular disease. Some epidemiological studies suggest a link with diabetes. How a single agent can cause such a wide variety of ailments has evaded a simple answer. A great variety of experimental systems, using wide-ranging exposures, have produced a mountain of published research. Despite all the published literature examining mode of action, there remains strong debate over how arsenic exerts its disease-causing effects, and no single unifying theme has emerged that can explain the diversity of diseases caused by chronic arsenic exposure. In my view, this lack of a unifying mechanism is at the heart of the issue.

The complexity of arsenic chemistry and biochemistry confounds many efforts to understand the mechanism of toxicity. Recent appreciation of the toxicity of trivalent metabolic intermediates has added to the problems of understanding toxicity and of mitigating toxicity by reducing exposure. We have attempted to address the complexity of the arsenic problem in *Arsenic: Exposure Sources, Health Risks, and Mechanisms of Toxicity* by compiling into a single-volume discussions of the exposure sources, exposure mitigation, chemistry, metabolism, the various diseases induced by arsenic exposure, and the variety of experimental models used to investigate arsenic toxicity.

The book is divided into four sections: Fundamentals of Arsenic Exposure and Metabolism, Epidemiology and Disease Manifestations of Arsenic Exposure, Mechanisms of Toxicity, and Models for Arsenic Toxicology and Risk Assessment. The chapters discuss a variety of topics including history of arsenic, sources of exposure, chemical and biochemical properties, molecular mechanisms, role in various diseases, genetics of susceptibility, and human health risk assessment, and concludes with a chapter discussing translation of experimental findings to human studies.

This book is offered as a resource for toxicologists, epidemiologists, risk assessors, environmental chemists, medical scientists, and other practicing professionals and researchers in academia, government, and industry. The book aims to provide a better understanding of the potential health problems posed by arsenic exposure and discuss ways that toxicological sciences can contribute to a characterization of how arsenic causes those problems and associated risks.

I am deeply indebted to the friends and colleagues who have contributed to this volume.

J. CHRISTOPHER STATES

PART I

FUNDAMENTALS OF ARSENIC EXPOSURE AND METABOLISM

1

HISTORY OF ARSENIC AS A POISON AND A MEDICINAL AGENT

MICHAEL F. HUGHES

Office of Research and Development, National Health and Environmental Effects Research Laboratory, US Environmental Protection Agency, Research Triangle Park, NC, USA

Disclaimer: This article has been reviewed in accordance with the policy of the National Health and Environmental Effects Research Laboratory, US Environmental Protection Agency, and approved for publication. Approval does not signify that the contents necessarily reflect the views and policies of the agency, nor does mention of trade names or commercial products constitute endorsement or recommendation for use.

1.1 INTRODUCTION

Arsenic is one of the most enigmatic elements known to humankind. For many centuries, arsenic has been used as an intentional human poison, for which it has generated much fear and interest. However, over this same time frame, arsenic has been used to benefit society, at least with good intentions, as a medicinal agent. The best example of this paradox is arsenic trioxide, which is also known as the white arsenic. This potent and lethal inorganic arsenical has not only been commonly used to commit homicide but also been used more recently as an effective cancer chemotherapeutic agent.

Arsenic is an insidious poison. Over the ages, arsenic came to be known as the "King of Poisons" because of its use to poison royalty [62]. Arsenic was the choice as a poison because it has no taste and could be discreetly mixed with food or drink.

Arsenic: Exposure Sources, Health Risks, and Mechanisms of Toxicity, First Edition.
Edited by J. Christopher States.
© 2016 John Wiley & Sons, Inc. Published 2016 by John Wiley & Sons, Inc.

The symptoms of arsenic poisoning were similar to those of common diseases (e.g., cholera) in the world at a time when hygienic practices were poor and safe drinking water was not readily available. Also, there was no chemical test to indicate that someone had been exposed to arsenic until the 1700s.

Although the poisonous nature of arsenic was well known, in 2010, in the United States alone, there were over 5000 cases of arsenic pesticide poisoning and 1000 cases of arsenic nonpesticidal poisoning [8]. Three deaths were noted in the arsenic non-pesticidal cases.

Arsenic is found in inorganic and organic forms as well as different valence or oxidation states (Fig. 1.1). The oxidation states of arsenic include –III, 0, III, and V. Examples of arsenicals in these states are arsine, elemental arsenic, arsenite, and arsenate, respectively. Arsine is a colorless, odorless gas and highly toxic [54]. Exposure to arsine is primarily occupational, so it will not be discussed in this chapter. The form and valence state of the arsenical is important in its potential toxic effects. In general terms (i) inorganic arsenicals are more potent than organic arsenicals; (ii) trivalent (III) arsenicals such as arsenite are more potent than pentavalent (V) arsenicals such as arsenate; and (iii) trivalent organic arsenicals are equally or more potent than trivalent inorganic arsenicals [30].

The clinical signs of acute oral arsenic toxicity are progressive and depend on the form, valence, and dose of the arsenical. In a human adult, the lethal range of inorganic arsenic is estimated at 1–3 mg As/kg [18]. The symptoms of acute arsenic poisoning are listed in Table 1.1 [25, 56]. Diarrhea is due to increased permeability of the blood vessels. Depending on the type and amount of arsenic consumed, death may occur within 24 h to 4 days. Death is usually due to massive fluid loss leading to dehydration, decreased blood volume, and circulatory collapse. Survivors of acute arsenic poisoning may develop peripheral neuropathy, which is displayed as severe ascending weakness. This effect may last for several years. Encephalopathy may also develop, potentially from the hemorrhage that can occur from the arsenic exposure.

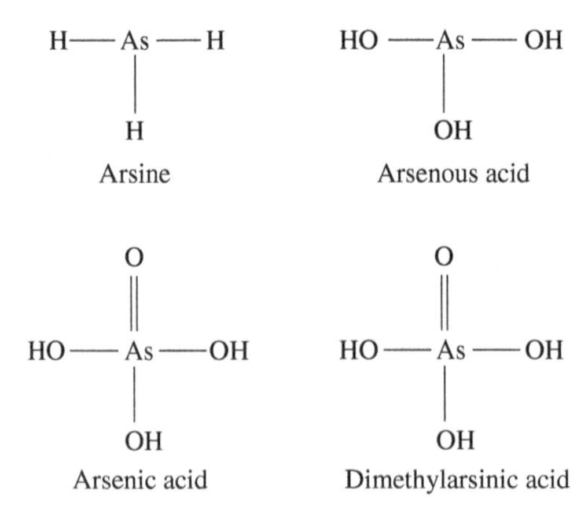

FIGURE 1.1 Structure of common arsenicals. The ionized forms of arsenous acid and arsenic acid are arsenite and arsenate, respectively. Dimethylarsinic acid is also named cacodylic acid.

TABLE 1.1 Acute and Chronic Clinical Effects of Inorganic Arsenic Exposure

Organ System	Acute Effects	Chronic Effects
Cardiac	Cardiomyopathy, hemorrhage, electrocardiographic changes	Hypertension, peripheral vascular disease, cardiomyopathy
Hematologic	Hemoglobinuria, bone marrow depression	Anemia, bone marrow hypoplasia
Gastrointestinal	Nausea, vomiting, diarrhea	Vomiting, diarrhea, weight loss
Hepatic	Fatty infiltration	Hepatomegaly, jaundice, cirrhosis, fibrosis, cancer
Neurologic	Peripheral neuropathy, ascending weakness, tremor encephalopathy, coma	Peripheral neuropathy, paresthesia, cognitive impairment
Pulmonary	Edema, respiratory failure	Cancer
Renal	Tubular and glomerular damage, oliguria, uremia	Nephritis, cancer
Skin	Alopecia	Hyperkeratosis, hypo- or hyperpigmentation, Mees' lines, cancer

Treatment for acute arsenic poisoning includes gastric lavage, administration of fluids and a chelator such as dimercaptopropanol, and hemodialysis.

In chronic arsenic poisoning, as in acute poisoning, essentially all the organs are affected [25, 56] (Table 1.1). The hallmark of chronic arsenic poisoning is the development of skin lesions. This includes hyper- or hypopigmentation and hyperkeratosis, particularly on the palms of the hands and soles of the feet. There is no known treatment for chronic arsenic poisoning that is of benefit to the individual. The best option is to minimize exposure to the source of arsenic and provide supportive care to the patient.

Arsenic is also a known human carcinogen, being classified as such by the International Agency for Research on Cancer [33] and the US Environmental Protection Agency (USEPA) [39]. Confirmed organs for cancerous development from chronic arsenic exposure include bladder, skin, and lung [51]. Potential target organs for cancer from arsenic exposure are liver, kidney, and prostate [33, 51].

1.2 INTENTIONAL POISONING BY ARSENIC

The poisonous nature of arsenic has been known for centuries, and thus, it has been used to commit homicide. It is so well known that poisoning by arsenic has been incorporated into the plots of literary works of Chaucer, Agatha Christie, and other writers, and even in the title of a 1940s Broadway play, "Arsenic and Old Lace" [6]. However, it is inconceivable that the victims would die so quickly from arsenic ingestion in that play. Their deaths would most likely be from the ingestion of cyanide and strychnine, which were also part of the poisonous concoction mixed with elderberry wine. But would the play have gained as much attention if it had been called "Cyanide, Strychnine and Old Lace"? No one knows for sure.

There have been many suspicious poisonings, potentially by arsenic, of powerful people in centuries long ago. It has been suggested that Alexander the Great and Britanicus were poisoned by arsenic [6, 12, 25]. The Greek physician Dioscorides, in the first century A.D., included arsenic as a poison in his five-volume publication *De Materia Medica* ("Regarding Medical Materials") [48]. During the Middle Ages and Renaissance periods, murder by poisoning reached its zenith [6, 12]. Noted individuals who poisoned others with arsenic for personal gain or profit during this time include the Italians Cesare Borgia, Giulia Toffana, and Hieronyma Spara, and the French woman Marie de Brinvillers [6, 12]. Some of the poisonings were politically motivated, particularly in the Catholic Church, as several senior clergymen were poisoned with arsenic over a 500-year period [12, 48].

An interesting and curious case of arsenic poisoning involved several elderly women of the village of Nagyrev in south-central Hungary [27]. In 1929, four women were brought to trial accused of murdering family members. Their basic plan was to call a doctor to the home of the intended victim. Many of the victims were chronically ill with tuberculosis or another debilitating disease. After the doctor departed, the victims were poisoned with arsenic. When the victims passed away, questions were not asked, because it was perceived that they died from complications of the noted illness. During this time, arsenic was easily available as arsenic acid as this agent was used as a rodenticide. Also, flypaper containing arsenic was commonly used. The arsenic was easily extracted from the flypaper and could be mixed with a drink, as one of the accused allegedly did with her husband's apricot brandy. There were other suspicious deaths at this time in this village, so 50 bodies from the town's cemetery were exhumed. Forty-six of the deceased had arsenic levels high enough to be lethal. Other women were brought to trial later and charged with the murders of husbands, fathers, sons, and mothers- and fathers-in-law. Several of the women were found guilty of murder and punished, while others were acquitted of the charges. It was alleged that the period of the murders in this village lasted over two decades and perhaps was even longer. It should be noted that even today, arsenic-containing flypaper is commercially available. The British newspaper, *The Guardian,* published an article in 2007 on poisons that could be purchased over the Internet [55]. The reporter was able to purchase from a company in Iowa flypaper with a packaging label indicating that it contained 2–4% metallic arsenic.

Intentional human poisoning with arsenic is not limited to Europe alone. In the state of North Carolina in the United States, from 1972 to 1982, there were 28 deaths attributed to arsenic exposure [47]. Of these deaths, 14 were declared homicides and 7 suicides. Four of the confirmed arsenic homicides were attributed to one woman, with the crimes occurring over a 4-year period. This woman may have been involved with additional arsenic poisonings.

Napoleon Bonaparte, the French military and political leader who died in 1821, may have been poisoned by arsenic [12]. Napoleon was exiled by the British to the south Atlantic island, St. Helena, in 1815. He appeared to be in good health upon arrival. Over time, he gained weight and had frequent illness. Several doctors on the island examined Napoleon and diagnosed hepatitis. Before his death Napoleon apparently lost weight. The official autopsy report indicated that Napoleon had a chronic stomach ulcer and died of stomach cancer. For political reasons, both the English and French accepted that stomach cancer was the cause of his death. However, Napoleon's personal physician, who actually performed the autopsy, maintained that Napoleon died from complications of hepatitis.

After 140 years, a Swedish dentist, Sten Forshufvud, became convinced that Napoleon's demise came from arsenic poisoning [23, 63]. Hair that was reportedly removed from the head of Napoleon after his death was analyzed for arsenic by neutron activation. The response was positive for arsenic [23]. As discussed in Chapter 13, trivalent arsenic readily binds to sulfhydryl groups. Keratin, the primary structural protein of hair, contains sulfhydryl groups, and thus arsenic will bind to it. Smith et al. [23, 63] analyzed another supposed portion of Napoleon's hair that was held by someone else, and it also tested positive for arsenic. These claims of high levels of arsenic in the hair of Napoleon brought about the theory that he had been intentionally poisoned. However, Lewin et al. [40] analyzed a different sample of Napoleon's hair and detected only background levels of arsenic. To complicate matters further, it has been suggested that Napoleon was treating himself with arsenic so that he could become tolerant to a lethal dose [6]. Although it is an interesting story, it is still not clear whether arsenic poisoning was the cause of Napoleon's death.

1.2.1 Chemical Warfare

Arsenic has a somewhat veiled but nonetheless wretched history of being a chemical warfare agent. There are writings of arsenic being utilized to provide smoke to cover advancing troops during battles in ancient Greece [12]. The use of arsenic in warfare became prominent in World War I by both the Germans and Allies (British, French, and United States). Arsenic was first employed in this war as an inactive agent by the French in 1916 [12]. Arsenic trichloride was mixed with phosgene in artillery shells. The arsenic minimized dissipation of the phosgene, a deadly gaseous agent, and also provided smoke so that observers could adjust the artillery to more accurately place the next round.

The Germans were the first to utilize arsenic as an active warfare agent in 1917 [12]. This particular agent, chlorodiphenylarsine, is a respiratory irritant, causing sneezing and mucous buildup in those exposed to it. The irritation develops into coughing, headache, and other detrimental effects for the soldiers. In some artillery shells, chlorodiphenylarsine was mixed with phosgene and diphosgene, both of which are deadly gases. Other arsenicals used as active agents in World War I by the Germans and Allies were dichloromethylarsine, dichlorophenylarsine, dibromophenylarsine, cyanodiphenylarsine, and others [12] (Fig. 1.2).

A very potent arsenical, β-chlorovinyldichloroarsine, which is a deadly blistering agent or vesicant, was developed late in World War I (Fig. 1.3). However, this chemical was never used in this war. This arsenical was synthesized by a US research team led by Captain Winford Lee Lewis. This agent became known as Lewisite [12]. There is some speculation that Lewisite was first synthesized by a priest in 1903 at

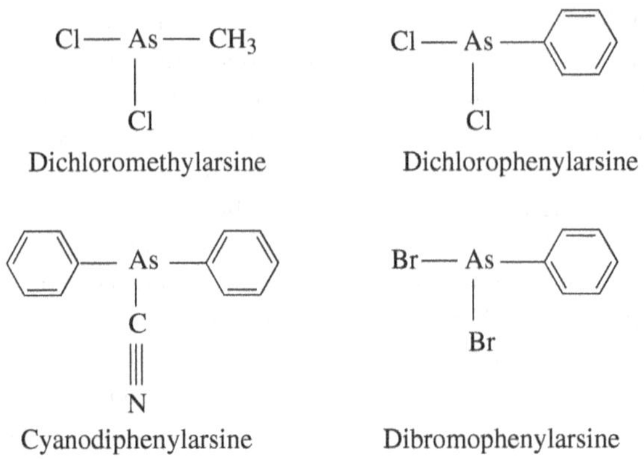

Dichloromethylarsine Dichlorophenylarsine

Cyanodiphenylarsine Dibromophenylarsine

FIGURE 1.2 Structure of some arsenical war gases used in World War I.

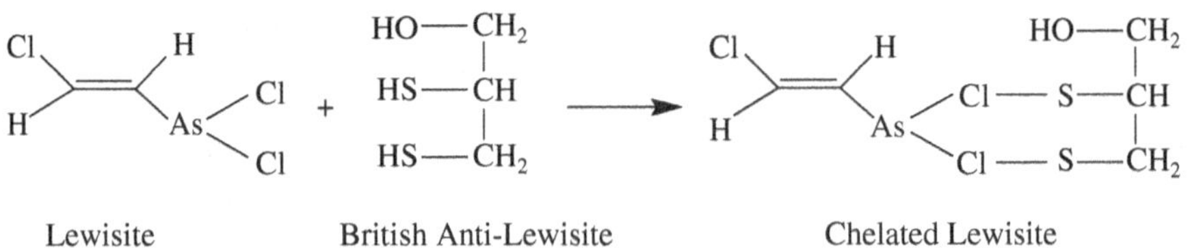

Lewisite British Anti-Lewisite Chelated Lewisite

FIGURE 1.3 Reaction of Lewisite with 2,3-dimercaptopropanol (BAL) to form chelated Lewisite.

the Catholic University of America, in Washington, D.C. [12]. For his dissertation studies, Father John Niewland was attempting to synthesize rubber using arsenic trichloride, acetylene, and aluminum chloride. During this synthetic work, Father Niewland became ill from a noxious odor that arose from the reaction of these chemicals. There were no further studies to determine the source of the noxious odor. Years later, Captain Lewis' research team, also based at the Catholic University of America, was alerted to the potential of the noxious compound from Father Niewland's experiments because they had access to his dissertation. Whether or not Father Niewland actually synthesized Lewisite is not clear, but it was eventually prepared by Captain Lewis' research team.

One significant outcome of the threat of the use of Lewisite (and other chemicals) in warfare was the research and development of chemicals that could prevent the effects of this toxic arsenical. This research on Lewisite also led to an understanding of a mode of action for arsenic. With the threat of war looming in Europe in the late 1930s, the British formed a research team to prepare antagonists for chemical warfare agents that the Germans might employ [52]. This team was led by biochemist Rudolph Peters. Through a series of *in vitro* experiments, it was determined that Lewisite bonded with two sulfhydryl groups that are in close proximity to one another, within an essential component of the pyruvate oxidase system. Several dithiols were synthesized and tested for their ability to antagonize the effects of Lewisite. Encouraging results were obtained with dithiols that could form a stable five- or six-membered ring with Lewisite. The most promising dithiols were those that bound more stably with Lewisite than the arsenical with the unknown biological receptor. It was determined that 2,3-dimercaptopropanol, which was eventually named British anti-Lewisite (BAL), was the best antagonist for Lewisite (Fig. 1.3). BAL penetrates cells and has low toxicity. BAL is effective both pre- and post-exposure of Lewisite.

While a major focus of chemical warfare is to directly incapacitate or kill the enemy, some effort may be directed to defoliate the surrounding landscape so the enemy is unable to conceal themselves in it as well as destroy their food supply. During the Vietnam War, the United States had a program, Operation Trail Dust, with a mission to defoliate the Vietnamese landscape with herbicides [12, 26]. Of the chemicals used in this program, the most famous was Agent Orange, which is a 1:1 mixture of the herbicides 2,4-D and 3,4,5-T. What is most known about Agent Orange is that 2,4,5-T was contaminated with dioxin. Another mission of Operation Trail Dust was to destroy the main food crop of the Viet Cong—rice. For this operation, a solution of dimethylarsinic acid, also known as cacodylic acid (Fig. 1.1), and its sodium salt was sprayed on the rice crops. The code name of this arsenical solution was Agent Blue (the names of the agents used were based on the color painted on the 55 gallon drums used to store the agent). This arsenical is a desiccant, drying out the crop it is intended to destroy. Cacodylic acid has been used in the United States as a cotton desiccant and a general weed killer but has recently been phased out of use [73]. The use of Agent Blue in the Vietnam War was approximately 10% of that of Agent Orange [12].

1.3 UNINTENTIONAL POISONING BY ARSENIC

Arsenic is an element found naturally in soil, air, and water. Because arsenic is pervasive, unintentional exposure to it has occurred in many unsuspecting populations. These exposures have led to adverse health effects, such as neuropathy, keratosis, peripheral vascular disease, malignancy, and death.

The unintentional poisoning of humans by arsenic most likely began during the first attempts to separate metals from other matter in the earth [75]. This is because arsenic is an impurity of the ores of copper, the first industrial metal, and iron. When humans discovered that metals could be cast by melting ores with fire, the exposure to arsenic and other metal contaminants most likely increased. Analysis of hair from the Tyrolean Neolithic mummy, the so-called "Ötzi-the-Iceman," who was discovered frozen in a glacier in the Alps along the Italian–Austrian border in 1991, contained copper and arsenic [7]. It is estimated that "Ötzi" died around 3300 B.C. It has been proposed that environmental arsenic pollution of cities from metallurgy caused the decline of the Etruscan civilization in Italy during the sixth century B.C. [29]. Even in more recent times, the exposure to arsenic from smelting of contaminated copper ores has continued to be problematic. Occupational exposure to arsenic by inhalation at copper smelters has led to increased deaths from lung cancer [20, 34].

1.3.1 Pigments and Dyes

Scheele's green is an arsenic- and copper-based pigment that was first prepared by Carl Schele in 1775. By its name it is green in color and was very popular in England in the 1800s. In fact, it was so fashionable during this time that Schele green was a common constituent in residential wallpaper, clothing, soap, paint, and other consumer products [5, 61]. Concerns were raised in England [5] and later in the United States [59] over the use of arsenic in wall paper and other products, in that people were being poisoned from arsenic exposure. There were reports of individuals becoming ill (e.g., fatigue and nausea) in their homes, which was attributed to arsenic in the wallpaper [5, 59]. There was even a suggestion that the wallpaper in the home of Napoleon, which contained arsenic, was the means of his alleged exposure to arsenic and his eventual death [35]. However, many of these illnesses were not well characterized, and the reports on morbidity and mortality from this type of exposure to arsenic were anecdotal.

Bartholomew Gosio, an Italian physician, found in 1892 that the mold *Scopulariopsis brevicaulis* was able to convert inorganic and organic arsenic into a gaseous arsenical [12, 13]. This arsenical has an odor of garlic and is known as "Gosio gas," and it has been suggested it was the toxic agent making people ill in their homes [35, 59]. Challenger et al. [10] determined the gas was trimethylarsine. However, trimethylarsine is about 1000-fold less toxic than arsine [13]. So it is unlikely that trimethylarsine had any significant role in the illnesses attributed to the arsenic-containing wallpaper. Sanger [59] also suggested the poisoning could have been due to absorption of inorganic arsenic dust from the wallpaper.

Efforts to ban arsenic-laden wallpaper began to arise in England [5]. Because studies on the actual reported illnesses were limited or nonexistent, it is too difficult to determine whether the arsenic in the wallpaper was actually the causative agent. By the late 1880s, the use of Schele green in wallpaper in England dramatically decreased, as different colors without arsenic became more popular [5].

In 1956, the US ambassador to Italy, Clare Boothe Luce, resigned from her post because of a chronic illness [12, 61]. This illness of unknown origin started in 1954 after she began her ambassadorship. Her symptoms included brittle fingernails, hair loss, anemia, and paresthesia in her lower right leg. It was finally determined that she had been poisoned by lead arsenate that was found in the painted bedroom ceiling of her residence in Rome. Over time, the ceiling plaster began to crumble into dust particles. Ambassador Luce was exposed to the lead arsenate tainted dust particles by inhalation and ingestion. She had noticed that the coffee she drank in her bedroom had a slight metallic taste. She returned to the United States and recovered from this chronic illness.

1.3.2 English Beer Drinkers

Cases of skin eruptions, erythema, keratosis, pigmentation, and "alcoholic paralysis" were reported in the summer of 1900 near and around Manchester, England [57]. The "alcoholic paralysis" was characterized as peripheral neuritis and paresthesia and numbness in the hands and feet. Ernest Reynolds, a local physician, saw a connection between the skin afflictions and neuritis in patients he was treating. The neuritis occurred mainly in drinkers of beer and not hard liquor. Reynolds obtained a sample of the beer that many of the sufferers had been consuming. With the beer he conducted the Reinsch test, and the results suggested the presence of arsenic. Further investigation by a brewery-sponsored commission established that the source of the arsenic was brewery sugar [24]. It was determined that a chemical company had prepared sulfuric acid from Spanish pyrite containing arsenic. The arsenic-contaminated sulfuric acid was sold to a company that prepared the brewery sugar. The arsenic-contaminated sugar was sold to 100 or more breweries in England and was subsequently used in the beer brewing process. Over 6000 people were affected at one point by the contaminated beer [17]. Dr. Reynolds treated about 500 people alone over a 2-month period during this epidemic. He noted that 13 people died from the contaminated beer during this time.

A Royal Commission was convened to investigate this arsenic-contaminated beer incident [12, 17]. One of their recommendations was that limits for the amount of arsenic in liquids such as beer (0.14 ppm) and foods (1.43 ppm) be established. These limits were later set into law in England in 1928.

1.3.3 Japanese Infant Milk

Outbreaks of skin rash, gastrointestinal distress, and in a few cases, death, were reported in infants in 1955 in western Japan. Overall, there were almost 12,000 affected infants, including 113 deaths. An inquiry into this illness determined that the infants were bottle-fed arsenic-contaminated powdered milk [71]. The source of the arsenic was sodium phosphate, which was added to the milk as a stabilizer. The sodium phosphate was a by-product of a process from which aluminum was extracted from bauxite. The company that manufactured the milk failed to adequately clean the sodium phosphate before adding it to the powdered milk. The sale of the milk was banned within a day following discovery of the arsenic contamination. The acute symptoms disappeared in many of the affected babies following this ban. However, follow-up studies of those poisoned showed that about 325 suffered from developmental retardation and 250 had other ailments (e.g., epilepsy) [14]. It was difficult to determine the dose of arsenic the infants received because different lots of milk were prepared, and the infants were exposed for various lengths of time. More recent analysis of cancer mortality data of an exposed birth cohort that was most severely affected by the arsenic milk contamination showed that mortality from skin, pancreatic, and liver cancer was increased [79].

1.3.4 Coal in China

Coal in the area of Guizhou, China, contains a high concentration of arsenic (>100 ppm) [41]. As wood used to heat homes and to cook became depleted in the 1960s in this rural region of China, coal began being used in its place. In many homes of this region, unvented coal-burning indoor stoves are used for residential heating and to dry food resulting in indoor arsenic air concentration that may reach 250 $\mu g/m^3$

or greater. These levels are 80–90 times higher than the China Air Quality Permission Standard of 3 μg/m^3 [43]. Arsenic in the air from burning the coal can coat and permeate the smoke-dried food. Approximately 17% of the residents of this area showed the dermal lesions of chronic arsenic intoxication. Malignancies were the most serious outcome of arsenic intoxication, accounting for 50% of deaths in the arsenicosis patients. The source of arsenic exposure to people with the coal-burning stoves is primarily food (50–80%) and air (10–20%) [43]. Other elements such as fluorine and antimony are also released into the air from burning this coal, and along with the poor nutrition of the population, may be additional factors in the overall toxic response [2, 41, 43].

1.3.5 Drinking Water

In the mid- to late 1950s, it became evident that the incidence of a unique malady, Blackfoot Disease, was increasing in Taiwan [67–69]. The hands, feet, or both of people with this disease become discolored; hence, the term "Blackfoot Disease." This disease occurred in an economically poor region of southwest Taiwan, affecting mainly farmers and fishermen. Gangrene of the digits also has been reported in West Bengal [49] and Bangladesh [1], another region of high endemic exposure. This disease is rarely, if at all, observed in other regions of the world.

Blackfoot Disease is a vaso-occlusive disease in the extremities of the body. Early symptoms include numbness or cold sensation in the extremities, particularly the feet. Other symptoms include burning sensations and intermittent claudication. This disease progresses to "shooting" pain, development of ulcers, and finally gangrene in the hands and feet. Of 1300 Blackfoot Disease cases examined by Tseng [69], 68% had spontaneous or surgical amputation of gangrenous hands or feet.

Investigations into the cause of this disease pointed to arsenic exposure [67–69]. In this region of Taiwan, with a population of approximately 100,000, the main source of drinking water was deep artesian wells, some over 200 m deep. These wells began to be used as a source of drinking water in the early 1900s because many of the shallow wells in the area had become tainted with saline. It was determined that these deep artesian wells were naturally contaminated with arsenic, with concentrations ranging from 0.01 to 1.2 ppm [68]. The severity of the disease was correlated with the concentration of arsenic in the artesian well water and the length of exposure. Individuals who lived in this region, but were able to drink water from the shallow wells, did not develop Blackfoot Disease. The arsenic concentration in the shallow wells ranged from 0.001 to 0.017 ppm. A low arsenic source of water became available to this region in the 1960s. After this time, no new cases of Blackfoot Disease developed in individuals who lived in the endemic area, used this low arsenic source of drinking water, and were less than 20 years old when they began to use the low arsenic water [68]. There were some suggestions that humic substances and chemicals other than arsenic present in the artesian well water also contributed to Blackfoot Disease [67]. But, this has never been positively determined.

In this same arsenic-exposed Taiwanese population, the risk of skin cancer and other dermal effects were also increased. Tseng et al. [70] reported that the overall prevalence rates for skin cancer, hyperpigmentation, and keratosis were 10.6/1000, 183.5/1000, and 71/1000, respectively. The prevalence rate for these afflictions increased with arsenic content of the artesian well water and with age (i.e., duration of exposure). These data have been used to determine the USEPA's oral reference dose (3×10^{-4} mg/kg/day) for arsenic and to support the classification of this metalloid as a human carcinogen [72].

A more recent calamity of unintentional arsenic exposure to a population has occurred in eastern India (West Bengal) [15, 16] and Bangladesh [38]. Like the Taiwanese, the exposure is from arsenic-contaminated drinking water. As India and Bangladesh are developing nations, surface water used as a source of drinking water in these countries is often contaminated with viruses, microorganisms, and pollutants. This has led to the occurrence of disease and mortality in people consuming this water, particularly infants. To alleviate this problem, several nongovernmental organizations (e.g., UNICEF and World Bank) in conjunction with the Bangladeshi government in the 1970s started to install tube-wells into the ground. These tube-wells were dug to provide ground water, which in theory would be less contaminated than surface water, and thus be a safer source of drinking water. Reports of dermal lesions in individuals that consumed this ground water began to emerge in the 1980s [9]. However, installation of the tube-wells in the region continued and millions are still in use. What has been determined is that the well water is naturally contaminated with arsenic due to the geological formations in the region. The water from many of these wells exceeds the World Health Organization (WHO) and USEPA arsenic drinking water standard of 10 μg/L. Tens of millions of people in this region of the world drink this contaminated water [62]. They are at risk of developing skin lesions, cardiovascular disease, diabetes, and cancer of skin, bladder, and other organs [62]. There has been a movement to test the water of each tube-well. Those wells with elevated levels are painted red to warn the user not to drink water from the well. However, this warning is not always heeded. Also, there are millions of wells, and to test all of them will take time.

Countries that have high arsenic levels in groundwater that is used as a source of drinking water include Argentina, Chile, Mexico, Thailand, Vietnam, and others [46]. Individuals in these countries who consume this contaminated water are also at risk for the development of several chronic diseases and cancer. In addition, mitigation of the high arsenic levels in the drinking water may not necessarily decrease the risk of cancer. Steinmaus et al. [64] have reported a high cancer risk in people in northern Chile 40 years after exposure cessation to high levels of arsenic in drinking water.

1.4 MEDICINAL USES OF ARSENIC

Although arsenic has been used with malicious intent over the ages, it has also been used benevolently to improve, or at least with good intentions, the health of man. There are writings of the use of arsenic sulfides in the form of orpiment (As_2S_3) and realgar (As_4S_4) by Hippocrates and Aristotle in the fourth century B.C. in pastes to treat ailments of the skin [6, 28, 45]. Dioscorides reported on the usefulness of orpiment in the first century A.D. as a depilatory [28]. Arsenic sulfide was used in the Middle East in the 1200s for the treatment of skin diseases, hemorrhoids, and syphilis [38].

Arsenic became more widely used in Europe in the late Middle Ages to the 1800s to treat an assortment of ailments [6, 28, 45]. The English physician William Withering advocated in the late eighteenth century the use of arsenic for medical purposes [66]. One of the most recognized arsenic treatments from this era was Fowler's solution [6, 12, 28, 37]. In the 1770s, a patented solution called "Tasteless Ague and Fever Drops" was used in English hospitals as an antiperiodic medicinal. Thomas Fowler, an English physician, asked the apothecarist Mr. Hughes, to determine the constituents of the patented solution. He found that it contained arsenic and prepared a new solution, by dissolving arsenic trioxide in alkali. Lavender was added to make the solution appear as a medicinal agent. This solution became to be known as Fowler's solution. It was used to treat many ailments including fever, asthma, syphilis, rheumatism, skin disorders, and leukemia [12, 45]. Fowler's solution and other arsenical solutions were listed as medicinal agents in the *Materia Medica* published in 1903 (Table 1.2). Fowler's solution was used up until the mid-1900s when other medicinal agents were found to be more efficacious with lower toxicity.

The medicinal use of arsenicals such as Fowler's solution is not without adverse effect. Sir Jonathan Hutchinson, an English surgeon, reported at the Pathological Society of London in 1887 that the long-term internal administration of arsenic at high doses could result in epithelial cancer [31]. Hutchinson presented a case of an American physician who had taken arsenic for psoriasis. Although the psoriasis on this patient had been cured, he had developed nodules on his palms and soles, which is a characteristic of chronic arsenic poisoning. The patient later died and was found to have epithelial cancer. This and other cases presented by Hutchinson were the first reports of the potential for inorganic arsenic to be a human carcinogen.

TABLE 1.2 Arsenical Preparations Used as Medicinal Agents in the Early 1900s[a]

Preparation	Arsenical	Formulation	Therapeutic Use
Acidum arsenosum	Arsenous acid	Pill; 1% solution with potassium carbonate (Fowler's solution)	Caustic to remove skin growth, anemia, malaria psoriasis, eczema
Arseni iodidum	Iodide of arsenic	1% solution	Tuberculosis, scrofula, bronchitis
Sodium cacodylas	Sodium cacodylate	Solution	Tuberculosis
Arsenii bromidi	Bromide of arsenic	1% solution (Clemen's solution)	Diabetes
Cupri arsenis	Arsenite of copper	Solution	Intestinal antiseptic for childhood diarrhea

[a] From Ref. [65].

In the latter part of the nineteenth century and early twentieth century, synthetic chemistry techniques had advanced to the point where chemicals were readily being synthesized to combat infectious diseases. Paul Ehrlich, a German physician, is generally credited to be the father of this new era, termed chemotherapy [44, 58, 78]. Ehrlich's goal was to synthesize the "Magic Bullet," a drug that would kill the infectious agent, but not harm the patient. Ehrlich and his coworker, Alfred Bertheim, were able to determine the structure of the arsenical atoxyl. Atoxyl was the first synthetic aromatic arsenical. It was prepared by Antoine Berchampe in 1863, when he heated aniline and arsenic together [44, 58]. Berchampe believed the structure to be an anilide. Forty years later, atoxyl began to be used in Europe to treat skin diseases and cancer. It had been found to be less acutely toxic than inorganic arsenic that was used for similar purposes. Ehrlich and Bertheim determined that atoxyl was actually an amino arsenic acid. This knowledge allowed them to synthesize over 900 derivatives of atoxyl. The most famous derivative, Compound 606 or arsphenamine, was found to be effective against syphilis. It was marketed as Salvarsan to treat syphilis in 1910 and was more successful than the standard mercury salt treatments used at that time. Salvarsan remained the most useful drug for syphilis until the emergence of penicillin in the 1940s.

The use of arsenicals in medicine has decreased over the years, primarily because of the potential for toxic effects of these compounds and because more effective medicinal agents are available. However, arsenic is being used for specific diseases (See Section 1.4.4). The trivalent organic arsenical melarsoprol is used to treat human African trypanosomiasis [50]. Melarsoprol is the drug of choice for the second stage of this disease because it can cross the blood–brain barrier and kill the parasites that reside in the cerebrospinal fluid. However, there are adverse effects with using this drug. About 20% of the patients treated with melarsoprol are affected by reactive encephalopathy. In addition, some strains of this parasite are becoming resistant to the therapeutic effects of melarsoprol.

1.4.1 Arsenic Eaters of Styria

A curious aspect of arsenic exposure is the "arsenic eaters of Styria" [28, 45, 53]. People living in the region near Graz, Austria, were reported to have begun eating small amounts of arsenic in the twelfth century. Both white and yellow arsenic (orpiment) were consumed. Consumption of arsenic in this region was for an assumed beneficial effect. Reasons given for eating arsenic were to enhance a woman's complexion, improve breathing during hiking at high altitudes, aid in digestion, increase courage and sexual potency, and as a preventative measure against infectious diseases. The use of arsenic in Styria in this manner is controversial because there is limited evidence that it occurred. If it happened, it was most likely in secret, and predominantly by poor people. But considering that other Europeans were using arsenic as a tonic during 1700 and 1800s, and there is use of the traditional Chinese and Indian Medicines (see below), the eating of arsenic in Styria could have been a frequent occurrence.

1.4.2 Traditional Medicines

Traditional medicines used in countries such as China and India contain metals such as arsenic, mercury, and lead [11, 36, 42, 60]. The WHO defines traditional medicine as health practices and approaches that use plant, animal, and mineral medicines as well as spiritual and physical techniques alone or in combination to diagnose and treat illness as well maintain an individual's health (WHO [80]). These practices were developed many hundreds if not thousands of years ago and are still in place. In many developing countries, the WHO estimates that up to 80% of the population relies on traditional medicine.

Chinese traditional medicines include the arsenical minerals orpiment, realgar, and arsenolite (essentially arsenic trioxide) [42]. Indian Ayurvedic metallic herbal preparations called Bhasmas can contain arsenic, mercury, silver, and other metals [36]. The metals are purposely added to the traditional medicines for therapeutic effect. For example, realgar is a constituent of *Hongling San* (15% realgar in seven components) and used to treat heatstroke, dizziness, headache, and nausea [42]. Most, if not all, of these traditional medicines are not regulated as pharmaceutical agents in the western world. However, these traditional medicines are available via the Internet and in specialized stores in developed countries [11, 60]. Poisonings have been noted from their use [21, 22]. As people emigrate from developing to developed countries, these traditional medicines may become more prevalent and care should be taken with their use.

1.4.3 Uses in Dentistry

Arsenic has had use in dentistry, primarily to relieve dental pain and for root canal therapy [32]. There are writings of the use of arsenic in China in 2700 B.C. for painful teeth to "kill a tooth worm." Two thousand years later, arsenic was used in the Middle East and Europe for tooth pulp devitalization, to treat dental fistula and tooth pain, and for root canal therapy. For the latter treatment, the roots of teeth were coated with yellow arsenic (orpiment) before extraction. To alleviate tooth pain, in some cases, a mixture of opium and arsenic was used.

John Roach Spooner of Canada is generally recognized as the first practitioner in North America to devitalize the dental pulp with arsenic. This occurred in the early 1800s. While arsenic was used by dentists in North America during the early to late 1800s, this was not without controversy. Several practitioners cautioned the use of arsenic because of its potent and acute toxic properties. In fact, there were several deaths of dental patients that had been treated with arsenic, although it was not completely determined whether the arsenic treatment was the cause of death. The use of arsenic in dentistry started to fade at the end of the 1800s. However, even today, homeopathic dentistry recommends the use of arsenicum album (diluted arsenic trioxide) for the treatment of tooth ache [74].

1.4.4 Treatment for Leukemia

One of the more interesting uses of arsenic is for the treatment of relapsed acute promyelocytic leukemia [19]. Treatment for this rare form of leukemia with arsenic trioxide reemerged in China in the early 1970s [3]. Arsenic had been used in China to treat leukemia up to the 1950s but became disfavored because of newly found radiation treatments and implementation of alkylating agents. However, because of political and cultural turmoil in China in the 1950s and 1960s, western-type anti-leukemia therapy became unavailable to the Chinese people. In 1971, researchers from Harbin Medical University learned that traditional Chinese medicine was being used in the village of Lindian in northeast China to treat cancer effectively. Analysis of the medicine showed it contained arsenic trioxide, mercuric oxide, and toad extract. TD Zhang of Harbin Medical University determined that arsenic trioxide was the active ingredient. In 1973, arsenic trioxide was successfully used in humans to treat chronic myelogenous and acute promyelocytic leukemia [3]. Word of its use as an anti-leukemia agent became known in the United States in the late 1990s. The Food and Drug Administration approved of the use of arsenic trioxide for treatment of refractive acute promyelocytic leukemia in 2001. Although arsenic trioxide is an effective cancer therapeutic agent, it has to be used cautiously, with careful monitoring of the patients because of the potential acute lethal toxicity of this arsenical [76].

Research is underway for candidates to supplement or replace arsenic trioxide as a treatment for leukemia. One of the candidates is realgar, which is a mineral found in ores removed from the ground. Realgar is a common constituent of traditional Chinese and Indian medicines. The acute toxicity of realgar is about 100-fold less than arsenic trioxide [42, 77]. Positive results using realgar in treating leukemia have been reported [77], but it currently is not approved for cancer treatment in the United States. Realgar is poorly soluble in water, which makes it less bioavailable than other arsenicals [4, 77]. Nanoparticles and quantum dots have been developed containing realgar in an attempt to increase its bioavailability, and animal studies suggest this occurs [4, 77]. Bioleaching of arsenic from realgar is another potential means for the dissolution of this arsenical for its medicinal use [81].

1.5 SUMMARY

Arsenic has a long history of being an intentional and unintentional human poison. For intentional exposure, there have been many unfortunate instances where arsenic was used for homicidal intent. However, with current analytical chemistry capabilities, the probability of determining the cause of the poisoning and perhaps the conviction of the accused is great. There is a great effort to reduce unintentional exposure, primarily in drinking water, by many world-wide governmental agencies and nongovernmental organizations. Adverse outcomes from this particular type of exposure, primarily in developing countries, may reach calamitous heights in the next few years. Arsenic has specific medicinal uses such as the treatment for acute promyelocytic leukemia and trypanosomiasis. However, its use needs to be carefully monitored because of the high probability of side effects. Research is underway to find less toxic forms of arsenic that will hopefully have a beneficial effect. Arsenic has had a long history of being a poison and a medicine. Human exposure to this metalloid will continue primarily because of its pervasiveness in the environment and less so from its use as a medicinal agent.

REFERENCES

[1] S. Ahamed, M.K. Sengupta, S.C. Mukherjee, S. Pati, A. Mukherjeel, M.M. Rahman, M.A. Hossain, B. Das, B. Nayakl, A. Pal, A. Zafar, S. Kabir, S.A. Banu, S. Morshed, T. Islam, M.M. Rahman, Q. Quamruzzaman, D. Chakraborti, An eight-year study report on arsenic contamination in groundwater and health effects in Eruani village, Bangladesh and an approach for its mitigation. The Journal of Health, Population and Nutrition 24 (2006) 129–141.

[2] D. An, Y.G. He, Q.X. Hu, Poisoning by coal smoke containing arsenic and fluoride. Fluoride 30 (1997) 29–32.

[3] W.Y. Au, A biography of arsenic and medicine in Hong Kong and China. Hong Kong Medical Journal 17 (2011) 507–512.

[4] P. Balaz, J. Sedlak, Arsenic in cancer treatment: Challenges for application of realgar nanoparticles (A mini review). Toxins 2 (2010) 1568–1581.

[5] P.W.J. Bartrip, How green was my valance? Environmental arsenic poisoning and the Victorian domestic ideal. English Historical Review 109 (1994) 891–913.

[6] R. Bentley, T.G. Chasteen, Arsenic curiosa and humanity. The Chemical Educator 7 (2002) 51–60.

[7] H.M. Bolt, Arsenic: An ancient toxicant of continuous public health impact, from iceman Otzi until now. Archives of Toxicology 86 (2012) 825–830.

[8] A.C. Bronstein, D.A. Spyker, L.R. Cantilene, J.L. Green, B.H. Rumack, R.C. Dart, 2010 Annual report of the American Association of Poison Control Centers' National Poison Data System (NPDS): 28th Annual Report. Clinical Toxicology 49 (2011) 910–941.

[9] A.K. Chakraborty, K.C. Saha, Arsenical dermatosis from tubewell water in West Bengal. The Indian Journal of Medical Research 85 (1987) 326–334.

[10] F. Challenger, C. Higginbottom, L. Ellis, The formation of organo-metalloidal compounds by microorganisms. Part I. Trimethylarsine and dimethylarsine. Journal of the Chemical Society 95 (1933) 101.

[11] K. Cooper, B. Noller, D. Connell, J. Yu, R. Sadler, H. Olszowy, G. Golding, U. Tinggi, M.R. Moore, S. Myers, Public health risks from heavy metals and metalloids present in traditional Chinese medicines. Journal of Toxicology and Environmental Health, Part A 70 (2007) 1694–1699.

[12] W.R. Cullen Is Arsenic an Aphrodisiac? The Sociochemistry of an Element, The Royal Society of Chemistry, Cambridge, MA, 2008.

[13] W.R. Cullen, R. Bentley, The toxicity of trimethylarsine: An urban myth. Journal of Environmental Monitoring 7 (2005) 11–15.

[14] M. Dakeishi, K. Murata, P. Grandjean, Long-term consequences of arsenic poisoning during infancy due to contaminated milk powder. Environmental Health: A Global Access Science Source 5 (2006) 1–7.

[15] D. Das, A. Chatterjee, B.K. Mandal, G. Samanta, D. Chakraborti, B. Chanda, Arsenic in ground water in six districts of West Bengal, India: The biggest arsenic calamity in the world. Part 2. Arsenic concentration in drinking water, hair, nails, urine, skin-scale and liver tissue (biopsy) of the affected people. The Analyst 120 (1995) 917–924.

[16] D. Das, A. Chatterjee, G. Samanta, B. Mandal, T.R. Chowdhury, G. Samanta, P.P. Chowdhury, C. Chanda, G. Basu, D. Lodh, Arsenic contamination in groundwater in six

districts of West Bengal, India: The biggest arsenic calamity in the world. The Analyst 119 (1994) 168N–170N.

[17] P. Dyer, The 1900 arsenic poisoning epidemic. Brewery History 130 (2009) 65–85.

[18] M.J. Ellenhorn Ellenhorn's Medical Toxicology: Diagnosis and Treatment of Human Poisoning, Williams & Wilkins, Baltimore, 1997.

[19] A. Emadi, S.D. Gore, Arsenic trioxide: An old drug rediscovered. Blood Reviews 24 (2010) 191–199.

[20] P.E. Enterline, R. Day, G.M. Marsh, Cancers related to exposure to arsenic at a copper smelter. Occupational and Environmental Medicine 52 (1995) 28–32.

[21] E. Ernst, Heavy metals in traditional Indian remedies. European Journal of Clinical Pharmacology 57 (2002) 891–869.

[22] E. Ernst, Toxic heavy metals and undeclared drugs in Asian herbal medicines. Trends in Pharmacological Sciences 23 (2002) 136–139.

[23] S. Forshufvud, H. Smith, A. Wassen, Arsenic content of Napoleon I's hair probably taken immediately after his death. Nature 192 (1961) 103–105.

[24] G. Gall, What's your poison. Brewery History 128 (2008) 49–53.

[25] M.S. Gorby, Arsenic poisoning. Western Journal of Medicine 149 (1988) 308–315.

[26] G. Greenfield, Agent Blue and the business of killing rice. *Ecologist* (2004). http://www.countercurrents.org/us-greenfield180604.htm. Accessed June 13, 2015.

[27] F. Gyorgyey, Arsenic and no lace. Caduceus 3 (1987) 40–65.

[28] J.S. Haller, Therapeutic mule: The use of arsenic in the nineteenth century materia medica. Pharmacy in History 17 (1975) 87–100.

[29] A.P. Harrison, I. Cattani, J.M. Turfa, Metallurgy, environmental pollution and the decline of Etruscan civilization. Environmental Science and Pollution Research International 17 (2010) 165–180.

[30] M.F. Hughes, Arsenic toxicity and potential mechanisms of action. Toxicology Letters 133 (2002) 1–16.

[31] J. Hutchinson, Arsenic cancer. British Medical Journal 2 (1887) 1281–1282.

[32] J.M. Hyson, A history of arsenic in dentistry. Journal of the California Dental Association 35 (2007) 135–139.

[33] IARC A Review of Human Carcinogens. C. Metals, Arsenic, Fibres and Dusts, International Agency for Research on Cancer, Lyon, France, 2012.

[34] L. Jarup, G. Pershagen, S. Wall, Cumulative arsenic exposure and lung cancer in smelter workers: A dose response study. American Journal of Industrial Medicine 15 (1989) 31–41.

[35] D.E.H. Jones, D.W.D. Ledingham, Arsenic in Napoleon's wallpaper. Nature 299 (1982) 626–627.

[36] A. Kumar, A.G. Nair, A.V. Reddy, A.N. Garg, Bhasmas: Unique Ayurvedic metallic-herbal preparations, chemical characterization. Biological Trace Element Research 109 (2006) 231–254.

[37] H.A. Langenhan, History of the arsenical solutions. Journal of the American Pharmaceutical Association 8 (1919) 189.

[38] E. Lev, Medicinal exploitation of inorganic substances in the Levant in the Medieval and Early Ottoman Periods. Adler Museum Bulletin 28 (2002) 11–16.

[39] T. Levine, A. Rispin, C. Chen, H. Gibb. Special Report on Ingested Inorganic Arsenic. Skin Cancer, Nutritional Essentiality, USEPA, Washington, DC, 1988. http://www.epa.gov/raf/publications/pdfs/EPA_625_3-87_013.PDF. Accessed April 09, 2015.

[40] P.K. Lewin, R.G.V. Hancock, P. Voynovich, Napoleon Bonaparte: No evidence of chronic arsenic poisoning. Nature 299 (1982) 622–628.

[41] D. Li, D. An, Y. Zhou, J. Liu, M.P. Waalkes, Current status and prevention strategy for coal-arsenic poisoning in Guizhou, China. Journal of Health, Population, and Nutrition 24 (2006) 273–276.

[42] J. Liu, Y. Lu, Q. Wu, R.A. Goyer, M.P. Waalkes, Mineral arsenicals in traditional medicines: Orpiment, realgar, and arsenolite. The Journal of Pharmacology and Experimental Therapeutics 326 (2008) 363–368.

[43] J. Liu, B. Zheng, H.V. Aposhian, Y. Zhou, M.L. Chen, A. Zhang, M.P. Waalkes, Chronic arsenic poisoning from burning high-arsenic-containing coal in Guizhou, China. Environmental Health Perspectives 110 (2002) 119–122.

[44] N.C. Lloyd, H.W. Morgan, B.L. Nichlson, R.S. Ronimus, R. Riethmiller, Salvarsan: The first chemotherapeutic compound. Chemistry in New Zealand 69 (2005) 24–27.

[45] A. Lykknes, L. Kvittingen, Arsenic: Not so evil after all? Journal of Chemical Education 80 (2003) 497–500.

[46] B.K. Mandal, K.T. Suzuki, Arsenic round the world: A review. Talanta 58 (2002) 201–235.

[47] E.W. Massey, D. Wold, A. Heyman, Arsenic: Homicidal intoxication. Southern Medical Journal 77 (1984) 848–851.

[48] W.J. Meek, The gentle art of poisoning. Journal of the American Medical Association 158 (1955) 335–339.

[49] S.C. Mukherjee, K.C. Saha, S. Pati, R.N. Dutta, M.M. Rahman, M.K. Sengupta, S. Ahamed, D. Lodh, B. Das, M.A. Hossain, B. Nayak, A. Mukherjee, D. Chakraborti, S.K. Dulta, S.K. Palit, I. Kaies, A.K. Barua, K.A. Asad, Murshidabad: One of the nine groundwater arsenic-affected districts of West Bengal, India. Part II: Dermatological, neurological, and obstetric findings. Clinical Toxicology 43 (2005) 835–848.

[50] A.J. Nok, Arsenicals (melarsoprol), pentamidine and suramin in the treatment of human African trypanosomiasis. Parasitology Research 90 (2003) 71–79.

[51] NRC Arsenic in Drinking Water, National Academy Press, Washington, DC, 1999.

[52] M.G. Ord, L.A. Stocken, A contribution to chemical defence in World War II. Trends in Biochemical Sciences 25 (2000) 253–256.

[53] G. Przygoda, J. Feldmann, W.R. Cullen, The arsenic eaters of Styria: A different picture of people who were chronically exposed to arsenic. Applied Organometallic Chemistry 15 (2001) 457–462.

[54] S. Pullen-James, S.E. Woods, Occupational arsine gas exposure. Journal of the National Medical Association 98 (2006) 1998–2001.

[55] J. Randerson, Lethal poisons for sale in the web marketplace. *The Guardian*. March 28, 2007. http://www.theguardian.com/technology/2007/mar/29/news.uknews. Accessed April 9, 2015.

[56] R.N. Ratnaike, Acute and chronic arsenic toxicity. Postgraduate Medical Journal 79 (2003) 391–396.

[57] E.S. Reynolds, An account of the epidemic outbreak of arsenical poisoning occurring in beer drinkers in the north of England and the Midland counties in 1900. Medico-Chirurgical Transactions 4 (1901) 409–452.

[58] S. Riethmiller, From Atoxyl to Salvarsan: Search for the Magic Bullet. Chemotherapy 51 (2005) 235–242.

[59] C.R. Sanger, On chronic arsenical poisoning from wall papers and fabrics. Proceedings of the American Academy of Arts and Sciences 29 (1893) 148–177.

[60] R.B. Saper, R.S. Phillips, A. Sehgal, N. Khouri, R.B. Davis, J. Paquin, V. Thuppil, S.N. Kales, Lead, mercury, and arsenic in US- and Indian-manufactured Ayurvedic medicines sold via the Internet. JAMA: The Journal of the American Medical Association 300 (2008) 915–923.

[61] S. Scheindlin, The duplicitous nature of inorganic arsenic. Molecular Interventions 5 (2005) 60–64.

[62] A.H. Smith, E.O. Lingas, M. Rahman, Contamination of drinking-water by arsenic in Bangladesh: A public health emergency. Bulletin of the World Health Organization 78 (2000) 1093–1103.

[63] H. Smith, S. Forshufvud, A. Wassen, Distribution of arsenic in Napoleon's hair. Nature 194 (1962) 725–726.

[64] C.M. Steinmaus, C. Ferreccio, J.A. Romo, Y. Yuan, S. Cortes, G. Marshall, L.E. Moore, J.R. Balmes, J. Liaw, T. Golden, A.H. Smith, Drinking water arsenic in northern Chile: High cancer risks 40 years after exposure cessation. Cancer Epidemiology, Biomarkers and Prevention 22 (2013) 623–630.

[65] A.A. Stevens Modern Materia Medica and Therapeutics, W.B. Saunders & Co., Philadelphia, PA, 1903.

[66] X. Thomas, J. Troncy, Arsenic: A beneficial therapeutic poison—a historical overview. Adler Museum Bulletin 35 (2009) 3–13.

[67] C.H. Tseng, Blackfoot disease and arsenic: A never-ending story. Journal of Environmental Science and Health. Part C, Environmental Carcinogenesis & Ecotoxicology Reviews 23 (2005) 55–74.

[68] W.P. Tseng, Effects and dose: Response relationships of skin cancer and blackfoot disease with arsenic. Environmental Health Perspectives 19 (1977) 109–119.

[69] W.P. Tseng, Blackfoot disease in Taiwan: A 30-year follow-up study. Angiology 40 (1989) 547–558.

[70] W.P. Tseng, H.M. Chu, S.W. How, J.M. Fong, C.S. Lin, S. Yeh, Prevalence of skin cancer in an endemic area of chronic arsenicism in Taiwan. Journal of the National Cancer Institute 40 (1968) 453–463.

[71] K. Tsuchiya, Various effects of arsenic in Japan depending on type of exposure. Environmental Health Perspectives 19 (1977) 35–42.

[72] US Environmental Protection Agency. Arsenic, inorganic (CASRN 7440-38-2). System, Integrated Risk Information. 2003. http://www.epa.gov/iris/subst/0278.htm. Accessed April 9, 2015.

[73] US Environmental Protection Agency. Organic Arsenicals. 2013. http://www.epa.gov/oppsrrd1/reregistration/organic_arsenicals_fs.html. Accessed April 9, 2015.

[74] P. Wander, Top 5 reasons we visit … the dentist. Health & Homeopathy (2010) 17–20.

[75] T.A. Wertime, Man's First Encounters With Metallurgy: Man's discovery of ores and metals helped to shape his sense of science, technology, and history. Science 146 (1964) 1257–1267.

[76] P. Westervelt, R.A. Brown, D.R. Adkins, H. Khoury, P. Curtin, D. Hurd, S.M. Luger, M.K. Ma, T.J. Ley, J.F. DiPersio, Sudden death among patients with acute promyelocytic leukemia treated with arsenic trioxide. Blood 98 (2001) 266–271.

[77] J. Wu, Y. Shao, J. Liu, G. Chen, P.C. Ho, The medicinal use of realgar (As4S4) and its recent development as an anticancer agent. Journal of Ethnopharmacology 135 (2011) 595–602.

[78] A. Yarnell, Salvarsan. Chemical and Engineering News 83 (2005) 116.

[79] T. Yorifuji, T. Tsuda, P. Grandjean, Unusual cancer excess after neonatal arsenic exposure from contaminated milk powder. Journal of the National Cancer Institute 102 (2010) 360–361.

[80] WHO Fact Sheet Number 134 (2008).

[81] J. Zhang, X. Zhang, Y. Ni, H. Li, Bioleaching of arsenic from medicinal realgar by pure and mixed cultures. Process Biochemistry 42 (2007) 1265–1271.

2

GEOGENIC AND ANTHROPOGENIC ARSENIC HAZARD IN GROUNDWATERS AND SOILS: DISTRIBUTION, NATURE, ORIGIN, AND HUMAN EXPOSURE ROUTES

David A. Polya and Michael Lawson

School of Earth, Atmospheric and Environmental Sciences and Williamson Research Centre for Molecular Environmental Science, The University of Manchester, Manchester, UK

2.1 INTRODUCTION

2.1.1 Scope and Structure

In this chapter, the distribution, nature, and origins of arsenic hazards in groundwaters and soils are outlined, along with the biogeochemical processes that give rise to these distributions. In addition, the major exposure routes from these reservoirs to humans are briefly summarized. Particularly given the importance of drinking groundwater as a major exposure route contributing to massive deleterious human health impacts [98, 99, 175, 179, 180, 195, 197, 201, 220], much of the focus of this chapter is on the biogeochemistry of arsenic in groundwater, particularly in, but not restricted to, shallow aquifers in the lowlands of circum-Himalayan Asia.

There are over 6000 articles to date published on the topic of "arsenic" and ("groundwater" or "soil")—in this chapter, we cite only a couple hundred of these articles, selected primarily on the basis of our experience, followed by "recommendation-by-volume-of-citations." Throughout this chapter, we have tried to emphasize more

Arsenic: Exposure Sources, Health Risks, and Mechanisms of Toxicity, First Edition.
Edited by J. Christopher States.
© 2016 John Wiley & Sons, Inc. Published 2016 by John Wiley & Sons, Inc.

generic theoretical concepts and then illustrate them with specific examples. For more detailed lists of species, minerals, compositions/ore grades, and the like, we refer the reader to specific reviews as outlined in the following.

2.1.2 Previous Reviews

Among the most prominent previous reviews of arsenic (bio-geo-)chemistry are those of Cullen and Reimer [56], a substantive landmark review of arsenic chemistry that details occurrences in geological and biological media; Matschullat [148], which provides a review of global cycling of arsenic; and Welch et al. [242], which reviews the occurrences of arsenic in US groundwaters. Mandal and Suzuki [143] as well as Smedley and Kinniburgh [219] provide detailed accounts of the distribution and origin of arsenic in key geological media in addition to summarizing selected studies of arsenic sorption on various mineral phases. More recently, Ravenscroft et al. [201] provides a substantive review of arsenic chemistry, geological occurrences, remediation options, health impacts, and human exposure routes, with a particular emphasis on drinking water. Henke [88] is a recent extensive review while, with the increasing recognition of the role of rice as a source of major exposure, Meharg and Zhao [154] represents an important and comprehensive account of arsenic and rice.

Summaries of thermodynamic data for arsenic species are notably provided by Nordstrom and Archer [176], Cleverley and Benning [50], Lu and Zhu [138], and Bessinger et al. [26], among others. The (bio)geochemistry of arsenic is often intimately linked with the strongly redox-sensitive [185] (bio)geochemistry of sulfur [227] and/or iron [183], so more general accounts of environmental, mineralogical, and geochemical controls on the transfer of inorganic chemical components between various geological reservoirs are helpful and are provided by Domenico and Schwartz [61], Appelo and Postma [11], Stumm and Morgan [222], Bethke [27], and Brown and Calas [34], among others. More arsenic-specific reviews of (bio-geo-)chemistry are provided in the 2006 special issue of *Elements* [44, 131, 164, 182, 237].

2.1.3 Changing State-of-the-Art

The groundwork for our collective understandings of the occurrence and behavior of arsenic in the environment is underpinned by science and arsenic-specific science that has been published over 20 years ago. As such, excellent "old" reviews, (e.g., [70, 56]) are, even today, valuable sources of information and understanding. Having said that, however, it is equally apparent that very substantial advances have been made over the last 20 years or so—particularly with respect to our understanding of the (i) detailed chemistry of S-bearing thioarsenite and thioarsenate species [188, 189, 244] and oligomeric [87] arsenic species, (ii) the role of microbes in mediating (bio)geochemical reactions controlling arsenic mobility [4, 101, 131, 184, 236], and (iii) the molecular scale nature of arsenic interactions with mineral surfaces [45, 107, 215] and organic matter [38, 92, 122]. In addition, the improvements in our collective ability to integrate empirical or *ab initio* and other thermodynamic [20, 86, 87] and kinetic models [26, 54, 97] of chemical reactivity with micro- to macro-scale models of fluid flow through porous media, whether they be soils, local aquifers, or basin-scale aquifer systems, have proven to be critical in the development of predictions on the location and extent of arsenic hazard. The rapid development of (i) more sensitive, more robust, and more species-specific analytical, including micron-scale spatially resolved, technologies, (ii) remote-sensing technologies, such as dispersed environmental sensors controlled and monitored by telemetry, and (iii) increased computational power and software sophistication indicates that further substantial advances will most likely be made over the next 20 years. Such advances will hopefully provide the improved level of understanding that is required to, among other things, inform the development of more sophisticated and accurate species-dependent health risk assessments [68].

In light of those advances, some reappraisal of currently published studies may also be required. While a detailed review of analytical and modeling methodologies is well beyond the scope of this chapter, we take the opportunity to highlight a number of significant reviews that may help the interested reader to understand better relevant technological developments as well as appreciating some of the limitations of studies published to date. A review of technologies utilized in studying microbial controls on metal(loid) mobilizing processes is provided by Lloyd et al. [130], while high spatial resolution imaging techniques have been reviewed by others [162, 171, 248].

Reviews of various technologies used for determining chemical speciation of arsenic in fluids are published [74, 140, 247]. Discussions of issues surrounding the preservation of arsenic speciation of collected samples are provided by several authors [22, 77, 81, 126, 140, 187, 196, 240], who emphasize not only the challenges of analyzing arsenic species that are either unstable or difficult to extract but also the paucity of stable speciation-certified reference materials. Of particular note is that the importance of aqueous thioarsenate species has only recently been fully recognized [188], and it is likely that many prior studies may have not only missed the determination of these moieties as species but also failed to have preserved them adequately before analyzing waters for total arsenic. Many organic arsenicals are yet to be fully characterized or perhaps even discovered. Solid-phase speciation techniques have been reviewed with regard to arsenic [239]. Powerful synchrotron-based techniques for determining the molecular environment of aqueous or solid-phase arsenic species are reviewed by Brown et al. [35]. The future of chemical speciation studies is considered by others [2, 72].

Software packages that are currently used worldwide for modeling chemical speciation and water–biomass–rock mass transfer processes include PHREEQC [230] and Geochemists Workbench [27]. These may be coupled with fluid flow modeling software (e.g., MODFLOW [231]) for the development of coupled 1D, 2D, or 3D contaminant transport models [37, 94, 156, 157]. Such models may be calibrated by ^{14}C and other groundwater dating/tracer techniques [28, 41, 63, 94].

2.2 BIOGEOCHEMISTRY OF ARSENIC

The distribution of arsenic in groundwaters and soils is strongly dependent on geological, biological, and, sometimes, anthropological processes. The relative importance of the processes in various contexts depends in part on (i) the nature of the biogeochemical environment and (ii) the chemical forms or species in which arsenic occurs and the subsequent biogeochemical reactivity of these species. These reactivities, in turn, depend upon atomic and molecular-scale properties.

As a Group VA p-block element in the periodic table, arsenic displays a wide range of chemistries. With a ground-state electronic configuration $[Ar]3d^{10}4s^24p^3$, multiple valence states, notably including −3, 0, +3, and +5, are exhibited by the range of naturally occurring arsenic-bearing molecules. With widely varying acid dissociation constants, arsenic species may act as acids, bases, or amphoteric substances in the Arrhenius sense. In combination, these lead to both oxidation potential and pH being important controls on arsenic speciation and consequent mobility and bioavailability. With intermediate ionization potentials between those of Group I alkali metals and those of Group VII halogens and with intermediate polarizabilities, arsenic also exhibits ambiguous soft/hard behavior. Arsenic demonstrates a tendency to soft behavior when in the III oxidation state but a tendency to hard behavior when in the V oxidation state. Consequently, environmentally significant molecular species of arsenic can be found that are oxygen-bonded or sulfur-bonded or indeed a combination of these. In addition, strong covalent As–C bonding gives rise, particularly in biological media, to a complex range of organic arsenicals, including phospholipids [76] and other oligomeric species [73].

Notwithstanding all these complexities, the inorganic chemistry of arsenic can be largely rationalized in terms of similarity of behavior of the various species to those of phosphate, silica, or chalcophiles. Indeed, many interactions of arsenic in *in vitro and in vivo* systems can be understood in the same way before the recourse to detailed thermodynamic-based pharmacokinetic modeling. Key examples of such behavior include the following:

1. Under oxidizing, near-neutral conditions, such as those found in many surface waters, arsenic often occurs predominately as the $HAsO_4^{2-}$ (aq) species, whose concentration in water may be controlled by sorption onto various Fe(III)–O–H phases, such as goethite or ferrihydrite. This behavior mimics that of HPO_4^{2-} (aq).
2. Under reducing slightly acidic conditions, such as those found in many reducing subsurface environments in which anaerobic metal-reducing bacteria are active, arsenic occurring predominately as H_3AsO_3 (aq) acts as a chalcophile, being highly reactive toward reduced sulfur, occurring as H_2S (aq) or HS^- (aq). This can result in the immobilization of arsenic as arsenic sulfides, such as orpiment (As_2S_3) or realgar (AsS), or arsenic-bearing iron sulfides, such as pyrite (FeS_2), marcasite (FeS_2), pyrrhotite ($Fe_{1-x}S$), greigite (Fe_3S_4), arsenopyrite (FeAsS), or iron monosulfide (FeS).
3. Within silicate minerals, arsenic typically occurs as As(V) in tetrahedral coordination substituting for SiO_4. For example, in serpentinite [85] and other silicates, such as smectite [186], hydrogarnet [47], and much more rarely as the arsenic feldspar, filatovite [71]; arguably of greater importance, the ingress of arsenic into rice grown in highly reducing flooded paddy fields has been shown to take place through the same aquaporins [139] that permit the transport of

silica. This substitution arises because of the great similarity—in terms of charge, size, molecular structure, and pK_a—of arsenous acid, H_3AsO_3 (aq), and silicic acid, H_4SiO_4 (aq) [251].

2.2.1 Arsenic Species

Arsenic species may occur in gas, solid, and liquid phases.

2.2.1.1 ***Gas Phase Species*** Gaseous arsenic species have been observed in natural geothermal gases, thermogenic hydrocarbon gases, emissions from sewage sludge, and emissions from soils and plants. Arsine (AsH_3) is a highly toxic inorganic arsenic species, but its reactivity to oxygen means that other species tend to be of greater importance in the environment—for example, (i) various methylated arsenicals, such as $(CH_3)_2AsCl$, $(CH_3)_3As$, $(CH_3)_2AsSCH_3$, and CH_3AsCl_2, have been observed as the predominant arsenic gas species in geothermal volatiles from the Yellowstone National Park [187]; (ii) the methylated arsenic species $(CH_3)_3As$ was found in higher concentrations than AsH_3 in incubated sewage sludge [158] and, in addition, CH_3AsH_2 and $(CH_3)_2AsH$ were found at similar concentrations to AsH_3 in the same environment; and (iii) $(CH_3)_3As$ is the dominant arsenic species observed to volatilize from rice plants and soils [106, 155].

2.2.1.2 ***Mineral Phase Species*** Minerals containing arsenic as a major component include both sulfide and oxide groups as well as a range of more complex phases, including sulfosalts. The more common minerals are listed here but there is a complex and wide variety of, particularly secondary low temperature formed, arsenic minerals, more details of which may be found in Bowles et al. [32], Smedley and Kinniburgh [219], and Vaughan [238] and references therein. The major arsenic-bearing minerals are as follows: sulfides—orpiment (As_2S_3), realgar (As_4S_4), nicolite (NiAs), gersdorffite (NiAsS), arsenopyrite (FeAsS), loellingite ($FeAs_2$); the sulfosalts—enargite (Cu_3AsS_4), and tennantite ($(Cu,Fe)_{12}As_4S_{13}$); and the oxides/oxyanion phases—scorodite ($FeAsO_4.2H_2O$), claudetite (As_2O_3), arsenolite (As_2O_3), erythrite ($Co_3(AsO_4)_2.8H_2O$), and pharmacosiderite ($Fe_3(AsO_4)_2(OH)_3.5H_2O$). In addition, arsenic occurs as a trace component in or on many other minerals, either in the form of molecular scale substitutions or as adsorbates. Arguably, the most important of these are arsenian pyrite and marcasite in which arsenic can occur in concentrations over 1 wt.%, and various Fe–O–H–(–C–S) bearing phases, such as goethite, ferrihydrite, green-rust, and jarosite, in which arsenic can commonly reach concentrations in the order of 0.1–1 wt.%. Further information on the nature of arsenic species physically or chemically sorbed onto various mineral surfaces is provided by Charlet et al. [45], which builds upon evidence provided by batch and flow through sorption experiments [60, 144], synchrotron-based X-ray absorption spectroscopy [107, 145, 215] and other spectroscopic studies [33], and/or *ab initio* or other modeling studies [91, 223].

2.2.1.3 Aqueous Phase Species In groundwaters, the predominant arsenic species are most commonly inorganic oxyanions of As(III), commonly referred to as arsenite, and As(V), commonly referred to as arsenate, with molecular formulas $H_xAsO_3^{(x-3)}$ with $x=3$ to 0, and $H_xAsO_4^{(x-3)}$ with $x=3$ to 0, respectively. In addition, in environments where there is appropriate microbial activity or industrial contamination, methylated arsenicals, notably monomethylarsonous acid (MMA(III)), monomethylarsonic acid (MMA(V)), dimethylarsinous acid (DMA(III)), and dimethylarsinic acid (DMA(V)), may also be important [10]. In the presence of high concentrations of reduced sulfur species, such as H_2S (aq) and HS^- (aq) and particularly under alkaline conditions, thioarsenites and to a greater extent thioarsenates may be important. Under some not so uncommon circumstances, thioarsenates may even represent the most abundant aqueous arsenic species [188, 189, 244]. In addition, the importance of aqueous arsenic carbonate complexes has also been proposed [112] but has been largely discredited [167].

2.2.1.4 Species in Biological Media Furthermore, a wide variety of arsenic species, in addition to arsenite, arsenate, and various methylated arsenicals, are found in biological media. These notably include arsenocholine, arsenobetaine, roxarsone, and numerous organoarsenic lipids, including phospholipids. However, these biological arsenicals are outside the scope of this chapter, are discussed elsewhere in this volume, and thus are not considered further.

2.2.2 Key Reactions of Arsenic Species

2.2.2.1 Acid Dissociation Reactions In aqueous solutions, inorganic arsenic species can undergo acid dissociation according to

$$H_xAs^{(z)}O_y^{(x-2y+z)}(aq) \leftrightarrow H_{x-1}As^{(z)}O_y^{(x-1-2y+z)}(aq) + H^+(aq). \quad (2.1)$$

This reaction is necessarily coupled to a base association reaction, most commonly the reaction below involving the solvent

$$H_2O(aq) + H^+(aq) \leftrightarrow H_3O^+(aq) \quad (2.2)$$

with the concentrations of the species involved being related, under equilibrium conditions, to the thermodynamic equilibrium constant, K_a, for reaction (2.1) by

$$K_a = \left[H^+(aq)\right]\left[H_{x-1}As^{(z)}O_y^{(x-1-2y+z)}(aq)\right] / \left[H_xAs^{(z)}O_y^{(x-2y+z)}(aq)\right], \quad (2.3)$$

where z is the valence state of the arsenic, x the number of hydrogen atoms, and y the number of oxygen atoms in the reactant arsenic species. Hence, the ratio of the two arsenic species is given by

$$\log\left\{\left[H_{x-1}As^{(z)}O_y^{(x-1-2y+z)}(aq)\right] / \left[H_xAs^{(z)}O_y^{(x-2y+z)}(aq)\right]\right\} = pH - pK_a, \quad (2.4)$$

where

$$pH \equiv -\log\left[H^+(aq)\right] \tag{2.5}$$

and

$$pK_a \equiv -\log K_a. \tag{2.6}$$

Accordingly, distinctive pH ranges in natural waters are characterized by the predominance of different variably protonated inorganic arsenic species. At ambient temperatures and pressures, for inorganic As(V) species, $H_3As^VO_4(aq)$ predominates at pH<2.30, $H_2As^VO_4^-(aq)$ in the pH range 2.3–7.0, $HAs^VO_4^{2-}(aq)$ in the pH range 7.0–11.8, and $As^VO_4^{3-}(aq)$ at pH>11.8; for inorganic As(III) species, $H_3As^VO_3(aq)$ predominates at pH<9.2, $H_2As^VO_3^-(aq)$ in the pH range 9.2–14.1, and $HAs^VO_3^{2-}(aq)$ or $As^VO_3^{3-}(aq)$ would only predominate at extremely alkaline pH not normally encountered in natural waters.

The differences in these ranges between inorganic As(V) and As(III) species have profound consequences for their reactivity and, in particular, the extent to which they adsorb onto mineral surfaces at different pHs. Notably, inorganic As(III) species are predominately neutral (i.e., zero-charged) for all pHs up to 9.2, whereas above pH 2.3, inorganic As(V) are negatively charged and consequently rather more likely to sorb on to positively charged mineral surfaces that are typically found in acidic and slightly acidic to neutral waters. The mobility of inorganic arsenic is therefore often strongly dependent upon its redox state.

2.2.2.2 Reduction/Oxidation of Oxyanions Notionally, coupled redox reactions between inorganic As(V) and As(III) can be expressed overall, for example, in the following form:

$$H_3As^{III}O_3(aq) + \tfrac{1}{2}O_2(g) \leftrightarrow H_2As^VO_4^-(aq) + H^+(aq). \tag{2.7}$$

However, in most groundwater systems, indeed even in many surface water systems, O_2 (g) or even O_2 (aq) is not the most abundant or chemically available electron acceptor. Accordingly, the oxidation of inorganic As(III) species is coupled to the reduction of various other electron acceptors, notably and most importantly Fe(III), for example, in the form of FeOOH (s); Mn(IV) in the form of MnO_2 (s); S(VI) in the form of $SO_4^{2-}(aq)$; or N(VII) in the form of $NO_3^-(aq)$. Conversely, the reduction of inorganic As(V) species is coupled to the oxidation of various electron donors, notably and most importantly Fe(II), for example, in the form of Fe^{++} (aq) or $FeCO_3$ (s); Mn(II), for example, in the form of Mn^{++} (aq); S(II) in the form of H_2S (aq) or HS^- (aq) or FeS (s); N(−3) in the form of $NH_4^+(aq)$ at lower pHs or NH_3 (aq) at higher pHs; H(0) as H_2 (aq); and C(−4), for example, in the form of CH_4 (aq); or C(0), for example, in the form of $CH_3COO^-(aq)$ or various forms of organic matter with chemical formula approximating CH_2O (s). Given these controls on the redox chemistry of arsenic, in near-surface soils/sediment/groundwater systems, the speciation of arsenic—in either the solid or the aqueous phase—is often highly correlated with depth-dependent [25, 40, 136, 222] and laterally [218] or temporally [206] variable trends in the relative concentrations of various electron acceptors and electron donors.

Notwithstanding the evident relationship of arsenic redox chemistry with observable redox gradients in natural systems, it is worth highlighting that, particularly at near-ambient temperatures, the kinetics of such systems are often slow, resulting in disequilibrium both between and within different redox couples. For example, despite Cherry et al.'s [48] proposal to use the As(III)/As(V) ratio as a redox indicator in natural waters, As(III) and As(V) are often not in equilibrium with each other nor with other redox couples in such systems [55]. Indeed, more broadly, the lack of equilibrium means that it is often not clear which redox couples are actually controlling measured Eh (or pE) and consequently the use of Eh to model redox speciation needs to be done with particular care or perhaps not at all [127].

At near-ambient temperatures, indeed even at higher temperatures, the sluggish kinetics of many heterogeneous and homogeneous redox reactions, including those involving arsenic species, combined with the large Gibbs free energies of the reactions, affords microbes the opportunity to derive energy from catalyzing these reactions. Indeed, the viability of extant indigenous life in most groundwater systems hinges critically on these reactions. A by-product of these microbial respiratory processes is that they control to a large extent the chemical speciation, solubility, and bioavailability of metals and metalloids in groundwater and soil water systems, either directly or indirectly.

2.2.2.3 Adsorption Sorption/desorption processes involving a variety of mineral and organic surfaces play a critical role in controlling the mobility of arsenic in many surface and groundwaters, particularly in the absence of reduced sulfur [44, 219]. There are several factors that can critically control these processes:

1. The predominately anionic form of arsenic oxyanions results in arsenic sorption on mineral surfaces being stronger under acidic conditions (under which mineral surfaces tend to be positively charged) than under alkaline conditions (under which mineral surfaces tend to be negatively charged).
2. The oxidation state of arsenic. Under acidic conditions, the sorption of anionic As(V) is often stronger than that of neutral As(III), and so microbial or abiotic reduction of As(V) to As(III) may lead to mobilization of arsenic. However, the often-misquoted work of Dixit and Hering [60] highlights that for amorphous iron oxide (HFO) and goethite substrates, at pHs above 7–8, the reverse is the case, with a cross-over in behavior at near-neutral pHs.
3. The nature of the arsenic–mineral surface complex, in particular whether it is an inner-sphere or an outer-sphere complex, or a unidentate or bidentate complex [44].
4. The specific surface area and the nature of the available mineral surfaces [60].
5. Competing anions, notably phosphate [12], have been proposed as being of importance to controlling arsenic mobility, as has passivating polymerized silica [225] and organics. Although carbonate sorption has been proposed to be significant [112], recent experimentation [33] suggests generally otherwise.

2.3 ARSENIC DISTRIBUTION IN ROCKS

Arsenic occurs ubiquitously in most rocks at concentrations in the range of 1–10 μg/g. Notable enrichments may also be found in loess, glacial tills, and peats, but most commonly in acid sulfate soils (up to 50 μg/g), shales (up to 200 μg/g), and rocks in certain mining districts (up to 10,000s μg/g) [219]. Arsenic is readily mobilized by hydrothermal fluids [89] and so is found enriched in a wide variety of ore deposit types, including skarns, granitoids associated Sn–W deposits [193], volcanogenic massive sulfide deposits, and continental epithermal Au–Ag deposits [212]. Indeed, its widespread association with many types of hydrothermal ore deposits means that it is often employed as a pathfinder element in geochemical exploration programs [80].

2.4 ARSENIC DISTRIBUTION IN GROUNDWATERS

Arsenic in groundwaters varies from much less than 1 μg/L up to 10,000 μg/L [219] or even 100,000 μg/L in highly polluted environments. Hazardous concentrations, variably defined by various authors [9, 219] as those greater than 10 μg/L or greater than 50 μg/L, are not uncommon and indeed are very widespread [9, 16, 219]. The major processes controlling arsenic in groundwater include (i) input from geothermal fluids; (ii) desorption of mineral-bound arsenic in oxidizing aquifers; (iii) microbially mediated reductive dissolution of arsenic-bearing iron host phases and of As(V) in reducing aquifers—these will be discussed in greater detail as it is associated with the most extensive human arsenic exposures—and (iv) input from anthropogenic sources (mining, industry, and pesticides). An indication of the global extent of high arsenic hazard groundwaters, based largely upon the recent compilation of Ravenscroft et al. [201] and references therein, is shown in Figure 2.1, while models by Amini et al. [9] predicting high arsenic groundwaters in both reducing and oxidized/alkaline waters are shown in Figure 2.2. Knowledge of the extent of high arsenic groundwaters is increasing year-on-year (e.g., [146]), so distribution maps such as Figure 2.1 are likely to change significantly in the future. In the absence of comprehensive global data coverage, geostatistics-based predictive tools [9, 120, 205] are useful indicators of where such further hazards may exist.

2.4.1 Geothermal Waters

High arsenic concentrations, up to 70,000 μg/L [133] but more typically within the 1,000–10,000 μg/L range, are often found in 100–400°C geothermal waters [241]. There is a strong plate tectonic control on the distribution of such geothermal areas, with the majority associated with recent volcanism located near either (i) destructive plate margins, and more notably in a circum-Pacific band including occurrences in and around the Andes (e.g., El Tatio, Chile), central America (e.g., Miravalles, Costa Rica; Cerro Prieto, Mexico), the western United States (e.g., Salton Sea), Japan (e.g., Otake), the Philippines (e.g., Mindanao), and New Zealand (e.g., Wairakei, Broadlands) or (ii) constructive plate margins/incipient intra-continental rifts (e.g., Hveragerdi, Iceland; Lake Naivasha, Kenya). In addition, notable geothermal waters are associated with intra-plate hot spots (e.g., Yellowstone, USA) [211, 241].

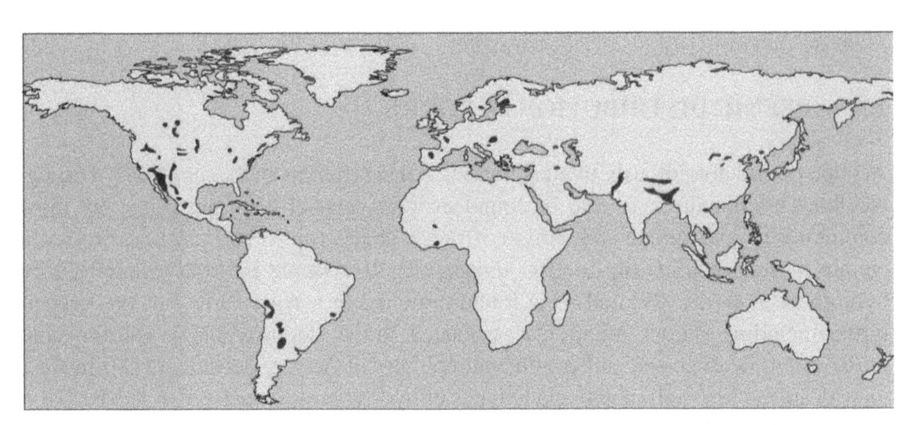

FIGURE 2.1 Some key known global occurrences of high hazard arsenic groundwaters. Plotted locations are indicative and approximate. Note that, not least of all due to widespread mesoscale heterogeneity, the map does not permit reliable prediction of groundwater arsenic in individual wells (From Polya et al. [197]. © Springer-Verlag).

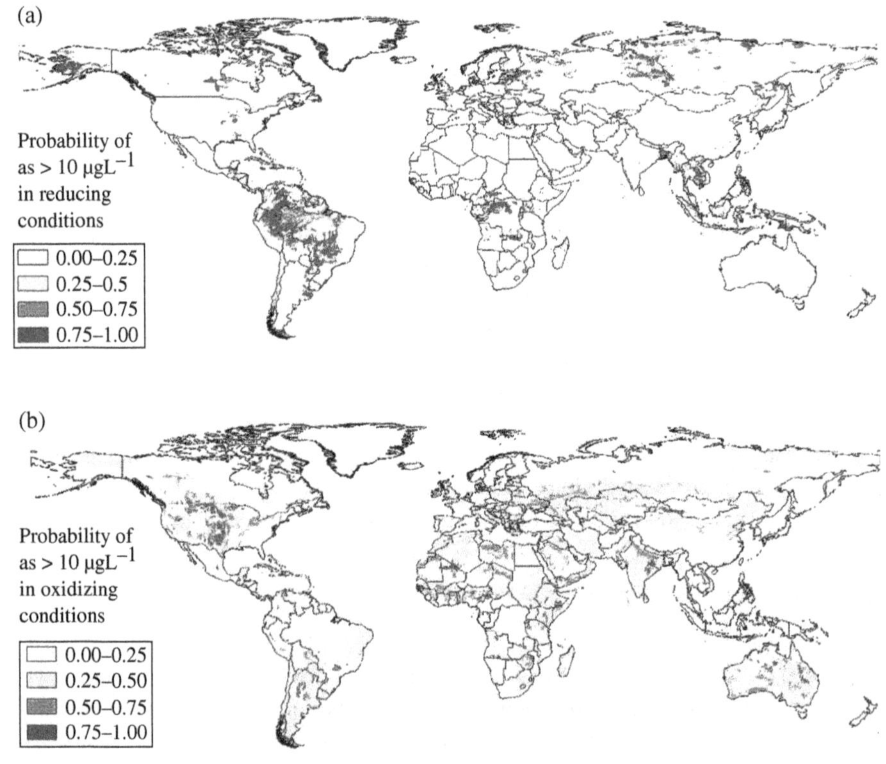

FIGURE 2.2 Modeled probability of geogenic arsenic in groundwater exceeding 10 μg/L in (a) reducing and (b) high pH/oxidizing conditions. These figures give a guide as to the likely widespread extent of the two major types geogenic arsenic in groundwaters, although (i) significant differences exist between modeled high hazard areas and actual high hazards, for example, notably in the eastern United States and (ii) geothermal high geogenic arsenic groundwaters and anthropogenically polluted and mineral deposit impacted groundwaters are not considered in the model (Reprinted with permission from Amini et al. [9]. © American Chemical Society). (*See insert for color representation of the figure.*)

At high temperatures, arsenic is solubilized in H_2S-rich geothermal fluids and is most likely stabilized as thioarsenate complexes [188]. In near-surface geothermal waters, Lord et al. [134] found the highest arsenic concentrations (~9000 μg/L) in alkali-chloride waters, representing the geothermal waters from which acid CO_2 had been exsolved by

$$HCO_3^-(aq) \rightarrow CO_2(g) - +OH^-(aq) \tag{2.8}$$

whereas acid sulfate waters formed by the interaction of exsolved geothermal gases with local groundwaters contained much lower arsenic concentrations, reflecting in part destabilization of thioarsenates and arsenite due to, respectively:

$$H_2S(g) - +2O_2(g) \rightarrow SO_4^{2-}(aq) + H_2O(aq) \tag{2.9}$$

and

$$H_3AsO_3(aq) + O_2(g) \rightarrow H_2AsO_4^-(aq) + H^+(aq). \tag{2.10}$$

In geothermal areas, further impacts on groundwaters and surface are frequently observed [188, 241, 243].

2.4.2 Oxidizing Groundwaters

In nonreducing groundwaters, there are two principal mechanisms by which arsenic is transferred from solid phases to the aqueous phase, viz., (i) oxidation of arsenic-bearing sulfides or sulfosalts, particularly in the ore field or mining environments (see Section 2.4.4.1) and (ii) desorption of arsenic by elevation of pH or competitive sorption processes.

Aquifers impacted by desorbed arsenic under nonreducing conditions are widespread with notable examples in Argentina, the United States, Spain, and China (Datong Basin) among other places [201]. The pH of such waters is often strongly alkaline and arsenic concentrations may reach as high as a 5000 μg/L in association with elevated concentrations of other chemicals, notably F, B, V, and U [218].

2.4.3 Reducing Groundwaters

2.4.3.1 Occurrence High arsenic-reducing groundwaters are typically found in flat-lying deltaic regions of large river systems draining the Himalaya [46] (see also Figs. 2.1 and 2.2). High concentrations of arsenic, up to 5000 μg/L, are almost universally restricted to the uppermost 150 m, with a few isolated occurrences at greater depths. Generally, low groundwater arsenic concentrations are found in relatively oxic Pleistocene aquifers, with the majority of the high arsenic aquifers being shallow and of Holocene age [23, 149, 200, 202]. Arsenic-rich groundwater in these regions are typically highly reducing, occasionally to the extent of methanogenesis, and with a distinct Ca–Mg–HCO_3 dominated chemical composition with relatively high concentrations of dissolved iron and phosphorus [24, 84, 124, 202, 207]. Sulfate concentrations are generally low, typically below 5000 μg/L. However, pockets of higher concentrations associated with elevated concentrations of sodium and chloride do occur, particularly in Cambodia, perhaps indicative of connate seawater trapped in the aquifer matrix. Nitrate concentrations are also generally low, except where locally impacted by pollution [202].

2.4.3.2 Source of Arsenic The arsenic in these groundwaters ultimately derives from the rapid weathering of As-bearing rocks in the upper Himalayan catchments [46]. This arsenic is then transported within or sorbed onto suspended particulates via the various great rivers of this region, and deposited and rapidly buried in the young, low-lying alluvial floodplains of the various river deltas. On deposition, initially under oxic conditions, this arsenic either sorbs to the surface of metal oxides that coat the surface of sedimentary grains, or under reducing conditions can be incorporated into the mineral structure of Fe(II) minerals such as magnetite [53] or Fe and/or As sulfides [114, 137]. In addition to the similar provenance of sediments associated with these arsenic-rich groundwaters, all of these aquifers have low topographically controlled hydraulic gradients that result in extremely slow aquifer flushing rates. Thus, generation of the high arsenic groundwaters in these regions is not a function of elevated concentrations in the host sediments, which typically have a concentration close to that of the world average [23, 29, 62, 117, 219]. Instead, the high arsenic groundwaters are the result of biogeochemical and hydrogeological processes that in turn are dependent upon tectonically and eustatically controlled patterns of sedimentation.

Several mechanisms have been postulated to describe the mobilization of arsenic from aquifer sediments to the groundwaters in these aquifers. The three most commonly invoked mechanisms responsible for arsenic release are (i) the reductive dissolution of Fe(III) phases, (ii) competitive exchange reactions, and (iii) the oxidative dissolution of arsenic-rich pyrite [201]. However, the most widely accepted theory [30, 172, 198] involves reductive dissolution of arsenic-bearing Fe(III)–O–H phases that exist in aquifers as coatings on sedimentary grains and are known to scavenge arsenic from solution [173]. The reductive dissolution of these phases in reducing groundwaters solubilizes Fe and its adsorbed load, including arsenic. The reductive dissolution of iron (oxy)hydroxides can be described by the following equation [150].

$$\begin{aligned}&4\mathrm{FeOOH}(\mathrm{s})+\mathrm{CH_2O}(\text{organic matter})+7\mathrm{H_2CO_3}(\mathrm{aq})\leftrightarrow\\&\qquad 4\mathrm{Fe}^{2+}(\mathrm{aq})+8\mathrm{HCO_3}^-(\mathrm{aq})+6\mathrm{H_2O}(\mathrm{aq}).\end{aligned}\tag{2.11}$$

The microbial oxidation of organic matter drives the biogeochemical processes that are responsible for establishing a redox gradient over the upper few meters of sediment [198]. Following the exhaustion of oxygen, a number of different electron acceptors are utilized until the onset of Fe-reducing conditions where Fe(III) is used as an electron acceptor in microbially mediated reactions. If sufficient organic matter is present to drive the process, sulfate reduction and even methanogenesis can result.

A series of carefully designed microcosm experiments demonstrated that this process of reductive dissolution is mediated by Fe(III)-respiring microbes [101] and is linked to the coupled oxidation of organic matter, which serves as an electron donor in this process. These authors also observed that Fe reduction occurred before arsenic release from sediments. The observed decoupled nature of arsenic and iron release from sediments has been used to suggest that Fe(III) phases may undergo transformation to secondary Fe(II) minerals without necessarily resulting in the release of iron to solution [95]. In fact, numerous studies have found that the reductive dissolution of Fe(III) phases leads to the precipitation of secondary Fe(II)/mixed valence phases such as magnetite [53, 82, 90, 116], vivianite [100, 102], and siderite [79], which are all capable of sequestering arsenic from solution either by incorporation into the mineral structure or through sorption to its surface. Coker et al. [53] argue that the order of reduction of iron and arsenic could be critical to determining whether or not aquifers become arsenic-prone or not—iron reduction followed by arsenic reduction would result in the effective sequestering of As(V) by Fe(II)-bearing secondary phases, whereas arsenic reduction followed by iron reduction would not.

In addition to the numerous secondary Fe phases that can form and sequester arsenic under Fe-reducing conditions, a number of studies have suggested that arsenic may be sequestered from solution under sulphate-reducing conditions [150]. However, while arsenic release has been shown to be rapid [90], sequestration of arsenic from solution is rate limited by thermodynamic and kinetic restrictions on the formation of mineral phases and the rate at which arsenic sorbs to their surface [113, 166]. To complicate matters further, the precipitation of arsenic bearing minerals such as As-sulfide may be inhibited by the presence of reactive Fe(III) or Fe(II) in solution [115].

Many studies suggest that reduction of Fe(III) alone is not sufficient to mobilize arsenic. Indeed, it is likely that multiple biogeochemical processes are occurring simultaneously that act both to release arsenic to solution and to sequester it from solution. Furthermore, it is possible that changes in the geochemical environment could result in the reductive dissolution, competitive exchange, and the oxidative dissolution mechanisms of arsenic release operating at different times at the same site to generate arsenic-rich groundwaters. Swartz et al. [224] suggest that high arsenic groundwater is associated with sediments whose sorption capacity has been exceeded and that the combined effect of the reductive dissolution of Fe(III)–O–H phases in the aquifer and competitive sorption contributes to maintain high As concentrations in the aquifer.

Others suggest that multiple mechanisms may be responsible for the release of arsenic from aquifer sediments [6, 190–192]. For instance, arsenic-rich sulfides in surface sediments are oxidized during periods of oxic conditions, forming Fe(III) oxides that sorb arsenic from solution. Subsequent reducing periods during periods of flooding, for example, drive the reductive dissolution of these secondary Fe(III) phases and releases arsenic to solution. However, because of the partial removal of sulfur via sulfate transport following oxidation, only a portion of the arsenic released can adsorb to the reprecipitated sulfide minerals. The remaining arsenic, present as arsenite because of the reducing conditions, is then easily transported to depth because of the paucity of sorbents in the underlying sediment. Such oscillating redox conditions are expected to be common, particularly in areas such as West Bengal and Bangladesh where seasonal monsoon flooding and subsequent extensive groundwater irrigation raise and lower the water table on a seasonal basis. Berg et al. [24] showed that the substantial groundwater abstraction for the water supply of Hanoi in Vietnam has led to dramatic lowering of groundwater levels, resulting in changing redox conditions. They suggest that this may lead to the partial oxidation of peat and the downward migration of natural organic matter-rich leachates that may be responsible for maintaining anoxic conditions in the aquifer below these peat layers, enhancing arsenic release through the reductive dissolution mechanism described earlier. They also note that the invading oxic conditions may release arsenic from sulfides and suggest that arsenic release may occur from both reductive and oxidative mechanisms.

Early studies focusing on the role of microorganisms in arsenic release reported that arsenic mobility could be enhanced by dissimilatory iron-reducing bacteria such as *Shewanella alga BrY* without the reduction of arsenate to arsenite (Cummings et al. [57]). Further studies showed that the iron-reducing bacteria, *Geobacter sulfurreducens* and *Geothrix fermentans*, were not able to directly reduce As(V) enzymatically or indirectly through Fe(II) in solution following reduction of Fe(III) [100, 102]. However, the reduction of Fe(III) alone may not be sufficient to result in arsenic release. Zobrist et al. [254] demonstrated the ability of the dissimilatory arsenate-respiring prokaryote *Sulfurospirillum barnesii* to reduce both Fe(III) and As(V), with reduction of As(V) to As(III) reported both in solution and when adsorbed to ferrihydrite. Oremland and Stolz [184] suggested that the ability of such dissimilatory metal-reducing microbes to reduce both iron and arsenic implicates such bacteria in the generation of the arsenic-contaminated aquifers of South and Southeast Asia.

Islam et al. [101] suggest that arsenic reduction may only occur after the majority of available Fe has been consumed. Indeed, at the time of writing this review, no obligate As(V)-reducing bacteria have been identified in any sediment taken from an arsenic-contaminated aquifer system, with all of the organisms capable of reducing arsenic also able to respire using alternative electron acceptors, including iron [184].

The catalytic role of microorganisms discussed earlier clearly highlights the importance of organic matter in driving arsenic mobilization in these sediments. However, organic matter is also implicated in As release through other processes. However, despite the obviously significant role played by organic matter, including in competitive sorption and in providing an electron shuttle between microorganisms and iron oxide surfaces [135, 159, 208], the source(s) of organic matter driving the biogeochemical processes associated with arsenic release remains poorly constrained and is currently a topic of intense debate. Indeed, this has been identified as one of the key limitations in developing models that are able to accurately predict the spatial distribution of As in these groundwaters [69].

Numerous potential sources of organic matter have been postulated to be driving As release in South and Southeast Asia. These can be broadly separated into subsurface/sedimentary sources and surface-derived sources. Subsurface sources of organic matter include the peat layers enriched in organic matter [149], and the migration of petroleum organics from depth [209] or petroleum organics reworked by sedimentary processes [232]. Conversely, organic matter could derive from the entrainment of dissolved organic matter with recharge from surface water [84]. It is perhaps more likely that there is not one single source responsible for the release of arsenic to groundwaters in reducing groundwaters, but that the natural environment and subsequent impacts of anthropogenic disturbance (through groundwater pumping and excavation of clay layers to introduce new sites of aquifer recharge) control the relative abundance of these different sources in any one location. Indeed, a recent study [125] demonstrates that both older (<10 kyrs) sedimentary and recent (<1000 years) surface-derived sources of organic matter are present in the groundwaters of arsenic contaminated study sites in West Bengal and Cambodia. In addition, Mailloux et al. [141] demonstrate, through the determination of [^{14}C] ages of DNA in groundwater, that the indigenous microbial community of an arsenic-contaminated aquifer system in Bangladesh consumed relatively young organic carbon, which was shown to be up to several thousands of years younger than the sedimentary organic matter. These studies suggest that any processes that change the distribution, abundance, and bioavailability of these different sources of organic matter may ultimately control the local redox environment, and hence the spatial distribution of arsenic in these shallow groundwaters.

2.4.4 Anthropogenic Inputs

Across the globe, both groundwaters and surface waters are contaminated in many places on local to regional scales either directly from anthropogenic inputs or indirectly through anthropogenic processes modifying environmental conditions to ones in which arsenic becomes mobile.

2.4.4.1 Direct Anthropogenic Inputs The key anthropogenic inputs are from (i) mining and metallurgical processes [18]; (ii) use and subsequent disposal of arsenical compounds as wood preservatives [31]; and (iii) application of arsenical pesticides [174]. Other arsenic pollution sources that may be locally important include chemical industries (e.g., in the Aberjona Watershed, [14]) and excretion of the growth-promotion agent roxarsone from poultry [210] or swine [142].

While background concentrations of arsenic in surface waters, including river, lake, and sea water, unimpacted by geothermal or anthropogenic inputs are typically below 1 μg/L, concentrations in anthropogenically impacted surface waters are highly variable and may reach 10 or even 1000 μg/L. A detailed discussion of the controls on arsenic mobility in such surface waters is beyond the scope of this chapter, but it is worth noting that under some circumstances, arsenic concentrations in such waters may exhibit strong cyclic variations, for example, related to seasonal overturn of seasonally stratified lakes [3, 14, 119], seasonal [93, 214], or diurnal [75] variations in the activities of microbes and other organisms [184], which either directly or indirectly impact arsenic cycling or seasonal variations in environmental parameters such as temperature [147], water saturation/redox [185, 206], or rainfall [208].

Mining and Metallurgical Processes The major commercial arsenic product from mining is arsenic trioxide, of which global production in 2011 was just under 46,000 tons, with over 90% of this originating in mines in China, Chile, Morocco, and Peru [229]. Contamination of ground and surface waters around such operations is a key environmental risk. In addition, many legacy sites, including Yellowknife (NWT, Canada) in which 237,000 tons of arsenic trioxide is currently stored [49], are also associated with elevated concentrations of arsenic in both mine waters and local groundwaters.

Arsenic is widespread in many types of ore deposit as an unwelcome penalty element. As such, it has tended to be disposed of in tailings dams along with other mined materials perceived to be too expensive or too difficult to treat economically. There is a long history of failures of such tailings dams, with a long-term integrated historical failure rate of around 1% [17], although failure rates in more recently constructed tailings dams are closer to around 0.1% [228]. High-profile failures that have caused widespread contamination of surface waters and to a lesser extent shallow groundwater systems include Chenzhou, P. R. C., Hunan, in 1985 [128] and Aznalcóllar, Spain, in 1998 [78, 96], while more recent tailings dam failures such as at Kolontár, Hungary, in 2010 [96, 103] confirm the ongoing risks from such construction.

Arsenical Pesticides and Herbicides These anthropogenic sources of arsenic are briefly outlined in Section 2.5.3 and particularly constitute a potential threat to shallow near-surface water-table aquifers and aquifers adjacent to incompletely lined landfills where arsenicals and wood treated with chromated copper arsenate (CCA) have been disposed.

2.4.4.2 Indirect Anthropogenic Controls In addition to direct anthropogenic inputs, arsenic in some groundwaters and surface waters becomes elevated indirectly as a result of human activity. Examples of this include (i) pollution of groundwaters with highly reducing organic wastes, which promote the mobilization of Fe(III)–O–H phase-incorporated arsenic by microbially mediated reductive dissolution processes; (ii) application of manure, silage, or other highly reducing amendments, typically aimed at immobilizing transition metal contaminants; and (iii) modification of the permeability structure of the ground, for example, by drilling uncased boreholes, permitting new groundwater flow pathways.

There is currently considerable debate [195] over whether or not groundwater management practices in already very seriously impacted regions, such as Bengal and similar settings, are likely to lead to exacerbation of groundwater arsenic mobilization in relatively shallow (<60 m depth) aquifers [84, 151, 168–170, 233] and also in deeper currently relatively arsenic-free aquifers [37, 64, 199, 246]. Irrespective of the processes occurring, it is difficult to argue that the shallow subsurface groundwater flow regime has not been influenced by groundwater pumping practices [157, 165]. Given that many deep tube wells tap relatively arsenic-free groundwater, many have touted the use of this deep reservoir as a sustainable source of safe potable water [5, 233, 234, 253]. However, recent studies suggest that the deep groundwater exhibits a similar vulnerability as the shallow groundwater to arsenic contamination, with irrigation abstraction of deep groundwater potentially resulting in the draw-down of arsenic-contaminated shallow groundwater [37, 44, 124, 156]. Indeed, in the Hanoi area of Vietnam, excessive abstraction of groundwater from the deep Pleistocene aquifer for public water supply has resulted in the vertical seepage of reduced groundwater rich in dissolved organic carbon, Fe, and ammonia into the Pleistocene aquifer [24], with similar observations reported in nearby Nam Du [177]. Deeper groundwater, it is suggested, may only represent a sustainable safe source of arsenic-free drinking water if wells are drilled in oxidized sediments, and its utilization is limited to domestic supply, with irrigation abstraction restricted to shallow groundwaters [37, 156]. Increases in the deep groundwater arsenic hazard may therefore only become apparent once the finite sorption capacity of deep sediments has been exceeded or once reducing conditions have become established to support arsenic release [235].

2.5 ARSENIC IN SOILS

Arsenic concentrations in soils are, on a global scale, largely controlled by climate-dependent weathering processes and the composition and mineralogy of the underlying rock. Locally, however, concentrations may deviate from geogenic concentrations by an order of magnitude as a result of contamination from anthropogenic processes, including applications of herbicides, mining, smelting, and other industrial activities.

2.5.1 Geogenic Rock and Weathering Control

Arsenic contents of soils are broadly similar to and largely reflect those of parent source rocks except where there are significant anthropogenic inputs. A recent review of arsenic distribution in European grazing and agricultural soils [226] summarizes results of the GEMAS dataset. Previously, the FOREGS project provided coverage of European soils for selected metals and metalloids, including arsenic—these values were interpolated using geostatistical methods as outlined [205].

2.5.2 Anthropogenic Inputs: Mining, Smelting, and Industrial

Contamination by mining beneficiation/smelting and other industrial activities can give rise to arsenic concentrations in soils that may exceed Clarke values by as much as a factor of 1000 [219, 239]. For example, arsenic concentrations as high as 50,000 mg/kg have been noted in mine-contaminated soils in South West England [109]. Such contamination tends be localized on a 100-m to 10-km scale, with the broad geographic distribution controlled largely by the distribution of ore fields [212] and for smelting/industrial operations by other geo-economic factors. It is noteworthy that efforts to remediate mining have sometimes inadvertently led to the mobilization of arsenic. Moreno-Jiménez et al. [163] and previously Hartley et al. [83]

warned of the likelihood of such mobilization when highly reducing organic composts, for example, are used in such remediation efforts.

2.5.3 Anthropogenic Inputs: Agrochemicals

Arsenic has been widely and extensively used a component of pesticides and herbicides. Compounds used are follows: the As(III) compounds—arsenic trioxide, calcium arsenite, copper arsenite, copper acetoarsenite; the $As^{(V)}$ compounds—arsenic pentoxide, calcium arsenate, lead arsenate, zinc arsenate; and the organoarsenicals—cacodylic acid (DMA), $MMA^{(III)}$, $MMA^{(V)}$, and various ammonium and calcium-substituted $MMA^{(V)}$ [178]. The usage of many of these is now banned, but contaminated soils arising from their legacy usage to combat prickly pear or plant pests in cotton fields, in apple orchards and on other crops [174] is widespread. More recent legacy usages have included that of chromated copper arsenate (CCA) as a wood preservative, which has been the largest use of arsenic in the United States in the latter part of the twentieth century with usage peaking at around 27,000 tons of CCA-As and 12,000,000 m^3 of treated wood per annum [31].

Concentrations of arsenic in contaminated agricultural soils in Louisiana, for example, were noted to be up to 10 times the norm for nonagricultural soils [245]. Soil arsenic can be nearly 200 mg/kg in lead arsenate-treated orchard soils and over 500 mg/kg in soils contaminated by CCA [51]. Arsenic soil concentrations of as high as 2000 mg/kg were reported in contaminated cattle dips in Australia [174].

2.5.4 Anthropogenic Inputs: Irrigation with High Arsenic Groundwaters

Massive scale irrigation of crops, particularly rice in paddy fields with high arsenic groundwaters, has led to concerns that this would lead to increases in soil arsenic in the impacted areas [153]. Until recently, these predicted cyclical increases have been difficult to discern because of heterogeneity of soil arsenic concentrations, heterogeneity of distribution of groundwater-derived added arsenic, and high seasonal variations in soil arsenic content, with a tendency to fluctuate between relatively high values immediately after premonsoon/dry-season irrigation and relatively low values immediately after flushing by monsoonal rains. However, year-after-year increases of soil arsenic of the order of a percent of so per year over a 3-year period suggest that such increases may be very significant over a period of decades [59, 203, 204].

2.5.5 Implications for Food Chain

Arsenic contamination of soils may lead, depending upon its chemical species-dependent and plant species-dependent bioavailability [109], to the accumulation of arsenic in plants [153] and subsequently in organisms, including humans, at higher trophic levels [65, 66]. Rice is particularly susceptible to accumulating arsenic in the edible grain because of the similarities of biogeochemical behavior of arsenic and silicon, which typically accumulates in rice at concentrations of as much as 20% [181] but arsenic may also accumulate at levels of concerns in other crops used for human consumption [13].

2.6 ARSENIC IN THE ATMOSPHERE

Arsenic concentrations in the atmosphere, as combined particulates and volatiles, vary approximately between 0.01 and 15 ng/m^3 [148] with a large contrast in concentrations between uncontaminated remote, oceanic (0.01–0.2 ng/m^3), and rural areas (0.01–1 ng/m^3) and contaminated urban areas (up to 15 ng/m^3) [148]. Mean atmospheric arsenic concentrations in the northern hemisphere are approximately five times those in the southern hemisphere, reflecting the greater concentration of both anthropogenic and geogenic surfaces in the northern hemisphere [148].

Anthropogenic fluxes of arsenic to the atmosphere are dominated by emissions from copper, lead, and zinc smelters, coal combustion, and herbicide use, with lesser contributions arising from deforestation and burning of grasslands [148]. Geogenic contributions arise largely from (high-temperature) volcanic activity and low-temperature release from soils. The relative importance of anthropogenic and geogenic emissions on a global scale is poorly known, particularly because of uncertainties in fluxes of arsenic volatilized from soils and plants. Maschullat [148] estimated anthropogenic fluxes to be at least 50% of total arsenic fluxes to the atmosphere; however, evidence of paleo-depositional profiles in peat bogs [216] suggests that anthropogenic fluxes are likely to have exceeded geogenic fluxes over the last 2000 years and this would be consistent with recent biovolatilization data [155] that indicate geogenic arsenic fluxes from soils may have been previously overestimated by a factor of around 20–60.

2.7 EXPOSURE ROUTES

Ultimate human exposure routes to environmental arsenic may be broadly classified as being through ingestion, inhalation, and dermal contact. Ingestion is typically primarily through diet—either drinking or eating—or may be through other behaviors, for example, ingestion of water during swimming or showering, or geophagia, which may result in exposures to soil arsenic, particularly in young children. Inhalation may involve exposure to dust particles, mists, or vapor. We exclude from consideration here other exposure routes, notably intravenous injection for medical or other purposes.

For each exposure route, the effective dose of arsenic received is dependent not only on the concentration of arsenic in the relevant medium and the relevant volume/mass/area of the medium, but also on the bioavailability, which is a function of the chemical and/or physical form of the arsenic in the relevant medium. For ingestion routes, the bioaccessibility refers to "the fraction of a soil [or other ingested material] contaminant that is soluble in the gastro-intestinal tract and is available for uptake into the circulatory system" [43]. Thus, bioaccessibility of inorganic arsenic in water is most commonly around 100%, that in rice in the order of 90% [108], while that of arsenic in soils may be considerably lower than these values, depending upon the chemical form of the arsenic. For example, the recently reported case [36] of a man who had ingested 84,000,000 μg of arsenic without any serious symptoms, other than nausea, highlights the importance of chemical form and bioaccessibility. In this case, the arsenic ingested was in the form of the highly insoluble sulfide mineral, orpiment, and (the authors speculate) was relatively coarse-grained, thereby ensuring a relatively slow rate of dissolution while in the gastrointestinal system. There are a wide range of methodologies being developed for determining bioaccessibilities for toxic contaminants, including arsenic, and standardization and interpretation of these remain an ongoing challenge as this chapter is written [118] in part because of the impact that gastrointestinal flora [121] and dietary factors [8] have on bioaccessibility.

2.7.1 Exposure from Ingesting "Water"

For the majority of those most seriously impacted by exposure to arsenic, water is the most important exposure route. Notably, in Bengal, typical water consumption rates of 3–6 L/day [160] combined with widespread arsenic concentration in water in the range of 50–500 μg/L give rise to exposures in the order of 150–3000 μg/day, with much of this range in excess of the 2.2 μg/kg-bw/day previously recommended as a provisional tolerable daily intake by FAO/WHO until they withdrew that value [104]. It is noteworthy, however, that even drinking 2 L/day of water at the WHO provisional guide value of 10 μg/L would result in daily arsenic exposures of 20 μg/day, an exposure level at which significant adverse cancer health outcomes have been modeled [180].

2.7.2 Exposure from Ingesting Food

Schoof et al. [213] were among the first researchers to conclude that dietary intake of arsenic, particularly through yams and rice, was an important exposure route, possibly contributing to "adverse health effects." Since then, many workers have highlighted the importance of rice as an exposure route [19, 21, 153, 161].

The relative importance of drinking water and rice as exposure routes obviously depends upon the relative concentrations of arsenic in drinking water and rice, and the relative consumption rates of drinking water and rice in different areas. For example, in rural West Bengal among populations consuming rice as a staple and with relatively uniform water and rice consumption rates, in areas with high (>100 μg/L) arsenic in groundwater-derived drinking water, rice was only a minor component of overall arsenic exposure, whereas in areas with low (<10 μg/L) groundwater arsenic, rice was the dominant route of dietary exposure [160]. However, notably, even in areas with intermediate (10–100 μg/L) groundwater-derived drinking water arsenic concentrations, rice was just as important an exposure route as water [161] consistent with earlier conclusions in studies in nearby Bangladesh [110, 111].

In countries, such as the United States and the United Kingdom, where rice is not a staple dietary item for the majority of the population, exposure to arsenic through consumption of food, notably rice, at relatively low levels is still clearly detectable through studies of urinary arsenic [42, 58] but nevertheless relatively low (<0.19 μg inorganic-As/kg-bw/day) for the vast majority (90%) of the population [249].

Of course, where there are particularly high exposures to a subpopulation, it would not be prudent to use a population mean value to base action to universally protect public health. In this regard, it is notable that Xue et al. [249] observed elevated exposures (>2 μg/kg-bw/day) from food (notably rice) for several percent of the US population and particularly in children younger than 5 years old.

Cooking methods and the arsenic content of cooking waters may also make a significant positive or negative impact on arsenic exposure. Approximately 10% of the arsenic exposure in a West Bengal-based population arose from cooking [161]. In addition, arsenic uptake from cooked rice depends upon the cooking method. For example, in parboiled rice arsenic was more bioaccessible than in nonparboiled rice, and arsenic bioaccessibility is arsenic species dependent [217].

The proportion of arsenic in rice occurring as inorganic arsenic (iAs), which is widely presumed to be a more toxic form than the remaining methylated (organic) species, varies widely between 20 and 95%, typically with lower proportions being found in high arsenic rice [154]. It has been suggested that methylation of arsenic is not a capability of higher plants such as rice and instead that methylated arsenic species are produced by the actions of soil microbes prior to being taken up by the plant [132, 252].

2.7.3 Exposure from Ingesting Soil

Ingestion of soil may take place through deliberate practices of geophagia, such as described in Bangladesh [7], or by accidental or exploratory ingestion by very young (>2 years) children, who may ingest as much as 200 mg of soil per day [129]. There are considerable uncertainties in estimates of such ingestion rates [1] as well as in effective bioaccessibilities [8, 118].

2.7.4 Exposure from Inhaling Dusts

Dusts are generally not to be considered to be an important exposure route, but there are notable exceptions, which are typically in rather specific occupational and environmental settings—for example:

1. Traffic pollution: association between mortality from lung cancer and elevated (up to 130 ng/m^3) concentrations of atmospheric arsenic, arising from traffic pollution in North West England and North Wales [221].
2. Near-mine waste dumps: elevated (900–26,000 μg/kg; $n = 8$) arsenic in the toe nails of human volunteers living near a former arsenic mine in Devon in the South West United Kingdom [39]; toe nail arsenic concentrations of 500 ± 90 μg/kg in environmentally exposed and 750 ± 350 μg/kg occupational-exposed volunteers compared to 220 ± 60 μg/kg in a control group from the near Minas da Panasqueira, a mine from which arsenic, largely in the form of arsenopyrite, is mined and discarded in tailings dams as a largely unwanted by-product of tin, tungsten, and silver abstraction [52]; arsenic is also commonly found in rocks up to 5 km away from the mine at concentrations of 50–200 mg/kg [193, 194].

3. Copper smelters: adverse health effects in workers exposed occupationally to greater than 500,000 ng/m^3 [15].
4. Indoor pollution: from burning of arsenic-bearing coals [250], although airborne coal-derived silica combined with volatile organic matter may be a major confounder in such studies [123].

2.8 CONCLUSIONS

Anthropogenic sources of arsenic are not uncommon, but are generally restricted to areas associated with current or former mining, smelting, or other industrial activity or to application of arsenic-bearing pesticides and herbicides.

Arsenic is very widespread in the environment but knowledge of this arguably is not. The conditions under which arsenic in soils and sediments is transferred into groundwater are common, particularly in highly populated flat-lying deltaic regions of major rivers draining the Himalayas but also in many other environments around the globe. Despite the large amount of work already carried out on delineating groundwater arsenic hazard, the next 10 years are likely to reveal an increasing number of occurrences of such arsenic hazard, with the concomitant risks to human health in areas where humans are exposed through its use for drinking, cooking, and irrigation.

Similarly, there is an increasing recognition that rice represents a significant exposure route for humans and one for which regulatory protection in most countries (China is a notable exception) lags behind that for drinking water [21, 67, 105, 152, 161, 195].

2.9 ACKNOWLEDGMENTS

The experience upon which this chapter is based includes field studies across Europe, the Indian subcontinent, Southeast Asia, the Far East, and Australasia, and has been funded over the past 10 years by grants from the European Commission (AquaTRAIN, EU ASIA-LINK CALIBRE), EPSRC, NERC, STFC, the former CCLRC, the British Council (UKIERI PRAMA), and the Royal Society, and enriched by discussions with many, including Chris Ballentine, Michael Berg, Ben Bostick, Steve Boult, Adrian Boyce, Charlotte Bryant, Willy Burgess, Andre Burnol, Benjamin Cancès, Mark Cave, Dipankar Chakraborti, Laurent Charlet, John Charnock, Raoul-Marie Couture, Jörg Feldmann, Scott Fendorf, Andrew Gault, Ashok Giri, Huaming Guo, Marina Héry, Karen Hudson-Edwards, Stephan Hug, Farhana Islam, Dave Kinniburgh, Jon Lloyd, Steve McGrath, Andy Meharg, Romain Millot, Debapriya Mondal, Luca Montanarella, Philippe Negrel, Kirk Nordstrom, Enoma Omoregie, Matt Polizzotto, Peter Ravenscroft, Keith Richards, Laura Richards, Helen Rowland, Joanne Santini, Tom Sato, Pauline Smedley, Allan Smith, Chansopheaktra Sovann, Chris States, Marie Vahter, Phillipe van Cappellen, Bart van Dongen, Lex van Geen, David Vaughan, Michael Watts, Jenny Webster, Bernhard Wehrli, Roy Wogelius, and Yong-Guan Zhu. The views expressed in this chapter do not necessarily reflect those of any of the funders or those individuals whose advice we acknowledge here.

REFERENCES

[1] P.W. Abrahams, Involuntary soil ingestion and geophagia: A source and sink of mineral nutrients and potentially harmful elements to consumers of earth materials. Applied Geochemistry 27 (2012) 954–968.

[2] F.C. Adams, Elemental speciation: Where do we come from? Where do we go? Journal of Analytical Atomic Spectrometry 19 (2004) 1090–1097.

[3] J. Aggett, G.A. Obrien, Detailed model for the mobility of arsenic in lacustrine sediments based on measurements in Lake Ohakuri. Environmental Science and Technology 19 (1985) 231–238.

[4] D. Ahmann, A.L. Roberts, L.R. Krumholz, F.M.M. Morel, Microbe grows by reducing arsenic. Nature 371 (1994) 749–750.

[5] M.F. Ahmed, S. Ahuja, M. Alauddin, S.J. Hug, J.R. Lloyd, A. Pfaff, T. Pichler, C. Saltikov, M. Stute, A. Van Geen, Epidemiology: Ensuring safe drinking water in Bangladesh. Science 314 (2006) 1687–1688.

[6] J. Akai, K. Izumi, H. Fukuhara, H. Masuda, S. Nakano, T. Yoshimura, H. Ohfuji, H. Md Anawar, K. Akai, Mineralogical and geomicrobiological investigations on groundwater arsenic enrichment in Bangladesh. Applied Geochemistry 19 (2004) 215–230.

[7] S.W. Al-Rmalli, R.O. Jenkins, M.J. Watts, P.I. Haris, Risk of human exposure to arsenic and other toxic elements from geophagy: Trace element analysis of baked clay using inductively coupled plasma mass spectrometry. Environmental Health 9 (2010) 79.

[8] P. Alava, G. du Laing, M. Odhiambo, A. Verliefde, F. Tack, T.R. Van de Wiele, Arsenic bioaccessibility upon gastrointestinal digestion is highly determined by its speciation and lipid-bile salt interactions. Journal of Environmental Science and Health, Part A: Toxic/Hazardous Substances and Environmental Engineering 48 (2013) 656–665.

[9] M. Amini, K.M. Abbaspour, M. Berg, L. Winkel, S.F. Hug, E. Hoehn, H. Yang, A. Johnson, Statistical modeling of global geogenic arsenic contamination in groundwater. Environmental Science and Technology 42 (2008) 3669–3675.

[10] L.C.D. Anderson, K.W. Bruland, Biogeochemistry of arsenic in natural-waters: The importance of methylated species. Environmental Science and Technology 25 (1991) 420–427.

[11] C.A.J. Appelo, D. Postma, Geochemistry Groundwater and Pollution, Balkema, New York, 1993.

[12] C.A.J. Appelo, M.J.J. Van der Weiden, C. Tournassat, L. Charlet, Surface complexation of ferrous iron and carbonate on ferrihydrite and the mobilization of arsenic. Environmental Science and Technology 36 (2002) 3096–3103.

[13] N.R. Atkinson, S.D. Young, A.M. Tye, N. Breward, E.H. Bailey, Does returning sites of historic peri-urban waste disposal to vegetable production pose a risk to human health? A case study near Manchester, UK. Soil Use and Management 28 (2012) 559–570.

[14] A.C. Aurillo, R.P. Mason, H.F. Hemond, Speciation and fate of arsenic in 3 lakes of the Aberjona watershed. Environmental Science and Technology 28 (1994) 577–585.

[15] O. Axelson, E. Dahlgren, C.-D. Jansson, S.O. Rehnlund, Arsenic exposure and mortality: A case-referent (sic) study from a Swedish copper smelter. British Journal of Industrial Medicine 35 (1978) 8–15.

[16] J.D. Ayotte, D.L. Montgomery, S.M. Flanagan, K.W. Robinson, Arsenic in groundwater in eastern New England: Occurrence, controls, and human health implications. Environmental Science and Technology 37 (2003) 2075–2083.

[17] S. Azam, Q. Li, Tailings dam failures: a review of the last one hundred years. Geotechnical News 28 (2010) 50–53.

[18] J.M. Azcue, J.O. Nriagu, Impact of abandoned mine tailings on the arsenic concentrations in Moira Lake, Ontario. Journal of Geochemical Exploration 52 (1995) 81–89.

[19] M. Bae, C. Watanabe, T. Inaoka, M. Sekiyama, N. Sudo, M.H. Bokul, R. Ohtsuka, Arsenic in cooked rice in Bangladesh. Lancet 360 (2002) 1839–1840.

[20] X. Bai, Q. Zhang, J. Yang, H. Ning, Probing the electronic structures and properties of neutral and charged monomethylated arsenic species ($CH_3As_n^{(-1,0,+1)}$, n = 1 – 7) using Gaussian-3 theory. Journal of Physical Chemistry 116 (2012) 9382–9390.

[21] M. Banerjee, N. Banerjee, P. Bhattacharjee, D. Mondal, P.R. Lythgoe, M. Martinez, J.X. Pan, D.A. Polya, A.K. Giri, High arsenic in rice is associated with elevated genotoxic effects in humans. Scientific Reports 3 (2013) 2195.

[22] A.J. Bednar, J.R. Garbarino, J.F. Ranville, T.R. Wildeman, Preserving the distribution of inorganic arsenic species in groundwater and acid mine drainage samples. Environmental Science and Technology 36 (2002) 2213–2218.

[23] M. Berg, H.C. Tran, T.C. Nguyen, H.V. Pham, R. Schertenleib, W. Giger, Arsenic contamination of groundwater and drinking water in Vietnam: A human health threat. Environmental Science and Technology 35 (2001) 2621–2626.

[24] M. Berg, P.T.K. Trang, C. Stengel, J. Buschmann, P.H. Viet, N. Van Dan, W. Giger, D. Stüben, Hydrological and sedimentary controls leading to arsenic contamination of groundwater in the Hanoi area, Vietnam: The impact of iron-arsenic ratios, peat, river bank deposits, and excessive groundwater abstraction. Chemical Geology 249 (2008) 91–112.

[25] R.A. Berner, A new geochemical classification of sedimentary environments. Journal of Sedimentary Petrology 51 (1981) 359–365.

[26] B.A. Bessinger, D. Vlassopoulos, S. Serrano, P.A. O'Day, Reactive transport modeling of subaqueous sediment caps and implications for the long-term of arsenic, mercury and methylmercury. Aquatic Geochemistry 18 (2012) 297–326.

[27] C.M. Bethke, Geochemical Reaction Modelling, Oxford University Press, New York and Oxford, 1996.

[28] C.M. Bethke, T.M. Johnson, Groundwater age and groundwater age dating. Annual Review of Earth and Planetary Sciences 36 (2008) 121–152.

[29] BGS, DPHE. Arsenic contamination of groundwater in Bangladesh. in: D.G. Kinniburgh and P.L. Smedley (Eds.), British geological survey report. WC/00/19. 2001. British Geological Survey, Keyworth, UK.

[30] P. Bhattacharya, D. Chatterjee, G. Jacks, Occurrence of arsenic contaminated groundwater in alluvial aquifers from delta plains, Eastern India: Options for safe drinking water supply. Water Resources Development 13 (1997) 79–92.

[31] D.I. Bleiwas. Arsenic and Old Waste. 2000. USGS. February 22, 2000.

[32] J.F.W. Bowles, D.J. Vaughan, R.A. Howie, J. Zussman, Rock-Forming Minerals Volume 5A: Non-Silicates: Oxides, Hydroxides and Sulphides, 2nd Edition, Geological Society, London, 2012.

[33] Y. Brechbühl, I. Christl, E.J. Elzinga, R. Kretzschmar, Competitive sorption of carbonate and arsenic to hematite: Combined ATR-FTIR and batch experiments. Journal of Colloid and Interface Science 377 (2012) 313–321.

[34] G.E. Brown, G. Calas, Environmental mineralogy: Understanding element behavior in ecosystems. Comptes Rendus Geoscience 343 (2011) 90–112.

[35] G.E. Brown Jr., J.G. Catalano, A.S. Templeton, T.P. Trainor, F. Farges, B.C. Bostick, T. Kendelewicz, C.S. Doyle, A.M. Spormann, K. Revill, G. Morin, F. Juillot, G. Calas, Environmental interfaces, heavy metals, microbes, and plants: Applications of XAFS spectroscopy and related synchrotron radiation methods to environmental science. Physica Scripta T115 (2005) 80.

[36] J.A. Buchanan, A. Everhardt, Z.D. Tebb, K. Heard, R.F. Wendlandt, M.J. Kosnett, Massive human ingestion of orpiment (arsenic trisulfide). The Journal of Emergency Medicine 44 (2013) 367–372.

[37] W.G. Burgess, M.A. Hoque, H.A. Michael, C.I. Voss, G.N. Breit, K.M. Ahmed, Vulnerability of deep groundwater in the Bengal Aquifer System to contamination by arsenic. Nature Geoscience 3 (2010) 83–87.

[38] J. Buschmann, A. Kappeler, U. Lindauer, D. Kistler, M. Berg, L. Sigg, Arsenite and arsenate binding to dissolved humic acids: Influence of pH, type of humic acid, and aluminum. Environmental Science and Technology 40 (2006) 6015–6020.

[39] M. Button, G.R.T. Jenkin, C.F. Harrington, M.J. Watts, Human toenails as biomarkers of exposure to elevated environmental arsenic. Journal of Environmental Monitoring 11 (2009) 610–617.

[40] B. Cancès, F. Juillot, G. Morin, V. Laperche, D. Polya, D.J. Vaughan, J.L. Hazemann, O. Proux, G.E. Brown Jr., G. Calas, Change in arsenic speciation through a contaminated soil profile: An XAS-based study. Science of the Total Environment 397 (2008) 178–189.

[41] I. Cartwright, Using groundwater geochemistry and environmental isotopes to assess the correction of ^{14}C ages in a silicate-dominated aquifer system. Journal of Hydrology 382 (2010) 174–187.

[42] C. Cascio, A. Raab, R.O. Jenkins, J. Feldmann, A.A. Mehard, P.I. Haris, The impact of rice based diet on urinary arsenic. Journal of Environmental Monitoring 13 (2011) 257–265.

[43] M.R. Cave, J. Wragg, H. Harrison, Measurement modelling and mapping of arsenic bio-accessibility in Northampton, United Kingdom. Journal of Environmental Science and Health, Part A: Toxic/Hazardous Substances and Environmental Engineering 48 (2013) 629–640.

[44] L. Charlet, S. Chakraborty, C.A.J. Appelo, G. Roman-Ross, B. Nath, A.A. Ansari, M. Lanson, D. Chatterjee, S.B. Mallik, Chemodynamics of an arsenic "hotspot" in a West Bengal aquifer: A field and reactive transport modeling study. Applied Geochemistry 22 (2007) 1273–1292.

[45] L. Charlet, G. Morin, J. Rose, Y.H. Wang, M. Auffan, A. Burnol, A. Fernandez-Martinez, Reactivity at (nano)particle-water interfaces, redox processes, and arsenic transport in the environment. Comptes Rendus Geoscience 343 (2011) 123–139.

[46] L. Charlet, D.A. Polya, Arsenic in shallow, reducing groundwaters in Southern Asia: An environmental health disaster. Elements 2 (2006) 91–96.

[47] J.M. Charnock, D. Polya, A.G. Gault, R.A. Wogelius, Direct EXAFS evidence for incorporation of As5+ in the tetrahedral site of natural andraditic garnet. American Mineralogist 92 (2007) 1856–1861.

[48] J.A. Cherry, A.U. Shaikh, D.E. Tallman, R.V. Nicholson, Arsenic species as an indicator of redox conditions in groundwater. Journal of Hydrology 43 (1979) 373–392.

[49] I.D. Clark, K.G. Raven, Sources and circulation of water and arsenic in the Giant Mine, Yellowknife, NWT, Canada. Isotopes in Environmental and Health Studies 40 (2004) 91–96.

[50] J.S. Cleverley, L.G. Benning, B.W. Mountain, Reaction path modelling in the As-S system: A case study for geothermal As transport. Applied Geochemistry 18 (2003) 1325–1345.

[51] E.E. Codling, Effect of flooding lead arsenate-contaminated orchard soil on growth and arsenic and lead accumulation in rice. Communications in Soil Science and Plant Analysis 40 (2009) 2800–2815.

[52] P. Coehlo, S. Costa, S. Silva, A. Walter, J. Ranville, A.C.A. Sousa, C. Costa, M. Coelho, J. Garcia-Leston, M.R. Pastorinho, B. Laffon, E. Pasaro, C.F. Harrington, A. Taylor, J.P. Teixeira, Metal(loid) levels in biological matrices from human populations exposed to mining contamination-Panasqueira Mine (Portugal). Journal of Toxicology and Environmental Health Part A: Current Issues 75 (2012) 893–908.

[53] V.S. Coker, A.G. Gault, C.I. Pearce, G. Van Der Laan, N.D. Telling, J.M. Charnock, D.A. Polya, J.R. Lloyd, XAS and XMCD evidence for species-dependent partitioning of arsenic during microbial reduction of ferrihydrite to Magnetite. Environmental Science and Technology 40 (2006) 7745–7750.

[54] R.M. Couture, B. Shafei, P. Van Cappellen, A. Tessier, C. Gobeil, Non-steady state modeling of arsenic diagenesis in lake sediments. Environmental Science and Technology 44 (2010) 197–203.

[55] E.A. Crecelius, N.S. Bloom, C.E. Cowan, E.A. Jenne. Speciation of Selenium and Arsenic in Natural Waters and Sediments. Volume 2: Arsenic Speciation EPRI EA-4641 Final Report. 1986. Electrical Power Research Institute, Palo Alto.

[56] W.R. Cullen, K.J. Reimer, Arsenic speciation in the environment. Chemical Reviews 89 (1989) 713–764.

[57] D.E. Cummings, F. Caccavo, S. Fendorf, R.F. Rosenzweig, Arsenic mobilization by the dissimilatory Fe(III)-reducing bacterium Shewanella alga BrY. Environmental Science & Technology 33 (1999) 723–729.

[58] M.A. Davis, T.A. Mackenzie, K.L. Cottingham, D. Gilbert-Diamond, T. Punshon, M.R. Karagas, Rice consumption and urinary arsenic concentrations in U.S. children. Environmental Health Perspectives 120 (2012) 1418–1424.

[59] J. Dittmar, A. Voegelin, L.C. Roberts, S.J. Hug, G.C. Saha, M.A. Ali, A.B.M. Badruzzaman, R. Kretzschmar, Spatial distribution and temporal variability of arsenic in irrigated rice fields in Bangladesh. 2. Paddy soil. Environmental Science and Technology 41 (2007) 5967–5972.

[60] S. Dixit, J.G. Hering, Comparison of arsenic(V) and arsenic(III) sorption onto iron oxide minerals: Implications for arsenic mobility. Environmental Science and Technology 37 (2003) 4182–4189.

[61] P.A. Domenico, F.W. Schwartz, Physical and Chemical Hydrogeology, John Wiley & Sons, Inc, New York, 1990.

[62] DPHE/MMI/BGS. Groundwater studies for arsenic contamination in Bangladesh. Rapid Investigation Phase. Final Report. 1999.

[63] W.M. Edmunds, P.L. Smedley, Residence time indicators in groundwater: The East Midlands Triassic sandstone aquifer. Applied Geochemistry 25 (2000) 737–752.

[64] L.E. Erban, S.M. Gorelick, H.A. Zebker, S. Fendorf, Release of arsenic to deep groundwater in the Mekong Delta, Vietnam, linked to pumping-induced land subsidence. Proceedings of the National Academy of Sciences of the United States of America 110 (2013) 13751–13756.

[65] B.V. Erry, M.R. Macnair, A.A. Meharg, R.F. Shore, Arsenic contamination in wood mice (Apodemus sylvaticus) and bank voiles (Clethrionomys glareolus) on abandoned mine sites in southwest Britain. Environmental Pollution 110 (2000) 179–187.

[66] B.V. Erry, M.R. Macnair, A.A. Meharg, R.F. Shore, I. Newton, Arsenic residues in predatory birds from an area of Britain with naturally and anthropogenically elevated arsenic levels. Environmental Pollution 106 (1999) 91–95.

[67] European Food Safety Authority, Panel on Contamination in the Food Chain: Scientific opinion on arsenic in food. European Food Safety Authority Journal 7 (2009) 1351.

[68] J. Feldmann, E.M. Krupp, Critical review or scientific opinion paper: Arsenosugars – a class of benign arsenic species or justification for developing partly speciated arsenic fractionation in foodstuffs? Analytical and Bioanalytical Chemistry 399 (2011) 1735–1741.

[69] S. Fendorf, H.A. Michael, A. Van Geen, Spatial and temporal variations of groundwater arsenic in south and southeast Asia. Science 328 (2010) 1123–1127.

[70] J.F. Ferguson, J. Gavis, Review of arsenic cycle in natural waters. Water Research 6 (1972) 1259–1274.

[71] S.K. Filatov, S.V. Krivovichev, P.C. Burns, L.P. Vergasova, Crystal structure of Flatovite, $K[(Al,Zn)_2(As,Si)_2O_8]$, the first arsenate of the feldspar group. European Journal of Mineralogy 16 (2004) 537–543.

[72] K.A. Francesconi, Complete extraction of arsenic species: A worthwhile goal? Applied Organometallic Chemistry 17 (2003) 682–683.

[73] K.A. Francesconi, J.S. Edmonds, Arsenic and marine organisms. Advances in Inorganic Chemistry 44 (1997) 147–189.

[74] K.A. Francesconi, D. Kuehnelt, Determination of arsenic species: A critical review of methods and applications, 2000–2003. Analyst 129 (2004) 373–395.

[75] C.C. Fuller, J.A. Davis, Influence of coupling of sorption and photosynthetic processes on trace-element cycles in natural-waters. Nature 340 (1989) 52–54.

[76] S. Garcia-Salgado, G. Rober, R. Raml, C. Magnes, K.A. Francesconi, Arsenosugar phospholipids and arsenic hydrocarbons in two species of brown algae. Environmental Chemistry 9 (2012) 63–66.

[77] A.G. Gault, J. Jana, S. Chakraborty, P. Mukherjee, M. Sarkar, B. Nath, D.A. Polya, D. Chatterjee, Preservation strategies for inorganic arsenic species in high iron, low Eh groundwater from West Bengal, India. Analytical and Bioanalytical Chemistry 381 (2005) 347–353.

[78] J.O. Grimalt, M. Ferrer, E. Macphearson, The mine tailing accident in Aznalcollar. Science of the Total Environment 242 (1999) 3–11.

[79] H. Guo, X. Tang, S. Yang, Z. Shen, Effect of indigenous bacteria on geochemical behavior of arsenic in aquifer sediments from the Hetao Basin, Inner Mongolia: Evidence from sediment incubations. Applied Geochemistry 23 (2008) 3267–3277.

[80] M. Hale, Pathfinder applications of arsenic, antimony and bismuth in geochemical exploration. Journal of Geochemical Exploration 15 (1981) 307–323.

[81] G.E.M. Hall, J.C. Pelchat, G. Gauthier, Stability of inorganic arsenic(III) and arsenic(V) in water samples. Journal of Analytical and Atomic Spectroscopy 14 (1999) 205–213.

[82] C.M. Hansel, S.G. Benner, S. Fendorf, Competing Fe(II)-induced mineralization pathways of ferrihydrite. Environmental Science and Technology 39 (2005) 7147–7153.

[83] W. Hartley, N.M. Dickinson, P. Riby, N.W. Lepp, Arsenic mobility in brownfield soils amended with green waste compost of biochar and planted with Miscanthus. Environmental Pollution 157 (2009) 2654–2662.

[84] C.F. Harvey, C.H. Swartz, A.B.M. Badruzzaman, N. Keon-Blute, W. Yu, M. Ashraf Ali, J. Jay, R. Beckie, V. Niedan, D. Brabander, P.M. Oates, K.N. Ashfaque, S. Islam, H.F. Hemond, M.F. Ahmed, Arsenic mobility and groundwater extraction in Bangladesh. Science 298 (2002) 1602–1606.

[85] K. Hattori, Y. Takahashi, S. Guillot, B. Johanson, Occurrence of As(V) in forearc mantle serpentinites based on X-ray absorption spectroscopy study. Geochimica et Cosmochimica Acta 69 (2005) 5585–5596.

[86] G.R. Helz, J.A. Tossell, Thermodynamic model for arsenic speciation in sulfidic waters: A noivel use of ab initio calculations. Geochimica et Cosmochimica Acta 72 (2008) 4457–4468.

[87] G.R. Helz, J.A. Tossell, J.M. Charnock, R.A.D. Pattrick, D.J. Vaughan, C.D. Garner, Oligomerization in As(III) sulfide solutions: Theoretical constraints and spectroscopic evidence. Geochimica et Cosmochimica Acta 59 (1995) 4604.

[88] K.R. Henke, Arsenic: Environmental Chemistry, Health Threats and Waste Treatment, John Wiley & Sons, Ltd, London, 2009.

[89] R.W. Henley, A.J. Ellis, Geothermal systems ancient and modern: A geochemical review. Earth Science Reviews 19 (1983) 1–50.

[90] M.J. Herbel, S. Fendorf, Biogeochemical processes controlling the speciation and transport of arsenic within iron coated sands. Chemical Geology 228 (2006) 16–32.

[91] T. Hiemstra, W.H. Van Riemsdijk, A surface structural approach to ion adsorption: The charge distribution model. Journal of Colloid and Interface Science 179 (1996) 488–508.

[92] M. Hoffman, C. Mikutta, R. Kretzschmar, Bisulfide reaction with natural organic matter enhances arsenite sorption: Insights from X-ray absorption spectroscopy. Environmental Science and Technology 46 (2012) 1788–1797.

[93] J.T. Hollibaugh, S. Carini, H. Gurleyuk, R. Jellison, S.B. Joye, G. LeCleir, C. Meile, L. Vasquez, D. Wallschlager, Arsenic speciation in Mono lake, California: Response to seasonal stratification and anoxia. Geochimica et Cosmochimica Acta 69 (2005) 1925–1937.

[94] M.A. Hoque, W.G. Burgess, C-14 dating of deep groundwater in the Bengal Aquifer System, Bangladesh: Implications for aquifer anisotropy, recharge sources and sustainability. Journal of Hydrology 444 (2012) 209–220.

[95] A. Horneman, A. Van Geen, D.V. Kent, P.E. Mathe, Y. Zheng, R.K. Dhar, S. O'Connell, M.A. Hoque, Z. Aziz, M. Shamsudduha, A.A. Seddique, K.M. Ahmed, Decoupling of As and Fe release to Bangladesh groundwater under reducing conditions. Part I: Evidence from sediment profiles. Geochimica et Cosmochimica Acta 68 (2004) 3459–3473.

[96] J.A. Hudson-Edwards, H.E. Jamieson, J.M. Charnock, M.G. Macklin, Arsenic speciation in waters and sediment of ephemeral floodplain pools, Rios Agrio-Guadiamar, Aznalcollar, Spain. Chemical Geology 219 (2005) 175–192.

[97] K.S. Hunter, Y.F. Wang, P. Van Cappellen, Kinetic modeling of microbially-driven redox chemistry of subsurface environments: coupling transport, microbial metabolism and geochemistry. Journal of Hydrology 209 (1998) 53–80.

[98] IARC, Some drinking-water disinfectants and contaminants, including arsenic. IARC Monograph Evaluation of Carcinogenic Risks to Humans 84 (2004) 1–477.

[99] IARC, Arsenic, metals, fibres and dusts. IARC Monographs 100-C (2012) 41–93.

[100] F.S. Islam, C. Boothman, A.G. Gault, D.A. Polya, J.R. Lloyd, Potential role of the Fe(III) reducing bacteria Geobacter and Geothrix in controlling arsenic solubility in Bengal delta sediments. Mineralogical Magazine 69 (2005) 865–875.

[101] F.S. Islam, A.G. Gault, C. Boothman, D.A. Polya, J.M. Charnock, D. Chatterjee, J.R. Lloyd, Role of metal-reducing bacteria in arsenic release from Bengal delta sediments. Nature 430 (2004) 68–71.

[102] F.S. Islam, R.L. Pederick, A.G. Gault, L.K. Adams, D.A. Polya, J.M. Charnock, J.R. Lloyd, Interactions between the Fe(III)-reducing bacterium Geobacter sulfurreducens and arsenate, and capture of the metalloid by biogenic Fe(II). Applied and Environmental Microbiology 71 (2005) 8642–8648.

[103] B. Jávor, The Kolontár Report: Causes and Lessons from the Red Mud Disaster, Budapest University of Technology and Economics, Budapest, 2011.

[104] JECFA (Joint FAO/WHO Expert Committee on Food Additives). Report of the seventy-second meeting, Rome, February 16–25, 2010. 2010.

[105] JECFA (Joint FAO/WHO Expert Committee on Food Additives). Proposed draft maximum levels for arsenic in rice (cx/cf 12/6/8), January 2012. 2012.

[106] Y. Jia, H. Huang, G.-X. Sun, F.-J. Zhao, Y.-G. Zhu, Pathways and relative contributions to arsenic volatilization from rice plants and soils. Environmental Science and Technology 46 (2012) 8090–8096.

[107] J. Jonsson, D.M. Sherman, Sorption of As(III) and As(V) to siderite, green rust (fougerite) and magnetite: Implications for arsenic release in anoxic groundwaters. Chemical Geology 255 (2008) 173–181.

[108] A.L. Juhasz, E. Smith, J. Weber, M. Rees, A. Rofe, T. Kuchel, L. Sansom, R. Naidu, In vivo assessment of arsenic bioavailability in rice and its significance for human health risk assessment. Environmental Health Perspectives 114 (2006) 1826–1831.

[109] P.J. Kavanagh, M.E. Farago, I. Thornton, R.S. Braman, Bioavailability of soil and mine wastes of the Tamar Valley, SW England. Chemical Speciation and Bioavailability 9 (1997) 77–81.

[110] M.L. Kile, E.A. Houseman, C.V. Breton, E. Smith, Q. Quamruzzaman, M. Rahman, G. Mahiuddin, D.C. Christiani, Association between total ingested arsenic and toenail arsenic concentrations. Journal of Environmental Science and Health Part A: Toxic/Hazardous Substances and Environmental Engineering 42 (2007) 1828–1834.

[111] M.L. Kile, E.A. Houseman, C.V. Breton, T. Smith, Q. Quamruzzaman, M. Rahman, G. Mahiuddin, D.C. Christiani, Dietary arsenic exposure in Bangladesh. Environmental Health Perspectives 115 (2007) 889–893.

[112] M.J. Kim, J.O. Nriagu, S. Haack, Carbonate ions and arsenic dissolution by groundwater. Environmental Science and Technology 34 (2000) 3094–3100.

[113] M.F. Kirk, E.E. Roden, L.J. Crossey, Experimental analysis of arsenic precipitation during microbial sulfate and iron reduction in model aquifer sediment reactors. Geochimica et Cosmochimica Acta 74 (2010) 2538–2555.

[114] M.F. Kirk, T.R. Holm, J. Park, Q. Jin, R.A. Sanford, B.W. Fouke, C.M. Bethke, Bacterial sulfate reduction limits natural arsenic contamination in groundwater. Geology 32 (2004) 953–956.

[115] B.D. Kocar, T. Borch, S. Fendorf, Arsenic repartitioning during biogenic sulfidization and transformation of ferrihydrite. Geochimica et Cosmochimica Acta 74 (2010) 980–994.

[116] B.D. Kocar, M.J. Herbel, K.J. Tufano, S. Fendorf, Contrasting effects of dissimilatory iron (III) and arsenic (V) reduction on arsenic retention and transport. Environmental Science and Technology 40 (2006) 6715–6721.

[117] B.D. Kocar, M.L. Polizzotto, S.G. Benner, S.C. Ying, M. Ung, K. Ouch, S. Samreth, B. Suy, K. Phan, M. Sampson, S. Fendorf, Integrated biogeochemical and hydrologic processes driving arsenic release from shallow sediments to groundwaters of the Mekong delta. Applied Geochemistry 23 (2008) 3059–3071.

[118] I. Koch, K.J. Reimer, M.I. Bakker, N.T. Basta, M.R. Cave, S. Denys, M. Dodd, B.A. Hale, R. Irwin, Y.W. Lowney, M.M. Moore, V. Paquin, P.E. Rasmussen, T. Repaso-Subang, G.L. Stephenson, S.D. Siciliano, J. Wragg, G.J. Zagury, Variability of bioaccessibility results using seventeen different methods on a standard reference material, NIST 2710. Journal of Environmental Science and Health, Part A: Toxic/ Hazardous Substances and Environmental Engineering 48 (2013) 641–655.

[119] A. Kuhn, L. Sigg, Arsenic cycling in eutrophic Lake Greifen, Switzerland: Influence of seasonal redox processes. Limnology and Oceanography 38 (1993) 1052–1059.

[120] L.R. Lado, D. Polya, L. Winkel, M. Berg, A. Hegan, Modelling arsenic hazard in Cambodia: A geostatistical approach using ancillary data. Applied Geochemistry 23 (2008) 3010–3018.

[121] B.D. Laird, K.J. James, T.R. Van de Wiele, M. Dodd, S.W. Casteel, M. Wickstrom, S.D. Siciliano, An investigation of the effect of gastrointestinal microbial activity on oral arsenic bioavailability. Journal of Environmental Science and Health, Part A: Toxic/ Hazardous Substances and Environmental Engineering 48 (2013) 612–619.

[122] P. Langner, C. Mikutta, R. Kretzschmar, Arsenic sequestration by organic sulphur in peat. Nature Geoscience 5 (2012) 66–73.

[123] D.J. Large, S. Kelly, B. Spiro, L.W. Tian, L.Y. Shao, R. Finkelman, M. Zhang, C. Somerfield, S. Plint, Y. Ali, Y.P. Zhou, Siliva-volatile interaction and the geological cause of the Xuan Wei lung cancer epidemic. Environmental Science and Technology 43 (2009) 9016–9021.

[124] M. Lawson, C.J. Ballentine, D.A. Polya, A.J. Boyce, D. Mondal, D. Chatterjee, S. Majumder, A. Biswas, The geochemical and isotopic composition of ground waters in West Bengal: Tracing ground-surface water interaction and its role in arsenic release. Mineralogical Magazine 72 (2008) 441–444.

[125] M. Lawson, D.A. Polya, A.J. Boyce, C. Bryant, D. Mondal, A. Shantz, C.J. Ballentine, Pond derived organic carbon driving changes in arsenic hazard found in Asian groundwaters. Environmental Science and Technology 34 (2013) 2342–2347.

[126] X.C. Le, S. Yalcin, M. Ma, Speciation of submicrogram per liter levels of arsenic in water: Om-site species separation integrated with sample collection. Environmental Science and Technology 34 (2000) 2342–2347.

[127] R.D. Linberg, D.D. Runnells, Ground water redox reactions: An analysis of equilibrium state applied to eh measurements and geochemical modelling. Science 225 (1984) 925–927.

[128] H.Y. Liu, A. Probst, B.H. Liao, Metal contamination of soils and crops affected by the Chenzhou lead/zinc mine spill (Hunan, China). Science of the Total Environment 339 (2005) 153–156.

[129] K. Ljung, O. Selinus, E. Otabbong, M. Berglund, Metal and arsenic distribution in soil particle sizes relevant to soil ingestion by children. Applied Geochemistry 21 (2006) 1613–1624.

[130] J.R. Lloyd, T.J. Beveridge, K. Morris, D.A. Polya Chapter 98: Techniques for studying microbial transformations of metals and radionuclides. in: Manual of Environmental Microbiology, 3rd Edition, C. J. Hurst, R. L. Crawford, J. L. Garland, D. A. Lipson, A. L. Mills, and L. D. Stetzenbach (Eds.), ASM Press, Washington, DC, 2006, pp. 1195–1213.

[131] J.R. Lloyd, R.S. Oremland, Microbial transformations of arsenic in the environment: From soda lakes to aquifers. Elements 2 (2006) 85–90.

[132] C. Lomax, W.J. Liu, L.Y. Wu, K. Xue, J.B. Xiong, J.Z. Zhou, S.P. McGrath, A.A. Meharg, A.J. Miller, F.J. Zhao, Methylated arsenic species in plants originate from soil microorganisms. New Phytologist 193 (2012) 665–672.

[133] D.L. Lopez, J. Bundschuh, P. Birkle, M.A. Armienta, L. Cumbal, O. Sracek, L. Cornejo, M. Ormachea, Arsenic in volcanic geothermal fluids of Latin America. Science of the Total Environment 429 (2012) 57–75.

[134] G. Lord, N. Kim, N.I. Ward, Arsenic speciation of geothermal waters of New Zealand. Journal of Environmental Monitoring 14 (2012) 3192–3201.

[135] D.R. Lovley, J.D. Coates, E.L. Blunt-Harris, E.J.P. Phillips, J.C. Woodward, Humic substances as electron acceptors for microbial respiration. Nature 382 (1996) 445–448.

[136] D.R. Lovley, S. Goodwin, Hydrogen concentrations as an indicator of the predominant terminal electron-accepting reactions in aquatic sediments. Geochimica et Cosmochimica Acta 52 (1988) 2993–3003.

[137] H.A. Lowers, G.N. Breit, A.L. Foster, J. Whitney, J. Yount, N. Uddin, A. Muneem, Arsenic incorporation into authigenic pyrite, Bengal Basin sediment, Bangladesh. Geochimica et Cosmochimica Acta 71 (2007) 2699–2717.

[138] P. Lu, C. Zhu, Arsenic Eh-pH diagrams at 25 C and 1 bar. Environmental Earth Science 62 (2011) 1673–1683.

[139] J.F. Ma, N. Yamaji, N. Mitani, X.Y. Xu, Y.H. Su, S.P. McGrath, F.J. Zhao, Transporters of arsenite in rice and their role in arsenic accumulation in rice grain. Proceedings of the National Academy of Sciences of the United States of America 105 (2008) 9931–9935.

[140] W. Maher, F. Krikowa, M. Ellwood, S. Foster, R. Jagtap, G. Raber, Overview of hyphenated techniques using an ICP-MS detector with an emphasis on extraction techniques for measurement of metalloids by HPLC-ICPMS. Microchemical Journal 105 (2012) 15–31.

[141] B.J. Mailloux, E. Trembath-Reichert, J. Cheung, M. Watson, M. Stute, G.A. Freyer, A.S. Ferguson, K.M. Ahmed, M.J. Alam, B.A. Buchholz, J. Thomas, A.C. Layton, Y. Zheng, B.C. Bostick, A. Van Geen, Advection of surface-derived organic carbon fuels microbial reduction in Bangladesh groundwater. Proceedings of the National Academy of Sciences of the United States of America 110 (2013) 5331–5335.

[142] K.C. Makris, M. Quazi, P. Punamiya, D. Sarkar, R. Datta, Fate of arsenic in swine waste from concentrated animal feeding operations. Journal of Environmental Quality 37 (2008) 1626–1633.

[143] B.K. Mandal, K.T. Suzuki, Arsenic round the world: A review. Talanta 58 (2002) 201–235.

[144] B.A. Manning, S.E. Fendorf, B.C. Bostick, D. Surez, Arsenic (III) oxidation and arsenic(V) adsorption reactions on synthetic birnessite. Environmental Science and Technology 36 (2002) 976–981.

[145] B.A. Manning, S.E. Fendorf, S. Goldberg, Surface structures and stability of arsenic(III) on goethite: Spectroscopic evidence for inner-sphere complexes. Environmental Science and Technology 32 (1998) 2383–2388.

[146] V.D. Martinez, E.A. Vucic, S. Lam, W.L. Lam, Emerging arsenic threat in Canada. Science 342 (2013) 559.

[147] M. Masson, J. Schafer, G. Blanc, A. Pierre, Seasonal variations and annual fluxes of arsenic in the Garonne, Dordogne and Isle Rivers, France. Science of the Total Environment 373 (2007) 196–207.

[148] J. Matschullat, Arsenic in the geosphere: A review. Science of the Total Environment 249 (2000) 297–312.

[149] J. McArthur, P. Ravenscroft, S. Safiulla, M.F. Thirwall, Arsenic in groundwater: Testing pollution mechanisms for sedimentary aquifers in Bangladesh. Water Resources Research 37 (2001) 109–117.

[150] J.M. McArthur, D.M. Banerjee, K.A. Hudson-Edwards, R. Mishra, R. Purohit, P. Ravenscroft, A. Cronin, R.J. Howarth, A. Chatterjee, T. Talukder, D. Lowry, S. Houghton, D.K. Chadha, Natural organic matter in sedimentary basins and its relation to arsenic in anoxic ground water: The example of West Bengal and its worldwide implications. Applied Geochemistry 19 (2004) 1255–1293.

[151] J.M. McArthur, P. Ravenscroft, O. Sracek, Aquifer arsenic source. Nature Geoscience 4 (2011) 655–656.

[152] A.A. Meharg, A. Raab, Getting to the bottom of arsenic standards and guidelines. Environmental Science and Technology 44 (2010) 4395–4399.

[153] A.A. Meharg, M. Rahman, Arsenic contamination of Bangladesh paddy field soils: Implications for rice contribution to arsenic consumption. Environmental Science and Technology 37 (2003) 229–234.

[154] A.A. Meharg, F.J. Zhao, Arsenic and Rice, Springer, New York, 2012.

[155] A. Mestrot, J. Feldmann, E.M. Krupp, M.S. Hossain, G. Roman-Ross, A.A. Meharg, Field fluxes and speciation of arsines emanating from soil. Environmental Science and Technology 45 (2011) 1798–1804.

[156] H.A. Michael, C.I. Voss, Evaluation of the sustainability of deep groundwater as an arsenic-safe resource in the Bengal Basin. Proceedings of the National Academy of Sciences of the United States of America 105 (2008) 8531–8536.

[157] H.A. Michael, C.I. Voss, Controls on groundwater flow in the Bengal Basin of India and Bangladesh: Regional modeling analysis. Hydrogeology Journal 17 (2009) 1561–1577.

[158] K. Michaelke, E.B. Wickenhesiter, M. Mehring, A.V. Hirner, R. Hensel, Product of volatile derivatives of metal(loid)s by microflora involved in anaerobic digestion of sewage sludge. Applied and Environmental Microbiology 66 (2000) 2791–2796.

[159] N. Mladenov, Y. Zheng, M.P. Miller, D.R. Nemergut, T. Legg, B. Simone, C. Hageman, M.M. Rahman, K.M. Ahmed, D.M. McKnight, Dissolved organic matter sources and consequences for iron and arsenic mobilization in Bangladesh aquifers. Environmental Science and Technology 44 (2010) 123–128.

[160] D. Mondal, M. Banerjee, M. Kundu, N. Banerjee, U. Bhattacharya, A.K. Giri, B. Ganguli, S. Sen Roy, D.A. Polya, Comparison of drinking water, raw rice and cooking of rice as arsenic exposure routes in three contrasting areas of West Bengal, India. Environmental Geochemistry and Health 32 (2010) 463–477.

[161] D. Mondal, D.A. Polya, Rice is a major exposure route for arsenic in Chakdaha block, Nadia district, West Bengal, India: A probabilistic risk assessment. Applied Geochemistry 23 (2008) 2987–2998.

[162] K.L. Moore, E. Lombi, F.J. Zhao, C.R.M. Grovenor, Elemental imaging at the nanoscale: NanoSIMS and complementary techniques for element localisation in plants. Analytical and Bioanalytical Chemistry 402 (2012) 3263–3273.

[163] E. Moreno-Jiménez, R. Clemente, A. Mestrot, A.A. Meharg, Arsenic and selenium mobilisation from organic matter treated mine spoil with and without inorganic fertilisation. Environmental Pollution 173 (2013) 238–244.

[164] G. Morin, G. Calas, Arsenic in soils, mine tailings, and former industrial sites. Elements 2 (2006) 97–101.

[165] A. Mukherjee, A.E. Fryar, H.D. Rowe, Regional-scale stable isotopic signatures of recharge and deep groundwater in the arsenic affected areas of West Bengal, India. Journal of Hydrology 334 (2007) 151–161.

[166] B. Nath, S. Chakraborty, A. Burnol, D. Stuben, D. Chatterjee, L. Charlet, Mobility of arsenic in the sub-surface environment: An integrated hydrogeochemical study and sorption model of the sandy aquifer materials. Journal of Hydrology 364 (2009) 236–248.

[167] C.S. Neuberger, G.R. Helz, Arsenic(III) carbonate complexing. Applied Geochemistry 20 (2005) 1218–1225.

[168] R.B. Neumann, K.N. Asfaque, B.M. Badruzzaman, M.A. Ali, J.K. Shoemaker, C.F. Harvey, Aquifer arsenic source reply. Nature Geoscience 4 (2011) 656.

[169] R.B. Neumann, K.N. Ashfaque, A.B.M. Badruzzaman, M.A. Ali, J.K. Shoemaker, C.F. Harvey, Anthropogenic influences on groundwater arsenic concentrations in Bangladesh. Nature Geoscience 3 (2010) 46–52.

[170] R.B. Neumann, A.P.S. Vincent, L.C. Roberts, B.M. Badruzzaman, M.A. Ali, C.F. Harvey, Rice field geochemistry and hydrology: An explanation for why groundwater irrigated fields in Bangladesh are net sinks of arsenic from groundwater. Environmental Science and Technology 45 (2011) 2072–2078.

[171] E.J. New, Tools to study distinct metal pools in biology. Dalton Transaction 42 (2013) 3210–3219.

[172] R.T. Nickson, J.M. McArthur, W.G. Burgess, K.M. Ahmed, P. Ravenscroft, M. Rahman, Arsenic poisoning of Bangladesh groundwater. Nature 395 (1998) 338.

[173] R.T. Nickson, J.M. McArthur, P. Ravenscroft, W.G. Burgess, K.M. Ahmed, Mechanism of arsenic release to groundwater, Bangladesh and West Bengal. Applied Geochemistry 15 (2000) 403–413.

[174] B.N. Noller, V. Diacomanolis, V.P. Matanitobua, J.C. Ng, H.H. Harris Arsenic from mining old and new: Legacies and challenges. in: Understanding the Geological and Medical Interface of Arsenic, J. C. Ng, B. N. Noller, R. Raidu, J. Bundschuh, and U. Bhattacharya (Eds.), Taylor & Francis Group, London, 2012, pp. 411–414.

[175] D.K. Nordstrom, Public health: Worldwide occurrences of arsenic in ground water. Science 296 (2002) 2143–2145.

[176] D.K. Nordstrom, D.G. Archer Arsenic thermodynamic data and environmental geochemistry. in: Arsenic in groundwater, A. H. Welch and K. G. Stollenwerk (Eds.), Springer, Berlin, 2002.

[177] J. Norrman, C.J. Sparrenbom, M. Berg, D.D. Nhan, P.Q. Nhan, H. Rosqvist, G. Jacks, E. Sigvardsson, D. Baric, J. Moreskog, P. Harms-Ringdahl, N.V. Hoan, Arsenic mobilisation in a new well field for drinking water production along the Red River, Nam Du, Hanoi. Applied Geochemistry 23 (2008) 3127–3142.

[178] NPIC (National Pesticide Information Centre). Arsenic. March 22, 2013. 2011.

[179] NRC, Arsenic in Drinking Water, National Academy Press, Washington, D.C., 1999.

[180] NRC, Arsenic in Drinking Water, 2001 Update, National Academy Press, Washington, D.C., 2001.

[181] J.O. Nriagu, T.-S. Lin, Trace metals in wild rice sold in the United States. Science of the Total Environment 172 (1995) 223–228.

[182] P.A. O'Day, Chemistry and mineralogy of arsenic. Elements 2 (2006) 77–83.

[183] P.A. O'Day, D. Vlassopoulos, R. Root, N. Rivera, The influence of sulfur and iron on dissolved arsenic concentrations in the shallow subsurface under changing redox conditions. Proceedings of the National Academy of Sciences of the United States of America 101 (2004) 13703–13708.

[184] R.S. Oremland, J.F. Stolz, The ecology of arsenic. Science 300 (2003) 939–944.

[185] C.T. Parsons, R.M. Couture, E.O. Omoregie, F. Bardelli, J.M. Greneche, G. Roman-Ross, L. Charlet, The impact of oscillating redox conditions: Arsenic immobilisation in contaminated calcareous floodplain soils. Environmental Pollution 178 (2013) 254–263.

[186] C. Pascua, J.M. Charnock, D. Polya, T. Sato, S. Yokoyama, M. Minato, Arsenic-bearing smectite from the geothermal environment. Mineralogical Magazine 69 (2005) 897–906.

[187] B. Planer-Friedrich, C. Lehr, J. Matschullat, B.J. Merkel, D.K. Nordstrom, M.W. Sandstrom, Speciation of volatile arsenic at geothermal features in Yellowstone National Park. Geochimica et Cosmochimica Acta 70 (2006) 2480–2491.

[188] B. Planer-Friedrich, J. London, R.B. McCleskey, D.K. Nordstrom, D. Wallschläger, Thioarsenates in geothermal waters of Yellowstone national park: Determination, preservation and geochemical importance. Environmental Science and Technology 41 (2007) 5245–5251.

[189] B. Planer-Friedrich, E. Suess, A.C. Schelnost, D. Wallschläger, Speciation in sulfidic waters: Reconciling contradictory spectroscopic and chromatographic evidence. Analytical Chemistry 82 (2010) 10228–10235.

[190] M.L. Polizzotto, C.F. Harvey, G. Li, A.B.M. Badruzzaman, A. Ali, M. Newville, S. Sutton, S. Fendorf, Solid-phases and desorption processes of arsenic within Bangladesh sediments. Chemical Geology 228 (2006) 97–111.

[191] M.L. Polizzotto, C.F. Harvey, S.R. Sutton, S. Fendorf, Processes conducive to the release and transport of arsenic into aquifers of Bangladesh. Proceedings of the National Academy of Sciences of the United States of America 102 (2005) 18819–18823.

[192] M.L. Polizzotto, B.D. Kocar, S.G. Benner, M. Sampson, S. Fendorf, Near-surface wetland sediments as a source of arsenic release to ground water in Asia. Nature 454 (2008) 505–508.

[193] D.A. Polya, Efficiency of hydrothermal ore formation and the Panasqueira W-Cu(Ag)-Sn vein deposit. Nature 333 (1988) 838–841.

[194] D.A. Polya, Chemistry of the main-stage ore-forming fluids of the Panasqueira W-CU(AG)-SN deposit, Portugal: Implications for models of ore genesis. Economic Geology 84 (1989) 1134–1152.

[195] D.A. Polya, L. Charlet, Rising arsenic risk? Nature Geoscience 2 (2009) 383–384.

[196] D.A. Polya, P.R. Lythgoe, F. Abou-Shakra, A.G. Gault, J.R. Brydie, J.G. Webster, K.L. Brown, M.K. Nimfopoulos, K.M. Michailidis, IC-ICP-MS and IC-ICP-HEX-MS

determination of arsenic speciation in surface and ground waters: Preservation and analytical issues. Mineralogical Magazine 67 (2003) 247–261.

[197] D.A. Polya, D. Mondal, A.K. Giri Quantification of deaths and DALYs arising from chronic exposure to arsenic in groundwaters utilized for drinking, cooking and irrigation of food crops. in: Handbook of Disease Burdens and Quality of Life Measures, V. R. Preedy and R. R. Watson (Eds.), Springer-Verlag, New York, 2010, pp. 701–728.

[198] D. Postma, F. Larsen, N.T. Minh Hue, M.T. Duc, P.H. Viet, P.Q. Nhan, S. Jessen, Arsenic in groundwater of the Red River floodplain, Vietnam: Controlling geochemical processes and reactive transport modeling. Geochimica et Cosmochimica Acta 71 (2007) 5054–5071.

[199] K.A. Radloff, Y. Zheng, H.A. Michael, M. Stute, B.C. Bostick, I. Mihajlov, M. Bounds, M.R. Huq, I. Choudhury, M.W. Rahman, P. Schlosser, K.M. Ahmed, A. Van Geen, Arsenic migration to deep groundwater in Bangladesh influenced by adsorption and water demand. Nature Geoscience 4 (2011) 793–798.

[200] P. Ravenscroft Distribution of groundwater arsenic in Bangladesh. in: G. Jacks, P. Bhattacharya, and A. A. Khan (Eds.), Groundwater Arsenic Contamination in the Bengal Delta Plain of Bangladesh: Proceedings of the KTH-Dhaka University of Dhaka, Bangladesh. February 7–8, 1999, 2001. KTH Special Publication, TRITA-AMI REPORT 3084.

[201] P. Ravenscroft, H. Brammer, K. Richards, Arsenic Pollution, Wiley-Richards, Oxford, 2009.

[202] P. Ravenscroft, W.C. Burgess, K.M. Ahmed, M. Burren, J. Perrin, Arsenic in groundwater of the Bengal Basin, Bangladesh: Distribution, field relations, and hydrogeological setting. Hydrogeology Journal 13 (2005) 727–751.

[203] L.C. Roberts, S.J. Hug, J. Dittmar, A. Voegelin, R. Kretzschmar, B. Wehrli, O.A. Cirpka, G.C. Saha, M.A. Ali, A.B.M. Badruzzaman, Arsenic release from paddy soils during monsoon flooding. Nature Geoscience 3 (2010) 53–59.

[204] L.C. Roberts, S.J. Hug, J. Dittmar, A. Voegelin, G.C. Saha, M.A. Ali, A.B.M. Badruzzaman, R. Kretzschmar, Spatial distribution and temporal variability of arsenic in irrigated rice fields in Bangladesh. 1. Irrigation water. Environmental Science and Technology 41 (2007) 5960–5966.

[205] L. Rodriguez-Lado, G.F. Sun, M. Berg, Q. Zhang, H.B. Xue, Q.M. Zheng, C.A. Johnson, Groundwater arsenic contamination throughout China. Science 341 (2013) 866–868.

[206] J.J. Rothwell, K.G. Taylor, E.L. Ander, M.G. Evans, S.M. Daniels, T.E.H. Allott, Arsenic retention and release in ombrotrophic peatlands. Science of the Total Environment 407 (2009) 1405–1417.

[207] H.A.L. Rowland, A.G. Gault, P.R. Lythgoe, D.A. Polya, Geochemistry of aquifer sediments and arsenic-rich groundwaters from Kandal Province, Cambodia. Applied Geochemistry 23 (2008) 3029–3046.

[208] H.A.L. Rowland, R.L. Pederick, D.A. Polya, R.D. Pancost, B.E. Van Dongen, A.G. Gault, D.J. Vaughan, C. Bryant, B. Anderson, J.R. Lloyd, The control of organic matter on microbially mediated iron reduction and arsenic release in shallow alluvial aquifers, Cambodia. Geobiology 5 (2007) 281–292.

[209] H.A.L. Rowland, D.A. Polya, J.R. Lloyd, R.D. Pancost, Characterisation of organic matter in a shallow, reducing, arsenic-rich aquifer, West Bengal. Organic Geochemistry 37 (2006) 1101–1114.

[210] D.W. Rutherford, A.J. Bednar, J.R. Garbarino, R. Needham, K.W. Staver, R.L. Wershaw, Environmental fate of roxarsone in poultry litter. Part II. Mobility of arsenic in soils amended with poultry litter. Environmental Science and Technology 37 (2003) 1515–1520.

[211] K. Saemundsson Geothermal systems in global perspective. Short Course IV on Exploration for Geothermal Resources, UNU-GTP, Lake Naivasha, Kenya, 2009.

[212] F.J. Sawkins, Metal Deposits in Relation to Plate Tectonics, Springer-Verlag, Berlin, 1984.

[213] R.A. Schoof, L.J. Yost, E.A. Crecelius, K. Irgolic, W. Goessler, H.R. Gui, H. Greene, Dietary arsenic intake in Taiwanese districts with elevated arsenic in drinking water. Human and Ecological Risk Assessment: An International Journal 4 (1998) 117–135.

[214] D.B. Senn, H.F. Hemond, Nitrate controls on iron and arsenic in an urban lake. Science 296 (2002) 2373–2376.

[215] D.M. Sherman, S.R. Randall, Surface complexation of arsenic(V) to iron(III) (hydr) oxides: Structural mechanism from ab initio molecular geometries and EXAFS spectroscopy. Geochimica et Cosmochimica Acta 67 (2003) 4223–4230.

[216] W. Shotyk, A.K. Cheburkin, P.G. Appleby, A. Fankhauser, J.D. Kramers, Two thousand years of atmospheric arsenic, antimony, and lead deposition recorded in an ombrotrophic peat bog profile, Jura Mountains, Switzerland. Earth and Planetary Science Letters 145 (1996) E1–E7.

[217] A.J. Signes-Pastor, S.W. Al-Rmalli, R.O. Jenkins, A.A. Carbonell-Barrachina, P.I. Haris, Arsenic bioaccessibility in cooked rice as affected by arsenic in cooking water. Journal of Food Science 77 (2012) T201–T206.

[218] P.L. Smedley, W.M. Edmunds, Redox patterns and trace-element behaviour in the East Midlands Triassic Sandstone Aquifer, U.K. Groundwater 40 (2002) 44–58.

[219] P.L. Smedley, D.G. Kinniburgh, A review of the source, behaviour and distribution of arsenic in natural waters. Applied Geochemistry 17 (2002) 517–568.

[220] A.H. Smith, E.O. Lingas, M. Rahman, Contamination of drinking-water by arsenic in Bangladesh: A public health emergency. Bulletin of the World Health Organisation 78 (2000) 1093–1103.

[221] P. Stocks, On the relations between atmospheric pollution in urban and rural localities and mortality from cancer, bronchitis and pneumonia, with particular reference to 3:4 benzopyrene, beryllium, molybdenum, vanadium and arsenic. British Journal of Cancer XIV (1960) 397–418.

[222] W. Stumm, J.J. Morgan, Aquatic Chemistry, 3rd Edition, John Wiley & Sons, Inc, New York, 1996.

[223] D.A. Sverjensky, K. Fukushi, A predictive model (ETLM) for As(III) adsorption and surface speciation on oxides consistent with spectroscopic data. Geochimica et Cosmochimica Acta 70 (2006) 3778–3802.

[224] C.H. Swartz, N.K. Blute, B. Badruzzman, A. Ali, D. Brabander, J. Jay, J. Besancon, S. Islam, H.F. Hemond, C.F. Harvey, Mobility of arsenic in a Bangladesh aquifer: Inferences from geochemical profiles, leaching data, and mineralogical characterization. Geochimica et Cosmochimica Acta 68 (2004) 4539–4557.

[225] P.J. Swedlund, J.G. Webster, Adsorption and polymerisation of silicic acid on ferrihydrite, and its effect on arsenic sorption. Water Resources Research 33 (1999) 3422.

[226] T. Tarvainen, S. Albanese, M. Birke, M. Ponavic, C. Reimass, Arsenic in agricultural and grazing land soils of Europe. Applied Geochemistry 28 (2013) 10.

[227] M.K. Ullrich, J.G. Pope, T.M. Seward, N. Wilson, B. Planer-Friedrich, Sulfur redox chemistry governs diurnal antimony and arsenic cycles at Champagne Pool, Waiotapu, New Zealand. Journal of Volcanology and Geothermal Research 262 (2013) 164–177.

[228] UNEP, Tailings Dams, Risk of Dangerous Occurrences, Lessons Learnt from Practical Experiences. UNEP Bulletin 121. 2001. ICOLD Committee on Tailings Dams and Waste Lagoons, Paris.

[229] USGS. Arsenic. http://minerals.usgs.gov/minerals/pubs/commodity/arsenic/mcs-2012-arsen.pdf. 2012. Accessed April 8, 2015.

[230] USGS. PHREEQC (Version 3) – A Computer Program for Speciation, Batch-Reaction, One-Dimensional Transport, and Inverse Geochemical Calculations. http://wwwbrr.cr.usgs.gov/projects/GWC_coupled/phreeqc/. 2013. Accessed March 21, 2013.

Ref Type: Electronic Citation

[231] USGS. USGS Ground-Water Software MODFLOW-2005 Version 1.10.00 Three-dimensional finite-difference ground-water model –2005 version. http://water.usgs.gov/nrp/gwsoftware/modflow2005/modflow2005.html. 2013. Accessed April 8, 2015.

[232] B.E. van Dongen, H.A.L. Rowland, A.G. Gault, D. Polya, C. Bryant, R.D. Pancost, Hopane, sterane and n-alkane distributions in shallow sediments hosting high arsenic groundwaters in Cambodia. Applied Geochemistry 23 (2008) 3047–3058.

[233] A. Van Geen, K.M. Ahmed, A.A. Seddique, M. Shamsudduha, Community wells to mitigate the arsenic crisis in Bangladesh. Bulletin of the World Health Organization 81 (2003) 632.

[234] A. Van Geen, Z. Cheng, Q. Jia, A.A. Seddique, M.W. Rahman, M.M. Rahman, K.M. Ahmed, Monitoring 51 deep community wells in Araihazar, Bangladesh, for up to 5 years: Implications for arsenic mitigation. Journal of Environmental Science and Health, Part A 42 (2007) 1729–1740.

[235] A. Van Geen, B.C. Bostick, T.K. Pham, M.L. Vi, M. Nguyen-Ngoc, D.M. Phu, H.V. Pham, K. Radloff, Z. Aziz, J.L. Mey, M.O. Stahl, C.F. Harvey, P. Oates, B. Weinman, C. Stengel, F. Frei, R. Kipfer, M. Berg, Retardation of arsenic transport through a Pleistocene aquifer. Nature 501 (2013) 204–207.

[236] A. Van Geen, S. Thoral, J. Rose, J.M. Garnier, Y. Zheng, J.Y. Bottero, Decoupling of As and Fe release to Bangladesh groundwater under reducing conditions. Part II: Evidence from sediment incubations. Geochim Cosmochim Acta 68 (2004) 3475–3486.

[237] D.J. Vaughan, D.A. Polya. Arsenic – the great poisoner revisited. Elements 9 (2013) 315–316.

[238] D.J. Vaughan, Sulfide mineralogy and geochemistry: Introduction and overview. Reviews in Mineralogy and Geochemistry 61 (2006) 1–5.

[239] S. Wang, C.N. Mulligan, Speciation and surface structure of inorganic arsenic in solid phases: A review. Environment International 34 (2008) 867–879.

[240] M.J. Watts, J. O'Reilly, A.L. Marcilla, R.A. Shaw, N.I. Ward, Field based speciation of arsenic in UK and Argentinean water samples. Environmental Geochemistry and Health 32 (2010) 479–490.

[241] J.G. Webster, D.K. Nordstrom Geothermal arsenic. in: Arsenic in Ground Water: Geochemistry and Occurrence, A. H. Welch and K. G. Stollenwerk (Eds.), Kluwer Academic, Boston, 2003, pp. 101–126.

[242] A.H. Welch, D.B. Westjohn, D.R. Helsel, R.B. Wanty, Arsenic in groundwater of the United States: Occurrence and geochemistry. Groundwater 38 (2000) 589–604.

[243] J.A. Wilkie, J.G. Hering, Rapid oxidation of geothermal arsenic(III) in streamwaters of the eastern Sierra Nevada. Environmental Science and Technology 32 (1998) 657–672.

[244] R.T. Wilkin, D. Wallschläger, R.G. Ford, Speciation of arsenic in sulfidic waters. Geochemical Transactions 4 (2003) 1–7.

[245] P.N. Williams, A. Raab, J. Feldmann, A.A. Meharg, Market basket survey shows elevated levels in As in South Central U.S. processed rice compared to California: Consequences for human dietary exposure. Environmental Science and Technology 41 (2007) 2178–2183.

[246] L.H.E. Winkel, P.T.K. Trang, V.M. Lan, C. Stengel, M. Amini, N.T. Ha, P.H. Viet, M. Berg, Arsenic pollution of groundwater in Vietnam exacerbated by deep aquifer exploitation for more than a century. Proceedings of the National Academy of Sciences of the United States of America 108 (2011) 1246–1251.

[247] B.A. Wood, J. Feldmann, Quantification of phytochelatins and their metal(loid) complexes: Critical assessment of current analytical methodology. Analytical and Bioanalytical Chemistry 402 (2012) 3299–3309.

[248] B. Wu, J.S. Becker, Imaging techniques for elements and element species in plany science. Metallomic 4 (2012) 403–416.

[249] J. Xue, V. Zartarian, S.-W. Wang, S.V. Liu, P. Georgopoulos, Probabilistic modeling of dietary arsenic exposure and dose and evaluation with 2003–2004 NHANES data. Environmental Health Perspectives 118 (2010) 345–350.

[250] J. Zhang, K.R. Smith, Household air pollution from coal and biomass fuels in China: Measurements, health impacts, and interventions. Environmental Health Perspectives 115 (2007) 848–855.

[251] F.J. Zhao, J.F. Ma, S.P. McGrath. Arsenic transport in plants. Annual Meeting of the Society for Experimental Biology 153A, S187. 2009. Glasgow, Scotland. Comparative Biochemistry and Physiology A Molecular & Integrative Physiology.

[252] F.J. Zhao, Y.G. Zhu, A.A. Meharg, Methylated arsenic species in rice: Geographical variation, origin, and uptake mechanisms. Environmental Science and Technology 47 (2013) 3957–3966.

[253] Y. Zheng, A. Van Geen, M. Stute, R. Dhar, Z. Mo, Z. Cheng, A. Horneman, I. Gavrieli, H.J. Simpson, R. Versteeg, M. Steckler, A. Grazioli-Venier, S. Goodbred, M. Shahnewaz, M. Shamsudduha, M.A. Hoque, K.M. Ahmed, Geochemical and hydrogeological contrasts between shallow and deeper aquifers in two villages of Araihazar, Bangladesh: Implications for deeper aquifers as drinking water sources. Geochimica et Cosmochimica Acta 69 (2005) 5203–5218.

[254] J. Zobrist, P.R. Dowdle, J.A. Davis, R.S. Oremland, Mobilization of arsenite by dissimilatory reduction of adsorbed arsenate. Environmental Science and Technology 34 (2000) 4747–4753.

3

REMEDIATION OF ARSENIC IN DRINKING WATER

CATHLEEN J. WEBB[1] AND ARDEN D. DAVIS[2]

[1]*Department of Chemistry, Western Kentucky University, Bowling Green, KY, USA*
[2]*Department of Geology and Geological Engineering, South Dakota School of Mines and Technology, Rapid City, SD, USA*

3.1 BACKGROUND

Arsenic contamination of groundwater and surface water is a problem of worldwide significance. The best way to prevent disease from chronic arsenic exposure is to remove the exposure. In areas where arsenic in drinking water is a health concern, remediation efforts have focused on providing an alternative source of water or treatment to remove arsenic from the water. The remediation choice depends on the availability of alternative water sources, technological and engineering infrastructure, economics, geographic location, cultural setting, geologic conditions, water chemistry, and other factors. The problems associated with and benefits of various water source substitutions and arsenic removal technologies will be discussed using specific examples from around the world.

3.2 PROVISION OF ALTERNATIVE WATER SOURCES

One of the most ready solutions for removing exposure to arsenic is to provide an alternative source of drinking water, if one is available. Examples of alternative sources are surface water, groundwater from a different aquifer, water from an existing piped supply nearby, or rainwater.

Arsenic: Exposure Sources, Health Risks, and Mechanisms of Toxicity, First Edition.
Edited by J. Christopher States.
© 2016 John Wiley & Sons, Inc. Published 2016 by John Wiley & Sons, Inc.

The maximum contaminant level for arsenic in drinking water varies from country to country. In much of the developed world, the standard has been set at 10 μg/L, as recommended by the World Health Organization in 1996 [25]; however, in some parts of the world, the standard currently ranges from 25 to 50 μg/L. Arsenic in drinking water is a problem of global scope [48] and has been described in diverse areas including Bangladesh [38], China [50], India [36], Mexico [15, 23], Vietnam [9], and the United States [29, 37, 49]. In 2006, the US Environmental Protection Agency (USEPA) lowered the maximum contaminant level for arsenic from 50 to 10 μg/L. Thus, some public water supplies in the United States that had been under the old limit were required to explore ways to meet the new, lower limit. An example is the southwestern United States, where some areas have groundwater with elevated arsenic concentrations.

The Arizona Department of Environmental Quality published its "Arsenic Master Plan" [5] in anticipation of the USEPA lowering of the maximum contaminant level from 50 to 10 μg/L. The plan stated that water systems could choose treatment options or nontreatment options to comply with arsenic standards. Nontreatment options included blending of water, replacing water sources with new sources, or connecting to another water system. The plan pointed out that nontreatment options tend to be more economical, easier to implement, and easier to manage than treatment options. Various alternatives have been considered and implemented by affected communities in the Phoenix and Tucson areas and in other parts of the state.

The New Mexico Environment Department Drinking Water Bureau published in 2004 its "Arsenic Compliance Strategy" [5, 40] for water supplies affected by the lowering of the USEPA maximum contaminant level. Options for water supplies included the following: (i) obtaining a new water source, (ii) blending water sources, (iii) modifying water sources, (iv) consolidation with another system, and (v) installing a treatment system for arsenic removal from water. Benefits and drawbacks were listed for each option. These included capital costs, water rights (much of New Mexico includes water-scarce regions), configurations of existing systems, monitoring infrastructure, availability of suitable water, water chemistry, waste handling, political factors, and social concerns.

3.2.1 Field Examples of Alternative Water Sources to Replace Arsenic-Contaminated Supplies

Provision of an alternative source of drinking water usually is preferable to treatment for removal of arsenic from contaminated supplies, primarily for reasons such as lower costs, ease of implementation, and decreased management requirements. Among the most practical and economical solutions for removing exposure to arsenic is to provide an alternative source of drinking water. Individual summaries for this option are described in the following sections for the Ganges Delta (Bangladesh), the Rosebud Indian Reservation in South Dakota (United States), Mount Rushmore National Memorial (United States), and the municipalities of Zacatecas and Guadalupe in the state of Zacatecas (Mexico).

3.2.1.1 Ganges Delta, Bangladesh In Bangladesh, drinking water supplies are based primarily on groundwater. Approximately 95% of the rural population in Bangladesh depends on tube wells for drinking water. More than 100 million persons rely on shallow tube wells. The shallow tube wells typically are small-diameter pipes, about 5 cm diameter, that were inserted at depths less than 200 m. Most of the tube wells are capped with cast iron or steel hand pumps [38]. It has been estimated that 35–77 million residents are at risk of exposure to arsenic-contaminated water from tube wells in Bangladesh [26]. The number of tube wells in Bangladesh has been estimated to be 6–11 million [28]; about 4.94 million were tested for arsenic between 2000 and 2003 [21]. Of these, about 1,400,000 (29% of the tested wells) had unsafe levels of arsenic.

Groundwater normally is free of harmful bacteria. Much of the surface water in Bangladesh, however, is contaminated with bacteria and viruses and should be treated by chlorination or other methods before domestic use. In the 1970s, high levels of infant and child mortality were common in Bangladesh; in response, about 95% of the population was systematically converted to groundwater for its domestic supplies, largely from shallow tube wells. At the time of the installation of tube wells, testing of water for arsenic levels was not routine. The tube wells provided an alternative to contaminated surface water, especially in the Ganges Delta, but problems with chronic arsenic poisoning from the shallow groundwater later became obvious. Although gastrointestinal diseases and deaths from diarrhea have declined dramatically in the area, the effects of the arsenic-contaminated water in Bangladesh have been described as the largest mass poisoning of a population in history [38].

In a study by Rahman et al. [33], in Bangladesh, symptoms of chronic arsenic poisoning had developed after 6 months to 2 years or more of exposure to water with high arsenic concentrations. Symptoms included melanosis, leukomelanosis, and keratosis, as well as problems relating to the gastrointestinal system; nervous system; lungs, liver, and other organs; and cardiovascular system. Problems of arsenic poisoning are intertwined with poverty and some of the cultural traditions of Bangladesh. UNICEF and various authors have stated that those with arsenic poisoning in Bangladesh suffer social stigma [1, 27]. For example, arsenicosis can cause darkening of the skin. In some parts of Bangladesh, lighter-skinned women are considered more attractive, which makes it more difficult for a woman to marry if she suffers from arsenic poisoning. In addition, it has been reported that many in Bangladesh believe arsenic poisoning is a curse or is contagious.

Potential sources of alternative supplies in Bangladesh include groundwater, surface water, and rainwater. In general, deeper wells within the country appear to have lower arsenic concentrations. The cost of new wells can be substantial, but typical methods of arsenic removal are not practical or affordable in much of the country. Site-specific conditions govern the most appropriate choice of alternatives. Groundwater options, surface water options, and some rainwater harvesting techniques have been used experimentally for alternative supply options in Bangladesh [2]. Where deeper wells are not an option because of factors such as economics, harvesting of rainwater is a potentially viable alternative, although water systems based completely on rainwater would be expensive.

The government of Bangladesh, along with national and international development partners, has developed mitigation technologies for arsenic [26]. However, most of the treatment methods for arsenic removal were discontinued within months of installation, mainly because of reasons such as high costs, efficiency, social problems, or other factors [26]. In 2004, the government published its "National Policy for Arsenic Mitigation and Its Implementation Plan." The policy emphasized alternative options such as improved dug wells, tube wells with acceptably low arsenic concentrations, pond sand filters, rainwater harvesting, and piped systems from safe water sources [26].

An investigation of alternative water supply options in Bangladesh [3, 42] described the four options of shallow dug wells, deep tube wells, rainwater harvesting, and pond sand filters. The authors found that user satisfaction and social acceptability of the options were area specific, depending on the quality and availability of water. Samples of water from dug wells, for example, showed microbial contamination in 95% of the wells that were tested. Although none of the dug wells that were sampled had an arsenic concentration greater than 50 μg/L, about 35% of these showed concentrations about 10 μg/L.

Clearly, a low-cost, low-technology solution is needed in order to provide safe drinking water to the affected population of Bangladesh.

3.2.1.2 Rosebud Indian Reservation, South Dakota, United States The Rosebud Indian Reservation (Fig. 3.1), in south-central South Dakota, is a rural area with scattered small towns. Todd County, in which the reservation is located, is one of the poorest counties in the United States, with a poverty rate of 49%. Unemployment levels on the reservation have been greater than 80% [42]. In many cases, housing and plumbing on the reservation are substandard. The area is underlain by the Arikaree

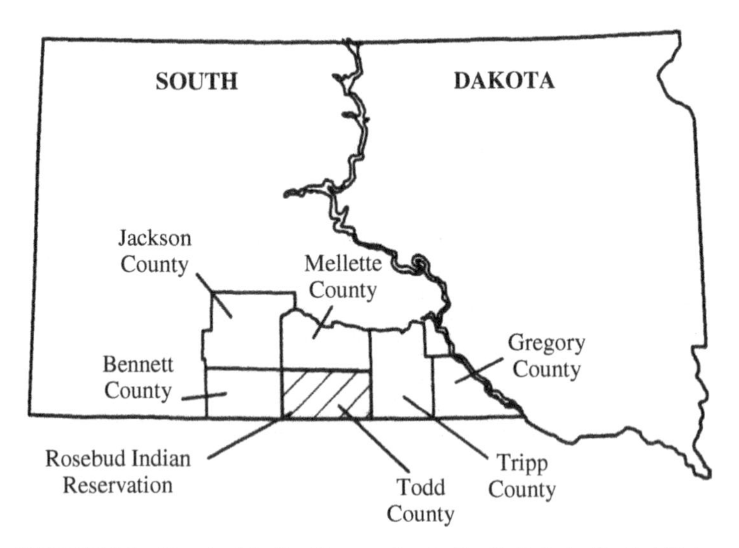

FIGURE 3.1 Rosebud Indian Reservation in South Dakota, United States.

aquifer, which generally supplies high-quality groundwater with reasonably low concentrations of total dissolved solids (TDS). However, the groundwater in certain areas of this sandstone aquifer contains arsenic concentrations that exceed the USEPA's maximum contaminant level of 10 μg/L. For example, a well that supplied the village of Two Strike near the reservation has shown arsenic concentrations as high as 21 μg/L, while a well at Parmelee has shown 19 μg/L and a well at Soldier Creek has shown 13 μg/L [39]. These concentrations were below the USEPA's old maximum contaminant level of 50 μg/L, which was in force until 2006, but are far above the current allowable limit. Numerous Arikaree wells in the area of Grass Mountain Community on the reservation have shown arsenic concentrations up to 0.110 μg/L, and a sample from a spring fed by the Arikaree aquifer had an arsenic concentration of 0.117 μg/L [11]. The City of Martin, South Dakota, about 30 miles west of the Rosebud Indian Reservation, has several wells in the Arikaree aquifer; samples from some of these have exceeded 10 μg/L.

The major source of arsenic in the shallow groundwater of the Rosebud Indian Reservation appears to be bedded ashfalls within sandstone and silt of the Arikaree Formation. The ash is likely to have originated from Tertiary-age volcanic eruptions farther west in Nevada or the Yellowstone area. Infiltrating recharge of slightly oxidized, mildly acidic rainwater and snowmelt then mobilized the arsenic and transported it laterally in dissolved form in groundwater of the Arikaree aquifer. A second possible source of arsenic in the Arikaree aquifer, at least locally, could be arsenic from poisoning of prairie dogs [11].

Because of the rural nature of the Rosebud Indian Reservation, conventional methods of arsenic removal from water are not practical. Instead, alternative sources of water usually are sought in cases where arsenic concentrations exceed the allowable limit. Wells in this area typically are several hundred feet deep, which can be a significant expense; however, small communities and rural homes in this area normally cannot afford the cost of a treatment system to remove arsenic from water, and the necessary infrastructure is not present. In some parts of the reservation, high-quality groundwater is available from the shallower Ogallala aquifer, but this aquifer is present only in the south-central part of the reservation and farther south in Nebraska.

Groundwater from deeper aquifers is not feasible in most cases because the costs of drilling a new well 1000 ft deep or greater are prohibitive, even if water is available from formations such as the low-permeability Precambrian basement rocks in this area. As mentioned, poverty levels are high in this area, with severe unemployment and few opportunities for jobs on the reservation. Deeper aquifers on the Rosebud reservation would not be productive enough, in most cases, to provide water for municipal supplies. In this semiarid region, harvesting of rainwater would not be likely to provide enough even for small municipal supplies.

Many of the water supplies of the smaller villages on the Rosebud Indian Reservation have recently connected to the Mni Wiconi pipeline, which brings water of relatively high chemical quality from the Missouri River, more than 100 miles to the northeast.

Arsenic concentrations in the Arikaree aquifer also vary laterally. The City of Martin, west of the reservation, has had some success in drilling Arikaree wells at different locations. Some of the city's wells provide water with arsenic concentrations below 10 μg/L; other wells with greater arsenic are not used currently.

3.2.1.3 Mount Rushmore National Memorial, South Dakota, United States

Drinking water at Mount Rushmore National Memorial is supplied by groundwater in Precambrian bedrock of the central Black Hills. The sculpture at Mount Rushmore (Fig. 3.2) is a major tourist attraction in the Black Hills. More than 2,000,000 persons per year have visited the monument each year since 2009. Tourist numbers typically are greatest during the summer months.

Groundwater that supplies wells and springs at Mount Rushmore National Memorial is contained in fractured Precambrian metamorphic and igneous rocks [32] as well as weathered colluvium [34]. Precambrian metamorphic rocks within Mount Rushmore National Memorial include arsenopyrite-bearing schist, phyllite, quartzite, and metagraywacke. The park's water supply formerly was obtained from Well 3, northeast of the sculpture. This well was completed in Precambrian phyllite and schist, which is recharged by infiltration through weathered colluvium.

FIGURE 3.2 The carving at Mount Rushmore National Memorial, United States, is within an intrusion of the Precambrian-age Harney Peak Granite. Underlying this is Precambrian phyllite and schist (visible below bust of Washington). In 2008, arsenic levels of 13–14 μg/L were detected in samples from the well that supplied water to the national memorial. The well was completed in Precambrian phyllite and schist.

Well 3 was drilled to a depth of 200 ft deep and was the major water supply for the park for several decades. The well can be pumped at 53 gallons per minute with 30 ft of drawdown. Fieldwork and related research [34] showed that the water has the lowest TDS of any listed in South Dakota public water supplies; one sample showed only 70 μg/L TDS. However, recent sampling has shown arsenic concentrations in the range of about 13 μg/L [10]. Deeper wells are not a practical option at Mount Rushmore National Memorial because existing wells are in Precambrian rock whose permeability decreases sharply with depth. Springs or shallower wells would be subject to surface contamination. Blending of the well's water with a planned new well could help bring the system into compliance. Current plans also include coagulation and low-pressure membrane treatment.

The nearby town of Keystone, approximately 2 miles from Mount Rushmore National Memorial, has had problems with arsenic in water from wells drilled in rock types that are similar to those at the memorial. The Precambrian rocks in this area are steeply dipping, and arsenic concentrations in wells can vary laterally, depending on the amount of mineralization and arsenopyrite present in the rocks at depth. Samples from one well showed arsenic concentrations of 36 μg/L [16]. That well currently is not used. Instead, a new well was drilled nearby and was brought into production when samples showed arsenic concentrations less than 10 μg/L. The City of Keystone depends heavily on tourism because of its proximity to Mount Rushmore National Memorial, and the town's population fluctuates from about 300 to 400 persons, in the winter, to several thousand during the summer months. Thus, water demands vary during the year and increase greatly during the tourist season. The cost of a new well is a substantial expense for the city's small population of year-round residents, but this must be weighed against the cost of a treatment system for arsenic in a small community without the infrastructure or tax base for expensive and technologically advanced arsenic removal.

The historic Keystone Mining District in the Black Hills has several inactive gold mines in which arsenopyrite was associated with the ore. For example, the Bullion Mine at Keystone produced 5630 ounces of gold between 1924 and 1931; in addition, nearly 100,000 pounds of white arsenic was produced [41]. At the nearby Holy Terror Mine in Keystone, a pumping test was conducted for planned dewatering of the mine. The pumping rate during the test was 600 gal/min; water samples showed arsenic concentrations of approximately 34 μg/L.

3.2.1.4 Zacatecas and Guadalupe Municipalities, Zacatecas State, Mexico

Groundwater is an important part of the economy and the national water supply in Mexico. Recent work [23] indicates that groundwater provides about 60% of Mexico's total water supply. Arsenic concentrations at levels above the national guideline of 25 μg/L have been detected in groundwater in several areas of Mexico, including the municipalities of Zacatecas and Guadalupe in the state of Zacatecas [23]. Zacatecas state is in north-central Mexico (Fig. 3.3) and is a region with more than 450 years of mining history, producing metallic ores such as silver and gold. Arsenic-bearing minerals are common in the ore [15]. Data from the US Geological Survey [43] show that Mexico was sixth in world arsenic production during 2010.

FIGURE 3.3 Map of Mexico showing Zacatecas state.

Accumulations of mine tailings are abundant in mining areas of Zacatecas state [15, 35], and soil contamination is severe in some areas. Particularly high levels of arsenic in agricultural soil were found in two irrigation zones of the Zacatecan valley where mine tailings were deposited by rivers [15]; associated with this were very high accumulations of arsenic in plants.

In 2010, the population of the Zacatecas municipality was about 138,000 persons, and the population of Guadalupe was about 160,000 [23]. Four aquifers supply water to the population in this area, with a combined extraction of almost 700 L/s (about 24.7 ft^3/s, or 11,000 gal/min). Recent information [23] shows that extraction of groundwater exceeds average annual recharge in the region. An associated risk is that overexploitation of groundwater containing dissolved metals could affect the population's health [4]. Arsenic concentrations in 48 wells in the aquifer system ranged from 1 to 492.5 μg/L. A study in which 10 extraction wells were sampled (Leal and Gelover [30]) showed that 80% of samples exceeded the Mexican guideline of 25 μg/L for arsenic at that time. Water supplies do not receive municipal treatment for removal of arsenic in Zacatecas and Guadalupe (González Dávila [23]); instead, water authorities blend water from contaminated wells with water from other wells that have a lower arsenic concentration.

A survey (González Dávila [23]) in the municipalities of Zacatecas and Guadalupe investigated arsenic levels in water supplies and associated health effects, as well as strategies and practices for avoiding ingestion of arsenic by residents. Questionnaires and personal interviews provided data on demographics, economic information, educational level, and expenditures for items such as food, medicine, bottled water, and household water, expressed in Mexican pesos. Average monthly expenses for households for tap water were reported to be $91 (as pesos). In contrast, average monthly household expenses for bottled water were $104. The research (González Dávila [23]) noted that this difference indicated a willingness to pay for clean, safe water, as well as indicated that households in the region did not trust the quality standards of their tap water. About 19% of the survey respondents said that tap water was their only drinking water supply, while the rest used a combination of filters, bottled water, and tap water. The study concluded by stressing the need for environmental and epidemiological studies in the Zacatecas and Guadalupe areas, along with a comprehensive public strategy to deal with the problem of arsenic contamination [23].

In cases where an alternative supply of water is not feasible, practical, or economical, treatment of water for arsenic removal must be considered. Treatment options for removing arsenic are described in the next section.

3.3 ARSENIC REMOVAL FROM DRINKING WATER

Remediation of arsenic in drinking water is constrained by the distribution and behavior of arsenic species. Inorganic arsenic exists mainly in the two oxidation states of As(III), or arsenite, and As(V), or arsenate. Arsenic speciation varies in water until equilibrium is reached between H_3AsO_3 (arsenite form) and H_3AsO_4 (arsenate form).

Arsenate in the form of H_3AsO_4 dissociates in three steps:

$$H_3AsO_4 \leftrightarrow H_2AsO_4^- + H^+ \quad K_1 = 6.31 \times 10^{-3} \tag{3.1}$$

$$H_2AsO_4^- \leftrightarrow H_2AsO_4^{2-} + H^+ \quad K_2 = 1.26 \times 10^{-7} \tag{3.2}$$

$$HAsO_4^{2-} \leftrightarrow AsO_4^{3-} + H^+ \quad K_3 = 3.16 \times 10^{-12} \tag{3.3}$$

Arsenite as H_3AsO_3 dissociates in the following steps:

$$H_3AsO_3 \leftrightarrow H_2AsO_3^- + H^+ \quad K_1 = 6.31 \times 10^{-10} \tag{3.4}$$

$$H_3AsO_3^- \leftrightarrow H_2AsO_3^{2-} + H^+ \quad K_2 = 7.41 \times 10^{-13} \tag{3.5}$$

$$HAsO_3^{2-} \leftrightarrow H_2AsO_3^{3-} + H^+ \quad K_3 = 5.01 \times 10^{-14} \tag{3.6}$$

Equilibrium constants shown above are from standard values for aqueous solutions [7].

Figure 3.4 shows fields of stability of various forms of dissolved arsenic in water. Under oxidizing conditions, As(V) is more abundant, as expected, but under reducing conditions, As(III) is more common. Current arsenic remediation methods for drinking water typically perform most effectively in removing arsenate. Preoxidation can be used, however, to convert arsenite to arsenate. Chlorine, ferric chloride, and potassium permanganate are effective, although chlorine can create undesirable byproducts. Ozone and hydrogen peroxide also can oxidize arsenite to arsenate. Performance of arsenic removal systems also can be improved or optimized by pH adjustment. For example, some iron-based processes have greater removal capacities

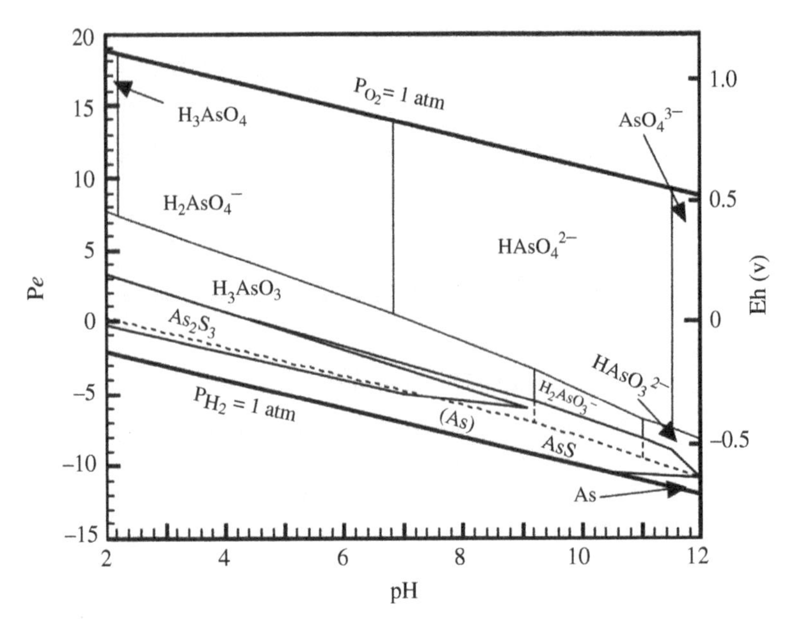

FIGURE 3.4 As–O–H_2O system at 25°C and 1 atm (Reprinted with permission from Drever [18]).

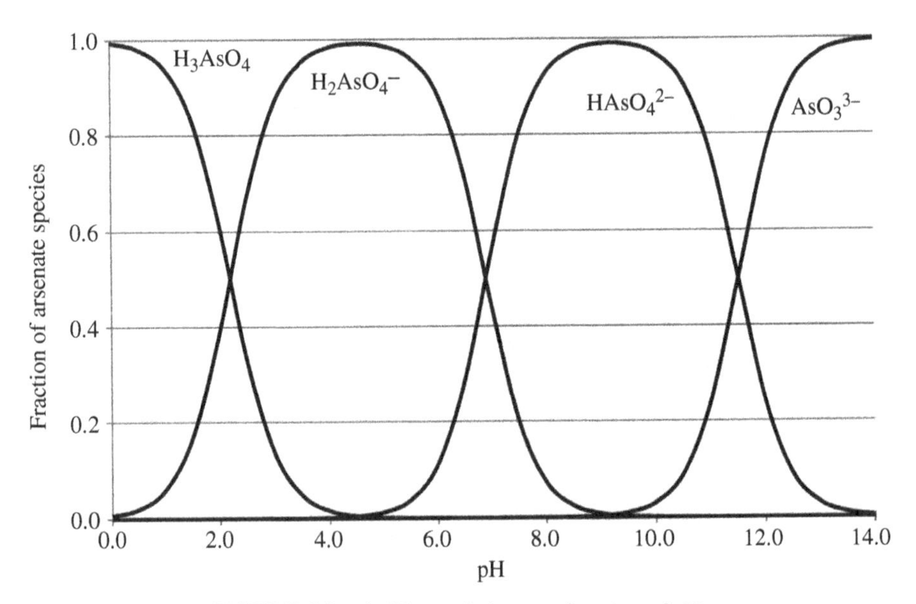

FIGURE 3.5 As(V) speciation as a function of pH.

below a pH value of about 8. Abundance of arsenic species as a function of pH is shown on Figures 3.5 and 3.6.

Arsenic removal technologies include the following: (i) precipitative processes such as coagulation/filtration, lime softening, and iron/manganese oxidation

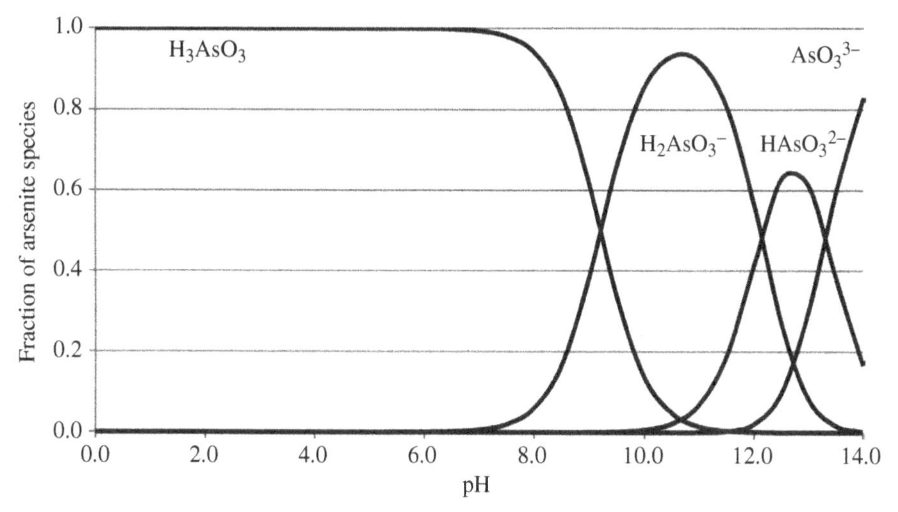

FIGURE 3.6 As(III) speciation as a function of pH.

(including greensand); (ii) adsorptive processes such as activated alumina; (iii) ion exchange; (iv) membrane processes such as reverse osmosis (RO), electrodialysis reversal, and filtration; and (v) alternative or emerging technologies such as granular ferric hydroxide, iron oxide-coated sand, titanium dioxide, photooxidation, and limestone-based material.

Blending of water also is a treatment option, especially if arsenic concentrations in the raw water are only slightly above the maximum contaminant level. Raw water can be blended with water from another source with lower arsenic concentrations, or it can be blended with water that has been treated previously for removal of arsenic. Blending methods reduce the costs of chemical usage and can extend the life of the treatment medium, as well as reduce overall operating costs.

The USEPA has listed the following as the best available technologies for arsenic removal: ion exchange, activated alumina, RO, enhanced lime softening, enhanced coagulation/filtration, and oxidation/filtration (iron removal) [44]. However, the USEPA also noted that these best available technologies often are not suited to smaller systems, especially in rural areas or developing countries where alternative or innovative, emerging technologies typically are more appropriate because of lower costs, smaller infrastructure requirements, or less complicated processes.

Specific arsenic removal technologies are discussed in more detail later.

3.3.1 Precipitative Processes

Precipitative processes for arsenic removal include coagulation/filtration, iron/manganese oxidation, and lime softening.

3.3.1.1 Coagulation/Filtration The coagulation/filtration process operates by creating a colloidal suspension, which then settles out by gravity or can be removed by filtering. In arsenic remediation, coagulants typically are aluminum sulfate, ferric chloride, or ferric sulfate, which can remove dissolved arsenic. A flocculent of larger particles is created, which is more readily filtered or can more easily settle for later removal.

The USEPA [46] has described demonstration projects for arsenic removal by coagulation/filtration at sites in the following areas: Three Forks, Montana; Pentwater, Michigan; Lidgerwood, North Dakota; Arnaudville, Louisiana; Okanogan, Washington; and Felton, Delaware.

3.3.1.2 Manganese Oxidation Fe/Mn oxidation is commonly used for removing arsenic from groundwater. Iron and manganese are treated during oxidation through the formation of hydroxides (e.g., $Fe(OH)_3$, or ferric hydroxide), onto which soluble arsenic species can adsorb. It has been reported [20] that removal of 2 μg/L of iron accomplished 92.5% removal of arsenic by adsorption. The USEPA [45] noted that arsenic removal during manganese precipitation is less effective than with iron.

Pilot projects in which iron removal methods were used have been described by the USEPA [46] at the following sites: Climax, Minnesota; Sabin, Minnesota; Stewart, Minnesota; Sauk Centre, Minnesota; Delavan, Wisconsin; Greenville, Wisconsin; Queen Anne's County, Maryland; Brown City, Michigan; Pentwater, Michigan; Sandusky, Michigan; Fountain City, Indiana; Alvin, Texas; Bruni, Texas; Wellman, Texas; Arnaudville, Louisiana; Goffstown, New Hampshire; Rollinsford, New Hampshire; Buckeye Lake, Ohio; Springfield, Ohio; Rimrock, Arizona; Sells, Arizona; Anthony, New Mexico; Nambe Pueblo, New Mexico; Taos, New Mexico; Three Forks, Montana; Reno, Nevada; and Vale, Oregon.

Greensand filtration also is used for arsenic removal. Greensand is so named because it contains glauconite, a green-colored, iron-rich mineral with ion exchange properties. In producing greensand, the glauconite-containing sand is treated with $KMnO_4$ to coat the particles with manganese dioxide. This treatment method embraces oxidation, ion exchange, and adsorption. Arsenic removal with greensand is more effective at the higher Fe/As ratios [47]. Greensand can be regenerated by treatment with $KMnO_4$ when its arsenic removal capacity is exhausted.

A pilot project using greensand has been described by the USEPA at Waynesville, Illinois [46].

3.3.1.3 Lime Softening Hardness of water is caused by divalent metallic cations, mainly calcium and magnesium. Lime softening can be used as a batch treatment to precipitate carbonates. For example, as lime is added to water, the pH rises, bicarbonate ions are converted to carbonate ions, and calcium carbonate is precipitated. Sodium carbonate also can be used for softening. These methods are used widely in large treatment systems. Lime softening methods are effective for removing arsenic, but the process is pH dependent and the efficiency of removal can be increased by oxidation of As(III) to As(V).

Large amounts of sludge are produced during lime softening, and disposal of the sludge is expensive. In almost all cases, construction of a lime softening plant for removing arsenic would not be practical unless the water also required removal of excessive hardness [44].

3.3.2 Adsorptive Processes

3.3.2.1 Activated Alumina Arsenic removal with activated alumina is accomplished by sorption of ions onto the oxidized activated alumina surface. Although the removal process with activated alumina is considered to be adsorption, the chemical reactions also involve ion exchange [31].

Activated alumina is produced by dehydration of aluminum hydroxide. In addition to its arsenic removal capacity, activated alumina also can remove selenium, fluoride, and some other contaminants at excessive levels. Regeneration of activated alumina is necessary when its adsorptive capacity is exhausted; this requires rinsing with a regenerant such as sodium hydroxide, flushing with water, and acid neutralization.

The optimum pH range for arsenic removal with activated alumina is from about 5.5 to 6, and the pH should always be held below 8.2 because activated alumina then has a net positive charge, which increases its effectiveness for removing anionic arsenic species. The oxidation state of arsenic also is important with activated alumina, which is far more effective in removing As(V), arsenate [44]. As(III), arsenite, is predominantly in its neutral form at this pH range (Fig. 3.6) and thus is less effectively removed.

Activated alumina exhibits ion preference, with increased preference for ions that are not removed as effectively by ion exchange. The addition of 360 μg/L of sulfate and about 1000 μg/L of TDS decreased the adsorption of As(V) by approximately 50%, compared to adsorption of dissolved arsenic in deionized water [13].

The USEPA [46] has described case studies of arsenic removal with activated alumina at Tucson, Arizona; Valley Vista, Arizona; Paramount, California; and Scottsdale, Arizona.

3.3.3 Ion Exchange

During ion exchange, ions on a solid phase are exchanged for ions in water. The solid phase can be a resin or a mineral such as zeolite, which is used to adsorb ions of the contaminant dissolved in the water that is being treated. When the exchange capacity of the medium is exhausted, the bed can be regenerated by rinsing with a concentrated solution of the ions that were originally on the medium before water treatment. Water quality parameters that can influence the effectiveness of ion exchange include pH, alkalinity, competing ions, and arsenic concentration. Also important is the type of resin or zeolite solid phase.

Factors that influence the suitability of ion exchange for a particular application are resin fouling, disposal of the regenerant solution, disposal of bed resin, and other design considerations [14].

Optimum pH values for arsenic removal by ion exchange are typically between about 6.5 and 9. Because arsenic-contaminated groundwater typically has a basic pH, ion exchange sometimes is a suitable technology.

Case studies of ion exchange for arsenic removal have been described by the USEPA [46] at Vale, Oregon, and at Fruitland, Idaho.

3.3.4 Membrane Processes

Membranes act as selective barriers in the removal of contaminants such as arsenic. The driving forces in membrane process can include pressure, concentration, electrical potential, and temperature [44]. Membrane processes also can be distinguished by pore sizes, for example, microfiltration, ultrafiltration, and nanofiltration.

Membrane processes for arsenic removal include RO, electrodialysis reversal, and filtration, which are discussed in the following.

3.3.4.1 RO RO has been used widely for desalination of water. It works by creating a pressure gradient across a membrane that exceeds the osmotic pressure of the treated water and can produce nearly pure water. The volume of discharge water and rejected salts can be substantial: from 10 to 50% of the influent flow [44].

The performance of RO is hindered by iron, manganese, silica, and turbidity. Performance also is strongly influenced by membrane type and operating conditions. RO is more effective in removing As(V) than As(III).

The USEPA [46] has described the results of a pilot project using RO at Carmel, Maine.

3.3.4.2 Electrodialysis Reversal In electrodialysis reversal, ions are transferred through selectively permeable membranes influenced by direct current electrical force. By this process, ions can be forced to migrate toward a solution of greater concentration. Membranes in the electrodialysis reversal process are placed in an array between opposing electrodes, with alternating cation and anion exchange membranes. The migration of cations or anions is governed by the direction of the negative or positive electrodes, in which periodic reversal of the electrodes is used to minimize the potential for fouling of electrodes [44], most likely because of the presence of the neutral form of arsenite at the standard pH range of drinking water.

Electrodialysis reversal can be effective in reducing 85% or more of TDS, with arsenic removal rates of 70% or more, although removal rates for As(III) are lower [44].

3.3.4.3 Filtration Filtration processes for arsenic removal can be grouped into the categories of microfiltration, ultrafiltration, and nanofiltration.

Microfiltration can remove particulate arsenic from water, but its pore size is too large to remove arsenic in dissolved or colloidal form. Microfiltration can be used for treating surface water with arsenic in particulate form. Because most arsenic in groundwater is present in dissolved form, microfiltration is not effective for treating groundwater sources. However, it can be combined with coagulation processes.

Ultrafiltration can remove particulate arsenic, some colloidal arsenic from water, but it typically is not used for removal of dissolved arsenic in groundwater.

Nanofiltration membranes, with their small pore sizes, can remove dissolved arsenic in water. However, a drawback of the small pore sizes is fouling of membranes, which can make pretreatment of water necessary. Removal of As(V) has reached 90% in testing, but removal rates for As(III) have been lower [44].

3.3.5 Alternative or Emerging Technologies

Alternative or innovative emerging technologies include granular ferric hydroxide, iron oxide-coated sand, titanium dioxide, photooxidation, and limestone-based material.

3.3.5.1 ***Granular Ferric Hydroxide*** Adsorption of arsenic on granular ferric hydroxide is a technique that appears to be efficient for treatment of contaminated water. In one study, a treatment capacity of 30,000–40,000 bed volumes was reported, with effluent concentrations less than 10 μg/L [19]. With granular ferric hydroxide, however, adsorption of arsenic decreases as pH increases, and competition from phosphate can hinder arsenic removal. The cost of granular ferric hydroxide can be quite high, and regeneration of the material might not be practical [44].

Case studies of arsenic removal with granular ferric hydroxide have been described by the USEPA [46] at Tucson, Arizona; Paramount, California; and Scottsdale, Arizona.

3.3.5.2 ***Iron Oxide-Coated Sand*** This process involves sand grains that are coated with ferric hydroxide and used in fixed-bed reactors. Studies have shown it is effective for removal of metals as well as arsenic. However, after exhaustion, the reactor bed must be regenerated by rinsing with a regenerant (typically sodium hydroxide), flushing with water, and neutralization with a strong acid such as sulfuric acid. The process is subject to factors such as pH, the oxidation state of arsenic, and ion competition, as well as empty bed contact time. In one study, an increase of pH from 5.5 to 8.5 decreased adsorption of As(V) by 30% (Benjamin et al. [8]). As with many other arsenic removal methods, iron oxide-coated sand is more effective in removing As(V) than AS(III). Dissolved organic carbon in water also reduces arsenic adsorption [44]. Iron oxide minerals also have been used for arsenic removal from water [22].

3.3.5.3 ***Titanium Dioxide*** Titanium dioxide is a relatively new technique for arsenic removal that is accomplished by adsorption of dissolved arsenate. Its performance can be improved by photocatalytic oxidation of As(III) to As(V) before treatment with titanium dioxide (Guan et al. [24]). Titanium dioxide-based material appears to have a large adsorptive capacity; in one study, approximately 45,000 bed volumes of groundwater with an influent concentration of 39 μg/L was treated before exhaustion of the material, with an effluent concentration of less than 10 μg/L (Bang et al. [6]).

Titanium dioxide has been used for arsenic removal in municipal supplies in Arizona. Potential drawbacks in its use include the need for backwashing and the disposal of backwash water.

Pilot projects using titanium dioxide for arsenic removal have been described by the USEPA [46] at Willard, Utah, and Woodstock, Connecticut.

3.3.5.4 Photooxidation Photooxidation primarily involves the oxidation of As(III) to As(V). This process can be accelerated greatly in the presence of light and naturally occurring materials that absorb light. After oxidation, the As(V) can be removed by coprecipitation or by various processes such as those described in this chapter. The photochemical process of oxidation can be accomplished with ultraviolet lamp reactors or by sunlight-assisted photooxidation [44]. Testing has shown substantial reduction of arsenic levels, even in the presence of high levels of dissolved ferrous iron content, relative to arsenic content. Ordinarily, dissolved Fe(II) would present additional chemical oxidant demand in an oxidation system. It also has been reported that the residual products from this process have passed the toxicity characteristic leaching procedure (TCLP) test, indicating that it could be declared as a nonhazardous waste for disposal in a landfill [44].

3.3.5.5 Limestone-Based Material The use of limestone-based material is an innovative process that shows potential for removal of both As(III) and As(V) from drinking water. Limestone in its untreated form is effective for reducing arsenic concentrations below the current limit of 10 μg/L for drinking water, typically resulting in final concentrations of about 4–6 μg/L [16].

The arsenic removal method with untreated limestone appears to be the formation of a low-solubility precipitate of calcium arsenate [16]. One such likely reaction involves the formation of a hydrated calcium arsenate, $Ca_5(AsO_4)_3OH$, known as arsenate apatite. The precipitated arsenate apatite has extremely low solubility, with a K_{sp} of approximately 10^{-40} [51]. The low-solubility products of arsenate apatite and other calcium arsenates are likely to account for the stability of the waste product after arsenic removal with limestone, which has passed the TCLP test, indicating that it is suitable for disposal as nonhazardous waste in an ordinary landfill [16]. In addition, the waste can be encapsulated in concrete, and leaching tests have shown that the concrete-encapsulated waste also passes the TCLP test [12].

The use of untreated limestone for arsenic removal is relatively inefficient, with a low adsorptive capacity before exhaustion is reached, so a large amount of waste is generated. Although untreated limestone is inexpensive, transportation and handling of materials could be drawbacks.

A modification of limestone-based material involves the precipitation of iron on the surface of limestone to produce a synthesized material. This method is effective at removing both arsenate and arsenite because both forms of arsenic adsorb readily to the surface of iron through ionic interactions, especially at elevated pH levels. Limestone as a substrate offers a heterogeneous surface for iron deposition and provides secondary binding sites, a very alkaline surface pH, and buffering capacities. In testing, this method has removed approximately 99% of dissolved arsenic [17]. It also appears to be an efficient process; in testing, 3800 bed volumes of water with an influent concentration of 100 μg/L was treated before exhaustion, with effluent concentrations of about 6 μg/L.

3.4 SUMMARY

Arsenic contamination of groundwater and surface water is a health problem in many parts of the world. In areas where arsenic-contaminated drinking water is a serious concern, efforts to remove exposure to arsenic have focused on alternative sources of water or treatment of the water to remove arsenic.

Provision of an alternative source of water depends on availability. Alternatives could include surface water, groundwater from a shallower or deeper aquifer, water from a piped supply nearby, or rainwater.

The second alternative for removing exposure to arsenic is treatment of water to remove arsenic. Current and potential arsenic removal technologies include (i) precipitative processes such as coagulation/filtration, lime softening, and iron/manganese oxidation; (ii) adsorptive processes such as activated alumina; (iii) ion exchange; (iv) membrane processes such as RO, electrodialysis reversal, and filtration; and (v) alternative or emerging technologies such as granular ferric hydroxide, iron oxide-coated sand, titanium dioxide, photooxidation, and limestone-based material.

Problems and benefits are associated with each option. There are problems and benefits associated with each of these approaches that are related to cultural, socioeconomic, and engineering aspects.

REFERENCES

[1] M.A. Abedin, R. Shaw, Community-level arsenic-mitigation practices in southwestern part of Bangladesh. Community, Environment and Disaster Risk Management 13 (2013) 51–73.

[2] M.F. Ahmed, C.M. Ahmed Arsenic mitigation in Bangladesh, Local Government Division, Ministry of Local Government Rural Development & Cooperatives, Government of the People's Republic of Bangladesh (2002). http://users.physics.harvard.edu/~wilson/arsenic/conferences/Feroze_Ahmed.html. Accessed June 13, 2015.

[3] M.A. Alam, M.M. Rahman, Assessment of dugwell as an alternative water supply option in arsenic affected areas of Bangladesh. International Journal of Civil & Environmental Engineering 11 (2011) 124–137.

[4] M.A. Armienta, N. Segovia, Arsenic and fluoride in the groundwater of Mexico. Environmental Geochemistry and Health 30 (2008) 345–353.

[5] Arsenic master plan summary report. 2001. Arizona Department of Environmental Quality. Online Source: http://www.rwaa.info/files/Arsenic_Master_Plan_Summary_Report.pdf.

[6] S. Bang, M. Patel, L. Lippincott, X. Meng, Removal of arsenic from groundwater by granular titanium dioxide adsorbent. Chemosphere 60 (2005) 389–397.

[7] A.J. Bard, R. Parsons, J. Jordan, Standard potentials in aqueous solution, CRC press, New York, (1985).

[8] M.M. Benjamin, R.S. Sletten, R.P. Bailey, T. Bennett, Sorption and filtration of metals using iron-oxide-coated sand. Water Research 30 (1996) 2609–2620.

[9] M. Berg, H.C. Tran, T.C. Nguyen, H.V. Pham, R. Schertenleib, W. Giger, Arsenic contamination of groundwater and drinking water in Vietnam: a human health threat. Environmental Science & Technology 35 (2001) 2621–2626.

[10] H. Betemariam, A. Davis, D. Dixon, M. Hansen, Arsenic contamination in the Mount Rushmore area and proposed removal techniques. Geological Society of America Abstracts with Programs 42 (2010) 44.

[11] J.M. Carter, S.K. Sando, T.S. Hayes. Source, occurrence, and extent of arsenic in the Grass Mountain area of the Rosebud Indian Reservation, South Dakota. Water Resources Investigation Report 97-4286, 1–90, (1998). U.S. Geological Survey.

[12] P.K. Chintalapati, A.D. Davis, M.R. Hansen, J.L. Sorensen, D. Dixon, Encapsulation of limestone waste in concrete after arsenic removal from drinking water. Environmental Earth Sciences 59 (2009) 185–190.

[13] D. Clifford, C.C. Lin. Arsenic removal from groundwater in Hanford, California-a preliminary report, Environmental Engineering, University of Houston, Department of Civil/Environmental Engineering, Houston, (1986).

[14] D.A. Clifford, G.L. Ghurye Metal-oxide adsorption, ion exchange, and coagulation-microfiltration for arsenic removal from water. in: W.T. Frankenberger, Jr. (Ed.), Environmental chemistry of arsenic, CRC Press, New York, (2002), pp. 217–245.

[15] O.G. Dávila, J.M. Gómez-Bernal, E.A. Ruíz-Huerta, Plants and soil contamination with heavy metals in agricultural areas of Guadalupe, Zacatecas, Mexico. in: J. Srivastava (Ed.), Environmental contamination, InTech, Rijeka, (2012), pp. 37–50.

[16] A. Davis, C. Webb, D. Dixon, J. Sorensen, S. Dawadi, Arsenic removal from drinking water by limestone-based material. Mining Engineering 59 (2007) 71.

[17] A.D. Davis, C.J. Webb, J.L. Sorensen, D.J. Dixon, H. Betemariam, Laboratory testing of trace metals removal from mine drainage and arsenic removal from groundwater in the Black Hills of South Dakota. Environmental Earth Sciences 72 (2014) 355–361.

[18] J.I. Drever. The geochemistry of natural waters: surface and groundwater environments (3rd edition), Prentice Hall, Upper Saddle River, NJ, (1997).

[19] W. Driehaus, M. Jekel, U. Hildebrandt, Granular ferric hydroxide—a new adsorbent for the removal of arsenic from natural water. Aqua 47 (1998) 30–35.

[20] M. Edwards, Chemistry of arsenic removal during coagulation and Fe-Mn oxidation. Journal-American Water Works Association 86 (1994) 64–78.

[21] S.V. Flanagan, R.B. Johnston, Y. Zheng, Arsenic in tube well water in Bangladesh: health and economic impacts and implications for arsenic mitigation. Bulletin of the World Health Organization 90 (2012) 839–846.

[22] M. Gallegos-Garcia, K. Ramírez-Muñiz, S. Song, Arsenic removal from water by adsorption using iron oxide minerals as adsorbents: a review. Mineral Processing and Extractive Metallurgy Review 33 (2012) 301–315.

[23] O. González Dávila, Water arsenic and fluoride contamination in Zacatecas Mexico: an exploratory study. 8th International Conference "Developments in Economic Theory and Policy." The University of the Basque Country, Spain, June 29–July 1, 2011 (2011).

[24] X. Guan, J. Du, X. Meng, Y. Sun, B. Sun, Q. Hu, Application of titanium dioxide in arsenic removal from water: a review. Journal of Hazardous Materials 215 (2012) 1–16.

[25] Guidelines for drinking-water quality: recommendations. 2004. World Health Organization. Online Source: http://www.who.int/water_sanitation_health/dwq/GDWQ2004web.pdf. Accessed April 16, 2015.

[26] B.A. Hoque, S. Yamaura, A. Sakai, S. Khanam, M. Karim, Y. Hoque, S. Hossain, S. Islam, O. Hossain, Arsenic mitigation for water supply in Bangladesh: appropriate technological and policy perspectives. Water Quality Research Journal of Canada 41 (2006) 226–234.

[27] M.Z.H. Khan, Managing the arsenic disaster in water supply: risk measurement, costs of illness, and policy choices for Bangladesh. South Asian Network for Development and Environmental Economics, Kathmandu, (2007).

[28] D.G. Kinniburgh, P.L. Smedley, Arsenic contamination of groundwater in Bangladesh, Volume 3: Hydrochemical atlas, British Geological Survey, Keyworth, (2001).

[29] A. Kumar, P. Adak, P.L. Gurian, J.R. Lockwood, Arsenic exposure in US public and domestic drinking water supplies: a comparative risk assessment. Journal of Exposure Science & Environmental Epidemiology 20 (2010) 245–254.

[30] M. Leal, S. Gelover. Evaluación de la calidad del agua subterranea de fuentes de abastecimiento en acuíferos prioritarios de la región Cuencas Centrales del Norte. Anuario IMTA 2002.

[31] F.W. Pontius, Water quality and treatment: a handbook of community water supplies, AWWA, New York, (1990).

[32] J.E. Powell, J.J. Norton, D.G. Adolphson, Water resources and geology of Mount Rushmore National Memorial, South Dakota, U.S. Government Printing Office, Washington, (1973).

[33] M.M. Rahman, U.K. Chowdhury, S.Ch. Mukherjee, B.K. Mondal, K. Paul, D. Lodh, B.K. Biswas, C.R. Chanda, G.K. Basu, K.C. Saha, Chronic arsenic toxicity in Bangladesh and West Bengal, India-a review and commentary. Clinical Toxicology 39 (2001) 683–700.

[34] P.H. Rahn, Ground-water recharge at Mount Rushmore. Proceedings of the South Dakota Academy of Science PSDAA 2 (1990) 69.

[35] M.A. Salas-Luevano, E. Manzanares-Acuña, C. Letechipía-de León, H.R. Vega-Carrillo, Tolerant and hyperaccumulators autochthonous plant species from mine tailing disposal sites. Asian Journal of Experimental Sciences 23 (2009) 27–32.

[36] S. Sarkar, L.M. Blaney, A. Gupta, D. Ghosh, A.K. SenGupta, Arsenic removal from groundwater and its safe containment in a rural environment: validation of a sustainable approach. Environmental Science & Technology 42 (2008) 4268–4273.

[37] M.E. Serfes, S.E. Spayd, G.C. Herman. Arsenic occurrence, sources, mobilization, and transport in groundwater in the Newark Basin of New Jersey, Advances in Arsenic Research, ACS Symposium Series, ACS Publications, Washington, DC, (2005), pp. 175–190.

[38] A.H. Smith, E.O. Lingas, M. Rahman, Contamination of drinking-water by arsenic in Bangladesh: a public health emergency. Bulletin of the World Health Organization 78 (2000) 1093–1103.

[39] South Dakota. Office of Drinking Water, Public water system data. Office of Drinking Water, South Dakota Department of Water and Natural Resources, Pierre, SD (1986).

[40] State of New Mexico Arsenic compliance strategy. 2004. New Mexico Environment Department. Online Source: http://www.nmenv.state.nm.us/dwb/contaminants/documents/StateArsenicStrategyAugust2004Final.pdf. Accessed April 16, 2015.

[41] U.S. Bureau of Mines, Bureau of Mines staff, Region V, Black Hills mineral atlas, South Dakota (in two parts) Part 2, United States. Bureau of Mines, Pittsburgh (1953).

[42] U.S. Department of the Interior, American Indian population and labor force report. (2005). Online Source: http://www.bia.gov/cs/groups/public/documents/text/idc-001719.pdf. Accessed April 16, 2015.

[43] U.S.G. Survey, Mineral commodity summaries, U.S. Department of the Interior, Bureau of Mines, U.S. Geological Survey, Reston, Virginia (2011).

[44] USEPA, Technologies and costs for removal of arsenic from drinking water. Agency, U.S. Environmental Protection. (2000). EPA 815-R-00-028, U.S. Environmental Protection Agency Office of Ground Water and Drinking Water. Online Source: http://water.epa.gov/drink/info/arsenic/upload/2005_11_10_arsenic_treatments_and_costs.pdf. Accessed April 16, 2015.

[45] USEPA, National primary drinking water regulations; arsenic and clarifications to compliance and new source contaminant monitoring; final rule. Agency, U.S. Environmental Protection. v. 66, 6981. (2001). Federal Register, U.S. Environmental Protection Agency. Online Source: http://www.epa.gov/rfa/drinking-water-arsenic.html. Accessed April 16, 2015.

[46] USEPA, Arsenic research: publications. U.S. Environmental Protection Agency. (2012). Online Source: http://www.epa.gov/nrmrl/wswrd/dw/arsenic/publications.html. Accessed April 16, 2015.

[47] T. Viraraghavan, K.S. Subramanian, J.A. Aruldoss, Arsenic in drinking water—problems and solutions. Water Science and Technology 40 (1999) 69–76.

[48] J.S. Wang, C.M. Wai, Arsenic in drinking water—a global environmental problem. Journal of Chemical Education 81 (2004) 207–213.

[49] A.H. Welch, D.R. Helsel, M.J. Focazio, S.A. Watkins Arsenic in ground water supplies of the United States. in: W.R. Chappel, C.O. Abernathy, R.L. Calderon (Eds.), Arsenic exposure and health effects III, Elsevier Ltd., Burlington, (1999), pp. 9–17.

[50] X. Xie, Y. Wang, M. Duan, H. Liu, Sediment geochemistry and arsenic mobilization in shallow aquifers of the Datong basin, northern China. Environmental Geochemistry and Health 31 (2009) 493–502.

[51] Y.N. Zhu, X.H. Zhang, Q.L. Xie, D.Q. Wang, G.W. Cheng, Solubility and stability of calcium arsenates at 25°C. Water, Air, and Soil Pollution 169 (2006) 221–238.

4

THE CHEMISTRY AND METABOLISM OF ARSENIC

David J. Thomas

Integrated Systems Toxicology Division, USEPA, Research Triangle Park, NC, USA

Disclaimer: This manuscript has been reviewed in accordance with the policy of the National Health and Environmental Effects Research Laboratory, U.S. Environmental Protection Agency, and approved for publication. Approval does not signify that the contents necessarily reflect the views and policies of the Agency, nor does mention of trade names or commercial products constitute endorsement or recommendation for use.

4.1 INTRODUCTION

A century of study of the process by which many organisms convert inorganic arsenic into an array of methylated metabolites has answered many questions and has posed some new ones. The capacity of microorganisms to form volatile arsenic compounds was first recognized in the nineteenth century [28]. This observation prompted Frederick Challenger and his colleagues to study arsenic methylation in microorganisms and to postulate a chemically plausible sequence of reactions in which pentavalent arsenic is reduced to trivalency and the resulting trivalent arsenical is oxidatively methylated (for a historical review see [19]). Identification of genes encoding arsenic methyltransferase in the three domains of the tree of life—Bacteria, Archaea, and Eukarya—provide an opportunity to test the chemical plausibility of Challenger's scheme and

Arsenic: Exposure Sources, Health Risks, and Mechanisms of Toxicity, First Edition.
Edited by J. Christopher States.
© 2016 John Wiley & Sons, Inc. Published 2016 by John Wiley & Sons, Inc.

other alternate schemes for arsenic methylation. Although our understanding of molecular aspects of enzymatically catalyzed methylation of arsenic remains in flux, data are sufficient to assert that conversion of inorganic arsenic to methylated species is a major determinant of the distribution and retention of arsenic among tissues and is an important factor in its actions as a toxicant and carcinogen.

In this chapter, attention first focuses on metabolic processes that convert inorganic arsenic into methylated oxyarsenical species; that is, compounds in which an arsenic atom is bound to one or more methyl groups and one or more oxygen atoms, and on the toxicological significance of the oxidation state of arsenic present in methylated oxyarsenicals. Attention then focuses on methylated thioarsenicals, a class of compounds in which an arsenic atom is bound to one or more methyl groups and one or more sulfur atoms, that are metabolites of inorganic arsenic and on the linkage between metabolic processes involved in formation of methylated oxyarsenicals and of methylated thioarsenicals. Finally, attention is focused on evidence linking the metabolism of complex organic arsenicals (e.g., arsenosugars, arsenobetaine, arsenolipids)[1] to the metabolism of methylated arsenicals. Integrating knowledge of these metabolic processes will provide a better understanding of aggregate exposure to arsenic in a variety of forms. Ultimately, this information will provide improved dosimetry for mode of action studies and for risk assessment and risk management decisions.

4.2 BIOMETHYLATION OF ARSENIC

4.2.1 Evolution of the Concept of Arsenic Biomethylation

Our understanding of arsenic biomethylation parallels growth and development of analytical chemistry and biochemistry that began in the nineteenth century. Recognition that the action of fungi resident on wallpaper colored with arsenic-containing pigments released a malodorous volatile compound ("Gosio gas") was a critical first step in the study of arsenic biomethylation (for a historical review see [28]). Identification of Gosio gas as trimethylarsine led to use of microorganisms to study of arsenic biomethylation. Work in microorganisms by Frederick Challenger and associates resulted in the so-called Challenger scheme for arsenic biomethylation [17, 18]. This eponymous scheme begins with an ionized trivalent arsenic atom that is oxidatively methylated by addition of a positively charged methyl group (CH_3^+) to yield a methylated product containing pentavalent arsenic.

[1] In this chapter, inorganic arsenic and its methylated metabolites are identified on the basis of the oxidation state of the arsenic atom in the molecule. For oxy-arsenicals, inorganic species are arsenate (As^V) and arsenite (As^{III}). Methylated species are monomethylarsonic acid (As^V), monomethylarsonous acid (As^{III}), dimethylarsinic acid (As^V), dimethylarsinous acid (As^{III}), trimethylarsine oxide (As^V), and trimethylarsine (As^{III}). Corresponding thioarsenical species are identified explicitly (e.g., thioarsenate or monomethylmonothioarsenate). Naming of complex organic arsenicals varies widely in the references cited in this chapter. As a general rule, terminology used in an original reference has been retained in this chapter.

Reduction of the pentavalent arsenic atom then precedes the next round of oxidative methylation [19, 30]. The sequence of reactions can be written as

$$\begin{aligned} As^{V}O_4^{3+} + 2e \rightarrow As^{III}O_3^{-3} + CH_3^{+} \rightarrow CH_3As^{V}O_4^{-2} + 2e \rightarrow CH_3As^{III}O_2^{-2} + CH_3^{+} \rightarrow \\ (CH_3)_2As^{V}O^{-1} + 2e \rightarrow (CH_3)_2As^{III}O^{-1} + CH_3^{+} \rightarrow (CH_3)_3As^{V}O \end{aligned} \tag{4.1}$$

As discussed in Section 4.3.2, this pathway of alternating methylation and reduction reactions is a chemically plausible pathway to produce an array of methylated arsenicals containing either trivalent or pentavalent arsenicals.

Improved analytical methods developed in the latter half of the twentieth century lead to detection of methylated arsenicals in environmental and biological samples. For example, hydride generation–cryotrapping-column chromatography-atomic absorption spectrometry detected inorganic and organic (methylated) arsenicals in natural waters and in human urine [8]. Perhaps the seminal work of this era was detection of methylated and dimethylated arsenic in urine of a volunteer after ingestion of wine or water that contained inorganic arsenic [27]. Notably, appearance of monomethylated arsenicals preceded appearance of dimethylated arsenicals in urine, a pattern broadly compatible with that predicted by the Challenger scheme depicted in Equation 4.1.

4.2.2 Arsenic Biomethylation in Humans and Other Organisms

Development and application of new analytical techniques that discriminate between arsenic in the pentavalent and trivalent oxidation state made possible the identification of methylated arsenicals containing trivalent arsenicals in biological samples. For example, monomethylarsonous and dimethylarsinous acid are found in urine from individuals who chronically ingest drinking water containing inorganic arsenic [4, 74, 85, 148] and in urine of patients who receive arsenic trioxide for treatment of acute promyelocytic leukemia [141]. The presence of monomethylarsonous acid in the bile of rats treated with arsenate or arsenite [57] and of dimethylarsinous acid and trimethylarsine oxide in rats treated with dimethylarsinic acid [83] provides additional evidence that methylated metabolites containing arsenic in the trivalent oxidation state are stable intermediates formed during biomethylation. Studies in mice treated with arsenite found the amount of dimethylarsinous acid in urine is dosage dependent, suggesting metabolic saturation at high dosage levels [55]. Monomethylarsonic acid and dimethylarsinous acid are detected in livers of mice treated with arsenite [32, 33]. Although the concentrations of methylated metabolites containing trivalent arsenic in urine have been used as biomarkers of exposure in epidemiological studies [34], their utility has been questioned due to concerns about stability of these compounds in urine [69]. Standardization of conditions for sample collection and processing as well as rigorous validation of analytical methods [31] would support their use in future population-based studies.

4.2.3 Toxic Consequences of Arsenic Methylation

Our understanding of the toxicological significance of arsenic methylation has evolved in recent years. Historically, formation of methylated metabolites has been regarded primarily as a process that detoxifies inorganic arsenic by producing less reactive and toxic methylated metabolites that are excreted more quickly than the parent compound [150]. In humans, ingested monomethylarsonic acid or dimethylarsinic acid are cleared more rapidly than arsenite, suggesting that methylation is a detoxification process [13, 14]. In rabbits and mice treated with periodate-oxidized adenosine, an inhibitor of S-adenosylmethionine-dependent methyltransferases, diminished capacity for conversion of arsenite into its methylated metabolites is associated with higher retention of arsenic in tissues [86, 87]. In pregnant mice treated with periodate-oxidized adenosine, reduced capacity for arsenic methylation enhances maternal toxicity and reduces fetal weights in mice treated with sodium arsenite or sodium arsenate [73].

The central role of methylation as a determinant of tissue distribution and retention of arsenic and as a factor affecting the toxicity of arsenic has been illuminated further by studies of the fate and effects of arsenicals in mice genetically engineered to minimize capacity for arsenic methylation. Knockout of the arsenic (+3 oxidation state) methyltransferase gene (*As3mt*) in mice produces animals with a greatly diminished capacity to convert inorganic arsenic into its methylated metabolites. This modification results in prolonged and elevated retention of inorganic arsenic in tissues [41, 66]. Altered tissue clearance of arsenic in *As3mt* knockout mice treated with inorganic arsenic or dimethylarsinous acid exacerbates damage to the uroepithelium [20, 39, 151]. These findings are consistent with the hypothesis that As3mt-catalyzed methylation of arsenic reduces inorganic arsenic concentrations in target tissues and that altered methylation capacity affects toxicity. In sum, these findings emphasize the importance of understanding metabolic processes that control formation of methylated arsenicals.

Studies in cellular systems demonstrate that some methylated derivatives of inorganic arsenic are more reactive and toxic than inorganic arsenic and can be linked to unique toxic effects. In a variety of test systems, mono- or dimethylated arsenicals that contain trivalent arsenic are more potent cytotoxins, genotoxins, and enzyme inhibitors than arsenite [81, 89, 97, 103, 104, 111, 125, 126]. Mono- and di-methylated arsenicals affect cell signaling pathways, including pathways involved in insulin-dependent glucose uptake and insulin secretion by pancreatic β-cells [40, 139]. A common mode of action of methylated arsenicals containing trivalent arsenic is the induction of a pro-oxidant state in cells. Monomethylarsonous acid treatment generates reactive oxygen species (ROS) in mitochondria of rat liver RLC-16 cells [101] and causes ARE-mediated activation of thioredoxin reductase gene expression in WI-38 lung fibroblasts [92]. Similarly, treatment of normal human epidermal keratinocytes with monomethylarsonous acid increases expression of genes involved in processes leading to arsenic-driven skin carcinogenesis, including genes involved in response to oxidative stress, in production of growth factors, and in modulation of cell signaling pathways [6].

The effects of monomethylarsonous acid on the proliferative properties of UROtsa cells, a SV40-immortalized human uroepithelial cell line, illustrate that a methylated arsenical can induce cellular transformation (see Chapter 8 for detailed discussion of arsenical transformation of bladder cells). Exposure of UROtsa cells to 50 nM monomethylarsonous acid reduces cell doubling time, causes anchorage-independent growth and produces cells that are tumorigenic in SCID mice [9, 146]. Transformation of UROtsa cells by monomethylarsonous acid exposure alters DNA methylation patterns, and produces a non-reverting phenotype in transformed cells [68]. Changes in DNA methylation induced by monomethylarsonous acid exposure are accompanied by changes in cytokine (e.g., IL1, IL6, IL8, TNF) levels consistent with a chronic inflammatory response [45]. Monomethylarsonous acid-induced changes in cytokine production in UROtsa cells are coupled with changes in gene expression commonly seen in cells exposed to oxidants, suggesting that monomethylarsonous acid exposure evokes a prooxidant state in cells [91]. At a molecular level, monomethylarsonous acid exposure of UROtsa cells affects expression and activity patterns for the DNA repair enzyme, poly(ADP-ribose) polymerase 1 (PARP1). Although monomethylarsonous acid exposure increases levels of apo-PARP1 in UROtsa cells, its DNA repair activity is reduced because monomethylarsonous acid displaces zinc from a zinc finger structure in the protein, preventing formation of the catalytically active holoenzyme [147]. A detailed discussion of arsenical inhibition of PARP1 is presented in Chapter 13.

In animal models methylated arsenicals can exert unique toxic or carcinogenic effects. In rats, exposure *in utero* to dimethylarsinic acid from gestational days 7 to 16 increases incidence of palatine rugae in the absence of overt maternal toxicity [114] and is a nephrotoxin in adult rats [96]. Methylated metabolites of inorganic arsenic can act as complete carcinogens or as promoters in rodents. For example, monomethylarsinous acid is a complete transplacental carcinogen in the mouse [135], dimethylarsinic acid is both a tumor promoter [80, 149] and a complete carcinogen in the rat [24], and trimethylarsine oxide is a liver carcinogen in the rat [119]. Arsenic carcinogenesis in animal models is discussed in detail in Chapter 19.

4.3 BIOCHEMICAL BASIS OF ARSENIC BIOMETHYLATION

4.3.1 Enzymology of Arsenic Methylation

As described in Section 4.2.2, identification of mono- and dimethylated arsenicals in the urine of humans exposed to ingested or inhaled inorganic arsenic [27, 122] sparked interest in biomethylation of this metalloid. Early studies of arsenic biomethylation using *in vitro* reaction mixtures that contained rat liver cytosol or with rat liver slices provided evidence of an enzymatically catalyzed reaction that used arsenite as the preferred substrate and S-adenosyl methionine as the methyl donor [11, 12, 56].

These results motivated studies to isolate, purify, and characterize arsenic methyltransferases. Functionally, any enzyme that catalyzes conversion of inorganic arsenic into mono-, di-, and tri-methylated products is an arsenic methyltransferase (EC # 2.1.1.137). One line of inquiry initially characterized arsenite methyltransferase and monomethylarsonous acid methyltransferase activities from rabbit liver cytosol and cultured human hepatocytes [153, 155] and from hamster liver [145]. Additional studies suggested that a monomethylarsonic acid reductase activity catalyzed the rate-limiting step in the Challenger scheme for arsenic methylation [154].

An arsenic methyltransferase activity purified from rat liver cytosol (molecular weight of about 42 kDa) catalyzes S-adenosyl methionine–dependent methylation of arsenite and monomethylarsonous acid [82]. Partial amino acid sequencing of this protein identified it as the predicted product of the *cyt19* gene previously annotated as a methyltransferase of unknown activity. The *cyt19* gene and its protein product have been given the systematic name of arsenic (+3 oxidation state) methyltransferase (*As3mt*). Although other enzymes (e.g., N-6 adenine-specific DNA methyltransferase 1, N6AMT1) may catalyze methylation of inorganic arsenic [61, 113], a preponderance of evidence [41–43] suggests that in higher organisms orthologous *As3mt* genes encode the primary arsenic methyltransferase.

Given differences in purification procedures and analytical methods, comparing the catalytic activity of rat As3mt and the rabbit liver arsenic methyltransferase is difficult. With monomethylarsonous acid as substrate, the kilometer is lower (250 nM *v.* 100 μM) and the Vmax is higher (68 pmol *v.* 4 pmol) for rat As3mt than for the rabbit liver enzyme. Thus, the activity of As3mt would be kinetically favored over that of the rabbit liver arsenic methyltransferase. In the absence of further sequence data on the rabbit liver enzyme, it is not possible to assess the linkage, if any, between the two enzymes.

4.3.2 Common Features of Arsenic Methyltransferases

In Eukarya, proteins encoded by orthologous As3mt genes are the primary catalysts of arsenic methylation [133]. In Bacteria and Archaea, proteins encoded by orthologous arsenic methyltransferase (ArsM) genes catalyze conversion of inorganic arsenic to methylated metabolites. Highly conserved amino acid sequences in ArsM and As3mt proteins from organisms across the three domains of life suggest that common molecular processes underlie arsenic methylation catalyzed by these enzymes [134]. Alignment of predicted amino acid sequences of human AS3MT and ArsM from a thermophilic eukaryotic alga *Cyanidioschyzo*n sp. 5508 shows extensive sequence conservation in organisms that diverged over one billion years ago (Fig. 4.1). Here, common sequence motifs for non-DNA methyltransferases (I, I′, II, and III) occur in both proteins with complete conservation of critical residues (e.g., DI in motif II and GGE in motif III). Site-directed mutagenesis shows that conservation of three cysteinyl residues is required for catalytic activity in each protein. In *Cyanidioschyzo*n ArsM, cysteinyl residues at positions 72, 174, and 224 are required for conversion of arsenite to methylated products; a cysteine to serine mutant at position 72 methylates monomethylarsonous acid [88]. In human AS3MT, cysteinyls at positions 156, 206, and 250 are required for catalytic activity [123]. Formation of intramolecular disulfides by these cysteinyl residues probably affects the structure and catalytic activity of the protein [47, 123].

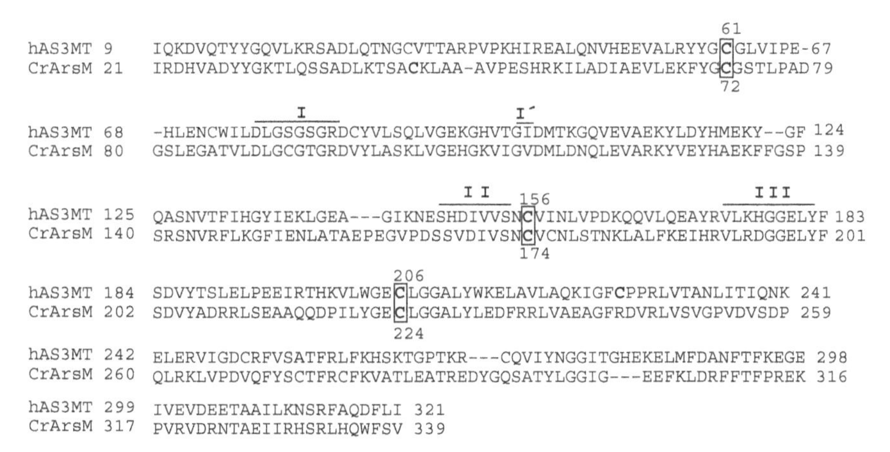

FIGURE 4.1 Conserved amino acid sequences in human AS3MT (hAS3MT) and Cyanidioschyzon sp. 5508 ArsM (CrArsM). Cysteinyl residues required for arsenic methyltransferase activity of each protein marked in boxes. Positions of common sequences motifs (I, I′, II, and III) of non-DNA methyltransferases are indicated.

4.3.3 Current Models of Enzymatically Catalyzed Arsenic Methylation

Studies of the kinetic properties of ArsM and As3mt provide insights into molecular processes underlying arsenic methylation. In particular, research on mechanisms of arsenic methylation has focused on two related issues. First, what physiological reductants are required for catalysis by arsenic methyltransferase? This issue reflects a long-standing interest in the roles of thiols in the reduction of pentavalent arsenicals and of arsenic-thiol complexes in arsenic metabolism. Second, do reactions catalyzed by arsenic methyltransferases proceed in accordance with the prediction of the Challenger scheme for biomethylation? The following paragraphs summarize the current state of research germane to these questions.

4.3.3.1 Role of Reductants The role of reductants in catalysis by arsenic methyltransferases has long been a topic of interest. Early studies showed that a monothiol such as glutathione (GSH) supports arsenic methylation capacity of cell homogenates and supernates [12] and that addition of GSH to buffers facilitates purification of arsenic methyltransferase activity from rat liver cytosol [82]. Surprisingly, GSH does not support recombinant rat As3mt (rAs3mt)-catalyzed methylation of arsenite to form mono-, di-, and tri-methylated metabolites. However, dithiol reductants strongly stimulate rAs3mt-catalyzed production of methylated arsenicals from arsenite [142]. In reaction mixtures that contained thioredoxin (Tr), glutaredoxin (Gr), or dihydrolipoic acid (DHLA) in coupled reaction systems with thioredoxin reductase (Trx) or GSH/glutathione reductase (Grx), and NADPH to regenerate the dithiol, each reductant supports arsenite methylation by rAs3mt. Addition of GSH to reaction mixtures containing rAs3mt and Tr/Trx/NADPH increases overall yield of methylated metabolites, especially dimethylated arsenic, but suppresses trimethylated arsenic formation [143]. Use of GSH as the sole reductant in hAS3MT-catalyzed reactions found relatively low rates of methylation of arsenite or monomethylarsonous acid [38]. By comparison, dimethylated arsenic is the predominant metabolite of either substrate in hAS3MT-catalyzed reactions with the Tr/Trx/NADPH-coupled system. In hAS3MT-catalyzed reactions with GSH as reductant, monomethylated arsenic is the predominant metabolite. In hAS3MT-catalyzed reactions, use of a non-thiol reductant, tris (2-carboxyethyl) phosphine (TCEP), yields a pattern of metabolites similar to that found in reactions with the Tr/Trx/NADPH-coupled system. Similarities in pattern and extent of methylation using a thiol and a non-thiol reductant may argue against a critical role of arsenic-thiol complexes as substrates for methylation. Kinetic analysis of hAS3MT-catalyzed reactions with arsenite and a monothiol reductant (GSH or L-cysteine) or a non-thiol reductant (TCEP) indicates a fully ordered sequence of reactions that begins with S-adenosyl methionine binding to the enzyme, continues with reduction of intramolecular disulfides, and concludes with transfer of a methyl group from S-adenosyl methionine to arsenite [140]. Data from hAS3MT-catalyzed reactions containing TCEP strongly suggest that formation of arsenic-thiol complexes is not required for catalysis but that the reductant may reduce critical disulfide bonds in the enzyme. Because GSH complexes of arsenite or monomethylarsonous acid bind faster to *Cyanidioschyzo*n sp. 5508 ArsM than do arsenite or monomethylarsonous acid, formation of GSH-arsenical complexes could facilitate methylation in the GSH-rich environment of the cell [88].

4.3.3.2 Molecular Processes in the Function of Arsenic Methyltransferases A central issue in the biomethylation of arsenic is the nature of reactions that transfer a methyl group from S-adenosyl methionine to an arsenical. As noted earlier, the Challenger scheme for arsenic biomethylation posits that addition of a methyl group to an arsenic atom in the trivalent oxidation state is an oxidative process yielding a methylated metabolite with an arsenic atom in the pentavalent oxidation state. A recent review examining the reactions that transfer a methyl group from a donor to an arsenic atom concludes that the Challenger scheme is the most plausible pathway for methylation [29]. However, earlier studies of enzymatically catalyzed methylation of arsenic in *in vitro* systems suggest that methylation occurs without a change in the oxidation state of the arsenic atom; that is, arsenic persists in the trivalent oxidation state through multiple rounds of methylation [63, 100]. Use of recombinant AS3MT and ArsM has led to evaluation of molecular processes involved in enzymatically catalyzed methylation of arsenic. Studies with *Cyanidioschyzo*n *merolae* ArsM suggest that arsenic remains in the trivalent oxidation state during initial binding to the protein and during transfer of a methyl group from S-adenosyl methionine [2]. Similarly, studies with human AS3MT suggest that a trivalent arsenical bound to the protein is not oxidized during methyl group addition [140]. For both enzymes, initial protein binding of substrate (a trivalent arsenical) depends on the presence of critical cysteinyl residues, and, as noted above, three cysteines in each protein are required for catalytic activity. If enzymatically catalyzed addition of a methyl group does not oxidize a trivalent arsenic atom to pentavalency, then appearance of methylated products containing pentavalent arsenic requires release of the methylated arsenical from enzyme and oxidation of arsenic in the liberated product. Reconciling these findings with recombinant enzymes with a chemically plausible scheme for transfer of a methyl group from S-adenosyl methionine to an arsenical substrate is a challenge for future research.

4.4 THIOLATION OF ARSENICALS

4.4.1 Detection of Thioarsenicals in Biological Systems

A thioarsenical is an arsenic-containing molecule that contains an arsenic-sulfur bond analogous to an arsenic-oxygen bond found in the corresponding oxyarsenical. For example, thioarsenate ($(OH)_3$-As(=S)) and arsenate ($(OH)_3$-As(=O)) are thioarsenical-oxyarsenical analogs. Interconversion of oxyarsenical and thioarsenical species is pH dependent [26] and arsenic in a thioarsenical readily oxidizes to pentavalency (Planer-Friedrich et al. [106]). Thus, in an oxygen-rich cellular environment, trivalent thioarsenicals are likely converted to pentavalent thioarsenicals. Detection of thioarsenicals in biological matrices is complicated by their instability in the presence of oxygen or reactive iron [128]. Until reliable procedures are developed to collect, store, and process biological samples for analysis of thioarsenicals, these metabolites may be of limited value in dosimetric studies or as biomarkers of exposure to inorganic arsenic [116]. Because thioarsenicals are relatively refractory to generation of arsines, their detection may be impaired in analytical methods that depend on hydride generation [64]. However, development of other analytical approaches such as mass spectrometry [44] has allowed rapid progress in detection and quantitation of thioarsenicals in biological samples.

Despite concerns about stability of these compounds, thioarsenicals have been detected in urine of mice and rats exposed to arsenate, arsenite, or dimethylarsinic acid [1, 41, 66, 99, 129]. A dimethylthioarsenical species was present in urines of Bangladeshi women who ingest drinking water contaminated with inorganic arsenic [110]. Urine from sheep that grazed on arsenic-rich seaweed on beaches in the Orkney Islands contained a variety of methylated oxy-arsenicals and a novel thioarsenical metabolite, 2-dimethylarsinothioyl acetic acid, derived from arsenosugar catabolism [46, 59]. This finding emphasizes a relation between the metabolism of arsenosugars, a class of complex organic arsenicals that occur in many foodstuffs, and the exposure to methylated thioarsenicals. This relation is considered in more detail is Section 4.5.2.

4.4.2 Origin of Thioarsenicals

The production of thioarsenicals from oxyarsenicals was demonstrated in the late nineteenth century in studies that exposed trialkyl arsine oxides to hydrogen sulfide (H_2S) [60]. H_2S is a “gasotransmitter” or signaling molecule in the cardiovascular, gastrointestinal, and nervous systems [79, 90, 102] and may regulate ATP production [51]. H_2S is produced in the gastrointestinal tract and in mammalian tissues. In the gastrointestinal tract, resident microbes produce H_2S by dissimilation of reduced organic compounds and by scavenging of H_2 formed by fermentation [36]. In rodents, microbial H_2S production rates are highest in cecum and colon, although H_2S production occurs from stomach to distal colon [36, 127]. Because H_2S is produced throughout the gastrointestinal tract, it can play a role in the preabsorptive metabolism of arsenicals which is discussed in Section 4.4.4.

In mammalian tissues, pyridoxal-5′-phosphate-dependent enzymes, cystathionine-β-synthase (CBS), cystathionine-γ-ligase (CSE), and cysteine aminotransferase (CAT) catalyze catabolism of sulfur-containing amino acids, cysteine and homocysteine to produce H_2S [120]. Relative contributions of these pathways to H_2S production are determined by levels of substrates and cofactors in tissues and by kinetic characteristics of relevant enzymes [121]. For example, homocysteine concentrations vary several-fold in different tissues of mice and tissue homocysteine concentrations correlate positively with levels of mRNAs for genes encoding enzymes involved in transsulfuration reactions and methionine regeneration [21]. Similarly, variation in tissue cysteine concentrations and capacity to metabolize this amino acid [124] may affect its role as a H_2S precursor. Estimated tissue H_2S concentrations vary widely [144] with some variability reflecting different analytical methods [70]. Rapid catabolism of and reversible binding of H_2S to proteins probably yields low nanomolar concentrations in tissues [53, 70, 78, 138].

Notably, the metabolic processes involved in H_2S production in mammalian tissues can be linked with the metabolic cycle for enzymatically catalyzed methylation of arsenic (Fig. 4.2). Both well-characterized arsenic methyltransferases, As3mt and

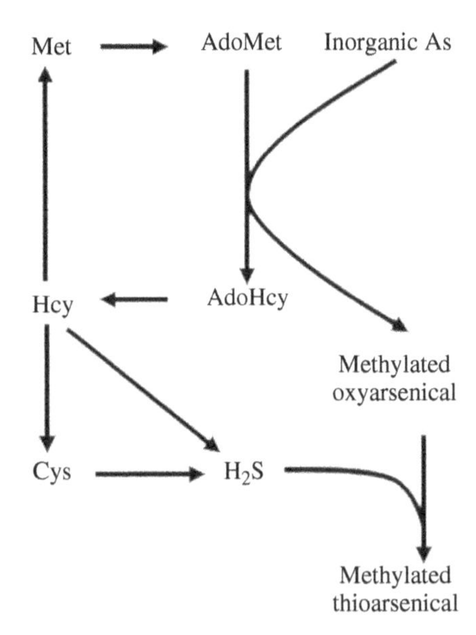

FIGURE 4.2 Linkage of metabolic pathways for methylation of arsenicals and production of hydrogen sulfide in mammalian tissues. Generation of S-adenosylhomocysteine (AdoHcy) during transfer of a methyl group to a methylated oxyarsenical is followed by production of homocysteine. Homocysteine can be recycled to form methionine, a precursor for S-adenosylmethionine (AdoMet) or serve as a direct or indirect precursor through cysteine for hydrogen sulfide (H_2S) that is used to convert a methylated oxyarsenical to a methylated thioarsenical.

ArsM, catalyze reactions that use S-adenosyl methionine as the methyl group donor. S-Adenosylhomocysteine produced in this reaction is enzymatically converted to homocysteine, the nexus for interaction of H_2S production and arsenic methylation. Homocysteine can be recycled to yield methionine, the direct precursor for S-adenosyl methionine, or serve as a substrate for H_2S production. Furthermore, as described in Section 4.4.4, S-adenosyl methionine-dependent production of a methylated oxyarsenical may be directly linked to the production of a homologous methylated thioarsenical.

4.4.3 Distribution and Toxicity of Thiolated Arsenicals

The acute cytotoxicity of thiolated arsenicals has been evaluated in a variety of test systems. In cultured human A549 lung adenocarcinoma cells, thiolated dimethylarsinic acid is a more potent cytotoxin than arsenite or a number of methylated compounds containing trivalent or pentavalent arsenic [7]. In human EJ1 urinary bladder cancer cells, dimethyldithioarsenate induces a pro-oxidant state and is cytotoxic [98]. In this cell line, dimethylarsinous acid and dimethyldithioarsenate are the most cytotoxic arsenicals [97]. Notably, exposure to these arsenicals results in similar cellular concentrations of arsenic. This study and another study [76] suggest that the cytotoxic potency of dimethyldithioarsenate might be related to facile accumulation in cells. A recent study found that among identified metabolites of arsenosugars, its thio-dimethylated arsenical metabolite is more cytotoxic in human uroepithelial UROtsa cells than is arsenite (LC_{30} of 2.3 μM *vs*. 3.6 μM) [75].

As recently summarized [112], the disposition and clearance of arsenic in rodents treated with dimethylmonothioarsenate approximates the pattern after treatment with dimethylarsinous acid. By comparision, dimethyldithioarsenate shows a pattern of tissue distribution and urinary clearance that approximates that of dimethylarsinic acid. Given the cytotoxic potency of dimethylarsinous acid, it is possible that dimethylmonothioarsenate distributed among tissues could produce an array of toxic effects.

4.4.4 Linkage of Thiolation and Methylation of Arsenic

As noted in Section 4.4.1, exposure to arsenate, arsenite, or dimethylarsinic acid is associated with the presence of thioarsenicals in urine of rodents or humans. These results suggest that processes involved in arsenic methylation are linked to processes which convert oxy-arsenicals to thio-arsenicals. Direct evidence of a linkage between methylation and thiolation has been obtained in *in vitro* systems which contain mouse cecal anaerobic microbiota. As noted above, microorganisms resident in the gastrointestinal tract produce H_2S and earlier work demonstrates that these organisms methylate arsenate and arsenite [58]. These factors suggest that examination of preabsorptive metabolism of arsenicals, that is, metabolism that occurs in the lumen of the gastrointestinal tract before absorption across the gastrointestinal barrier, might provide insights into the linkage between thiolation and methylation processes. In anaerobic reaction mixtures containing mouse cecal microbiota, arsenate is converted into a variety of methylated oxy- and thio-arsenical species [105], suggesting

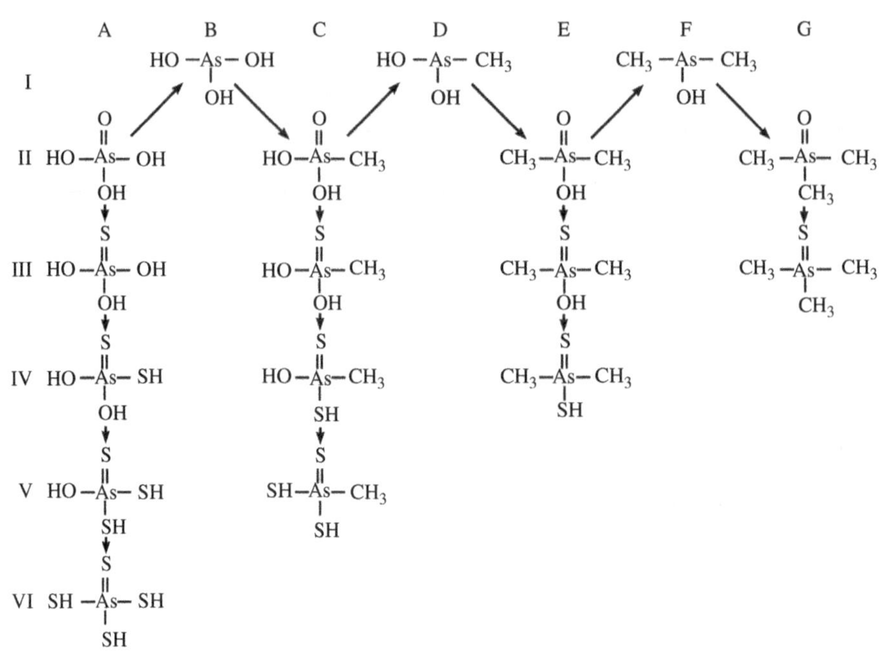

FIGURE 4.3 The linkage of pathways for formation of thio- and oxy-arsenicals. Arsenate (II A) is reduced to arsenite (I B) and converted to monomethylarsonic acid (II C). After reduction to monomethylarsonous acid (I D), it is converted to dimethylarsinic acid (II E). This species is reduced to dimethylarsinous acid (I F) and converted to trimethylarsine oxide (II G). Arsenate (II A) undergoes sequential reactions to form monothioarsenate (III A), dithioarsenate (IV A), trithioarsenate (V A), and tetrathioarsenate (VI A). Monomethylarsonic acid (II C) undergoes sequential reactions to form monomethylmonothioarsenate (III C), monomethyldithioarsenate (IV C), and monomethyltrithioarsenate (V C). Dimethyarsinic acid (II E) undergoes sequential reactions to form dimethylmonothioarsenate (III E) and dimethyldithioarsenate (IV E). Trimethylarsine oxide (II G) is converted to trimethylarsine sulfide (III G).

a linkage between methylation and thiolation pathways. Figure 4.3 shows a scheme for linkage of methylation of oxy-arsenical and formation of either inorganic or methylated thio-arsenicals. This scheme posits that methylation of oxy-arsenicals follows the Challenger scheme for biomethylation and that thio-arsenicals originate from oxy-arsenicals containing pentavalent arsenic. Additional studies with mouse cecal anaerobic microbiota have found that methylated thio-arsenicals, like methylated oxy-arsenicals, are substrates for methylation [71, 72]. Studies with [^{34}S]-labeled dimethylmonothioarsinic acid as substrate found [^{34}S]-labeled trimethylarsine sulfide as a product, suggesting that a methylated thio-arsenical is directly converted to a trimethylated thio-arsenical. This finding is surprising as it implies that a methylated thio-arsenical containing pentavalent arsenic or an unknown derivative is a substrate for methylation. To date, it is uncertain whether any arsenic methyltransferase can accept a methylated thio-arsenical as substrate. In terms of dosimetry and exposure assessment, preabsorptive metabolism of ingested arsenicals could result in concurrent exposure to an array of methylated oxy- and thio-arsenicals that display different patterns of systemic distribution, metabolism and toxicities.

4.5 METABOLISM OF COMPLEX ORGANIC ARSENICALS

4.5.1 Occurrence of Complex Organic Arsenicals in Biological Systems

Although the study of arsenic as a toxicant and a carcinogen has largely focused on inorganic arsenic and its methylated metabolites, a variety of complex organic arsenicals occur in biological materials, including many foodstuffs. Among these compounds arsenobetaine, arsenocholine, and arsenosugars are probably better characterized; arsenolipids have been recognized and studied only in recent years. Structures of these complex organic arsenicals are shown in Figure 4.4.

4.5.1.1 Arsenobetaine and Arsenocholine These complex organic arsenicals are the products of biosynthetic processes in a variety of organisms. Arsenobetaine does not occur in marine phytoplankton; its presence in marine zooplankton suggests that this compound is an osmolyte in these organisms [16]. Consumption of marine phytoplankton introduces this compound into higher trophic levels of the food chain so that much of the arsenic present in a variety of marine organisms is in the form of arsenobetaine. Hence, arsenobetaine makes a large contribution to arsenic intake in humans. For example, in a Spanish population, arsenobetaine in seafoods accounts for more than 90% of the total arsenic intake [48].

Studies of the metabolism of arsenobetaine have yielded divergent results. In mice, orally administered [^{73}As]-labeled arsenobetaine is highly bioavailable with most of the radioarsenic recovered in urine as the administered compound [136]. In contrast, studies of the fate of arsenobetaine administered orally to rats found that urine contained the parent compound, trimethylarsine oxide, dimethylarsinic acid, monomethylarsonic acid, tetramethyarsonium ion, arsenate, arsenite, and an unknown arsenical [152]. Studies in humans provide some evidence on the kinetic behavior of arsenobetaine in humans. An early study in volunteers who consumed [^{74}As]-labeled arsenobetaine in a fish meal found rapid distribution and clearance of the radioarsenic but provides no data on metabolic transformation [10]. A study of the kinetics of clearance of arsenic from blood of volunteers after ingestion of a fish meal in which arsenobetaine is stated (although unproven) to be the major source of arsenic found a biphasic pattern of clearance of arsenic from blood with half-life values of 7.1 and 63 h [77]. The authors suggest that the slower component of clearance from blood is related to ingestion of arsenic in the form of arsenobetaine.

Early work with cultured primary rat hepatocytes or with subcellular fractions from rat liver found that arsenocholine is converted to arsenobetaine aldehyde; the latter compound is degraded to trimethylarsine oxide and trimethylarsine [23]. Choline dehydrogenase or a similar enzyme likely catalyses conversion of

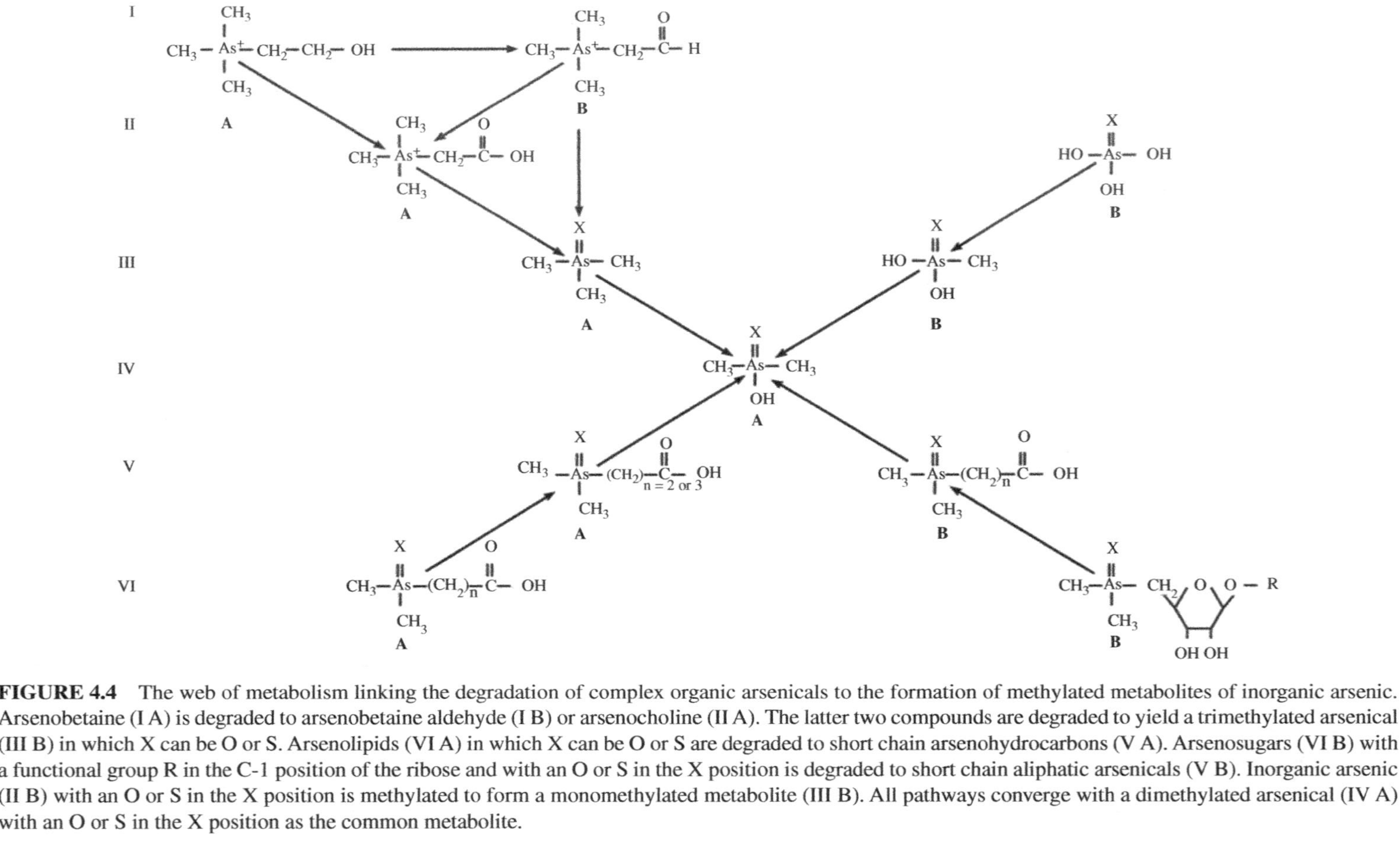

FIGURE 4.4 The web of metabolism linking the degradation of complex organic arsenicals to the formation of methylated metabolites of inorganic arsenic. Arsenobetaine (I A) is degraded to arsenobetaine aldehyde (I B) or arsenocholine (II A). The latter two compounds are degraded to yield a trimethylated arsenical (III B) in which X can be O or S. Arsenolipids (VI A) in which X can be O or S are degraded to short chain arsenohydrocarbons (V A). Arsenosugars (VI B) with a functional group R in the C-1 position of the ribose and with an O or S in the X position is degraded to short chain aliphatic arsenicals (V B). Inorganic arsenic (II B) with an O or S in the X position is methylated to form a monomethylated metabolite (III B). All pathways converge with a dimethylated arsenical (IV A) with an O or S in the X position as the common metabolite.

arsenocholine to arsenobetaine and trimethylarsine oxide, although the steps in this pathway are not well understood [22]. More recent research on the metabolism of arsenobetaine in higher organisms has focused on the role of commensal microbes with two general schemes proposed for arsenobetaine degradation [37]. In the first, the carboxymethyl-arsenic bond of arsenobetaine is cleaved to generate trimethylarsenic oxide with subsequent cleavage of methyl-arsenic bonds producing di- and mono-methylated arsenicals. In the second, cleavage of methyl-arsenic bonds of arsenobetaine generates dimethylarsenoyl acetate. Microorganisms isolated from the mussel *Mytilus edulis* degrade arsenobetaine to a dimethylated product [67]. Studies with aerobic microbiota isolated from human gastrointestinal tract found that arsenobetaine is converted to tri-, di-, and mono-methylated species, including dimethylarsenoyl acetate [62]. Microbes isolated from tissues of the freshwater crayfish *Procambarus clarkii* also convert arsenobetaine to tri-, di-, and mono-methylated species. Isolates of *Pseudomonas putida* have been identified as the organisms responsible for degradation of arsenobetaine [37]. A recent study in volunteers who ingested a variety of seafoods rich in arsenicals provides evidence of net production of arsenobetaine and dimethylarsinic acid based on the balance between the amount of these species ingested and the amount of these species excreted in urine [95]. Production of arsenobetaine could be linked to use of a trimethylarsenic moiety released during phosphatidyarsenocholine degradation [52].

4.5.1.2 Arsenosugars Arsenosugars are a class of compounds in which a dimethylated oxy-arsenic ($(CH_3)_2AsO$) or a dimethylated thio-arsenic moiety ($(CH_3)_2AsS$) is bound at the C-5 position of a D-riboside derivative (e.g. compound VI B in Fig. 4.4). Arsenosugars are notable for the wide variety of derivatives formed through moieties added at the C-1 position of the ribose. It is likely that these compounds are produced in organisms at the lowest level of aquatic food chains with transfer and concentration in organisms at higher trophic levels. Edible species (e.g., fish, shellfish, and algae) can attain part per million concentrations of arsenosugars [65]. Because arsenosugars in seafoods are highly bioaccessible, potential exposure to arsenic from ingestion of arsenosugar-rich foods is substantial [3]. Although arsenosugars are weakly or non-cytotoxic in test systems [75], the products formed from these compounds may be active as toxicants or carcinogens.

The pathway for arsenosugar synthesis has not been fully elucidated. Because a dimethylarsenic moiety is a prominent component of arsenosugars, enzymatically catalyzed methylation of arsenic to form a dimethylated species is probably a step in arsenosugar synthesis. In the unicellular green algae *Chlamydomonas reinhardtii* and freshwater cyanobacteria *Synechocystis* sp. PCC 6803 and *Nostoc* sp. PCC 7120, culture in the presence of arsenate results in formation of mono- and di-methylated arsenicals and accumulation of arsenosugars containing a glycerol or phosphate moiety at C-1 of the ribose [93, 94]. In these organisms, ArsM may catalyze conversion of inorganic arsenic to dimethylated arsenicals as the first step in arsenosugar synthesis, although this hypothesis has not been confirmed experimentally.

Studies in volunteers have examined the fate of ingested arsenosugars. Notably, these studies are largely descriptive, providing evidence on the pattern of arsenic-containing metabolites in urine following consumption of arsenosugars, but not necessarily providing evidence concerning underlying metabolic processes. In addition, studies that used foodstuffs (e.g., seaweed) as the arsenosugar source pose particular difficulties because these foods contain significant amounts of other arsenicals (e.g., dimethylarsinic acid). Concurrent metabolism of these other arsenicals may diminish the study's value in elucidating arsenosugar metabolism. An early study of arsenosugar metabolism in four volunteers after ingestion of commercial seaweed products containing arsenosugars found dimethylarsinic acid and several unknown arsenicals in urine [84]. Ingestion of a synthetic arsenosugars, 2, 3′-dihydroxypropyl 5-deoxy-5 dimethylarsinoyl-β-D-riboside, yields at least 12 arsenic-containing metabolites in urine [50]. Among these metabolites, dimethylarsinic acid predominates, accounting for about 67% of arsenic excreted in urine; both dimethylarsinoylethanol and trimethylarsine oxide are also found in urine. Later studies identified dimethylarsinoylethanol, dimethylarsinic acid, and methylarsonic acid in urine of volunteers who ingested the seaweed *Laminaria* which contains three arsenosugars [137]. After ingestion of 2, 3′-dihydroxypropyl 5-deoxy-5 dimethylarsinoyl-β-D-riboside, urines from volunteers contain thiolated metabolites, including thio-dimethylarsenoacetate, thio-dimethylarsenoethanol, and a thio-arsenosugar, as well as oxo-dimethylarsenoacetate [108]. Considerable variation has been found among volunteers in capacity to metabolize ingested arsenosugars [109]. Over 4 days after ingestion, the percentage of arsenic ingested as an arsenosugar that is recovered in urine ranged from 4 to 95%. Notably, the extent of clearance of ingested arsenic correlates with the complexity of the profile of arsenicals in urine. Individuals excreting the lowest percentage of the oral dose in urine had the lowest levels of the metabolites and the highest levels of the parent compound and its thio-arsenosugar analog in urine. In this study, detection of both thio-dimethylarsenoethanol and thio-dimethylarsenoacetate in serum after ingestion of arsenosugars suggests that these metabolites originate at some point during pre- or post-absorption processing and are not metabolites formed during the residence of urine in the urinary bladder before micturition. Arsenosugars are relatively resistant to degradation under *in vitro* conditions that mimic the environment of the stomach and upper gastrointestinal tract [3]. Treatment of four arsenosugars isolated from seaweed extract with simulated gastric juice at pH 1.1 hydrolyzed all compounds at the C-1 carbon on the ribose ring, yielding an arsenosugar with molecular mass of 254 [54]. Notably, these acidic conditions do not result in the formation of other arsenic containing products. In an anaerobic reaction mixture containing mouse cecal microbiota that produce H_2S, 3-[5′-deoxy-5-(dimethylarsinoyl)-beta-ribofuranosyloxy]-2-hydroxypropanesulfonic acid is quantitatively converted to its thiolated analog [25]. Similarly, reaction mixtures containing lamb liver cytosol convert oxy-arsenosugars extracted from the seaweed *Laminaria digitata* into thiolated analogs of these compounds, suggesting that H_2S generated in liver can provide active S for transformation under aerobic conditions [59]. Thus, although currently available data are not sufficient to identify the site at which arsenosugars are converted to the variety of urinary metabolites found after ingestion of these compounds, significant conversion may occur during preabsorptive metabolism in the gastrointestinal tract as well as during postabsorptive systemic metabolism.

4.5.1.3 Arsenolipids Arsenolipids are structural analogs of neutral lipids that occur in microorganisms, plants, and a wide variety of marine animals [35]. Phosphatidylarsenocholine, a major member of this group, is formed in a reaction catalyzed by phospholipase D [52]. Phosphatidylarsenocholine administered orally to mice is highly bioavailable; arsenic derived from the administered compound is excreted in urine as arsenobetaine as the major metabolite and arsenocholine as a minor metabolite. Neither the site of, nor steps in, phosphatidylarsenocholine metabolism have been determined. Because human intestinal microbiota convert phosphatidylcholine to trimethylamine-N-oxide [132], phosphatidylarsenocholine may be converted to trimethylarsine oxide in the gastrointestinal tract. In cod liver oil, common arsenolipids are analogs of myristic, palmitic, stearic, and arachidic acids in which the methyl group of the fatty acid is replaced with a dimethylarsinoyl moiety [115]. Dimethylarsinoyl-arsenolipids are abundant in fish muscle; about one-half of the 5.9 µg of arsenic per gram of sashimi grade tuna is fat soluble 1-dimethylarsinoylpentadecane and 1-dimethylarsinoyl-all cis 4, 7, 10, 13, 16, 19 docosahexane [130]. Arsenolipids are found in oil prepared from the capelin, *Mallotus villoneus* [107, 131] and in cod liver oil [5]. The metabolic origin of dimethylarsinoyl-arsenolipids is unknown. These compounds might result from biosynthetic infidelity [115]. Synthesis of arsenolipids with even numbers of C atoms cannot use dimethylarsinic acid as the starting material. However, dimethylarsinoylpropionic acid could be the precursor for production of even-numbered arsenolipids. Metabolic studies in volunteers after ingestion of arsenolipid-rich cod liver and cod liver oil found that the peak of urinary excretion of arsenicals occurs within 15 h of ingestion [117, 118]. The major urinary metabolite of arsenolipids is dimethylarsinic acid. Other urinary metabolites include arseno-fatty acids, oxy-dimethylarsenopropanoic acid, thio-dimethylarsenopropanoic acid, oxo-dimethylbutanoic acid, and thio-dimethyarsenobutanoic acid. Although these studies do not identify the sites or the pathways for conversion of arsenolipids to these urinary metabolites, they demonstrate that ingested arsenolipids are bioavailable and can be converted to water-soluble metabolites.

4.5.2 The Web of Metabolism

Dimethylated arsenic is the common metabolite that links degradation of the major classes of complex organic arsenicals to the metabolic process that converts inorganic arsenic into methylated metabolites (Fig. 4.4). Identification of dimethylated arsenic as a common metabolite in part reflects use of urinary metabolites to characterize patterns of arsenic metabolism. Many studies of disposition of dimethylated arsenicals including studies that compared retention of arsenic in humans following ingestion of inorganic, mono-, or di-methylated arsenicals found that arsenic appears in urine most quickly following ingestion of dimethylarsinic acid [14]. Early work noted conversion of complex organic arsenicals into dimethylated arsenicals excreted in urine from seafood-consuming volunteers [15, 84]. Notably, dimethylated arsenicals exert unique toxic effects, including teratogenic effects in mice and carcinogenicity in rats. Thus, evaluation of risk of exposure to either inorganic arsenic or to complex organic arsenicals should consider production of dimethylated metabolites. This approach may be most useful in assessment of risk in situations in which aggregate exposure to arsenic from all sources and in all forms is under consideration. For example, evaluation of aggregate exposure to complex organic arsenicals in food (e.g., seafoods) and to inorganic arsenic in food and drinking water should evaluate risk posed by dimethylated arsenic derived from all sources. In sum, given the evidence of linkage in the metabolic fate of many arsenical species, there is an impetus to reconsider risk assessments for this metalloid in media including water and foods [49].

4.6 SUMMARY AND FUTURE DIRECTIONS

Improved analytical techniques have led to the identification of a wide range of metabolites of inorganic arsenic and of complex arsenicals. In the past decade, it has become possible to determine the oxidation state of arsenic in inorganic arsenicals and their methylated metabolites and to identify a spectrum of thiolated arsenicals. These improved methods make possible dosimetric studies and are at the heart of studies of dose-effect and dose-response relations. Continued work on analytical methods should be complemented with development of procedures that preserve the native oxidation state of arsenic and protects thio-arsenicals from oxidation during sample collection and handling.

Although arsenic has long been identified as a carcinogen in humans, a growing body of research suggests that exposure to this metalloid is also associated with a variety of other adverse health effects. Understanding the mode of action of arsenic in the induction of disease is intimately tied to understanding the array of arsenic containing metabolites that are formed after exposure to inorganic arsenic or complex organic arsenicals. The scope of research on the role of metabolites of inorganic arsenic in disease induction must be expanded to evaluate the potential role of thioarsenicals in mediation of adverse biological effects. This aspect of research will require a better understanding of the linkage of processes that produce methylated oxy- and thio-arsenicals.

4.7 ACKNOWLEDGEMENT

This chapter honors Professor William R. Cullen of the Department of Chemistry, University of British Columbia, on his 80th birthday. Bill's many contributions to understanding the chemical basis of arsenic methylation have influenced a generation of researchers. I gratefully acknowledge his friendship and sage advice over the past two decades.

REFERENCES

[1] B.M. Adair, T. Moore, S.D. Conklin, J.T. Creed, D.C. Wolf, D.J. Thomas, Tissue distribution and urinary excretion of dimethylated arsenic and its metabolites in dimethylarsinic acid- or arsenate-treated rats. Toxicol. Appl. Pharmacol. 222 (2007) 235–242.

[2] A.A. Ajees, K. Marapakala, C. Packianathan, B. Sankaran, B.P. Rosen, Structure of an As(III) S-adenosylmethionine methyltransferase: insights into the mechanism of arsenic biotransformation. Biochemistry 51 (2012) 5476–5485.

[3] C. Almela, J.M. Laparra, D. Velez, R. Barbera, R. Farre, R. Montoro, Arsenosugars in raw and cooked edible seaweed: characterization and bioaccessibility. J. Agric. Food Chem. 53 (2005) 7344–7351.

[4] H.V. Aposhian, E.S. Gurzau, X.C. Le, A. Gurzau, S.M. Healy, X. Lu, M. Ma, L. Yip, R.A. Zakharyan, R.M. Maiorino, R.C. Dart, M.G. Tircus, D. Gonzalez-Ramirez, D.L. Morgan, D. Avram, M.M. Aposhian, Occurrence of monomethylarsonous acid in urine of humans exposed to inorganic arsenic. Chem. Res. Toxicol. 13 (2000) 693–697.

[5] U. Arroyo-Abad, J. Mattusch, S. Mothes, M. Moder, R. Wennrich, M.P. Elizalde-Gonzalez, F.M. Matysik, Detection of arsenic-containing hydrocarbons in canned cod liver tissue. Talanta 82 (2010) 38–43.

[6] K.A. Bailey, S.D. Hester, G.W. Knapp, R.D. Owen, S.F. Thai, Gene expression of normal human epidermal keratinocytes modulated by trivalent arsenicals. Mol. Carcinog. 49 (2010) 981–998.

[7] M. Bartel, F. Ebert, L. Leffers, U. Karst, T. Schwerdtle, Toxicological characterization of the inorganic and organic arsenic metabolite thio-DMA in cultured human lung cells. J. Toxicol. 2011 (2011) 373141.

[8] R.S. Braman, C.C. Foreback, Methylated forms of arsenic in the environment. Science 182 (1973) 1247–1249.

[9] T.G. Bredfeldt, B. Jagadish, K.E. Eblin, E.A. Mash, A.J. Gandolfi, Monomethylarsonous acid induces transformation of human bladder cells. Toxicol. Appl. Pharmacol. 216 (2006) 69–79.

[10] R.M. Brown, D. Newton, C.J. Pickford, J.C. Sherlock, Human metabolism of arsenobetaine ingested with fish. Hum. Exp. Toxicol. 9 (1990) 41–46.

[11] J.P. Buchet, R. Lauwerys, Study of inorganic arsenic methylation by rat liver in vitro: relevance for the interpretation of observations in man. Arch. Toxicol. 57 (1985) 125–129.

[12] J.P. Buchet, R. Lauwerys, Role of thiols in the in-vitro methylation of inorganic arsenic by rat liver cytosol. Biochem. Pharmacol. 37 (1988) 3149–3153.

[13] J.P. Buchet, R. Lauwerys, H. Roels, Comparison of the urinary excretion of arsenic metabolites after a single oral dose of sodium arsenite, monomethylarsonate, or dimethylarsinate in man. Int. Arch. Occup. Environ. Health 48 (1981) 71–79.

[14] J.P. Buchet, R. Lauwerys, H. Roels, Urinary excretion of inorganic arsenic and its metabolites after repeated ingestion of sodium metaarsenite by volunteers. Int. Arch. Occup. Environ. Health 48 (1981) 111–118.

[15] J.P. Buchet, J. Pauwels, R. Lauwerys, Assessment of exposure to inorganic arsenic following ingestion of marine organisms by volunteers. Environ. Res. 66 (1994) 44–51.

[16] G. Caumette, I. Koch, K.J. Reimer, Arsenobetaine formation in plankton: a review of studies at the base of the aquatic food chain. J. Environ. Monit. 14 (2012) 2841–2853.

[17] F. Challenger, Biological methylation. Sci. Prog. 35 (1947) 396–416.

[18] F. Challenger, Biological methylation. Adv. Enzymol. Relat. Subj. Biochem. 12 (1951) 429–491.

[19] T.G. Chasteen, R. Bentley, Historic review: Frederick Challenger, 1887–1983: chemist and biochemist. Appl. Organomet. Chem. 17 (2003) 201–211.

[20] B. Chen, L.L. Arnold, S.M. Cohen, D.J. Thomas, X.C. Le, Mouse arsenic (+3 oxidation state) methyltransferase genotype affects metabolism and tissue dosimetry of arsenicals after arsenite administration in drinking water. Toxicol. Sci. 124 (2011) 320–326.

[21] N.C. Chen, F. Yang, L.M. Capecci, Z. Gu, A.I. Schafer, W. Durante, X.F. Yang, H. Wang, Regulation of homocysteine metabolism and methylation in human and mouse tissues. FASEB J. 24 (2010) 2804–2817.

[22] M.K. Chern, R. Pietruszko, Evidence for mitochondrial localization of betaine aldehyde dehydrogenase in rat liver: purification, characterization, and comparison with human cytoplasmic E3 isozyme. Biochem. Cell Biol. 77 (1999) 179–187.

[23] A. Christakopoulos, H. Norin, M. Sandstrom, H. Thor, P. Moldeus, R. Ryhage, Cellular metabolism of arsenocholine. J. Appl. Toxicol. 8 (1988) 119–127.

[24] S.M. Cohen, T. Ohnishi, L.L. Arnold, X.C. Le, Arsenic-induced bladder cancer in an animal model. Toxicol. Appl. Pharmacol. 222 (2007) 258–263.

[25] S.D. Conklin, A.H. Ackerman, M.W. Fricke, P.A. Creed, J.T. Creed, M.C. Kohan, K. Herbin-Davis, D.J. Thomas, In vitro biotransformation of an arsenosugar by mouse anaerobic cecal microflora and cecal tissue as examined using IC-ICP-MS and LC-ESI-MS/MS. Analyst 131 (2006) 648–655.

[26] S.D. Conklin, M.W. Fricke, P.A. Creed, J.T. Creed, Investigation of the pH effects on the formation of methylated thio-arsenicals, and the effects of pH and temperature on their stability. J. Anal. At. Spectrom. 23 (2008) 711–716.

[27] E.A. Crecelius, Changes in the chemical speciation of arsenic following ingestion by man. Environ. Health Perspect. 19 (1977) 147–150.

[28] W.R. Cullen, Is Arsenic an Aphrodisiac? The Sociochemistry of an Element, The Royal Society of Chemistry, Cambridge, 2008.

[29] W.R. Cullen, Chemical mechanism of arsenic biomethylation. Chem. Res. Toxicol. 27 (2014) 457–461.

[30] W.R. Cullen, B.C. McBride, J. Reglinski, The reduction of trimethylarsine oxide to trimethylarsine by thiols: a mechanistic model for the biological reduction of arsenicals. J. Inorg. Biochem. 21 (1984) 45–60.

[31] J. Currier, R.J. Saunders, L. Ding, W. Bodnar, P. Cable, T. Matousek, J.T. Creed, M. Styblo, Comparative oxidation state specific analysis of arsenic species by high-performance liquid chromatography-inductively coupled plasma-mass spectrometry and hydride generation-cryotrapping-atomic absorption spectrometry. J. Anal. At. Spectrom. 28 (2013) 843–852.

[32] J.M. Currier, M. Svoboda, D.P. de Moraes, T. Matousek, J. Dedina, M. Styblo, Direct analysis of methylated trivalent arsenicals in mouse liver by hydride generation-cryotrapping-atomic absorption spectrometry. Chem. Res. Toxicol. 24 (2011) 478–480.

[33] J.M. Currier, M. Svoboda, T. Matousek, J. Dedina, M. Styblo, Direct analysis and stability of methylated trivalent arsenic metabolites in cells and tissues. Metallomics 3 (2011) 1347–1354.

[34] L.M. Del Razo, G.G. Garcia-Vargas, O.L. Valenzuela, E.H. Castellanos, L.C. Sanchez-Pena, J.M. Currier, Z. Drobna, D. Loomis, M. Styblo, Exposure to arsenic in drinking water is associated with increased prevalence of diabetes: a cross-sectional study in the Zimapan and Lagunera regions in Mexico. Environ. Health 10 (2011) 73.

[35] V.M. Dembitsky, D.O. Levitsky, Arsenolipids. Prog. Lipid Res. 43 (2004) 403–448.

[36] B. Deplancke, K. Finster, W.V. Graham, C.T. Collier, J.E. Thurmond, H.R. Gaskins, Gastrointestinal and microbial responses to sulfate-supplemented drinking water in mice. Exp. Biol. Med. (Maywood) 228 (2003) 424–433.

[37] V. Devesa, A. Loos, M.A. Suner, D. Velez, A. Feria, A. Martinez, R. Montoro, Y. Sanz, Transformation of organoarsenical species by the microflora of freshwater crayfish. J. Agric. Food Chem. 53 (2005) 10297–10305.

[38] L. Ding, R.J. Saunders, Z. Drobna, F.S. Walton, P. Xun, D.J. Thomas, M. Styblo, Methylation of arsenic by recombinant human wild-type arsenic (+3 oxidation state) methyltransferase and its methionine 287 threonine (M287T) polymorph: role of glutathione. Toxicol. Appl. Pharmacol. 264 (2012) 121–130.

[39] P.R. Dodmane, L.L. Arnold, K.L. Pennington, D.J. Thomas, S.M. Cohen, Effect of dietary treatment with dimethylarsinous acid (DMA(III)) on the urinary bladder epithelium of arsenic (+3 oxidation state) methyltransferase (As3mt) knockout and C57BL/6 wild type female mice. Toxicology 305 (2013) 130–135.

[40] C. Douillet, J. Currier, J. Saunders, W.M. Bodnar, T. Matousek, M. Styblo, Methylated trivalent arsenicals are potent inhibitors of glucose stimulated insulin secretion by murine pancreatic islets. Toxicol. Appl. Pharmacol. 267 (2013) 11–15.

[41] Z. Drobna, H. Naranmandura, K.M. Kubachka, B.C. Edwards, K. Herbin-Davis, M. Styblo, X.C. Le, J.T. Creed, N. Maeda, M.F. Hughes, D.J. Thomas, Disruption of the arsenic (+3 oxidation state) methyltransferase gene in the mouse alters the phenotype for methylation of arsenic and affects distribution and retention of orally administered arsenate. Chem. Res. Toxicol. 22 (2009) 1713–1720.

[42] Z. Drobna, S.B. Waters, V. Devesa, A.W. Harmon, D.J. Thomas, M. Styblo, Metabolism and toxicity of arsenic in human urothelial cells expressing rat arsenic (+3 oxidation state)-methyltransferase. Toxicol. Appl. Pharmacol. 207 (2005) 147–159.

[43] Z. Drobna, W. Xing, D.J. Thomas, M. Styblo, shRNA silencing of AS3MT expression minimizes arsenic methylation capacity of HepG2 cells. Chem. Res. Toxicol. 19 (2006) 894–898.

[44] J.L. Ellis, S.D. Conklin, C.M. Gallawa, K.M. Kubachka, A.R. Young, P.A. Creed, J.A. Caruso, J.T. Creed, Complementary molecular and elemental detection of speciated thioarsenicals using ESI-MS in combination with a xenon-based collision-cell ICP-MS with application to fortified NIST freeze-dried urine. Anal. Bioanal. Chem. 390 (2008) 1731–1737.

[45] C. Escudero-Lourdes, M.K. Medeiros, M.C. Cardenas-Gonzalez, S.M. Wnek, J.A. Gandolfi, Low level exposure to monomethyl arsonous acid-induced the over-production of inflammation-related cytokines and the activation of cell signals associated with tumor progression in a urothelial cell model. Toxicol. Appl. Pharmacol. 244 (2010) 162–173.

[46] J. Feldmann, K. John, P. Pengprecha, Arsenic metabolism in seaweed-eating sheep from Northern Scotland. Fresenius J. Anal. Chem. 368 (2000) 116–121.

[47] D.E. Fomenko, W. Xing, B.M. Adair, D.J. Thomas, V.N. Gladyshev, High-throughput identification of catalytic redox-active cysteine residues. Science 315 (2007) 387–389.

[48] M. Fontcuberta, J. Calderon, J.R. Villalbi, F. Centrich, S. Portana, A. Espelt, J. Duran, M. Nebot, Total and inorganic arsenic in marketed food and associated health risks for the Catalan (Spain) population. J. Agric. Food Chem. 59 (2011) 10013–10022.

[49] K.A. Francesconi, Arsenic species in seafood: origin and human health implications. Pure Appl. Chem. 82 (2010) 373–381.

[50] K.A. Francesconi, R. Tanggaar, C.J. McKenzie, W. Goessler, Arsenic metabolites in human urine after ingestion of an arsenosugar. Clin. Chem. 48 (2002) 92–101.

[51] M. Fu, W. Zhang, L. Wu, G. Yang, H. Li, R. Wang, Hydrogen sulfide (H2S) metabolism in mitochondria and its regulatory role in energy production. Proc. Natl. Acad. Sci. U. S. A. 109 (2012) 2943–2948.

[52] S. Fukuda, M. Terasawa, K. Shiomi, Phosphatidylarsenocholine, one of the major arsenolipids in marine organisms: synthesis and metabolism in mice. Food Chem. Toxicol. 49 (2011) 1598–1603.

[53] J. Furne, A. Saeed, M.D. Levitt, Whole tissue hydrogen sulfide concentrations are orders of magnitude lower than presently accepted values. Am. J. Physiol. Regul. Integr. Comp. Physiol. 295 (2008) R1479–R1485.

[54] B.M. Gamble, P.A. Gallagher, J.A. Shoemaker, X. Wei, C.A. Schwegel, J.T. Creed, An investigation of the chemical stability of arsenosugars in simulated gastric juice and acidic environments using IC-ICP-MS and IC-ESI-MS/MS. Analyst 127 (2002) 781–785.

[55] E.A. Garcia-Montalvo, O.L. Valenzuela, L.C. Sanchez-Pena, A. Albores, L.M. Del Razo, Dose-dependent urinary phenotype of inorganic arsenic methylation in mice with a focus on trivalent methylated metabolites. Toxicol. Mech. Methods 21 (2011) 649–655.

[56] B. Georis, A. Cardenas, J.P. Buchet, R. Lauwerys, Inorganic arsenic methylation by rat tissue slices. Toxicology 63 (1990) 73–84.

[57] Z. Gregus, A. Gyurasics, I. Csanaky, Biliary and urinary excretion of inorganic arsenic: monomethylarsonous acid as a major biliary metabolite in rats. Toxicol. Sci. 56 (2000) 18–25.

[58] L.L. Hall, S.E. George, M.J. Kohan, M. Styblo, D.J. Thomas, In vitro methylation of inorganic arsenic in mouse intestinal cecum. Toxicol. Appl. Pharmacol. 147 (1997) 101–109.

[59] H.R. Hansen, M. Jaspars, J. Feldmann, Arsinothioyl-sugars produced by in vitro incubation of seaweed extract with liver cytosol analysed by HPLC coupled simultaneously to ES-MS and ICP-MS. Analyst 129 (2004) 1058–1064.

[60] H.R. Hansen, R. Pickford, J. Thomas-Oates, M. Jaspars, J. Feldmann, 2-Dimethylarsinothioyl acetic acid identified in a biological sample: the first occurrence of a mammalian arsinothioyl metabolite. Angew. Chem. Int. Ed. Engl. 43 (2004) 337–340.

[61] F. Harari, K. Engstrom, G. Concha, G. Colque, M. Vahter, K. Broberg, N-6-adenine-specific DNA methyltransferase 1 (N6AMT1) polymorphisms and arsenic methylation in Andean women. Environ. Health Perspect. 121 (2013) 797–803.

[62] C.F. Harrington, E.I. Brima, R.O. Jenkins, Biotransformation of arsenobetaine by microorganisms from the human gastrointestinal tract. Chem. Speciat. Bioavailab. 20 (2008) 173–180.

[63] T. Hayakawa, Y. Kobayashi, X. Cui, S. Hirano, A new metabolic pathway of arsenite: arsenic-glutathione complexes are substrates for human arsenic methyltransferase Cyt19. Arch. Toxicol. 79 (2005) 183–191.

[64] A. Hernandez-Zavala, T. Matousek, Z. Drobna, D.S. Paul, F. Walton, B.M. Adair, D. Jiri, D.J. Thomas, M. Styblo, Speciation analysis of arsenic in biological matrices by automated hydride generation-cryotrapping-atomic absorption spectrometry with multiple microflame quartz tube atomizer (multiatomizer). J. Anal. At. Spectrom. 23 (2008) 342–351.

[65] S. Hirata, H. Toshimitsu, Determination of arsenic species and arsenosugars in marine samples by HPLC-ICP-MS. Anal. Bioanal. Chem. 383 (2005) 454–460.

[66] M.F. Hughes, B.C. Edwards, K.M. Herbin-Davis, J. Saunders, M. Styblo, D.J. Thomas, Arsenic (+3 oxidation state) methyltransferase genotype affects steady-state distribution and clearance of arsenic in arsenate-treated mice. Toxicol. Appl. Pharmacol. 249 (2010) 217–223.

[67] R.O. Jenkins, A.W. Ritchie, J.S. Edmonds, W. Goessler, N. Molenat, D. Kuehnelt, C.F. Harrington, P.G. Sutton, Bacterial degradation of arsenobetaine via dimethylarsinoylacetate. Arch. Microbiol. 180 (2003) 142–150.

[68] T.J. Jensen, P. Novak, S.M. Wnek, A.J. Gandolfi, B.W. Futscher, Arsenicals produce stable progressive changes in DNA methylation patterns that are linked to malignant transformation of immortalized urothelial cells. Toxicol. Appl. Pharmacol. 241 (2009) 221–229.

[69] D.A. Kalman, R.L. Dills, C. Steinmaus, M. Yunus, A.F. Khan, M.M. Prodhan, Y. Yuan, A.H. Smith, Occurrence of trivalent monomethyl arsenic and other urinary arsenic species in a highly exposed juvenile population in Bangladesh. J. Expo. Sci. Environ. Epidemiol. 24 (2014) 113–120.

[70] H. Kimura, Metabolic turnover of hydrogen sulfide. Front. Physiol. 3 (2012) 101.

[71] K.M. Kubachka, M.C. Kohan, S.D. Conklin, K. Herbin-Davis, J.T. Creed, D.J. Thomas, In vitro biotransformation of dimethylarsinic acid and trimethylarsine oxide by anaerobic microflora of mouse cecum analyzed by HPLC-ICP-MS and HPLC-ESI-MS. J. Anal. At. Spectrom. 24 (2009) 1062–1068.

[72] K.M. Kubachka, M.C. Kohan, K. Herbin-Davis, J.T. Creed, D.J. Thomas, Exploring the in vitro formation of trimethylarsine sulfide from dimethylthioarsinic acid in anaerobic microflora of mouse cecum using HPLC-ICP-MS and HPLC-ESI-MS. Toxicol. Appl. Pharmacol. 239 (2009) 137–143.

[73] C.A. Lammon, X.C. Le, R.D. Hood, Pretreatment with periodate-oxidized adenosine enhances developmental toxicity of inorganic arsenic in mice. Birth Defects Res. B Dev. Reprod. Toxicol. 68 (2003) 335–343.

[74] X.C. Le, M. Ma, W.R. Cullen, H.V. Aposhian, X. Lu, B. Zheng, Determination of monomethylarsonous acid, a key arsenic methylation intermediate, in human urine. Environ. Health Perspect. 108 (2000) 1015–1018.

[75] L. Leffers, F. Ebert, M.S. Taleshi, K.A. Francesconi, T. Schwerdtle, In vitro toxicological characterization of two arsenosugars and their metabolites. Mol. Nutr. Food Res. 57 (2013) 1270–1282.

[76] L. Leffers, M. Unterberg, M. Bartel, C. Hoppe, I. Pieper, J. Stertmann, F. Ebert, H.U. Humpf, T. Schwerdtle, In vitro toxicological characterisation of the S-containing arsenic metabolites thio-dimethylarsinic acid and dimethylarsinic glutathione. Toxicology 305 (2013) 109–119.

[77] B. Lehmann, E. Ebeling, C. Alsen-Hinrichs, Kinetics of arsenic in human blood after a fish meal. Gesundheitswesen 63 (2001) 42–48.

[78] M.D. Levitt, M.S. Abdel-Rehim, J. Furne, Free and acid-labile hydrogen sulfide concentrations in mouse tissues: anomalously high free hydrogen sulfide in aortic tissue. Antioxid. Redox Signal. 15 (2011) 373–378.

[79] L. Li, P. Rose, P.K. Moore, Hydrogen sulfide and cell signaling. Annu. Rev. Pharmacol. Toxicol. 51 (2011) 169–187.

[80] W. Li, H. Wanibuchi, E.I. Salim, S. Yamamoto, K. Yoshida, G. Endo, S. Fukushima, Promotion of NCI-Black-Reiter male rat bladder carcinogenesis by dimethylarsinic acid an organic arsenic compound. Cancer Lett. 134 (1998) 29–36.

[81] S. Lin, L.M. Del Razo, M. Styblo, C. Wang, W.R. Cullen, D.J. Thomas, Arsenicals inhibit thioredoxin reductase in cultured rat hepatocytes. Chem. Res. Toxicol. 14 (2001) 305–311.

[82] S. Lin, Q. Shi, F.B. Nix, M. Styblo, M.A. Beck, K.M. Herbin-Davis, L.L. Hall, J.B. Simeonsson, D.J. Thomas, A novel S-adenosyl-L-methionine:arsenic(III) methyltransferase from rat liver cytosol. J. Biol. Chem. 277 (2002) 10795–10803.

[83] X. Lu, L.L. Arnold, S.M. Cohen, W.R. Cullen, X.C. Le, Speciation of dimethylarsinous acid and trimethylarsine oxide in urine from rats fed with dimethylarsinic acid and dimercaptopropane sulfonate. Anal. Chem. 75 (2003) 6463–6468.

[84] M. Ma, X.C. Le, Effect of arsenosugar ingestion on urinary arsenic speciation. Clin. Chem. 44 (1998) 539–550.

[85] B.K. Mandal, Y. Ogra, K.T. Suzuki, Identification of dimethylarsinous and monomethylarsonous acids in human urine of the arsenic-affected areas in West Bengal, India. Chem. Res. Toxicol. 14 (2001) 371–378.

[86] E. Marafante, M. Vahter, The effect of methyltransferase inhibition on the metabolism of [74As]arsenite in mice and rabbits. Chem. Biol. Interact. 50 (1984) 49–57.

[87] E. Marafante, M. Vahter, J. Envall, The role of the methylation in the detoxication of arsenate in the rabbit. Chem. Biol. Interact. 56 (1985) 225–238.

[88] K. Marapakala, J. Qin, B.P. Rosen, Identification of catalytic residues in the As(III) S-adenosylmethionine methyltransferase. Biochemistry 51 (2012) 944–951.

[89] M.J. Mass, A. Tennant, B.C. Roop, W.R. Cullen, M. Styblo, D.J. Thomas, A.D. Kligerman, Methylated trivalent arsenic species are genotoxic. Chem. Res. Toxicol. 14 (2001) 355–361.

[90] M. Medani, D. Collins, N.G. Docherty, A.W. Baird, P.R. O'Connell, D.C. Winter, Emerging role of hydrogen sulfide in colonic physiology and pathophysiology. Inflamm. Bowel Dis. 17 (2011) 1620–1625.

[91] M. Medeiros, X. Zheng, P. Novak, S.M. Wnek, V. Chyan, C. Escudero-Lourdes, A.J. Gandolfi, Global gene expression changes in human urothelial cells exposed to low-level monomethylarsonous acid. Toxicology 291 (2012) 102–112.

[92] S.R. Meno, R. Nelson, K.J. Hintze, W.T. Self, Exposure to monomethylarsonous acid (MMA(III)) leads to altered selenoprotein synthesis in a primary human lung cell model. Toxicol. Appl. Pharmacol. 239 (2009) 130–136.

[93] S. Miyashita, S. Fujiwara, M. Tsuzuki, T. Kaise, Rapid biotransformation of arsenate into oxo-arsenosugars by a freshwater unicellular green alga, *Chlamydomonas reinhardtii*. Biosci. Biotechnol. Biochem. 75 (2011) 522–530.

[94] S.i. Miyashita, S. Fujiwara, M. Tsuzuki, T. Kaise, Cyanobacteria produce arsenosugars. Environ. Chem. 9 (2012) 474–484.

[95] M. Molin, S.M. Ulven, L. Dahl, V.H. Telle-Hansen, M. Holck, G. Skjegstad, O. Ledsaak, J.J. Sloth, W. Goessler, A. Oshaug, J. Alexander, D. Fliegel, T.A. Ydersbond, H.M. Meltzer, Humans seem to produce arsenobetaine and dimethylarsinate after a bolus dose of seafood. Environ. Res. 112 (2012) 28–39.

[96] T. Murai, H. Iwata, T. Otoshi, G. Endo, S. Horiguchi, S. Fukushima, Renal lesions induced in F344/DuCrj rats by 4-weeks oral administration of dimethylarsinic acid. Toxicol. Lett. 66 (1993) 53–61.

[97] H. Naranmandura, M.W. Carew, S. Xu, J. Lee, E.M. Leslie, M. Weinfeld, X.C. Le, Comparative toxicity of arsenic metabolites in human bladder cancer EJ-1 cells. Chem. Res. Toxicol. 24 (2011) 1586–1596.

[98] H. Naranmandura, Y. Ogra, K. Iwata, J. Lee, K.T. Suzuki, M. Weinfeld, X.C. Le, Evidence for toxicity differences between inorganic arsenite and thioarsenicals in human bladder cancer cells. Toxicol. Appl. Pharmacol. 238 (2009) 133–140.

[99] H. Naranmandura, K. Rehman, X.C. Le, D.J. Thomas, Formation of methylated oxyarsenicals and thioarsenicals in wild-type and arsenic (+3 oxidation state) methyltransferase knockout mice exposed to arsenate. Anal. Bioanal. Chem. 405 (2013) 1885–1891.

[100] H. Naranmandura, N. Suzuki, K.T. Suzuki, Trivalent arsenicals are bound to proteins during reductive methylation. Chem. Res. Toxicol. 19 (2006) 1010–1018.

[101] H. Naranmandura, S. Xu, T. Sawata, W.H. Hao, H. Liu, N. Bu, Y. Ogra, Y.J. Lou, N. Suzuki, Mitochondria are the main target organelle for trivalent monomethylarsonous acid (MMA(III))-induced cytotoxicity. Chem. Res. Toxicol. 24 (2011) 1094–1103.

[102] K.R. Olson, Hydrogen sulfide: both feet on the gas and none on the brake? Front. Physiol. 4 (2013) 2.

[103] J.S. Petrick, F. Ayala-Fierro, W.R. Cullen, D.E. Carter, A.H. Vasken, Monomethylarsonous acid (MMA(III)) is more toxic than arsenite in Chang human hepatocytes. Toxicol. Appl. Pharmacol. 163 (2000) 203–207.

[104] J.S. Petrick, B. Jagadish, E.A. Mash, H.V. Aposhian, Monomethylarsonous acid (MMA(III)) and arsenite: LD(50) in hamsters and in vitro inhibition of pyruvate dehydrogenase. Chem. Res. Toxicol. 14 (2001) 651–656.

[105] T.S. Pinyayev, M.J. Kohan, K. Herbin-Davis, J.T. Creed, D.J. Thomas, Preabsorptive metabolism of sodium arsenate by anaerobic microbiota of mouse cecum forms a variety of methylated and thiolated arsenicals. Chem. Res. Toxicol. 24 (2011) 475–477.

[106] B. Planer-Friedrich, E. Suess, A.C. Scheinost, D. Wallschlager, Arsenic speciation in sulfidic waters: reconciling contradictory spectroscopic and chromatographic evidence. Anal. Chem. 82 (2010) 10228–10235.

[107] G. Raber, S. Khoomrung, M.S. Taleshi, J.S. Edmonds, K.A. Francesconi, Identification of arsenolipids with GC/MS. Talanta 78 (2009) 1215–1218.

[108] R. Raml, W. Goessler, P. Traar, T. Ochi, K.A. Francesconi, Novel thioarsenic metabolites in human urine after ingestion of an arsenosugar, 2′,3′-dihydroxypropyl 5-deoxy-5-dimethylarsinoyl-beta-D-riboside. Chem. Res. Toxicol. 18 (2005) 1444–1450.

[109] R. Raml, G. Raber, A. Rumpler, T. Bauernhofer, W. Goessler, K.A. Francesconi, Individual variability in the human metabolism of an arsenic-containing carbohydrate, 2′,3′-dihydroxypropyl 5-deoxy-5-dimethylarsinoyl-beta-D-riboside, a naturally occurring arsenical in seafood. Chem. Res. Toxicol. 22 (2009) 1534–1540.

[110] R. Raml, A. Rumpler, W. Goessler, M. Vahter, L. Li, T. Ochi, K.A. Francesconi, Thiodimethylarsinate is a common metabolite in urine samples from arsenic-exposed women in Bangladesh. Toxicol. Appl. Pharmacol. 222 (2007) 374–380.

[111] K. Rehman, Z. Chen, W.W. Wang, Y.W. Wang, A. Sakamoto, Y.F. Zhang, H. Naranmandura, N. Suzuki, Mechanisms underlying the inhibitory effects of arsenic compounds on protein tyrosine phosphatase (PTP). Toxicol. Appl. Pharmacol. 263 (2012) 273–280.

[112] K. Rehman, H. Naranmandura, Arsenic metabolism and thioarsenicals. Metallomics 4 (2012) 881–892.

[113] X. Ren, M. Aleshin, W.J. Jo, R. Dills, D.A. Kalman, C.D. Vulpe, M.T. Smith, L. Zhang, Involvement of N-6 adenine-specific DNA methyltransferase 1 (N6AMT1) in arsenic biomethylation and its role in arsenic-induced toxicity. Environ. Health Perspect. 119 (2011) 771–777.

[114] E.H. Rogers, N. Chernoff, R.J. Kavlock, The teratogenic potential of cacodylic acid in the rat and mouse. Drug Chem. Toxicol. 4 (1981) 49–61.

[115] A. Rumpler, J.S. Edmonds, M. Katsu, K.B. Jensen, W. Goessler, G. Raber, H. Gunnlaugsdottir, K.A. Francesconi, Arsenic-containing long-chain fatty acids in cod-liver oil: a result of biosynthetic infidelity? Angew. Chem. Int. Ed. Engl. 47 (2008) 2665–2667.

[116] J. Scheer, S. Findenig, W. Goessler, K.A. Francesconi, B. Howard, J.G. Umans, J. Pollak, M. Tellez-Plaza, E.K. Silbergeld, E. Guallar, A. Navas-Acien, Arsenic species and selected metals in human urine: validation of HPLC/ICPMS and ICPMS procedures for a long-term population-based epidemiological study. Anal. Methods 4 (2012) 406–413.

[117] E. Schmeisser, W. Goessler, K.A. Francesconi, Human metabolism of arsenolipids present in cod liver. Anal. Bioanal. Chem. 385 (2006) 367–376.

[118] E. Schmeisser, A. Rumpler, M. Kollroser, G. Rechberger, W. Goessler, K.A. Francesconi, Arsenic fatty acids are human urinary metabolites of arsenolipids present in cod liver. Angew. Chem. Int. Ed. Engl. 45 (2005) 150–154.

[119] J. Shen, H. Wanibuchi, E.I. Salim, M. Wei, A. Kinoshita, K. Yoshida, G. Endo, S. Fukushima, Liver tumorigenicity of trimethylarsine oxide in male Fischer 344 rats—association with oxidative DNA damage and enhanced cell proliferation. Carcinogenesis 24 (2003) 1827–1835.

[120] S. Singh, R. Banerjee, PLP-dependent H(2)S biogenesis. Biochim. Biophys. Acta 1814 (2011) 1518–1527.

[121] S. Singh, D. Padovani, R.A. Leslie, T. Chiku, R. Banerjee, Relative contributions of cystathionine beta-synthase and gamma-cystathionase to H2S biogenesis via alternative trans-sulfuration reactions. J. Biol. Chem. 284 (2009) 22457–22466.

[122] T.J. Smith, E.A. Crecelius, J.C. Reading, Airborne arsenic exposure and excretion of methylated arsenic compounds. Environ. Health Perspect. 19 (1977) 89–93.

[123] X. Song, Z. Geng, J. Zhu, C. Li, X. Hu, N. Bian, X. Zhang, Z. Wang, Structure-function roles of four cysteine residues in the human arsenic (+3 oxidation state) methyltransferase (hAS3MT) by site-directed mutagenesis. Chem. Biol. Interact. 179 (2009) 321–328.

[124] M.H. Stipanuk, J.E. Dominy, Jr., J.I. Lee, R.M. Coloso, Mammalian cysteine metabolism: new insights into regulation of cysteine metabolism. J. Nutr. 136 (2006) 1652S–1659S.

[125] M. Styblo, L.M. Del Razo, L. Vega, D.R. Germolec, E.L. LeCluyse, G.A. Hamilton, W. Reed, C. Wang, W.R. Cullen, D.J. Thomas, Comparative toxicity of trivalent and pentavalent inorganic and methylated arsenicals in rat and human cells. Arch. Toxicol. 74 (2000) 289–299.

[126] M. Styblo, S.V. Serves, W.R. Cullen, D.J. Thomas, Comparative inhibition of yeast glutathione reductase by arsenicals and arsenothiols. Chem. Res. Toxicol. 10 (1997) 27–33.

[127] F. Suarez, J. Furne, J. Springfield, M. Levitt, Production and elimination of sulfur-containing gases in the rat colon. Am. J. Physiol. 274 (1998) G727–G733.

[128] E. Suess, D. Wallschlager, B. Planer-Friedrich, Stabilization of thioarsenates in iron-rich waters. Chemosphere 83 (2011) 1524–1531.

[129] S. Suzuki, L.L. Arnold, K.L. Pennington, B. Chen, H. Naranmandura, X.C. Le, S.M. Cohen, Dietary administration of sodium arsenite to rats: relations between dose and urinary concentrations of methylated and thio-metabolites and effects on the rat urinary bladder epithelium. Toxicol. Appl. Pharmacol. 244 (2010) 99–105.

[130] M.S. Taleshi, J.S. Edmonds, W. Goessler, M.J. Ruiz-Chancho, G. Raber, K.B. Jensen, K.A. Francesconi, Arsenic-containing lipids are natural constituents of sashimi tuna. Environ. Sci. Technol. 44 (2010) 1478–1483.

[131] M.S. Taleshi, K.B. Jensen, G. Raber, J.S. Edmonds, H. Gunnlaugsdottir, K.A. Francesconi, Arsenic-containing hydrocarbons: natural compounds in oil from the fish capelin, *Mallotus villosus*. Chem. Commun. (Camb.) 2008 (2008) 4706–4707.

[132] W.H. Tang, Z. Wang, B.S. Levison, R.A. Koeth, E.B. Britt, X. Fu, Y. Wu, S.L. Hazen, Intestinal microbial metabolism of phosphatidylcholine and cardiovascular risk. N. Engl. J. Med. 368 (2013) 1575–1584.

[133] D.J. Thomas, J. Li, S.B. Waters, W. Xing, B.M. Adair, Z. Drobna, V. Devesa, M. Styblo, Arsenic (+3 oxidation state) methyltransferase and the methylation of arsenicals. Exp. Biol. Med. (Maywood) 232 (2007) 3–13.

[134] D.J. Thomas, B.P. Rosen, Arsenic methyltransferases. in: R.H. Kretsinger, V.N. Uversky, E.A. Permyakov (Eds.), Encyclopedia of Metalloproteins, Springer, Berlin, 2013, pp. 138–143.

[135] E.J. Tokar, B.A. Diwan, D.J. Thomas, M.P. Waalkes, Tumors and proliferative lesions in adult offspring after maternal exposure to methylarsonous acid during gestation in CD1 mice. Arch. Toxicol. 86 (2012) 975–982.

[136] M. Vahter, E. Marafante, L. Dencker, Metabolism of arsenobetaine in mice, rats and rabbits. Sci. Total Environ. 30 (1983) 197–211.

[137] M. Van Hulle, C. Zhang, B. Schotte, L. Mees, F. Vanhaecke, R. Vanholder, X.R. Zhang, R. Cornelis, Identification of some arsenic species in human urine and blood after ingestion of Chinese seaweed Laminaria. J. Anal. At. Spectrom. 19 (2004) 58–64.

[138] V. Vitvitsky, O. Kabil, R. Banerjee, High turnover rates for hydrogen sulfide allow for rapid regulation of its tissue concentrations. Antioxid. Redox Signal. 17 (2012) 22–31.

[139] F.S. Walton, A.W. Harmon, D.S. Paul, Z. Drobna, Y.M. Patel, M. Styblo, Inhibition of insulin-dependent glucose uptake by trivalent arsenicals: possible mechanism of arsenic-induced diabetes. Toxicol. Appl. Pharmacol. 198 (2004) 424–433.

[140] S. Wang, X. Li, X. Song, Z. Geng, X. Hu, Z. Wang, Rapid equilibrium kinetic analysis of arsenite methylation catalyzed by recombinant human arsenic (+3 oxidation state) methyltransferase (hAS3MT). J. Biol. Chem. 287 (2012) 38790–38799.

[141] Z. Wang, J. Zhou, X. Lu, Z. Gong, X.C. Le, Arsenic speciation in urine from acute promyelocytic leukemia patients undergoing arsenic trioxide treatment. Chem. Res. Toxicol. 17 (2004) 95–103.

[142] S.B. Waters, V. Devesa, L.M. Del Razo, M. Styblo, D.J. Thomas, Endogenous reductants support the catalytic function of recombinant rat cyt19, an arsenic methyltransferase. Chem. Res. Toxicol. 17 (2004) 404–409.

[143] S.B. Waters, V. Devesa, M.W. Fricke, J.T. Creed, M. Styblo, D.J. Thomas, Glutathione modulates recombinant rat arsenic (+3 oxidation state) methyltransferase-catalyzed formation of trimethylarsine oxide and trimethylarsine. Chem. Res. Toxicol. 17 (2004) 1621–1629.

[144] N.L. Whitfield, E.L. Kreimier, F.C. Verdial, N. Skovgaard, K.R. Olson, Reappraisal of H2S/sulfide concentration in vertebrate blood and its potential significance in ischemic preconditioning and vascular signaling. Am. J. Physiol. Regul. Integr. Comp. Physiol. 294 (2008) R1930–R1937.

[145] E. Wildfang, R.A. Zakharyan, H.V. Aposhian, Enzymatic methylation of arsenic compounds. VI. Characterization of hamster liver arsenite and methylarsonic acid methyltransferase activities in vitro. Toxicol. Appl. Pharmacol. 152 (1998) 366–375.

[146] S.M. Wnek, T.J. Jensen, P.L. Severson, B.W. Futscher, A.J. Gandolfi, Monomethylarsonous acid produces irreversible events resulting in malignant transformation of a human bladder cell line following 12 weeks of low-level exposure. Toxicol. Sci. 116 (2010) 44–57.

[147] S.M. Wnek, C.L. Kuhlman, J.M. Camarillo, M.K. Medeiros, K.J. Liu, S.S. Lau, A.J. Gandolfi, Interdependent genotoxic mechanisms of monomethylarsonous acid: role of ROS-induced DNA damage and poly(ADP-ribose) polymerase-1 inhibition in the malignant transformation of urothelial cells. Toxicol. Appl. Pharmacol. 257 (2011) 1–13.

[148] R. Xie, W. Johnson, S. Spayd, G.S. Hall, B. Buckley, Arsenic speciation analysis of human urine using ion exchange chromatography coupled to inductively coupled plasma mass spectrometry. Anal. Chim. Acta 578 (2006) 186–194.

[149] S. Yamamoto, Y. Konishi, T. Matsuda, T. Murai, M.A. Shibata, I. Matsui-Yuasa, S. Otani, K. Kuroda, G. Endo, S. Fukushima, Cancer induction by an organic arsenic compound, dimethylarsinic acid (cacodylic acid), in F344/DuCrj rats after pretreatment with five carcinogens. Cancer Res. 55 (1995) 1271–1276.

[150] H. Yamauchi, B.A. Fowler in: J. O. Nriagu (Ed.), Arsenic in the Environment, Part II: Human Health and Ecosystem Effects, John Wiley & Sons, Inc., New York, 1994, pp. 35–43.

[151] M. Yokohira, L.L. Arnold, K.L. Pennington, S. Suzuki, S. Kakiuchi-Kiyota, K. Herbin-Davis, D.J. Thomas, S.M. Cohen, Effect of sodium arsenite dose administered in the drinking water on the urinary bladder epithelium of female arsenic (+3 oxidation state) methyltransferase knockout mice. Toxicol. Sci. 121 (2011) 257–266.

[152] K. Yoshida, K. Kuroda, Y. Inoue, H. Chen, H. Wanibuchi, S. Fukushima, G. Endo, Metabolites of arsenobetaine in rats: does decomposition of arsenobetaine occur in mammals? Appl. Organomet. Chem. 15 (2001) 271–276.

[153] R. Zakharyan, Y. Wu, G.M. Bogdan, H.V. Aposhian, Enzymatic methylation of arsenic compounds: assay, partial purification, and properties of arsenite methyltransferase and monomethylarsonic acid methyltransferase of rabbit liver. Chem. Res. Toxicol. 8 (1995) 1029–1038.

[154] R.A. Zakharyan, H.V. Aposhian, Enzymatic reduction of arsenic compounds in mammalian systems: the rate-limiting enzyme of rabbit liver arsenic biotransformation is MMA(V) reductase. Chem. Res. Toxicol. 12 (1999) 1278–1283.

[155] R.A. Zakharyan, F. Ayala-Fierro, W.R. Cullen, D.M. Carter, H.V. Aposhian, Enzymatic methylation of arsenic compounds. VII. Monomethylarsonous acid (MMAIII) is the substrate for MMA methyltransferase of rabbit liver and human hepatocytes. Toxicol. Appl. Pharmacol. 158 (1999) 9–15.

PART II

EPIDEMIOLOGY AND DISEASE MANIFESTATIONS OF ARSENIC EXPOSURE

5

HUMAN POPULATION STUDIES AND NUTRITIONAL INTERVENTION

Yu Chen and Fen Wu

Departments of Population Health and Environmental Medicine, New York University School of Medicine, New York, NY, USA

5.1 INTRODUCTION

There is wide variation in the susceptibility to arsenic toxicity, and nutrition is believed to be an important susceptibility factor. A number of studies have suggested that malnutrition may increase the prevalence or severity of arsenic-induced health effects [1]. In areas with severe arsenic-related health effects due to ingestion of drinking water with high arsenic concentrations, that is, southwestern Taiwan, inhabitants were reported to have a poor nutritional status and a low socioeconomic level [2]; their reported diet was adequate in calories and high in carbohydrates but low in protein and extremely low in fat. In another study on the same population, Hsueh et al. [3] reported that undernourishment—indexed by a high consumption of dried sweet potatoes as a staple food—was associated with an increased prevalence of arsenic-induced skin cancer. It has been suggested that low intake of micronutrients in this population may have led to a greater susceptibility to carcinogenesis [4]. More detailed nutritional data, although still lacking quantities of food consumed, were presented in a study in Taiwan that found increased risks of blackfoot disease (a peripheral vascular disease attributed to arsenic in drinking water) associated with undernourishment [5]. Poor nutritional status might indicate an increased susceptibility to arsenic toxicity, leading to reduced methylation of arsenic and therefore increased tissue retention of arsenic. Since these earlier studies suggesting a role of

Arsenic: Exposure Sources, Health Risks, and Mechanisms of Toxicity, First Edition.
Edited by J. Christopher States.
© 2016 John Wiley & Sons, Inc. Published 2016 by John Wiley & Sons, Inc.

nutrition in arsenic toxicity, numerous studies have set forth to investigate the influence of specific nutrients. The remainder of this chapter discusses epidemiologic data on the various nutritional factors that influence arsenic metabolism in humans and nutritional factors in relation to arsenic-induced health effects.

5.2 NUTRITION AND ARSENIC METABOLISM

5.2.1 Nutrients Contributing to One-Carbon Metabolism

Arsenic in drinking water is present as inorganic arsenic (iAs), namely arsenite (As^{III}) and arsenate (As^{V}). As discussed in Chapter 4, methylation of iAs, first generates monomethylarsonic acid (MMA^{V}). After the reduction of MMA^{V} to monomethylarsonous acid (MMA^{III}), a second methylation can occur to generate dimethylarsinic acid (DMA^{V}) [6]. It is unclear to what extent DMA^{V} is reduced to dimethylarsinous acid (DMA^{III}) in vivo, as DMA^{III} is an unstable intermediate [7]. In these reactions of one-carbon metabolism, methyl groups are transferred from *S*-adenosylmethionine (SAM) [8] to As^{III}. The reactions require availability of dietary methyl groups for the formation of SAM and the presence of reduced glutathione or other thiols for reduction of As^{V} [9]. Full functioning of one-carbon metabolism also requires adequate intakes of folate, vitamin B12, and vitamin B6 to remethylate homocysteine back to methionine [10]. Other nutrients that are not required per se, such as protein, choline, betaine, and cysteine also contribute to the availability of methyl groups ultimately used in SAM biosynthesis. For instance, dietary protein intake provides a source of methionine in the one-carbon metabolism. The amino acid cysteine is produced from homocysteine as an intermediate in glutathione biosynthesis. Like glutathione, cysteine is capable of reducing pentavalent As^{V} to As^{III} *in vitro* [11], and this reduction is a prerequisite to methylation. Choline is oxidized to betaine, which can donate its methyl group to homocysteine to form methionine. This reaction, catalyzed by betaine:homocysteine methyltransferase, is the sole alternate route to the folate-dependent methionine synthase–catalyzed homocysteine remethylation [12]. Epidemiologic studies on nutritional factors that influence the one-carbon metabolism are discussed below.

The earliest human data implicating nutritional influences on methylation and toxicity of arsenic came from a case study which reported that a girl with methylenetetrahydrofolate reductase (MTHFR) deficiency developed severe clinical signs and symptoms of arsenic poisoning upon exposure to an arsenic-containing pesticide, whereas no other exposed family members developed symptoms [13]. In a study of 11 families in Chile, Smith's group reported intra-family associations in arsenic methylation. The father/mother correlation for iAs/(MMA+DMA) was low ($r=0.18$). However, adjustment for plasma folate or homocysteine substantially increased the correlations ($r=0.33$ and 0.55, respectively) [14]. Although the authors did not conclude that there was a significant effect of nutritional factors on arsenic methylation, the data were highly suggestive. A few studies reported lowered arsenic methylation in people with low protein intake. Assessment of dietary intakes and urinary arsenic methylation patterns in 87 subjects from two arsenic-exposed regions in the western United States showed that subjects in the lower quartile of protein intake had higher %MMA and lower %DMA in urine than did subjects in the upper quartile [15]. No associations were found for folate, but the study was conducted several years after mandatory folic acid fortification of the USA food supply and therefore all of the study subjects were essentially folate supplemented. In a cross-sectional study with more than 1000 Bangladeshi adults exposed to arsenic in drinking water, higher intakes of cysteine, methionine, protein, and vitamin B12 were associated with lower %iAs and higher ratios of MMA to iAs in urine; higher intakes of choline (beta=0.10, $P=0.02$) were associated with higher DMA-to-MMA ratios, after adjustment for age, sex, smoking, total urinary arsenic, and total energy intake [16]. However, folate intake was not associated with urinary arsenic metabolites.

More recently, Vahter's group analyzed plasma concentrations of folate, vitamin B12, zinc, ferritin and selenium in a cross-sectional study of 442 women in early pregnancy from Matlab, Bangladesh. In their analyses, they first stratified by arsenic exposure (tertiles of urinary arsenic) and then compared %iAs, %MMA, and %DMA across tertiles of plasma micronutrient concentrations. In a multivariate adjusted model, only %iAs was found to be lower with increasing plasma folate concentrations and only among the highest arsenic exposure subgroup [17]. In a subset of 324 women who had urine samples available at gestational weeks 8, 14, and 30, Vahter et al. examined changes in arsenic methylation during the course of pregnancy. There were no observed associations between plasma folate or vitamin B12 and the change in urinary %iAs, %MMA, and %DMA over the course of pregnancy [18]. Although the authors concluded from these studies that nutritional status had little influence on arsenic methylation, this is not surprising given that plasma folate concentrations change dramatically over the course of pregnancy, introducing noise to the variable. Also, all of these women had been given prenatal folic acid (400 µg) starting at week 14 in addition to other vitamin and/or mineral supplements. It is also possible that other nutrients which were not examined in this study, such as choline and betaine, play a more substantial role in arsenic methylation during pregnancy; choline biosynthesis is significantly upregulated by estrogen, concentrations of which rise dramatically during pregnancy [19].

Gamble's group conducted a series of studies in Bangladesh on nutritional influences on arsenic metabolism and toxicity. They first evaluated the underlying prevalence of folate and B12-deficiency and hyperhomocysteinemia in a random sample of 1650 Bangladeshi adults. This survey revealed that the study population has an extremely high prevalence of hyperhomocysteinemia, particularly among males: 63% of males and 26% of females were found to have hyperhomocysteinemia (using cutoffs of ≥11.4 and 10.4 µmol/l, for men and women, respectively as defined in National Health and Nutrition Examination Survey [NHANES]) [20]. The data are consistent with a study of healthy males which reported that plasma total homocysteine concentrations were higher in Indian Asian men residing in the UK than their white European counterparts [21]. The survey also revealed modest but statistically significant negative correlations between water arsenic and plasma folate concentrations ($r=-0.13$, $P>0.0001$), suggesting that arsenic may in some way negatively impact folate nutritional status. The group subsequently selected a subset of 300 participants for a cross-sectional study on the associations between folate, total homocysteine, and arsenic methylation [22], after excluding those who were vitamin B12 deficient. There was a moderate but significant positive correlations between plasma folate and %DMA in urine and negative correlations between folate and both %iAs and %MMA in urine (Spearman correlations −0.12, −0.12, and 0.14, for %iAs, %MMA and %DMA, respectively; $P<0.05$ for all). Concentrations of total homocysteine were positively correlated with %MMA ($r=0.21$, $P<0.001$) and negatively correlated with %DMA ($r=-0.14$, $P<0.001$).

The strongest evidence supporting a specific role for folate in arsenic metabolism comes from a double-blind, placebo-controlled, 12-week folic acid supplementation trial conducted also by Gamble's group in 200 adults recruited from rural Bangladesh [23]. All participants, known to have low plasma folate concentrations at the beginning of the study, were randomly assigned to receive either folic acid supplementation (400 µg/day) or a placebo; compliance with the study regimen was confirmed through direct observation. Urinary arsenic metabolites were measured at enrollment, after 1 week and after 12 weeks. Folic acid supplementation resulted in an increase in %DMA (72% before and 79% after) that was significantly ($P<0.0001$) greater than that in the placebo group, as was the reduction in %MMA (13% before and 10% after, $P<0.0001$) and %iAs (15% before and 11% after, $P<0.001$). A subsequent study measured arsenic metabolites in blood for 130 participants, that is, those participants from the previous trial who had detectable levels of all arsenic metabolites. The study results revealed that folic acid supplementation resulted in a decline in total blood arsenic of 13.6±2.9% as compared to 2.5±3.2% for the placebo group ($P=0.01$). The decline in blood arsenic was largely due to the decline in MMA in blood. Whereas total blood arsenic (i.e., iAs+MMA+DMA) declined, on average, by 1.7, 1.1 µg/l of this was MMA; MMA declined by 22% from baseline [24]. These findings further support a role for folate in arsenic metabolism, and they suggest that folic acid supplementation can favorably alter arsenic metabolism in this population.

Gamble's group also conducted a cross-sectional study of 778 Bangladeshi adults in which vitamin B12-deficient participants were oversampled. This study found that vitamin B12 was inversely associated with %iAs [unstandardized regression coefficient (b)=−0.10; 95% CI, −0.17 to −0.02; $P=0.01$] and positively associated with %MMA ($b=0.12$; 95% CI, 0.05–0.20; $P=0.001$). Both of these associations were stronger among folate-sufficient participants (%iAs: $b=-0.17$; 95% CI, −0.30 to −0.03; $P=0.02$. %MMA: $b=0.20$; 95% CI, 0.11 to 0.30; $P<0.0001$), and the differences by folate status were statistically significant. They concluded that in contrast to folate, vitamin B12 appeared to facilitate the first arsenic methylation step among folate-sufficient individuals and may not favorably influence arsenic metabolism given the toxicity of MMA^{III} [25].

5.2.2 Selenium

Animal studies have shown an interaction between selenium and arsenic, such that uptake of one of these elements causes release, redistribution, or elimination of the other element by urinary and/or biliary routes [26, 27]. Chen et al. observed an inverse association between blood selenium level and urinary arsenic indicating that selenium-induced biliary excretion may occur in humans. A study of 252 subjects from Lanyang Basin, Taiwan, reported that urinary selenium levels increased the metabolism or methylation of arsenic, indicated by a higher percentage of DMA over total urinary arsenic [28]. Another study in Taiwan reported that plasma vitamin E levels were significantly inversely related to urinary total arsenic concentrations and %iAs, and significantly positively related to %DMA [29].

5.2.3 Creatinine

Gamble et al. observed that urinary creatinine is negatively correlated with %iAs and positively correlated with %DMA in urine ($r=-0.32$ and 0.30, respectively, $P>0.0001$); the correlations remain equally robust with and without control for covariates including body weight, age, and water or urine arsenic concentrations. They confirmed this observation in several subsequent studies in Bangladesh [23, 30–33]. Smith's group subsequently reported similar findings in West Bengal [34]. Urinary creatinine is influenced by dietary intake of creatine (derived from meat), which downregulates endogenous creatine biosynthesis. Because creatine biosynthesis is the major consumer of methyl groups [35, 36], downregulation by dietary sources may lower total homocysteine [37–39] and thereby increase the pool of methyl groups for methylation of arsenic and other substrates.

5.2.4 Other Nutrients

In the U.S. study discussed above [15], subjects in the lower quartile of iron, zinc, and vitamin B3 had higher %MMA and lower %DMA than did subjects in the higher quartile [15]. In one study, oral administration of iron reduced arsenic-caused DNA damage in mice, although it is unknown whether this effect is related to impacts on arsenic methylation [40]. Zinc has been linked to decreased arsenic toxicity in some studies [41, 42] but not in others [43–45]. No animal or human studies have identified a similar association for dietary vitamin B3. In the Bangladeshi study [16], higher intakes of calcium were associated with lower urinary %iAs and higher ratios of MMA to iAs, while higher intakes of vitamin B3 (beta=0.22, $P=0.02$) were associated with higher DMA-to-MMA ratios [16]. However, the mechanisms underlying these associations are not clear.

5.3 NUTRITION AND ARSENIC-INDUCED HEALTH EFFECTS

5.3.1 Skin Lesions, Blackfoot Disease, and Skin Cancer

In 2004, Smith's group conducted a dietary recall study in West Bengal, India, of 192 skin lesion cases and 192 age- and sex-matched controls. The study reported a modest increase in the risk of arsenic-induced skin lesions for persons in the lowest quintiles of animal protein (OR 1.94; 95% CI 1.05–3.59), calcium (OR 1.89; 95% CI 1.04–3.43), fiber (OR 2.20; 95% CI 1.15–4.21), and folate (OR 1.67; 95% CI 0.87–3.20) [43].

In 2006, this group reported results of a study of plasma concentrations of a series of 17 metabolites (including a series of micronutrients and cholesterol, glucose, glutathione, homocysteine, and transthyretin) and risk for arsenic-induced skin lesions; plasma analyses were done on a subset of 180 of the original 192 cases. No statistically significant odds ratios were observed for any of the parameters studied, including folate and homocysteine [46].

Using data from a large population-based, prospective cohort study, the Health Effects of Arsenic Longitudinal Study (HEALS) in Araihazar, Bangladesh, Ahsan's group first conducted a cross-sectional study of 10,628 subjects with nonmissing dietary data. This study reported that vitamins B2 and B6, folate, and vitamins A, C, and E significantly modified risk of arsenic-related skin lesions. The deleterious effect of ingested arsenic, at a given exposure level, was significantly reduced (ranging from 46% reduction for pyridoxine to 68% for vitamin C) for persons in the highest quintiles of vitamin intake [47]. The same group prospectively evaluated the association of nutrient intake with incident skin lesions [48]. The study observed significant associations between low intakes of various nutrients (retinol, calcium, fiber, folate, iron, vitamins B1 and B2 and vitamins A, C, and E) and skin lesion incidence, particularly for keratotic skin lesions. Associations for vitamins C and E showed significant linear trends. Intracellular antioxidants such as vitamins A, C, and E decrease arsenic toxicity by reversing disturbances in lipid peroxidation, generation of nitric oxide, reactive oxygen species, and apoptosis initiated by arsenic metabolites [49]. Though the explanation for the association with fiber is yet to be determined, fiber potentially plays a role via the absorption of arsenic through the gastrointestinal tract [43]. In a nested-case control study of 274 skin-lesion cases individually matched to controls for gender and age (within 5 years) and frequency matched for water arsenic (within 100 μg/l), Pilsner et al. reported that folate deficiency and hyperhomocysteinemia were both associated with increased risk for skin lesions [50], as is low urinary creatinine [OR (95% CI) were 1.8 (1.1–2.9) for plasma folate <9 nmol/l, 1.7 (1.1–2.6) for hyperhomocysteinemia, and 0.7 (0.5–0.8) for a fold increase in urinary creatinine]. A study in southwest Taiwan, where the prevalence of blackfoot disease is the highest, reported that skin cancer cases had a significantly lower serum β-carotene level than matched healthy controls [51].

Chen et al. did a case-cohort analysis to prospectively evaluate the association between arsenic-related premalignant skin lesions and prediagnostic blood selenium levels in 303 cases of skin lesions and 849 subcohort members with available baseline blood and urine samples [52]. Incidence rate ratios for skin lesions in increasing blood selenium quintiles were 1.00 (reference), 0.68 (0.39–1.18), 0.51 (0.29–0.87), 0.52 (0.30–0.91), and 0.53 (0.31–0.90). Effect estimates remained similar with adjustments for age, sex, body mass index, smoking status, excessive sunlight exposure (in men), well water arsenic concentration at baseline, and nutritional intakes of folate, iron, protein, vitamin E, and B vitamins. At any given arsenic exposure, the risk of premalignant skin lesions was consistently greater among participants with blood selenium lower than average [52]. The findings support the hypothesis that dietary selenium intake may reduce the incidence of arsenic-related premalignant skin lesions among populations exposed to arsenic in drinking water.

However, findings from previous studies were mostly inconclusive on the relationship between selenium intake and arsenic toxicity. A case-control study in Taiwan found that patients with blackfoot disease had lower blood selenium levels than controls, whereas a similar case-control study found that blood selenium was higher in patients with late-stage blackfoot disease compared with that in controls [45, 53]. In another case-control study in West Bengal, odds ratios for arsenic-related skin lesions did not differ by blood selenium levels [46]. It is unclear, however, whether the blood selenium levels observed in cases were a consequence or a contributing factor to blackfoot disease or arsenic-related skin lesions in these case-control analyses. A placebo-controlled trial in Inner Mongolia found that selenium supplementation significantly improved skin lesions [54]. However, the trial was neither randomized nor double blind, and the dropout rates in both the placebo and the treatment groups were high. A pilot randomized, placebo-controlled, double-blind trial conducted by Ahsan's group among 121 men and women found that selenium supplementation slightly improved skin lesion status; however, the sample size of the study was small and the improvement was not significant [55]. This trial also found that supplementation with vitamin E, either alone or in combination with selenium, slightly improved skin lesion status, although the improvement was not significant [55].

5.3.2 Urothelial Carcinoma

One of the strongest studies to date on the impact of nutritional status on risk for arsenic-induced health effects is the recent case-control study of 177 urothelial carcinoma cases and 488 controls in a population in Taiwan exposed to low concentrations of arsenic in drinking water. This study found that higher %DMA in urine and higher plasma folate concentrations were associated with an decreased risk. In a multivariate-adjusted model, the odds ratios (95% CI) for increasing quartiles of plasma folate concentrations were 1.0 (referent), 0.33 (0.20–0.54), 0.22 (0.13–0.38), and 0.09 (0.04–0.19), P trend<0.0001. Furthermore, a significant interaction was observed between urinary arsenic profiles and plasma folate in affecting urothelial carcinoma risk [56]. In another case-control study of 170 urothelial carcinoma cases and 402 healthy controls in the same population, Chung et al. showed that study participants with lower vitamin E, higher urinary total arsenic, higher %iAs, higher %MMA, and lower %DMA had a higher urothelial carcinoma risk than those with higher vitamin E, lower urinary total arsenic, lower %iAs, lower %MMA, and higher %DMA [29].

5.3.3 Cardiovascular Effects

Synergistic effect of low serum carotene level and duration of consuming arsenic-contaminated well water on ischemic heart disease was observed in a case-control study conducted in southwest Taiwan [57]. The OR was higher (OR=5.0, 95% CI 1.5–16.8) for those who had both long duration of consuming artesian well water and decreased serum level of α- and β- carotene, compared to ORs for those with either one alone (OR=2.5, 95% CI 0.8–7.9; and OR=2.9, 95% CI 0.5–16.8, respectively)

[57]. Chen et al. performed a cross-sectional analysis to evaluate the influence of dietary intakes of B vitamins and folate on the association between arsenic exposure from drinking water and blood pressure [58]. The study found that among participants with a lower than average dietary intake level of B vitamins and folate, the odds ratios for high pulse pressure by increasing quintiles of time-weighted average arsenic were 1.00 (referent), 1.84 (95% CI: 1.07, 3.16), 1.89 (95% CI: 1.11, 3.20), 1.83 (95% CI: 1.09, 3.07), and 1.89 (95% CI: 1.12, 3.20). These findings indicate that the effect of moderate-level arsenic exposure on blood pressure may be more pronounced in persons with lower intake of nutrients related to arsenic metabolism and cardiovascular health.

5.4 CONCLUSIONS

These findings support the hypothesis that nutrients relevant to arsenic metabolism and antioxidant nutrients may modify the risk of arsenic-related health effects. Nutrients influencing the susceptibility to or expression of arsenic-associated cancer and noncancer effects need to be better characterized. Particular attention should be given to the cardiovascular effects under various exposure scenarios, especially in low-income countries such as South Asia, where the populations are undergoing an epidemiologic transition in which the burden of CVD is increasing. On the other hand, some methodological issues may have limited the effect estimates in the above-mentioned studies. For instance, the use of food-frequency questionnaire (FFQ) introduced inevitable measurement errors. FFQ is not an ideal methodology for quantifying intake of food or various nutrients [59]. Certain nutrients have been shown to have poor concordance between FFQ-measured intake and intake based on diary or recall methods, including folate, fiber, protein, potassium, retinol, and vitamin E [60, 61]. However, although the FFQ is limited in its ability to quantify exact intake of nutrients, it is a useful and efficient methodology for ranking individuals from highest to lowest based on their nutrient intake. Future studies with biochemical measures of nutrient intakes are needed to further evaluate the effects of these nutrients. It is of note that limitations also exist for the use of biochemical measures such as serum or plasma levels of nutrients. For instance, the accuracy of one-spot evaluation of nutrients may be in doubt, although the values might be reliable if all subjects reported no accompanying changes in lifestyle. In addition, if the specimen collection and measurement of nutrients were completed after disease diagnosis, the association between nutrients and disease risk might be underestimated if patients improved their dietary habits after diagnosis. Finally, different nutrients vary in their responses to sample handling procedures. It has been reported that folate is highly unstable and would likely be degraded if not frozen in time after collection, but homocysteine continues to be released into plasma by red blood cells after sample collection [62]. The inherent difficulty of accurately measuring specific nutrients using FFQ and isolating the effects of individual nutrients in the context of a complex diet of correlated foods and nutrients has motivated Pierce et al. to examine the associations among FFQ-derived dietary patterns [63–65] and arsenical skin lesion incidence in a large prospective cohort [66]. Using the dietary pattern approach, the authors did not measure specific nutrients, but rather focused on broad patterns derived from the correlation structure of the FFQ data. They found that an increasing gourd and root factor score showed a clear association with decreasing skin lesion risk. This score also modified the association between water arsenic exposure and skin lesions risk, with decreased arsenic-related risk for individuals with high gourd and root scores [66]. This study provides additional support for the hypothesis that diet plays a critical role in susceptibility to arsenic-related toxicity.

The findings of a close interaction between arsenic exposure and nutrients in disease risk and the increased excretion of arsenic after folate supplementation have prompted some researchers to believe that nutritional intervention may provide an additional approach for lessening the burden of disease resulting from long-term arsenic exposure. Although numerous observational studies have indicated a role of nutrition in arsenic toxicity, intervention studies, especially those that focus on clinical endpoints, are limited. For instance, the clinical trial of folate shows that folate supplementation can improve arsenic methylation capacity. However, to what extent this effect can translate to risk reduction of arsenic-related disease is unknown. Similar evidence is largely lacking for other nutrients. Therefore, dietary recommendation of specific diet or nutrients for the exposed population is probably not advisable at present. Thus, more intervention studies may be needed. Meanwhile, it is clear that the discontinuation of drinking arsenic-contaminated water and using "safe" drinking water remain of primary importance. The observations of individuals with varying susceptibility due to nutrition help the characterization of disease occurrence in the population in terms of etiologic heterogeneity. The data may be used to suggest priority of clean water provision for certain population subgroups that are more susceptible to arsenic toxicity than others.

REFERENCES

[1] National Research Council, *Arsenic in Drinking Water*, National Academy of Sciences, Washington, DC, 2001.

[2] T.H. Yang, R.Q. Blackwell, Nutritional and environmental conditions in the endemic Blackfoot area, Formosan Science 15 (1961) 101–129.

[3] Y.M. Hsueh, G.S. Cheng, M.M. Wu, H.S. Yu, T.L. Kuo, C.J. Chen, Multiple risk factors associated with arsenic-induced skin cancer: effects of chronic liver disease and malnutritional status, British Journal of Cancer 71 (1995) 109–114.

[4] R.R. Engel, O. Receveur, Arsenic ingestion and internal cancers: a review, American Journal of Epidemiology 138 (1993) 896–897.

[5] C.J. Chen, M.M. Wu, S.S. Lee, J.D. Wang, S.H. Cheng, H.Y. Wu, Atherogenicity and carcinogenicity of high-arsenic artesian well water. Multiple risk factors and related malignant neoplasms of blackfoot disease, Arteriosclerosis 8 (1988) 452–460.

[6] M. Vahter, E. Marafante, Effects of low dietary intake of methionine, choline or proteins on the biotransformation of arsenite in the rabbit, Toxicology Letters 37 (1987) 41–46.

[7] H.R. Hansen, A. Raab, M. Jaspars, B.F. Milne, J. Feldmann, Sulfur-containing arsenical mistaken for dimethylarsinous acid [DMA(III)] and identified as a natural metabolite in urine: major implications for studies on arsenic metabolism and toxicity, Chemical Research in Toxicology 17 (2004) 1086–1091.

[8] E. Marafante, M. Vahter, The effect of methyltransferase inhibition on the metabolism of [^{74}As]arsenite in mice and rabbits, Chemico-Biological Interactions 50 (1984) 49–57.

[9] S.B. Waters, V. Devesa, L.M. Del Razo, M. Styblo, D.J. Thomas, Endogenous reductants support the catalytic function of recombinant rat cyt19, an arsenic methyltransferase, Chemical Research in Toxicology 17 (2004) 404–409.

[10] J. Selhub, Folate, vitamin B12 and vitamin B6 and one carbon metabolism, The Journal of Nutrition, Health & Aging 6 (2002) 39–42.

[11] A. Celkova, J. Kubova, V. Stresko, Determination of arsenic in geological samples by HG AAS, Analytical and Bioanalytical Chemistry 355 (1996) 150–153.

[12] P.M. Ueland, P.I. Holm, S. Hustad, Betaine: a key modulator of one-carbon metabolism and homocysteine status, Clinical Chemistry and Laboratory Medicine: CCLM FESCC 43 (2005) 1069–1075.

[13] O.F. Brouwer, W. Onkenhout, P.M. Edelbroek, J.F. de Kom, F.A. de Wolff, A.C. Peters, Increased neurotoxicity of arsenic in methylenetetrahydrofolate reductase deficiency, Clinical Neurology and Neurosurgery 94 (1992) 307–310.

[14] J.S. Chung, D.A. Kalman, L.E. Moore, M.J. Kosnett, A.P. Arroyo, M. Beeris, D.N. Mazumder, A.L. Hernandez, A.H. Smith, Family correlations of arsenic methylation patterns in children and parents exposed to high concentrations of arsenic in drinking water, Environmental Health Perspectives 110 (2002) 729–733.

[15] C. Steinmaus, K. Carrigan, D. Kalman, R. Atallah, Y. Yuan, A.H. Smith, Dietary intake and arsenic methylation in a U.S. population, Environmental Health Perspectives 113 (2005) 1153–1159.

[16] J.E. Heck, M.V. Gamble, Y. Chen, J.H. Graziano, V. Slavkovich, F. Parvez, J.A. Baron, G.R. Howe, H. Ahsan, Consumption of folate-related nutrients and metabolism of arsenic in Bangladesh, The American Journal of Clinical Nutrition 85 (2007) 1367–1374.

[17] L. Li, E.C. Ekstrom, W. Goessler, B. Lonnerdal, B. Nermell, M. Yunus, A. Rahman, S. El Arifeen, L.A. Persson, M. Vahter, Nutritional status has marginal influence on the metabolism of inorganic arsenic in pregnant Bangladeshi women, Environmental Health Perspectives 116 (2008) 315–321.

[18] R.M. Gardner, B. Nermell, M. Kippler, M. Grander, L. Li, E.C. Ekstrom, A. Rahman, B. Lonnerdal, A.M. Hoque, M. Vahter, Arsenic methylation efficiency increases during the first trimester of pregnancy independent of folate status, Reproductive Toxicology 31 (2011) 210–218.

[19] S.H. Zeisel, Importance of methyl donors during reproduction, The American Journal of Clinical Nutrition 89 (2009) 673S–677S.

[20] M.V. Gamble, H. Ahsan, X. Liu, P. Factor-Litvak, V. Ilievski, V. Slavkovich, F. Parvez, J.H. Graziano, Folate and cobalamin deficiencies and hyperhomocysteinemia in Bangladesh, The American Journal of Clinical Nutrition 81 (2005) 1372–1377.

[21] J.C. Chambers, O.A. Obeid, H. Refsum, P. Ueland, D. Hackett, J. Hooper, R.M. Turner, S.G. Thompson, J.S. Kooner, Plasma homocysteine concentrations and risk of coronary heart disease in UK Indian Asian and European men, Lancet 355 (2000) 523–527.

[22] M.V. Gamble, X. Liu, H. Ahsan, R. Pilsner, V. Ilievski, V. Slavkovich, F. Parvez, D. Levy, P. Factor-Litvak, J.H. Graziano, Folate, homocysteine, and arsenic metabolism in arsenic-exposed individuals in Bangladesh, Environmental Health Perspectives 113 (2005) 1683–1688.

[23] M.V. Gamble, X. Liu, H. Ahsan, J.R. Pilsner, V. Ilievski, V. Slavkovich, F. Parvez, Y. Chen, D. Levy, P. Factor-Litvak, J.H. Graziano, Folate and arsenic metabolism: a double-blind, placebo-controlled folic acid-supplementation trial in Bangladesh, The American Journal of Clinical Nutrition 84 (2006) 1093–1101.

[24] M.V. Gamble, X. Liu, V. Slavkovich, J.R. Pilsner, V. Ilievski, P. Factor-Litvak, D. Levy, S. Alam, M. Islam, F. Parvez, H. Ahsan, J.H. Graziano, Folic acid supplementation lowers blood arsenic, The American Journal of Clinical Nutrition 86 (2007) 1202–1209.

[25] M.N. Hall, X. Liu, V. Slavkovich, V. Ilievski, Z. Mi, S. Alam, P. Factor-Litvak, H. Ahsan, J.H. Graziano, M.V. Gamble, Influence of cobalamin on arsenic metabolism in Bangladesh, Environmental Health Perspectives 117 (2009) 1724–1729.

[26] O.A. Levander, C.A. Baumann, Selenium metabolism. VI. Effect of arsenic on the excretion of selenium in the bile, Toxicology and Applied Pharmacology 9 (1966) 106–115.

[27] J. Gailer, G.N. George, I.J. Pickering, R.C. Prince, H.S. Younis, J.J. Winzerling, Biliary excretion of [(GS)(2)AsSe](-) after intravenous injection of rabbits with arsenite and selenate, Chemical Research in Toxicology 15 (2002) 1466–1471.

[28] Y.M. Hsueh, Y.F. Ko, Y.K. Huang, H.W. Chen, H.Y. Chiou, Y.L. Huang, M.H. Yang, C.J. Chen, Determinants of inorganic arsenic methylation capability among residents of the Lanyang Basin, Taiwan: arsenic and selenium exposure and alcohol consumption, Toxicology Letters 137 (2003) 49–63.

[29] C.J. Chung, Y.S. Pu, Y.T. Chen, C.T. Su, C.C. Wu, H.S. Shiue, C.Y. Huang, Y.M. Hsueh, Protective effects of plasma alpha-tocopherols on the risk of inorganic arsenic-related urothelial carcinoma, The Science of the Total Environment 409 (2011) 1039–1045.

[30] J.R. Pilsner, X. Liu, H. Ahsan, V. Ilievski, V. Slavkovich, D. Levy, P. Factor-Litvak, J.H. Graziano, M.V. Gamble, Folate deficiency, hyperhomocysteinemia, low urinary creatinine, and hypomethylation of leukocyte DNA are risk factors for arsenic-induced skin lesions, Environmental Health Perspectives 117 (2009) 254–260.

[31] D.B. Barr, L.C. Wilder, S.P. Caudill, A.J. Gonzalez, L.L. Needham, J.L. Pirkle, Urinary creatinine concentrations in the U.S. population: implications for urinary biologic monitoring measurements, Environmental Health Perspectives 113 (2005) 192–200.

[32] M.N. Hall, X. Liu, V. Slavkovich, V. Ilievski, J.R. Pilsner, S. Alam, P. Factor-Litvak, J.H. Graziano, M.V. Gamble, Folate, cobalamin, cysteine, homocysteine, and arsenic metabolism among children in Bangladesh, Environmental Health Perspectives 117 (2009) 825–831.

[33] M.V. Gamble, X. Liu, Urinary creatinine and arsenic metabolism, Environmental Health Perspectives 113 (2005) A442; author reply A442–443.

[34] A. Basu, S. Mitra, J. Chung, D.N. Guha Mazumder, N. Ghosh, D. Kalman, O.S. von Ehrenstein, C. Steinmaus, J. Liaw, A.H. Smith, Creatinine, diet, micronutrients, and arsenic methylation in West Bengal, India, Environmental Health Perspectives 119 (2011) 1308–1313.

[35] S.H. Mudd, J.R. Poole, Labile methyl balances for normal humans on various dietary regimens, Metabolism, Clinical and Experimental 24 (1975) 721–735.

[36] L.M. Stead, J.T. Brosnan, M.E. Brosnan, D.E. Vance, R.L. Jacobs, Is it time to reevaluate methyl balance in humans? The American Journal of Clinical Nutrition 83 (2006) 5–10.

[37] W.J. Korzun, Oral creatine supplements lower plasma homocysteine concentrations in humans, Clinical Laboratory Science: Journal of the American Society for Medical Technology 17 (2004) 102–106.

[38] L.M. Stead, K.P. Au, R.L. Jacobs, M.E. Brosnan, J.T. Brosnan, Methylation demand and homocysteine metabolism: effects of dietary provision of creatine and guanidinoacetate, American Journal of Physiology. Endocrinology and Metabolism 281 (2001) E1095–E1100.

[39] Y.E. Taes, J.R. Delanghe, A.S. De Vriese, R. Rombaut, J. Van Camp, N.H. Lameire, Creatine supplementation decreases homocysteine in an animal model of uremia, Kidney International 64 (2003) 1331–1337.

[40] S. Poddar, P. Mukherjee, G. Talukder, A. Sharma, Dietary protection by iron against clastogenic effects of short-term exposure to arsenic in mice in vivo, Food and Chemical Toxicology: An International Journal Published for the British Industrial Biological Research Association 38 (2000) 735–737.

[41] A.G. Milton, P.D. Zalewski, R.N. Ratnaike, Zinc protects against arsenic-induced apoptosis in a neuronal cell line, measured by DEVD-caspase activity, Biometals: An International Journal on the Role of Metal Ions in Biology, Biochemistry, and Medicine 17 (2004) 707–713.

[42] G.H. Rabbani, S.K. Saha, M. Akhtar, F. Marni, A.K. Mitra, S. Ahmed, M. Alauddin, M. Bhattacharjee, S. Sultana, A.K. Chowdhury, Antioxidants in detoxification of arsenic-induced oxidative injury in rabbits: preliminary results, Journal of Environmental Science and Health, Part A: Toxic/Hazardous Substances & Environmental Engineering 38 (2003) 273–287.

[43] S.R. Mitra, D.N. Mazumder, A. Basu, G. Block, R. Haque, S. Samanta, N. Ghosh, M.M. Smith, O.S. von Ehrenstein, A.H. Smith, Nutritional factors and susceptibility to arsenic-caused skin lesions in West Bengal, India, Environmental Health Perspectives 112 (2004) 1104–1109.

[44] M. Shimizu, J.F. Hochadel, B.A. Fulmer, M.P. Waalkes, Effect of glutathione depletion and metallothionein gene expression on arsenic-induced cytotoxicity and c-myc expression in vitro, Toxicological Sciences: An Official Journal of the Society of Toxicology 45 (1998) 204–211.

[45] C.T. Wang, Concentration of arsenic, selenium, zinc, iron and copper in the urine of blackfoot disease patients at different clinical stages, European Journal of Clinical Chemistry and Clinical Biochemistry: Journal of the Forum of European Clinical Chemistry Societies 34 (1996) 493–497.

[46] J.S. Chung, R. Haque, D.N. Guha Mazumder, L.E. Moore, N. Ghosh, S. Samanta, S. Mitra, M.M. Hira-Smith, O. von Ehrenstein, A. Basu, J. Liaw, A.H. Smith, Blood concentrations of methionine, selenium, beta-carotene, and other micronutrients in a case-control study of arsenic-induced skin lesions in West Bengal, India, Environmental Research 101 (2006) 230–237.

[47] L.B. Zablotska, Y. Chen, J.H. Graziano, F. Parvez, A. van Geen, G.R. Howe, H. Ahsan, Protective effects of B vitamins and antioxidants on the risk of arsenic-related skin lesions in Bangladesh, Environmental Health Perspectives 116 (2008) 1056–1062.

[48] S. Melkonian, M. Argos, Y. Chen, F. Parvez, B. Pierce, A. Ahmed, T. Islam, H. Ahsan, Intakes of several nutrients are associated with incidence of arsenic-related keratotic skin lesions in bangladesh, The Journal of Nutrition 142 (2012) 2128–2134.

[49] S. Chattopadhyay, S. Bhaumik, M. Purkayastha, S. Basu, A. Nag Chaudhuri, S. Das Gupta, Apoptosis and necrosis in developing brain cells due to arsenic toxicity and protection with antioxidants, Toxicology Letters 136 (2002) 65–76.

[50] J.R. Pilsner, X. Liu, H. Ahsan, V. Ilievski, V. Slavkovich, D. Levy, P. Factor-Litvak, J.H. Graziano, M.V. Gamble, Genomic methylation of peripheral blood leukocyte DNA: influences of arsenic and folate in Bangladeshi adults, The American Journal of Clinical Nutrition 86 (2007) 1179–1186.

[51] Y.M. Hsueh, H.Y. Chiou, Y.L. Huang, W.L. Wu, C.C. Huang, M.H. Yang, L.C. Lue, G.S. Chen, C.J. Chen, Serum beta-carotene level, arsenic methylation capability, and incidence of skin cancer, Cancer Epidemiology, Biomarkers & Prevention: A Publication of the American Association for Cancer Research, Cosponsored by the American Society of Preventive Oncology 6 (1997) 589–596.

[52] Y. Chen, M. Hall, J.H. Graziano, V. Slavkovich, A. van Geen, F. Parvez, H. Ahsan, A prospective study of blood selenium levels and the risk of arsenic-related premalignant skin lesions, Cancer Epidemiology, Biomarkers & prevention: A Publication of the American Association for Cancer Research, Cosponsored by the American Society of Preventive Oncology 16 (2007) 207–213.

[53] S.M. Lin, M.H. Yang, Arsenic, selenium, and zinc in patients with blackfoot disease, Biological Trace Element Research 15 (1988) 213–221.

[54] L. Yang, W. Wang, S. Hou, P.J. Peterson, W.P. Williams, Effects of selenium supplementation on arsenism: an intervention trial in inner Mongolia, Environmental Geochemistry and Health 24 (2002) 359–374.

[55] W.J. Verret, Y. Chen, A. Ahmed, T. Islam, F. Parvez, M.G. Kibriya, J.H. Graziano, H. Ahsan, A randomized, double-blind placebo-controlled trial evaluating the effects of vitamin E and selenium on arsenic-induced skin lesions in Bangladesh, Journal of Occupational and Environmental Medicine/American College of Occupational and Environmental Medicine 47 (2005) 1026–1035.

[56] Y.K. Huang, Y.S. Pu, C.J. Chung, H.S. Shiue, M.H. Yang, C.J. Chen, Y.M. Hsueh, Plasma folate level, urinary arsenic methylation profiles, and urothelial carcinoma susceptibility, Food and Chemical Toxicology 46 (2008) 929–938.

[57] Y.M. Hsueh, W.L. Wu, Y.L. Huang, H.Y. Chiou, C.H. Tseng, C.J. Chen, Low serum carotene level and increased risk of ischemic heart disease related to long-term arsenic exposure, Atherosclerosis 141 (1998) 249–257.

[58] Y. Chen, P. Factor-Litvak, G.R. Howe, J.H. Graziano, P. Brandt-Rauf, F. Parvez, A. van Geen, H. Ahsan, Arsenic exposure from drinking water, dietary intakes of B vitamins and folate, and risk of high blood pressure in Bangladesh: a population-based, cross-sectional study, American Journal of Epidemiology 165 (2007) 541–552.

[59] A.R. Kristal, U. Peters, J.D. Potter, Is it time to abandon the food frequency questionnaire? Cancer Epidemiology, Biomarkers & Prevention: A Publication of the American Association for Cancer Research, Cosponsored by the American Society of Preventive Oncology 14 (2005) 2826–2828.

[60] A.F. Malekshah, M. Kimiagar, M. Saadatian-Elahi, A. Pourshams, M. Nouraie, G. Goglani, A. Hoshiarrad, M. Sadatsafavi, B. Golestan, A. Yoonesi, N. Rakhshani, S. Fahimi, D. Nasrollahzadeh, R. Salahi, A. Ghafarpour, S. Semnani, J.P. Steghens, C.C. Abnet, F. Kamangar, S.M. Dawsey, P. Brennan, P. Boffetta, R. Malekzadeh, Validity and reliability of a new food frequency questionnaire compared to 24 h recalls and biochemical measurements: pilot phase of Golestan cohort study of esophageal cancer, European Journal of Clinical Nutrition 60 (2006) 971–977.

[61] R.L. Prentice, Dietary assessment and the reliability of nutritional epidemiology reports, Lancet 362 (2003) 182–183.

[62] M.N. Hall, M.V. Gamble, Nutritional manipulation of one-carbon metabolism: effects on arsenic methylation and toxicity, Journal of Toxicology 2012 (2012) 595307.

[63] K.B. Michels, M.B. Schulze, Can dietary patterns help us detect diet-disease associations? Nutrition Research Reviews 18 (2005) 241–248.

[64] A.K. Kant, Dietary patterns and health outcomes, Journal of the American Dietetic Association 104 (2004) 615–635.

[65] F.B. Hu, Dietary pattern analysis: a new direction in nutritional epidemiology, Current Opinion in Lipidology 13 (2002) 3–9.

[66] B.L. Pierce, M. Argos, Y. Chen, S. Melkonian, F. Parvez, T. Islam, A. Ahmed, R. Hasan, P.J. Rathouz, H. Ahsan, Arsenic exposure, dietary patterns, and skin lesion risk in Bangladesh: a prospective study, American Journal of Epidemiology 173 (2011) 345–354.

6

SKIN MANIFESTATIONS OF CHRONIC ARSENICOSIS

NILENDU SARMA
Department of Dermatology, NRS Medical College, Kolkata, India

6.1 INTRODUCTION

Arsenic (As) is classified by the United States Environmental Protection Agency (USEPA) and International Agency for Research on Cancer (IARC) as a group 1 human carcinogen. Arsenicosis is defined by the World Health Organization (WHO) as a chronic health condition resulting from consumption of arsenic above safe limit for more than 6 months usually manifested by characteristic skin lesions including melanosis and keratosis [3].

Trace amounts of arsenic are ubiquitously detected in nature. Drinking ground water with high level of arsenic, much above safe limit has affected millions of people in many countries worldwide. In addition to exposure via arsenic contaminated drinking water, human exposure to arsenic also has occurred through herbal medicines or occupationally. Skin involvement is a most distinctive symptom of chronic arsenic exposure and is useful to diagnose arsenicosis. However, skin symptoms usually take a long time to manifest.

Skin manifestations of arsenicosis include melanosis, hyperkeratosis, and both basal cell and squamous cell carcinoma. These skin changes as a consequence of chronic arsenic exposure will be discussed in this chapter based upon the author's experience with the situation in West Bengal, one of the regions of the world experiencing severe arsenic exposure.

Arsenic: Exposure Sources, Health Risks, and Mechanisms of Toxicity, First Edition.
Edited by J. Christopher States.
© 2016 John Wiley & Sons, Inc. Published 2016 by John Wiley & Sons, Inc.

6.2 MECHANISM OF ARSENIC TOXICITY IN SKIN AND OTHER TISSUES

The pathogenesis of arsenic-induced diseases is yet to be elucidated. Skin is particularly sensitive to the toxic effect of arsenic. Epidermal stem cells might be the prime target for arsenic induced damage.

As discussed in Chapter 4, after entry into the body, inorganic arsenic is reduced from the pentavalent form to trivalent form. Subsequently, methylation occurs with formation of tri- and pentavalent methylated arsenicals. This methylation has been considered a detoxification process until recently. However, methylation, particularly to trivalent methylated arsenicals, is now considered to be metabolic activation contributing to the carcinogenic and genotoxic mechanisms of arsenic. Methylated trivalent arsenic adversely affects DNA repair, generates reactive oxygen species, inhibits cellular proliferation and induces chromosomal abnormality.

Arsenic 3 methyl transferase (AS3MT) uses *S*-adenosylmethionine (SAM) as the primary methyl group donor. Depletion of SAM by AS3MT may inhibit various protective cellular processes that require methyl groups. This interaction with methyl transfer pathways may also contribute to the ill effects of arsenic.

In rodent studies, arsenic acted as cocarcinogen with arsenic in drinking water potentiating the carcinogenic effect of ultraviolet light on skin [17]. These studies suggested that skin cancer in arsenicosis patients may arise through cocarcinogenic action of arsenic in conjunction with sun exposure. However, arsenicosis patients exhibit a limited set of skin cancers and these occur not only in sun-exposed but also in sun-protected parts of the body. Thus, the role of potential interaction of arsenic with ultraviolet light in skin carcinogenesis is not clear.

6.3 CUTANEOUS MANIFESTATIONS

Skin is the most sensitive organ and visible alteration is noted in a high percentage of arsenic exposed population. Cutaneous manifestation of arsenicosis is the most distinctive change associated with arsenicosis. Thus, the skin changes are the most useful tool for clinical diagnosis of arsenicosis. Molecular basis for the universal involvement of skin in arsenicosis lies in the high affinity of the arsenic for the thiol groups in the keratin rich environment of skin, hair and nails.

When exposure occurs through drinking water, it requires many years for clinically detectable signs to develop. Pigmentary changes may develop after 6 months to 10 years of continuous exposure [19]. Pigmentation changes may be of several types including diffuse melanosis or variable pigmentary change in a “raindrop” pattern. The classic cutaneous changes of chronic arsenicosis, pigmentary alterations and keratosis, are discussed in detail in Section 6.3.

Generally, pigmentation change develops earlier than keratosis. The keratoses may give rise to basal cell and squamous cell carcinomas. However, it may take more than two decades of exposure before carcinoma develops. Internal malignancy may take even longer period, may be 30 years or more. However, in the case of exposure to higher concentration of arsenic, as may occur through medications or other sources, manifestation may be much earlier. Keratosis is reported to occur as early as 2.5 years of continuous exposure through Fowler's solution. Non-malignant cutaneous changes may regress after withdrawal from exposure but such duration may be many years and not known [27].

Due to long incubation period and also due to lesser number of published studies, little is known about the cutaneous changes in children. Possibly the changes are less common in children. Cutaneous changes however may occur in small children as early as young as 2 years of age [4, 11]. Children may have higher incidence of skin lesion than adults when exposed through sources with higher arsenic content like Chinese herbal medicines.

6.3.1 Pigmentary Alterations

Pigmentary alterations in arsenicosis may be spotted or diffuse. Size of the smaller pigmentary changes generally varies from pinpoint to 1 cm. Spotted pigmentation larger than 2 cm is rare.

Pigmentary alteration may be composed of hyper (darker) or hypo (lighter) regions. Well defined small hyperpigmented macules in arsenicosis are often called "raindrop" pigmentation (Fig. 6.1, Panel a). Pigmentation can be diffuse and may involve larger body surfaces. Pigmentary alteration may develop at any site but are common over trunk, thighs and arms, and are clearly distinct in sun protected areas. It has been reported that hyperpigmentation has a tendency to develop particular at normally pigmented sites like areola or groin [19]. Diffuse hyperpigmentation is fairly common in palms (Fig. 6.1, Panel b) and soles and often aids diagnosis [18].

Small hypopigmented macules are also present scattered over the entirebody and are called leukomelanosis (Fig. 6.1, Panel c). Diffuse pigmentation, when present in the background makes the smaller hypopigmented macules stand out.

Affected area may have both hypopigmentation and hyperpigmentation simultaneously. This is also called dyschromia. Generally there are numerous lesions and large areas are affected and in advanced cases, innumerable lesion may merge. In very severe cases the area appears xerodermoid (Fig. 6.1, Panel d).

Hyperpigmentation has been reported to be the most common pigmentary alteration [15]. In a population with arsenical skin cancers, 89.7% had hyperpigmentation [15]. However, this co-presentation of hyperpigmentation and skin cancer is not found to be common among populations other than those of Oriental descent [15]. There is controversy regarding the relative prevalence of keratosis and pigmentary alteration. Some studies suggested that keratoses were more common [16, 22, 23]. Others found pigmentary alterations to be more common [15, 22]. However, the disagreement could be due to differences in the definition of the keratosis.

Mucosal pigmentation such as on the undersurface of tongue or buccal mucosa is reported to occur and may be diffuse or spotted. Exact prevalence of this type of pigmentation change is unknown. Mild to moderate diffuse pigmentation may develop

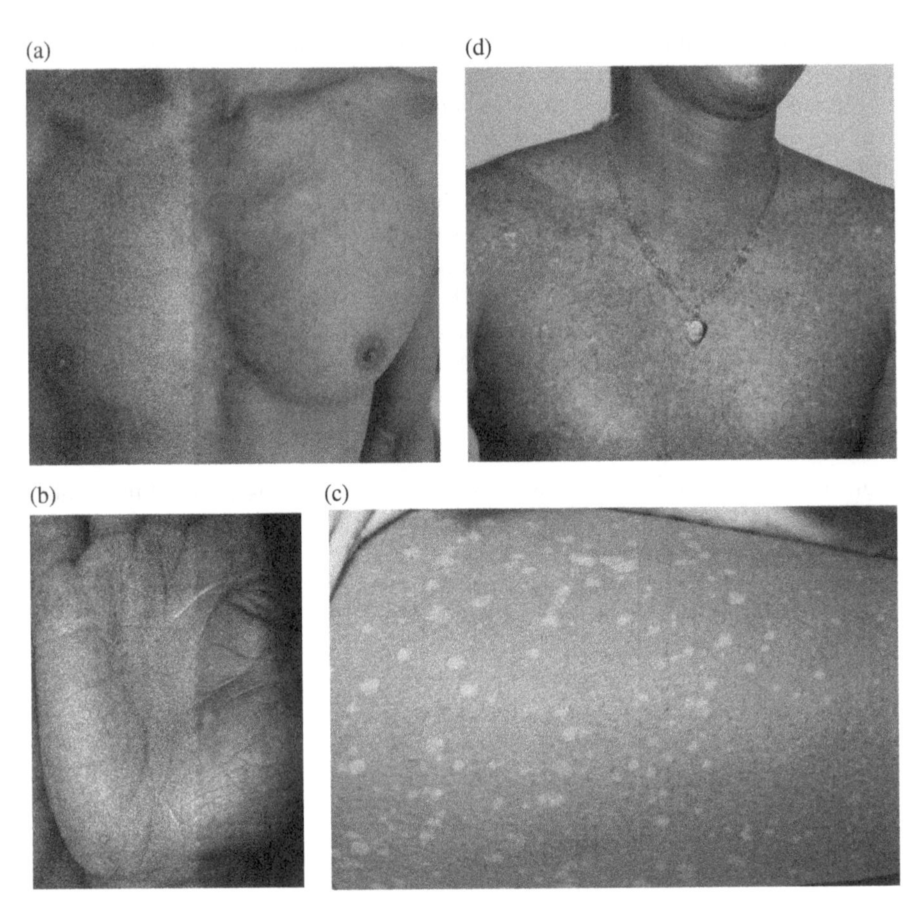

FIGURE 6.1 Pigmentary changes induced by arsenicosis. Small, distinct hyperpigmented macules on chest due to arsenicosis (a). Diffuse pigmentation on palm in arsenicosis (b). Well defined hypopigmented macules on thigh due to arsenicosis (c). Extensive xerodermoid dyschromia on chest in arsenicosis (d). (*See insert for color representation of the figure.*)

over palms, trunks and limbs. Exposed areas often show darker pigmentation. Increased melanin is detected high in the epidermis as well as in the dermis [24].

Pigmentary alterations seen in arsenicosis need to be differentiated from many other similar appearing conditions. Pityriasis versicolor, post inflammatory hypopigmentation, pityriasis lichenoides chronica (PLC), leprosy, idiopathic guttate hypomelanosis (IGH), all may mimic arsenical hypopigmentation.

Hyperpigmented spots may resemble freckles, lichen planus and post inflammatory hyperpigmentation. Differentiation of diffuse arsenical pigmentation over exposed skin may be difficult from chronic photo darkening.

Diffuse pigmentation is particularly common among people exposed to arsenic through herbal medicines [8, 21]. Diffuse pigmentation should be differentiated from ashy dermatoses, lichen planus pigmentosus (LPP) and macular amyloidosis.

6.3.2 Keratoses

Arsenical "keratosis" is a term used to define typical palpable skin changes. Keratoses are the most distinctive cutaneous changes in arsenicosis. "Keratosis" is otherwise a less commonly used term in clinical dermatology. Arsenical "keratosis" is used for "papules" (≤0.5 mm) and "plaques" (>0.5 cm) that are the standard dermatological terms for palpable skin changes.

Classical sites for arsenical keratosis are palms (Fig. 6.2, Panel a) and soles (Fig. 6.2, Panel b). Palmoplantar keratosis are considered diagnostic hallmark of the disease. However, other body areas like trunk and limbs are also frequently involved. Face is only rarely affected.

On palms and soles, the keratoses have a tendency to develop on friction or trauma prone areas like thenar eminences, lateral borders, roots or lateral surfaces of fingers and soles, heels and toes of feet. Usually they are darker but may be of skin color. They have a firm, spiny, or "gritty" feel when touched. They are generally small (<1 cm) but can be bigger. Based on the sizes of palmo-plantar keratosis, disease

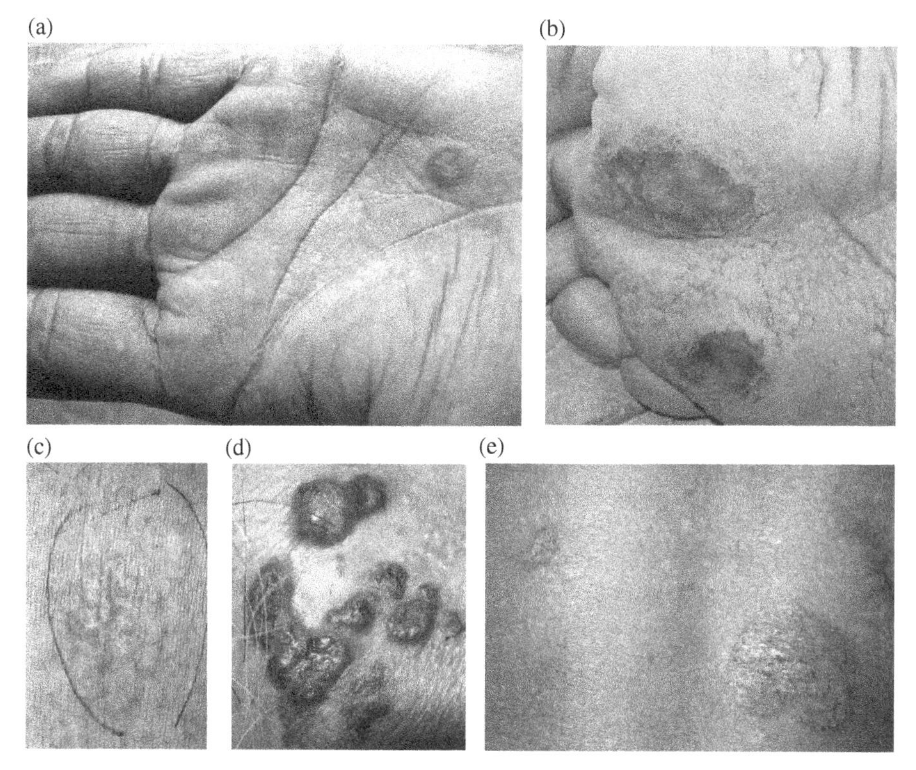

FIGURE 6.2 Premalignant and malignant lesions induced by arsenicosis. Arsenical keratosis on palm (a). Large arsenical keratosis on sole mimicking callosity (b). Classical Bowen's disease in arsenicosis (c). Pigmented nodular basal cell carcinoma in arsenicosis (d). Multiple superficial basal cell carcinoma on back with typical arsenical pigmentary changes in the background (e). (*See insert for color representation of the figure.*)

severity has been graded into mild (size of <2 mm), moderate (2–5 mm) and severe (>5 mm). Larger lesions appear as warty nodules. They are commonly discrete but may be confluent in severe cases. Entire palmo-plantar surface may be thickened. Regular activities often dislodge these keratotic skin lesions leading to depressed areas (pits or crater). Based on the size of the keratosis, size of the crater also varies. Other changes in palms and soles may be diffuse fissuring without any keratoses.

They should be differentiated from viral wart, occupational keratosis, corn/callus, pitted keratolysis, epidermodysplasia of Hopf and Darier's disease. With the constellation of clinical appearance, arsenical palmoplantar keratosis appears distinctive and diagnosis is generally not difficult.

Non-palmoplantar keratosis distributed over trunk and limbs may mimic lichen planus, viral wart, actinic keratosis and seborrheic keratosis.

6.3.3 Other Non-malignant Cutaneous Changes

Vascular compromisation of the limbs may lead to ischemic changes and gangrene of the toes and feet. This is often called Blackfoot disease. In some cases, ulcer may be noted. Ulcers usually have high malignant potential. Finger nails may occasionally show transverse white bands, called Mee's lines.

6.3.4 Malignant Changes

Chronic exposure to arsenic induces various types of skin cancers like Bowen's disease (Fig. 6.2, Panel c), basal cell carcinoma (Fig. 6.2, Panel d), and squamous cell carcinoma. Bowen's disease, which is considered to be squamous cell carcinoma in situ [26], is reported to be the most common type of cutaneous malignancy in chronic arsenicosis [25]. Superficial basal cell carcinomas (Fig. 6.2, Panel e) are also extremely common. Merkel cell carcinoma may develop from arsenicosis [13]. Multiple and different types of skin cancers in a single individual is common. A characteristic finding is development of multiple Bowen's disease or basal cell carcinoma in the sun protected areas which is otherwise very unusual. Bowen's disease appears as erythematosus scaly eczematous patch or plaque with well-defined margin. It is notable that although melanotic changes are common, malignant melanoma is not observed with arsenicosis. Malignant changes may appear in internal organs like lung, bladder, prostate, kidney and liver as discussed elsewhere in this volume.

6.4 CLINICAL ISSUES

6.4.1 Histopathology

In arsenical hyperkeratosis, there is presence of prominent hyperkeratosis and acanthosis. Different grades of nuclear atypicality like hyperchromasia or clumping has been reported to be present occasionally especially in the deeper sections of palmoplantar keratosis. Disordered arrangement of the squamous cells, a feature of Bowen's disease has also been detected in such cases. Bowen's disease and superficial basal cell carcinoma are very common in arsenicosis. Histological differentiation may be difficult in some cases. Possibly these cases feature squamous metaplasia in basal cell carcinoma or true combined form, referred to as basosquamous carcinoma.

In pigmented skin lesions, most part of epidermis shows presence of increased melanin within keratinocytes and in the dermis, melanophages are detected in increased number.

6.4.2 Case Definition, Diagnosis and Severity Grading

As per case definition proposed by WHO, arsenicosis is a chronic health condition resulting from consumption of arsenic above safe limit more than 6 months, and is usually manifested by melanosis and keratosis. However, the skin manifestations, the gold standard for diagnosis of the condition generally may not become apparent with only 6 months exposure. Thus it is understandable that as per this definition, a case may be defined as arsenicosis before clinically detectable manifestations appear. Laboratory confirmation of the clinical diagnosis requires measurement of arsenic body burden. Measurement of biomarkers in hair and nails above a cut off limit may indicate chronic intoxication but does not accurately define exposure duration over 6 months.

Diagnosis is relatively easier when the cutaneous manifestations are apparent. Cutaneous manifestations are considered sine qua non for clinical diagnosis of chronic arsenicosis. Typical pigmentary alterations and palmo-planatar keratosis are considered the classical changes for the diagnosis. However, there are so far no well accepted diagnostic criteria for diagnosis of arsenicosis. Various workers have used different criteria in the studies making the results incomparable. The inclusion criteria used in a study by Guha Mazumder [6] were as follows: typical raindrop pigmentation and/or depigmentation and/or keratosis of skin of body and limbs, and arsenic level above permissible limit (>0.05 mg/L) in the water consumed by these people. Thus high drinking water arsenic level was mandatory in that criterion. Keratosis on the other hand was an optional component. A study by Kadono et al. [10] arbitrarily set five nodules or pits as diagnostic criterion for keratosis, a criterion that may not be accepted by many.

Criteria proposed by WHO published in a field guide [3] requires ruling out alternative causes of both keratosis and pigmentary alterations. This field guide presents an algorithmic approach for diagnosis of "suspected", "probable", "clinically confirmed" and "clinically and laboratory confirmed" cases of arsenicosis in a step wise manner based on the clinical manifestations, differentiation from other clinical simulators, and results of investigations. It has been mentioned that the algorithm for case definition has acceptable sensitivity and specificity (both >80%). Although there could have been more simplicity in the algorithm, the diagnostic criteria are reasonably simple and easy to use.

Different severity grading systems have been proposed. Saha described four stages, 7 grades and 22 sub grades [18]. Kadono et al. [10] developed a different grading system for keratosis (sole-5 grades, palms- 4 grades) and melanosis (3 grades). These grading systems were complicated, difficult to reproduce, not suitable for use in the field and not validated. In the WHO filed guide, keratoses were graded into mild, moderate and severe based on the size and not the total number. Melanosis was not graded.

A trained dermatologist or an individual who has expertise in examining cases of arsenicosis should be able to diagnose cases clinically with minimal error. In fact, cutaneous manifestations of arsenicosis are distinct enough to diagnose the cases and differentiate from many other clinical simulators. When there is a definite history of chronic exposure to arsenic, diagnosis can be even easier. Problems may arise when clinical manifestation is less apparent as may occur in early cases or in some people even after many years. Incomplete manifestation includes either pigmentary alteration or keratosis, scattered and few lesions and limited extent of involvement.

Laboratory confirmation of increased tissue level of biomarkers can confirm clinical suspicion. If available, laboratory confirmation should be attempted. However this may not be feasible in some situations especially epidemiological surveys. In many areas facilities for quality laboratory measurement for tissue arsenic level are not widely available. Clinical diagnosis can satisfactorily be used for diagnosis of cases in such conditions.

6.4.3 Management

Stopping intake of arsenic containing water and switching to safer water is the most important step in management of patients suffering from chronic arsenicosis. Many systemic illnesses like weakness, anaemia and neuropathy and even non-malignant skin manifestations [5] may improve on permanent discontinuation of contaminated water. Approaches to reducing arsenic in drinking water or providing alternative sources are discussed in Chapter 3 of this volume.

Severe anemia, chronic bronchitis with or without obstruction, dyspepsia, gastrointestinal varices, peripheral vascular disease including gangrene and peripheral neuropathy are the predominant causes of morbidity. Some of these are also important causes of mortality. All these should be taken care. If required, patients should be referred to specialists if available for specific management. There are basic steps that can be employed effectively to improve the general condition of the patient. High protein diet is recommended in these patients. It has additional benefit of reducing arsenical toxicity by enhanced methylation of toxic inorganic arsenic.

Management of palmoplantar keratosis includes keratolytics like topical salicylic acids (3–20%), 10–20% of urea and dermabrasion. All the suspected and confirmed skin cancers are immediately excised.

Retinoids are known to have antikeratinizing effects. They are used in many disorders of keratinization. Thus retinoids are used in chronic arsenicosis also. Additionally, they may have some role in the chemoprevention of arsenic-related cancers in various organs [9, 14]. Efficacy of retinoids like isotretinoinand acitretin in the management of arsenical keratosis is however infrequently studied. Considering the limited number of available drugs, they should be used in large scale population study for proper documentation and assessment of their efficacy.

DMSA, DMPS, dimercaprol, D-penicillamine are metal chelators sometimes used in acute and chronic arsenic poisoning [7]. The first three are more useful in acute arsenic poisoning. These should be used immediately, preferably within minutes after the exposure and the efficacy is greatly reduced if their use is delayed. Their use in chronic arsenic poisoning is limited because their efficacy lacks sufficient evidence. Moreover, their use may be associated with significant toxicity.

D-penicillamine is possibly also effective but is costly and has many side effects. DMSA and DMPS are water soluble analogues of dimercaprol. DMSA and DMPS are better than dimercaprol.

Urine excretion of arsenic increases many folds after therapy with DMPS [1]. Guha Mazumder did not find any clinical improvement of chronic arsenicosis after two courses of therapy with DMSA [7]. However, the same author found some clinical improvement with DMPS. Experience by other workers yielded mixed results with DMSA and DMPS [12, 20]. There is a report of significant improvement in neuropathy after 6 weeks of therapy with d-penicillamine [2]. Overall, the reported efficacy of these chelators is less than necessary to justify wide usage.

6.5 SUMMARY

The skin is a primary target organ for arsenic toxicity. Non-malignant skin lesions and pigmentary changes are early signs of arsenic toxicity and are primary diagnostic criteria for arsenicosis. Malignant neoplasms including basal cell carcinoma and squamous cell carcinoma often occur with prolonged exposure. Notable is the frequent occurrence of Bowen's disease which is squamous cell carcinoma in situ, as well as the absence of malignant melanoma. Remediation of arsenic contaminated drinking water with uncontaminated water may reduce the non-malignant skin changes.

REFERENCES

[1] H.V. Aposhian, A. Arroyo, M.E. Cebrian, L.M. del Razo, K.M. Hurlbut, R.C. Dart, D. Gonzalez-Ramirez, H. Kreppel, H. Speisky, A. Smith, M.E. Gonsebatt, P. Ostrosky-Wegman, M.M. Aposhian, DMPS-arsenic challenge test. I: increased urinary excretion of monomethylarsonic acid in humans given dimercaptopropane sulfonate. J. Pharmacol. Exp. Ther. 282 (1997) 192–200.

[2] S.K. Bansal, N. Haldar, U.K. Dhand, J.S. Chopra, Phrenic neuropathy in arsenic poisoning. Chest 100 (1991) 878–880.

[3] D. Caussy. WHO Technical Bulletin No. 31: Detection, Management and Surveillance of Arsenicosis in South-East Asia Region. 2005. New Delhi, World Health Organization.

[4] U. Fierz, Catamnestic studies on the side-effects of therapy of skin diseases with arsenic. Arch. Klin. Exp. Dermatol. 227 (1966) 286–290.

[5] D.V. Frost, Arsenicals in biology—retrospect and prospect. Fed. Proc. 26 (1967) 194–208.

[6] D.N. Guha Mazumder, Chronic arsenic toxicity: clinical features, epidemiology, and treatment: experience in West Bengal. J. Environ. Sci. Health A Tox. Hazard. Subst. Environ. Eng. 38 (2003) 141–163.

[7] D.N. Guha Mazumder, U.C. Ghoshal, J. Saha, A. Santra, B.K. De, A. Chatterjee, S. Dutta, C.R. Angle, J.A. Centeno, Randomized placebo-controlled trial of 2,3-dimercaptosuccinic acid in therapy of chronic arsenicosis due to drinking arsenic-contaminated subsoil water. J. Toxicol. Clin. Toxicol. 36 (1998) 683–690.

[8] N.M. Hanjani, A.B. Fender, M.G. Mercurio, Chronic arsenicism from Chinese herbal medicine. Cutis 80 (2007) 305–308.

[9] W.K. Hong, M.B. Sporn, Recent advances in chemoprevention of cancer. Science 278 (1997) 1073–1077.

[10] T. Kadono, T. Inaoka, N. Murayama, K. Ushijima, M. Nagano, S. Nakamura, C. Watanabe, K. Tamaki, R. Ohtsuka, Skin manifestations of arsenicosis in two villages in Bangladesh. Int. J. Dermatol. 41 (2002) 841–846.

[11] M.M. Khan, F. Sakauchi, T. Sonoda, M. Washio, M. Mori, Magnitude of arsenic toxicity in tube-well drinking water in Bangladesh and its adverse effects on human health including cancer: evidence from a review of the literature. Asian Pac. J. Cancer Prev. 4 (2003) 7–14.

[12] K. Lenz, K. Hruby, W. Druml, A. Eder, A. Gaszner, G. Kleinberger, M. Pichler, M. Weiser, 2,3-Dimercaptosuccinic acid in human arsenic poisoning. Arch. Toxicol. 47 (1981) 241–243.

[13] H.C. Lien, T.F. Tsai, Y.Y. Lee, C.H. Hsiao, Merkel cell carcinoma and chronic arsenicism. J. Am. Acad. Dermatol. 41 (1999) 641–643.

[14] R. Lotan, Retinoids in cancer chemoprevention. FASEB J. 10 (1996) 1031–1039.

[15] M.E. Maloney, Arsenic in dermatology. Dermatol. Surg. 22 (1996) 301–304.

[16] Y. Miki, T. Kawatsu, K. Matsuda, H. Machino, K. Kubo, Cutaneous and pulmonary cancers associated with Bowen's disease. J. Am. Acad. Dermatol. 6 (1982) 26–31.

[17] T.G. Rossman, A.N. Uddin, F.J. Burns, Evidence that arsenite acts as a cocarcinogen in skin cancer. Toxicol. Appl. Pharmacol. 198 (2004) 394–404.

[18] K.C. Saha, Diagnosis of arsenicosis. J. Environ. Sci. Health A Tox. Hazard. Subst. Environ. Eng. 38 (2003) 255–272.

[19] R.L. Shannon, D.S. Strayer, Arsenic-induced skin toxicity. Hum. Toxicol. 8 (1989) 99–104.

[20] S. Shum, J. Whitehead, L. Vaughn, S. Shum, T. Hale, Chelation of organoarsenate with dimercaptosuccinic acid. Vet. Hum. Toxicol. 37 (1995) 239–242.

[21] C.H. Tay, Cutaneous manifestations of arsenic poisoning due to certain Chinese herbal medicine. Australas. J. Dermatol. 15 (1974) 121–131.

[22] W.P. Tseng, H.M. Chu, S.W. How, J.M. Fong, C.S. Lin, S. Yeh, Prevalence of skin cancer in an endemic area of chronic arsenicism in Taiwan. J. Natl. Cancer Inst. 40 (1968) 453–463.

[23] S.S. Wong, K.C. Tan, C.L. Goh, Cutaneous manifestations of chronic arsenicism: review of seventeen cases. J. Am. Acad. Dermatol. 38 (1998) 179–185.

[24] A. Woollons, R. Russell-Jones, Chronic endemic hydroarsenicism. Br. J. Dermatol. 139 (1998) 1092–1096.

[25] S. Yeh, S.W. How, C.S. Lin, Arsenical cancer of skin. Histologic study with special reference to Bowen's disease. Cancer 21 (1968) 312–339.

[26] H.S. Yu, W.T. Liao, C.Y. Chai, Arsenic carcinogenesis in the skin. J. Biomed. Sci. 13 (2006) 657–666.

[27] R. Zaldivar, A. Guillier, Environmental and clinical investigations on endemic chronic arsenic poisoning in infants and children. Zentralbl. Bakteriol. Orig. B 165 (1977) 226–234.

7

LUNG CANCER AND OTHER PULMONARY DISEASES

CARA L. SHERWOOD[1] AND R. CLARK LANTZ[2]

[1] *Arizona Respiratory Center, University of Arizona, Tucson, AZ, USA*
[2] *Cellular and Molecular Medicine, University of Arizona, Tucson, AZ, USA*

7.1 INTRODUCTION

Over the past 30 years, it has become evident that the lung is a target organ for adverse health outcomes following exposure to arsenic. Exposure to high levels of arsenic has been associated with lung cancer, nonmalignant lung diseases such as bronchiectasis and chronic obstructive pulmonary disease (COPD), and other respiratory effects (e.g., chronic cough, chest sounds) [1–7]. Individuals exposed *in utero* and/or chronically to arsenic can have increased respiratory infection morbidity and mortality [8–11]. Despite the clear connection between arsenic exposure and adverse respiratory effects, the cellular and molecular mechanisms behind these effects have not been fully elucidated.

7.2 ARSENIC AS A LUNG CARCINOGEN

While exposure to arsenic can occur through a number of routes, including ingestion in water, foods, and soil, early work focused on the ability of arsenic to increase the risk of lung cancer through inhalation, especially in occupational settings. As early as 1879, Harting and Hess determined that 75% of Saxony miners were dying from lung cancer attributed to occupational inhalation of arsenic [12]. Beginning in the

Arsenic: Exposure Sources, Health Risks, and Mechanisms of Toxicity, First Edition.
Edited by J. Christopher States.
© 2016 John Wiley & Sons, Inc. Published 2016 by John Wiley & Sons, Inc.

1970s and through the 1980s, a number of investigations reported on increased risk of lung cancer among smelter workers, showing a positive correlation between exposures measured either through work history or urine analysis [13–20]. In 1980, these data provided the International Agency for Research on Cancer (IARC) with sufficient evidence to support that inorganic arsenic was a human lung carcinogen based on studies involving exposure through inhalation [21].

In the 1980s and 1990s, it became apparent through studies conducted in Taiwan, Argentina, Japan, and Chile that ingestion of high levels of arsenic (>100 ppb) through drinking water can also lead to an increased risk of developing lung cancer (reviewed in [22]). Some of the first studies correlating lung cancer with ingestion of arsenic (350–1140 ppb) from artesian well water were conducted in an area of Taiwan endemic for arsenic-induced blackfoot disease [23, 24]. A dose–response relationship was seen between standard mortality ratios (SMRs) from cancer and levels of arsenic in drinking water. The highest SMRs were observed in areas where only artesian wells were used as opposed to areas that used a combination of artesian and shallow wells. In 2007, Allan Smith and colleagues published a 50-year study on lung and bladder cancer mortality based on a unique environmental arsenic exposure population from northern Chile (Region II) [25]. Residents of Antofagasta, Chile (Region II), experienced a dramatic increase in concentrations of arsenic (>500 ppb) in their drinking water from 1958 until remediation efforts began in the 1970s. Marshall et al. reported three to fourfold rate ratios for lung cancer in this exposed population. There are limited studies on lower levels of arsenic exposure (<100 ppb) through contaminated drinking water and elevated lung cancer risks [26–28]; therefore, further evidence is needed to explore this possibility. The accumulation of evidence for a causal relationship between arsenic-contaminated drinking water and lung cancer led the IARC [29] to designate arsenic as a pulmonary toxicant through ingestion in addition to inhalation [29]. This makes arsenic the first substance to be classified as a carcinogen where exposure occurs through two different routes.

Smith et al. have suggested that it is the absorbed dose rather than the route of exposure that determines the risk of developing cancer [30]. However, there may be differences in the histological types of lung cancer between inhalation and ingestion. Guo et al. were the first to examine systematically lung cancer histological types following arsenic ingestion [31]. These studies showed an increased percentage of squamous cell and small cell carcinomas. Similar results were also found in a US population [32]. However, previous reports studying cancer types following inhalation of arsenic in smelter workers found an increase in adenocarcinomas [17]. The sites of action for induction of lung cancers may be different depending on whether the exposures are through ingestion or inhalation.

7.3 ARSENIC AND NONCARCINOGENIC LUNG DISEASE

The detrimental effects of arsenic on the lung are not limited to cancer development. Chronic exposure to arsenic can lead to development of noncarcinogenic lung disease and other adverse respiratory effects. In separate studies in West Bengal and Bangladesh, direct measurements of lung function (forced expiratory volume in 1 s (FEV_1), forced vital capacity (FVC), and peak expiratory flow rate (PEFR)) were shown to be reduced by chronic arsenic exposure [6, 33, 34], while symptoms associated with respiratory disease (i.e., cough, chest sounds, and shortness of breath) were all increased in a dose-dependent manner with arsenic concentration in the drinking water [4]. In Bangladesh populations, the prevalence of chronic bronchitis and chronic cough increased as arsenic concentrations in well water were increased from less than 600 to 1000 ppb [5] and over 63% of subjects in a separate study with mean arsenic exposure of 216 ± 211 ppb (compared to 11 ± 20 ppb in controls) displayed increased respiratory complications that ranged from shortness of breath to bronchitis [35]. In Antofagasta, Chile, Smith et al. found that *in utero* and early childhood exposure to high levels (>500 ppb) of arsenic-contaminated drinking water increased SMRs for the obstructive lung disease bronchiectasis [7]. Bronchiectasis prevalence was also associated with adulthood exposure to chronic arsenic-contaminated drinking water in a population from West Bengal, India [4]. Despite the strong evidence for drinking water arsenic-induced lung disease at high levels of exposure, the chronic effects at lower levels (e.g., <100 ppb), which are prevalent in the United States and near the current EPA MCL of 10 ppb, are only now beginning to be studied.

7.4 ARSENIC AS A SYNERGISTIC LUNG TOXICANT

The molecular and cellular mechanisms by which arsenic acts as a toxicant have not been fully elucidated although, as a carcinogen, it appears to act through nonmutagenic mechanisms. Arsenic has been implicated in promoting alterations in multiple cellular pathways. These pathways include suppression of cell cycle checkpoint proteins, altered expression of growth factors, resistance to apoptosis, inhibition of DNA repair, decreased immunosurveillance, alteration of DNA methylation, and increased oxidative stress (reviewed in [36, 37]).

Synergism between exposure to arsenic and other carcinogens, predominantly cigarette smoke, has been appreciated for some time. Increased incidence of lung cancer was seen in smelter workers occupationally exposed to arsenic, who were also smokers [38, 39]. Similar synergistic effects have also been reported following ingestion of arsenic. There was a significant dose–response trend of ingested arsenic on lung cancer risk, which was more prominent among cigarette smokers [27, 40]. Using animal models, several researchers have investigated the mechanisms behind this synergy. Application of benzo[*a*]pyrene (BaP), a known carcinogen found in cigarette smoke, to animals ingesting arsenic in their drinking water resulted in increased BaP/DNA binding in the lung [41]. This result suggests that arsenic may be inhibiting the recognition and removal of these adducts. Similarly, Lantz and Hays [42] and more recently Wang et al. [43] have shown that combined inhalation of arsenic and cigarette smoke can reduce protective antioxidants in the lung, resulting in increased levels of DNA oxidation in the airway epithelium [42, 43].

Not only does arsenic act synergistically with cigarette smoke to enhance the risk of developing lung cancer, but arsenic and cigarette smoke together can also exacerbate the development of emphysema [43]. At the levels administered in their study, neither arsenic nor cigarette smoke resulted in increased alveolar size. However, when administered together, changes in alveolar size and destruction of alveolar septa as would occur in emphysema were evident.

The interaction between arsenic exposure and other agents that can produce adverse noncancerous diseases also has been reported. Arsenic exposure appears to be synergistic with infectious agents. In an animal study reported by Kozul et al., arsenic ingestion through drinking water resulted in increased morbidity and viral titers following infection with flu virus [44]. This report was the impetus for epidemiological studies linking arsenic ingestion to reduced airway immunity. In the same population of northern Chile (Region II) that was reported to have increased bronchiectasis, Smith et al. found a causal relationship between ingestion of arsenic-contaminated drinking water and higher mortality rates from tuberculosis [11]. Increased risk for lower respiratory tract infections and depressed immunity during infancy was found in a population in Bangladesh exposed to arsenic *in utero* [9, 10]. Liao et al. explored the link between arsenic ingestion and reduced immune response to influenza A (H1N1) in Taiwan, West Bengal, India, and the United States. They found that chronic exposure to arsenic-contaminated drinking water was correlated with increased respiratory effects following infection with H1N1 virus [8]. These studies are significant in identifying the respiratory tract immune system as a target for arsenic. Taken together, there is clear evidence that arsenic causes adverse respiratory effects at high levels of exposure (i.e., 200–1000 ppb) when combined with other adverse events such as inhalation of cigarette smoke.

7.5 DIFFERENTIAL LUNG EFFECTS OF ARSENIC BASED ON AGE AT TIME OF EXPOSURE

In utero and early life exposures appear to be particularly sensitive times for the development of both arsenic-induced carcinogenic and noncarcinogenic diseases. Residents of northern Chile (Region II) experienced a 13-year period of high (>500 ppb) arsenic concentrations in their drinking water supply [45]. This unique exposure scenario provided an opportunity to evaluate the long-term effects of high arsenic levels during *in utero* and early postnatal developmental periods. Individuals who were exposed *in utero* or during early childhood had an increased risk of dying from lung cancers and chronic obstructive lung disease as young adults. Exposures in early childhood led to a SMR for lung cancer of 7.0 [95% confidence interval (CI), 5.4–8.9; $p<0.001$] and a SMR for bronchiectasis of 12.4 (95% CI, 3.3–31.7; $p<0.001$). For those individuals exposed both *in utero* and during early childhood, the SMRs were 6.1 (95% CI, 3.5–9.9; $p<0.001$) for lung cancer and 46.2 (95% CI, 21.1–87.7; $p<0.001$) for bronchiectasis [7]. Additional studies from this same population indicate that early developmental exposures to high levels of arsenic in drinking water result in decrement in lung function in adults [1]. Exposure to high levels of arsenic prior to age 10 resulted in decreased FEV_1 and FVC and increased breathlessness (prevalence odds ratio = 5.94, 95% CI 1.36–26.0). The magnitude of the responses was associated with the level of exposure. These findings suggest that exposure to arsenic in drinking water during early childhood and/or *in utero* has pronounced pulmonary effects, greatly increasing subsequent mortality in young adults from both carcinogenic and noncarcinogenic lung disease.

7.6 ANIMAL STUDIES OF ARSENIC EXPOSURE THAT MODEL HUMAN EFFECTS

While arsenic had been characterized as a human carcinogen, it was difficult to show carcinogenic effects in rodent model systems until Waalkes et al. [46] developed a transplacental model of exposure. Using high doses in the pregnant female drinking water (40–100 ppm), Waalkes and his coworkers were able to demonstrate increased formation of pulmonary tumors in mice exposed only to arsenic *in utero* when the arsenic was administered between embryonic day 8 and 18 alone. Prenatally exposed female C_3H offspring showed a dose-related increase in lung carcinoma [46]. Analysis of gene expression in the lung tumors indicated that aberrant estrogen receptor signaling could play a role in the development of tumors later in life [47].

The ability of arsenic to increase lung tumors following *in utero* exposures has led to the idea that arsenic is affecting the stem cell populations [48]. More detail of this proposed pathway for development of arsenic-induced cancers is presented in Chapter 17 of this book. It is hypothesized that arsenic blocks progression of stem cells into more differentiated cells and that chronic exposure may result in reactivation of the blocked pathways with an aberrant self-renewal pathway. This disruption of normal differentiation then can lead to development of cancer stem cells that can be the precursors for dysregulated self-renewal and differentiation. It is interesting that an early event is for arsenic to block the activity and expansion of stem cells. This could also affect noncancerous endpoints, such as wound repair by inhibition of expansion and differentiation of stem cells and stem-like cells following epithelial injury.

Animal studies have also proved useful in modeling noncarcinogenic effects of arsenic exposure. Kozul et al. [49] showed that mice chronically exposed (5–6 weeks) to arsenic in their drinking water at environmentally relevant levels (10 and 100 ppb) had alterations in gene expression, including genes mediating innate immune functions in the lung [49]. In a second study, the same group showed mice administered H1N1 virus that had been exposed chronically (5 weeks) to 100 ppb arsenic in their drinking water suffered increased morbidity and lung viral titers compared to untreated controls [44]. These data indicate that mechanisms of airway innate immunity are compromised after arsenic exposure.

Lantz et al. used a mouse model of environmentally relevant levels (0–100 ppb) of arsenic in drinking water to investigate arsenic effects and the molecular targets in the lung. Continuous exposure from conception to adulthood resulted in increased airway reactivity accompanied by alterations in airway smooth muscle and collagen content [50]. *In utero* and early postnatal exposures to arsenic were found to alter airway reactivity to methacholine challenge in 28-day-old pups. Removal of mice from arsenic exposure 28 days after birth did not reverse the alterations in sensitivity to methacholine. In addition, adult mice exposed to similar levels of arsenic in drinking water did not show alterations. Therefore, alterations in airway reactivity were irreversible and specific to exposures during lung development. The observed functional changes were correlated with protein and gene expression changes as well as morphological structural changes around the airways. Arsenic increased the whole lung levels of smooth muscle actin in a dose-dependent manner. The level of smooth muscle mass around airways was increased with arsenic exposure, especially around airways smaller than 100 μm in diameter. This increase in smooth muscle was associated with alterations in extracellular matrix (ECM) (collagen, elastin) expression. This model system demonstrates that *in utero* and postnatal exposure to environmentally relevant levels of arsenic can irreversibly alter pulmonary structure and function in the adult animals. These results combined with the epidemiological evidence from Chile are indicative that early life exposure to arsenic in drinking water may have irreversible respiratory effects.

7.7 ARSENIC MECHANISMS OF ACTION IN LUNG DISEASE DEVELOPMENT

Mechanisms involved in the development of bronchiectasis, COPD (chronic bronchitis, emphysema), and lung cancer in the absence of arsenic are still being elucidated, but what is clear is that lung effects from arsenic exposure share many of the same tissue/cellular characteristics of these diseases. In a recent review of the molecular mechanisms involved in COPD, Fischer and colleagues refer to a pathogenic triad in COPD that includes oxidative stress, protease–antiprotease imbalance, and inflammation [51]. Arsenic has been implicated in all three of these pathological events in multiple organ systems, including the lung, and a physiologic outcome common to these effects is aberrant airway remodeling. The focus of the remainder of this chapter will be on the role of arsenic in airway remodeling and how that relationship might lead to both carcinogenic and noncarcinogenic lung disease.

Aberrant airway remodeling is a hallmark of many respiratory diseases including emphysema, asthma, idiopathic pulmonary fibrosis, tuberculosis, and bronchiectasis [52–55]. It consists of an array of persistent tissue structural changes that occur through a process of injury and dysregulated repair that can lead to airway chronic inflammation and altered ECM deposition in the airway wall, resulting in airflow obstruction [52, 56–59]. In addition to matrix alterations, cell migration is a critical step in tissue remodeling [60], and abnormalities in cell migration have the potential to lead to respiratory disease.

Dysregulated wound repair may also contribute to the occurrence of lung cancer. It is well established that malignant tumors can arise from sites of chronic tissue wounding (reviewed in [61, 62]). The similarities in the signaling pathways between cellular repair/remodeling and initiation of cancer following a wound are strikingly similar. These similarities suggest that common molecular and cellular pathways are active in both wound repair and cancer tissues. In his seminal publication, Harold Dvorak postulated, "tumors are wounds that do not heal" [63]. These observations were based on similarities in the stroma between the two processes, with the resulting cancer being caused by an absence of appropriate regulation of the healing process. Therefore, aberrant wound healing can lead to both carcinogenic and noncarcinogenic lung disease.

The airway epithelium is the first line of defense from inhaled particulates, pathogens, allergens, and toxicants (reviewed in [64]). In addition to providing a physical barrier, the airway epithelium coordinates innate immune defense by responding to inhaled insults with a dynamic range of cellular mediators. Along with the epithelium, the dense basement membrane is essential for proper epithelial function. The basement membrane anchors the epithelium, establishes and maintains epithelial polarity, provides survival signals to the epithelium, and acts as a barrier between the epithelium and the underlying mesenchyme. Because of the exposure to pollutants and infectious agents, the epithelium is constantly undergoing cellular renewal. If there is an injury to the epithelium, wound repair mechanisms are initiated to reestablish the epithelial barrier. The epithelial repair process in the airway involves spreading and migration of wound-adjacent cells and cellular proliferation and differentiation (reviewed in [65]). Some of the molecules involved in epithelial wound repair include cytokines, growth factors, ECM components, and matrix metalloproteinases (MMPs). A denuded airway epithelium results in secretion of a provisional ECM by wound-adjacent epithelial cells. Surviving epithelial cells also secrete matrix-remodeling MMPs, in which MMP-7 and MMP-9 are known to be involved in wound repair and cellular differentiation (reviewed in [66]). Both *in vitro* and *in vivo* epithelial repair models have shown that the initial and most important response in airway epithelial reconstitution is the spreading and migration of wound-adjacent cells that occurs up to 24 h prior to proliferation. Therefore, efficient directed cell migration in epithelial wound repair is critical and requires precise coordination of cytoskeletal dynamics.

Purinergic signaling is essential in the coordination of wound repair in multiple epithelial cell types, including those that line the airway [67–70]. Through mechanisms that are not fully defined, ATP is released into the extracellular space following airway injury and binds purinergic receptors. The G-protein-coupled receptor $P2Y_2$ appears to be the principal purinergic receptor involved in cell migration [69, 71, 72]. Klepeis et al. [69] showed that P2Y activation following scratch wounding of an *in vitro* corneal epithelial monolayer was essential for complete wound closure. They further showed that ATP-induced P2Y activation resulted in increased phosphorylation of paxillin, important in focal adhesions and migration, as well as caused the translocation of RhoA to the cell membrane.

We have outlined lung cancer and noncancer adverse health outcomes following arsenic exposures that have been reported in human populations and/or animal studies. Those adverse outcomes suggest that the ECM and aberrant cell motility and wound repair are targets of arsenic, leading to the chronic lung disease phenotypes seen in populations exposed to high levels of arsenic. Multiple model systems have been used to validate such hypotheses. Evidence for arsenic-induced alterations in expression and organization of ECM and the components involved in matrix remodeling has been shown. Aberrant wound repair and signaling mechanisms involved in cellular migration, as well as changes in airway epithelial barrier structure and function, have been demonstrated following arsenic exposure as well. The data presented in the following sections summarize the findings from those studies.

7.8 ARSENIC EFFECTS ON THE ECM

Lantz and Hays [42] investigated the molecular targets involved in long-term, low-dose, arsenic-associated respiratory toxicity in adult male C57Bl/6 mice exposed to either 10 or 50 ppb arsenic in their drinking water for up to 8 weeks. This exposure paradigm resulted in a very small number of arsenic-induced changes in gene expression; however, of particular interest was the disproportionate number of ECM genes that were altered by the exposure. The ECM genes that were downregulated included several collagen types (i.e., Col1A1, Col1A2, Col3a1, Col6a2, Col6a3), elastin, fibronectin, and procollagen C-endopeptidase enhancer (PColce). Collagen type 1 is a fibrillar form of collagen found in most connective tissues. Collagen type 3 is also a fibrillar form found in extensible connective tissues in the lung, skin, blood vessels, and intestine. Type 6 collagens are beaded filament-type collagens and are matrix organizers by binding other matrix proteins. Elastin is a component of elastic fibers found extensively around blood vessels and in the distal lung alveolar septa. Fibronectin is a basement membrane protein involved in cell adhesion and migration processes including embryogenesis, wound healing, blood coagulation, host defense, and metastasis. Finally, PColce is an enzyme necessary for cleavage of procollagen type 1. A reduction would lead to a decrease in the active form of type 1 collagens. In general, alterations were dose dependent with greater suppression of expression of the matrix genes in animals exposed to the higher arsenic dose (i.e., 50 ppb). Clearly, alterations in these genes suggest that chronic low-dose arsenic is causing alterations in the levels of important matrix proteins in the lung and is also affecting genes that control matrix organization and cell motility. Staining of lung tissues from animals exposed to 50 ppb arsenic for 8 weeks showed what would be expected from the decreased expression of matrix-related genes. Compared to the control organization, arsenic exposure resulted in a more disorganized and expanded matrix around airways and blood vessels in the lung.

It is interesting to note that arsenic-induced changes in the matrix after exposure in adult animals were not limited to changes in the lung. Similar alterations were also seen in the extravascular matrix of small arteries in the heart following AsIII exposure (50 ppb) [73]. Hearts were also harvested, and real-time RT-PCR demonstrated an AsIII-associated decrease in the levels of expression of many of the same ECM genes found to be decreased in the lung. This result was also supported *in vitro* by using a murine embryonic fibroblast cell line (NIH3T3) exposed to AsIII. Direct *in vitro* application of arsenic (10–50 ppb) resulted in decreased expression of Col1A1, Col1A2, Col3A1, Col4A1, fibronectin, and hyaluronan acid synthase 2. These data show that arsenic can directly alter matrix production in embryonic fibroblasts.

Because studies in adult mice had identified ECM as a potential target for arsenic, it would be expected that similar arsenic-induced changes in matrix during lung development would lead to structural and functional alterations in the adult. As discussed in Section 7.5, developmental exposures in humans have been linked to increased risks of dying from lung cancers and other chronic lung diseases in adults [7]. During fetal and early postnatal lung development, ECM gene expression is necessary for proper development of lung and blood vessels. The level of expression of the matrix genes depends on the gene and on the developmental age. Agents that can alter this expression during these critical times are known to cause long-term morphological and functional alterations in the lung and blood vessels. Examples are inhibition of elastin expression following maternal cigarette smoking [74] or postnatal exposure to hyperoxia [75], viral infections [76], or dexamethasone [77]. These exposures lead to permanent alveolar enlargement.

Arsenic-induced alterations in these important matrix genes have been reported in mice following *in utero* and early postnatal exposure. Arsenic induced increases in the expression of Col1a2, Col3a1, and elastin [50, 78]. The degree of the alterations was both developmental time and dose dependent. Rather than suppression of expression of these matrix genes as was seen in adult exposures, arsenic exposure resulted in increases in mRNA levels of expression for matrix genes. Exposure to either 50 or 100 ppb arsenic resulted in increased Col1a2 expression on postnatal day 7. Col3a1 expression was increased on postnatal day 12 by 100 ppb arsenic. The pattern of elastin expression was more complex but, in general, resulted in increased elastin expression at early postnatal times. Interestingly, levels of expression of each of these matrix genes were decreased in arsenic-exposed animals on postnatal day 28, similar to what they had previously seen in adults. Arsenic did not alter the expression pattern of Col1a1 during the early postnatal periods.

Changes in the levels of matrix protein gene expression suggest that arsenic is interacting with normal developmental processes to alter expression of these matrix genes. While levels of gene expression were altered by arsenic, whole lung levels of matrix protein were only marginally altered. Analysis of Western blot intensities, normalized to GAPDH expression levels, showed increased protein expression of Col1a2 on day 12, decreased protein expression of Col3a1 on day 12, and decreased protein expression of elastin on day 7. While these values show trends, only the arsenic-induced decrease in Col3a1 on day 12 reached significance. Whole lung protein levels for Col1a1, Col1a2, Col3a1, and elastin from arsenic-exposed animals on day 28 were not different from controls. One explanation of this observed result is that mRNA levels of expression may be increasing to compensate for losses of proteins, so that normal levels of matrix protein required for developmental processes are maintained. Increases in mRNA expression could be a compensatory response. For example, arsenic-induced increases in MMP-9 during early postnatal time periods, as has been seen in a mouse model, would degrade matrix, requiring increased matrix gene expression to maintain appropriate protein levels.

While whole lung levels of matrix proteins were unchanged, regional decreases in total collagen in adventitia around airways was seen in the 28-day-old mice exposed to arsenic during development. This arsenic-induced localized regional decrease in collagen was also associated with dose-dependent arsenic-induced increased levels of smooth muscle actin staining around airways, particularly the small airways.

Whole lung protein analysis demonstrated significant arsenic-induced increases in both smooth muscle actin mRNA and protein levels. Exposure in adults also resulted in an apparent alteration in the amount of smooth muscle surrounding airways and blood vessels. Increased thickness and extent of staining for smooth muscle actin were detectable in animals that had been exposed to 50 ppb arsenic for 8 weeks. These alterations were most apparent in the larger airways, although some increase was also seen in the terminal airways. It is possible that arsenic-induced alterations in the matrix are responsible at least in part for the alterations in smooth muscle. Interactions between matrix and expression of smooth muscle cell genes have been directly demonstrated in matrix protein knockout mice. This localized regional decrease in collagen could contribute to the increased smooth muscle around airways [79, 80], similar to collagen knockout mice. Elastin and collagen knockout mice both show hyperproliferation of smooth muscle around blood vessels [81, 82]. This demonstrates that arsenic-induced changes may be specific for particular structural regions of the lung.

Another aspect of airway ECM regulation affected by arsenic is matrix remodeling. Proteins of the MMP family are involved in the breakdown of ECM in normal physiological processes, such as embryonic development, reproduction, and tissue remodeling, as well as in disease processes, such as asthma, COPD, arthritis, and metastasis (reviewed in [83]). The enzyme encoded by this gene degrades type IV and V collagens. Type IV collagens are found in the basement membranes and the alveolar septa of the lung. Type V collagen is important for fibril formation and matrix organization. Alterations in MMP-9 and its inhibitor, tissue inhibitor of metalloproteinase-1 (TIMP-1), have been seen across numerous arsenic model systems. Levels of MMP-9 and TIMP-1 were measured in human populations with low drinking water exposures of arsenic (<20 ppb) [84]. Levels of MMP-9/TIMP-1 ratio in induced sputum were significantly positively associated with levels of total urinary arsenic. In the same study, levels of α1-antitrypsin (A1AT), the protein deficiency known to contribute to genetic emphysema, were found to be lower in individuals exposed to arsenic. Model systems of arsenic exposure have differed in the ratios of MMP-9/TIMP-1 (in the human populations, decreases in the ratio were due predominantly to decreases in TIMP-1), but the underlying effect is the same, that is, increased degradation and alterations of matrix.

The cellular pathways affected by arsenic that lead to alterations in the ECM are obscure. The primary responses to arsenic exposure in model systems are altered matrix expression, increased levels of smooth muscle around airways, increased smooth muscle actin in whole lung, and alterations in MMP-9/TIMP-1 ratios. One growth factor that plays a role in all of these processes is TGF-β [85, 86]. In mice, arsenic-induced increases in TGF-β1 have been reported for both acute and chronic exposures to arsenite. Subcutaneous injection of arsenite in mice resulted in increased levels of TGF-β1 in kidney [87]. Mice exposed chronically (10 months) to arsenic (200 ppm) in their drinking water had increased TGF-β1 gene expression in the liver [88]. In this case, mRNAs for procollagen 1 and 3 and smooth muscle actin were also increased. Therefore, arsenic-induced changes in TGF-β1 signaling may contribute to the alterations during development. Petrick et al. [78] previously identified the β-catenin pathway as an important target of arsenic in the lung during *in utero* development. Wnt and β-catenin signaling, which can be modulated by TGF-β [89], can also lead to increases in airway smooth muscle [90]. In the *in utero* exposure model, arsenic-induced increases in TGF-β1 are seen beginning on day 12 after birth. These increases persisted in 28-day-old animals.

Another potential arsenic mechanism of action in aberrant airway remodeling is through a secondary effect following activation of inflammatory mediators. Of particular interest in airway hyperresponsiveness is the cytokine interleukin-13 (IL-13). IL-13 is a well-known mediator of chronic allergic inflammation, sufficient to induce most if not all of the key features of experimental asthma [91]. IL-13 promotes airway hyperresponsiveness [92], epithelial cell damage, increased airway smooth muscle mass, increased collagen deposition, decreased epithelial barrier function, altered wound healing, and increased MMP-9 expression [93, 94]. Furthermore, IL-13 stimulates airway fibrosis, largely through the ability of MMPs to activate TGF-β1, thus favoring the accumulation of eosinophils and macrophages in the lung [95, 96]. Data show that arsenic can induce all of these same alterations, suggesting that IL-13 may play a role in arsenic-induced alterations in lung structure and function [50, 97].

Analysis of gene expression in the lungs from 12-day-old mice that were exposed to 50 ppb arsenic during development supports the role of IL-13 in mediating arsenic-induced alterations. Whole lung mRNA expression data collected from 50 ppb arsenic-exposed mice on day 12 after birth shows increased levels of expression of IL-13 and other markers. Specifically, IL-13, GATA3 (a major transcription factor for regulation of IL-13 and IL-4), and macrophage secretion product markers FIZZ1 and YM2 are upregulated. FIZZ1 and YM2 have been shown to be secreted by macrophages in response to IL-13. Classical macrophage activation genes interferon gamma (IFNγ) and IL-1β were downregulated. These data support the hypothesis that *in utero* and early postnatal exposure to arsenic can alter expression of IL-13 and other alternative macrophage activation-associated genes. Figure 7.1 shows that arsenic exposure did not result in any changes in IL-13 expression in the lungs of 1- and 7-day-old pups. However, IL-13 levels in arsenic-exposed animals were increased in 12-day-old animals. These levels were further increased in 28-day-old animals. These data indicate that arsenic can induce IL-13 in the lungs during development. This induction appears to occur with postnatal exposure because changes are not seen until animals reach 12 days of age. Similar increases in IL-13 and TGF-β have been seen following inhalation of arsenic-containing dusts [98].

7.9 ARSENIC AFFECTS CELL MOTILITY AND WOUND REPAIR

Epithelial cell barriers, such as those that line the lung, are constantly exposed to foreign insults. These compounds, either directly or indirectly, can lead to a denuded epithelium that must be repaired to reestablish normal function. As part of this repair process, cells responsible for the repair must be able to migrate to damaged areas, differentiate into the cells distinctive for those found in the damaged area, and

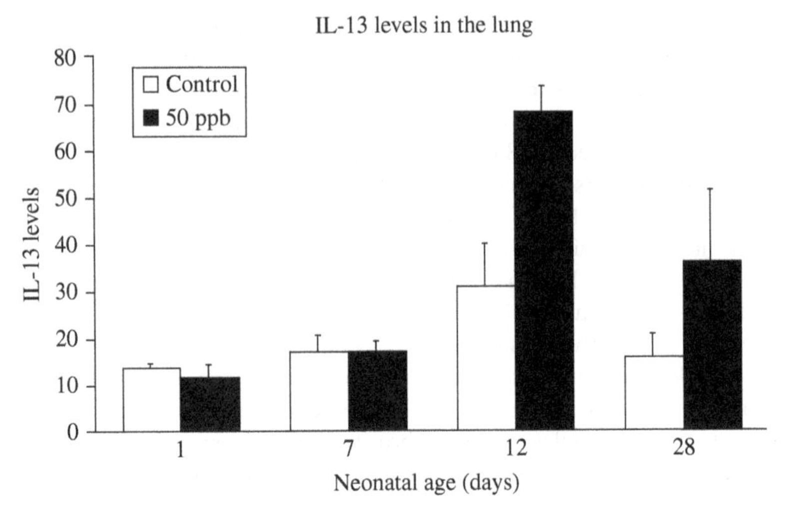

FIGURE 7.1 IL-13 protein levels in postnatal lung. Continuous *in utero* and postnatal exposures to 50 ppb arsenic (as sodium arsenite) in drinking water resulted in increased levels of IL-13 in 12- and 28-day-old animals.

repopulate the epithelium (reviewed in [65]). If toxicants affect any of these processes, then timely wound repair will be affected, increasing the risk for loss of function and increased adverse health outcomes.

Pathway analysis using gene and protein expression data from multiple model systems suggests that wound repair and cell motility are two of the more probable processes affected by exposure to environmentally relevant levels of arsenic [50, 73, 78, 99]. In addition to gene expression analysis of whole lungs, proteomic analysis has also indicated that chronic exposure of mice to 50 ppb arsenic in their drinking water can lead to alterations in proteins that are important in cell motility and can affect ECM. Differential expression of approximately 1000 proteins was characterized for proteins extracted from whole lungs of mice exposed for 8 weeks to 50 ppb arsenic in their water using a BD PowerBlot [42]. The analysis showed that arsenic led to alterations in levels of expression of 18 different proteins. Of these 18, nine were associated with cell motility and adhesion. These include increased expression of Rac1, Rab4, and 4.1N and decreased expression of ß-catenin, CapZ, gelsolin, Hic-5, nucleoporin, p62, and Stat3. In general, proteins that enhanced cell motility were suppressed, while those that inhibit motility were induced. Lung lavage fluid proteomic analysis also showed alterations in the expression of proteins that are associated with altered cell migration and wound repair (soluble receptor for advanced glycation end products (sRAGE) and A1AT). The sRAGE was also decreased in the sputum of a human population exposed to 20 ppb arsenic in drinking water) [84].

Inhibition of appropriate airway epithelial repair has been seen in an *in vivo* animal model, using naphthalene-induced lung injury method (Fig. 7.2) (Lantz unpublished results). Naphthalene selectively damages the distal airway epithelium, but the epithelium fully recovers 2 weeks after exposure. Mice treated with

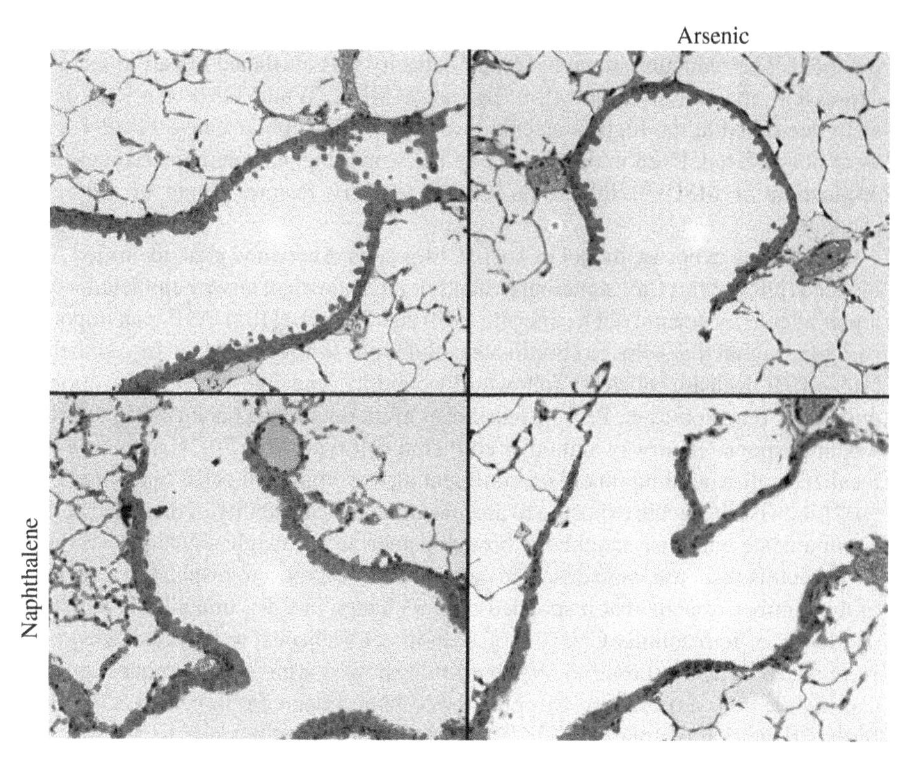

FIGURE 7.2 Lung pathology induced by combined exposure to arsenic and naphthalene. Representative images of the bronchiolar epithelium of male C57Bl/6 mice 14 days after intraperitoneal injection with either corn oil (top row) or 200 mg/kg naphthalene (bottom row). Four weeks prior to naphthalene or corn oil injection, animals were placed on either control water (0 ppb arsenic, left column) or water containing 50 ppb arsenic (as sodium arsenite) (right column). Airways from animals exposed to only arsenic (upper right panel) or naphthalene (lower left panel) appeared to be similar to control animals (upper left panel). However, animals that had received naphthalene while ingesting arsenic displayed a much more squamous bronchiolar epithelial layer, with reduced cell–cell contacts (lower right panel).

naphthalene in the presence of arsenic (50 ppb arsenic in drinking water beginning 4 weeks prior to naphthalene exposure) showed an aberrant wound repair response to the naphthalene-induced epithelial damage compared to untreated controls. The airway epithelium in the arsenic-exposed animals was either not appropriately replaced or was replaced by squamous epithelium. The ability of arsenic to affect tissue remodeling is not limited to the airway. Arsenic has also been shown to cause vascular remodeling of liver sinusoids [100, 101].

Arsenic effects on cellular migration were tested in an *in vitro* model of wound repair in human bronchial epithelial cells (16HBE14o-) [97]. In that study, a 24 h arsenic exposure (30–290 ppb Na arsenite) significantly increased the time needed to close a scrape wound in a confluent monolayer of cells. Arsenic at concentrations as low as 30 ppb inhibited reformation of epithelial monolayers following scrape wounding. At concentrations of 290 ppb arsenic, wound repair was almost completely inhibited. The reduction in wound repair capacity was attributed in part to increased expression and activity of MMP-9, but not MMP-2. While TIMP-1 mRNA levels were decreased at the highest arsenic level tested (290 ppb arsenic), TIMP-1 levels were not affected. Even in the absence of the wounding, arsenic induced significant production of MMP-9. Inhibiting MMP-9 activity restored some of the repair capability.

In the same exposure model of 16HBE14o- cells, Sherwood et al. identified ATP-induced (purinergic) Ca^{2+} signaling as another mechanism of airway epithelial wound repair altered by arsenic (24 h exposure to 60 and 290 ppb) [102]. ATP is an important paracrine signal that helps to coordinate cellular physiology in the airway epithelium [103–107]. Release of ATP following wounding and subsequent activation of purinergic receptors (i.e., $P2Y_2$ discussed in more detail later) contributes to proper wound response in airway and other epithelial cell types [69, 70]. Using an *in vitro* localized cell wounding model on confluent airway epithelial cells, Sherwood et al. [102] showed that acute exposure to arsenic attenuates the ability of damaged cells to communicate with their neighbors through paracrine purinergic signaling. The alterations in this response were dose dependent. Arsenic exposure resulted in a decrease in the number of cells that responded with a change in Ca^{2+}. In addition, the shape and levels of intracellular Ca^{2+} $[Ca^{2+}]_i$ seen in cells adjacent to the localized wound were altered. Cells exposed to 60 ppb arsenic showed a more rapid sequestration of Ca^{2+}, while cells exposed to 290 ppb showed both a decrease in peak $[Ca^{2+}]_i$ and a further reduction of total $[Ca^{2+}]_i$. These alterations were not due to a lower ATP released from the damaged cell, but rather to alterations in the level and function of the purinergic receptors, as arsenic-exposed cells responded with lower $[Ca^{2+}]_i$ signaling in response to an exogenous application of ATP.

Extracellular nucleotides act as agonists of P2 receptors that are categorized into the metabotropic P2Y receptors and the ionotropic P2X receptors [104, 108, 109]. P2Y receptors are G-protein-coupled receptors that are activated by a variety of nucleotides and include at least eight family members [110]. P2X receptors are cation-selective ion channels activated by ATP with seven family members [111]. Activation of both receptor types results in increased intracellular Ca^{2+} concentration ($[Ca^{2+}]_i$). Sherwood et al. found that the arsenic-induced reduction in Ca^{2+} influx after wounding was due in part to alterations in function of the $P2Y_2$ and $P2X_4$ receptors. However, the mode of inhibition differed between the receptor types. Arsenic reduced $[Ca^{2+}]_i$ release by suppressing $P2Y_2$ receptor expression in a dose-dependent manner. Arsenic appeared to block the extracellular Ca^{2+} influx through the $P2X_4$ receptor by interacting directly to inhibit the receptor activation. The $P2X_4$ receptor contains a number of thiol binding sites as well as a zinc activation site reminiscent of a zinc finger binding motif [112]. Zinc finger binding sites have been shown to be sensitive sites for arsenic binding [113, 114] (see also Chapter 13). Thus, interaction with zinc binding site also could be occurring with the $P2X_4$ receptor.

While acute higher doses of arsenic have been shown to alter wound repair signaling, what are more relevant are the effects of a low-dose chronic exposure. Sherwood et al. have also examined the effects of a 4–5-week low-dose (10 and 25 ppb) arsenic exposure on wound repair signaling in 16HBE14o- cells [115]. As seen in

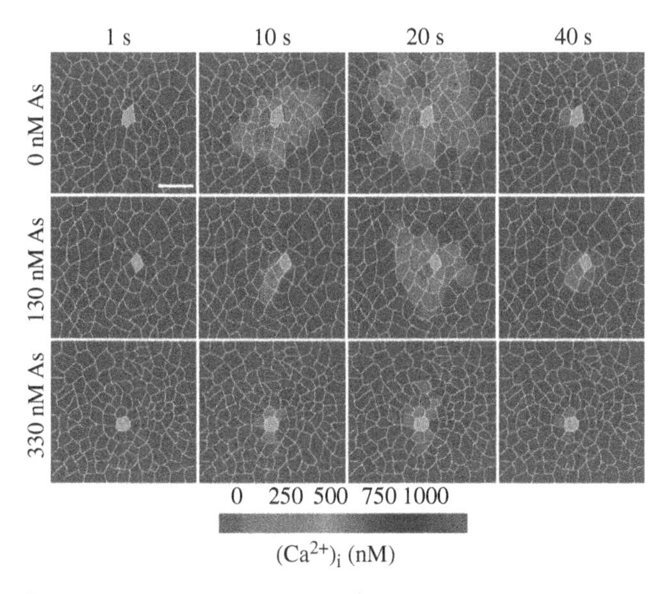

FIGURE 7.3 Localized wound-induced Ca^{2+} signaling dynamics is suppressed in 16HBE14o- cells by chronic low-dose arsenic exposure. Representative images of 16HBE14o- cell monolayers subjected to a localized wound (1–2 cells; gray area) at approximately 3 s and monitored for $[Ca^{2+}]_i$ over 60 s. Color scale at bottom indicates approximate $[Ca^{2+}]_i$ (nM); cells are outlined with white borders; white bar represents 50 μm (From Sherwood et al. [115]. © Oxford University Press). (*See insert for color representation of the figure.*)

acute studies, chronic low-dose exposures also led to decreased ability to close a scrape wound as well as alterations in ATP-induced Ca^{2+} signaling following localized damage (Fig. 7.3). These effects of arsenic on wound-induced Ca^{2+} signaling were also seen in a mouse model under conditions of natural exposure. Cultured tracheal epithelial cells were obtained from mice exposed to control or 50 ppb arsenic-supplemented drinking water for 4 weeks. Tracheal epithelial cells from arsenic-exposed mice displayed attenuated wound-induced Ca^{2+} signaling dynamics similar to the *in vitro* chronic exposure. These findings demonstrate that chronic arsenic exposure at levels that are commonly found in drinking water (i.e., 10–50 ppb) alters cellular mechanisms critical to airway epithelial wound repair.

7.10 ARSENIC AFFECTS AIRWAY EPITHELIAL BARRIER PROPERTIES

The airway epithelium acts as a barrier to protect the underlying submucosa from inhaled particulates and pathogens. An important aspect of this barrier is the formation and maintenance of tight junctions that are the apical most junctions between adjacent cells (reviewed in [116]). Breaches in this barrier can lead to pathogenic infiltration, chronic infection, and inflammation. Proteins within the tight junction complex also regulate paracellular ion conductance that can impact levels of airway surface liquid that subsequently can impact innate immune functions such as mucociliary clearance. When injury occurs to the epithelium, resolution of the wound requires appropriate reformation of the tight junctions. Alterations in the formation and maintenance of the tight junctions have been associated with chronic lung disease and increased risk of respiratory infections (reviewed in [117]).

Arsenic has been shown to alter the function and structure of the tight junction complex in primary mouse and immortalized human bronchial epithelial cells (16HBE14o-) [118]. In that study, a 5-day exposure to 60 or 290 ppb arsenic reduced transepithelial resistance, a measure of barrier function, in well-differentiated primary mouse tracheal epithelial (MTE) cell cultures. Alterations in transepithelial resistance can be indicative of increased permeability to macromolecules and viruses. Immunofluorescent staining of arsenic-treated MTE cells showed altered patterns of localization of the transmembrane tight junction proteins claudin (Cl) Cl-1, Cl-4, and Cl-7 and occludin at cell–cell contacts when compared with untreated controls. In experiments with 16HBE14o- cells under the same exposure time and concentrations of arsenic, protein levels were increased for Cl-4, Cl-5, Cl-7, and occludin as well as the mRNA levels of Cl-4 and Cl-7 in these cells. Additionally, cultures treated with 290 ppb arsenic had altered phosphorylation of occludin. In summary, exposure to environmentally relevant levels of arsenic can alter both the function and structure of airway epithelial barrier constituents. The consequences of these changes likely contribute to the observed arsenic-induced loss in basic innate immune defense and increased infection in the airway.

It is reasonable to think that MMP-9 plays a role in the arsenic-induced alterations observed in the airway tight junctions. Arsenic exposure at the levels reported in the tight junction study were shown by Olsen et al. [97] in the same cell type (i.e., 16HBE14o-) to lead to increased secretion and activity of MMP-9, even in the absence of wounding. Apical administration of MMP-9 in 16HBE14o- cell monolayers reduces transepithelial resistance and changes localization Cl-1 and occludin [119]. Other cytokines and growth factors may also be involved in the observed changes in barrier properties induced by arsenic. In the human lung epithelial cell line CaluIII, IL-13 caused an increase in epithelial permeability and a decrease in expression of ZO-1 and occludin [93]. These factors also are altered by arsenic exposure.

7.11 SUMMARY AND CONCLUSION

Aberrant tissue remodeling is characteristic of several chronic lung diseases that are associated with arsenic exposure. The airway epithelium is under constant bombardment by inhaled pollutants, allergens, toxicants, and pathogens. Although inflammatory cells and subepithelial fibroblasts contribute to repair processes, it is the airway epithelial cells that largely orchestrate the response to cellular damage and inflammation initiated by these inhaled insults (depicted in Fig. 7.4). Restoration of the airway epithelium following injury involves multiple steps including recognition of damage by adjacent cells, breakdown of cellular junctions, cellular dedifferentiation, migration to the damaged area, cellular proliferation, cellular redifferentiation,

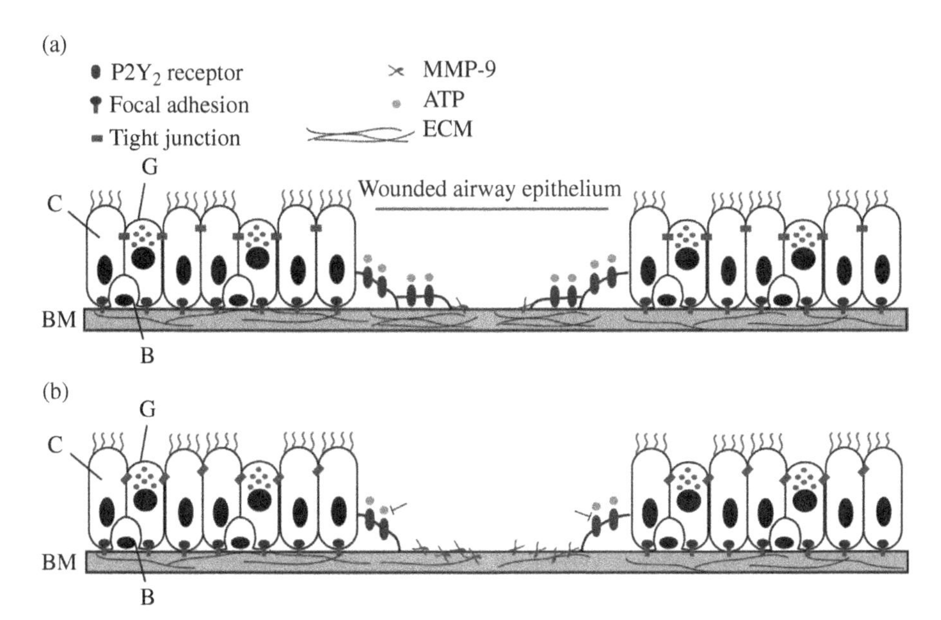

FIGURE 7.4 Aberrant tissue repair under arsenic exposure. (a) Normal epithelial wound repair involves the following from wound-adjacent cells: (1) recognition of a wound (e.g., paracrine ATP signaling), (2) secretion of MMPs to degrade cell–matrix attachments, (3) secretion of new ECM components, and (4) cell spreading and migration across the new provisional matrix to close the wound. A wounded airway epithelium under arsenic exposure (b) fails to repair the wound due to the following altered mechanisms: (1) loss of ATP recognition from wound-adjacent cells, (2) overactive MMP secretion, and (3) alterations in ECM constituents. B, basal cell; BM, basement membrane; C, ciliated cell; G, goblet cell.

and finally reestablishment of epithelial junctions (reviewed in [120]). Delayed and aberrant airway epithelial repair can lead to increased access of pathogens and damage to the underlying tissues.

Damage to the airway epithelium results in the release of ATP from damaged cells that in turn activates purinergic receptors in a paracrine fashion—a requirement for directed cellular migration [67, 69–71]. One downstream pathway of the purinergic receptors is to activate Duox1, an NADPH oxidase homolog, which along with EGF signaling leads to the production of MMP-9 [70]. MMP-9 can serve multiple roles, including assisting in the breakdown of tight junctions and detachment of migrating cells [119, 121–123]. Appropriate migration also requires ECM components. Epithelial cells autologously secrete a provisional matrix to aid in wound repair. An important ECM component in wound repair is fibronectin that enhances cell migration through its interaction with $\alpha5/\beta1$ integrins on the cell surface [124]. Fibronectin also exhibits chemotactic activity as well as collagens I and IV [125]. Various inflammatory mediators and growth factors can also affect wound repair, and this appears to be dependent on cellular background (reviewed in [65]). Cytokines TNFα, IL-1β, IFNγ, IL-4, and IL-6 and growth factors EGF and IGF-1 all enhance wound repair in the airway [93, 126–128]. Nitric oxide, IL-4, IL-13, and TGF-β inhibit airway cellular migration [93, 129, 130].

A number of environmental agents can delay or inhibit migration, leading to delayed wound repair. These agents include cigarette smoke, various farm dusts, air pollution, and ozone (reviewed in [131]). Multiple studies ranging from cellular to human have demonstrated arsenic-induced changes in proteins and pathways integral to the development and maintenance of the airway epithelium. Arsenic decreases β-catenin levels, attenuates purinergic signaling, alters MMP-9 expression and important ECM proteins such as collagens and fibronectin, and increases the expression of inhibitory cytokines and growth factors [50, 73, 78, 97, 102, 115]. Additionally, arsenic alters the molecular expression and localization of tight junction proteins critical in providing a physical barrier (Sherwood et al., unpublished results). For noncancerous endpoints, disruption and dysregulation of these airway epithelial homeostasis processes can certainly result in the development of chronic diseases. Continued chronic exposure to arsenic will delay or even inhibit the repair processes, leading to a chronic wound. In these situations, while proteins like MMP-9 are necessary for appropriate repair, chronic production of MMP-9 can lead to dysregulation by digestion of fibronectin and inhibition of the reformation of tight junctions. This continual damage can lead to chronic inflammation and airway remodeling including squamous metaplasia, mucous hyperplasia, fibrosis, and tissue destruction [132].

Dysregulated wound repair can also promote the development of cancer (reviewed in [63]). Chronic wounding and repetitive activation of repair mechanisms can result in chronic inflammation producing an underlying increased oxidative stress. This can result in an increased risk for DNA oxidation. Continuous repair also means that the epithelium will be undergoing continual proliferation. Should transformed cells exist in the epithelium or in the stem cell population, then expansion of those cells during the repair process could develop into cancer. Finally, there is a large amount of similarity in signaling pathways and processes between wound healing and development of cancer. Typically, epithelial wounds will resolve and heal, while cancers exhibit amplified repair responses that do not resolve. Since arsenic is producing a chronic wound environment, appropriate resolution of the wounds will not occur, predisposing the lung to carcinogenesis.

In conclusion, environmental levels of ingested arsenic alter a number of processes involved in the development and maintenance of airway epithelial architecture. These alterations have been seen in several model systems and in humans. Arsenic-induced pathologies vary based on the exposure platform (e.g., developmental vs. adult) and the level and length of exposure, thereby affecting functional and structural phenotypic changes. Additional research needs to be carried out in humans to validate these changes seen in model systems. A shift in focus to lower-level arsenic exposures in model systems and human studies is also warranted. Included in future studies should be the association of environmental arsenic levels with alterations in lung function, especially in children, with the determination of the incidence of lower respiratory infections (a prediction from the results from model systems) and analysis of arsenic-induced changes in additional mediators (e.g., TGF-β) that can affect wound repair and airway remodeling.

REFERENCES

[1] D.C. Dauphine, C. Ferreccio, S. Guntur, Y. Yuan, S.K. Hammond, J. Balmes, A.H. Smith, C. Steinmaus, Lung function in adults following in utero and childhood exposure to arsenic in drinking water: preliminary findings, Int Arch Occup Environ Health 84 (2011) 591–600.

[2] J.X. Guo, L. Hu, P.Z. Yand, K. Tanabe, M. Miyatalre, Y. Chen, Chronic arsenic poisoning in drinking water in Inner Mongolia and its associated health effects, J Environ Sci Health A Tox Hazard Subst Environ Eng 42 (2007) 1853–1858.

[3] D.N. Mazumder, R. Haque, N. Ghosh, B.K. De, A. Santra, D. Chakraborti, A.H. Smith, Arsenic in drinking water and the prevalence of respiratory effects in West Bengal, India, Int J Epidemiol 29 (2000) 1047–1052.

[4] D.N. Mazumder, C. Steinmaus, P. Bhattacharya, O.S. von Ehrenstein, N. Ghosh, M. Gotway, A. Sil, J.R. Balmes, R. Haque, M.M. Hira-Smith, A.H. Smith, Bronchiectasis in persons with skin lesions resulting from arsenic in drinking water, Epidemiology 16 (2005) 760–765.

[5] A.H. Milton, M. Rahman, Respiratory effects and arsenic contaminated well water in Bangladesh, Int J Environ Health Res 12 (2002) 175–179.

[6] F. Parvez, Y. Chen, P.W. Brandt-Rauf, A. Bernard, X. Dumont, V. Slavkovich, M. Argos, J. D'Armiento, R. Foronjy, M.R. Hasan, H.E. Eunus, J.H. Graziano, H. Ahsan, Nonmalignant respiratory effects of chronic arsenic exposure from drinking water among never-smokers in Bangladesh, Environ Health Perspect 116 (2008) 190–195.

[7] A.H. Smith, G. Marshall, Y. Yuan, C. Ferreccio, J. Liaw, O. von Ehrenstein, C. Steinmaus, M.N. Bates, S. Selvin, Increased mortality from lung cancer and bronchiectasis in young adults after exposure to arsenic in utero and in early childhood, Environ Health Perspect 114 (2006) 1293–1296.

[8] C.M. Liao, C.P. Chio, Y.H. Cheng, N.H. Hsieh, W.Y. Chen, S.C. Chen, Quantitative links between arsenic exposure and influenza A (H1N1) infection-associated lung function exacerbations risk, Risk Anal 31 (2011) 1281–1294.

[9] A. Rahman, M. Vahter, E.C. Ekstrom, L.A. Persson, Arsenic exposure in pregnancy increases the risk of lower respiratory tract infection and diarrhea during infancy in Bangladesh, Environ Health Perspect 119 (2011) 719–724.

[10] R. Raqib, S. Ahmed, R. Sultana, Y. Wagatsuma, D. Mondal, A.M. Hoque, B. Nermell, M. Yunus, S. Roy, L.A. Persson, S.E. Arifeen, S. Moore, M. Vahter, Effects of in utero arsenic exposure on child immunity and morbidity in rural Bangladesh, Toxicol Lett 185 (2009) 197–202.

[11] A.H. Smith, G. Marshall, Y. Yuan, J. Liaw, C. Ferreccio, C. Steinmaus, Evidence from Chile that arsenic in drinking water may increase mortality from pulmonary tuberculosis, Am J Epidemiol 173 (2011) 414–420.

[12] O. Neubauer, Arsenical cancer: a review, Br J Cancer 1 (1947) 192–251.

[13] P.E. Enterline, G.M. Marsh, Mortality studies of smelter workers, Am J Ind Med 1 (1980) 251–259.

[14] P.E. Enterline, G.M. Marsh, Cancer among workers exposed to arsenic and other substances in a copper smelter, Am J Epidemiol 116 (1982) 895–911.

[15] A. Lee-Feldstein, Arsenic and respiratory cancer in humans: follow-up of copper smelter employees in Montana, J Natl Cancer Inst 70 (1983) 601–610.

[16] A. Lee-Feldstein, Cumulative exposure to arsenic and its relationship to respiratory cancer among copper smelter employees, J Occup Med 28 (1986) 296–302.

[17] G. Pershagen, F. Bergman, J. Klominek, L. Damber, S. Wall, Histological types of lung cancer among smelter workers exposed to arsenic, Br J Ind Med 44 (1987) 454–458.

[18] G. Pershagen, N.E. Bjorklund, On the pulmonary tumorigenicity of arsenic trisulfide and calcium arsenate in hamsters, Cancer Lett 27 (1985) 99–104.

[19] G. Pershagen, C.G. Elinder, A.M. Bolander, Mortality in a region surrounding an arsenic emitting plant, Environ Health Perspect 19 (1977) 133–137.

[20] S.S. Pinto, V. Henderson, P.E. Enterline, Mortality experience of arsenic-exposed workers, Arch Environ Health 33 (1978) 325–331.

[21] IARC, Monographs of the Evaluation of Carcinogenic Risk to Humans: some metals and metallic compounds. International Agency for Research on Cancer, Lyon 23 (1980).

[22] I. Celik, L. Gallicchio, K. Boyd, T.K. Lam, G. Matanoski, X. Tao, M. Shiels, E. Hammond, L. Chen, K.A. Robinson, L.E. Caulfield, J.G. Herman, E. Guallar, A.J. Alberg, Arsenic in drinking water and lung cancer: a systematic review, Environ Res 108 (2008) 48–55.

[23] C.J. Chen, Y.C. Chuang, T.M. Lin, H.Y. Wu, Malignant neoplasms among residents of a blackfoot disease-endemic area in Taiwan: high-arsenic artesian well water and cancers, Cancer Res 45 (1985) 5895–5899.

[24] C.J. Chen, C.J. Wang, Ecological correlation between arsenic level in well water and age-adjusted mortality from malignant neoplasms, Cancer Res 50 (1990) 5470–5474.

[25] G. Marshall, C. Ferreccio, Y. Yuan, M.N. Bates, C. Steinmaus, S. Selvin, J. Liaw, A.H. Smith, Fifty-year study of lung and bladder cancer mortality in Chile related to arsenic in drinking water, J Natl Cancer Inst 99 (2007) 920–928.

[26] C. Ferreccio, C. Gonzalez, V. Milosavjlevic, G. Marshall, A.M. Sancha, A.H. Smith, Lung cancer and arsenic concentrations in drinking water in Chile, Epidemiology 11 (2000) 673–679.

[27] Y. Chen, H. Ahsan, Cancer burden from arsenic in drinking water in Bangladesh, Am J Public Health 94 (2004) 741–744.

[28] J.P. Buchet, D. Lison, Mortality by cancer in groups of the Belgian population with a moderately increased intake of arsenic, Int Arch Occup Environ Health 71 (1998) 125–130.

[29] IARC, Monographs on the Evaluation of Carcinogenic Risks to Humans. International Agency for Research on Cancer, Lyon 84 (2004).

[30] A.H. Smith, A. Ercumen, Y. Yuan, C.M. Steinmaus, Increased lung cancer risks are similar whether arsenic is ingested or inhaled, J Expo Sci Environ Epidemiol 19 (2009) 343–348.

[31] H.R. Guo, N.S. Wang, H. Hu, R.R. Monson, Cell type specificity of lung cancer associated with arsenic ingestion, Cancer Epidemiol Biomarkers Prev 13 (2004) 638–643.

[32] J.E. Heck, A.S. Andrew, T. Onega, J.R. Rigas, B.P. Jackson, M.R. Karagas, E.J. Duell, Lung cancer in a U.S. population with low to moderate arsenic exposure, Environ Health Perspect 117 (2009) 1718–1723.

[33] B.K. De, D. Majumdar, S. Sen, S. Guru, S. Kundu, Pulmonary involvement in chronic arsenic poisoning from drinking contaminated ground-water, J Assoc Physicians India 52 (2004) 395–400.

[34] O.S. von Ehrenstein, D.N. Mazumder, Y. Yuan, S. Samanta, J. Balmes, A. Sil, N. Ghosh, M. Hira-Smith, R. Haque, R. Purushothamam, S. Lahiri, S. Das, A.H. Smith, Decrements in lung function related to arsenic in drinking water in West Bengal, India, Am J Epidemiol 162 (2005) 533–541.

[35] L.N. Islam, A.H. Nabi, M.M. Rahman, M.S. Zahid, Association of respiratory complications and elevated serum immunoglobulins with drinking water arsenic toxicity in human, J Environ Sci Health A Tox Hazard Subst Environ Eng 42 (2007) 1807–1814.

[36] C.O. Abernathy, Y.P. Liu, D. Longfellow, H.V. Aposhian, B. Beck, B. Fowler, R. Goyer, R. Menzer, T. Rossman, C. Thompson, M. Waalkes, Arsenic: health effects, mechanisms of actions, and research issues, Environ Health Perspect 107 (1999) 593–597.

[37] T.G. Rossman, Mechanism of arsenic carcinogenesis: an integrated approach, Mutat Res 533 (2003) 37–65.

[38] I. Hertz-Picciotto, A.H. Smith, D. Holtzman, M. Lipsett, G. Alexeeff, Synergism between occupational arsenic exposure and smoking in the induction of lung cancer, Epidemiology 3 (1992) 23–31.

[39] L. Jarup, G. Pershagen, Arsenic exposure, smoking, and lung cancer in smelter workers: a case-control study, Am J Epidemiol 134 (1991) 545–551.

[40] S.K. Wadhwa, T.G. Kazi, N.F. Kolachi, H.I. Afridi, S. Khan, A.A. Chandio, A.Q. Shah, G.A. Kandhro, S. Nasreen, Case-control study of male cancer patients exposed to arsenic-contaminated drinking water and tobacco smoke with relation to non-exposed cancer patients, Hum Exp Toxicol 30 (2011) 2013–2022.

[41] C.D. Evans, K. LaDow, B.L. Schumann, R.E. Savage, Jr., J. Caruso, A. Vonderheide, P. Succop, G. Talaska, Effect of arsenic on benzo[*a*]pyrene DNA adduct levels in mouse skin and lung, Carcinogenesis 25 (2004) 493–497.

[42] R.C. Lantz, A.M. Hays, Role of oxidative stress in arsenic-induced toxicity, Drug Metab Rev 38 (2006) 791–804.

[43] C.K. Wang, H.L. Lee, H. Chang, M.H. Tsai, Y.C. Kuo, P. Lin, Enhancement between environmental tobacco smoke and arsenic on emphysema-like lesions in mice, J Hazard Mater 221–222 (2012) 256–263.

[44] C.D. Kozul, K.H. Ely, R.I. Enelow, J.W. Hamilton, Low-dose arsenic compromises the immune response to influenza A infection in vivo, Environ Health Perspect 117 (2009) 1441–1447.

[45] A.H. Smith, M. Goycolea, R. Haque, M.L. Biggs, Marked increase in bladder and lung cancer mortality in a region of Northern Chile due to arsenic in drinking water, Am J Epidemiol 147 (1998) 660–669.

[46] M.P. Waalkes, J. Liu, B.A. Diwan, Transplacental arsenic carcinogenesis in mice, Toxicol Appl Pharmacol 222 (2007) 271–280.

[47] J. Shen, J. Liu, Y. Xie, B.A. Diwan, M.P. Waalkes, Fetal onset of aberrant gene expression relevant to pulmonary carcinogenesis in lung adenocarcinoma development induced by in utero arsenic exposure, Toxicol Sci 95 (2007) 313–320.

[48] E.J. Tokar, W. Qu, M.P. Waalkes, Arsenic, stem cells, and the developmental basis of adult cancer, Toxicol Sci 120 Suppl 1 (2011) S192–S203.

[49] C.D. Kozul, T.H. Hampton, J.C. Davey, J.A. Gosse, A.P. Nomikos, P.L. Eisenhauer, D.J. Weiss, J.E. Thorpe, M.A. Ihnat, J.W. Hamilton, Chronic exposure to arsenic in the drinking water alters the expression of immune response genes in mouse lung, Environ Health Perspect 117 (2009) 1108–1115.

[50] R.C. Lantz, B. Chau, P. Sarihan, M.L. Witten, V.I. Pivniouk, G.J. Chen, In utero and post-natal exposure to arsenic alters pulmonary structure and function, Toxicol Appl Pharmacol 235 (2009) 105–113.

[51] B.M. Fischer, E. Pavlisko, J.A. Voynow, Pathogenic triad in COPD: oxidative stress, protease-antiprotease imbalance, and inflammation, Int J Chron Obstruct Pulmon Dis 6 (2011) 413–421.

[52] D.E. Davies, J. Wicks, R.M. Powell, S.M. Puddicombe, S.T. Holgate, Airway remodeling in asthma: new insights, J Allergy Clin Immunol 111 (2003) 215–225; quiz 226.

[53] K. Dheda, H. Booth, J.F. Huggett, M.A. Johnson, A. Zumla, G.A. Rook, Lung remodeling in pulmonary tuberculosis, J Infect Dis 192 (2005) 1201–1209.

[54] A. Niimi, A. Torrego, A.G. Nicholson, B.G. Cosio, T.B. Oates, K.F. Chung, Nature of airway inflammation and remodeling in chronic cough, J Allergy Clin Immunol 116 (2005) 565–570.

[55] H.Y. Reynolds, D.B. Gail, J.P. Kiley, Interstitial lung diseases: where we started from and are now going, Sarcoidosis Vasc Diffuse Lung Dis 22 (2005) 5–12.

[56] S.T. Holgate, D.E. Davies, S. Rorke, J. Cakebread, G. Murphy, R.M. Powell, J.W. Holloway, ADAM 33 and its association with airway remodeling and hyperresponsiveness in asthma, Clin Rev Allergy Immunol 27 (2004) 23–34.

[57] B.E. McParland, P.T. Macklem, P.D. Pare, Airway wall remodeling: friend or foe? J Appl Physiol 95 (2003) 426–434.

[58] Y. Nakano, N.L. Muller, G.G. King, A. Niimi, S.E. Kalloger, M. Mishima, P.D. Pare, Quantitative assessment of airway remodeling using high-resolution CT, Chest 122 (2002) 271S–275S.

[59] D. Ramos-Barbon, M.S. Ludwig, J.G. Martin, Airway remodeling: lessons from animal models, Clin Rev Allergy Immunol 27 (2004) 3–21.

[60] A. Dosanjh, B. Zuraw, Endothelin-1 (ET-1) decreases human bronchial epithelial cell migration and proliferation: implications for airway remodeling in asthma, J Asthma 40 (2003) 883–886.

[61] E.N. Arwert, E. Hoste, F.M. Watt, Epithelial stem cells, wound healing and cancer, Nat Rev Cancer 12 (2012) 170–180.

[62] M. Schafer, S. Werner, Cancer as an overhealing wound: an old hypothesis revisited, Nat Rev Mol Cell Biol 9 (2008) 628–638.

[63] H.F. Dvorak, Tumors: wounds that do not heal. Similarities between tumor stroma generation and wound healing, N Engl J Med 315 (1986) 1650–1659.

[64] M. Vareille, E. Kieninger, M.R. Edwards, N. Regamey, The airway epithelium: soldier in the fight against respiratory viruses, Clin Microbiol Rev 24 (2011) 210–229.

[65] L.M. Crosby, C.M. Waters, Epithelial repair mechanisms in the lung, Am J Physiol Lung Cell Mol Physiol 298 (2010) L715–L731.

[66] C. Coraux, J. Roux, T. Jolly, P. Birembaut, Epithelial cell-extracellular matrix interactions and stem cells in airway epithelial regeneration, Proc Am Thorac Soc 5 (2008) 689–694.

[67] A.U. Dignass, A. Becker, S. Spiegler, H. Goebell, Adenine nucleotides modulate epithelial wound healing in vitro, Eur J Clin Invest 28 (1998) 554–561.

[68] V.E. Klepeis, A. Cornell-Bell, V. Trinkaus-Randall, Growth factors but not gap junctions play a role in injury-induced Ca^{2+} waves in epithelial cells, J Cell Sci 114 (2001) 4185–4195.

[69] V.E. Klepeis, I. Weinger, E. Kaczmarek, V. Trinkaus-Randall, P2Y receptors play a critical role in epithelial cell communication and migration, J Cell Biochem 93 (2004) 1115–1133.

[70] U.V. Wesley, P.F. Bove, M. Hristova, S. McCarthy, A. van der Vliet, Airway epithelial cell migration and wound repair by ATP-mediated activation of dual oxidase 1, J Biol Chem 282 (2007) 3213–3220.

[71] S. Bagchi, Z. Liao, F.A. Gonzalez, N.E. Chorna, C.I. Seye, G.A. Weisman, L. Erb, The P2Y2 nucleotide receptor interacts with alphav integrins to activate Go and induce cell migration, J Biol Chem 280 (2005) 39050–39057.

[72] L. Yang, D. Cranson, V. Trinkaus-Randall, Cellular injury induces activation of MAPK via P2Y receptors, J Cell Biochem 91 (2004) 938–950.

[73] A.M. Hays, R.C. Lantz, L.S. Rodgers, J.J. Sollome, R.R. Vaillancourt, A.S. Andrew, J.W. Hamilton, T.D. Camenisch, Arsenic-induced decreases in the vascular matrix, Toxicol Pathol 36 (2008) 805–817.

[74] M.H. Collins, A.C. Moessinger, J. Kleinerman, J. Bassi, P. Rosso, A.M. Collins, L.S. James, W.A. Blanc, Fetal lung hypoplasia associated with maternal smoking: a morphometric analysis, Pediatr Res 19 (1985) 408–412.

[75] M.C. Bruce, C. Honaker, P. Karathanasis, Postnatal age at onset of hyperoxic exposure influences developmentally regulated tropoelastin gene expression in the neonatal rat lung, Am J Respir Cell Mol Biol 14 (1996) 177–185.

[76] W.L. Castleman, R.L. Sorkness, R.F. Lemanske, G. Grasee, M.M. Suyemoto, Neonatal viral bronchiolitis and pneumonia induces bronchiolar hypoplasia and alveolar dysplasia in rats, Lab Invest 59 (1988) 387–396.

[77] L.N. Blanco, L. Frank, The formation of alveoli in rat lung during the third and fourth postnatal weeks: effect of hyperoxia, dexamethasone, and deferoxamine, Pediatr Res 34 (1993) 334–340.

[78] J.S. Petrick, F.M. Blachere, O. Selmin, R.C. Lantz, Inorganic arsenic as a developmental toxicant: in utero exposure and alterations in the developing rat lungs, Mol Nutr Food Res 53 (2009) 583–591.

[79] B.G. Dekkers, D. Schaafsma, S.A. Nelemans, J. Zaagsma, H. Meurs, Extracellular matrix proteins differentially regulate airway smooth muscle phenotype and function, Am J Physiol Lung Cell Mol Physiol 292 (2007) L1405–L1413.

[80] K. Parameswaran, A. Willems-Widyastuti, V.K. Alagappan, K. Radford, A.R. Kranenburg, H.S. Sharma, Role of extracellular matrix and its regulators in human airway smooth muscle biology, Cell Biochem Biophys 44 (2006) 139–146.

[81] S.K. Karnik, B.S. Brooke, A. Bayes-Genis, L. Sorensen, J.D. Wythe, R.S. Schwartz, M.T. Keating, D.Y. Li, A critical role for elastin signaling in vascular morphogenesis and disease, Development 130 (2003) 411–423.

[82] Y. Liu, P. Choudhury, C.M. Cabral, R.N. Sifers, Intracellular disposal of incompletely folded human alpha1-antitrypsin involves release from calnexin and post-translational trimming of asparagine-linked oligosaccharides, J Biol Chem 272 (1997) 7946–7951.

[83] S. Loffek, O. Schilling, C.W. Franzke, Series "matrix metalloproteinases in lung health and disease": biological role of matrix metalloproteinases: a critical balance, Eur Respir J 38 (2011) 191–208.

[84] A.B. Josyula, G.S. Poplin, M. Kurzius-Spencer, H.E. McClellen, M.J. Kopplin, S. Sturup, R. Clark Lantz, J.L. Burgess, Environmental arsenic exposure and sputum metalloproteinase concentrations, Environ Res 102 (2006) 283–290.

[85] H. Ohbayashi, K. Shimokata, Matrix metalloproteinase-9 and airway remodeling in asthma, Curr Drug Targets Inflamm Allergy 4 (2005) 177–181.

[86] S. Xie, M.B. Sukkar, R. Issa, N.M. Khorasani, K.F. Chung, Mechanisms of induction of airway smooth muscle hyperplasia by transforming growth factor-beta, Am J Physiol Lung Cell Mol Physiol 293 (2007) L245–L253.

[87] A. Kimura, Y. Ishida, T. Hayashi, T. Wada, H. Yokoyama, T. Sugaya, N. Mukaida, T. Kondo, Interferon-gamma plays protective roles in sodium arsenite-induced renal injury by up-regulating intrarenal multidrug resistance-associated protein 1 expression, Am J Pathol 169 (2006) 1118–1128.

[88] J. Wu, J. Liu, M.P. Waalkes, M.L. Cheng, L. Li, C.X. Li, Q. Yang, High dietary fat exacerbates arsenic-induced liver fibrosis in mice, Exp Biol Med (Maywood) 233 (2008) 377–384.

[89] F. Caraci, E. Gili, M. Calafiore, M. Failla, C. La Rosa, N. Crimi, M.A. Sortino, F. Nicoletti, A. Copani, C. Vancheri, TGF-beta1 targets the GSK-3beta/beta-catenin pathway via ERK activation in the transition of human lung fibroblasts into myofibroblasts, Pharmacol Res 57 (2008) 274–282.

[90] R.O. Nunes, M. Schmidt, G. Dueck, H. Baarsma, A.J. Halayko, H.A. Kerstjens, H. Meurs, R. Gosens, GSK-3/beta-catenin signaling axis in airway smooth muscle: role in mitogenic signaling, Am J Physiol Lung Cell Mol Physiol 294 (2008) L1110–L1118.

[91] L. Cohn, J.A. Elias, G.L. Chupp, Asthma: mechanisms of disease persistence and progression, Annu Rev Immunol 22 (2004) 789–815.

[92] D.M. Walter, J.J. McIntire, G. Berry, A.N. McKenzie, D.D. Donaldson, R.H. DeKruyff, D.T. Umetsu, Critical role for IL-13 in the development of allergen-induced airway hyperreactivity, J Immunol 167 (2001) 4668–4675.

[93] M. Ahdieh, T. Vandenbos, A. Youakim, Lung epithelial barrier function and wound healing are decreased by IL-4 and IL-13 and enhanced by IFN-gamma, Am J Physiol Cell Physiol 281 (2001) C2029–C2038.

[94] P. Nath, S.Y. Leung, A.S. Williams, A. Noble, S. Xie, A.N. McKenzie, K.F. Chung, Complete inhibition of allergic airway inflammation and remodelling in quadruple IL-4/5/9/13-/- mice, Clin Exp Allergy 37 (2007) 1427–1435.

[95] S. Lanone, T. Zheng, Z. Zhu, W. Liu, C.G. Lee, B. Ma, Q. Chen, R.J. Homer, J. Wang, L.A. Rabach, M.E. Rabach, J.M. Shipley, S.D. Shapiro, R.M. Senior, J.A. Elias, Overlapping and enzyme-specific contributions of matrix metalloproteinases-9 and -12 in IL-13-induced inflammation and remodeling, J Clin Invest 110 (2002) 463–474.

[96] J.H. Lee, N. Kaminski, G. Dolganov, G. Grunig, L. Koth, C. Solomon, D.J. Erle, D. Sheppard, Interleukin-13 induces dramatically different transcriptional programs in three human airway cell types, Am J Respir Cell Mol Biol 25 (2001) 474–485.

[97] C.E. Olsen, A.E. Liguori, Y. Zong, R.C. Lantz, J.L. Burgess, S. Boitano, Arsenic upregulates MMP-9 and inhibits wound repair in human airway epithelial cells, Am J Physiol Lung Cell Mol Physiol 295 (2008) L293–L302.

[98] S. Tao, Y. Zheng, A. Lau, M.C. Jaramillo, B.T. Chau, R.C. Lantz, P.K. Wong, G.T. Wondrak, D.D. Zhang, Tanshinone I activates the Nrf2-dependent antioxidant response and protects against As(III)-induced lung inflammation in vitro and in vivo, Antioxid Redox Signal 19 (2013) 1647–1661.

[99] R.C. Lantz, B.J. Lynch, S. Boitano, G.S. Poplin, S. Littau, G. Tsaprailis, J.L. Burgess, Pulmonary biomarkers based on alterations in protein expression after exposure to arsenic, Environ Health Perspect 115 (2007) 586–591.

[100] A.C. Straub, K.A. Clark, M.A. Ross, A.G. Chandra, S. Li, X. Gao, P.J. Pagano, D.B. Stolz, A. Barchowsky, Arsenic-stimulated liver sinusoidal capillarization in mice requires NADPH oxidase-generated superoxide, J Clin Invest 118 (2008) 3980–3989.

[101] A.C. Straub, L.R. Klei, D.B. Stolz, A. Barchowsky, Arsenic requires sphingosine-1-phosphate type 1 receptors to induce angiogenic genes and endothelial cell remodeling, Am J Pathol 174 (2009) 1949–1958.

[102] C.L. Sherwood, R.C. Lantz, J.L. Burgess, S. Boitano, Arsenic alters ATP-dependent Ca(2)+ signaling in human airway epithelial cell wound response, Toxicol Sci 121 (2011) 191–206.

[103] T. Lieb, C.W. Frei, J.I. Frohock, R.J. Bookman, M. Salathe, Prolonged increase in ciliary beat frequency after short-term purinergic stimulation in human airway epithelial cells, J Physiol 538 (2002) 633–646.

[104] E.M. Schwiebert, A. Zsembery, Extracellular ATP as a signaling molecule for epithelial cells, Biochim Biophys Acta 1615 (2003) 7–32.

[105] R.E. Bucheimer, J. Linden, Purinergic regulation of epithelial transport, J Physiol 555 (2004) 311–321.

[106] A. van der Vliet, NADPH oxidases in lung biology and pathology: host defense enzymes, and more, Free Radic Biol Med 44 (2008) 938–955.

[107] E.R. Lazarowski, R.C. Boucher, Purinergic receptors in airway epithelia, Curr Opin Pharmacol 9 (2009) 262–267.

[108] L. Erb, Z. Liao, C.I. Seye, G.A. Weisman, P2 receptors: intracellular signaling, Pflugers Arch 452 (2006) 552–562.

[109] G. Burnstock, Purine and pyrimidine receptors, Cell Mol Life Sci 64 (2007) 1471–1483.

[110] M.P. Abbracchio, G. Burnstock, J.M. Boeynaems, E.A. Barnard, J.L. Boyer, C. Kennedy, G.E. Knight, M. Fumagalli, C. Gachet, K.A. Jacobson, G.A. Weisman, International Union of Pharmacology LVIII: update on the P2Y G protein-coupled nucleotide receptors: from molecular mechanisms and pathophysiology to therapy, Pharmacol Rev 58 (2006) 281–341.

[111] R.A. North, Molecular physiology of P2X receptors, Physiol Rev 82 (2002) 1013–1067.

[112] E.M. Schwiebert, L. Liang, N.L. Cheng, C.R. Williams, D. Olteanu, E.A. Welty, A. Zsembery, Extracellular zinc and ATP-gated P2X receptor calcium entry channels: new zinc receptors as physiological sensors and therapeutic targets, Purinergic Signal 1 (2005) 299–310.

[113] W. Ding, W. Liu, K.L. Cooper, X.J. Qin, P.L. de Souza Bergo, L.G. Hudson, K.J. Liu, Inhibition of poly(ADP-ribose) polymerase-1 by arsenite interferes with repair of oxidative DNA damage, J Biol Chem 284 (2009) 6809–6817.

[114] K.T. Kitchin, K. Wallace, Arsenite binding to synthetic peptides based on the Zn finger region and the estrogen binding region of the human estrogen receptor-alpha, Toxicol Appl Pharmacol 206 (2005) 66–72.

[115] C.L. Sherwood, R.C. Lantz, S. Boitano, Chronic arsenic exposure in nanomolar concentrations compromises wound response and intercellular signaling in airway epithelial cells, Toxicol Sci 132 (2013) 222–234.

[116] J.M. Anderson, C.M. Van Itallie, Physiology and function of the tight junction, Cold Spring Harb Perspect Biol 1 (2009) a002584.

[117] Y. Soini, Claudins in lung diseases, Respir Res 12 (2011) 70.

[118] C.L. Sherwood, A.E. Liguori, C.E. Olsen, R.C. Lantz, J.L. Burgess, S. Boitano. Arsenic compromises conducting airway epithelial barrier properties in primary mouse and immortalized human cell cultures, PLoS ONE 8 (2013) e82970.

[119] P.D. Vermeer, J. Denker, M. Estin, T.O. Moninger, S. Keshavjee, P. Karp, J.N. Kline, J. Zabner, MMP9 modulates tight junction integrity and cell viability in human airway epithelia, Am J Physiol Lung Cell Mol Physiol 296 (2009) L751–L762.

[120] E. Puchelle, J.M. Zahm, J.M. Tournier, C. Coraux, Airway epithelial repair, regeneration, and remodeling after injury in chronic obstructive pulmonary disease, Proc Am Thorac Soc 3 (2006) 726–733.

[121] H. Ichiyasu, J.M. McCormack, K.M. McCarthy, D. Dombkowski, F.I. Preffer, E.E. Schneeberger, Matrix metalloproteinase-9-deficient dendritic cells have impaired migration through tracheal epithelial tight junctions, Am J Respir Cell Mol Biol 30 (2004) 761–770.

[122] A.C. Buisson, J.M. Zahm, M. Polette, D. Pierrot, G. Bellon, E. Puchelle, P. Birembaut, J.M. Tournier, Gelatinase B is involved in the in vitro wound repair of human respiratory epithelium, J Cell Physiol 166 (1996) 413–426.

[123] C. Legrand, C. Gilles, J.M. Zahm, M. Polette, A.C. Buisson, H. Kaplan, P. Birembaut, J.M. Tournier, Airway epithelial cell migration dynamics. MMP-9 role in cell-extracellular matrix remodeling, J Cell Biol 146 (1999) 517–529.

[124] A.L. Herard, D. Pierrot, J. Hinnrasky, H. Kaplan, D. Sheppard, E. Puchelle, J.M. Zahm, Fibronectin and its alpha 5 beta 1-integrin receptor are involved in the wound-repair process of airway epithelium, Am J Physiol 271 (1996) L726–L733.

[125] R.M. Senior, A. Hinek, G.L. Griffin, D.J. Pipoly, E.C. Crouch, R.P. Mecham, Neutrophils show chemotaxis to type IV collagen and its 7S domain and contain a 67 kD type IV collagen binding protein with lectin properties, Am J Respir Cell Mol Biol 1 (1989) 479–487.

[126] T. Geiser, K. Atabai, P.H. Jarreau, L.B. Ware, J. Pugin, M.A. Matthay, Pulmonary edema fluid from patients with acute lung injury augments in vitro alveolar epithelial repair by an IL-1beta-dependent mechanism, Am J Respir Crit Care Med 163 (2001) 1384–1388.

[127] T. Geiser, P.H. Jarreau, K. Atabai, M.A. Matthay, Interleukin-1beta augments in vitro alveolar epithelial repair, Am J Physiol Lung Cell Mol Physiol 279 (2000) L1184–L1190.

[128] S.R. White, B.M. Fischer, B.A. Marroquin, R. Stern, Interleukin-1beta mediates human airway epithelial cell migration via NF-kappaB, Am J Physiol Lung Cell Mol Physiol 295 (2008) L1018–L1027.

[129] C. Neurohr, S.L. Nishimura, D. Sheppard, Activation of transforming growth factor-beta by the integrin alphavbeta8 delays epithelial wound closure, Am J Respir Cell Mol Biol 35 (2006) 252–259.

[130] P.F. Bove, U.V. Wesley, A.K. Greul, M. Hristova, W.R. Dostmann, A. van der Vliet, Nitric oxide promotes airway epithelial wound repair through enhanced activation of MMP-9, Am J Respir Cell Mol Biol 36 (2007) 138–146.

[131] A. Tam, S. Wadsworth, D. Dorscheid, S.F. Man, D.D. Sin, The airway epithelium: more than just a structural barrier, Ther Adv Respir Dis 5 (2011) 255–273.

[132] M.F. Beers, E.E. Morrisey, The three R's of lung health and disease: repair, remodeling, and regeneration, J Clin Invest 121 (2011) 2065–2073.

8

BLADDER CANCER AND ARSENIC

MATTHEW K. MEDEIROS AND A. JAY GANDOLFI

Department of Pharmacology and Toxicology, College of Pharmacy, University of Arizona, Tucson, AZ, USA

8.1 INTRODUCTION

8.1.1 Demographics

In the United States, over 70,000 bladder cancer cases are diagnosed, and more than 14,000 deaths reported annually [46]. It is the second most common malignancy of the genitourinary tract after prostate cancer. The risk of bladder cancer increases with age with about 66% of bladder cancers diagnosed in patients 65 years or older. Men have a higher risk of bladder cancer than women, a ratio of about 7:2. Occupational exposure to aromatic amines was once shown to be strongly associated with bladder cancer. Presently, the greatest environmental risk factor in developing bladder cancer is active smoking, contributing to more than 50% of cases [58, 75, 103]. However, there is strong evidence showing a link between exposure to arsenic in the drinking water at concentrations exceeding 300–500 μg/L and the development of bladder cancer [69]. At arsenic concentrations of 10–100 μg/L, health effects are not obvious, but synergistic effects may also exist between smoking, nutritional factors, and arsenic.

8.1.2 Bladder Tumors

About 90% of bladder cancer cases originate from the transitional cells of the bladder urothelium and exhibit a great deal of heterogeneity with respect to gene expression profiling and patient response to treatments [66, 71]. These cancers can present as

Arsenic: Exposure Sources, Health Risks, and Mechanisms of Toxicity, First Edition.
Edited by J. Christopher States.
© 2016 John Wiley & Sons, Inc. Published 2016 by John Wiley & Sons, Inc.

noninvasive papillary tumors that protrude into the bladder lumen and usually do not metastasize. These tumors are easily removed but have high relapse frequency. About one-third of bladder cancers present as invasive nonpapillary tumors. These tumors can aggressively invade the bladder wall and metastasize to other sites. A small percentage (10–15%) of low-grade papillary tumors can progress to this aggressive form of bladder cancer. Tumor staging of bladder cancer is determined relative to the extent of penetration of the tumor within the bladder wall. Superficial bladder cancer, which has not invaded the muscle layer, is designated Ta. T1 is a tumor that has invaded the subepithelial connective tissue layer. Stage 2 (T2a and T2b) tumors are defined as more aggressive and invasive and penetrate into the muscle layer (Fig. 8.1). Rapid progression and metastasis will occur once tumors invade the muscle layer. The development of chemotherapy resistance with invasive tumors is a severe problem leading to rapid disease progression and mortality [21]. Identification of the

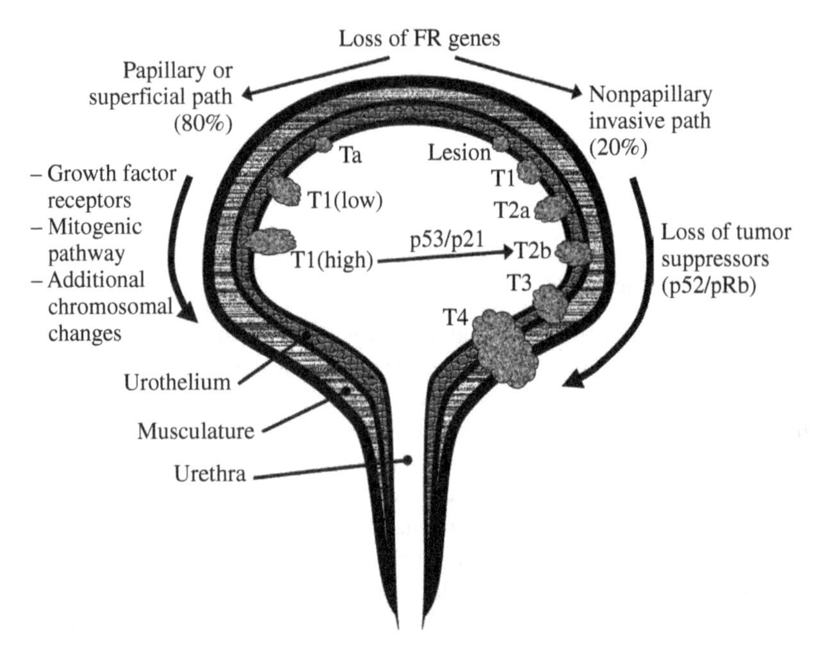

FIGURE 8.1 Illustration of proposed pathways in the development of bladder cancer. Superficial and nonpapillary pathways are defined by distinct cellular changes that follow the loss of forerunner (FR) genes, which impart a growth advantage. Superficial or papillary tumors display hyperplasia but are noninvasive. These tumors are commonly driven by mutations in a growth factor receptor or in a mitogen-activated protein kinase pathway. Progression from the low-grade T1 stage to a high-grade stage is thought to involve additional chromosomal changes and changes in cell cycle control. The invasive nonpapillary transitional cell carcinoma typically evolves from a genetically unstable population. These carcinomas are characterized by changes in p53 function with subsequent genetic alterations and instability, inactivation of pRb, and continued chromosomal changes. A small proportion of high-grade T1 papillary tumors will form invasive tumors if alterations in p53 and p21 (CDKN1A) and/or additional chromosomal changes occur.

initiating events that lead to invasion, metastasis, and resistance is of great importance to researchers and patients who present with this form of cancer. Although early detection is an important factor in yielding a favorable prognosis, the advancement in chemotherapeutics has helped to significantly reduce the mortality rates from bladder cancer in the past two decades.

8.1.3 Molecular Aspects of Bladder Cancers

Distinct differences in morphology and progression between invasive nonpapillary tumors and superficial papillary tumors suggest separate carcinogenic pathways [66, 78]. Where p53 mutations occur early, invasive carcinomas typically arise *in situ*. These tumors display genetic instability with further genetic alterations accumulating as the disease progresses, including Rb and p16 inactivation from chromosomal segment 8p loss. In superficial papillary tumors, there are typically constitutively active components of the MAPK pathway, that is, H-Ras, preceded by activating mutations in FGFR3. Additional events such as overexpression of cyclin D1, loss of chromosomal segment 11p, and inactivation of p16 characterize the transition from the low-grade papillary tumor to the high-grade invasive carcinoma. Similarly, the transition into an invasive tumor has been observed when superficial tumors acquire alterations in p53, cyclin-dependent kinase inhibitor 1A (CDKN1A, p21), and/or loss of the 8p chromosomal segment. Irrespective of the path taken, the shared molecular event for almost all bladder cancers is the allelic loss on chromosome 9 [103]. The 9p region contains the p16/ARF gene, which is frequently inactivated by loss of heterozygosity or methylation. Additionally, located in this region is the gene for interferon alpha, which can induce apoptosis in the bladder cancer cells [118]. Therefore, inactivation of this gene may also contribute to disease progression.

8.1.4 Epigenetics and Bladder Cancer

Epigenetic alterations are typically associated with bladder cancer progression and prognosis [64, 65, 112]. As studies suggest the irreversible nature of epigenetic changes, an understanding of epigenetic processes may be useful in the clinical setting in helping to restore gene function in the diseased bladder [23]. DNA hypermethylation is observed in patients with bladder cancer and can be detected in the blood and urine [53, 82]. Such epigenetic patterns can be used to discriminate between subtypes of bladder cancer and allow for the proper selection of treatments [7]. Accumulation of epigenetic changes may be related to aging and increased duration of exposure to carcinogens with concomitant increase in genomic instability and DNA damage. Both smoking and arsenic exposure are associated with promoter hypermethylation at specific genes in bladder cancer, that is, p16 and RASSF1A [103, 111]. Globally, about 17% of sequences in normal urothelial cells have aberrant methylation patterns in bladder cancer. Genes involved in cell cycle control, DNA damage repair, cell differentiation, and adhesion are typically altered by aberrant methylation. Such epigenetic changes can possibly serve to predict tumor progression since promoter hypermethylation will most likely affect an increasing number of genes with each advancing stage of the disease. Some typical bladder cancer markers with promoter hypermethylation include DBC1, p16, p14, APC, and CDKN2A [35, 40]. The hypermethylation status of, for example, DAPK, p14, TIMP3, SOCS-1, and STAT-1 was positively correlated with the recurrence of the cancer [32, 45].

Another mechanism of epigenetic regulation involves the covalent posttranslational acetylation, methylation, or ubiquitination of histone proteins [20]. These modifications can alter the overall chromatin state and thus the level of ease or difficulty of transcribing genes. DNA methylation and histone modifications are closely related, and often, the cooperative interaction of both mechanisms of regulation is necessary for proper gene expression. Since acetylation is a potentially reversible modification, the application of epigenetic therapy using HDAC inhibitors is showing some promise as an effective antitumor regimen for bladder cancer treatment [51, 96].

8.1.5 Role of p53 Alterations in Cancer

The role of altered p53 in human cancer is critical, since half of human cancers harbor p53 gene mutations [31]. In 60% of invasive bladder cancers, loss-of-heterozygosity studies have shown that the TP53 locus is lost in chromosome 17p [26, 70, 99]. This event is rare in noninvasive tumors. The main responses following the activation of p53 include regulation of cell cycle, DNA repair, and apoptosis. In general, p53 is activated by specific stressors (i.e., hypoxia) and DNA damage. Therefore, p53 mutations that alter p53 function can confer genomic instability and uncontrolled cell cycle properties. Disease progression, decreased patient survival, and disease stage are generally correlated to p53 mutations [79]. p53 status is important, more so, as a prognostic marker in bladder cancer than as a predictor of how well the patient will respond to therapies. A number of studies support the role of p53 in the regulation of apoptosis [73]. Dysregulation of this pathway plays an important role in tumorigenesis, allowing the survival and continuation of genetically damaged tumor-forming cells.

8.2 ARSENIC AND BLADDER CANCER

The International Agency for Research on Cancer (IARC) concluded in 2002 that there was sufficient evidence for an increased risk of bladder cancer in individuals exposed to arsenic in drinking water, implicating arsenic as a human carcinogen. Although recognized as a human bladder carcinogen, the concentration necessary to promote carcinogenesis is not certain. Epidemiological studies conducted in Taiwan, Chile, Argentina, the United States, Japan, England, and Finland have provided evidence for an association between bladder cancer when there are high-concentrations and, in one case, low arsenic exposures [41, 55, 98, 116]. In these regions, arsenic concentrations have been reported to exceed 500 μg/L. As much as 40 million people are exposed to concentrations above 50 μg/L in Bangladesh; West Bengal, India; and Vietnam [50, 80]. Elevated mortality and incidence rates of bladder cancer have been investigated in regions of Taiwan where blackfoot disease is prevalent. These studies suggest that the correlation between the bladder cancer and arsenic exposure is strong when concentrations of arsenic exceed 170 μg/L. In another Taiwan study, researchers found that an increased risk of incidence of transitional cell carcinoma of the bladder occurred at an average rate of 0.45 per 100,000 person-years with a 1% increase in the proportion of wells with arsenic above 640 μg/L [34]. A Finnish study found evidence of a link between arsenic exposure and bladder cancer risk even at a very low concentration (0.5 μg/L) and that smoking and nutritional factors had a synergistic effect [55]. Other studies involving cohorts in the United States, Japan, and England have given some support to the association, but the weaknesses in these studies include lack of specific information regarding individual lifestyles, water consumption behavior, smoking habits, and occupational history.

A recent study of bladder cancer incidence in Antofagasta, northern Chile, where a significant contamination of drinking water was recorded (as high as 17 times what is recommended by WHO), revealed elevations in bladder cancer mortality rates for the regional population compared to the total Chilean population [30]. Records show that the incorporation of a new water source in 1958 led to an abrupt jump in arsenic concentration from 90 μg/L to between 800 and 900 μg/L. Thus, the entire population of Antofagasta was exposed to arsenic until 1970s when treatment plants were implemented. The study showed that mortality rates started to increase about 10–15 years after the start of exposure, and the rates were persistent until the end of the study in 2000. The peak mortality rate with bladder cancer for men in Antofagasta in 1991 was 26.7 per 100,000 compared to 5.8 per 100,000 for Chile. For women, peak mortality was 18.6 per 100,000 in 2001. Obviously, with a clear risk associated with bladder cancer, arsenic-exposed populations should be informed of the risk, and early detection of arsenic-contaminated water supplies should be acted upon.

8.3 ARSENIC BIOTRANSFORMATION PRODUCTS AND BLADDER EXPOSURE

Arsenic usually enters the body in the trivalent form [18]. Arsenic is biotransformed via oxidation/reduction and methylation processes to inorganic (arsenite [As(III)], arsenate [As(V)]) and methylated metabolites. The methylated metabolites include monomethylarsonous acid [MMA(III)], monomethylarsonic acid [MMA(V)], dimethylarsinic acid [MMA(V)], and dimethylarsinous acid [DMA(III)] (see Chapter 4). Elimination occurs by urinary excretion in both organic and inorganic forms, with organic arsenic being eliminated more rapidly [3]. Although the liver is an important site for metabolism following arsenic ingestion, biotransformation can occur in other tissues, including the testes, kidney, and lung [38]. However, considerable variation in the biotransformation of inorganic arsenic between mammalian species has been observed, as well as between persons in human populations [97]. As these biotransformation products are generated, they are distributed via the blood to nearly all tissues. The production of urine by the kidneys from filtration, reabsorption, and secretion generally concentrates soluble substances, such as the arsenic biotransformation products. Such high concentration of arsenicals increases the insult to the bladder when the urine is stored. Based on mice exposed to inorganic arsenic, the total tissue distribution of arsenicals was greatest in the kidney followed by the lungs and then urinary bladder, and in these tissues, arsenic levels exceeded levels in the skin, blood, and liver tissue [52].

MMA(V) and DMA(V) have long been detoxification products formed in mammals after inorganic arsenic exposure [97]. The methylation efficiency can be evaluated by measuring the proportions of, for example, MMA(V) and DMA(V) excreted in the urine. Humans excrete appreciable amounts of MMA(V) in the urine compared to other mammals. Mice, dogs, and rats all methylate arsenic to DMA(V) with high efficiency and have varying rates of excretion. Irrespective of the type of exposure, the average relative distribution of arsenic metabolites in urine for human populations is consistent with inorganic arsenic (arsenate + arsenite) measured at 10–30%, MMA (III + V) at 10–20%, and DMA (III + V) at 60–80% [63] (Fig. 8.2). Only very rare situations exist in small exposed populations and cultures around the world where significant deviations in the proportions of metabolites occur compared to the general population. Additionally, variations in the production in MMA are related to the presence of genetic polymorphisms in enzymes involved in arsenic biotransformation [101]. Interindividual variation in the metabolism of arsenic can be influenced by dose, route of exposure, and nutritional status as determined by animal studies, but little is known on what factors influence arsenic biotransformation in humans other than genetic factors.

Genetic polymorphisms in arsenic biotransformation enzymes could provide individuals with varying degrees of susceptibility to arsenic exposure or development of cancer. Polymorphisms in genes involved in methylation of arsenic such as GSTO1-1, GSTM1, and GSTT1 can influence proportions of MMA(III) and DMA(III) in urine that are in contact with bladder urothelium [42, 91]. Since arsenic and its metabolites are in continuous contact with the bladder urothelium during the process of excretion, the bladder may be exposed to higher concentrations of arsenicals due to the bioconcentration of urine after kidney filtration.

The capacity for arsenic methylation can be assessed from urinary metabolites, MMA(V) and DMA(V) [100]. These metabolites represent the activity of first and second methylation stages. The primary methylation index (PMI) is calculated from the ratio of MMA(V) to total inorganic arsenic ingestion or intake. Secondary methylation index (SMI) is determined from the ratio of DMA(V) to MMA(V). The susceptibility of humans to arsenic-induced health effects exemplifies the importance of complete secondary methylation, as humans excrete a higher percentage of MMA(V) compared to rodents. Chen et al. [13] assessed the risks with the development of bladder cancer associated from cumulative arsenic exposure and urinary arsenic methylation status (using PMI and SMI) in Taiwan. Working from 49 newly diagnosed bladder cancer cases between 1996 and 1999 and control groups, measurements of arsenic speciation in the urine, and publications on arsenic exposure from 1974 to 1976, Chen et al. found that subjects with a low SMI (<4.8) had an increased risk for bladder cancer especially when coupled with a high cumulative arsenic exposure level (>12 mg/L × year). Risks were higher for men (smoking or

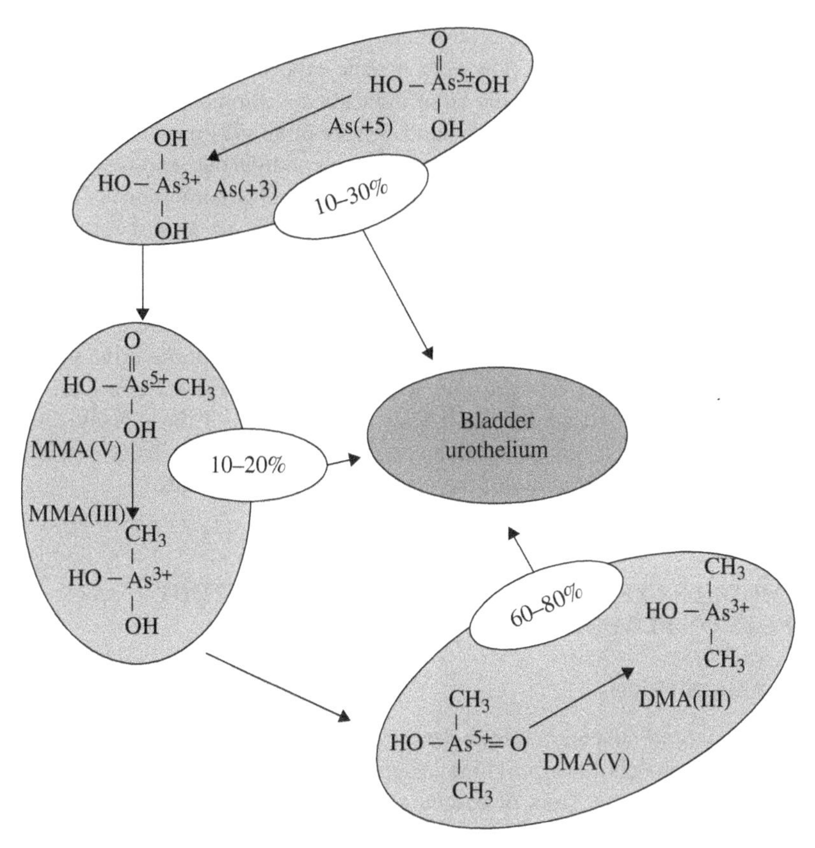

FIGURE 8.2 Biotransformation and metabolite exposure. Exposure of either inorganic species of arsenic, arsenite [As(+3)], or arsenate [As(+5)] increases levels of all arsenic species in the urine including MMA(V), MMA(III), DMA(V), and DMA(III). For a detailed discussion of arsenic metabolism, the reader is directed to Chapter 4. All of these various species can be detected in the urine of individuals exposed to arsenic, and levels will depend on time, route, dose, and individual metabolic differences. The bladder urothelium, therefore, is exposed to both inorganic and organic arsenicals. Relative proportions of the various species typically detected in the urine are given in the diagram.

nonsmoking) than for women. In another study [108], an investigation of micronucleation in exfoliated bladder cells and buccal cells was conducted and revealed a positive association between the frequency of micronucleated bladder cells and the urinary concentration of inorganic arsenic and its methylated metabolites. The study utilized a group of 18 individuals in Nevada, United States, exposed chronically to arsenic-contaminated water (average 1312 μg/L of arsenic) and compared with a matched group with low exposure (16 μg/L).The SMI in the exposed group was about 2. However, micronuclei in buccal cells failed to yield an association, suggesting that the bladder is particularly sensitive to the genotoxic effects of arsenic. Since micronuclei are acentric fragments or whole chromosomes found in the cell cytoplasm after abnormal cell division, the assessment of micronuclei formation serves as an indicator for bladder toxicity to low-level arsenic exposure. An abundance of micronuclei within a population of cells is an indicator of chromosomal damage and/or precancerous tissue. Utilizing this method, Moore et al. [72] measured micronuclei formation in exfoliated bladder cells of Chilean individuals exposed to 50 μg/L arsenic. The study demonstrated a dose-dependent increase in micronuclei formation.

In Taiwan, increased MMA(V) output in the urine or decrease excretion of DMA(V) has been correlated with peripheral vascular disease, skin cancer, and bladder cancer. These studies demonstrated that the variations in biotransformation rates of arsenic play a significant role in the severity in arsenic-induced health effects in individuals exposed to arsenic. The increase of MMA(V) in the urine may be an indicator for high MMA(III) exposure. As the trivalent forms are considered more toxic, the presence of disease is most likely associated with these arsenicals. The studies, which show chromosomal damage in exfoliated cells, clearly demonstrate the sensitivity and susceptibility of the bladder to arsenic exposure.

8.4 MODELS USED IN ARSENIC RESEARCH TO STUDY EFFECTS ON BLADDER

8.4.1 Animal Models

Animal models of arsenic-induced cancer have been developed using primarily rodents. These models have provided insight into the biotransformation process of arsenic and disposition. Tests of carcinogenicity of arsenic have been performed. At dietary concentration as high as 400 ppm arsenate (and 250 ppm arsenite) administered to mice and 416 ppm of arsenite in rats, standard carcinogenicity bioassays were negative [94]. Although adult exposure to inorganic arsenic is not directly carcinogenic in rodents, DMA(V) has been shown to be either a promoter or a complete carcinogen in the rat bladder [19]. The earliest report on the sensitivity of the rat bladder to DMA(V) was through the work by Yamamoto et al. [117]. However, in these studies, several initiators of carcinogenesis (i.e., nitrosamines, nitrosoureas) were applied prior to the administration of DMA(V). Wei et al. [109] would later demonstrate complete carcinogenicity of DMA at concentrations of 50 and 200 ppm in the drinking water. In a 2-year bioassay where 50 and 200 ppm of DMA(V) in drinking water was administered to male rats, a significantly increased incidence of bladder tumors was observed. At a concentration of 12.5 ppm, there was no evidence of cytotoxicity, hyperproliferation, or tumors. At a concentration of 40 ppm DMA(V), a significantly increased incidence of hyperplasia was observed. Additional studies with MMA(V) and DMA(V) administered to the diets of mice in 2-year bioassays did not reveal any significant changes in the bladder, nor tumorigenesis, indicating species-specific effects [5, 6]. However, the accumulation of arsenite, and to a lesser extent DMA, arsenate, and MMA, has been observed in the bladder tissue of mice when administered arsenate or arsenite, and short-term studies have shown hyperproliferative effects in the bladder epithelium associated with arsenic-induced activator protein 1 (AP-1) activation [44, 52, 62, 87, 88]. Urinary excretion has been shown to be mostly in the form of MMA and DMA. Based on these rodent models, the cytotoxicity and potential for carcinogenicity of arsenic may be due to a combined effect from the presence of methylated forms of arsenic and the more reactive trivalent intermediates, accumulation of arsenicals in the bladder, and regenerative cell proliferation.

Waalkes et al. [104] have developed a mouse model showing induced cancers in the offspring of mouse dams that were administered high doses of inorganic arsenic *in utero* (see Chapter 19). Following transplacental exposures of 85 ppm of As(III) during gestation days 8 through 18, multiple organ carcinogenesis in the offspring occurred with exposure to As(III). Increases in papilloma and carcinomas of the bladder of female offspring were observed and only occurred in males when tamoxifen or diethylstilbestrol was administered postpartum [105]. The relevance of these studies has not been determined in human carcinogenesis, as the mechanisms of the model have not been completely elucidated. However, the model does implicate arsenic as a transplacental agent, acting to sensitize target tissues *in utero*, such as the bladder, to carcinogen exposure in the adult stage.

Given that DMA(V) has been shown to be carcinogenic in rats, considerable effort has gone into the determination of the mode of action for this arsenic metabolite. Several modes of action have been considered in the DMA(V)-induced bladder tumors in rats. These modes of action include the production of reactive intermediates, indirect genotoxicity, inhibition of DNA repair, direct simulation of proliferation, and/or cytotoxicity with mitogenic regeneration [19] (Fig. 8.3). Follow-up shorter-term (10 weeks) studies with rats have been done to investigate the cytotoxicity of DMA(V) [4]. It was clearly shown using urinalysis, scanning electron microscope, and BrdU pulse administration that changes in urinary composition (i.e., volume, calcium concentration), cytotoxicity of the urothelium, and hyperplasia were observed at dietary concentrations above 2 ppm DMA(V). These findings established that a mode of action of DMA(V)-induced carcinogenesis involved a cytotoxicity followed by regenerative proliferation and hyperplasia.

DMA(V) is a stable chemical and only causes toxicity at a very high concentration; thus, a potentially reactive intermediate may have been the major contributor to urothelial cytotoxicity. To test this hypothesis, Cohen et al. [17] coadministered 2,3-dimercaptopropane-1-sulfonate (DMPS), a scavenger of trivalent arsenicals, with 100 ppm DMA(V) in the rat diets, and demonstrated that cytotoxicity was significantly reduced, providing evidence of the involvement of a trivalent arsenical. Subsequent urinary analysis from the rats expose to DMA(V) showed the presence of MMA(III), DMA(V), and trimethylarsine oxide. With DMPS coadministration, both MMA(III) and trimethylarsine oxide were significantly reduced. As a follow-up, coadministration for 4 weeks of sodium arsenate at 100 ppm in the diet and DMPS yielded significant reduction in cytotoxicity in superficial cells and the incidence of simple hyperplasia [94]. Wang et al. [107] attributed DMA(V) toxicity to dose-dependent damage to cellular organelles with particular emphases on the damage to mitochondria.

Given that much of the published literature on arsenic exposure has been performed with relatively high concentrations, actual arsenic exposure through the food and water and the chronic health issues seen in exposed populations are a result of

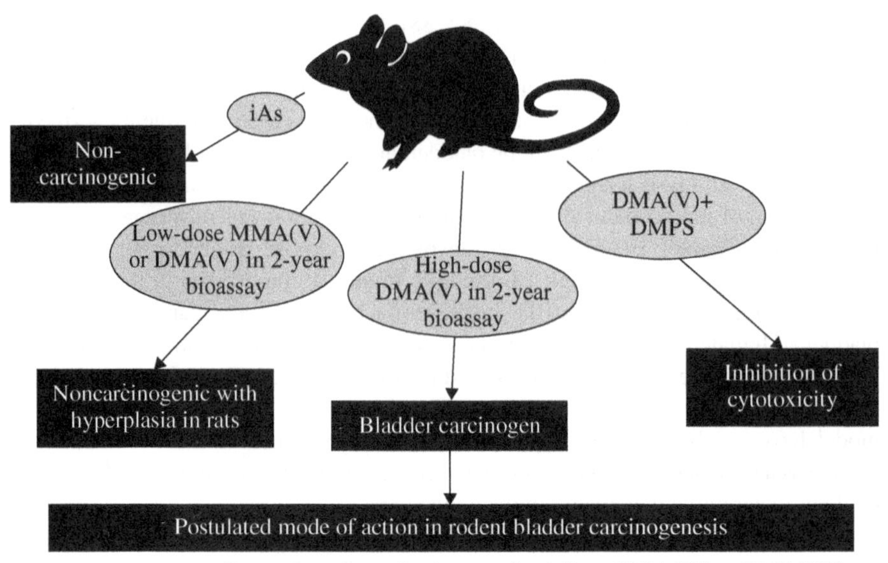

- Formation of reactive intermediates [i.e., MMA(III) or DMA(III)]
- DNA repair inhibition
- Cytotoxicity with regenerative mitogenesis and hyperplasia
- Changes in urine composition

FIGURE 8.3 Rodent model of arsenic-induced carcinogenesis in the bladder. Inorganic arsenic (iAs) administration in the diet has not been found to be carcinogenic. Low-dose administration of MMA(V) and DMA(V) was shown to promote hyperplasia in rat bladders. At higher doses of DMA(V), bladder tumors formed. Coadministration of 2,3-dimercapto-1-propanesulfonic acid (DMPS), a trivalent arsenical scavenger, reduced overall toxicity of DMA(V), suggesting that a reactive intermediate may be playing a role in toxicity and carcinogenesis. Postulated modes of action of DMA(V)-induced carcinogenesis are given.

subtoxic, low-dose exposures. The character and magnitude of responses to low-dose exposures are much different than to high-level toxic doses, with biphasic responses that are indicative of hormesis. Such a biphasic effect was recently demonstrated in mice exposed to 0.5, 2, 10, or 50 mg As/l over 1- or 12-week period [16]. Gene expression changes in the bladders of these mice occurred primarily at the lower concentrations and at the two highest concentrations, compared to the intermediate concentrations used. The time-dependent effects were characterized by a general downregulation of gene expression in the 1-week exposure versus the 12-week exposure. As carcinogenicity studies have failed to show tumor formation in mice exposed to 50 mg·As/L, Clewell et al. [16] have suggested that high concentrations of arsenic preferentially promote apoptosis and that carcinogenicity of arsenic in mice may actually peak at a fivefold lower concentration.

Although useful, the rat model of arsenic-induced carcinogenesis presents some issues. Arsenic methylation varies in different species of animals [100]. New World animals such as the monkey, tamarin, guinea pig, and marmoset do not methylate inorganic arsenic. However, Old World animals such as the hamster, rabbit, and pigeon have arsenic methylation capacity. This difference in metabolic capability can present challenges when extrapolating results from animals to humans, especially in regard to the susceptibility of the urothelium and the toxicokinetics of arsenicals. Humans excrete an appreciable amount of MMA(V) compared to other species with a great deal of interindividual variations. Rats can efficiently methylate arsenic, yet slow urinary excretion occurs since the methylated metabolites bind to hemoglobin. This is not known to occur in other species. Additionally, trimethylarsine oxide is another terminal metabolite of arsenic biotransformation in rats, mice, and to a small extent other species. Generally, trimethylarsine oxide is not formed in humans or only at very low levels. Another issue is that arsenic alone is not sufficient to cause cancer in most mammalian species, which contrasts with the strong association between arsenic exposure and the development of cancer in human populations. With the work by Waalkes et al. (see Chapter 19), the effects of arsenic at critical stages of development in the rodent model suggest that transplacental exposures may have a greater impact on health later in life.

8.4.2 *In Vitro* Models

UROtsa cells are derived from the urothelium of the ureter from a 12-year-old female and immortalized by the introduction of the SV40 large T antigen [76]. This cell line displays normal characteristics of the urothelium, and, although immortalized, the cell line follows normal biological behavior including contact-inhibited growth and nontumorigenic behavior, that is, does not form colonies in soft agar and does not form tumors in immunocompromised mice when injected subcutaneously. Although other urothelial cell lines exist such as the SV-HUC-1 and SW780, none of these lines meet the requirements for long-term carcinogenesis studies [25]. SV-HUC-1 cells are normal human urothelial cells (HUC) immortalized with the SV40 virus. However, SV-HUC-1 loses important normal urothelial characteristics and exhibits abnormalities such as altered growth, irregularly shaped nuclei, and loss of the glycocalyx on the cell surface. Additionally, SV-HUC-1 is able to grow in soft agar, and this phenotype increases with subsequent passages. After 50 passages, SV-HUC-1 obtains the ability to form tumors in athymic mice. SW780 is an established tumor cell line derived from the bladder of an elderly female. It displays characteristics of a transitional cell carcinoma and can also form tumors in athymic mice [56]. Conversely, the UROtsa cell line has been characterized and shown to have similarities between cytoplasmic structures, stratification, and expression level of specific stress response proteins with *in situ* bladder urothelium [83]. Although UROtsa lack uroplakin and are not fully differentiated, the UROtsa cell line has been considered an excellent model for the study of transitional epithelium of the bladder [25]. As such, UROtsa cells provide a valuable model to study arsenic toxicity and chronic exposure to the metabolites.

Cytotoxicity assays of inorganic arsenic and MMA(III) have been performed in UROtsa cells. IC50 for inorganic arsenic was determined to be 100 μM, whereas MMA(III) was 20-fold more toxic [10]. These results support earlier work by Cohen et al. [17] where MMA(III) and DMA(III) were demonstrated to be highly toxic to immortalized rat (MYP3) and human (1T1) urothelial cells and had IC50 values in the micromolar range compared to the pentavalent forms that had IC50 values in the millimolar range. In 24 h exposures of pentavalent arsenicals, up to 20 μM, the viability of UROtsa cells was not affected [92]. The established IC50 values for the various arsenicals have given researchers a guide on the overall level of toxicity of arsenicals and a foundation for deciding which concentrations to use for studying prolonged low-level exposures to urothelial cells.

With epidemiological studies demonstrating the association between chronic low-level arsenic exposure and the incidence of chronic diseases including cancer, the determination of the mechanisms involved in arsenic-induced health effects has been explored using the UROtsa model. Sens et al. [85] first demonstrated the carcinogenic potential of arsenate in UROtsa with a low-level chronic exposure. Initial observation of continuous exposure to 4 or 8 μM As(III) severely reduced the survival of UROtsa cells, whether in serum-free or serum-enriched media. When treated with 1 μM, continuously for a year, UROtsa cells exhibited increased proliferation rate with concomitant ability to form colonies in soft agar, indicating the cells had undergone malignant transformation. With subsequent transplantation in immunocompromised mice, the chronically exposed UROtsa cells developed tumors. The results were the first to demonstrate a direct malignant transformation of urothelial cells after chronic 52-week exposure to arsenate.

Since humans excrete a large percentage of arsenic as MMA(III), the increased toxicity of the trivalent arsenical may play a part in the overall toxicity from arsenic exposure. Therefore, an examination of the potential for MMA(III) to induce transformation was conducted [10]. After a 52-week continuous exposure to 50 nM MMA(III), UROtsa cells underwent a malignant transformation, with characteristics similar to the As(III)-transformed cell line generated by Sens et al. [85]. Isolated tumors in the immunocompromised mice xenotransplanted with MMA(III)-induced malignant cells were highly proliferative and had extensive Ki-67 staining, a marker of cell proliferation. However, harvested tumors exhibited the characteristics of moderately differentiated squamous cell carcinoma, whereas transitional cell carcinoma accounts for 90% of human bladder cancer. In subsequent studies, transformation of UROtsa cells with chronic exposure to 50 nM MMA(III) was possible in as few as 24 weeks [114]. At 24 weeks of exposure, doubling time of the treated cultures fell from 42 to 27 h, in addition to acquiring anchorage-independent growth. With a clear carcinogenesis model established from chronic exposure to low-level arsenicals, mechanisms of initiation, transformation, and progress can now be studied.

The UROtsa cell line was developed by the introduction of a temperature-sensitive SV40 large T antigen construct. The large T antigen is a multifunctional protein, which serves as a DNA helicase, recruiter of the cellular replication apparatus, transcriptional repressor, and transcriptional activator [1]. Large T antigen will prevent p53-dependent gene transcription by binding directly to p53 and blocking the ability of p53 to bind DNA. The negative regulation of cell proliferation is attributed to the tumor suppressor function of p53 [74]. In addition, large T antigen can bind to Rb, which in effect abolishes control of cell cycle progression. Due to the crucial roles of Rb and p53, manipulation of the functions of these proteins by viral oncoproteins allows for the virus to gain control of cell growth regulation to permit viral growth and replication. With previous studies showing that a SV40-transformed cell line, SV-HUC-1, resulted in genetic instability within a few passages, it was believed that the SV40 temperature-sensitive construct would allow modulation of proliferation and generate a model, which would more closely resemble HUC, *in vitro* [68]. However, UROtsa cells invariably grow at permissive and nonpermissive temperatures. The similarity between biological events leading to malignant transformation of a normal urothelial cell *in vitro* and the genetic changes observed in bladder cancer biopsies highlights the utility of the model in studying carcinogenesis [37].

8.4.3 Studies with Primary Cells

As of this writing, there are no published studies utilizing primary urothelial cells in arsenic research. We initiated a study attempting to replicate the MMA(III)-induced transformation of UROtsa cells in a primary urothelial cell culture. However, a 6-month chronic exposure to MMA(III) failed to transform primary bladder epithelial cells (unpublished results). In primary urothelial cells, p53 expression was found to be elevated twofold by MMA(III) exposure or UV radiation. The cytotoxic profiles of As(III) and MMA(III) were determined to be similar to those measured in UROtsa cells. The failure to transform primary urothelial cells may inherently revolve around the stringent mechanisms controlling replicative senescence [90]. A proposed scheme of carcinogenesis suggests that induced immortalization is an essential first step in carcinogenesis and a requirement for complete transformation. Due to the innate mechanisms controlling replicative senescence, it is difficult to induce transformation in primary cells using chemical carcinogens, radiation, or oncogenes. These outcomes suggest that arsenic-induced carcinogenesis may require an initiating event (i.e., p53 abnormality, viral interaction) or a cocarcinogen that compromises antitumor or apoptotic pathways.

8.5 MECHANISMS OF ARSENIC-INDUCED BLADDER CANCER

In 2001, Kitchin proposed several mechanisms by which arsenic can induce carcinogenesis [54]. These proposed mechanisms include the generation of oxidative stress, DNA damage, inhibition of DNA repair processes, alterations in the expression of growth factors, suppression of p53, enhanced cell proliferation, and changes in chromosomal properties. To date, each of these mechanisms has been well characterized in the involvement with arsenic-induced carcinogenesis. However, the potential for the development of bladder cancer through these mechanisms may very well depend on the coordinate actions of each.

8.5.1 Oxidative Stress

The overproduction of reactive oxygen species (ROS) has been implicated in the pathogenesis of a number of diseases including neurological disorders, cardiovascular disease, and cancer [102]. Although ROS can serve a role in normal biological functions, if an imbalance occurs where ROS generation overwhelms cellular antioxidant systems, damage will occur to cellular structures such as membranes, proteins, and nucleic acids [77]. Exposure to arsenic increases intracellular ROS. Direct DNA damage does not occur with arsenic exposure, but indirectly through the formation of DNA adducts like 8-hydroxy-2′-deoxyguanosine (8-OHdG) from, for example, hydroxyl radical formation. Continuous low-level exposure to MMA(III) increases ROS in UROtsa cells [113]. Coupled with the binding of MMA(III) to PARP-1, the combined effects of ROS generation and inhibition of DNA repair components suggest that arsenic is acting on multiple fronts in promoting DNA damage and inducing carcinogenesis in bladder cells [115]. The generated ROS will also alter lipid structures, inhibit enzymes, and decrease antioxidant levels. Regulatory pathways of cell growth such MAPK pathways and apoptosis are also affected by the presence of ROS. The damage that occurs to DNA from oxidative stress is considered a primary event in the initiation of carcinogenesis, coupled with perturbations in cellular signaling.

Investigating the protective effects of antioxidants on rat bladder urothelium, Wei et al. [110] demonstrated reduced cytotoxicity in MYP3 rat bladder epithelial cells when cotreatments of arsenicals and antioxidants *N*-acetylcysteine or vitamin C were performed. However, Takahashi et al. [95] found that coadministration of *N*-acetylcysteine and DMA(V) induced bladder injury in rats but attributed the effect to increased oxidative stress from thiyl radical formation or generated reactive metabolites from the interaction between DMA(V) and *N*-acetylcysteine at the doses used in the study. Increases in ubiquitin-conjugated proteins, a marker for acute arsenic-induced stress, were observed in UROtsa treated with 0.5–25 μM $NaAsO_2$ [9]. Treatment with buthionine sulfoximine, a compound that reduces available GSH, led to a substantial increase in ubiquitinated proteins. This result suggested that the bladder was more susceptible to toxicity with a depletion of antioxidants. Induction of ROS in UROtsa exposed to 1–10 μM $NaAsO_2$ and 50–500 nM MMA(III) was measured by the nonspecific oxidative stress probe DCFDA. Treatments with superoxide dismutase or catalase inhibited the induction of ROS. The level of toxicity of As(III) and MMA(III) in UROtsa could be reduced when Nrf2 activation was induced by sulforaphane and tert-butylhydroquinone [106]. In UROtsa with compromised Nrf2 activity, the sensitivity to arsenicals was increased. In human populations exposed to arsenic, urinary 8-OHdG concentrations have been correlated with total urinary arsenic levels [15, 59]. Given the direct and indirect evidence for ROS generation from arsenic exposure, it is clear that ROS is playing a critical role in the damage of internal cell structures and the promotion of aberrant signaling.

8.5.2 Changes in Intracellular Signaling

Perturbations or dysregulation of key signaling pathways that are involved in growth, proliferation, and survival play an important role in cancer initiation and progression [36]. Since intracellular signaling serves to control the expression of a variety of genes, any defects within the signaling cascade would alter cellular processes and cellular homeostasis. The process of carcinogenesis may be enhanced by aberrant mitogenic signals or a consequence of regeneration following cytotoxic damage.

With exposure to arsenic, multiple signaling pathways are perturbed and include the MAPK pathway, pathways that tie into nuclear factor kappa B (NF-κB) and AP-1 transcription factors, and growth factor receptors (i.e., epidermal growth factor receptor (EGFR)).

Both As(III) and MMA(III) in urothelial cells have been shown to enhance cellular proliferation and activate the MAPK signaling pathway, a major mediator of proliferation, differentiation, and cell death [24]. It has been demonstrated that the cellular hyperproliferation is linked to an activated MAPK signaling from peroxide overproduction. With acute exposure to arsenite, most cell types will show activation of JNK and p38. However, differential activation is observed, since ERK activation has been shown to differ between cell types. Additionally, the MAPK pathway appears to be affected by the dose of arsenic. At low concentrations of trivalent arsenic ($<10\,\mu M$), cell survival and proliferation are enhanced through the PI3K pathway and ERK and p38 [86]. At higher concentrations ($>10\,\mu M$), arsenic stimulates stress-related kinases such as JNK that promote cell death [43]. Short-term studies have examined the activation of EGFR signaling pathway by As(III), MMA(III), and DMA(III) with similar observations in mouse urinary bladder exposed to As(III) [89].

8.5.3 Perturbations in Gene Expression Patterns from Acute and Chronic Arsenic Exposure

Several modes of action have been proposed for carcinogenicity of the organoarsenical metabolites. These metabolites (see Chapter 4) have been shown to induce oxidative stress, enhance or suppress DNA repair activity, alter cell cycle control, activate mitogenic responses, and both induce and suppress apoptosis. These proposed events suggest that dose-dependent transitions in gene expression with concomitant changes in biological adaptation, proliferation, and cell fate are likely responsible for arsenic-induced toxicity and carcinogenicity [33]. A general overview of arsenic research has given some insight in the behavior of cells on a dose–response basis. The earliest events at low-level exposure involve a cascade of biological responses that shifts cell state from an adaptive condition to a proliferative condition. At higher concentrations of arsenic, the induction of genes related to cell proliferation and cellular toxicity occurs, and with even higher concentration, toxicity-related cell death and anticarcinogenic effects govern.

8.5.4 Gene Expression Changes in Urothelial Cells Associated with Arsenic Metabolites

To address more specifically the gene expression changes that occur in human bladder cells, the application of urothelial cell models has provided some insight in the progressive changes that occur with acute and chronic exposure to arsenicals. A summary of the studies to be reviewed is presented in Table 8.1. Drobna et al. [22] showed that $0.1–5\,\mu M$ $NaAsO_2$ increased the expression and activity of AP-1 in UROtsa cells exposed acutely. With acute exposures to micromolar concentrations of

TABLE 8.1 Studies of Arsenic-Induced Changes in Gene Expression Involving Human Tissues or Human Urothelial Cells

Tissue	Arsenical[a]	Patterns of Changes in Gene Expression	Notable Gene(s) or Protein(s) Expression Altered	References
Human studies				
Urine	As	—	↓beta-defensin-1	Hegedus et al. [39]
Bladder tumors	As	Hedgehog signaling activated	↑GLI1	Fei et al. [29]
Urothelial carcinoma	As	—	↑caveolin-1, ↓eNOS, ↑IKK-beta, ↑COX-2	Liu et al. [61]
In vitro studies				
UROtsa	As(III)	Induction of cellular stress and growth arrest genes, mitogenic pathway, altered antiapoptotic and cell adhesion genes	↑GADD45, ↑GADD153, ↑c-fos, ↑EGR-1; total of 13 genes altered at low concentration, additional nine genes at fivefold higher	Simeonova et al. [87]
UROtsa	As(III) Methylarsine oxide Iododimethylarsine	—	↑AP-1	Drobna et al. [22]
SV-HUC-1	As(III) MMA(III) DMA(III)	Distinct patterns of expression for each arsenical tested, suppression of genes common to each arsenical; some effects are epigenetic	↑IL1R2; 11 suppressed genes common to each arsenical; about 5.7% of genes altered from 2000 genes probed after 3 months of exposure	Su et al. [93]
UROtsa	MMA(III)	—	↑COX-2	Eblin et al. [24]
Normal human bladder cells and UROtsa	As(III)	Modest changes in the expression of autophagy proteins	↑beclin-1	Larson et al. [57]

				Chai et al. [12]
UROtsa (transformed)	As(III)	—	↑metallothionein isoform 3	Zhou et al. [119]
UROtsa (transformed)	MMA(III)	—	↑WNT5A	Jensen et al. [49]
UROtsa (transformed)	As(III)	Changes in keratin expression indicative of squamous cell differentiation in some tumor heterotransplants	↑keratin-6	Cao et al. [11]
UROtsa	MMA(III)	Changes occurring early in the exposure period appear to involve inflammatory genes	Roughly 10% of genes altered from 28,000 genes probed after 3 months of exposure	Medeiros et al. [67]
SV-HUC-1	As(III)	—	↑caveolin-1, ↓eNOS, ↑IKK-beta, ↑COX-2	Liu et al. [61]
UROtsa	MMA(III)	Overexpression of inflammatory cytokines and chemokines	↑IL-8	Escudero-Lourdes et al. [28]
				Escudero-Lourdes et al. [27]

[a] Arsenical abbreviations: As, arsenic; As(III), arsenite; DMA(III), dimethylarsinous acid; MMA(III), monomethylarsonous acid.

inorganic arsenic and methylated trivalent arsenicals, there was a significant increase in the nuclear levels of phosphorylated c-jun and DNA binding activity of AP-1. The methylated arsenicals were noted to be more potent inducers, and the induction pathway appeared to proceed through the ERK pathway as opposed to the JNK pathway. To explore the effects of the methylated metabolites further, Eblin et al. measured induction of COX-2 in UROtsa, mediated through the ERK1/2 pathway, after chronic exposure to low-level MMA(III) [24]. Selective inhibitors of the ERK1/2 pathway reduced the colony-forming potential of transformed UROtsa cells. A more recent study by Liu et al. [61] showed elevated levels of eNOS, IKK-beta, and COX-2 downstream of arsenic-induced attenuation of caveolin-1 in SV-HUC-1 cells. These observations were supported by immunocytochemical staining of urothelial carcinomas from patients of blackfoot disease-endemic areas. Overall, these arsenic-induced effects, the activation of the ERK mitogenic pathway, AP-1 transcriptional activity, and COX-2 elevation have been considered as important events in arsenic-induced carcinogenesis.

Simeonova et al. performed a gene expression study on UROtsa cells exposed to 10 or 50 μM $NaAsO_2$ [89].Of the 588 genes probed, 13 genes were differentially expressed when cells were exposed at the lower concentration of arsenite, and an additional 9 were differentially expressed at the higher concentration. A consistent induction of stress response genes (i.e., growth arrest and DNA damage-inducible genes [GADD45, GADD153], early growth response protein 1 [EGR1], and cyclin-dependent kinase inhibitor 1A [CDKN1A, p21/WAF1]) was observed. Additionally, increased expression of c-fos and c-jun occurred within 1 h of arsenite treatment, followed by a rapid decline. These results are consistent with the induction of the AP-1 complex and proliferative effects from mitogens (i.e., epidermal growth factor [EGF]) through the overexpression of EGR1.

Utilizing the SV-HUC-1 cell line, Su et al. showed that a 3-month chronic exposure to the trivalent metabolites of arsenic (As(III), MMA(III), DMA(III)) induced distinct gene expression profiles [93]. Using a cDNA microarray analysis of 2000 genes, about 5.7% of the genes were differentially expressed when compared to passage-matched controls. Unlike the UROtsa, the long-term exposure to these metabolites was not sufficient to transform the cell. However, significant morphological changes were observed. Gene ontology analysis indicated that a large group of differentially expressed genes regulate cell growth and maintenance processes, cell death, motility, and transport. The second group included genes involved with metabolism, protein phosphorylation, RNA processing, proteolysis, and transcriptional regulation. Common to each of the metabolites used was the enhanced expression of IL1R2. Eleven genes were suppressed by all three arsenicals, and these genes are involved with the regulation of integrins, transmembrane receptors, and signal transduction in the IP3 and NF-κB pathways. As arsenic is known to induce oxidative stress, the study did find changes in heme oxygenase-1, SOD1, and heat shock proteins. Medeiros et al., focusing mainly on the effects of low-level MMA(III) and MMA(III)-induced transformation of UROtsa, performed a comprehensive gene expression study [67]. This analysis revealed only minor changes in gene expression at 1 and 2 months of exposure, contrasting with substantial changes observed at

3 months of exposure when the cells undergo malignant transformation. The gene expression changes at 3 months were analyzed showing distinct alterations in biological processes and pathways such as response to oxidative stress, enhanced cell proliferation, antiapoptosis, MAPK signaling, as well as inflammation. Although the substantial changes that occur at 3 months of exposure may be a consequence of the transformation, there are common occurrences of altered biological processes between the first 2 months of exposure and the third, which may be pivotal in driving transformation. These changes complement the findings of Su et al. [93] and, based on gene ontology analysis, include alterations in transcriptional activity, changes in the regulation of protein stability, suppression of immune-related genes, and alterations in proinflammatory genes just prior to the transformation.

The role of inflammation in the processes of malignant transformation in UROtsa has been explored. Chronic inflammation has been linked with cancer development mainly because of a variety of proinflammatory cytokines, growth factors, and angiogenic chemokines that are associated with tumors [8, 60, 81]. Escudero-Lourdes et al. found early indications of the overproduction of proinflammatory cytokines when UROtsa cells were exposed to low-level MMA(III) for as short as 12 h [27]. These changes persisted around the third month of exposure with elevated expression of interleukins IL-1, IL-6, and IL-8, coinciding with NF-κB and c-Jun activation, and activation of autocrine chemokine receptors (CXCR1, CXCR2). IL-8 gene silencing in UROtsa cells exposed for 3 months to low-level MMA(III) attenuated proliferation rate and colony formation in soft agar [27]. These studies suggested that a sustained response to arsenic exposure includes the activation of inflammatory mediators, which may play a critical role in fostering malignant transformation.

Studies of gene expression changes in human bladder tissue within human populations exposed to arsenic are limited. Only one study is known regarding gene expression changes of the urinary cells/fluid acquired from exposed human populations. While examining the composition of urine specimens from exposed individuals of Nevada and Chile, Hegedus et al. found that beta-defensin-1 (HBD-1), a constitutively expressed peptide, was reduced in the arsenic-exposed populations [39]. HBD-1 is an antimicrobial peptide expressed in urothelial and respiratory tissues. Recent studies have suggested that HBD-1 may serve as a tumor suppressor. Additionally, as HBD-1 is involved with innate immunity, this result would support an immune-suppressing or immune-toxicant effect of arsenic. The relationship of immune suppression from toxicant exposures and the increased risk of infectious and/or neoplastic disease were proposed by Selgrade [84], although no specific link has been demonstrated between arsenic-induced immune suppression and susceptibility of tumorigenesis in rodents. However, Andres [2] proposed that the damage of the immune system would impair the response to transformed cells, and this assertion would have implications in arsenic-exposed human populations. Earliest changes observed with MMA(III) exposure in the UROtsa prior to malignant transformation, although not well supported by other studies with urothelial cells, were the alterations of immune and inflammatory responses characterized by a general suppression of proinflammatory cytokines (i.e., interleukins, STATs, TNF) and chemokine transcripts [67].

8.5.5 Epigenetic Effects of Arsenicals in Urothelial Cells

Arsenic-induced epigenetic modifications have been observed in a number of studies with human cell lines and tumor specimens from human populations. Studies by Marsit et al. [64, 65] and Chen et al. [14] on urothelial and bladder cancer specimens have revealed DNA hypermethylation in the promoters of genes such as Ras association (RalGDS/AF-6) domain family member 1 (RASSF1a), protease, serine, 3 (PRSS3), and death-associated protein kinase (DAPK) 1. These specimens were acquired from arsenic-exposed populations from the United States, Taiwan, China, and India. The specificity of the epigenetic modification examined in these studies could prove to be a key molecular event contributing to the malignant phenotype of cancer arising in individuals from arsenic-contaminated environments. This observation also suggests that arsenic toxicity and induced carcinogenesis may be tied to epigenetic remodeling as opposed to genetic damage.

The epigenetic effects of arsenicals have been studied in the UROtsa cell line. The models of arsenic-induced malignant transformation developed by Sens et al. [85] and Bredfelt et al. [10] were characterized by irreversible changes even after a withdrawal of the arsenic insult. Jensen et al. [47] profiled the acetylation state of lysine-9 and lysine-14 of histone H3 in the As(III)-induced and MMA(III)-induced tumorigenic phenotypes. It was shown that a majority of significant promoter alterations were nonrandomly hypoacetylated histones, whether the exposure was to As(III) or MMA(III). Functional consequences of these histone alterations appeared to result in a decrease in the expression of associated genes DBC1 (now named "cell cycle and apoptosis regulator 2" (CCAR2)), family with sequence similarity 83, member A (FAM83A), zinc finger- and SCAN domain-containing 12 (ZSCAN12), and C1q and tumor necrosis factor-related protein 6 (C1QTNF6). Patterns of increased DNA methylation in these promoter regions corresponded to decreased histone acetylation. Prevalent DNA hypermethylation was observed in select promoter regions that increased in a time-dependent manner through malignant transformation [48]. DNA hypermethylation was also evident in potential tumor suppressors such as deleted in bladder cancer chromosome region candidate 1 (DBCCR1) (now known as bone morphogenetic protein/retinoic acid-inducible neural-specific 1 (BRINP1)). DBCCR1 has commonly been observed as deleted or silenced in human bladder cancers [74]. Epigenetic activation of WNT5A, a glycoprotein found elevated in a number of cancers including bladder cancer, was correlated with permissive histone modifications and likely plays a role in arsenic-induced transformation [49]. Furthermore, the histone modifications were stable in the malignant phenotype even after the arsenicals were withdrawn. Such studies link chronic arsenical exposure to aberrant epigenetic remodeling that is characterized by a prevalent genome-wide hypomethylation coupled with hypermethylation in specific promoters (Fig. 8.4).

8.6 CONCLUSIONS

Epidemiological studies have established an association between arsenic exposure and the development of bladder cancer. Studies in arsenic-endemic areas such Taiwan, Chile, and Argentina have led the IARC to list arsenic as an environmental

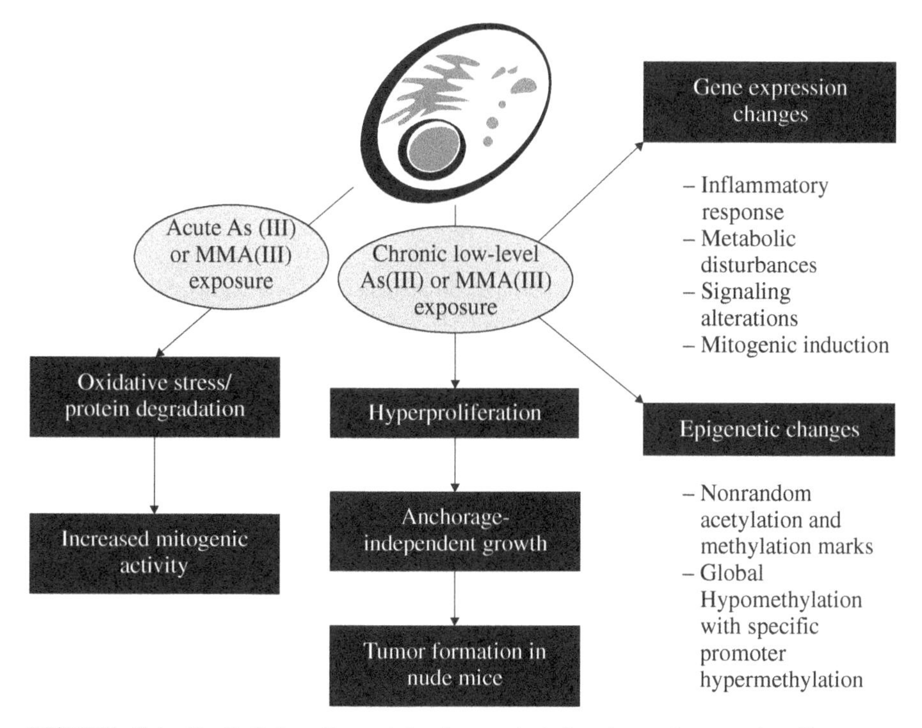

FIGURE 8.4 Urothelial cell model of arsenic-induced carcinogenesis. Short-term exposures to arsenite [As(III)] or monomethylarsonous acid [MMA(III)] are characterized by oxidative stress and increased cell proliferation. With chronic low-level exposures to As(III) or MMA(III), the cell dynamics change and are unbalanced toward carcinogenesis. Gene expression changes are noted as well as concomitant epigenetic alterations.

contaminant and human carcinogen. The susceptibility of individuals is possibly dependent on the biotransformation capacity and the proportion of arsenicals in circulation. The arsenic biotransformation process, which serves to improve arsenic removal, can generate reactive intermediates. As the bladder's physiological property is to store and remove concentrated urine and other metabolites, it is clear that the bladder is a susceptible organ to arsenic-induced toxicity and carcinogenesis.

To determine the biological effects of and mechanisms involved with arsenic exposure in humans, *in vivo* and *in vitro* models have been developed. Despite differences in the response of animals to arsenic when compared to humans, rodent models have revealed that arsenic promotes its effects through the biotransformation-generated metabolites and exposures *in utero* may impact the health of adults. Methylated trivalent arsenicals have now been implicated in the process of cytotoxicity and carcinogenesis. With *in vitro* models such as the HUC line, UROtsa, researchers have now found that arsenical toxicity can vary with oxidation state and methylation, arsenic impacts the intracellular signaling of bladder cells by promoting proliferation and growth, arsenic can change the redox state of the cells, and arsenical interactions can lead to alterations in normal gene expression patterns with concomitant epigenetic changes (Fig. 8.4).

ACKNOWLEDGMENTS

The authors and portions of the work described were supported by NIH grants ES04940, ES07091, and ES06694.

REFERENCES

[1] D. Ahuja, M.T. Saenz-Robles, J.M. Pipas, SV40 large T antigen targets multiple cellular pathways to elicit cellular transformation. Oncogene 24 (2005) 7729–7745.

[2] A. Andres, Cancer incidence after immunosuppressive treatment following kidney transplantation. Crit. Rev. Oncol. Hematol. 56 (2005) 71–85.

[3] H.V. Aposhian, E.S. Gurzau, X.C. Le, A. Gurzau, S.M. Healy, X. Lu, M. Ma, L. Yip, R.A. Zakharyan, R.M. Maiorino, R.C. Dart, M.G. Tircus, D. Gonzalez-Ramirez, D.L. Morgan, D. Avram, M.M. Aposhian, Occurrence of monomethylarsonous acid in urine of humans exposed to inorganic arsenic. Chem. Res. Toxicol. 13 (2000) 693–697.

[4] L.L. Arnold, M. Cano, J.M. St, M. Eldan, M. van Germert, S.M. Cohen, Effects of dietary dimethylarsinic acid on the urine and urothelium of rats. Carcinogenesis 20 (1999) 2171–2179.

[5] L.L. Arnold, M. Eldan, M. van Germert, C.C. Capen, S.M. Cohen, Chronic studies evaluating the carcinogenicity of monomethylarsonic acid in rats and mice. Toxicology 190 (2003) 197–219.

[6] L.L. Arnold, M. Eldan, A. Nyska, M. van Germert, S.M. Cohen, Dimethylarsinic acid: results of chronic toxicity/oncogenicity studies in F344 rats and in B6C3F1 mice. Toxicology 223 (2006) 82–100.

[7] R. Baffa, J. Letko, C. McClung, J. LeNoir, A. Vecchione, L.G. Gomella, Molecular genetics of bladder cancer: targets for diagnosis and therapy. J. Exp. Clin. Cancer Res. 25 (2006) 145–160.

[8] F. Balkwill, A. Mantovani, Inflammation and cancer: back to Virchow? Lancet 357 (2001) 539–545.

[9] T.G. Bredfeldt, M.J. Kopplin, A.J. Gandolfi, Effects of arsenite on UROtsa cells: low-level arsenite causes accumulation of ubiquitinated proteins that is enhanced by reduction in cellular glutathione levels. Toxicol. Appl. Pharmacol. 198 (2004) 412–418.

[10] T.G. Bredfeldt, B. Jagadish, K.E. Eblin, E.A. Mash, A.J. Gandolfi, Monomethylarsonous acid induces transformation of human bladder cells. Toxicol. Appl. Pharmacol. 216 (2006) 69–79.

[11] L. Cao, X.D. Zhou, M.A. Sens, S.H. Garrett, Y. Zheng, J.R. Dunlevy, D.A. Sens, S. Somji, Keratin 6 expression correlates to areas of squamous differentiation in multiple independent isolates of As(+3)-induced bladder cancer. J. Appl. Toxicol. 30 (2010) 416–430.

[12] C.Y. Chai, Y.C. Huang, W.C. Hung, W.Y. Kang, W.T. Chen, Arsenic salts induced autophagic cell death and hypermethylation of DAPK promoter in SV-40 immortalized human uroepithelial cells. Toxicol. Lett. 173 (2007) 48–56.

[13] Y.C. Chen, H.J. Su, Y.L. Guo, Y.M. Hsueh, T.J. Smith, L.M. Ryan, M.S. Lee, D.C. Christiani, Arsenic methylation and bladder cancer risk in Taiwan. Cancer Causes Control 14 (2003) 303–310.

[14] W.T. Chen, W.C. Hung, W.Y. Kang, Y.C. Huang, C.Y. Chai, Urothelial carcinomas arising in arsenic-contaminated areas are associated with hypermethylation of the gene promoter of the death-associated protein kinase. Histopathology 51 (2007) 785–792.

[15] C.J. Chung, C.J. Huang, Y.S. Pu, C.T. Su, Y.K. Huang, Y.T. Chen, Y.M. Hsueh, Urinary 8-hydroxydeoxyguanosine and urothelial carcinoma risk in low arsenic exposure area. Toxicol. Appl. Pharmacol. 226 (2008) 14–21.

[16] H.J. Clewell, R.S. Thomas, E.M. Kenyon, M.F. Hughes, B.M. Adair, P.R. Gentry, J.W. Yager, Concentration- and time-dependent genomic changes in the mouse urinary bladder following exposure to arsenate in drinking water for up to 12 weeks. Toxicol. Sci. 123 (2011) 421–432.

[17] S.M. Cohen, L.L. Arnold, E. Uzvolgyi, M. Cano, J.M. St, S. Yamamoto, X. Lu, X.C. Le, Possible role of dimethylarsinous acid in dimethylarsinic acid-induced urothelial toxicity and regeneration in the rat. Chem. Res. Toxicol. 15 (2002) 1150–1157.

[18] S.M. Cohen, L.L. Arnold, M. Eldan, A.S. Lewis, B.D. Beck, Methylated arsenicals: the implications of metabolism and carcinogenicity studies in rodents to human risk assessment. Crit. Rev. Toxicol. 36 (2006) 99–133.

[19] S.M. Cohen, T. Ohnishi, L.L. Arnold, X.C. Le, Arsenic-induced bladder cancer in an animal model. Toxicol. Appl. Pharmacol. 222 (2007) 258–263.

[20] M.A. Dawson, T. Kouzarides, Cancer epigenetics: from mechanism to therapy. Cell 150 (2012) 12–27.

[21] R.M. Drayton, J.W. Catto, Molecular mechanisms of cisplatin resistance in bladder cancer. Expert. Rev. Anticancer Ther. 12 (2012) 271–281.

[22] Z. Drobna, I. Jaspers, D.J. Thomas, M. Styblo, Differential activation of AP-1 in human bladder epithelial cells by inorganic and methylated arsenicals. FASEB J. 17 (2003) 67–69.

[23] E. Dudziec, J.R. Goepel, J.W. Catto, Global epigenetic profiling in bladder cancer. Epigenomics 3 (2011) 35–45.

[24] K.E. Eblin, T.G. Bredfeldt, S. Buffington, A.J. Gandolfi, Mitogenic signal transduction caused by monomethylarsonous acid in human bladder cells: role in arsenic-induced carcinogenesis. Toxicol. Sci. 95 (2007) 321–330.

[25] K.E. Eblin, T.G. Bredfeldt, A.J. Gandolfi, Immortalized human urothelial cells as a model of arsenic-induced bladder cancer. Toxicology 248 (2008) 67–76.

[26] N. Erill, A. Colomer, M. Verdu, R. Roman, E. Condom, N. Hannaoui, J.M. Banus, C. Cordon-Cardo, X. Puig, Genetic and immunophenotype analyses of TP53 in bladder cancer: TP53 alterations are associated with tumor progression. Diagn. Mol. Pathol. 13 (2004) 217–223.

[27] C. Escudero-Lourdes, M.K. Medeiros, M.C. Cardenas-Gonzalez, S.M. Wnek, J.A. Gandolfi, Low level exposure to monomethyl arsonous acid-induced the over-production of inflammation-related cytokines and the activation of cell signals associated with tumor progression in a urothelial cell model. Toxicol. Appl. Pharmacol. 244 (2010) 162–173.

[28] C. Escudero-Lourdes, T. Wu, J.M. Camarillo, A.J. Gandolfi, Interleukin-8 (IL-8) over-production and autocrine cell activation are key factors in monomethylarsonous acid [MMA(III)]-induced malignant transformation of urothelial cells. Toxicol. Appl. Pharmacol. 258 (2012) 10–18.

[29] D.L. Fei, H. Li, C.D. Kozul, K.E. Black, S. Singh, J.A. Gosse, J. DiRenzo, K.A. Martin, B. Wang, J.W. Hamilton, M.R. Karagas, D.J. Robbins, Activation of Hedgehog signaling by the environmental toxicant arsenic may contribute to the etiology of arsenic-induced tumors. Cancer Res. 70 (2010) 1981–1988.

[30] M.I. Fernandez, J.F. Lopez, B. Vivaldi, F. Coz, Long-term impact of arsenic in drinking water on bladder cancer health care and mortality rates 20 years after end of exposure. J. Urol. 187 (2012) 856–861.

[31] W.A. Freed-Pastor, C. Prives, Mutant p53: one name, many proteins. Genes Dev. 26 (2012) 1268–1286.

[32] M.G. Friedrich, S. Chandrasoma, K.D. Siegmund, D.J. Weisenberger, J.C. Cheng, M.I. Toma, H. Huland, P.A. Jones, G. Liang, Prognostic relevance of methylation markers in patients with non-muscle invasive bladder carcinoma. Eur. J. Cancer 41 (2005) 2769–2778.

[33] P.R. Gentry, T.B. McDonald, D.E. Sullivan, A.M. Shipp, J.W. Yager, H.J. Clewell, III, Analysis of genomic dose-response information on arsenic to inform key events in a mode of action for carcinogenicity. Environ. Mol. Mutagen. 51 (2010) 1–14.

[34] H.R. Guo, H.S. Chiang, H. Hu, S.R. Lipsitz, R.R. Monson, Arsenic in drinking water and incidence of urinary cancers. Epidemiology 8 (1997) 545–550.

[35] T. Habuchi, M. Luscombe, P.A. Elder, M.A. Knowles, Structure and methylation-based silencing of a gene (DBCCR1) within a candidate bladder cancer tumor suppressor region at 9q32-q33. Genomics 48 (1998) 277–288.

[36] D. Hanahan, R.A. Weinberg, Hallmarks of cancer: the next generation. Cell 144 (2011) 646–674.

[37] F. He, L. Mo, X.Y. Zheng, C. Hu, H. Lepor, E.Y. Lee, T.T. Sun, X.R. Wu, Deficiency of pRb family proteins and p53 in invasive urothelial tumorigenesis. Cancer Res. 69 (2009) 9413–9421.

[38] S.M. Healy, E.A. Casarez, F. Ayala-Fierro, H. Aposhian, Enzymatic methylation of arsenic compounds. V. Arsenite methyltransferase activity in tissues of mice. Toxicol. Appl. Pharmacol. 148 (1998) 65–70.

[39] C.M. Hegedus, C.F. Skibola, M. Warner, D.R. Skibola, D. Alexander, S. Lim, N.L. Dangleben, L. Zhang, M. Clark, R.M. Pfeiffer, C. Steinmaus, A.H. Smith, M.T. Smith, L.E. Moore, Decreased urinary beta-defensin-1 expression as a biomarker of response to arsenic. Toxicol. Sci. 106 (2008) 74–82.

[40] A.M. Hoffman, P. Cairns, Epigenetics of kidney cancer and bladder cancer. Epigenomics 3 (2011) 19–34.

[41] C. Hopenhayn-Rich, M.L. Biggs, A. Fuchs, R. Bergoglio, E.E. Tello, H. Nicolli, A.H. Smith, Bladder cancer mortality associated with arsenic in drinking water in Argentina. Epidemiology 7 (1996) 117–124.

[42] L.I. Hsu, W.P. Chen, T.Y. Yang, Y.H. Chen, W.C. Lo, Y.H. Wang, Y.T. Liao, Y.M. Hsueh, H.Y. Chiou, M.M. Wu, C.J. Chen, Genetic polymorphisms in glutathione S-transferase (GST) superfamily and risk of arsenic-induced urothelial carcinoma in residents of southwestern Taiwan. J. Biomed. Sci. 18 (2011) 51.

[43] C. Huang, W.Y. Ma, J. Li, A. Goranson, Z. Dong, Requirement of Erk, but not JNK, for arsenite-induced cell transformation. J. Biol. Chem. 274 (1999) 14595–14601.

[44] M.F. Hughes, E.M. Kenyon, B.C. Edwards, C.T. Mitchell, L.M. Razo, D.J. Thomas, Accumulation and metabolism of arsenic in mice after repeated oral administration of arsenate. Toxicol. Appl. Pharmacol. 191 (2003) 202–210.

[45] Z. Jablonowski, E. Reszka, J. Gromadzinska, W. Wasowicz, M. Sosnowski, Hypermethylation of p16 and DAPK promoter gene regions in patients with non-invasive urinary bladder cancer. Arch. Med. Sci. 7 (2011) 512–516.

[46] A. Jemal, R. Siegel, E. Ward, Y. Hao, J. Xu, M.J. Thun, Cancer statistics, 2009. CA Cancer J. Clin. 59 (2009) 225–249.

[47] T.J. Jensen, P. Novak, K.E. Eblin, A.J. Gandolfi, B.W. Futscher, Epigenetic remodeling during arsenical-induced malignant transformation. Carcinogenesis 29 (2008) 1500–1508.

[48] T.J. Jensen, P. Novak, S.M. Wnek, A.J. Gandolfi, B.W. Futscher, Arsenicals produce stable progressive changes in DNA methylation patterns that are linked to malignant transformation of immortalized urothelial cells. Toxicol. Appl. Pharmacol. 241 (2009) 221–229.

[49] T.J. Jensen, R.J. Wozniak, K.E. Eblin, S.M. Wnek, A.J. Gandolfi, B.W. Futscher, Epigenetic mediated transcriptional activation of WNT5A participates in arsenical-associated malignant transformation. Toxicol. Appl. Pharmacol. 235 (2009) 39–46.

[50] K. Jomova, Z. Jenisova, M. Feszterova, S. Baros, J. Liska, D. Hudecova, C.J. Rhodes, M. Valko, Arsenic: toxicity, oxidative stress and human disease. J. Appl. Toxicol. 31 (2011) 95–107.

[51] W.K. Kelly, V.M. Richon, O. O'Connor, T. Curley, B. MacGregor-Curtelli, W. Tong, M. Klang, L. Schwartz, S. Richardson, E. Rosa, M. Drobnjak, C. Cordon-Cordo, J.H. Chiao, R. Rifkind, P.A. Marks, H. Scher, Phase I clinical trial of histone deacetylase inhibitor: suberoylanilide hydroxamic acid administered intravenously. Clin. Cancer Res. 9 (2003) 3578–3588.

[52] E.M. Kenyon, M.F. Hughes, B.M. Adair, J.H. Highfill, E.A. Crecelius, H.J. Clewell, J.W. Yager, Tissue distribution and urinary excretion of inorganic arsenic and its methylated metabolites in C57BL6 mice following subchronic exposure to arsenate in drinking water. Toxicol. Appl. Pharmacol. 232 (2008) 448–455.

[53] W.J. Kim, Y.J. Kim, Epigenetic biomarkers in urothelial bladder cancer. Expert. Rev. Mol. Diagn. 9 (2009) 259–269.

[54] K.T. Kitchin, Recent advances in arsenic carcinogenesis: modes of action, animal model systems, and methylated arsenic metabolites. Toxicol. Appl. Pharmacol. 172 (2001) 249–261.

[55] P. Kurttio, E. Pukkala, H. Kahelin, A. Auvinen, J. Pekkanen, Arsenic concentrations in well water and risk of bladder and kidney cancer in Finland. Environ. Health Perspect. 107 (1999) 705–710.

[56] A.A. Kyriazis, A.P. Kyriazis, W.B. McCombs, III, W.D. Peterson, Jr., Morphological, biological, and biochemical characteristics of human bladder transitional cell carcinomas grown in tissue culture and in nude mice. Cancer Res. 44 (1984) 3997–4005.

[57] J.L. Larson, S. Somji, X.D. Zhou, M.A. Sens, S.H. Garrett, D.A. Sens, J.R. Dunlevy, Beclin-1 expression in normal bladder and in Cd2+ and As3+ exposed and transformed human urothelial cells (UROtsa). Toxicol. Lett. 195 (2010) 15–22.

[58] S. Letasiova, A. Medve'ova, A. Sovcikova, M. Dusinska, K. Volkovova, C. Mosoiu, A. Bartonova, Bladder cancer, a review of the environmental risk factors. Environ. Health 11 Suppl 1 (2012) S11.

[59] X. Li, J. Pi, B. Li, Y. Xu, Y. Jin, G. Sun, Urinary arsenic speciation and its correlation with 8-OHdG in Chinese residents exposed to arsenic through coal burning. Bull. Environ. Contam. Toxicol. 81 (2008) 406–411.

[60] W.W. Lin, M. Karin, A cytokine-mediated link between innate immunity, inflammation, and cancer. J. Clin. Invest 117 (2007) 1175–1183.

[61] X.P. Liu, Y.C. Huang, W.C. Hung, W.T. Chen, H.S. Yu, C.Y. Chai, Sodium arsenite-induced abnormalities in expressions of Caveolin-1, eNOS, IKKbeta, and COX-2 in SV-40 immortalized human uroepithelial cells and in urothelial carcinomas. Toxicol. In Vitro 26 (2012) 1098–1105.

[62] M.I. Luster, P.P. Simeonova, Arsenic and urinary bladder cell proliferation. Toxicol. Appl. Pharmacol. 198 (2004) 419–423.

[63] N. Marchiset-Ferlay, C. Savanovitch, M.P. Sauvant-Rochat, What is the best biomarker to assess arsenic exposure via drinking water? Environ. Int. 39 (2012) 150–171.

[64] C.J. Marsit, M.R. Karagas, H. Danaee, M. Liu, A. Andrew, A. Schned, H.H. Nelson, K.T. Kelsey, Carcinogen exposure and gene promoter hypermethylation in bladder cancer. Carcinogenesis 27 (2006) 112–116.

[65] C.J. Marsit, M.R. Karagas, A. Schned, K.T. Kelsey, Carcinogen exposure and epigenetic silencing in bladder cancer. Ann. N. Y. Acad. Sci. 1076 (2006) 810–821.

[66] D.J. McConkey, S. Lee, W. Choi, M. Tran, T. Majewski, S. Lee, A. Siefker-Radtke, C. Dinney, B. Czerniak, Molecular genetics of bladder cancer: emerging mechanisms of tumor initiation and progression. Urol. Oncol. 28 (2010) 429–440.

[67] M. Medeiros, X. Zheng, P. Novak, S.M. Wnek, V. Chyan, C. Escudero-Lourdes, A.J. Gandolfi, Global gene expression changes in human urothelial cells exposed to low-level monomethylarsonous acid. Toxicology 291 (2012) 102–112.

[68] L.F. Meisner, S.Q. Wu, B.J. Christian, C.A. Reznikoff, Cytogenetic instability with balanced chromosome changes in an SV40 transformed human uroepithelial cell line. Cancer Res. 48 (1988) 3215–3220.

[69] J.R. Meliker, R.L. Wahl, L.L. Cameron, J.O. Nriagu, Arsenic in drinking water and cerebrovascular disease, diabetes mellitus, and kidney disease in Michigan: a standardized mortality ratio analysis. Environ. Health 6 (2007) 4.

[70] A.P. Mitra, R.H. Datar, R.J. Cote, Molecular pathways in invasive bladder cancer: new insights into mechanisms, progression, and target identification. J. Clin. Oncol. 24 (2006) 5552–5564.

[71] A.P. Mitra, C.C. Bartsch, R.J. Cote, Strategies for molecular expression profiling in bladder cancer. Cancer Metastasis Rev. 28 (2009) 317–326.

[72] L.E. Moore, A.H. Smith, C. Hopenhayn-Rich, M.L. Biggs, D.A. Kalman, M.T. Smith, Micronuclei in exfoliated bladder cells among individuals chronically exposed to arsenic in drinking water. Cancer Epidemiol. Biomarkers Prev. 6 (1997) 31–36.

[73] S.K. Nayak, P.S. Panesar, H. Kumar, Non-genotoxic p53-activators and their significance as antitumor therapy of future. Curr. Med. Chem. 18 (2011) 1038–1049.

[74] H. Nishiyama, J.H. Gill, E. Pitt, W. Kennedy, M.A. Knowles, Negative regulation of G(1)/S transition by the candidate bladder tumour suppressor gene DBCCR1. Oncogene 20 (2001) 2956–2964.

[75] S.E. Patton, M.C. Hall, H. Ozen, Bladder cancer. Curr. Opin. Oncol. 14 (2002) 265–272.

[76] J.L. Petzoldt, I.M. Leigh, P.G. Duffy, C. Sexton, J.R. Masters, Immortalisation of human urothelial cells. Urol. Res. 23 (1995) 377–380.

[77] G. Poli, G. Leonarduzzi, F. Biasi, E. Chiarpotto, Oxidative stress and cell signalling. Curr. Med. Chem. 11 (2004) 1163–1182.

[78] I. Proctor, K. Stoeber, G.H. Williams, Biomarkers in bladder cancer. Histopathology 57 (2010) 1–13.

[79] K.N. Qureshi, J. Lunec, D.E. Neal, Molecular biological changes in bladder cancer. Cancer Surv. 31 (1998) 77–97.

[80] M.M. Rahman, J.C. Ng, R. Naidu, Chronic exposure of arsenic via drinking water and its adverse health impacts on humans. Environ. Geochem. Health 31 Suppl 1 (2009) 189–200.

[81] J. Rodriguez-Vita, T. Lawrence, The resolution of inflammation and cancer. Cytokine Growth Factor Rev. 21 (2010) 61–65.

[82] P.H. Roos, N. Jakubowski, Methods for the discovery of low-abundance biomarkers for urinary bladder cancer in biological fluids. Bioanalysis 2 (2010) 295–309.

[83] M.R. Rossi, J.R. Masters, S. Park, J.H. Todd, S.H. Garrett, M.A. Sens, S. Somji, J. Nath, D.A. Sens, The immortalized UROtsa cell line as a potential cell culture model of human urothelium. Environ. Health Perspect. 109 (2001) 801–808.

[84] M.K. Selgrade, Use of immunotoxicity data in health risk assessments: uncertainties and research to improve the process. Toxicology 133 (1999) 59–72.

[85] D.A. Sens, S. Park, V. Gurel, M.A. Sens, S.H. Garrett, S. Somji, Inorganic cadmium- and arsenite-induced malignant transformation of human bladder urothelial cells. Toxicol. Sci. 79 (2004) 56–63.

[86] P.P. Simeonova, M.I. Luster, Arsenic carcinogenicity: relevance of c-Src activation. Mol. Cell Biochem. 234–235 (2002) 277–282.

[87] P.P. Simeonova, S. Wang, W. Toriuma, V. Kommineni, J. Matheson, N. Unimye, F. Kayama, D. Harki, M. Ding, V. Vallyathan, M.I. Luster, Arsenic mediates cell proliferation and gene expression in the bladder epithelium: association with activating protein-1 transactivation. Cancer Res. 60 (2000) 3445–3453.

[88] P.P. Simeonova, S. Wang, M.L. Kashon, C. Kommineni, E. Crecelius, M.I. Luster, Quantitative relationship between arsenic exposure and AP-1 activity in mouse urinary bladder epithelium. Toxicol. Sci. 60 (2001) 279–284.

[89] P.P. Simeonova, S. Wang, T. Hulderman, M.I. Luster, c-Src-dependent activation of the epidermal growth factor receptor and mitogen-activated protein kinase pathway by arsenic. Role in carcinogenesis. J. Biol. Chem. 277 (2002) 2945–2950.

[90] M.R. Stampfer, P. Yaswen, Human epithelial cell immortalization as a step in carcinogenesis. Cancer Lett. 194 (2003) 199–208.

[91] C. Steinmaus, L.E. Moore, M. Shipp, D. Kalman, O.A. Rey, M.L. Biggs, C. Hopenhayn, M.N. Bates, S. Zheng, J.K. Wiencke, A.H. Smith, Genetic polymorphisms in MTHFR 677 and 1298, GSTM1 and T1, and metabolism of arsenic. J. Toxicol. Environ. Health A 70 (2007) 159–170.

[92] M. Styblo, L.M. Del Razo, L. Vega, D.R. Germolec, E.L. LeCluyse, G.A. Hamilton, W. Reed, C. Wang, W.R. Cullen, D.J. Thomas, Comparative toxicity of trivalent and pentavalent inorganic and methylated arsenicals in rat and human cells. Arch. Toxicol. 74 (2000) 289–299.

[93] P.F. Su, Y.J. Hu, I.C. Ho, Y.M. Cheng, T.C. Lee, Distinct gene expression profiles in immortalized human urothelial cells exposed to inorganic arsenite and its methylated trivalent metabolites. Environ. Health Perspect. 114 (2006) 394–403.

[94] S. Suzuki, L.L. Arnold, T. Ohnishi, S.M. Cohen, Effects of inorganic arsenic on the rat and mouse urinary bladder. Toxicol. Sci. 106 (2008) 350–363.

[95] N. Takahashi, T. Yoshida, A. Ohnuma, H. Horiuchi, K. Ishitsuka, Y. Kashimoto, M. Kuwahara, N. Nakashima, T. Harada, The enhancing effect of the antioxidant N-acetylcysteine on urinary bladder injury induced by dimethylarsinic acid. Toxicol. Pathol. 39 (2011) 1107–1114.

[96] N. Tanji, A. Ozawa, T. Kikugawa, N. Miura, T. Sasaki, K. Azuma, M. Yokoyama, Potential of histone deacetylase inhibitors for bladder cancer treatment. Expert. Rev. Anticancer Ther. 11 (2011) 959–965.

[97] C.H. Tseng, Arsenic methylation, urinary arsenic metabolites and human diseases: current perspective. J. Environ. Sci. Health C. Environ. Carcinog. Ecotoxicol. Rev. 25 (2007) 1–22.

[98] T. Tsuda, A. Babazono, E. Yamamoto, N. Kurumatani, Y. Mino, T. Ogawa, Y. Kishi, H. Aoyama, Ingested arsenic and internal cancer: a historical cohort study followed for 33 years. Am. J. Epidemiol. 141 (1995) 198–209.

[99] T. Uchida, C. Wada, H. Ishida, C. Wang, S. Egawa, E. Yokoyama, T. Kameya, K. Koshiba, p53 mutations and prognosis in bladder tumors. J. Urol. 153 (1995) 1097–1104.

[100] M. Vahter, Methylation of inorganic arsenic in different mammalian species and population groups. Sci. Prog. 82 (Pt 1) (1999) 69–88.

[101] M. Vahter, Genetic polymorphism in the biotransformation of inorganic arsenic and its role in toxic. Toxicol. Lett. 112–113 (2000) 209–217.

[102] M. Valko, D. Leibfritz, J. Moncol, M.T. Cronin, M. Mazur, J. Telser, Free radicals and antioxidants in normal physiological functions and human disease. Int. J. Biochem. Cell Biol. 39 (2007) 44–84.

[103] D. Volanis, T. Kadiyska, A. Galanis, D. Delakas, S. Logotheti, V. Zoumpourlis, Environmental factors and genetic susceptibility promote urinary bladder cancer. Toxicol. Lett. 193 (2010) 131–137.

[104] M.P. Waalkes, J.M. Ward, J. Liu, B.A. Diwan, Transplacental carcinogenicity of inorganic arsenic in the drinking water: induction of hepatic, ovarian, pulmonary, and adrenal tumors in mice. Toxicol. Appl. Pharmacol. 186 (2003) 7–17.

[105] M.P. Waalkes, J. Liu, J.M. Ward, B.A. Diwan, Enhanced urinary bladder and liver carcinogenesis in male CD1 mice exposed to transplacental inorganic arsenic and postnatal diethylstilbestrol or tamoxifen. Toxicol. Appl. Pharmacol. 215 (2006) 295–305.

[106] X.J. Wang, Z. Sun, W. Chen, K.E. Eblin, J.A. Gandolfi, D.D. Zhang, Nrf2 protects human bladder urothelial cells from arsenite and monomethylarsonous acid toxicity. Toxicol. Appl. Pharmacol. 225(2) (2007) 206–213

[107] A. Wang, D.C. Wolf, B. Sen, G.W. Knapp, S.D. Holladay, W.R. Huckle, T. Caceci, J.L. Robertson, Dimethylarsinic acid in drinking water changed the morphology of urinary bladder but not the expression of DNA repair genes of bladder transitional epithelium in F344 rats. Toxicol. Pathol. 37 (2009) 425–437.

[108] M.L. Warner, L.E. Moore, M.T. Smith, D.A. Kalman, E. Fanning, A.H. Smith, Increased micronuclei in exfoliated bladder cells of individuals who chronically ingest arsenic-contaminated water in Nevada. Cancer Epidemiol. Biomarkers Prev. 3 (1994) 583–590.

[109] M. Wei, H. Wanibuchi, S. Yamamoto, W. Li, S. Fukushima, Urinary bladder carcinogenicity of dimethylarsinic acid in male F344 rats. Carcinogenesis 20 (1999) 1873–1876.

[110] M. Wei, L. Arnold, M. Cano, S.M. Cohen, Effects of co-administration of antioxidants and arsenicals on the rat urinary bladder epithelium. Toxicol. Sci. 83 (2005) 237–245.

[111] H. Wei, A.M. Kamat, S. Aldousari, Y. Ye, M. Huang, C.P. Dinney, X. Wu, Genetic variations in the transforming growth factor beta pathway as predictors of bladder cancer risk. PLoS One 7 (2012) e51758.

[112] C.S. Wilhelm-Benartzi, D.C. Koestler, E.A. Houseman, B.C. Christensen, J.K. Wiencke, A.R. Schned, M.R. Karagas, K.T. Kelsey, C.J. Marsit, DNA methylation profiles delineate etiologic heterogeneity and clinically important subgroups of bladder cancer. Carcinogenesis 31 (2010) 1972–1976.

[113] S.M. Wnek, M.K. Medeiros, K.E. Eblin, A.J. Gandolfi, Persistence of DNA damage following exposure of human bladder cells to chronic monomethylarsonous acid. Toxicol. Appl. Pharmacol. 241 (2009) 202–209.

[114] S.M. Wnek, T.J. Jensen, P.L. Severson, B.W. Futscher, A.J. Gandolfi, Monomethylarsonous acid produces irreversible events resulting in malignant transformation of a human bladder cell line following 12 weeks of low-level exposure. Toxicol. Sci. 116 (2010) 44–57.

[115] S.M. Wnek, C.L. Kuhlman, J.M. Camarillo, M.K. Medeiros, K.J. Liu, S.S. Lau, A.J. Gandolfi, Interdependent genotoxic mechanisms of monomethylarsonous acid: role of ROS-induced DNA damage and poly(ADP-ribose) polymerase-1 inhibition in the malignant transformation of urothelial cells. Toxicol. Appl. Pharmacol. 257 (2011) 1–13.

[116] M.M. Wu, T.L. Kuo, Y.H. Hwang, C.J. Chen, Dose-response relation between arsenic concentration in well water and mortality from cancers and vascular diseases. Am. J. Epidemiol. 130 (1989) 1123–1132.

[117] S. Yamamoto, Y. Konishi, T. Matsuda, T. Murai, M.A. Shibata, I. Matsui-Yuasa, S. Otani, K. Kuroda, G. Endo, S. Fukushima, Cancer induction by an organic arsenic compound, dimethylarsinic acid (cacodylic acid), in F344/DuCrj rats after pretreatment with five carcinogens. Cancer Res. 55 (1995) 1271–1276.

[118] H. Zhang, Z.H. Chen, T.M. Savarese, Codeletion of the genes for p16INK4, methylthioadenosine phosphorylase, interferon-alpha1, interferon-beta1, and other 9p21 markers in human malignant cell lines. Cancer Genet. Cytogenet. 86 (1996) 22–28.

[119] X.D. Zhou, M.A. Sens, S.H. Garrett, S. Somji, S. Park, V. Gurel, D.A. Sens, Enhanced expression of metallothionein isoform 3 protein in tumor heterotransplants derived from As+3- and Cd+2-transformed human urothelial cells. Toxicol. Sci. 93 (2006) 322–330.

9

NEUROLOGICAL EFFECTS OF ARSENIC EXPOSURE

DOMINIC B. FEE
Department of Neurology, Medical College of Wisconsin in Milwaukee, WI, USA

9.1 INTRODUCTION

Arsenic is a ubiquitous chemical, occurring naturally and due to industrial activity, both in obtaining/processing raw materials and in final products. Arsenic exposure and toxicity (arsenic causing a clinical deficit regardless of severity) is not uniform. Because of these issues, arsenic is underrecognized as the cause of medical problems, especially in areas of low environmental levels and in situations of chronic, low-dose exposure.

As described in other chapters, arsenic is a metalloid that can occur in inorganic and organic forms. In general, organic arsenic exposure is less toxic and, in the United States, more common. Organic arsenic exposures typically occur through ingestion of water-based organisms, for example, shellfish such as shrimp, scallops, crab, and lobster, but can also occur through ingestion other foodstuffs, for example, meat from animals fed arsenic-containing feed [1, 76, 77, 127, 134, 148, 151, 158]. Inorganic exposure occurs both naturally, for example, present in groundwater and some foodstuffs such as rice and some vegetables and fruits, and due to various industrial activities and products, for example, mining, cosmetics, herbal supplements, contaminated recreational drugs, etc. [1, 24, 27, 38, 76, 77, 95, 127, 134, 138, 142, 148, 151, 158, 170].

Arsenic: Exposure Sources, Health Risks, and Mechanisms of Toxicity, First Edition.
Edited by J. Christopher States.
© 2016 John Wiley & Sons, Inc. Published 2016 by John Wiley & Sons, Inc.

Arsenic toxicity is dependent on factors beyond whether it is in organic or inorganic forms [1, 77, 127, 134, 142, 158]. Arsenic has different valence states; however, clinically, the two most relevant forms are trivalent (3+) arsenite and pentavalent (5+) arsenate [76, 127, 151, 158]. Inorganic arsenite is 60 times more toxic than inorganic arsenate; arsenate is reduced to arsenite soon after entering the body [76, 127, 151, 158]. Arsenic is typically eliminated by excretion predominately in urine but also in bile; some inorganic arsenic is methylated after absorption (see Chapter 4), with methylated arsenite potentially more toxic than inorganic arsenite [76, 142, 151, 158]. Route of exposure, oral (most common) versus inhalation (e.g., breathing in smoke from burning arsenic ("CCA") treated wood) versus transcutaneous also influence toxicity [1, 76]. Arsenate compounds are more water soluble and, henceforth, are better absorbed through mucosal membranes; arsenite compounds are more lipophilic [142]. Another significant factor in the clinical presentation of arsenic toxicity is age of the individual exposed. Adults, children, and fetuses will have different clinical manifestations to the same exposure [35, 158, 161]. Toxicity is related to amount, duration, and type of arsenic exposure and is influenced by one's own biochemistry; individuals with the same exposure can have different clinical presentations [3, 15, 25, 35, 76, 77, 102, 144].

Although arsenic toxicity is multifactorial, clinical assessment is limited. Taking a patient's history gives clues to duration and amount of exposure but rarely describes the actual exposure [91]. Specific testing for arsenic is limited to total levels that can be subdivided into organic and inorganic components, usually on special request only. Valence levels are not reported. Urine and blood levels are most commonly tested. If the testing is done remote from the exposure, it may be falsely negative; sending hair or nails for assessment minimizes this. A false-positive situation can occur in endemic areas of arsenic exposure; just because elevated levels are present, it does not prove that arsenic is the cause of the symptoms.

9.2 MEDICAL OVERVIEW

To diagnose arsenic-induced damage to the body, one must have a high index of suspicion. There is no uniform presentation. Amount of exposure, duration of exposure, and age of the patient are all variables that influence symptoms at presentation [76, 91, 142]. Again, a group of individuals with a similar exposure can present with different symptoms; additionally, different species and different ages process arsenic differently [3, 15, 35, 158, 161].

As indicated previously, there are multiple variables contributing to arsenic toxicity. The main variables are amount of exposure and duration of exposure [91, 127, 142]. Typically, patients are seen with "high-dose" acute/short-term exposure and "low-dose" chronic/long-term exposure [127]. There is significant biologic variability between species but also within species on how arsenic is metabolized [158, 161]. "High dose" chronic is typically fatal; "low dose" short term may only produce minimal and/or transient symptoms that are not clinically relevant. Some individuals with elevated arsenic levels are asymptomatic [76]. This does not mean that the person may not develop symptoms later, but highlights the biological variability present. A certain amount of damage needs to occur over a certain time frame for symptoms to be apparent.

History is the key to all medical diagnoses. Frequently, in arsenic-induced disorders, the exposure is not obvious [91, 142]. Accidental exposure/toxicity is the most common etiology; hence, since accidental, it may not be recognized [127]. In cases of chronic exposure/toxicity, since occurring over a long period of time, nothing different is appreciated [142]. Similarly, in acute toxicity, the source may not be deemed unusual by the patient, but it may be a change from a normal behavior, for example, inhalation of smoke from burning arsenic-treated lumber in a bonfire, using rat poison purchased decades ago for a new infestation, and ingesting a newly purchased dietary supplement contaminated with arsenic.

To further complicate making a diagnosis of arsenic-induced nervous system injury, no symptoms are specific solely to arsenic. In areas of low environmental levels of arsenic, the symptoms can be caused by many etiologies more common than arsenic [47, 91]. The body has a limited number of ways that it can show that it is injured. When a patient presents with a symptom or set of symptoms, a differential diagnosis is created. A differential diagnosis is essentially a list of what could be injured to explain the symptoms and what could be causing that injury. Unless there is an obvious exposure reported when the patient presents to medical attention, arsenic as the cause will not be investigated until more common etiologies are excluded, if it is even assessed [142]. On physical exam, Mees' lines, transverse white bands in the nail beds, suggest arsenic exposure, but this is not always present and can take weeks to occur after acute exposure [35, 40, 43, 48, 91, 99, 108, 169].

Even if arsenic is assessed in situations in which it is causing the symptoms, proof of arsenic toxicity may not be found. Since the onset of many neurologic disorders due to arsenic is delayed by weeks from exposure, the assessment for arsenic in urine or blood may not show it. For example, inorganic arsenic has a half life of 1–2 h in blood; 40–70% of arsenic is absorbed, metabolized, and excreted within 48 h of exposure in humans with a residual fraction still present a week or two later [25, 77, 91]. Thus, lack of elevated arsenic in urine or blood may show a false-negative result, not because of a problem with the sensitivity of the test but because the arsenic is "out of the system." Assessing for arsenic incorporated into hair or nails helps to minimize this potential problem [19, 40, 41, 91, 142]. Conversely, in many parts of the world, high arsenic levels occur in water and foods or due to occupational exposure. Just because elevated arsenic levels are present in the patient or an exposure has occurred, it does not "prove" causality for the patient's symptoms; a false-positive test result can be present [41, 47, 150]. Another form of false positivity occurs when testing is done soon after the patient eats shellfish high in organic arsenic. Having the laboratory differentiate whether a high level is due to inorganic or organic forms (not always done routinely) helps assess for this type of false-positive result as well. This is a typical cause of a high arsenic level in the United States, and if further history is taken from the patient, a history of eating water-based food known to be high in organic arsenic, typically shellfish, prior to the testing will be found.

Treatment in the acute setting is geared toward restoring bodily functioning; typically, this involves correcting dehydration. Gastric lavage and activated charcoal help remove arsenic remaining in the gastrointestinal tract [142]. For arsenic absorbed, chelation therapy, such as British anti-Lewisite (BAL, dimercaprol), *D*-penicillamine, 2,3-dimercapto-1-propanesulfonate (DMPS), or dimercaptosuccinic acid (DMSA), is used; hemodialysis can also be tried [142, 158, 167]. Even with early treatment, neurologic recovery can be very delayed, and permanent, severe deficits may still result [35, 108].

9.3 NEUROLOGIC OVERVIEW

There are many ways to classify neurologic problems. The main is by anatomic localization. The central nervous system (CNS) is the brain and spinal cord. The peripheral nervous system (PNS) is the nerves that are outside of the CNS; these are the nerves that course throughout the body. This is an artificial distinction because many neurons have projections in both. However, there are differences in supporting cells and milieu, for example, the brain and spinal cord float in a filtrate of the blood called the cerebrospinal fluid (CSF), oligodendrocytes versus Schwann cells for myelin, etc., that validates this distinction. There are many diseases such as multiple sclerosis that only affect one but not the other. Arsenic can affect both. Most reported cases though are of either CNS or PNS involvement, but both can occur simultaneously.

Classification schemes used in toxicity include rate of exposure (acute vs. chronic), age of the patient, and if the damage is directly due to the agent or is secondary to other damage/dysfunction induced by the agent. Classification schemes help us to understand and explain. Nature does not care about them; there will always be exceptions or issues that do not fit. Frequently, classification schemes will be superimposed. Arsenic has been used for millennia; its toxicity has been studied for nearly as long and by different scientists with different approaches and agendas. As explained later, arsenic toxicity has different presentations and different mechanisms of injury; there is no simple explanation for arsenic toxicity.

9.3.1 Indirect Involvement of the Nervous System

Frequently, derangements in one organ system will cause secondary dysfunction in the nervous system. It is joked that the role of the body is to provide for the brain. This is not to minimize the importance of other organs; the body is highly interrelated. But it recognizes that a very strict microenvironment must be maintained for the brain, and to a slightly lesser degree the rest of the nervous system, to function. Weakness is an example of a symptom occurring because of the interdependence of the body and nervous system. Certainly, all of the issues mentioned under indirect and direct effects of arsenic on the nervous system can cause/contribute to weakness; however, in the setting of acute arsenic toxicity, weakness is most likely due to the dehydration resultant from the severe diarrhea.

9.3.1.1 CNS Encephalopathy, also termed delirium, is due to poorly coordinated or disrupted brain activity. Affected individuals are not appropriately awake, alert, attentive, or interactive. This is also described as altered or clouded consciousness. The layperson would define it as inappropriately sleepy, being confused, or not thinking properly. Infections and medications/drugs are the most common causes in general, and certainly, these can be present in patients with arsenic toxicity. However, in arsenic toxicity, arsenic-induced liver and/or kidney damage with resultant buildup of waste products is more likely. For example, in a case of arsenic exposure in a ship's cargo hold with subsequent toxicity, the encephalopathy seen in some patients was due to hemolysis with concomitant anoxia and renal failure, not due to the arsenic toxicity itself [169]. Encephalopathy is very common in the hospitalized patient, and usually, there is more than one potential etiology present.

9.3.1.2 PNS Polyneuropathy is a condition in which something is damaging the peripheral nerves. It has many synonyms with the most common being peripheral nerve disease, peripheral neuropathy, and neuropathy usually with the etiology preceding it, for example, diabetic neuropathy. Polyneuropathy symptoms are a combination of numbness and tingling, pain frequently burning in character, weakness, and/or coordination problems. Axonal length is a significant risk factor for a nerve cell to be damaged; therefore, there is a length-dependent aspect to polyneuropathy. The symptoms typically start in the feet and progress up the leg. Once the symptoms reach mid-calf or below the knees, the hands start developing symptoms—termed stocking–glove distribution.

Polyneuropathy symptoms are heavily influenced by duration and severity of the disease process [91]. A hyperacute, very severe process may result in symptoms present throughout the body within hours to days of onset. A chronic, mildly severe process may have minimal sensory symptoms, and the presenting complaint frequently is balance or walking problems. Subacute and acute processes of mild to moderate severity typically will have a combination of sensory and motor symptoms, but the sensory, that is, numbness, tingling, and burning pain, usually predominate.

Many conditions can cause polyneuropathy; arsenic can do it directly as detailed in the next section. However, just being sick can cause it—termed neuropathy of critical illness or of chronic illness. Arsenic toxicity can cause gastrointestinal, liver, and kidney damage that can subsequently cause polyneuropathy, for example, nutritional deficiency polyneuropathy and uremic polyneuropathy [91, 158]. These would be unlikely in the acute setting. However, these and other conditions caused by arsenic, such as diabetes mellitus in chronic exposure, could cause a subacute or chronic polyneuropathy. Coexisting conditions make the diagnosis of polyneuropathy due solely to arsenic difficult in any toxicity setting [41]. Prior records and testing should be reviewed to see if polyneuropathy was present prior to the arsenic exposure.

9.3.2 Direct Involvement of the Nervous System

9.3.2.1 CNS Experimentally, in rodents, both inorganic arsenite and arsenate and the organic metabolites cross the immature fetal and mature adult blood–brain barrier and get into the brain with increased brain arsenic concentration as exposure/toxicity increases [7, 31, 44, 52, 65, 73, 78, 92, 94, 98, 103, 113, 122, 130, 133, 135, 140, 145, 152, 159, 163, 164, 175, 183, 184]. At autopsy of humans who died from arsenic toxicity, arsenic was detected in the brain with, in one of the cases, arsenite and its methylated species prominent [10, 57]. Arsenite and arsenate exposure *in utero* has been shown to cause neural tube deficits and neuronal migrational abnormalities [42, 71, 105, 161]. Arsenite has been shown to alter Purkinje cell (cerebellar neuron) migration, delay their maturation, and result in them having increased size [34]. Arsenic (III) methyltransferase (AS3MT) and methylated metabolites of arsenite are found throughout the adult rodent brain suggesting that arsenite is metabolized in brain tissues [52, 132, 140, 175]. Ingestion of organic arsenate and inorganic arsenite can lead to detectable levels of arsenite in the CSF; arsenic also accumulates in the choroid plexus, site of CSF production, potentially in a "protective" manner by resulting in lower arsenic levels in the CSF versus blood [4, 142, 184]. Arsenite instilled into the lateral ventricle, that is, increased CSF arsenic concentration, results in CNS-mediated effects including induced hyperthermia thought to be mediated by arsenic effects on the hypothalamus [141]. Others have also shown that arsenite affects the hypothalamic–pituitary axis and alters hormonal expression [31, 74, 97, 98]. Arsenic is also detectable in rodent and human spinal cord after exposure and toxicity [52, 57].

Age at time of exposure is a significant factor when discussing arsenic affecting the brain. Arsenic exposure during brain development may end up being its biggest public health problem; however, this has not been completely settled. Arsenic exposure early in life has been associated with cognitive deficits, usually reported as lower IQ scores, and behavioral abnormalities in school age [63]. The studies typically are cross sectional, assessing current arsenic levels in urine, hair, nearby well water, etc., with current cognitive ability, usually IQ score or performance on neuropsychological testing, in toddlers and adolescents. These studies are severely limited by not having an assessment of arsenic exposure early in life/throughout life nor being able to take into account confounding variables, for example, nutrition over time. Most are able to demonstrate an association [16, 101, 121, 136, 153, 160, 162, 165, 166, 172]. However, not all are able to show an association or not with all parameters assessed [137, 160, 165, 166]. Again, there are significant limitations to this study design, and, therefore, results can be both overemphasized or inappropriately dismissed. A similar association of arsenic and poor cognitive ability in a cross-sectional study has been seen in adults who would have been exposed to arsenic throughout their life [55, 115]. Ishii *et al.* [72] reported adult and infant cerebellar and brain stem signs and symptoms in patients whose water supply was contaminated by organic arsenite. Cross-sectional study in children and toddlers has also shown an association with arsenic and behavioral changes, for example, oppositional, attention deficit, and hyperactivity [137]. Similar to cognitive ability studies, not all variables assessed had an association [137]. In rodents, exposure of pups *in utero* and adults to arsenite causes behavioral changes, locomotor deficits, and learning difficulties [7, 20, 21, 26, 73, 94, 96–98, 110, 124, 130, 131, 135, 143, 161, 164, 176, 177]. Similar behavioral/locomotor/learning abnormalities were seen in zebrafish exposed to arsenate [5, 29].

Arsenic can directly affect the brain typically causing an encephalopathy, which usually resolves over time, but permanent brain damage can occur [3, 8, 13, 43, 48, 49, 67, 72, 106, 125, 134]. One example of an arsenic-induced encephalopathy is reactive arsenical encephalopathy, which occurs due to an arsenic-containing treatment for trypanosomiasis, "sleeping sickness" [2, 68, 134]. These cases are associated with signs and symptoms of brain injury, for example, seizures and stroke. However, given that trypanosomiasis can directly affect the brain, it is difficult to state which symptoms are due solely to arsenic, but arsenic-induced changes to cerebral blood vessels appear to be present [2, 68]. Chronic arsenic exposure is felt to be a risk factor for stroke, but not all studies demonstrate an association [23, 150, 173]. Experimentally, arsenite and organic arsenate alter expression of vasoactive compounds in the rodent brain [114, 129, 183].

In a case of arsenic-induced encephalopathy due to industrial exposure, neuropsychological testing revealed impaired concentration, new learning, and short-term memory, which improved as his condition improved [106]. In another case of encephalopathy due to industrial exposure, repeat neuropsychological testing over the patient's recovery revealed deficits most prominent in verbal learning and verbal memory with no significant deficits in visual memory, general intelligence, or language [13]. O'Bryant *et al.* [115] did extensive neuropsychological testing in a cross-sectional study of arsenic exposure and cognitive ability in adults [115]. Acute exposure seemed associated with impairment in visuospatial skills, language, and executive function. Chronic, low-dose exposure was associated with impairments in global cognition, language, executive functioning, and memory [115]. In another study of chronic arsenic exposure, the arsenic-exposed group had deficits in a wide range of neurobehavioral tests [55].

Arsenic exposure and toxicity causes other CNS signs and symptoms. Seizures have been associated with arsenic toxicity [2, 3, 80, 128]. Long-term follow-up to the 1955 arsenic-tainted powdered milk outbreak in Japan revealed increased mortality in the exposed individuals, especially due to neurologic causes with epilepsy being the main cause of this [150]. Psychiatric symptoms, mainly depression and anxiety, nonspecific mood changes, sleep disturbances/somnolence, and headaches, have also been reported [13, 14, 48–50, 99, 106, 107, 125, 149]. Arsenate exposure in utero in rodents leads to adult-onset depression-like behaviors [97]. Arsenic toxicity also has been associated with vision loss via optic neuropathy and hearing loss [9, 142].

9.3.2.2 PNS Experimentally, arsenic can be found in rodent and porcine peripheral nerves, but was not seen in human peripheral nerves to a significant degree at autopsy; potentially, the discrepancy could be due to the time from exposure to assessment of the tissue [52, 54, 57, 89, 156]. Polyneuropathy is the most recognized and reported neurologic problem associated with arsenic toxicity in the neurologic literature. In the acute setting, its presentation mimics Guillain–Barre syndrome (GBS); therefore, it is important to remind providers frequently to investigate for arsenic toxicity in the setting of GBS that does not present or respond as anticipated [35, 40, 60, 75]. Typically, it is not diagnosed until a situation of proposed recurrent GBS occurs.

The polyneuropathy associated with acute arsenic toxicity can present days to weeks after exposure [19, 35, 61, 81, 88, 99, 108, 169]. It has a distal-to-proximal spread of sensory symptoms and weakness. In all aspects, it can mimic GBS; there is significant overlap in clinical presentation as well as test results such as CSF and nerve conduction studies/electromyography (NCS/EMG) [35, 40, 60, 64]. GBS is an autoimmune condition in which the body's immune system attacks the myelin sheath (Schwann cells) that surrounds the axons of the peripheral nerves. The initial damage though seems to be at the level of the nerve roots, the part of the peripheral nerve as they enter/exit the spinal cord. GBS is frequently a consequence of a gastrointestinal illness due to *Campylobacter jejuni* but can be due to other bacterial and viral infections.

Incidence of arsenic-induced polyneuropathy is difficult to determine. In areas of endemic arsenic exposure, cross-sectional studies assessing arsenic levels in urine, hair, nails, or drinking water have shown an association with subclinical neuropathy [66, 69, 90, 101, 119, 120, 154, 155]. However, not every study shows an association [51, 83]. Again, there are significant limitations in this study design. Rahman *et al.* [125] assessed patients from different villages in West Bengal, an area of high endemic arsenic exposure, for clinical neuropathy [107, 125]. To be assessed for neuropathy, patients had to have arsenic-induced skin lesions and have no conditions known to cause neuropathy. One group had an incidence rate for clinical neuropathy of 37%; the other was 87%. In both groups, the findings were predominately sensory and mild. To highlight the difficulties in determining true incidence, the cases in the second group (87% incidence) were associated with a well that had elevated, but not remarkedly so, arsenic levels and occurred despite not intaking arsenic-contaminated water for 5 months prior to the study. The first group was felt to have chronic exposure, the second subacute. NCS were abnormal in about a third of the patients with clinical neuropathy [107, 125]. Gerr *et al.* [55] assessed individuals exposed to arsenic-contaminated dust in a town in Georgia, United States. Of the exposed group, 15.3% had clinical polyneuropathy, whereas only 3.4% of the unexposed group had it. Severity of exposure was not associated with increased incidence nor was there significant difference between the groups via NCS. Patients with known causes of polyneuropathy were excluded [55]. Feldman *et al.* [41] examined arsenic exposure in workers in a smelting factory in which arsenic trioxide was a by-product. Forty-two percent of high exposure and 39% of low exposure had clinical polyneuropathy versus 12% of controls. Arsenic levels in tissue (nails and hair) correlated with these findings. Individuals with other potential causes of polyneuropathy were excluded. Although there was a definite trend of worsening NCS in the exposed group and a couple of values were statistically significant, overall, NCS differences between control and exposed groups were not clinically relevant [41]. Assessment of Swedish smelter factory workers had essentially the same NCS findings as in Feldman *et al.* [41]; since clinical examination was similar between exposed and control groups, it was felt that arsenic predisposed to a subclinical polyneuropathy [12, 87]. In a clinical trial of arsenic trioxide for treatment of acute promyelocytic leukemia, 42.5% developed clinical neuropathy [149].

Individuals can have symptoms persisting for years after a single exposure [61, 81, 88, 108]. In polyneuropathy due to chronic arsenic exposure, there usually is incomplete recovery [125]. As previously emphasized, even with early treatment, neurologic recovery can be delayed, and permanent, severe deficits may still result [35, 43, 108].

9.4 PATHOLOGIC MECHANISMS

9.4.1 Clinical Assessments

9.4.1.1 Electrodiagnostic Testing NCS/EMG are two studies, both of which assess the PNS. They are done at the same visit and complement each other. Simplistically, NCS involve shocking a nerve and measuring the response at a distance from the stimulus [39, 91, 168]. The speed of the conduction, that is, distal latency and conduction velocity, correlates with myelin integrity. Size of the response, that is, amplitude, correlates with axonal integrity. NCS preferentially assess large, myelinated fibers. NCS are most useful for polyneuropathy and focal nerve damage. Depending on placement of the electrodes or stimulator, sensory and motor aspects of peripheral nerves can be assessed. EMG involves placing a needle electrode into the muscles [28, 91]. The muscle is assessed at rest for spontaneous firing of individual muscle fibers and for spontaneous firing of individual axons. The muscle is also tested during voluntary activity, looking for reinnervation changes and altered firing patterns. EMG preferentially assesses nerve root damage, that is, radiculopathy, and primary muscle damage, that is, myopathy.

Typically, in both acute and chronic arsenic toxicity, the NCS findings are of a mixed involvement, both sensory and motor aspects of the peripheral nerves involving both demyelinating and axonal features; the sensory conduction studies appear to be more significantly affected than the motor conduction studies [6, 35, 40, 43, 57, 60, 64, 88, 99, 100, 108, 116, 117]. EMG findings are consistent with the distal-to-proximal axonal changes seen in polyneuropathy, same sites. Obviously, some reported NCS/EMG in arsenic toxicity will have a different pattern. Many articles reporting acute arsenic exposures emphasize the early demyelinating features to highlight the similarity to GBS, but axonal changes are also present.

For single-dose toxicity, the NCS/EMG changes usually are noticed a week or so after the exposure, typically worsen over a few weeks to months, and then improve [35, 57, 61, 88, 108, 116, 117]. The NCS/EMG may be unremarkable for days after an exposure because enough Wallerian degeneration will not have occurred. During chronic exposures and multiple acute exposures, the NCS/EMG worsen overtime; axonal changes predominate [35, 60, 64]. In both acute and chronic exposures, improvement in the NCS/EMG is delayed from when the exposure ceases but may never normalize [43, 60, 88, 108].

Another set of electrodiagnostic testing are evoked potentials (EVPs). Simplistically, EVPs are a recording of the time from providing a sensory stimulus and it reaching and/or being processed through the CNS, for example, with visual EVPs, measuring the time from providing an alternating checkerboard pattern visual stimulus and recording when the occipital lobe initially processes this input. When assessed in arsenic exposure/toxicity, the EVPs, visual EVPs, brain stem auditory EVPs, and somatosensory EVPs are either unremarkable or appropriate for the patients' symptoms [43, 125].

9.4.1.2 Biopsies At times, biopsies of peripheral nerves are performed to confirm/exclude diagnoses. Typically, the sural nerve is biopsied; it is a sensory-only nerve that runs on the lateral aspect of the ankle. Those reported in the literature were done weeks to months after symptom onset and similarly remote from arsenic exposure/toxicity [35, 61, 88, 100, 117, 118]. Large, myelinated fibers are preferentially involved, similar to GBS. However, the changes seen in GBS that indicate an active autoimmune response, that is, segmental demyelination and inflammatory cell infiltrates, are not present. The changes are that of degenerating axons and their associated myelin; this degeneration represents distal changes of proximal damage, that is, Wallerian degeneration. Gherardi *et al.* [57] reported a superficial peroneal biopsy; the superficial peroneal nerve is a sensory-only nerve on the anterior aspect of the ankle. It had similar findings to those reported for the sural nerve. In an experimental model of arsenic toxicity, the rodent sural nerve had histologic findings similar to those reported for human biopsies—axonal loss, decrease in axonal size of myelinated fibers, and thin myelin sheath [54].

Muscle biopsies have shown no direct muscle damage, but there are changes indicating that the nerves supplying the muscles are damaged [57, 61]. At autopsy of a patient that died from arsenic toxicity, arsenic levels were "normal" in the muscles [57].

9.4.1.3 Autopsy/Necropsy Gherardi *et al.* reported a case of a lady with trypanosomiasis, "sleeping sickness," who died from progressive weakness due to arsenic toxicity 2 months after initiation of a treatment containing arsenic, melarsoprol [57]. It does not appear to be a fatal case of reactive arsenical encephalopathy; there was no gross autopsy evidence for trypanosomiasis involvement in the CNS. The main neuropathological abnormality was severe alterations of anterior horn cells; the findings included cytoplasmic vacuolation, central chromatolysis, and cellular degeneration with poorly reactive astrocytes. There were axoplasmic accumulations of neurofilaments; it was implied to be present in the anterior horn cells, but this was not definitively stated, nor was it definitively stated that these were not present in other neurons. The dorsal root ganglion had evidence for sensory neuronal loss. Arsenic levels were elevated in the spinal cord (highest for the nervous system) and brain as well as other nonneurologic tissues; arsenic levels were "normal" in muscle and undetectable in distal peripheral nerves. This pattern is a typical of changes reported in autopsies from arsenic toxicity causing weakness reported in the non-English literature or performed over 40 years ago in the English literature [35, 57].

Experimentally, in rodents exposed to arsenite, there were changes in anterior horn cells, cervical more so than lumbar, without vacuolation, chromatolysis, or degenerative changes; this pattern may have resulted because the arsenic exposure in the study was nonfatal [33]. In pigs given toxic dosages of an organic arsenic compound, 3-nitro-4-hydroxyphenylarsonic acid (roxarsone), there was vacuolation of the myelin and axonal degeneration of the white matter of the spinal cord but with minimal changes in peripheral, cranial, and optic nerves [80]. Unfortunately, anterior horn cell involvement was not definitively stated, but the authors felt that the changes seen were at the level of the nerve fibers with myelin changes being initial. In another study of toxicity in pigs exposed to 4-aminophenylarsonic acid (arsanilic acid), severe changes in myelinated peripheral nerves and the optic nerve/tract were seen, no significant degenerative changes were seen in the anterior horn of the spinal cord nor in the dorsal root ganglion [89]. The initial damage was felt to be at the level of the myelin. The differences reported in this paragraph may be due to how the pathophysiology of cellular injury occurs—organic arsenate versus inorganic arsenite in the rodent studies versus organic arsenite in the autopsy study by Gherardi *et al.* [57]. With arsenite, there is evidence that it may exert its toxicity at the level of the perikaryon, that is, neuronal cell body or very proximal neuronal process such as the axonal hillock, of the anterior horn cell. The two organic arsenicals used in the pig toxicity studies have very similar chemical structures, there are a couple of side chain differences on the phenyl ring bound to the arsenic and that may preferentially affect myelin.

In those cases of death due to reactive arsenical encephalopathy, autopsy revealed evidence for meningoencephalitis, most likely due to the trypanosomiasis infection, and stroke-like damage [2, 68]. A hemorrhagic leukoencephalopathy with features of perivascular lymphoplasmic infiltration, diffuse microglia hyperplasia, large reactive astrocytes in the white matter (myelin), necrosis of small blood vessels, ischemic brain injury, and associated petechial hemorrhages can be seen in some cases [2, 68, 134]. Again, given that there is an active infection involving the CNS, it is difficult to determine which features are due solely to arsenic. In rodents exposed to arsenite, the cortex of the brain had hyperchromatic cells and disordered fibers, correlating to the degree of exposure [129]. The axons appeared discontinuous with variation in fiber size and with undulated fibers; these observations suggest problems with the myelin sheath [129]. Others have reported similar findings of axonal loss, shrunken/hyperchromatic cells, cytoplasmic vacuolar degeneration, and nuclear pyknosis in the rodent cerebellar and cerebral cortex [79, 122, 123, 147]. Electron microscopy has shown shrunken neurons with condensed nuclei and decreased synaptic vesicles; mitochondria were swollen and vacuolated [94].

9.4.1.4 Summary of Clinical Assessments The clinical presentation of acute arsenic toxicity mimicking GBS suggests that arsenic damages the peripheral nerves in a manner similar to GBS; however, although anatomically similar structures are affected, the mechanism of damage is different. In arsenic toxicity, there is no evidence for an autoimmune response. GBS predominately affects the myelin present in Schwann cells that wrap around the peripheral nerve axons; the nerve root, the most proximal part of the PNS, is the site of initial damage but with the entire peripheral nerve soon being affected. Electrophysiologically, the NCS/EMG findings in arsenic toxicity can be consistent with this type of damage. However, the mechanism of damage appears to be different. Distal sensory nerve biopsies do not show the inflammation and segmental demyelination that is seen in GBS; only degenerative changes are seen. This presentation simply indicates that the damage is more proximal. Autopsy data indicate that the damage is at the level of the anterior horn (neuronal) cell body or very proximal projections (e.g., axonal hillock). CSF analysis in these cases also supports this location as a site of active damage, but does not suggest a mechanism. The elevated CSF protein solely indicates that an abnormal process is present in close approximation to the CSF. Similar to GBS, nerve biopsy shows that large, myelinated nerve fibers are preferentially affected in arsenic toxicity.

In cases of fatal reactive arsenical encephalopathy, there is felt to be direct brain involvement of arsenic, leading to death. Obviously, there are issues of more than one process present given direct CNS involvement of the trypanosomiasis. However, it most likely highlights that arsenic can have different presentations/mechanisms of toxicity, that is, direct brain involvement in these cases but predominately spinal cord involvement in the cases presenting like GBS. Similarly, arsenic may just predispose to other pathologic conditions to develop. For example, Yip *et al.* reported a case of encephalopathy consistent with Wernicke–Korsakoff syndrome, which is caused by thiamine deficiency, associated with arsenic-containing therapy for acute promyelocytic leukemia; it was felt that the arsenic exposure did not cause the Wernicke's encephalopathy, but did predispose to it [180]. Freeman and Couch [49] reported a case with psychotic symptoms suggesting Korsakoff psychosis of Wernicke–Korsakoff syndrome in addition to a polyneuropathy. The patient was an alcoholic with coexisting nutritional deficiency, which is the typical underlying cause of the thiamine deficiency causing this syndrome. However, she had minimal response to thiamine replacement, but once elevated arsenic levels were discovered and chelation therapy with BAL was initiated, she had a rapid improvement in her mental state and a slower improvement in her polyneuropathy [49]. Arsenic alters acetyl-CoA and succinyl-CoA production from thiamine that can result in a biochemical state similar to that seen in thiamine deficiency; this can be a mechanism predisposing to, if not causing, Wernicke–Korsakoff syndrome [142].

9.4.2 Cellular Mechanisms

The majority of arsenic absorbed into the body is arsenite (trivalent form); absorbed arsenate (pentavalent form) is reduced to arsenite [76, 134, 142]. In general, arsenic causes cellular damage by either interacting with thiol (sulfhydryl) bonds or by substituting for phosphorus. Arsenite interaction with thiol bonds can lead to enzymatic inactivation or misfolding/inappropriate aggregation of proteins [76, 77, 134, 158]. Potentially, over 200 enzymes are inactivated by arsenite, predominately in cellular energy pathways and DNA synthesis and repair [1, 127]. Arsenate can inappropriately substitute for phosphorus [134]. The arsenate bonds are less stable and can lead to disruption of high energy bonds, for example, bonds typically using phosphorus in ATP [76, 157]. Given the number of proteins affected by arsenic, differences in the cells involved, and that arsenite and arsenate have different mechanisms of toxicity, there are probably multiple means by which arsenic damages cells of the nervous system.

Arsenic can affect the oxidative phosphorylation pathway. The main mechanism is by arsenite inactivation of sulfhydral groups in multiple enzymes in this pathway with the end result being prevention of acetyl coenzyme A formation [134, 142, 158]. Arsenate can also disrupt energy production by substituting for phosphorus in ATP leading to ADP-arsenate, which rapidly and irreversibly degrades [134, 142]. Disruption of oxidative phosphorylation pathways can affect cells in a couple of manners: less energy production that induces significant cell stress and production of reactive oxygen species [142, 151, 158]. In rodents exposed to arsenite, but not at lethal levels, alterations in enzymes in the oxidative phosphorylation pathways in anterior horn cells suggested alterations in energy production [33]. Arsenate exposure has been shown to alter ectonucleotide activity and transcription; ectonucleotides are involved in the hydrolysis of ATP through to AMP [5].

In rodent and human brain explants and neuronal cell cultures, arsenite exposure causes oxidative stress and apoptosis through the Bcl-2–caspase pathway, potentially by activation of c-Jun N-terminal kinase 3 (JNK3) and p38 mitogen-activated protein (MAP) kinases [20, 21, 46, 111, 112]. Similar findings were seen for a dimethylated arsenate compound but with concentrations 1000-fold higher than those needed for arsenite [112]. Cultured rodent astrocytes exposed to arsenite also revealed DNA damage [18]. In rodent brains, *in utero* and adult exposure to arsenite causes oxidative stress with reduced antioxidant enzyme activity and to induce reactive oxygen species and reactive nitrogen species, with these effects being partially reversed by antioxidant or chelation therapy; resultant lipid peroxidation has been shown to cause neuronal cell membrane damage in multiple areas of the brain [7, 11, 22, 36, 37, 44, 45, 53, 58, 65, 79, 104, 122, 126, 132, 135, 139, 145–147, 152, 161, 175, 177, 179, 183]. Similar oxidative stress with resultant lipid peroxidation due to arsenite toxicity was seen in the rodent spinal cord and peripheral nerves [52]. Zebrafish brains exposed to arsenite had increased oxidized proteins but no change in antioxidant capacity [29]. Another proposed mechanism of reactive oxygen species damage besides lipid peroxidation is DNA damage [151]. In rodents, arsenite induces oxidative DNA damage in the cerebral cortex and cerebellum; these areas also had neuronal loss with evidence of lysis and nuclear vacuolation [123]. Apoptosis has been seen in the brains of arsenite-exposed rodents with increased phosphorylation of p38 MAPK and decreased phosphorylation of ERK 1/2 [179]. RNA expression of antioxidant genes in cultured neurons and different brain regions is altered following exposure to inorganic or organic arsenicals [70, 114, 122, 135, 179]. However, it is still to be determined if this response is due to arsenic affecting factors promoting gene expression or due to direct DNA damage.

The cytoskeleton is a highly organized intracellular network of filaments comprised of microfilaments, intermediate filaments, and microtubules. It is critical for proper cell functioning. Cytoskeletal proteins are vital to axonal functioning, and damage to them, by any cause, is felt to contribute to development of polyneuropathy and other neurologic disorders [76, 134]. Arsenic exposure has been shown to affect cytoskeletal proteins [76, 134, 158]. Autopsy in a case of arsenic toxicity revealed neurofilament (an intermediate filament) aggregation in anterior horn cell bodies [57]. Experimentally, high-dose arsenite toxicity in rodents decreases neurofilament light and medium subunits in a time-delayed manner with some changes in the heavy subunit at the highest dosage and latest time point assessed, 9 h; however, long-term low-dose exposure did not affect any neurofilament subunit expression [156]. There were also changes in protein expression in supporting fibroblasts of the sciatic nerve [156]. Arsenite does decrease neurofilament transport and results in an increase in phosphorylated neurofilament in the perikaryon possibly by increased JNK and glycogen synthase kinase-3-beta activity [30]. It inhibits neurite, neuronal outgrowth, and formation at low concentrations [171]. Neurite outgrowth inhibited by arsenite is related to deficient activation of adenosine monophosphate-activated kinase (AMPK) and the serine/threonine kinase 11 (STK11 *aka* LKB1); antioxidants antagonized this effect [163]. The LKB1–AMPK signaling pathway is involved in regulating cell structure, integrity, division, and growth. Arsenite also causes hyperphosphorylation of tau, a microtubule-associated protein whose deficits are implicated in Alzheimer's disease, at multiple amino acid residues [59]. This hyperphosphorylation is suggested either to inhibit binding to microtubules or to make microtubules unstable. Given that arsenic exposure results in hyperphosphorylated tau and activates JNK3 and p38 MAP kinases in a manner similar to that seen in Alzheimer's disease, it has been proposed that arsenic exposure is a risk factor for Alzheimer's disease [56, 62]. Arsenite and dimethylarsinous acid (DMA^{III}) exposure stimulates beta-amyloid precursor protein expression in cell culture and alters its processing, for example, cleavage; beta-amyloid precursor protein overexpression and accumulation is suggested to cause Alzheimer's disease [32, 182]. Interestingly, in neuroblastoma, CNS-derived glial tumor, and schwannoma, PNS-derived Schwann cell tumor, cell lines, arsenite, and arsenate had no-to-minimal effect on neurofilament light (NEFL), medium (NEF3), heavy (NEFH), or microtubule-associated tau (MAPT) gene expression, but the chemicals were toxic to the cultured cells; however, monomethylarsonic acid (MMA^V) and dimethylarsinic acid (DMA^V) were less toxic to the cells and significantly affected NEFL, NEF3, and NEFH expression [157]. DMA^V induced more alterations than MMA^V in the schwannoma-derived cells, and MMA^V induced more in the neuroblastoma-derived cells. MAPT expression was only increased by MMA^V in the neuroblastoma cell line [157]. Thus, there appears to be arsenic species specificity in adverse effects on cytoskeleton in neurologic diseases.

In rodent studies, inorganic arsenite altered neurotransmitter levels and appeared to affect all neurotransmitter systems, for example, monoaminergic, cholinergic, GABAergic, and glutamatergic [31, 73, 74, 78, 82, 97, 103, 109, 110, 130, 134, 135, 152, 159, 174, 176, 178, 181]. However, not all studies reveal abnormalities, and not all neurotransmitters assessed had detectable changes [7, 31, 133]. These studies for the most part attempt to correlate neurotransmitter changes in certain brain regions with a phenotype, for example, dopamine with locomotion. However, these changes are not assessed in specific nuclei so it is hard to determine significance. Age appears to be a significant variable, with young versus old rodents having different levels and distribution of neurotransmitter changes [110, 159]. Arsenite alters spontaneously generated action potentials in snail central neurons: low concentrations decreased the frequency of action potentials, and higher concentrations initially lead to depolarization and then irreversible bursts [93]. This affect appears to be mediated by the phospholipase C pathway [93]. In young and adult rodent hippocampal slices, high-dose arsenite exposure leads to a reduction in amplitude of evoked excitatory postsynaptic potential (EPSP) and inhibited long-term potentiation [84]. MMA^{III} and DMA^{III} depressed EPSP at high concentration at baseline and with high frequency stimulation and enhanced them at low concentration with high frequency stimulation; MMA^{V} and DMA^{V} had no effect [85, 86]. MMA^{III} and DMA^{III} action on postsynaptic glutamatergic receptors was proposed to mediate this response [85, 86]. Arsenite was shown to alter glutamate decarboxylase mRNA expression in brains of rodent pups exposed in utero; GABA-T mRNA expression was unchanged [175]. In neuronal cell culture, glutamate transport was decreased by arsenite exposure by altering glutamate/aspartate transporter (GLAST) and excitatory amino acid transporter 1 (EAAT1) activity via protein kinase A, protein kinase C, and MAPK [17]. Other receptor gene expression has been shown to be altered by arsenite-decreased *N*-methyl-*D*-aspartate (NMDA) receptor gene expression in the hippocampus, and decreased calcium/calmodulin-dependent protein kinase IV (camk4) [94, 164]. Thus, arsenite and its metabolites have demonstrated adverse effects on a variety of processes essential to normal neurological function.

9.5 SUMMARY

Arsenic affects on the nervous system are varied. Inorganic and organic forms of arsenite and arsenate probably have different mechanisms and sites of damage. Experimental studies focus on inorganic arsenite, which skews the literature; even so, inorganic arsenite appears to have different affects, for example, disruption of oxidative phosphorylation, cytoskeleton formation, etc. Dosage and organism susceptibility are significant variables. Arsenic-induced peripheral neuropathy, which mimics GBS, is probably its most well-known effect on the nervous system. Currently, the biggest issue is arsenic effects on the developing nervous system. Specifically, are arsenic-contaminated water and foodstuffs significantly impairing cognitive and other human neurodevelopment?

REFERENCES

[1] C.O. Abernathy, Y.P. Liu, D. Longfellow, H.V. Aposhian, B. Beck, B. Fowler, R. Goyer, R. Menzer, T. Rossman, C. Thompson, M. Waalkes, Arsenic: health effects, mechanisms of actions, and research issues. Environ Health Perspect 107 (1999) 593–597.

[2] J.H. Adams, L. Haller, F.Y. Boa, F. Doua, A. Dago, K. Konian, Human African trypanosomiasis (T.b. gambiense): a study of 16 fatal cases of sleeping sickness with some observations on acute reactive arsenical encephalopathy. Neuropathol Appl Neurobiol 12 (1986) 81–94.

[3] C.W. Armstrong, R.B. Stroube, T. Rubio, E.A. Siudyla, G.B. Miller, Jr., Outbreak of fatal arsenic poisoning caused by contaminated drinking water. Arch Environ Health 39 (1984) 276–279.

[4] W.Y. Au, S. Tam, B.M. Fong, Y.L. Kwong, Determinants of cerebrospinal fluid arsenic concentration in patients with acute promyelocytic leukemia on oral arsenic trioxide therapy. Blood 112 (2008) 3587–3590.

[5] L.A. Baldissarelli, K.M. Capiotti, M.R. Bogo, G. Ghisleni, C.D. Bonan, Arsenic alters behavioral parameters and brain ectonucleotidases activities in zebrafish (Danio rerio). Comp Biochem Physiol C Toxicol Pharmacol 155 (2012) 566–572.

[6] S.K. Bansal, N. Haldar, U.K. Dhand, J.S. Chopra, Phrenic neuropathy in arsenic poisoning. Chest 100 (1991) 878–880.

[7] U. Bardullas, J.H. Limon-Pacheco, M. Giordano, L. Carrizales, M.S. Mendoza-Trejo, V.M. Rodriguez, Chronic low-level arsenic exposure causes gender-specific alterations in locomotor activity, dopaminergic systems, and thioredoxin expression in mice. Toxicol Appl Pharmacol 239 (2009) 169–177.

[8] W.S. Beckett, J.L. Moore, J.P. Keogh, M.L. Bleecker, Acute encephalopathy due to occupational exposure to arsenic. Br J Ind Med 43 (1986) 66–67.

[9] V. Bencko, K. Symon, V. Chladek, J. Pihrt, Health aspects of burning coal with a high arsenic content. II. Hearing changes in exposed children. Environ Res 13 (1977) 386–395.

[10] L. Benramdane, M. Accominotti, L. Fanton, D. Malicier, J.J. Vallon, Arsenic speciation in human organs following fatal arsenic trioxide poisoning: a case report. Clin Chem 45 (1999) 301–306.

[11] V.K. Bharti, R.S. Srivastava, B. Sharma, J.K. Malik, Buffalo (*Bubalus bubalis*) epiphyseal proteins counteract arsenic-induced oxidative stress in brain, heart, and liver of female rats. Biol Trace Elem Res 146 (2012) 224–229.

[12] S. Blom, B. Lagerkvist, H. Linderholm, Arsenic exposure to smelter workers. Clinical and neurophysiological studies. Scand J Work Environ Health 11 (1985) 265–269.

[13] K. Bolla-Wilson, M.L. Bleecker, Neuropsychological impairment following inorganic arsenic exposure. J Occup Med 29 (1987) 500–503.

[14] J. Brinkel, M.H. Khan, A. Kraemer, A systematic review of arsenic exposure and its social and mental health effects with special reference to Bangladesh. Int J Environ Res Public Health 6 (2009) 1609–1619.

[15] O.F. Brouwer, W. Onkenhout, P.M. Edelbroek, J.F. de Kom, F.A. de Wolff, A.C. Peters, Increased neurotoxicity of arsenic in methylenetetrahydrofolate reductase deficiency. Clin Neurol Neurosurg 94 (1992) 307–310.

[16] J. Calderon, M.E. Navarro, M.E. Jimenez-Capdeville, M.A. Santos-Diaz, A. Golden, I. Rodriguez-Leyva, V. Borja-Aburto, F. Diaz-Barriga, Exposure to arsenic and lead and neuropsychological development in Mexican children. Environ Res 85 (2001) 69–76.

[17] Y. Castro-Coronel, L.M. Del Razo, M. Huerta, A. Hernandez-Lopez, A. Ortega, E. Lopez-Bayghen, Arsenite exposure downregulates EAAT1/GLAST transporter expression in glial cells. Toxicol Sci 122 (2011) 539–550.

[18] I. Catanzaro, G. Schiera, G. Sciandrello, G. Barbata, F. Caradonna, P. Proia, I. Di Liegro, Biological effects of inorganic arsenic on primary cultures of rat astrocytes. Int J Mol Med 26 (2010) 457–462.

[19] D. Chakraborti, S.C. Mukherjee, K.C. Saha, U.K. Chowdhury, M.M. Rahman, M.K. Sengupta, Arsenic toxicity from homeopathic treatment. J Toxicol Clin Toxicol 41 (2003) 963–967.

[20] S. Chattopadhyay, S. Bhaumik, A. Nag Chaudhury, S. Das Gupta, Arsenic induced changes in growth development and apoptosis in neonatal and adult brain cells in vivo and in tissue culture. Toxicol Lett 128 (2002) 73–84.

[21] S. Chattopadhyay, S. Bhaumik, M. Purkayastha, S. Basu, A. Nag Chaudhuri, S. Das Gupta, Apoptosis and necrosis in developing brain cells due to arsenic toxicity and protection with antioxidants. Toxicol Lett 136 (2002) 65–76.

[22] A.N. Chaudhuri, S. Basu, S. Chattopadhyay, S. Das Gupta, Effect of high arsenic content in drinking water on rat brain. Indian J Biochem Biophys 36 (1999) 51–54.

[23] H.Y. Chiou, W.I. Huang, C.L. Su, S.F. Chang, Y.H. Hsu, C.J. Chen, Dose-response relationship between prevalence of cerebrovascular disease and ingested inorganic arsenic. Stroke 28 (1997) 1717–1723.

[24] U.K. Chowdhury, B.K. Biswas, T.R. Chowdhury, G. Samanta, B.K. Mandal, G.C. Basu, C.R. Chanda, D. Lodh, K.C. Saha, S.K. Mukherjee, S. Roy, S. Kabir, Q. Quamruzzaman, D. Chakraborti, Groundwater arsenic contamination in Bangladesh and West Bengal, India. Environ Health Perspect 108 (2000) 393–397.

[25] S.M. Cohen, L.L. Arnold, M. Eldan, A.S. Lewis, B.D. Beck, Methylated arsenicals: the implications of metabolism and carcinogenicity studies in rodents to human risk assessment. Crit Rev Toxicol 36 (2006) 99–133.

[26] M.T. Colomina, M.L. Albina, J.L. Domingo, J. Corbella, Influence of maternal stress on the effects of prenatal exposure to methylmercury and arsenic on postnatal development and behavior in mice: a preliminary evaluation. Physiol Behav 61 (1997) 455–459.

[27] W.R. Cullen, K.J. Reimer, Arsenic speciation in the environment. Chem Rev 89 (1989) 713–764.

[28] J.R. Daube, D.I. Rubin, Needle electromyography. Muscle Nerve 39 (2009) 244–270.

[29] M.R. de Castro, J.V. Lima, D.P. de Freitas, S. Valente Rde, N.S. Dummer, R.B. de Aguiar, L.C. dos Santos, L.F. Marins, L.A. Geracitano, J.M. Monserrat, D.M. Barros, Behavioral and neurotoxic effects of arsenic exposure in zebrafish (*Danio rerio*, Teleostei: Cyprinidae). Comp Biochem Physiol C Toxicol Pharmacol 150 (2009) 337–342.

[30] J. DeFuria, T.B. Shea, Arsenic inhibits neurofilament transport and induces perikaryal accumulation of phosphorylated neurofilaments: roles of JNK and GSK-3beta. Brain Res 1181 (2007) 74–82.

[31] J.M. Delgado, L. Dufour, J.I. Grimaldo, L. Carrizales, V.M. Rodriguez, M.E. Jimenez-Capdeville, Effects of arsenite on central monoamines and plasmatic levels of adrenocorticotropic hormone (ACTH) in mice. Toxicol Lett 117 (2000) 61–67.

[32] N.N. Dewji, C. Do, R.M. Bayney, Transcriptional activation of Alzheimer's beta-amyloid precursor protein gene by stress. Brain Res Mol Brain Res 33 (1995) 245–253.

[33] P. Dhar, M. Jaitley, M. Kalaivani, R.D. Mehra, Preliminary morphological and histochemical changes in rat spinal cord neurons following arsenic ingestion. Neurotoxicology 26 (2005) 309–320.

[34] P. Dhar, N. Mohari, R.D. Mehra, Preliminary morphological and morphometric study of rat cerebellum following sodium arsenite exposure during rapid brain growth (RBG) period. Toxicology 234 (2007) 10–20.

[35] P.D. Donofrio, A.J. Wilbourn, J.W. Albers, L. Rogers, V. Salanga, H.S. Greenberg, Acute arsenic intoxication presenting as Guillain-Barre-like syndrome. Muscle Nerve 10 (1987) 114–120.

[36] N. Dwivedi, A. Mehta, A. Yadav, B.K. Binukumar, K.D. Gill, S.J. Flora, MiADMSA reverses impaired mitochondrial energy metabolism and neuronal apoptotic cell death after arsenic exposure in rats. Toxicol Appl Pharmacol 256 (2011) 241–248.

[37] F.M. El-Demerdash, M.I. Yousef, F.M. Radwan, Ameliorating effect of curcumin on sodium arsenite-induced oxidative damage and lipid peroxidation in different rat organs. Food Chem Toxicol 47 (2009) 249–254.

[38] E.O. Espinoza, M.J. Mann, B. Bleasdell, Arsenic and mercury in traditional Chinese herbal balls. N Engl J Med 333 (1995) 803–804.

[39] B. Falck, E. Stalberg, Motor nerve conduction studies: measurement principles and interpretation of findings. J Clin Neurophysiol 12 (1995) 254–279.

[40] H. Feit, R.S. Tindall, M. Glasberg, Sources of error in the diagnosis of Guillain-Barre syndrome. Muscle Nerve 5 (1982) 111–117.

[41] R.G. Feldman, C.A. Niles, M. Kelly-Hayes, D.S. Sax, W.J. Dixon, D.J. Thompson, E. Landau, Peripheral neuropathy in arsenic smelter workers. Neurology 29 (1979) 939–944.

[42] V.H. Ferm, D.P. Hanlon, Arsenate-induced neural tube defects not influenced by constant rate administration of folic acid. Pediatr Res 20 (1986) 761–762.

[43] R.M. Fincher, R.M. Koerker, Long-term survival in acute arsenic encephalopathy. Follow-up using newer measures of electrophysiologic parameters. Am J Med 82 (1987) 549–552.

[44] S.J. Flora, Arsenic-induced oxidative stress and its reversibility following combined administration of *N*-acetylcysteine and meso 2,3-dimercaptosuccinic acid in rats. Clin Exp Pharmacol Physiol 26 (1999) 865–869.

[45] S.J. Flora, S. Bhadauria, S.C. Pant, R.K. Dhaked, Arsenic induced blood and brain oxidative stress and its response to some thiol chelators in rats. Life Sci 77 (2005) 2324–2337.

[46] S.J. Flora, K. Bhatt, A. Mehta, Arsenic moiety in gallium arsenide is responsible for neuronal apoptosis and behavioral alterations in rats. Toxicol Appl Pharmacol 240 (2009) 236–244.

[47] B.A. Fowler, J.B. Weissberg, Arsine poisoning. N Engl J Med 291 (1974) 1171–1174.

[48] A. Franzblau, R. Lilis, Acute arsenic intoxication from environmental arsenic exposure. Arch Environ Health 44 (1989) 385–390.

[49] J.W. Freeman, J.R. Couch, Prolonged encephalopathy with arsenic poisoning. Neurology 28 (1978) 853–855.

[50] Y. Fujino, X. Guo, J. Liu, L. You, M. Miyatake, T. Yoshimura, Mental health burden amongst inhabitants of an arsenic-affected area in Inner Mongolia, China. Soc Sci Med 59 (2004) 1969–1973.

[51] Y. Fujino, X. Guo, K. Shirane, J. Liu, K. Wu, M. Miyatake, K. Tanabe, T. Kusuda, T. Yoshimura, Arsenic in drinking water and peripheral nerve conduction velocity among residents of a chronically arsenic-affected area in Inner Mongolia. J Epidemiol 16 (2006) 207–213.

[52] E. Garcia-Chavez, I. Jimenez, B. Segura, L.M. Del Razo, Lipid oxidative damage and distribution of inorganic arsenic and its metabolites in the rat nervous system after arsenite exposure: influence of alpha tocopherol supplementation. Neurotoxicology 27 (2006) 1024–1031.

[53] E. Garcia-Chavez, A. Santamaria, F. Diaz-Barriga, P. Mandeville, B.I. Juarez, M.E. Jimenez-Capdeville, Arsenite-induced formation of hydroxyl radical in the striatum of awake rats. Brain Res 976 (2003) 82–89.

[54] E. Garcia-Chavez, B. Segura, H. Merchant, I. Jimenez, L.M. Del Razo, Functional and morphological effects of repeated sodium arsenite exposure on rat peripheral sensory nerves. J Neurol Sci 258 (2007) 104–110.

[55] F. Gerr, R. Letz, P.B. Ryan, R.C. Green, Neurological effects of environmental exposure to arsenic in dust and soil among humans. Neurotoxicology 21 (2000) 475–487.

[56] S. Gharibzadeh, S.S. Hoseini, Arsenic exposure may be a risk factor for Alzheimer's disease. J Neuropsychiatry Clin Neurosci 20 (2008) 501.

[57] R.K. Gherardi, P. Chariot, M. Vanderstigel, D. Malapert, J. Verroust, A. Astier, C. Brun-Buisson, A. Schaeffer, Organic arsenic-induced Guillain-Barre-like syndrome due to melarsoprol: a clinical, electrophysiological, and pathological study. Muscle Nerve 13 (1990) 637–645.

[58] A. Ghosh, A.K. Mandal, S. Sarkar, N. Das, Hepatoprotective and neuroprotective activity of liposomal quercetin in combating chronic arsenic induced oxidative damage in liver and brain of rats. Drug Deliv 18 (2011) 451–459.

[59] B.I. Giasson, D.M. Sampathu, C.A. Wilson, V. Vogelsberg-Ragaglia, W.E. Mushynski, V.M. Lee, The environmental toxin arsenite induces tau hyperphosphorylation. Biochemistry 41 (2002) 15376–15387.

[60] M.J. Goddard, J.L. Tanhehco, P.C. Dau, Chronic arsenic poisoning masquerading as Landry-Guillain-Barre syndrome. Electromyogr Clin Neurophysiol 32 (1992) 419–423.

[61] H.H. Goebel, P.F. Schmidt, J. Bohl, B. Tettenborn, G. Kramer, L. Gutmann, Polyneuropathy due to acute arsenic intoxication: biopsy studies. J Neuropathol Exp Neurol 49 (1990) 137–149.

[62] G. Gong, S.E. O'Bryant, The arsenic exposure hypothesis for Alzheimer disease. Alzheimer Dis Assoc Disord 24 (2010) 311–316.

[63] P. Grandjean, P.J. Landrigan, Developmental neurotoxicity of industrial chemicals. Lancet 368 (2006) 2167–2178.

[64] S.A. Greenberg, Acute demyelinating polyneuropathy with arsenic ingestion. Muscle Nerve 19 (1996) 1611–1613.

[65] R. Gupta, S.J. Flora, Effect of *Centella asiatica* on arsenic induced oxidative stress and metal distribution in rats. J Appl Toxicol 26 (2006) 213–222.

[66] D.M. Hafeman, H. Ahsan, E.D. Louis, A.B. Siddique, V. Slavkovich, Z. Cheng, A. van Geen, J.H. Graziano, Association between arsenic exposure and a measure of subclinical sensory neuropathy in Bangladesh. J Occup Environ Med 47 (2005) 778–784.

[67] A.H. Hall, Chronic arsenic poisoning. Toxicol Lett 128 (2002) 69–72.

[68] L. Haller, H. Adams, F. Merouze, A. Dago, Clinical and pathological aspects of human African trypanosomiasis (T. b. gambiense) with particular reference to reactive arsenical encephalopathy. Am J Trop Med Hyg 35 (1986) 94–99.

[69] J. Hindmarsh, O. McLetchie, L. Heffernan, O. Hayne, H. Ellenberger, R. McCurdy, H. Thiebaux, Electromyographic abnormalities in chronic environmental arsenicalism. J Anal Toxicol 1 (1977) 270–276.

[70] Y. Hong, F. Piao, Y. Zhao, S. Li, Y. Wang, P. Liu, Subchronic exposure to arsenic decreased Sdha expression in the brain of mice. Neurotoxicology 30 (2009) 538–543.

[71] R.D. Hood, S.L. Bishop, Teratogenic effects of sodium arsenate in mice. Arch Environ Health 24 (1972) 62–65.

[72] K. Ishii, A. Tamaoka, F. Otsuka, N. Iwasaki, K. Shin, A. Matsui, G. Endo, Y. Kumagai, T. Ishii, S. Shoji, T. Ogata, M. Ishizaki, M. Doi, N. Shimojo, Diphenylarsinic acid poisoning from chemical weapons in Kamisu, Japan. Ann Neurol 56 (2004) 741–745.

[73] T. Itoh, Y.F. Zhang, S. Murai, H. Saito, H. Nagahama, H. Miyate, Y. Saito, E. Abe, The effect of arsenic trioxide on brain monoamine metabolism and locomotor activity of mice. Toxicol Lett 54 (1990) 345–353.

[74] K. Jana, S. Jana, P.K. Samanta, Effects of chronic exposure to sodium arsenite on hypothalamo-pituitary-testicular activities in adult rats: possible an estrogenic mode of action. Reprod Biol Endocrinol 4 (2006) 9.

[75] S. Jha, A.K. Dhanuka, M.N. Singh, Arsenic poisoning in a family. Neurol India 50 (2002) 364–365.

[76] K. Jomova, Z. Jenisova, M. Feszterova, S. Baros, J. Liska, D. Hudecova, C.J. Rhodes, M. Valko, Arsenic: toxicity, oxidative stress and human disease. J Appl Toxicol 31 (2011) 95–107.

[77] F.T. Jones, A broad view of arsenic. Poult Sci 86 (2007) 2–14.

[78] G.M. Kannan, N. Tripathi, S.N. Dube, M. Gupta, S.J. Flora, Toxic effects of arsenic (III) on some hematopoietic and central nervous system variables in rats and guinea pigs. J Toxicol Clin Toxicol 39 (2001) 675–682.

[79] K. Kato, M. Mizoi, Y. An, M. Nakano, H. Wanibuchi, G. Endo, Y. Endo, M. Hoshino, S. Okada, K. Yamanaka, Oral administration of diphenylarsinic acid, a degradation product of chemical warfare agents, induces oxidative and nitrosative stress in cerebellar Purkinje cells. Life Sci 81 (2007) 1518–1525.

[80] S. Kennedy, D.A. Rice, P.F. Cush, Neuropathology of experimental 3-nitro-4-hydroxyphenylarsonic acid toxicosis in pigs. Vet Pathol 23 (1986) 454–461.

[81] Y. Kishi, H. Sasaki, H. Yamasaki, K. Ogawa, M. Nishi, K. Nanjo, An epidemic of arsenic neuropathy from a spiked curry. Neurology 56 (2001) 1417–1418.

[82] H. Kobayashi, A. Yuyama, M. Ishihara, N. Matsusaka, Effects of arsenic on cholinergic parameters in brain in vitro. Neuropharmacology 26 (1987) 1707–1713.

[83] K. Kreiss, M.M. Zack, P.J. Landrigan, R.G. Feldman, C.A. Niles, J. Chirico-Post, D.S. Sax, M.H. Boyd, D.H. Cox, Neurologic evaluation of a population exposed to arsenic in Alaskan well water. Arch Environ Health 38 (1983) 116–121.

[84] K. Kruger, N. Binding, H. Straub, U. Musshoff, Effects of arsenite on long-term potentiation in hippocampal slices from young and adult rats. Toxicol Lett 165 (2006) 167–173.

[85] K. Kruger, H. Repges, J. Hippler, L.M. Hartmann, A.V. Hirner, H. Straub, N. Binding, U. Musshoff, Effects of dimethylarsinic and dimethylarsinous acid on evoked synaptic potentials in hippocampal slices of young and adult rats. Toxicol Appl Pharmacol 225 (2007) 40–46.

[86] K. Kruger, H. Straub, A.V. Hirner, J. Hippler, N. Binding, U. Musshoff, Effects of monomethylarsonic and monomethylarsonous acid on evoked synaptic potentials in hippocampal slices of adult and young rats. Toxicol Appl Pharmacol 236 (2009) 115–123.

[87] B.J. Lagerkvist, B. Zetterlund, Assessment of exposure to arsenic among smelter workers: a five-year follow-up. Am J Ind Med 25 (1994) 477–488.

[88] P.M. Le Quesne, J.G. McLeod, Peripheral neuropathy following a single exposure to arsenic. Clinical course in four patients with electrophysiological and histological studies. J Neurol Sci 32 (1977) 437–451.

[89] A.E. Ledet, J.R. Duncan, W.B. Buck, F.K. Ramsey, Clinical, toxicological, and pathological aspects of arsanilic acid poisoning in swine. Clin Toxicol 6 (1973) 439–457.

[90] Y. Li, Y. Xia, L. He, Z. Ning, K. Wu, B. Zhao, X.C. Le, R. Kwok, M. Schmitt, T. Wade, J. Mumford, D. Otto, Neurosensory effects of chronic exposure to arsenic via drinking water in Inner Mongolia: I. Signs, symptoms and pinprick testing. J Water Health 4 (2006) 29–37.

[91] Z. London, J.W. Albers, Toxic neuropathies associated with pharmaceutic and industrial agents. Neurol Clin 25 (2007) 257–276.

[92] C. Lu, F. Zhao, D. Sun, Y. Zhong, X. Yu, G. Li, X. Lv, G. Sun, Y. Jin, Comparison of speciated arsenic levels in the liver and brain of mice between arsenate and arsenite exposure at the early life. Environ Toxicol 29 (2014) 797–803.

[93] G.L. Lu, Y.C. Chang, Y.H. Chen, C.L. Tsai, M.C. Tsai, Arsenic trioxide modulates the central snail neuron action potential. J Formos Med Assoc 108 (2009) 683–693.

[94] J.H. Luo, Z.Q. Qiu, W.Q. Shu, Y.Y. Zhang, L. Zhang, J.A. Chen, Effects of arsenic exposure from drinking water on spatial memory, ultra-structures and NMDAR gene expression of hippocampus in rats. Toxicol Lett 184 (2009) 121–125.

[95] E. Lynch, R. Braithwaite, A review of the clinical and toxicological aspects of "traditional" (herbal) medicines adulterated with heavy metals. Expert Opin Drug Saf 4 (2005) 769–778.

[96] V.P. Markowski, E.A. Reeve, K. Onos, M. Assadollahzadeh, N. McKay, Effects of prenatal exposure to sodium arsenite on motor and food-motivated behaviors from birth to adulthood in C57BL6/J mice. Neurotoxicol Teratol 34 (2012) 221–231.

[97] E.J. Martinez, B.L. Kolb, A. Bell, D.D. Savage, A.M. Allan, Moderate perinatal arsenic exposure alters neuroendocrine markers associated with depression and increases depressive-like behaviors in adult mouse offspring. Neurotoxicology 29 (2008) 647–655.

[98] E.J. Martinez-Finley, A.M. Ali, A.M. Allan, Learning deficits in C57BL/6J mice following perinatal arsenic exposure: consequence of lower corticosterone receptor levels? Pharmacol Biochem Behav 94 (2009) 271–277.

[99] E.W. Massey, D. Wold, A. Heyman, Arsenic: homicidal intoxication. South Med J 77 (1984) 848–851.

[100] L. Mathew, A. Vale, J.E. Adcock, Arsenical peripheral neuropathy. Pract Neurol 10 (2010) 34–38.

[101] D.N. Mazumder, J. Das Gupta, A.K. Chakraborty, A. Chatterjee, D. Das, D. Chakraborti, Environmental pollution and chronic arsenicosis in south Calcutta. Bull World Health Organ 70 (1992) 481–485.

[102] M.N. Mead, Arsenic: in search of an antidote to a global poison. Environ Health Perspect 113 (2005) A378–A386.

[103] J.J. Mejia, F. Diaz-Barriga, J. Calderon, C. Rios, M.E. Jimenez-Capdeville, Effects of lead-arsenic combined exposure on central monoaminergic systems. Neurotoxicol Teratol 19 (1997) 489–497.

[104] D. Mishra, S.J. Flora, Differential oxidative stress and DNA damage in rat brain regions and blood following chronic arsenic exposure. Toxicol Ind Health 24 (2008) 247–256.

[105] R.E. Morrissey, N.K. Mottet, Arsenic-induced exencephaly in the mouse and associated lesions occurring during neurulation. Teratology 28 (1983) 399–411.

[106] W.E. Morton, G.A. Caron, Encephalopathy: an uncommon manifestation of workplace arsenic poisoning? Am J Ind Med 15 (1989) 1–5.

[107] S.C. Mukherjee, M.M. Rahman, U.K. Chowdhury, M.K. Sengupta, D. Lodh, C.R. Chanda, K.C. Saha, D. Chakraborti, Neuropathy in arsenic toxicity from groundwater arsenic contamination in West Bengal, India. J Environ Sci Health A Tox Hazard Subst Environ Eng 38 (2003) 165–183.

[108] M.J. Murphy, L.W. Lyon, J.W. Taylor, Subacute arsenic neuropathy: clinical and electrophysiological observations. J Neurol Neurosurg Psychiatry 44 (1981) 896–900.

[109] T.N. Nagaraja, T. Desiraju, Regional alterations in the levels of brain biogenic amines, glutamate, GABA, and GAD activity due to chronic consumption of inorganic arsenic in developing and adult rats. Bull Environ Contam Toxicol 50 (1993) 100–107.

[110] T.N. Nagaraja, T. Desiraju, Effects on operant learning and brain acetylcholine esterase activity in rats following chronic inorganic arsenic intake. Hum Exp Toxicol 13 (1994) 353–356.

[111] U. Namgung, Z. Xia, Arsenite-induced apoptosis in cortical neurons is mediated by c-Jun N-terminal protein kinase 3 and p38 mitogen-activated protein kinase. J Neurosci 20 (2000) 6442–6451.

[112] U. Namgung, Z. Xia, Arsenic induces apoptosis in rat cerebellar neurons via activation of JNK3 and p38 MAP kinases. Toxicol Appl Pharmacol 174 (2001) 130–138.

[113] H. Naranmandura, N. Suzuki, J. Takano, T. McKnight-Whitford, Y. Ogra, K.T. Suzuki, X.C. Le, Systemic distribution and speciation of diphenylarsinic acid fed to rats. Toxicol Appl Pharmacol 237 (2009) 214–220.

[114] T. Negishi, M. Takahashi, Y. Matsunaga, S. Hirano, T. Tashiro, Diphenylarsinic acid increased the synthesis and release of neuroactive and vasoactive peptides in rat cerebellar astrocytes. J Neuropathol Exp Neurol 71 (2012) 468–479.

[115] S.E. O'Bryant, M. Edwards, C.V. Menon, G. Gong, R. Barber, Long-term low-level arsenic exposure is associated with poorer neuropsychological functioning: a Project FRONTIER study. Int J Environ Res Public Health 8 (2011) 861–874.

[116] E. O'Shaughnessy, G.H. Kraft, Arsenic poisoning: long-term follow-up of a nonfatal case. Arch Phys Med Rehabil 57 (1976) 403–406.

[117] S.J. Oh, Electrophysiological profile in arsenic neuropathy. J Neurol Neurosurg Psychiatry 54 (1991) 1103–1105.

[118] M. Ota, Ultrastructure of sural nerve in a case of arsenical neuropathy. Acta Neuropathol 16 (1970) 233–242.

[119] D. Otto, L. He, Y. Xia, Y. Li, K. Wu, Z. Ning, B. Zhao, H.K. Hudnell, R. Kwok, J. Mumford, A. Geller, T. Wade, Neurosensory effects of chronic exposure to arsenic via drinking water in Inner Mongolia: II. Vibrotactile and visual function. J Water Health 4 (2006) 39–48.

[120] D. Otto, Y. Xia, Y. Li, K. Wu, L. He, J. Telech, H. Hundell, J. Prah, J. Mumford, T. Wade, Neurosensory effects of chronic human exposure to arsenic associated with body burden and environmental measures. Hum Exp Toxicol 26 (2007) 169–177.

[121] F. Parvez, G.A. Wasserman, P. Factor-Litvak, X. Liu, V. Slavkovich, A.B. Siddique, R. Sultana, T. Islam, D. Levy, J.L. Mey, A. van Geen, K. Khan, J. Kline, H. Ahsan, J.H. Graziano, Arsenic exposure and motor function among children in Bangladesh. Environ Health Perspect 119 (2011) 1665–1670.

[122] F. Piao, S. Li, Q. Li, J. Ye, S. Liu, Abnormal expression of 8-nitroguanine in the brain of mice exposed to arsenic subchronically. Ind Health 49 (2011) 151–157.

[123] F. Piao, N. Ma, Y. Hiraku, M. Murata, S. Oikawa, F. Cheng, L. Zhong, T. Yamauchi, S. Kawanishi, K. Yokoyama, Oxidative DNA damage in relation to neurotoxicity in the brain of mice exposed to arsenic at environmentally relevant levels. J Occup Health 47 (2005) 445–449.

[124] G.T. Pryor, E.T. Uyeno, H.A. Tilson, C.L. Mitchell, Assessment of chemicals using a battery of neurobehavioral tests: a comparative study. Neurobehav Toxicol Teratol 5 (1983) 91–117.

[125] M.M. Rahman, U.K. Chowdhury, S.C. Mukherjee, B.K. Mondal, K. Paul, D. Lodh, B.K. Biswas, C.R. Chanda, G.K. Basu, K.C. Saha, S. Roy, R. Das, S.K. Palit, Q. Quamruzzaman, D. Chakraborti, Chronic arsenic toxicity in Bangladesh and West Bengal, India: a review and commentary. J Toxicol Clin Toxicol 39 (2001) 683–700.

[126] M.V. Rao, G. Avani, Arsenic induced free radical toxicity in brain of mice. Indian J Exp Biol 42 (2004) 495–498.

[127] R.N. Ratnaike, Acute and chronic arsenic toxicity. Postgrad Med J 79 (2003) 391–396.

[128] D.A. Rice, C.H. McMurray, R.M. McCracken, D.G. Bryson, R. Maybin, A field case of poisoning caused by 3-nitro-4-hydroxy phenyl arsonic acid in pigs. Vet Rec 106 (1980) 312–313.

[129] R. Rios, S. Zarazua, M.E. Santoyo, J. Sepulveda-Saavedra, V. Romero-Diaz, V. Jimenez, F. Perez-Severiano, G. Vidal-Cantu, J.M. Delgado, M.E. Jimenez-Capdeville, Decreased nitric oxide markers and morphological changes in the brain of arsenic-exposed rats. Toxicology 261 (2009) 68–75.

[130] V.M. Rodriguez, L. Carrizales, M.E. Jimenez-Capdeville, L. Dufour, M. Giordano, The effects of sodium arsenite exposure on behavioral parameters in the rat. Brain Res Bull 55 (2001) 301–308.

[131] V.M. Rodriguez, L. Carrizales, M.S. Mendoza, O.R. Fajardo, M. Giordano, Effects of sodium arsenite exposure on development and behavior in the rat. Neurotoxicol Teratol 24 (2002) 743–750.

[132] V.M. Rodriguez, L.M. Del Razo, J.H. Limon-Pacheco, M. Giordano, L.C. Sanchez-Pena, E. Uribe-Querol, G. Gutierrez-Ospina, M.E. Gonsebatt, Glutathione reductase inhibition and methylated arsenic distribution in Cd1 mice brain and liver. Toxicol Sci 84 (2005) 157–166.

[133] V.M. Rodriguez, L. Dufour, L. Carrizales, F. Diaz-Barriga, M.E. Jimenez-Capdeville, Effects of oral exposure to mining waste on in vivo dopamine release from rat striatum. Environ Health Perspect 106 (1998) 487–491.

[134] V.M. Rodriguez, M.E. Jimenez-Capdeville, M. Giordano, The effects of arsenic exposure on the nervous system. Toxicol Lett 145 (2003) 1–18.

[135] V.M. Rodriguez, J.H. Limon-Pacheco, L. Carrizales, M.S. Mendoza-Trejo, M. Giordano, Chronic exposure to low levels of inorganic arsenic causes alterations in locomotor activity and in the expression of dopaminergic and antioxidant systems in the albino rat. Neurotoxicol Teratol 32 (2010) 640–647.

[136] J.L. Rosado, D. Ronquillo, K. Kordas, O. Rojas, J. Alatorre, P. Lopez, G. Garcia-Vargas, M. Del Carmen Caamano, M.E. Cebrian, R.J. Stoltzfus, Arsenic exposure and cognitive performance in Mexican schoolchildren. Environ Health Perspect 115 (2007) 1371–1375.

[137] A. Roy, K. Kordas, P. Lopez, J.L. Rosado, M.E. Cebrian, G.G. Vargas, D. Ronquillo, R.J. Stoltzfus, Association between arsenic exposure and behavior among first-graders from Torreon, Mexico. Environ Res 111 (2011) 670–676.

[138] E.L. Sainio, R. Jolanki, E. Hakala, L. Kanerva, Metals and arsenic in eye shadows. Contact Dermatitis 42 (2000) 5–10.

[139] S. Samuel, R. Kathirvel, T. Jayavelu, P. Chinnakkannu, Protein oxidative damage in arsenic induced rat brain: influence of DL-alpha-lipoic acid. Toxicol Lett 155 (2005) 27–34.

[140] L.C. Sanchez-Pena, P. Petrosyan, M. Morales, N.B. Gonzalez, G. Gutierrez-Ospina, L.M. Del Razo, M.E. Gonsebatt, Arsenic species, AS3MT amount, and AS3MT gene expression in different brain regions of mouse exposed to arsenite. Environ Res 110 (2010) 428–434.

[141] P.N. Saxena, S.S. Raza, A. Attri, R. Agarwal, S. Gupta, M. Saksena, Central hyperthermic effect of arsenic in rabbits. Indian J Med Res 94 (1991) 241–245.

[142] W.L. Schoolmeester, D.R. White, Arsenic poisoning. South Med J 73 (1980) 198–208.

[143] H. Schulz, L. Nagymajtenyi, L. Institoris, A. Papp, O. Siroki, A study on behavioral, neurotoxicological, and immunotoxicological effects of subchronic arsenic treatment in rats. J Toxicol Environ Health A 65 (2002) 1181–1193.

[144] M. Sharpe, Deadly waters run deep: the global arsenic crisis. J Environ Monit 5 (2003) 81N–85N.

[145] S. Shila, V. Kokilavani, M. Subathra, C. Panneerselvam, Brain regional responses in antioxidant system to alpha-lipoic acid in arsenic intoxicated rat. Toxicology 210 (2005) 25–36.

[146] S. Shila, M. Subathra, M.A. Devi, C. Panneerselvam, Arsenic intoxication-induced reduction of glutathione level and of the activity of related enzymes in rat brain regions: reversal by DL-alpha-lipoic acid. Arch Toxicol 79 (2005) 140–146.

[147] M. Sinha, P. Manna, P.C. Sil, Protective effect of arjunolic acid against arsenic-induced oxidative stress in mouse brain. J Biochem Mol Toxicol 22 (2008) 15–26.

[148] P.L. Smedley, D.G. Kinniburgh, A review of the source, behaviour and distribution of arsenic in natural waters. Appl Geochem 17 (2002) 517–568.

[149] S.L. Soignet, S.R. Frankel, D. Douer, M.S. Tallman, H. Kantarjian, E. Calleja, R.M. Stone, M. Kalaycio, D.A. Scheinberg, P. Steinherz, E.L. Sievers, S. Coutre, S. Dahlberg, R. Ellison, R.P. Warrell, Jr., United States multicenter study of arsenic trioxide in relapsed acute promyelocytic leukemia. J Clin Oncol 19 (2001) 3852–3860.

[150] H. Tanaka, H. Tsukuma, A. Oshima, Long-term prospective study of 6104 survivors of arsenic poisoning during infancy due to contaminated milk powder in 1955. J Epidemiol 20 (2010) 438–445.

[151] D.J. Thomas, M. Styblo, S. Lin, The cellular metabolism and systemic toxicity of arsenic. Toxicol Appl Pharmacol 176 (2001) 127–144.

[152] N. Tripathi, G.M. Kannan, B.P. Pant, D.K. Jaiswal, P.R. Malhotra, S.J. Flora, Arsenic-induced changes in certain neurotransmitter levels and their recoveries following chelation in rat whole brain. Toxicol Lett 92 (1997) 201–208.

[153] S.Y. Tsai, H.Y. Chou, H.W. The, C.M. Chen, C.J. Chen, The effects of chronic arsenic exposure from drinking water on the neurobehavioral development in adolescence. Neurotoxicology 24 (2003) 747–753.

[154] C.H. Tseng, Abnormal current perception thresholds measured by neurometer among residents in blackfoot disease-hyperendemic villages in Taiwan. Toxicol Lett 146 (2003) 27–36.

[155] H.P. Tseng, Y.H. Wang, M.M. Wu, H.W. The, H.Y. Chiou, C.J. Chen, Association between chronic exposure to arsenic and slow nerve conduction velocity among adolescents in Taiwan. J Health Popul Nutr 24 (2006) 182–189.

[156] A. Vahidnia, F. Romijn, M. Tiller, G.B. van der Voet, F.A. de Wolff, Arsenic-induced toxicity: effect on protein composition in sciatic nerve. Hum Exp Toxicol 25 (2006) 667–674.

[157] A. Vahidnia, R.J. van der Straaten, F. Romijn, J. van Pelt, G.B. van der Voet, F.A. de Wolff, Arsenic metabolites affect expression of the neurofilament and tau genes: an in-vitro study into the mechanism of arsenic neurotoxicity. Toxicol In Vitro 21 (2007) 1104–1112.

[158] A. Vahidnia, G.B. van der Voet, F.A. de Wolff, Arsenic neurotoxicity: a review. Hum Exp Toxicol 26 (2007) 823–832.

[159] S. Valkonen, H. Savolainen, J. Jarvisalo, Arsenic distribution and neurochemical effects in peroral sodium arsenite exposure of rats. Bull Environ Contam Toxicol 30 (1983) 303–308.

[160] O.S. von Ehrenstein, S. Poddar, Y. Yuan, D.G. Mazumder, B. Eskenazi, A. Basu, M. Hira-Smith, N. Ghosh, S. Lahiri, R. Haque, A. Ghosh, D. Kalman, S. Das, A.H. Smith, Children's intellectual function in relation to arsenic exposure. Epidemiology 18 (2007) 44–51.

[161] A. Wang, S.D. Holladay, D.C. Wolf, S.A. Ahmed, J.L. Robertson, Reproductive and developmental toxicity of arsenic in rodents: a review. Int J Toxicol 25 (2006) 319–331.

[162] S.X. Wang, Z.H. Wang, X.T. Cheng, J. Li, Z.P. Sang, X.D. Zhang, L.L. Han, X.Y. Qiao, Z.M. Wu, Z.Q. Wang, Arsenic and fluoride exposure in drinking water: children's IQ and growth in Shanyin county, Shanxi province, China. Environ Health Perspect 115 (2007) 643–647.

[163] X. Wang, D. Meng, Q. Chang, J. Pan, Z. Zhang, G. Chen, Z. Ke, J. Luo, X. Shi, Arsenic inhibits neurite outgrowth by inhibiting the LKB1-AMPK signaling pathway. Environ Health Perspect 118 (2010) 627–634.

[164] Y. Wang, S. Li, F. Piao, Y. Hong, P. Liu, Y. Zhao, Arsenic down-regulates the expression of Camk4, an important gene related to cerebellar LTD in mice. Neurotoxicol Teratol 31 (2009) 318–322.

[165] G.A. Wasserman, X. Liu, F. Parvez, H. Ahsan, P. Factor-Litvak, J. Kline, A. van Geen, V. Slavkovich, N.J. Loiacono, D. Levy, Z. Cheng, J.H. Graziano, Water arsenic exposure and intellectual function in 6-year-old children in Araihazar, Bangladesh. Environ Health Perspect 115 (2007) 285–289.

[166] G.A. Wasserman, X. Liu, F. Parvez, H. Ahsan, P. Factor-Litvak, A. van Geen, V. Slavkovich, N.J. Loiacono, Z. Cheng, I. Hussain, H. Momotaj, J.H. Graziano, Water arsenic exposure and children's intellectual function in Araihazar, Bangladesh. Environ Health Perspect 112 (2004) 1329–1333.

[167] P.M. Wax, C.A. Thornton, Recovery from severe arsenic-induced peripheral neuropathy with 2,3-dimercapto-1-propanesulphonic acid. J Toxicol Clin Toxicol 38 (2000) 777–780.

[168] A.J. Wilbourn, Sensory nerve conduction studies. J Clin Neurophysiol 11 (1994) 584–601.

[169] S.P. Wilkinson, P. McHugh, S. Horsley, H. Tubbs, M. Lewis, A. Thould, M. Winterton, V. Parsons, R. Williams, Arsine toxicity aboard the Asiafreighter. Br Med J 3 (1975) 559–563.

[170] P.N. Williams, M.R. Islam, E.E. Adomako, A. Raab, S.A. Hossain, Y.G. Zhu, J. Feldmann, A.A. Meharg, Increase in rice grain arsenic for regions of Bangladesh irrigating paddies with elevated arsenic in groundwaters. Environ Sci Technol 40 (2006) 4903–4908.

[171] A.J. Windebank, Specific inhibition of myelination by lead in vitro; comparison with arsenic, thallium, and mercury. Exp Neurol 94 (1986) 203–212.

[172] R.O. Wright, C. Amarasiriwardena, A.D. Woolf, R. Jim, D.C. Bellinger, Neuropsychological correlates of hair arsenic, manganese, and cadmium levels in school-age children residing near a hazardous waste site. Neurotoxicology 27 (2006) 210–216.

[173] M.M. Wu, T.L. Kuo, Y.H. Hwang, C.J. Chen, Dose-response relation between arsenic concentration in well water and mortality from cancers and vascular diseases. Am J Epidemiol 130 (1989) 1123–1132.

[174] S. Xi, L. Guo, R. Qi, W. Sun, Y. Jin, G. Sun, Prenatal and early life arsenic exposure induced oxidative damage and altered activities and mRNA expressions of neurotransmitter metabolic enzymes in offspring rat brain. J Biochem Mol Toxicol 24 (2010) 368–378.

[175] S. Xi, Y. Jin, X. Lv, G. Sun, Distribution and speciation of arsenic by transplacental and early life exposure to inorganic arsenic in offspring rats. Biol Trace Elem Res 134 (2010) 84–97.

[176] R.S. Yadav, L.P. Chandravanshi, R.K. Shukla, M.L. Sankhwar, R.W. Ansari, P.K. Shukla, A.B. Pant, V.K. Khanna, Neuroprotective efficacy of curcumin in arsenic induced cholinergic dysfunctions in rats. Neurotoxicology 32 (2011) 760–768.

[177] R.S. Yadav, M.L. Sankhwar, R.K. Shukla, R. Chandra, A.B. Pant, F. Islam, V.K. Khanna, Attenuation of arsenic neurotoxicity by curcumin in rats. Toxicol Appl Pharmacol 240 (2009) 367–376.

[178] R.S. Yadav, R.K. Shukla, M.L. Sankhwar, D.K. Patel, R.W. Ansari, A.B. Pant, F. Islam, V.K. Khanna, Neuroprotective effect of curcumin in arsenic-induced neurotoxicity in rats. Neurotoxicology 31 (2010) 533–539.

[179] C.C. Yen, T.J. Ho, C.C. Wu, C.F. Chang, C.C. Su, Y.W. Chen, T.R. Jinn, T.H. Lu, P.W. Cheng, Y.C. Su, S.H. Liu, C.F. Huang, Inorganic arsenic causes cell apoptosis in mouse cerebrum through an oxidative stress-regulated signaling pathway. Arch Toxicol 85 (2011) 565–575.

[180] S.F. Yip, Y.M. Yeung, E.Y. Tsui, Severe neurotoxicity following arsenic therapy for acute promyelocytic leukemia: potentiation by thiamine deficiency. Blood 99 (2002) 3481–3482.

[181] M.I. Yousef, F.M. El-Demerdash, F.M. Radwan, Sodium arsenite induced biochemical perturbations in rats: ameliorating effect of curcumin. Food Chem Toxicol 46 (2008) 3506–3511.

[182] S. Zarazua, S. Burger, J.M. Delgado, M.E. Jimenez-Capdeville, R. Schliebs, Arsenic affects expression and processing of amyloid precursor protein (APP) in primary neuronal

cells overexpressing the Swedish mutation of human APP. Int J Dev Neurosci 29 (2011) 389–396.

[183] S. Zarazua, F. Perez-Severiano, J.M. Delgado, L.M. Martinez, D. Ortiz-Perez, M.E. Jimenez-Capdeville, Decreased nitric oxide production in the rat brain after chronic arsenic exposure. Neurochem Res 31 (2006) 1069–1077.

[184] W. Zheng, D.F. Perry, D.L. Nelson, H.V. Aposhian, Choroid plexus protects cerebrospinal fluid against toxic metals. FASEB J 5 (1991) 2188–2193.

10

DIABETES MELLITUS

Miroslav Stýblo and Christelle Douillet

Department of Nutrition, University of North Carolina Chapel Hill, Chapel Hill, NC, USA

10.1 INTRODUCTION

Growing evidence from epidemiological and laboratory studies suggests that exposures to naturally occurring environmental chemicals or the environmental contaminants produced by man contribute to the current worldwide epidemic of diabetes mellitus [74, 76, 105]. Diabetes is a metabolic disease characterized by the disruption of mechanisms that regulate carbohydrate metabolism, resulting in an impaired utilization of glucose by liver and peripheral tissues and by increased glucose levels in blood (hyperglycemia). Insulin-dependent (type 1) diabetes is caused by severe deficiency of circulating insulin (hypoinsulinemia) due to the autoimmune destruction of insulin-producing pancreatic β-cells. Although often called juvenile diabetes, type 1 diabetes strikes both children and adults. In addition to hyperglycemia and hypoinsulinemia, the untreated type 1 diabetes is characterized by increased levels of ketone bodies in the blood (ketoacidosis). Ketone bodies are products of the breakdown of fatty acids that, in case of type 1 diabetes, replace glucose as the major source of energy in the body. Hyperglycemia is also characteristic of noninsulin-dependent (type 2) diabetes. However, in this case, the impairment of glucose metabolism is associated with insulin resistance, that is, with the decreased sensitivity of the liver and peripheral tissues to insulin signal, which is required to stimulate cellular glucose uptake and metabolism. The compensatory increase of insulin production by pancreas ultimately leads to an impairment of β-cell function, resulting in a decreased insulin secretion. More than two thirds of type 2 diabetes cases are represented by overweight or obese individuals [29].

Arsenic: Exposure Sources, Health Risks, and Mechanisms of Toxicity, First Edition.
Edited by J. Christopher States.
© 2016 John Wiley & Sons, Inc. Published 2016 by John Wiley & Sons, Inc.

Type 2 diabetes was originally characterized as an adult onset disease. However, with growing obesity of children, type 2 diabetes is now being diagnosed also in early stages of life [74]. The pathogenesis and clinical symptoms of gestational (type 3) diabetes are similar to those of type 2 diabetes. Type 3 diabetes occurs in women during pregnancy and usually improves or disappears after childbirth.

10.2 DIABETES ASSOCIATED WITH EXPOSURE TO ARSENIC

Association between exposure to arsenic (As) and diabetes has been extensively examined in both laboratory and population-based studies. Results of these studies have been thoroughly reviewed in several recent publications focusing mainly on exposure to inorganic As (iAs) and on drinking water as the major source of iAs exposure for human populations worldwide [4, 12, 26, 42, 73]. One of the most comprehensive reviews was published by members of a National Toxicology Program expert panel who met in January 2011 to evaluate current evidence linking iAs exposure to the risk of diabetes [64]. Although causality of the association remains unclear, the experts agreed that existing human data support a link between chronic exposure to iAs and diabetes in populations with high exposure levels (≥150 μg As/L in drinking water), while the evidence was viewed as insufficient for the association between diabetes and low iAs exposures (<150 μg As/L drinking water) [64]. This chapter summarizes results of the published research and provides insights into potential cellular and molecular mechanisms by which iAs exposure could disrupt glucose homeostasis in a manner consistent with diabetes.

10.2.1 Epidemiological Studies

The association between iAs exposure and increased prevalence of diabetes was first reported in arsenicosis-endemic areas of Taiwan [51]. This study found a twofold increase in the prevalence of diabetes among local residents drinking water with 700–930 μg As/L as compared to residents in nonendemic areas. A significant dose–response relationship between cumulative exposure to iAs and the prevalence of diabetes was also found. The link between diabetes and chronic exposure to high levels of iAs in drinking water was later confirmed by several prospective and cross-sectional studies from Taiwan [110, 111] and Bangladesh [90, 91]. These studies used fasting blood glucose, oral glucose tolerance test (OGTT), glucosuria, and self-reported diagnosis or medication to identify diabetic individuals. Another study from high-exposure areas of Taiwan [108] linked iAs exposure to increased mortality attributed to diabetes. Other early studies focused on individuals exposed to iAs in occupational settings, including copper smelter, glass worker, and taxidermists [48, 89, 92]. These studies were retrospective in nature and primarily used death certificates to establish association between iAs exposure and diabetes as the cause of death.

Lately, more than two dozen of original papers have been published that examined the association between environmental exposures to iAs and diabetes in various countries, including Bangladesh [15, 46, 70], China [114], Cyprus [63], Mexico

[19, 21, 25], Pakistan [2, 50], Serbia [49], Spain [94], Taiwan [115, 116], Turkey [96], and the United States [28, 39, 54, 67, 71, 72, 102, 103, 106, 126]. Most of the recent studies were cross-sectional, case–control, or retrospective. Some relied on death certificates, self-reported diabetes, doctor's diagnosis, or insurance claims to identify diabetes cases. Other studies used glucosuria that may be indicative of diabetes but is not an accepted diagnostic tool. More recent studies used clinical indicators that are recommended for diabetes diagnosis, including fasting blood or plasma glucose levels or glucose levels determined during OGTT. Only a few studies also measured plasma insulin levels and calculated homeostasis model assessment of insulin resistance (HOMA-IR), which is the generally accepted measure of insulin resistance. Some studies only compared As levels in urine of diabetic and nondiabetic individuals and provided no quantitative data on diabetes risk [2, 50, 96]. In general, almost all studies carried out in moderate- or high-exposure areas have reported a significant association between chronic exposure to iAs and risk of diabetes. Results of studies in low-exposure areas were rather inconsistent.

10.2.1.1 Reports Linking Diabetes to High iAs Exposure The strongest associations of diabetes with iAs exposure have consistently been reported from high-exposure areas of Bangladesh and Taiwan (Table 10.1) [46, 51, 70, 90, 91, 108, 110, 111, 116]. Here, the levels of iAs exposure ranged from <10 to 2100 μg As/L of drinking water, and some of the study subjects displayed skin lesions, a typical sign of chronic iAs poisoning. The odds ratio (OR) for risk of diabetes reported in these studies for the highest exposure levels was as high as 10.05 with the corresponding 95% confidence interval (95% CI) of 1.3–77.9 [51]. The diabetes diagnosed in the high-exposure areas of Taiwan and Bangladesh was characterized as type 2 diabetes because of the adult onset of the disease and the lack of ketoacidosis, typically associated with type 1 diabetes. None of these studies, however, assessed blood insulin level, insulin resistance, or beta-cell function to further characterize the disease phenotype.

10.2.1.2 Reports Linking Diabetes to Moderate or Low iAs Exposure The reports of significant associations between iAs exposure and diabetes are not limited only to the high-exposure areas. Two cross-sectional studies in the Coahuila state [19] and in the Zimapan and Lagunera regions [21] of Mexico examined diabetes prevalence in populations with low to moderately high levels of exposure to iAs in drinking water—up to 400 and 215 μg As/L, respectively. In both studies, diabetes was diagnosed using the established diagnostic tools, fasting blood glucose or OGTT. The Coahuila study found a significantly increased risk of diabetes (OR = 2.16; 95% CI, 1.23–3.79) for subjects with urinary As levels of 63.5–104 μg/g of creatinine and even higher risk (OR = 2.84; 95% CI, 1.64–4.92) for subjects with urinary As levels >104 μg/g of creatinine. In Zimapan and Lagunera, the risk of diabetes was examined for four exposure categories, including <10, 10–49.9, 50–124.9, and ≥125 μg As/L of drinking water. The risk was significantly associated with the level of As in drinking water: OR = 1.13 (95% CI, 1.05–1.22) for an incremental increase of 10 μg As/L. Here, the ORs of 15.8 (95% CI 1.69–147.6) and

TABLE 10.1 Epidemiologic Studies of the Association between Arsenic Exposure and Diabetes

Reference	Study Area	Subjects (Average Age or Age Range)	Average Arsenic Exposure/Exposure Categories	Diabetes Diagnosis	Statistically Significant Association or Difference
Afridi et al. [2]	Pakistan	Men nonsmokers, $n=225$ Men smokers $n=209$ Age range: 31–60	Nonsmokers: urine As 4.7 μg/L (healthy) versus 5.59 μg/L (diabetic) Smokers: urine As 5.41 μg/L (healthy) versus 7.27 μg/L (diabetic)	FBG HbA1c Self-reported	Yes
Chen et al. [13]	Central Taiwan	$N=1{,}043$ Age range: 35–64	Urine As: >200 μg/g versus ≤35 μg/g creatinine	FBG	Yes AdjOR = 2.22
Chen et al. [14]	Southwest Taiwan	Controls, $n=136$, age = 60.4 Patients with MetS, $n=111$, age = 64.3	As in water: 684 (MetS) μg/L versus 570 (controls) μg/L CAE: 17.71 (MetS) mg/L/y versus 13.96 (controls) mg/L/y	Not reported (average FBG for MetS = 127.4 mg/dL)	Yes For MetS: OR = 1.51 for As in water >768 μg/L versus <700 μg/L OR = 2.71 for CAE >18.9 mg/L/y versus <12.6 mg/L/y
Chen et al. [15]	Bangladesh	$N=11{,}319$ Age range: 25–56	As in water: 176–864 (Q5) μg/L versus 0.1–8 (Q1) μg/L	Self-reported	No AdjOR for = 1.1
Coronado-Gonzalez et al. [19]	Mexico	$N=400$ Age range: 30–87	Urine As: <63.5 (T1) μg/g versus >104 (T3) μg/g creatinine	FBG Treatment	Yes OR = 2.84
Del Razo et al. [21]	Mexico (Zimapan, Lagunera)	$N=258$ Age for diabetics = 50.4 Age for nondiabetics = 32.3	As in water: 42.9 μg/L (range 3.1–215.2 ppb)	FBG ≥ 126 mg/dL 2HBG ≥ 200 mg/dL Self-reported	Yes OR increases by 13% per 10 ppb As in water

Ettinger et al. [28]	Oklahoma	N = 456 pregnant women Age = 24.5 (range: 14–36+)	Blood As: 0.2–0.9 (Q1) μg/L versus 2–24 (Q4) μg/L	OGTT	Yes AdjOR = 2.79
Flores et al. [31]	Mexico (Guanajuato)	N = 76 Age = 52	Serum As (0.5–1.8 μg/L) and urine As (<0.5–48.2 μg/L)	Previous diagnosis (FBG)	No Higher urine As in diabetics versus controls
Gribble et al. [39]	American Indians (AZ, OK, ND, SD)	N = 3,925 Age range: 45–74	Urine As median = 14.1 μg/L (interquartile range: 7.9–24.2 μg/L)	FBG ≥ 126 mg/dL 2HBG ≥ 200 mg/dL HbA1c ≥ 6.5% Treatment	Yes For subjects with poor glycemic control: PR = 1.55 Q4 versus Q1 urinary As
Islam et al. [46]	Bangladesh	N = 1,004 Age = 44.9	As in water = 159 μg/L (range: 10–1401 μg/L)	FBG ≥ 126 mg/dL Self-reported	Yes OR = 1.9 for As >50 μg/L of water
Jovanovic et al. [49]	Serbia	Exposed, n = 541 Unexposed, n = 2,746 Age for Men = 60 Women = 63	As in water = 56.1 μg/L (range: 1–349 μg/L) versus 2.0 μg/L (range: 0.5–4 μg/L)	FBG OGTT	Yes OR = 1.22–1.56 for the period of 2006–2009
Kolachi et al. [50]	Pakistan	Healthy women, n = 68 Diabetic women, n = 76 Age range: 30–40	Hair As = 1.48 μg/g Blood As = 2.37 μg/L Urine As = 4.13 μg/L	FBG ≥ 140 mg/ dL 2HBG ≥ 200 mg/dL	Yes As in hair, blood, and urine from healthy mothers and their infants lower as compared to diabetic mothers and their infants
Lai et al. [51]	Taiwan	N = 891 Age range: 30–60+	CAE: ≥15 ppm-y versus 0 ppm-y	OGTT Self-reported	Yes AdjOR = 10.05

(continued)

TABLE 10.1 *(Continued)*

Reference	Study Area	Subjects (Average Age or Age Range)	Average Arsenic Exposure/Exposure Categories	Diabetes Diagnosis	Statistically Significant Association or Difference
Lewis et al. [54]	Utah	N=961 female deaths N=1,242 male deaths	As in water range of median=14–166 μg/L (overall range: 3.5–620 μg/L)	Death certificate	Yes Women SMR=1.23 No Men SMR=0.79
Makris et al. [63]	Cyprus	N=317 Age range: 18–60+	CLAEX median=613 mg (range: 237–1305 mg)	Self-reported	Yes OR=1.78 Higher CLAEX in diabetes versus nondiabetics (999 mg vs. 573 mg) No No longer significant after adjusting for age, sex, and smoking
Meliker et al. [67]	Michigan	N=38,722 female deaths N=41,282 male deaths	As in water=7.58 μg/L (range: 1.27–11.98 μg/L)	Death certificate	Yes Women SMR=1.28 Men SMR=1.27
Nabi et al. [70]	Bangladesh	N=235 Age range: 35.4	As in water: 218 μg/L versus 11.3 μg/L (range: 3–875)	Glycemia	Yes OR=2.95
Navas-Acien et al. [71]	United States (NHANES 2003–2004)	N=788 Age≥20	Urine As: 18 (≥80th) μg/L versus 3.5 (≤20th percentile) μg/L	FBG Self-reported Treatment	Yes AdjOR=3.58

Navas-Acien et al. [72]	United States (NHANES 2003–2006)	N = 1,279 (with arsenobetaine < LOD) Age ≥ 20	Urine As: 7.4 (≥80th) μg/L versus 1.6 (≤20th percentile) μg/L	FBG Self-reported Treatment	Yes AdjOR = 2.60
Rahman et al. [90]	Bangladesh	N = 1,107 with keratosis Age range: 30–60+	As in water: <10–2100 μg/L	Self-reported Glucosuria + FBG and 2HBG	Yes AdjPR = 5.2
Rahman et al. [91]	Bangladesh	N = 430 with skin lesion Age range: 30–60+	As in water: <500 μg/L versus >1000 μg/L Cumulative exposure index: <1 mg-y/L versus >10 mg-y/L	Glucosuria (by glucometric strips)	Yes AdjPR = 2.9
Ruiz-Navarro et al. [94]	Spain	N = 87	75th versus 25th percentile microgram per liter urine Urine As = 3.68 μg/L	Not reported	No RR = 0.87
Serdar et al. [96]	Turkey	N = 87 Age for nondiabetics = 52 Age for diabetics = 59	Plasma As: 1.22 (diabetics) μg/L versus 0.86 (nondiabetics) μg/L	Treatment	No
Steinmaus et al. [102]	United States (NHANES 2003–2004)	N = 795 Age ≥ 20	Urine As: 12 (≥80th) μg/L versus 2.7 (≤20th percentile) μg/L urine not adjusted for creatinine [urine As = total As-(arsenobetaine + arsenocholine)]	FBG Self-reported Treatment	No AdjOR = 1.15
Steinmaus et al. [103]	United States (NHANES 2003–2006)	N = 1,280 (with arsenobetaine < LOD) Age ≥ 20	Urine As: ≥80th versus ≤20th percentile microgram per liter not adjusted for creatinine	FBG Self-reported Treatment	No AdjOR = 1.03
Tollestrup et al. [106]	Washington	N = 1,074 deaths (lived near smelter as children)	≥10 years residence within 1.6 km of smelter versus <1 year residence	Death certificate	No RR = 1.6

(continued)

TABLE 10.1 *(Continued)*

Reference	Study Area	Subjects (Average Age or Age Range)	Average Arsenic Exposure/Exposure Categories	Diabetes Diagnosis	Statistically Significant Association or Difference
Tsai et al. [108]	Taiwan	$N = 19{,}536$	As in water median = 780 μg/L (range: 250–1140 μg/L)	Death certificate	Yes SMR = 1.46
Tseng et al. [110, 111]	Taiwan	$N = 446$ Age = 47.4	As in water: <17 mg/L-y versus ≥17 mg/L-y	FBG ≥ 7.8 mM 2HBG ≥ 11.1 mM	Yes OR = 2.1
Yassine et al. [123]	Arizona	$N = 43$ Age for controls = 51 Age for diabetics = 58	Urine As: 5.3 ppb (controls) versus 6.6 ppb (diabetics)	Prior history 2HBG ≥ 200 mg/dL	Yes Total iAs and DMAs adjusted to creatinine increases in diabetics versus nondiabetics No Unadjusted or specific gravity adjusted iAs not different in diabetic versus control
Wang et al. [116]	Taiwan	$N = 706{,}314$ Age range: 25–65+	Endemic (>0.35 mg/L water) versus nonendemic regions As in water median: 780 μg/L (range: 350–1140 μg/L)	National Health Insurance claims	Yes OR = 2.69

Wang et al. [115]	Taiwan	$N = 660$ Age range: 35–64	Range of As in water: ~6–15 μg/L water High versus low microgram per gram in hair	Metabolic syndrome (FBG)	Yes AdjOR = 2.35
Wang et al. [114]	China (Xinjiang)	$N = 235$ Age > 30	Range of As in water: 21–272 (endemic) μg/L versus 16–38 (control) μg/L	Hospital records, exam (BG)	Yes RR = 1.098
Ward and Pim 1984 [118]	England	$N = 117$ Age range for controls = 18–64 Age range for diabetics = 38–78	Plasma As: 75th versus 25th percentile	Not reported	Yes RR = 1.09
Zierold et al. [126]	WI	$N = 1,185$ Age = 62	As in water: >10 μg/L versus <2 μg/L As in water median = 2 μg/L (range: 0–2389 μg/L)	Self-reported	No AdjOR = 1.02

2HBG, 2-hour blood/plasma glucose recorded during OGTT; AdjOR, adjusted odds ratio; AdjPR, adjusted prevalence ratio; BG, blood/plasma glucose; CAE, cumulative arsenic exposure; CLAEX, cumulative lifetime As exposure; FBG, fasting blood/plasma glucose; LOD, limit of detection; MetS, metabolic syndrome; NHANES, National Health and Nutrition Examination Survey; NS, nonsignificant; OGTT, oral glucose tolerance test; OR, odds ratio; ppb, parts per billion; ppm-y, parts per million year; PR, prevalence ratio; Q1, Q3, Q4, Q5, Quartiles 1, 3, 4, and 5; RR, relative risk; SMR, standardized mortality ratio; T1, T3, Tertile 1 and 3.

TABLE 10.2 Association of Diabetes Classified by FBG or 2HBG with Exposure To iAs in Zimapan and Lagunera, Mexico (Adjusted for Age, Sex, Obesity, and Hypertension)[a]

iAs in water	FBG					
(μg/L)	Cases	Noncases	OR	95% CI		p^b
<50	8	186	1.00			
50–124.9	9	50	3.3	1.11	9.85	0.03
≥125	6	22	7.40	2.13	25.65	<0.01
iAs in water	**2HBG**					
(μg/L)	Cases	Noncases	OR	95% CI		p^b
<50	9	177	1.00			
50–124.9	9	41	2.89	0.97	8.56	0.06
≥125	6	16	7.01	2.02	24.44	<0.01

2HBG, 2 hours blood glucose; CI, confidence interval; FBG, fasting blood glucose; OR, odds ratio.
[a] Reanalysis of data from Ref. [21].
[b] *p*-value for comparison of cases to noncases.

5.01 (95% CI, 1.02–24.7) were found for As levels ≥125 μg/L based on the fasting blood glucose level (FBG ≥ 126 mg/dL) and the glucose level recorded 2 h into OGTT (2HBG ≥ 200 mg/dL), respectively. However, a reanalysis of data from this study using only three exposure categories (<50, 50–124.9, and ≥125 μg As/L) shows that the risk increases significantly already for exposures at or above 50 μg As/L (Table 10.2). Notably, 50 μg As/L was the US EPA maximum contaminant level for As in drinking water before 2001. The Zimapan and Lagunera study also found a positive statistically significant association between iAs exposure and the blood levels of glycated hemoglobin (HbA1c), an indicator of chronic hyperglycemia. This was also one of the few studies that measured fasting plasma insulin levels and assessed insulin resistance. It found statistically significant negative correlations between iAs exposure and both the fasting plasma insulin level and HOMA-IR. The fact that in this study iAs exposure increased the risk of diabetes, as demonstrated by high fasting blood glucose levels and impaired glucose tolerance, but was associated with lower plasma insulin and lower HOMA-IR challenges the generally accepted view that iAs-induced diabetes is type 2 diabetes, which is typically characterized by high plasma insulin levels (hyperinsulinemia) and insulin resistance (i.e., high HOMA-IR values).

A significant increase in the risk of diabetes has also been reported from several other regions with low to moderate levels of iAs in drinking water (Table 10.1), including low-exposure areas of central Taiwan (6–15 μg As/L of drinking water) where fasting plasma glucose levels correlated with As concentration in hair [115] and the Middle Banat region of Serbia with the average concentration of As in drinking water of 56 μg/L [49]. Consistent with these findings are reports from Pakistan [2, 50] where higher concentrations of As were found in urine, blood, or hair of diabetics, including diabetic mothers and their newborns, as compared to nondiabetic controls. Similarly, Makris and associates [63] reported a statistically significant association between cumulative lifetime exposure to As and risk of diabetes in Mammari, Cyprus, where As levels in drinking water ranged from <10 to 70 μg As/L. However, after adjusting for age, sex, smoking, education, and fish consumption, the effect of As on diabetes was not statistically significant, raising questions about the methodology used for the assessment of iAs exposure and for the evaluation of associations between this exposure and health outcomes.

10.2.1.3 Reports Linking Diabetes to iAs Exposure in the United States Several US studies found significant association between exposure to iAs and risk of diabetes (Table 10.1). One of these studies used blood As levels to characterize As exposure in pregnant women from Tar Creek Superfund site, a former lead and zinc mining area in Oklahoma [28]. This study found impaired glucose tolerance to be associated with higher As levels in blood (2–24 μg/L). Another study linked iAs exposure from drinking water to diabetes as the cause of death among residents of six counties of southeastern Michigan where groundwater As levels typically range from 10 to 100 μg/L [67]. A statistically significant association between As exposure (total As level in urine) and diabetes classified by fasting blood glucose was also reported for American Indian adults from Arizona, Oklahoma, and North and South Dakota who participated in the Strong Heart Study (a population-based study of cardiovascular disease and diabetes in American Indian communities) [39]. However, this association was restricted to participants with poor diabetes control, that is, with HbA1c levels exceeding 8% of the total blood hemoglobin. Notably, urinary As was not associated with HbA1c or with insulin resistance (HOMA-IR) in participants without diabetes.

Of particular interest are three studies that analyzed data from the 2005 to 2008 National Health and Nutrition Examination Survey (NHANES) that was designed to assess the health and nutritional status of adults and children in the United States. Two of these studies reported a statistically significant association between the risk of diabetes and urinary As levels after adjusting for fish consumption [71] or after excluding individuals with fish As (arsenobetaine) levels in urine above the limit of detection [72]. Arsenobetaine, which is commonly found in marine fish and other sea organisms, is relatively nontoxic, and there are no reports linking exposure to this arsenical to adverse health effects. A low level or absence of arsenobetaine in urine indicates that little or no seafood was consumed in several days before urine collection and that As species in urine are associated with As exposure from drinking water or from other sources. Arsenobetaine, as well as some of the other arsenicals found in seafood (e.g., arsenocholine or arsenosugars), could be partially metabolized in human body to dimethylarsenic, which is also produced in the course of iAs metabolism [64]. Thus, accounting for seafood consumption or correcting for biomarkers of the exposure to arsenicals from seafood is essential for proper characterization of the exposure to iAs. This is particularly important for studies involving populations with low levels of iAs in drinking water for which seafood is often the major source of As exposure and arsenobetaine or other arsenicals originated from seafood represent most of the As species excreted in urine [57, 72, 102].

10.2.1.4 Studies Reporting No Association between iAs Exposure and Diabetes Studies that did not find linkage between iAs exposure and risk of diabetes come from both low- and high-exposure areas (Table 10.1). One of these studies used the NHANES data and found no association between diabetes and As concentration in urine after subtracting urinary arsenobetaine [102, 103]. However, this approach is likely to misclassify iAs exposure because it does not account for other arsenicals of seafood origin, and specifically for the part of dimethylarsenic produced by metabolism of these arsenicals [64]. Two other studies in the United States examined diabetes as the cause of death using death certificates [54, 106], which are known to have low sensitivity and specificity for diabetes [16]. Two studies from Spain [94] and Turkey [96] only examined As levels in urine and plasma of diabetic and nondiabetic individuals and reported no statistically significant differences. Another US study examined links between As exposure from drinking water and self-reported chronic illnesses in 19 townships in Wisconsin [126]. An increased risk was found for depression, high blood pressure, circulatory problems, and bypass surgery for residents drinking water with As levels higher that 10 μg/L; however, no statistically significant association was found for diabetes.

One of the most recent and the largest studies in high-exposure areas of Bangladesh also found no association with diabetes [15], thus, contradicting result of all previous studies carried out in this region. This cross-sectional study used urinary As concentration and time-weighted water As to characterize As exposure and self-reported physician diagnosis, glucosuria, or Hb1Ac to identify diabetic individuals. However, several aspects of this study raise concerns. First of all, an unusually low prevalence of diabetes (~2%) was found at the baseline in this study cohort. In comparison, prevalence reported by other studies for As-exposed residents of the endemic areas of Bangladesh was substantially higher: 9–14.8% [46, 90, 91]. In addition, although the HbA1c level in blood is now an accepted and validated measure for diagnosis of diabetes in the United States [1], it is unclear whether it was also validated for the Bangladeshi population. Finally, the glucose level in urine does not necessarily reflect blood glucose level; therefore, glucosuria is not a reliable indicator of diabetes [36]. Thus, using the approved diagnostic tools, including FBG, 2HBG, and HbA1c (if validated for the study population), could prevent misclassification of the associations between iAs exposure and diabetes.

10.2.2 Laboratory Studies

Information about the effects of iAs and organic As species on glucose metabolism can be found in a large number of laboratory studies. However, many of these studies focused on endpoints other than diabetes or glucose homeostasis. Arsenic has frequently been used as an acute stressor in animal and tissue culture studies examining stress-mediated cell signaling or stress-induced responses in various metabolic pathways, including the pathways of carbohydrate metabolism. These types of studies have typically examined glucose metabolism or insulin secretion in cultured cells or laboratory animals exposed to acutely toxic concentrations of As, which are generally incompatible with chronic exposures in humans [83]. Therefore, data from these studies are of little value for research of the diabetogenic effects of environmental exposures to iAs. However, some of the recent studies using better designs and models are supportive of the association between iAs exposure and diabetes and provide data about potential mechanisms by which iAs could disrupt glucose homeostasis in a manner that is consistent with diabetes.

10.2.2.1 Animal Studies The earlier *in vivo* laboratory studies measured insulin or glucose levels in fasted or nonfasted blood or urine of rats, mice, or goats after exposures to iAs or methylated As species in food, drinking water, or oral capsules or via gavage and subcutaneous or intraperitoneal injections [3, 5, 9, 10, 18, 35, 44, 68, 77–79]. The doses, duration of the exposure, and chemical forms of As used in these studies have varied greatly, producing conflicting results and confusing data interpretation with respect to the environmentally induced diabetes. However, results of recent studies that were designed to assess effects of chronic exposure to iAs on relevant diabetes indicators have brought more consistent results. Notably, most of these studies are supportive of the link between diabetes and iAs exposure.

Thus, iAs species (arsenite, arsenate, or arsenic trioxide) provided in drinking water for up to 30 weeks or administered daily by IP injection or gavage have been shown to impair fasting blood glucose or glucose tolerance in adult mice [41, 83, 84, 124] and adult rats [18, 35, 47, 97, 114]. Impaired glucose tolerance was also found in rat offspring after a combined prenatal and postnatal exposure to iAs in drinking water [20]. Oxidative stress in pancreas, pancreatitis, or β-cell damage have been reported in both mice and rats exposed to iAs [10, 11, 20, 47, 69, 124]. Several studies also examined the effects of iAs on plasma insulin and HOMA-IR. The combined prenatal and postnatal exposure to iAs (3 ppm As as arsenite) has been shown to increase insulin resistance (HOMA-IR) in rat offspring [20]. Consistent with these findings, rats receiving arsenite by gavage (1.7 mg As/kg body weight every 12 h) for 90 days developed hyperglycemia, hyperinsulinemia, and insulin resistance [47]. Similar effects were described for pregnant mice injected (IP) on gestational days 1.5 and 8.5 with a high dose of arsenate (9.6 mg/kg body weight), which is expected to cause exencephaly in offspring [41]. These mice developed fasting hyperglycemia and hyperinsulinemia, glucose intolerance, and insulin resistance. Contradicting data on insulin secretion and resistance were reported in two other studies in which mice were exposed to trivalent iAs species in drinking water. Specifically, exposure to arsenic trioxide (10 ppm As for 5 or 7 weeks) alone or combined with exposure to humic acid (500 mg/L) decreased plasma insulin levels in male CD-1 mice [124]. Similarly, exposure of C57BL/6 male mice to arsenite (25 or 50 ppm As for 20 weeks) combined with high-fat diet decreased fasting plasma insulin, HOMA-IR, and insulin response to glucose challenge during OGTT in a dose-dependent manner [84]. Notably, the arsenite-exposed mice fed with the obesogenic high-fat diet were glucose intolerant but were not obese, suggesting that iAs exposure produced diabetes while suppressing fat accumulation.

These contradicting data suggest that while chronic exposure to iAs can produce diabetes in laboratory rodents, the disease phenotype and particularly the effects of exposure on insulin production and β-cell function vary as different animal or As species and treatment protocols were used in different studies. Susceptibility to the diabetogenic effects of iAs exposure may also differ between animal species and between animals and humans. Specifically, mice are thought to be less susceptible than humans to iAs toxicity due to faster metabolism and lower internal dose of iAs and its metabolites [81, 82]. Rats, unlike mice or humans, sequester As (specifically dimethyl-As) in erythrocytes [58–60], thus prolonging iAs clearance and changing the distribution of iAs metabolites between target organs. Generally, rats are not recommended as an animal model for assessing As metabolism or toxicity [75].

10.2.2.2 Tissue Culture Studies and Modes of Action Diabetes associated with iAs exposure has been generally referred to as type 2 diabetes. This type of diabetes is characterized primarily by insulin resistance due to disruption in insulin signaling, which activates glucose uptake and metabolism and thus helps to maintain the blood glucose levels. Results of laboratory studies suggest that disruption of insulin-stimulated glucose uptake (ISGU) could be a mechanism responsible for glucose intolerance associated with chronic exposures to iAs. Early studies showed that subtoxic concentrations of phenylarsine oxide (an aromatic derivative of trivalent As not commonly found in the environment) inhibit insulin signaling and ISGU by cultured murine adipocytes and by isolated rat skeletal muscle [32, 33, 40, 99]. In contrast, high cytotoxic concentrations of phenylarsine oxide or arsenite stimulated the basal, insulin-independent glucose uptake in various cell types [7, 8, 30, 37, 65, 80, 104, 119, 121]. This effect is associated with activation of glucose metabolism as a part of the prosurvival response to an acute stress, which is not consistent with chronic environmental exposure to iAs [83].

Recent studies using physiologically relevant As species and cell culture models have provided relatively detailed information on mechanisms by which subtoxic exposure to iAs inhibits insulin signal and ISGU. These studies focused on mechanisms that mediate insulin signal in peripheral tissues or regulate differentiation of preadipocytes and myoblasts to form the insulin-sensitive adipocytes and myotubes, respectively. In these cells, binding of insulin to the insulin receptor activates the IRS-1, IRS-2, PI3K, PDK, PKB/Akt, and/or PKC-ζ/PKC-λ signaling cascade [17, 100], which in turn triggers the translocation of GLUT4 transporters from the perinuclear compartment to the plasma membrane to increase glucose uptake [45, 52]. Exposures to subtoxic concentrations of arsenite or its methylated trivalent metabolites, methylarsonous and dimethylarsinous acids, have been shown to inhibit the insulin-dependent translocation of GLUT4 to the plasma membrane of differentiated murine adipocytes (Fig. 10.1) [113]. Arsenite and methylarsonous acid inhibited the insulin-stimulated phosphorylation of protein kinase B (PKB/Akt) by phosphoinositide-dependent kinase (PDK), while dimethylarsinous acid inhibited GLUT4 translocation downstream from PKB/Akt [82, 113]. Notably, both methylated trivalent arsenicals were more potent than arsenite as inhibitors of insulin signaling, suggesting that methylation may activate iAs as a diabetogen. Additional evidence suggests that the inhibition of insulin-stimulated PKB/Akt phosphorylation may be associated with an adaptive antioxidant response in adipocytes chronically exposed to low levels of arsenite [122]. However, exposures to arsenite have also been shown to inhibit the differentiation of preadipocytes and myoblasts by disrupting the expression of genes

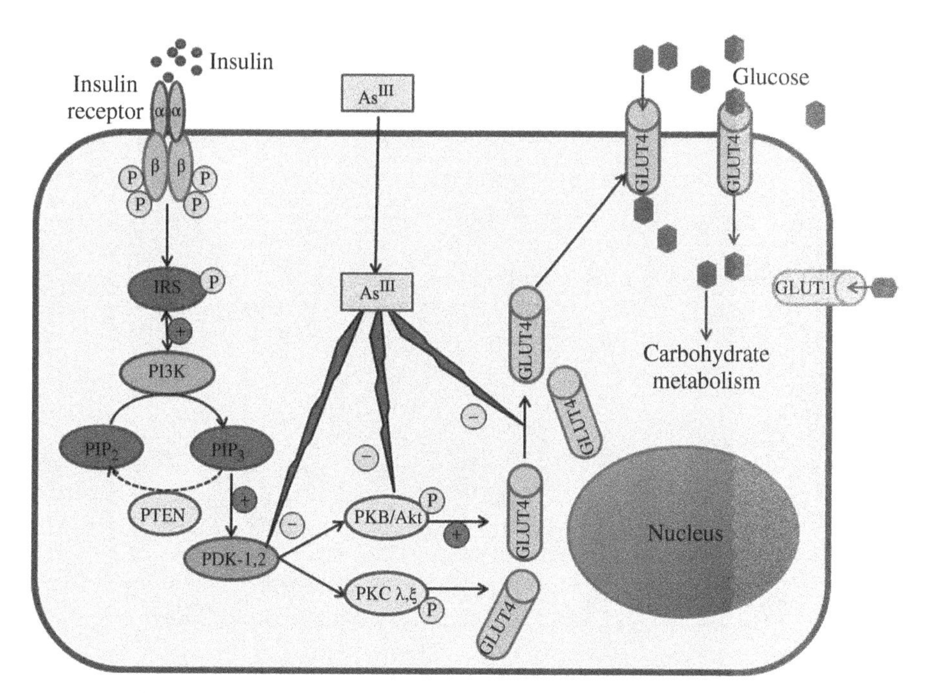

FIGURE 10.1 Arsenic inhibition of insulin signaling in murine adipocyte.

and transcription factors that regulate this process [95, 101, 107, 117, 120, 125]. Thus, both the insulin signal disruption and the inhibition of myogenic and adipogenic differentiation by iAs or by its methylated metabolites may contribute to the diabetogenic effects of iAs exposure and may also explain the inhibition of fat accumulation in the mouse studies described above.

An alternative approach to study the diabetogenic effects of iAs exposure has involved cultured pancreatic islets or insulinoma (β-cell) lines. This work has examined effects of As exposure on glucose-stimulated insulin expression and secretion. Similar to the effects on insulin signaling, effects of As on β-cell function differed between experiments using different exposure levels or different As species. Several earlier studies examined the effects of iAs on pancreatic/duodenal homeobox-1 (PDX-1, insulin promoter factor 1), a transcription factor that activates preproinsulin transcription in response to elevated blood glucose concentrations [66]. Here, a highly cytotoxic concentration of arsenite (1 mM), which is inconsistent with chronic exposure to iAs, was shown to promote PDX-1 translocation to the nucleus [27, 61] and to stimulate PDX-1 DNA binding in murine insulinoma (MIN6) cells and in isolated human pancreatic islets [62]. However, the opposite effect was reported in a study in which isolated rat islets were exposed to a moderately toxic concentration of 5 μM arsenite for 72 h [23]. This treatment inhibited glucose-stimulated expression of insulin. In recent studies, exposures to even lower concentrations of arsenite or methylated trivalent arsenicals had little or no effects on insulin expression, but inhibited glucose-stimulated insulin secretion (GSIS) [24, 34].

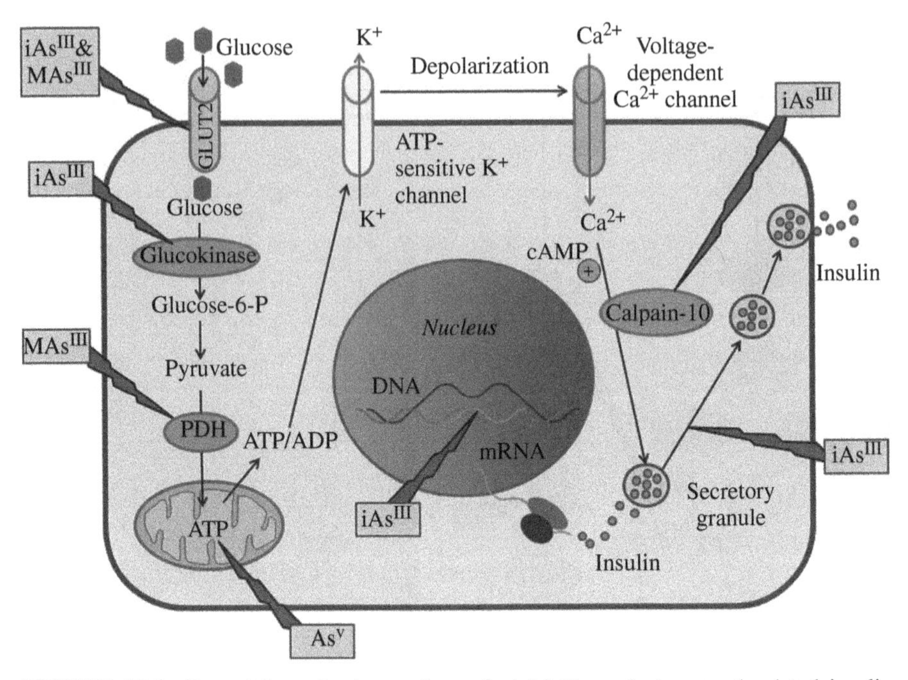

FIGURE 10.2 Potential mechanisms of arsenic inhibition of glucose-stimulated insulin secretion proposed in the literature.

GSIS is regulated by a complex mechanism involving the uptake and oxidative degradation of glucose by β-cells, followed by an increase in ATP production, which results in depolarization of plasma membrane and influx of Ca^{2+} cations, which in turn activate the assembly of insulin secretory vesicles [93]. In addition, reactive oxygen species (ROS) produced during degradation of glucose and ATP production in mitochondria may play a stimulating role in the regulation of GSIS [53, 86, 87]. Recent work by Fu and associates [34] showed that exposure to subtoxic concentrations of arsenite that have no effect on insulin expression inhibits GSIS by cultured rat insulinoma (INS-1) cells, possibly by provoking an adaptive oxidative stress response that increases antioxidant levels and dampens the essential signaling involving ROS. In addition to arsenite, the methylated trivalent arsenicals, methylarsonous and dimethylarsinous acids, were also found to be potent inhibitors of GSIS. Low submicromolar concentrations of these arsenicals inhibited GSIS by isolated murine pancreatic islets with only minor effects on insulin expression [24]. In fact, both methylarsonous and dimethylarsinous acids were more potent than arsenite as GSIS inhibitors. Notably, inhibition of GSIS by arsenite, methylarsonous acid, or dimethylarsinous acid could be almost completely reversed by incubation of the islets in As-free medium.

Although the suppression of the essential ROS signaling may be the mechanism by which exposures to low concentrations of trivalent arsenicals inhibit GSIS, other mechanisms involving multiple steps or components of the GSIS could be proposed based on the current literature (Fig. 10.2). GLUT2 transporters are responsible for the uptake of glucose by β-cells in the first step of GSIS. However, arsenite is also transported across the plasma membrane by GLUT transporters [56] and thus, could compete with glucose uptake. Arsenite has also been shown to modulate the expression of hexokinase, the enzyme that controls the flux of glucose-6-phosphate into glycolysis, the pathway for oxidative degradation of glucose producing pyruvate [88]. Arsenite and methylarsonous acid are inhibitors of pyruvate dehydrogenase [85, 109], the enzyme that channels carbons from pyruvate into Krebs cycle, which in turn generates substrates for ATP synthesis in mitochondria. Arsenite has also been reported to inhibit calpain-10 [22], the Ca^{2+}-dependent protease that participates in the assembly of insulin secretory vesicles in β-cells [112]. Finally, a pentavalent iAs, arsenate, is known to replace phosphate in phosphorylation reactions, including the reactions of ATP synthesis in mitochondria [38].

Thus, being multitarget toxins, iAs and its metabolites can interfere with multiple processes that regulate both insulin signaling and β-cell function in a manner that is consistent but not necessarily identical with type 2 or type 1 diabetes.

10.3 CONCLUSIONS

Results of epidemiologic and recent animal studies are generally supportive of the association between chronic exposures to moderate-to-high levels of iAs in drinking water. Some of the recent human population studies suggest that even relatively low exposures (≤50 μg As/L) may increase the risk of developing diabetes. Diabetes associated with iAs exposure is characterized by hyperglycemia and impaired glucose tolerance, but not necessarily by hyperinsulinemia or insulin resistance. In fact, some data suggest impairment in insulin production by pancreas. Consistent with the population-based and animal studies are results of the *in vitro* experiments that suggest that both insulin signaling and insulin production by β-cells are impaired by exposure to iAs or to its reactive methylated metabolites, methylarsonous and dimethylarsinous acids. Notably, while submicromolar concentrations of the methylated arsenicals can effectively inhibit GSIS by pancreatic islets [24], it takes micromolar concentrations to inhibit insulin signaling and ISGU in adipocytes [82]. Taken together, these data suggest that β-cells may be the primary targets for chronic exposure to iAs and that the basic mechanism and phenotype of diabetes associated with this exposure may combine the mechanisms and phenotypes typically associated with type 1 and type 2 diabetes.

Future epidemiological research should focus on prospective studies to examine causality of the association between iAs exposure and diabetes while providing additional information on the diabetes phenotype, including data on insulin resistance and β-cell function. Characterization of the diabetes phenotype associated with iAs exposure will help to design effective strategies for treatment and possibly for prevention of this disease. Obesity is the number one cause of type 2 diabetes worldwide. However, little is known about interactions between obesity and iAs exposure. Future work should examine if obesity or obesogenic diet can modify the diabetogenic effects of iAs exposure or if the exposure and obesity can act in a synergistic manner to produce diabetes. More information is also needed on the association between duration of iAs exposure and the risk of developing diabetes, as well as on the diabetogenic effects of iAs exposure at different periods of life. The effects of prenatal and early-life exposures are of particular interest. Recent work has shown that chronic exposure to iAs is associated with changes in gene promoter methylation in circulating leukocytes; genes involved in both tumor suppression [98] and in type 1 and 2 diabetes [6] have been identified as targets. These findings suggest that epigenetic mechanisms, including epigenetic changes in fetuses exposed to iAs *in utero*, may be in part responsible for diabetes associated with iAs exposure.

Future laboratory studies using animal and tissue culture models should be designed to examine the mechanisms of iAs-induced diabetes at levels of exposure that are relevant to the environmental exposure levels. The goal should be to achieve the internal As doses (i.e., the concentrations and speciation of As in animal tissues or cells) that are consistent with the concentrations of As metabolites in blood and tissues of human subjects exposed to iAs from drinking water. Further development of methods for quantitative, oxidation state-specific analysis of As species in complex biological matrices, including blood and tissues, will be needed to achieve this goal. Particular attention should be given to the role of As metabolism, including enzymatic methylation in modulation of the diabetogenic effects of iAs exposure. A recently established strain of C57Bl/6 mice knocked out for As (+3 oxidation state) methyltransferase [43], the key enzyme in the pathway for iAs metabolism [55], provides an excellent model for this type of studies.

REFERENCES

[1] ADA, Standards of medical care in diabetes—2011. Diabetes Care 34 Suppl 1 (2011) S11–S61.

[2] H.I. Afridi, T.G. Kazi, N. Kazi, M.K. Jamali, M.B. Arain, N. Jalbani, J.A. Baig, R.A. Sarfraz, Evaluation of status of toxic metals in biological samples of diabetes mellitus patients. Diabetes Research and Clinical Practice 80 (2008) 280–288.

[3] M.V. Aguilar, M.C. Martinez-Para, M.J. Gonzalez, Effects of arsenic (V)-chromium (III) interaction on plasma glucose and cholesterol levels in growing rats. Annals of Nutrition & Metabolism 41 (1997) 189–195.

[4] S.S. Andra, K.C. Makris, C.A. Christophi, A.S. Ettinger, Delineating the degree of association between biomarkers of arsenic exposure and type-2 diabetes mellitus. International Journal of Hygiene and Environmental Health 216 (2013) 35–49.

[5] L.L. Arnold, M. Eldan, M. van Gemert, C.C. Capen, S.M. Cohen, Chronic studies evaluating the carcinogenicity of monomethylarsonic acid in rats and mice. Toxicology 190 (2003) 197–219.

[6] K.A. Bailey, M. Wu, W.O. Ward, L. Smeester, J.E. Rager, G. Garcia-Vargas, L.M. Del Razo, Z. Drobna, M. Styblo, R.C. Fry, Arsenic and the Epigenome: inter-individual differences in arsenic metabolism related to distinct patterns of DNA methylation. Journal of Biochemical and Molecular Toxicology 27 (2013) 106–115.

[7] M. Bazuine, F. Carlotti, M.J. Rabelink, J. Vellinga, R.C. Hoeben, J.A. Maassen, The p38 mitogen-activated protein kinase inhibitor SB203580 reduces glucose turnover by the glucose transporter-4 of 3T3-L1 adipocytes in the insulin-stimulated state. Endocrinology 146 (2005) 1818–1824.

[8] M. Bazuine, D.M. Ouwens, D.S. Gomes de Mesquita, J.A. Maassen, Arsenite stimulated glucose transport in 3T3-L1 adipocytes involves both Glut4 translocation and p38 MAPK activity. European Journal of Biochemistry/FEBS 270 (2003) 3891–3903.

[9] U. Biswas, S. Sarkar, M.K. Bhowmik, A.K. Samanta, S. Biswas, Chronic toxicity of arsenic in goats: clinicobiochemical changes, pathomorphology and tissue residues. Small Ruminant Research : The Journal of the International Goat Association 38 (2000) 229–235.

[10] L. Boquist, S. Boquist, I. Ericsson, Structural beta-cell changes and transient hyperglycemia in mice treated with compounds inducing inhibited citric acid cycle enzyme activity. Diabetes 37 (1988) 89–98.

[11] D. Chakraborty, A. Mukherjee, S. Sikdar, A. Paul, S. Ghosh, A.R. Khuda-Bukhsh, [6]-Gingerol isolated from ginger attenuates sodium arsenite induced oxidative stress and plays a corrective role in improving insulin signaling in mice. Toxicology Letters 210 (2012) 34–43.

[12] C.J. Chen, S.L. Wang, J.M. Chiou, C.H. Tseng, H.Y. Chiou, Y.M. Hsueh, S.Y. Chen, M.M. Wu, M.S. Lai, Arsenic and diabetes and hypertension in human populations: a review. Toxicology and Applied Pharmacology 222 (2007) 298–304.

[13] J.W. Chen, H.Y. Chen, W.F. Li, S.H. Liou, C.J. Chen, J.H. Wu, S.L. Wang, The association between total urinary arsenic concentration and renal dysfunction in a community-based population from central Taiwan. Chemosphere 84 (2011) 17–24.

[14] J.W. Chen, S.L. Wang, Y.H. Wang, C.W. Sun, Y.L. Huang, C.J. Chen, W.F. Li, Arsenic methylation, GSTO1 polymorphisms, and metabolic syndrome in an arseniasis endemic area of southwestern Taiwan. Chemosphere 88 (2012) 432–438.

[15] Y. Chen, H. Ahsan, V. Slavkovich, G.L. Peltier, R.T. Gluskin, F. Parvez, X. Liu, J.H. Graziano, No association between arsenic exposure from drinking water and diabetes mellitus: a cross-sectional study in Bangladesh. Environmental Health Perspectives 118 (2010) 1299–1305.

[16] W.S. Cheng, D.L. Wingard, D. Kritz-Silverstein, E. Barrett-Connor, Sensitivity and specificity of death certificates for diabetes: as good as it gets? Diabetes Care 31 (2008) 279–284.

[17] K. Choi, Y.B. Kim, Molecular mechanism of insulin resistance in obesity and type 2 diabetes. The Korean Journal of Internal Medicine 25 (2010) 119–129.

[18] J.M. Cobo, M. Castineira, Oxidative stress, mitochondrial respiration, and glycemic control: clues from chronic supplementation with Cr3+ or As3+ to male Wistar rats. Nutrition 13 (1997) 965–970.

[19] J.A. Coronado-González, L.M. Del Razo, G. Garcia-Vargas, F. Sanmiguel-Salazar, J. Escobedo-de la Peña, Inorganic arsenic exposure and type 2 diabetes mellitus in Mexico. Environmental Research 104 (2007) 383–389.

[20] M.E. Davila-Esqueda, J.M. Morales, M.E. Jimenez-Capdeville, E. De la Cruz, R. Falcon-Escobedo, E. Chi-Ahumada, S. Martin-Perez, Low-level subchronic arsenic exposure from prenatal developmental stages to adult life results in an impaired glucose homeostasis. Experimental and Clinical Endocrinology & Diabetes : Official Journal, German Society of Endocrinology [and] German Diabetes Association 119 (2011) 613–617.

[21] L.M. Del Razo, G.G. Garcia-Vargas, O.L. Valenzuela, E.H. Castellanos, L.C. Sanchez-Pena, J.M. Currier, Z. Drobna, D. Loomis, M. Styblo, Exposure to arsenic in drinking water is associated with increased prevalence of diabetes: a cross-sectional study in the Zimapan and Lagunera regions in Mexico. Environmental Health 10 (2011) 73.

[22] A. Diaz-Villasenor, A.L. Burns, A.M. Salazar, M. Sordo, M. Hiriart, M.E. Cebrian, P. Ostrosky-Wegman, Arsenite reduces insulin secretion in rat pancreatic beta-cells by decreasing the calcium-dependent calpain-10 proteolysis of SNAP-25. Toxicology and Applied Pharmacology 231 (2008) 291–299.

[23] A. Diaz-Villasenor, M.C. Sanchez-Soto, M.E. Cebrian, P. Ostrosky-Wegman, M. Hiriart, Sodium arsenite impairs insulin secretion and transcription in pancreatic beta-cells. Toxicology and Applied Pharmacology 214 (2006) 30–34.

[24] C.D. Douillet, J.M. Currier, J. Saunders, W.M. Bodnar, T. Matoušek, M. Stýblo, Methylated trivalent arsenicals are potent inhibitors of glucose stimulated insulin secretion by murine pancreatic islets. Toxicology and Applied Pharmacology 267 (2013) 11–15.

[25] Z. Drobna, L.M. Del Razo, G.G. Garcia-Vargas, L.C. Sanchez-Pena, A. Barrera-Hernandez, M. Styblo, D. Loomis, Environmental exposure to arsenic, AS3MT polymorphism and prevalence of diabetes in Mexico. Journal of Exposure Science & Environmental Epidemiology 23 (2013) 151–155.

[26] EFSA, European Food Safety Authority panel on contaminants in the food chain (CONTAM); Scientific opinion on arsenic in food. EFSA Journal. http://www.efsa.europa.eu/en/scdocs/scdoc/1351.htm 7 (2009) (accessed April 9, 2015).

[27] L.J. Elrick, K. Docherty, Phosphorylation-dependent nucleocytoplasmic shuttling of pancreatic duodenal homeobox-1. Diabetes 50 (2001) 2244–2252.

[28] A.S. Ettinger, A.R. Zota, C.J. Amarasiriwardena, M.R. Hopkins, J. Schwartz, H. Hu, R.O. Wright, Maternal arsenic exposure and impaired glucose tolerance during pregnancy. Environmental Health Perspectives 117 (2009) 1059–1064.

[29] H. Eyre, R. Kahn, R.M. Robertson, Preventing cancer, cardiovascular disease, and diabetes: a common agenda for the American Cancer Society, the American Diabetes Association, and the American Heart Association. CA: a Cancer Journal for Clinicians 54 (2004) 190–207.

[30] C. Fladeby, G. Serck-Hanssen, Stress-induced glucose uptake in bovine chromaffin cells: a comparison of the effect of arsenite and anisomycin. Biochimica et Biophysica Acta 1452 (1999) 313–321.

[31] C.R. Flores, M.P. Puga, K. Wrobel, M.E. Garay Sevilla, Trace elements status in diabetes mellitus type 2: possible role of the interaction between molybdenum and copper in the progress of typical complications. Diabetes Research and Clinical Practice 91 (2011) 333–341.

[32] S.C. Frost, R.A. Kohanski, M.D. Lane, Effect of phenylarsine oxide on insulin-dependent protein phosphorylation and glucose transport in 3T3-L1 adipocytes. The Journal of Biological Chemistry 262 (1987) 9872–9876.

[33] S.C. Frost, M.D. Lane, Evidence for the involvement of vicinal sulfhydryl groups in insulin-activated hexose transport by 3T3-L1 adipocytes. The Journal of Biological Chemistry 260 (1985) 2646–2652.

[34] J. Fu, C.G. Woods, E. Yehuda-Shnaidman, Q. Zhang, V. Wong, S. Collins, G. Sun, M.E. Andersen, J. Pi, Low-level arsenic impairs glucose-stimulated insulin secretion in

pancreatic beta cells: involvement of cellular adaptive response to oxidative stress. Environmental Health Perspectives 118 (2010) 864–870.

[35] T. Ghafghazi, J.W. Ridlington, B.A. Fowler, The effects of acute and subacute sodium arsenite administration on carbohydrate metabolism. Toxicology and Applied Pharmacology 55 (1980) 126–130.

[36] D.E. Goldstein, R.R. Little, R.A. Lorenz, J.I. Malone, D. Nathan, C.M. Peterson, D.B. Sacks, Tests of glycemia in diabetes. Diabetes Care 27 (2004) 1761–1773.

[37] G.W. Gould, G.E. Lienhard, L.I. Tanner, E.M. Gibbs, Phenylarsine oxide stimulates hexose transport in 3T3-L1 adipocytes by a mechanism other than an increase in surface transporters. Archives of Biochemistry and Biophysics 268 (1989) 264–275.

[38] M.J. Gresser, ADP-arsenate. Formation by submitochondrial particles under phosphorylating conditions. The Journal of Biological Chemistry 256 (1981) 5981–5983.

[39] M.O. Gribble, B.V. Howard, J.G. Umans, N.M. Shara, K.A. Francesconi, W. Goessler, C.M. Crainiceanu, E.K. Silbergeld, E. Guallar, A. Navas-Acien, Arsenic exposure, diabetes prevalence, and diabetes control in the strong heart study. American Journal of Epidemiology 176 (2012) 865–874.

[40] E.J. Henriksen, J.O. Holloszy, Effects of phenylarsine oxide on stimulation of glucose transport in rat skeletal muscle. The American Journal of Physiology 258 (1990) C648–C653.

[41] D.S. Hill, B.J. Wlodarczyk, L.E. Mitchell, R.H. Finnell, Arsenate-induced maternal glucose intolerance and neural tube defects in a mouse model. Toxicology and Applied Pharmacology 239 (2009) 29–36.

[42] C.F. Huang, Y.W. Chen, C.Y. Yang, K.S. Tsai, R.S. Yang, S.H. Liu, Arsenic and diabetes: current perspectives. The Kaohsiung Journal of Medical Sciences 27 (2011) 402–410.

[43] M.F. Hughes, B.C. Edwards, K.M. Herbin-Davis, J. Saunders, M. Styblo, D.J. Thomas, Arsenic (+3 oxidation state) methyltransferase genotype affects steady-state distribution and clearance of arsenic in arsenate-treated mice. Toxicology and Applied Pharmacology 249 (2010) 217–223.

[44] M.F. Hughes, M. Menache, D.J. Thompson, Dose-dependent disposition of sodium arsenate in mice following acute oral exposure. Fundamental and Applied Toxicology : Official Journal of the Society of Toxicology 22 (1994) 80–89.

[45] T. Imamura, J. Huang, I. Usui, H. Satoh, J. Bever, J.M. Olefsky, Insulin-induced GLUT4 translocation involves protein kinase C-lambda-mediated functional coupling between Rab4 and the motor protein kinesin. Molecular and Cellular Biology 23 (2003) 4892–4900.

[46] R. Islam, I. Khan, S.N. Hassan, M. McEvoy, C. D'Este, J. Attia, R. Peel, M. Sultana, S. Akter, A.H. Milton, Association between type 2 diabetes and chronic arsenic exposure in drinking water: a cross sectional study in Bangladesh. Environmental Health : A Global Access Science Source 11 (2012) 38.

[47] J.A. Izquierdo-Vega, C.A. Soto, L.C. Sanchez-Pena, A. De Vizcaya-Ruiz, L.M. Del Razo, Diabetogenic effects and pancreatic oxidative damage in rats subchronically exposed to arsenite. Toxicology Letters 160 (2006) 135–142.

[48] G.E. Jensen, M.L. Hansen, Occupational arsenic exposure and glycosylated haemoglobin. The Analyst 123 (1998) 77–80.

[49] D. Jovanovic, Z. Rasic-Milutinovic, K. Paunovic, B. Jakovljevic, S. Plavsic, J. Milosevic, Low levels of arsenic in drinking water and type 2 diabetes in Middle Banat region, Serbia. International Journal of Hygiene and Environmental Health 216 (2013) 50–55.

[50] N.F. Kolachi, T.G. Kazi, H.I. Afridi, N. Kazi, S. Khan, G.A. Kandhro, A.Q. Shah, J.A. Baig, S.K. Wadhwa, F. Shah, M.K. Jamali, M.B. Arain, Status of toxic metals in biological samples of diabetic mothers and their neonates. Biological Trace Element Research 143 (2011) 196–212.

[51] M.S. Lai, Y.M. Hsueh, C.J. Chen, M.P. Shyu, S.Y. Chen, T.L. Kuo, M.M. Wu, T.Y. Tai, Ingested inorganic arsenic and prevalence of diabetes mellitus. American Journal of Epidemiology 139 (1994) 484–492.

[52] J.O. Lee, S.K. Lee, J.H. Jung, J.H. Kim, G.Y. You, S.J. Kim, S.H. Park, K.O. Uhm, H.S. Kim, Metformin induces Rab4 through AMPK and modulates GLUT4 translocation in skeletal muscle cells. Journal of Cellular Physiology 226 (2011) 974–981.

[53] C. Leloup, C. Tourrel-Cuzin, C. Magnan, M. Karaca, J. Castel, L. Carneiro, A.L. Colombani, A. Ktorza, L. Casteilla, L. Penicaud, Mitochondrial reactive oxygen species are obligatory signals for glucose-induced insulin secretion. Diabetes 58 (2009) 673–681.

[54] D.R. Lewis, J.W. Southwick, R. Ouellet-Hellstrom, J. Rench, R.L. Calderon, Drinking water arsenic in Utah: a cohort mortality study. Environmental Health Perspectives 107 (1999) 359–365.

[55] S. Lin, Q. Shi, F.B. Nix, M. Styblo, M.A. Beck, K.M. Herbin-Davis, L.L. Hall, J.B. Simeonsson, D.J. Thomas, A novel S-adenosyl-L-methionine:arsenic(III) methyltransferase from rat liver cytosol. Journal of Biological Chemistry 277 (2002) 10795–10803.

[56] Z. Liu, M.A. Sanchez, X. Jiang, E. Boles, S.M. Landfear, B.P. Rosen, Mammalian glucose permease GLUT1 facilitates transport of arsenic trioxide and methylarsonous acid. Biochemical and Biophysical Research Communications 351 (2006) 424–430.

[57] M.P. Longnecker, On confounded fishy results regarding arsenic and diabetes. Epidemiology 20 (2009) 821–823; discussion e821–e822.

[58] M. Lu, H. Wang, X.F. Li, L.L. Arnold, S.M. Cohen, X.C. Le, Binding of dimethylarsinous acid to cys-13alpha of rat hemoglobin is responsible for the retention of arsenic in rat blood. Chemical Research in Toxicology 20 (2007) 27–37.

[59] M. Lu, H. Wang, X.F. Li, X. Lu, W.R. Cullen, L.L. Arnold, S.M. Cohen, X.C. Le, Evidence of hemoglobin binding to arsenic as a basis for the accumulation of arsenic in rat blood. Chemical Research in Toxicology 17 (2004) 1733–1742.

[60] M. Lu, H. Wang, Z. Wang, X.F. Li, X.C. Le, Identification of reactive cysteines in a protein using arsenic labeling and collision-induced dissociation tandem mass spectrometry. Journal of Proteome Research 7 (2008) 3080–3090.

[61] W.M. Macfarlane, C.M. McKinnon, Z.A. Felton-Edkins, H. Cragg, R.F. James, K. Docherty, Glucose stimulates translocation of the homeodomain transcription factor PDX1 from the cytoplasm to the nucleus in pancreatic beta-cells. The Journal of Biological Chemistry 274 (1999) 1011–1016.

[62] W.M. Macfarlane, S.B. Smith, R.F. James, A.D. Clifton, Y.N. Doza, P. Cohen, K. Docherty, The p38/reactivating kinase mitogen-activated protein kinase cascade mediates the activation of the transcription factor insulin upstream factor 1 and insulin gene transcription by high glucose in pancreatic beta-cells. The Journal of Biological Chemistry 272 (1997) 20936–20944.

[63] K.C. Makris, C.A. Christophi, M. Paisi, A.S. Ettinger, A preliminary assessment of low level arsenic exposure and diabetes mellitus in Cyprus. BMC Public Health 12 (2012) 334.

[64] E.A. Maull, H. Ahsan, J. Edwards, M.P. Longnecker, A. Navas-Acien, J. Pi, E.K. Silbergeld, M. Styblo, C.H. Tseng, K.A. Thayer, D. Loomis, Evaluation of the association between arsenic and diabetes: a National Toxicology Program workshop review. Environmental Health Perspectives 120 (2012) 1658–1670.

[65] H.E. McDowell, T. Walker, E. Hajduch, G. Christie, I.H. Batty, C.P. Downes, H.S. Hundal, Inositol phospholipid 3-kinase is activated by cellular stress but is not required for the stress-induced activation of glucose transport in L6 rat skeletal muscle cells. European Journal of Biochemistry/FEBS 247 (1997) 306–313.

[66] C.M. McKinnon, K. Docherty, Pancreatic duodenal homeobox-1, PDX-1, a major regulator of beta cell identity and function. Diabetologia 44 (2001) 1203–1214.

[67] J.R. Meliker, R.L. Wahl, L.L. Cameron, J.O. Nriagu, Arsenic in drinking water and cerebrovascular disease, diabetes mellitus, and kidney disease in Michigan: a standardized mortality ratio analysis. Environmental Health : A Global Access Science Source 6 (2007) 4.

[68] R.D. Mitchell, F. Ayala-Fierro, D.E. Carter, Systemic indicators of inorganic arsenic toxicity in four animal species. Journal of Toxicology and Environmental Health. Part A 59 (2000) 119–134.

[69] S. Mukherjee, D. Das, M. Mukherjee, A.S. Das, C. Mitra, Synergistic effect of folic acid and vitamin B12 in ameliorating arsenic-induced oxidative damage in pancreatic tissue of rat. The Journal of Nutritional Biochemistry 17 (2006) 319–327.

[70] A.H. Nabi, M.M. Rahman, L.N. Islam, Evaluation of biochemical changes in chronic arsenic poisoning among Bangladeshi patients. International Journal of Environmental Research and Public Health 2 (2005) 385–393.

[71] A. Navas-Acien, E.K. Silbergeld, R. Pastor-Barriuso, E. Guallar, Arsenic exposure and prevalence of type 2 diabetes in US adults. JAMA 300 (2008) 814–822.

[72] A. Navas-Acien, E.K. Silbergeld, R. Pastor-Barriuso, E. Guallar, Rejoinder: arsenic exposure and prevalence of type 2 diabetes: updated findings from the National Health Nutrition and Examination Survey, 2003–2006. Epidemiology 20 (2009) 816–820; discussion e811–e812.

[73] A. Navas-Acien, E.K. Silbergeld, R.A. Streeter, J.M. Clark, T.A. Burke, E. Guallar, Arsenic exposure and type 2 diabetes: a systematic review of the experimental and epidemiological evidence. Environmental Health Perspectives 114 (2006) 641–648.

[74] NIDDK, Diabetes research strategic plan. http://www2.niddk.nih.gov/AboutNIDDK/ReportsAndStrategicPlanning/DiabetesPlan/PlanPosting.htm (2011) (accessed April 9, 2015).

[75] NRC, Arsenic in drinking water. Subcommittee on arsenic in drinking water, National Research Council. The National Academies Press. http://www.nap.edu/catalog.php?record_id=6444 (1999) (accessed April 9, 2015).

[76] NTP, Role of environmental chemicals in the development of diabetes and obesity. http://ntp.niehs.nih.gov/pubhealth/hat/noms/diabetesobesity/index.html (2011) (accessed May 5, 2011).

[77] S. Pal, A.K. Chatterjee, Protective effect of methionine supplementation on arsenic-induced alteration of glucose homeostasis. Food and Chemical Toxicology 42 (2004) 737–742.

[78] S. Pal, A.K. Chatterjee, Protective effect of N-acetylcysteine against arsenic-induced depletion in vivo of carbohydrate. Drug and Chemical Toxicology 27 (2004) 179–189.

[79] S. Pal, A.K. Chatterjee, Prospective protective role of melatonin against arsenic-induced metabolic toxicity in Wistar rats. Toxicology 208 (2005) 25–33.

[80] C.A. Pasternak, J.E. Aiyathurai, V. Makinde, A. Davies, S.A. Baldwin, E.M. Konieczko, C.C. Widnell, Regulation of glucose uptake by stressed cells. Journal of Cellular Physiology 149 (1991) 324–331.

[81] D.S. Paul, V. Devesa, A. Hernandez-Zavala, B.M. Adair, F.S. Walton, Z. Drobna, D.J. Thomas, M. Styblo, Environmental arsenic as a disruptor of insulin signaling. Metal Ions in Biology and Medicine 10 (2008) 1–7.

[82] D.S. Paul, A.W. Harmon, V. Devesa, D.J. Thomas, M. Styblo, Molecular mechanisms of the diabetogenic effects of arsenic: inhibition of insulin signaling by arsenite and methylarsonous acid. Environmental Health Perspectives 115 (2007) 734–742.

[83] D.S. Paul, A. Hernandez-Zavala, F.S. Walton, B.M. Adair, J. Dedina, T. Matousek, M. Styblo, Examination of the effects of arsenic on glucose homeostasis in cell culture and animal studies: development of a mouse model for arsenic-induced diabetes. Toxicology and Applied Pharmacology 222 (2007) 305–314.

[84] D.S. Paul, F.S. Walton, R.J. Saunders, M. Styblo, Characterization of the impaired glucose homeostasis produced in C57BL/6 mice by chronic exposure to arsenic and high-fat diet. Environmental Health Perspectives 119 (2011) 1104–1109.

[85] J.S. Petrick, B. Jagadish, E.A. Mash, H.V. Aposhian, Monomethylarsonous acid (MMA(III)) and arsenite: LD(50) in hamsters and in vitro inhibition of pyruvate dehydrogenase. Chemical Research in Toxicology 14 (2001) 651–656.

[86] J. Pi, Y. Bai, Q. Zhang, V. Wong, L.M. Floering, K. Daniel, J.M. Reece, J.T. Deeney, M.E. Andersen, B.E. Corkey, S. Collins, Reactive oxygen species as a signal in glucose-stimulated insulin secretion. Diabetes 56 (2007) 1783–1791.

[87] J. Pi, Q. Zhang, J. Fu, C.G. Woods, Y. Hou, B.E. Corkey, S. Collins, M.E. Andersen, ROS signaling, oxidative stress and Nrf2 in pancreatic beta-cell function. Toxicology and Applied Pharmacology 244 (2010) 77–83.

[88] M.D. Pysher, J.J. Sollome, S. Regan, T.R. Cardinal, J.B. Hoying, H.L. Brooks, R.R. Vaillancourt, Increased hexokinase II expression in the renal glomerulus of mice in response to arsenic. Toxicology and Applied Pharmacology 224 (2007) 39–48.

[89] M. Rahman, O. Axelson, Diabetes mellitus and arsenic exposure: a second look at case-control data from a Swedish copper smelter. Occupational and Environmental Medicine 52 (1995) 773–774.

[90] M. Rahman, M. Tondel, S.A. Ahmad, O. Axelson, Diabetes mellitus associated with arsenic exposure in Bangladesh. American Journal of Epidemiology 148 (1998) 198–203.

[91] M. Rahman, M. Tondel, I.A. Chowdhury, O. Axelson, Relations between exposure to arsenic, skin lesions, and glucosuria. Occupational and Environmental Medicine 56 (1999) 277–281.

[92] M. Rahman, G. Wingren, O. Axelson, Diabetes mellitus among Swedish art glass workers—an effect of arsenic exposure? Scandinavian Journal of Work, Environment & Health 22 (1996) 146–149.

[93] P. Rorsman, L. Eliasson, E. Renstrom, J. Gromada, S. Barg, S. Gopel, The cell physiology of biphasic insulin secretion. News in Physiological Sciences 15 (2000) 72–77.

[94] M.L. Ruiz-Navarro, M. Navarro-Alarcon, H. Lopez Gonzalez-de la Serrana, V. Perez-Valero, M.C. Lopez-Martinez, Urine arsenic concentrations in healthy adults as

indicators of environmental contamination: relation with some pathologies. The Science of the Total Environment 216 (1998) 55–61.

[95] B. Salazard, L. Bellon, S. Jean, M. Maraninchi, C. El-Yazidi, T. Orsiere, A. Margotat, A. Botta, J.L. Berge-Lefranc, Low-level arsenite activates the transcription of genes involved in adipose differentiation. Cell Biology and Toxicology 20 (2004) 375–385.

[96] M.A. Serdar, F. Bakir, A. Hasimi, T. Celik, O. Akin, L. Kenar, O. Aykut, M. Yildirimkaya, Trace and toxic element patterns in nonsmoker patients with noninsulin-dependent diabetes mellitus, impaired glucose tolerance, and fasting glucose. International Journal of Diabetes in Developing Countries 29 (2009) 35–40.

[97] N. Singh, S.V. Rana, Effect of insulin on arsenic toxicity in diabetic rats-liver function studies. Biological Trace Element Research 132 (2009) 215–226.

[98] L. Smeester, J.E. Rager, K.A. Bailey, X. Guan, N. Smith, G. Garcia-Vargas, L.M. Del Razo, Z. Drobna, H. Kelkar, M. Styblo, R.C. Fry, Epigenetic changes in individuals with arsenicosis. Chemical Research in Toxicology 24 (2011) 165–167.

[99] M.O. Sowell, K.A. Robinson, M.G. Buse, Phenylarsine oxide and denervation effects on hormone-stimulated glucose transport. The American Journal of Physiology 255 (1988) E159–E165.

[100] M.L. Standaert, G. Bandyopadhyay, L. Perez, D. Price, L. Galloway, A. Poklepovic, M.P. Sajan, V. Cenni, A. Sirri, J. Moscat, A. Toker, R.V. Farese, Insulin activates protein kinases C-zeta and C-lambda by an autophosphorylation-dependent mechanism and stimulates their translocation to GLUT4 vesicles and other membrane fractions in rat adipocytes. The Journal of Biological Chemistry 274 (1999) 25308–25316.

[101] A.A. Steffens, G.M. Hong, L.J. Bain, Sodium arsenite delays the differentiation of C2C12 mouse myoblast cells and alters methylation patterns on the transcription factor myogenin. Toxicology and Applied Pharmacology 250 (2011) 154–161.

[102] C. Steinmaus, Y. Yuan, J. Liaw, A.H. Smith, Low-level population exposure to inorganic arsenic in the United States and diabetes mellitus: a reanalysis. Epidemiology 20 (2009) 807–815.

[103] C. Steinmaus, Y. Yuan, J. Liaw, A.H. Smith, On arsenic, diabetes, creatinine, and multiple regression modeling: a response to the commentaries on our reanalysis [Editorial]. Epidemiology 20 (2009) e1–e2.

[104] E.V. Sviderskaya, E. Jazrawi, S.A. Baldwin, C.C. Widnell, C.A. Pasternak, Cellular stress causes accumulation of the glucose transporter at the surface of cells independently of their insulin sensitivity. The Journal of Membrane Biology 149 (1996) 133–140.

[105] K.A. Thayer, J.J. Heindel, J.R. Bucher, M.A. Gallo, Role of environmental chemicals in diabetes and obesity: a National Toxicology Program workshop review. Environmental Health Perspectives 120 (2012) 779–789.

[106] K. Tollestrup, F.J. Frost, L.C. Harter, G.P. McMillan, Mortality among children residing near the American Smelting and Refining Company (ASARCO) copper smelter in Ruston, Washington. Archives of Environmental Health 58 (2003) 683–691.

[107] K.J. Trouba, E.M. Wauson, R.L. Vorce, Sodium arsenite inhibits terminal differentiation of murine C3H 10T1/2 preadipocytes. Toxicology and Applied Pharmacology 168 (2000) 25–35.

[108] S.M. Tsai, T.N. Wang, Y.C. Ko, Mortality for certain diseases in areas with high levels of arsenic in drinking water. Archives of Environmental Health: An International Journal 54 (1999) 186–193.

[109] C.H. Tseng, The potential biological mechanisms of arsenic-induced diabetes mellitus. Toxicology and Applied Pharmacology 197 (2004) 67–83.

[110] C.H. Tseng, C.K. Chong, L.T. Heng, C.P. Tseng, T.Y. Tai, The incidence of type 2 diabetes mellitus in Taiwan. Diabetes Research and Clinical Practice 50 Suppl 2 (2000) S61–S64.

[111] C.H. Tseng, T.Y. Tai, C.K. Chong, C.P. Tseng, M.S. Lai, B.J. Lin, H.Y. Chiou, Y.M. Hsueh, K.H. Hsu, C.J. Chen, Long-term arsenic exposure and incidence of non-insulin-dependent diabetes mellitus: a cohort study in arseniasis-hyperendemic villages in Taiwan. Environmental Health Perspectives 108 (2000) 847–851.

[112] M.D. Turner, Coordinated control of both insulin secretion and insulin action through calpain-10-mediated regulation of exocytosis? Molecular Genetics and Metabolism 91 (2007) 305–307.

[113] F.S. Walton, A.W. Harmon, D.S. Paul, Z. Drobna, Y.M. Patel, M. Styblo, Inhibition of insulin-dependent glucose uptake by trivalent arsenicals: possible mechanism of arsenic-induced diabetes. Toxicology and Applied Pharmacology 198 (2004) 424–433.

[114] J.P. Wang, S.L. Wang, Q. Lin, L. Zhang, D. Huang, J.C. Ng, Association of arsenic and kidney dysfunction in people with diabetes and validation of its effects in rats. Environment International 35 (2009) 507–511.

[115] S.L. Wang, F.H. Chang, S.H. Liou, H.J. Wang, W.F. Li, D.P. Hsieh, Inorganic arsenic exposure and its relation to metabolic syndrome in an industrial area of Taiwan. Environment International 33 (2007) 805–811.

[116] S.L. Wang, J.M. Chiou, C.J. Chen, C.H. Tseng, W.L. Chou, C.C. Wang, T.N. Wu, L.W. Chang, Prevalence of non-insulin-dependent diabetes mellitus and related vascular diseases in southwestern arseniasis-endemic and nonendemic areas in Taiwan. Environmental Health Perspectives 111 (2003) 155–159.

[117] Z.X. Wang, C.S. Jiang, L. Liu, X.H. Wang, H.J. Jin, Q. Wu, Q. Chen, The role of Akt on arsenic trioxide suppression of 3T3-L1 preadipocyte differentiation. Cell Research 15 (2005) 379–386.

[118] N.I. Ward, B. Pim, Trace element concentrations in blood plasma from diabetic patients and normal individuals. Biological Trace Element Research 6 (1984) 469–487.

[119] A.P. Warren, M.H. James, D.E. Menzies, C.C. Widnell, P.A. Whitaker-Dowling, C.A. Pasternak, Stress induces an increased hexose uptake in cultured cells. Journal of Cellular Physiology 128 (1986) 383–388.

[120] E.M. Wauson, A.S. Langan, R.L. Vorce, Sodium arsenite inhibits and reverses expression of adipogenic and fat cell-specific genes during in vitro adipogenesis. Toxicological Sciences : An Official Journal of the Society of Toxicology 65 (2002) 211–219.

[121] C.C. Widnell, S.A. Baldwin, A. Davies, S. Martin, C.A. Pasternak, Cellular stress induces a redistribution of the glucose transporter. FASEB Journal : Official Publication of the Federation of American Societies for Experimental Biology 4 (1990) 1634–1637.

[122] P. Xue, Y. Hou, Q. Zhang, C.G. Woods, K. Yarborough, H. Liu, G. Sun, M.E. Andersen, J. Pi, Prolonged inorganic arsenite exposure suppresses insulin-stimulated AKT S473 phosphorylation and glucose uptake in 3T3-L1 adipocytes: involvement of the adaptive antioxidant response. Biochemical and Biophysical Research Communications 407 (2011) 360–365.

[123] H. Yassine, M.J. Kimzey, M.A. Galligan, A.J. Gandolfi, C.S. Stump, S.S. Lau, Adjusting for urinary creatinine overestimates arsenic concentrations in diabetics. Cardiorenal Medicine 2 (2012) 26–32.

[124] C.C. Yen, F.J. Lu, C.F. Huang, W.K. Chen, S.H. Liu, S.Y. Lin-Shiau, The diabetogenic effects of the combination of humic acid and arsenic: in vitro and in vivo studies. Toxicology Letters 172 (2007) 91–105.

[125] Y.P. Yen, K.S. Tsai, Y.W. Chen, C.F. Huang, R.S. Yang, S.H. Liu, Arsenic inhibits myogenic differentiation and muscle regeneration. Environmental Health Perspectives 118 (2010) 949–956.

[126] K.M. Zierold, L. Knobeloch, H. Anderson, Prevalence of chronic diseases in adults exposed to arsenic-contaminated drinking water. American Journal of Public Health 94 (2004) 1936–1937.

11

HEPATOTOXICITY

GAVIN E. ARTEEL
Department of Pharmacology and Toxicology, University of Louisville, Louisville, KY, USA

11.1 THE LIVER AS A TARGET ORGAN OF ARSENIC TOXICITY

The liver plays multiple roles in the intact organism. It is critical for maintaining metabolic homeostasis and for synthesis of lipids and carbohydrates; it is also the major site of synthesis of several key proteins (e.g., albumin and clotting factors), and it synthesizes and excretes bile acids, which is critical for normal uptake of vitamins, lipids, and excretion of many xenobiotics (including metals). Additionally, the strategic location of the liver between the intestinal tract and the rest of the body makes it a critical organ for clearance of xenobiotics and toxins that enter the portal blood. It is therefore not surprising that liver has a very high capacity for phase I and II metabolic processes, which is responsible for the well-known "first-pass effect" in xenobiotic metabolism.

As the main detoxifying organ in the body, the liver has a high likelihood of toxic injury. It is therefore not surprising that the liver has tremendous regenerative capacity. This capacity distinguishes it from other vital organs (e.g., the brain, heart, and lungs) that are far less able to replace functional tissue once it has been destroyed. Due to its regenerative properties, however, the liver is able to restore to full size and ensure survival. In experimental animals (e.g., mice), the liver can fully regenerate after surgical removal of 2/3 of the organ within 7–10 days [75]. Although hepatocytes rarely proliferate in the healthy adult liver, virtually all surviving hepatocytes replicate at least once after partial hepatectomy (PHx). Residual hepatocytes upregulate both proliferative and liver-specific gene expression in order to preserve

Arsenic: Exposure Sources, Health Risks, and Mechanisms of Toxicity, First Edition.
Edited by J. Christopher States.
© 2016 John Wiley & Sons, Inc. Published 2016 by John Wiley & Sons, Inc.

tissue-specific function. In addition to hepatocyte proliferation, there is a tightly coordinated response to complement the regenerative process (e.g., angiogenesis, extracellular matrix (ECM) metabolism), so that the entire organ can be reconstituted within days. During liver regeneration, a complex network of cytokines, growth factors, kinases, and transcription factors drive hepatocytes out of the G_0 phase to enter and progress through replication [66]. The complex and synchronized regenerative response in liver can be perturbed, and thereby can impact normal tissue recovery from injury or damage. Indeed, it is now clear that impaired or altered regeneration and/or restitution is critical to the chronicity of numerous hepatic diseases. Indeed, it is hypothesized that hepatic fibrosis, which is collagenous scarring of the liver, is initiated predominantly by a failure of the liver to sufficiently reconstitute itself after chronic injury.

The liver is a well-known target organ of arsenic exposure. This is not surprising given that the major route of arsenic exposure is via ingestion and that the liver plays a key role in the metabolism of ingested arsenic (see Chapter 4). Hepatic abnormalities caused by arsenic exposure include hepatomegaly, non-cirrhotic portal fibrosis, cirrhosis, and portal hypertension [49, 64, 65]. Furthermore, arsenic exposure has been linked to hepatic malignancies, namely hepatic angiosarcoma and hepatocellular carcinoma in both humans and in animal models [70, 81]. These hepatic abnormalities can be in response to pharmacologic exposure; for example, Fowler's solution (1% arsenic trioxide), which was often prescribed as a remedy or "tonic" ([41]; see Chapter 1), will cause hepatic damage with chronic use [30, 52]. Furthermore, hepatic injury is common in areas in which high levels (ppm) of arsenic drinking water contamination are endemic; for example, hepatomegaly was observed in approximately 90% of members of families who drank water contaminated with high (ppm) levels of arsenic in West Bengal [33].

Whereas exposure to high arsenic levels, be it pharmacologic or environmental, clearly causes hepatic injury, the relative risks of lower arsenic exposure levels are unclear. For example, arsenic levels in municipal wells in the United States are generally well below the established threshold for direct hepatic damage caused by arsenic [49]. However, arsenic may cause more subtle changes to the liver at lower concentrations that do not translate to overt pathology. For example, Straub et al. [73] demonstrated that lower concentrations (ppb) of arsenic cause hepatic endothelial cell capillarization and vessel remodelling in mouse liver. The functional impact of such changes is incompletely understood. Furthermore, as discussed in the following text, the potential interaction between arsenic and other hepatotoxicities has not been fully explored. It is therefore unclear at this time if environmental arsenic exposure at the levels observed in the United States and other countries is a risk for liver disease.

11.2 MECHANISMS OF LIVER INJURY CAUSED BY ARSENIC

The mechanism(s) by which arsenic causes liver injury is incompletely understood. However, several viable mechanisms have been identified experimentally that may explain, at least in part, hepatic injury caused by arsenic exposure (see Fig. 11.1).

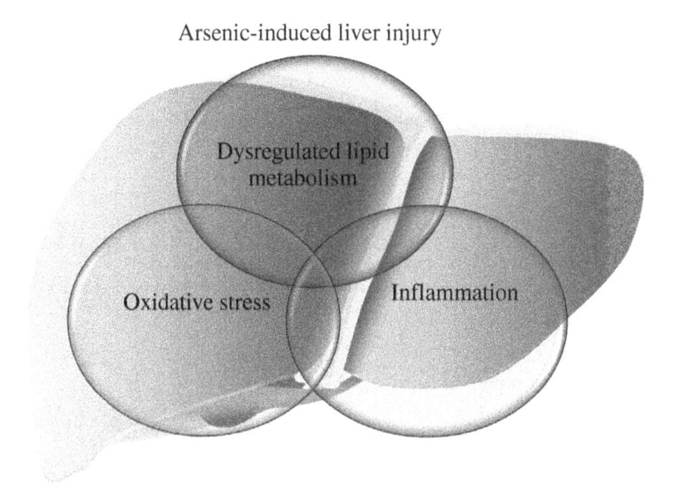

FIGURE 11.1 Overlapping mechanisms of arsenic-induced liver injury. The mechanism(s) by which arsenic causes liver injury is incompletely understood, but several have been identified experimentally that may explain hepatic injury caused by arsenic exposure. Specifically, arsenic-induced liver injury is hypothesized to involve dysregulated lipid metabolism, oxidative stress, and altered cytokine profiles and enhanced inflammation. Importantly, these mechanisms are not mutually exclusive and have several overlapping components that may contribute to overall hepatic injury. See text for details.

Specifically, arsenic-induced liver injury is hypothesized to involve dysregulated lipid metabolism [65], oxidative stress [28], and altered cytokine profiles and enhanced inflammation [17]. Each of these mechanisms will be discussed in detail.

11.2.1 Dysregulated Lipid Metabolism

As mentioned earlier, the liver plays a key role in the uptake, synthesis, metabolism, and excretion of lipids and lipid products. Furthermore, the liver plays a central role in synthesizing glucose from glycerol, which is released by peripheral lipolysis, to protect glucose-dependent organs (e.g., the brain) during periods of fasting. Any alterations of the flux of lipids through the liver can cause lipids to accumulate in the hepatocytes as triglycerides (i.e., steatosis; Fig. 11.2). Steatosis is a common pathologic response to a myriad of hepatic insults, including alcohol [4], drugs (e.g., tamoxifen, amiodarone, and glucocorticoids; [25]), obesity, metabolic syndrome and/or diabetes (see later [48]), and hepatitis virus infection [63], as well as environmental exposure to toxic chemicals (e.g., vinyl chloride [11]). Steatosis can be characterized as macrovesicular (having one large fat droplet per hepatocyte and lateral displacement of the nucleus) or microvesicular (many small fat droplets per hepatocyte), the latter of which is associated with more severe liver disease [38].

Steatosis is generally considered a benign pathologic change that resolves once the insult is removed; for example, alcoholic steatosis readily reverses with abstinence [8]. Indeed, at the level of the organism, hepatic steatosis may be viewed as

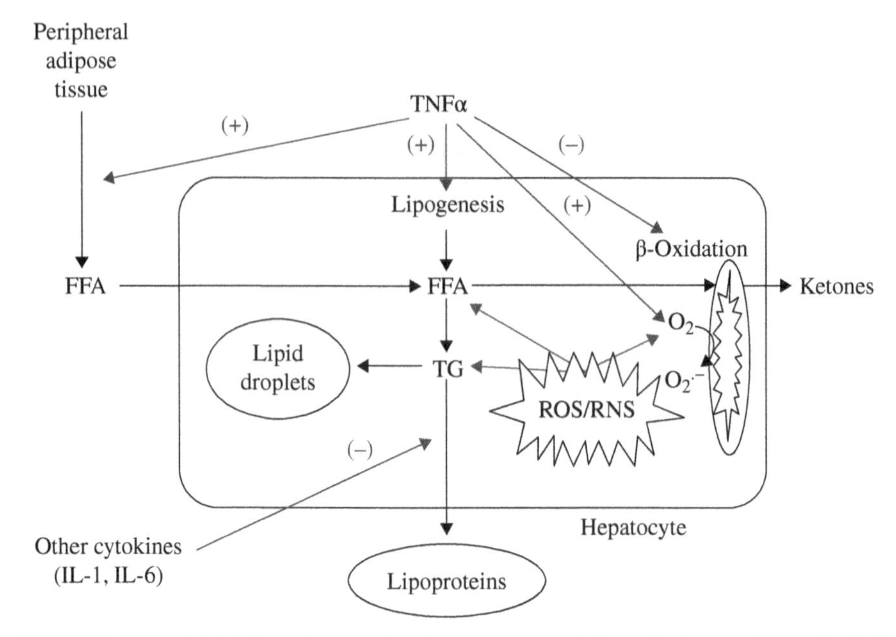

FIGURE 11.2 Lipid flux through the liver and potential mechanisms of hepatic lipid accumulation. The liver plays a key role in the uptake, synthesis, metabolism, and excretion of lipids and lipid products. Furthermore, cytokines released during inflammation can modulate lipid metabolism. Direct effects of TNFα include increased free fatty acid (FFA) release from adipocytes in the periphery, increased lipogenesis in hepatocytes, and inhibition of β-oxidation of fatty acids. Taken together, the net amount of FFA in hepatocytes is increased. Further, TNFα can directly increase ROS formation by impairing mitochondrial electron flow, leading formation of $O_2^{\cdot -}$. Indirectly, the oxidation of lipids by ROS/RNS can further impair β-oxidation of fatty acids, and further damage to mitochondria. Other cytokines (e.g., IL-1 and IL-6), may impair transport and secretion of triglycerides (TG). Any alterations of the flux of lipids through the liver can cause lipids to accumulate in the hepatocytes as triglycerides.

potentially a protective measure, as it partitions lipids away from the blood and stores them for potential later use [78]. However, it is thought that this "thrifty gene" response plays a causal role in hepatic pathology [56, 78]; numerous studies have established that steatosis can enhance the hepatotoxic response caused by a second agent [19]. For example, Yang et al. [87] demonstrated that livers from genetically obese (fa/fa) rats are exquisitely sensitive to hepatotoxicity caused by the injection of bacterial lipopolysaccharide (LPS) compared to their lean littermates; this exacerbation of liver damage was characterized by a more robust inflammatory response and enhanced cell death. These points in the context of arsenic exposure will be discussed later in this chapter.

Previous studies have shown that prolonged (>9 months) exposure to arsenic in mice causes hepatic steatosis (e.g., [65]). The mechanism(s) by which arsenic exposure mediates these effects is currently unclear. However, recent studies have suggested that arsenic may exacerbate glucose dyshomeostasis when coupled with a high-fat diet that induces insulin resistance and diabetes-like symptoms in mice ([59]; see Chapter 10). Furthermore, the metabolism of arsenic has been shown to deplete methyl donor pools [43, 58]; methyl donor deficiencies cause hepatic steatosis by impairing normal lipid metabolism (e.g., [62]). Lastly, as discussed in the following text, low-grade inflammation can also alter hepatic lipid metabolism, favoring lipid accumulation.

11.2.2 Oxidative Stress

Reactive oxygen species (ROS) and reactive nitrogen species (RNS), respectively, are products of normal cellular metabolism and have beneficial effects. For example, ROS mediates a variety of cellular signaling pathways. However, due to the potential of these molecules also to damage normal tissue, the balance between pro-oxidants and antioxidants is critical for the survival and function of aerobic organisms. If the balance is tipped to favor overproduction of these species, oxidative stress can occur (Fig. 11.3; [68]). Oxidative stress has been proposed to be critically involved in arsenic toxicity [17, 43, 58]. Although important advancements have been made in understanding the role of pro-oxidants in arsenic-induced liver injury, the mechanisms and implications of arsenic-induced oxidative stress are incompletely understood.

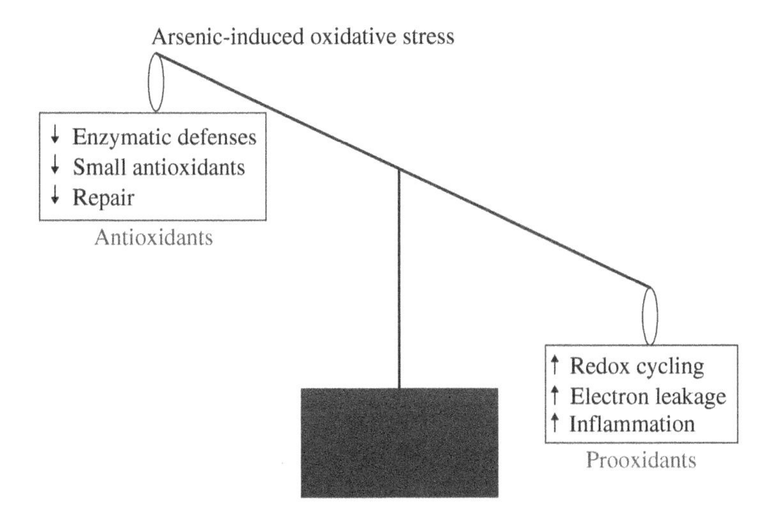

FIGURE 11.3 Imbalance between antioxidant and pro-oxidants, favoring oxidative stress, by arsenic exposure. Arsenic can directly and indirectly (e.g., via stimulating release of free iron) can increase redox cycling, which will generate pro-oxidants. Arsenic has also been shown to increase electron leakage from normal biochemical processes (e.g., mitochondria). Lastly, arsenic exposure induces low-grade inflammation in liver, which will increase production of pro-oxidants from inflammatory cells. The net effect is an increase in pro-oxidant production. Arsenic also impairs critical defenses against oxidative stress, be they enzymatic (e.g., SOD and catalase) or small molecules (e.g., GSH). Lastly, mechanisms designed to repair oxidative damage (e.g., DNA repair processes) are impaired by arsenic exposure.

*11.2.2.1 **Arsenic and Increased Pro-oxidant Production*** Arsenic exposure has clearly been shown to increase pro-oxidant production in cell culture models, and that pro-oxidant production is critical for several of it cytotoxic effects (Fig. 11.3). Although arsenic is a redox-active metal, it is unclear if redox cycling of arsenic, *per se*, will directly produce potent oxidant species. However, some metabolites of arsenic (e.g., dimethylarsine) have been shown to generate reactive oxygen intermediates [86]. In addition to directly forming reactive species, arsenic exposure can also mediate changes that indirectly increase the production of pro-oxidants. For example, methylated arsenic species can release redox-active iron from ferritin [1], which in turn then catalyzes Fenton reactions and the production of potent oxygen metabolites (e.g., OH). Arsenic exposure can also increase the leakage of free electrons from normal biologic processes. For example, the reduction of O_2 to H_2O by the mitochondrion is not completely controlled; it is estimated that 1–2% of O_2 consumption by mitochondria leads to the formation of O_2^- [9]. In other words, an 80-kg human would produce 215–430 mmol O_2^- a day from mitochondria [10]. Several lines of evidence support the hypothesis that arsenic increases the production of O_2^- from the mitochondria, most likely via damaging proteins involved in the electron transport chain (see [28] for review). Arsenic may activate oxidant-producing enzymes that increase production of reactive species; for example, endothelial NADPH oxidase (Nox-2) is activated by arsenic exposure in mice and plays a key role in hepatic endothelial capillarization [72]. Lastly, as summarized later, arsenic induces inflammation, which can indirectly increase oxidative stress in the tissue. The net effect is an increase in pro-oxidant production after arsenic exposure.

*11.2.2.2 **Arsenic and Decreased Anti-oxidant Defenses*** As mentioned earlier, oxidative stress is an imbalance between pro-oxidant production and antioxidant defenses, favoring the former [69]. As such, oxidative stress can occur owing to impaired antioxidant defenses (Fig. 11.3), even in the absence of an increase in pro-oxidant production, *per se*. Chronic arsenic exposure is known to decrease the activity of key antioxidant enzymes in the cell, such as superoxide dismutase (SOD), catalase (CAT), glutathione peroxidase (GPx), GST, and glutathione reductase (GR). Arsenic metabolism also depletes key nonprotein antioxidants in the cell, such as glutathione (GSH). Furthermore, within the cell, there exists a host of proteins and systems involved in the "antioxidant network." This network does not directly intercept pro-oxidants, but serves instead to repair or remove oxidatively modified biomolecules in response to oxidative stress. For example, arsenic inhibits systems designed to repair oxidatively modified DNA (see [20] for review).

*11.2.2.3 **How Does Arsenic-Induced Oxidative Stress Mediate Damage?*** One mechanism by which oxidative stress is proposed to cause cellular injury is via chemical modification of biologic molecules. These chemical modifications can alter and/or interfere with normal biologic processes and be directly toxic to the cell. Pro-oxidants can also confer highly specific changes in a cell at concentrations well below observable chemical damage [39]. It is now clear that pro-oxidants can mediate and/or amplify their signal by modifying signaling cascades within the cell [2, 22, 29, 42]. Oxidant-sensitive signaling cascades include small molecules (e.g., intracellular Ca^{++} [24]), stress-activated protein kinases (SAPKs; e.g., JNK, ERK 1/2, and p38), transcription factors (e.g., AP-1, HIF-1, and NFκB) [16]), and modulators of apoptosis signaling (e.g., caspases, Bad, and Bcl-2 [36]). Further, the hypothesis that redox modification of thiols is an important post-translational modification within the cell is receiving wider acceptance [29, 67]. Arsenic exposure has been shown to activate many of these signaling cascades in experimental models (see [28] for review). However, whether oxidative stress is the dominant pathway of activation of these kinase cascades by arsenic remains unclear and requires further study.

11.2.3 Altered Cytokine Profiles and Enhanced Inflammation

11.2.3.1 Arsenic Exposure and Priming of the Inflammatory Response in Liver

Chronic hepatic diseases, regardless of etiology, share a similar pathologic progression. A key step in these disease processes is an increase in low-grade inflammation. As noted earlier, chronic arsenic exposure is characterized by chronic inflammation in the liver, and involves an increase in proinflammatory cytokines. Low-grade hepatic inflammation is often not only due to an increase in stimuli but also due to an exacerbated immune/inflammatory response to stimuli. The two-hit hypothesis described earlier in the context of steatosis is mirrored on a molecular level by the concept of "priming." Here, "priming" refers to the ability of a pre-exposure to cause inflammatory cells of the liver (e.g., Kupffer cells) to more robustly release pro-inflammatory cytokines in response to a second stimulus, such as LPS. Although first defined for steatosis, the "two-hit" hypothesis can be expanded to include any pathologically benign changes to liver that enhance the hepatotoxic response caused by a second agent.

This group tested the hypothesis that arsenic at subhepatotoxic doses may sensitize the liver to a second hepatotoxin (i.e., serve as a first hit in a "two-hit" paradigm [3]). This hypothesis was tested in mice exposed to lipopolysaccharide (LPS), a Gram-negative bacterial wall product that is often elevated in systemic blood during liver disease [45], and is employed in basic research as a model hepatotoxicant. Liver damage caused by LPS at this dose is characterized by early inflammation, followed by hepatocyte death. Male C57Bl/6J mice (4–6 weeks) were exposed to arsenic (49 ppm) in drinking water. After 7 months of exposure, animals were injected with LPS (10 mg/kg i.p.) and sacrificed 24 h later. Arsenic alone caused no overt hepatotoxicity under these conditions, as determined by plasma enzymes and histology. In contrast, arsenic exposure dramatically enhanced liver damage caused by LPS, increasing the number and size of necroinflammatory foci. This effect of arsenic was coupled with increases in indices of oxidative stress (4-HNE adducts, depletion of GSH, and methionine pools). Taken together, these results suggest that arsenic, at doses that are not overtly hepatotoxic *per se*, "primes" the liver for inflammatory damage due to a second hit. These results further suggest that arsenic levels in the drinking water may be a risk modifier for the development of chronic liver diseases (see later).

Altered cytokine profiles may not only cause inflammatory cell damage but may also contribute to other pathologies associated with arsenic exposure. For example, TNFα (and other cytokines) influence lipid metabolism both in the liver and in the periphery (see Fig. 11.2; [61]). For example, TNFα increases free fatty acid release from adipocytes in the periphery [34], increases lipogenesis in hepatocytes [26], and inhibits β-oxidation of fatty acids [53]. Moreover, pro-oxidant production stimulated by TNFα in hepatocytes could impair mitochondrial electron flow and cause lipid peroxidation, processes that could also slow the metabolism of fat by mitochondria. Other cytokines (e.g., IL-1 and IL-6) may also impair transport and secretion of triglycerides [55]. The net consequence is that these proinflammatory cytokines increase the supply of fatty acids to liver while simultaneously impairing the ability of the hepatocytes to metabolize and secrete them.

11.3 ARSENIC AS A POTENTIAL RISK FACTOR FOR UNDERLYING LIVER DISEASE

There are many gaps in our understanding of the relative safety of arsenic to the human population. Nearly 4000 wells providing community water in the United States have arsenic levels greater than the current WHO recommended maximum contaminant level (MCL) of 10 ppb [23, 31]. Furthermore, even higher arsenic concentrations may be found in private artesian water supplies, which are not regulated by the Safe Drinking Water Act. As mentioned earlier, although it is well known that arsenic can be directly hepatotoxic in humans, these effects are associated with relatively high (ppm) contaminant levels in the drinking water supply in select areas of the world. In contrast, arsenic concentrations in US water supplies have not been directly linked to direct liver damage, *per se*. Most studies to date have focused on the effect of arsenic alone and not taken into consideration risk-modifying factors. It is now clear that the risk for developing a human disease derived from environmental exposure is not based solely on that environmental exposure, but is rather modified by other mitigating conditions, such as other environmental or genetic factors. Indeed, it has been suggested that diet may contribute to the risk of developing arsenic toxicity [70]. However, whether arsenic modifies the risk of developing liver damage owing to other insults has not been determined.

11.3.1 Obesity-Induced Liver Disease: A Growing Epidemic in the Developed World

Another major health concern for the US population is obesity, the prevalence of which is increasing at an alarming rate [27, 57]. Among the myriad of health complications associated with obesity (e.g., diabetes and cardiovascular risk) is nonalcoholic fatty liver disease (NAFLD). The moniker "nonalcoholic" is derived from the fact that the pathology of NAFLD is indistinguishable from that of alcoholic fatty liver disease, but occurs in the absence of significant alcohol consumption [48]. NAFLD is a spectrum of liver diseases, ranging from simple steatosis, to active inflammation, to advanced fibrosis and cirrhosis [18]. Risk factors for primary NAFLD (i.e., not secondary to other proximate causes) are analogous to those of metabolic syndrome (e.g., obesity, type II diabetes, and dyslipidemia [15]). It is however also clear that there are likely other unidentified risk factors that contribute to the development of disease.

11.3.2 Overlap between Arsenic and NAFLD?

A striking feature of arsenic-induced liver disease and NAFLD is that there is significant demographic and mechanistic overlap between arsenic exposure and NAFLD. Arsenic "hot-spots" in the US water supplies often colocalize with areas that have high incidences of obesity and its sequelae (e.g., NAFLD [51, 83]). For example, states with clusters of municipal wells with high levels of arsenic (e.g., Michigan, Texas, West Virginia, and Oklahoma [77]) also have high incidences of obesity and diabetes [12]. Furthermore, as stated earlier, arsenic levels in private artesian water supplies is not regulated by the USEPA and is often much higher than in municipal wells; the rate of obesity in the United States in rural areas tends to be even higher than in municipalities [54].

In addition to the potential demographic overlap of arsenic exposure and risk factors for NAFLD, there is also significant overlap in potential mechanisms by which arsenic and NAFLD cause liver damage. Specifically, like arsenic (see Section II, earlier), mechanisms hypothesized to contribute to the initiation and progression of NAFLD include impaired lipid metabolism and lipotoxicity [13], oxidative stress [13, 50], and altered cytokine profiles and inflammation [14, 21]. There is also the potential that arsenic exposure may be a risk factor for insulin resistance and type II diabetes [47, 80] (see Chapter 10), which may exacerbate the impact of obesity on the liver. It is therefore possible that arsenic exposure is an unidentified environmental risk factor in the development of NAFLD.

11.3.3 Arsenic Enhances Experimental NAFLD

The possibility that arsenic exposure, at exposure levels that are not overtly hepatotoxic, will exacerbate underlying liver disease has significant ramifications for environmental exposure. If such an interaction between arsenic and liver disease indeed exists, then the lower levels of arsenic observed in developed countries (such as the United States), which are generally considered below the threshold for hepatic injury [49], may need to be re-evaluated for risk. This concern is especially true given the increasing incidence of NAFLD (and other liver diseases) in these countries. As mentioned earlier, the concept that arsenic exposure will enhance inflammatory liver damage was already demonstrated in a model of acute hepatic inflammation caused by LPS [3]. However, whether or not arsenic exposure will exacerbate injury in a model of chronic liver disease had never been determined.

To test the hypothesis that low arsenic exposure sensitizes the liver to chronic injury; the effect of arsenic exposure on liver damage in a mouse model of NAFLD was determined [74]. Accordingly, male C57Bl/6J mice were exposed to either low-fat diet (LFD; 13% calories as fat) or high-fat diet (HFD; 42% calories as fat) for

10 weeks; this model is a model of the "Western diet" and causes obesity, insulin resistance and fatty liver injury (e.g., [84]), with pathology similar to that found in NAFLD. Animals were given either control tap water or arsenic-containing (4.9 ppm as sodium arsenite) water during the feeding period. Biochemical and histologic indices of liver damage were determined. HFD (± arsenic) significantly increased body weight gain in mice compared with low-fat controls. As expected, HFD under these conditions caused obesity, insulin resistance and glucose dyshomeostasis, hepatomegaly, and steatohepatitis [74]. Although arsenic exposure had no effect on indices of liver damage in LFD-fed animals, it significantly increased the liver damage caused by HFD. These data support the hypothesis that arsenic may exacerbate underlying liver disease, even at exposure levels that are not overtly hepatotoxic, *per se*.

The main effect of arsenic under these conditions appeared to be to enhance the inflammatory response caused by experimental NAFLD, analogous to previous studies with LPS [3]. This effect correlated with an increase in hepatic expression of pro-inflammatory cytokines (e.g., TNFα), as well as a decrease in anti-inflammatory cytokines (e.g., IL-10) compared to animals fed LFD or HFD + tap water [74]. This imbalance between pro- and anti-inflammatory is common to chronic liver diseases (e.g., NAFLD [76]), and is hypothesized to be a critical component of the priming of the inflammatory response that was discussed earlier (see Section II).

11.3.4 Does Plasminogen Activator Inhibitor-1 (PAI-1) Play a Key Role in Arsenic-Enhanced NAFLD?

A key finding in the study of the interaction between arsenic and experimental NAFLD was that arsenic dramatically enhanced the increase in hepatic expression of PAI-1 caused by HFD feeding in mice (see Fig. 11.4; [74]). PAI-1 is an acute phase protein that plays an important role in injury and inflammation, at least in part via regulation of fibrinolysis. PAI-1 prevents the activation of plasmin and subsequent degradation of fibrin into fibrin degradation products (FDP) by inhibiting both the urokinase-type activators (uPAs) and tissue-type plasminogen activators (tPAs; Fig. 11.4; [44]). The net effect of PAI-1 induction is therefore a decrease in fibrin degradation and a subsequent accumulation of fibrin ECM. Indeed, this effect of arsenic correlated with an increase in perisinusoidal deposition of fibrin ECM in the liver (Fig. 11.4) in HFD-fed mice. The role of PAI-1 and fibrin accumulation in vascular disease is well understood to contribute to endothelial dysfunction and inflammation. Although the liver can produce large amounts of PAI-1 in response to stress, the role of PAI-1 in liver diseases is not well understood [79].

Previous work by this group and others has shown a correlation between fibrin ECM and inflammation in models of hepatic injury (e.g., [5–7]). There are several potential mechanisms by which fibrin ECM is proinflammatory (see Fig. 11.4). For example, fibrin clots disrupt the flow of blood within the hepatic parenchyma (i.e., hemostasis); the subsequent microregional hypoxia and hepatocellular death may directly and indirectly increase a proinflammatory response [32, 35, 60, 82]. Furthermore, fibrin matrices have been shown to be permissive to chemotaxis and activation of monocytes and leukocytes [37, 46]. Therefore, the increase in PAI-1

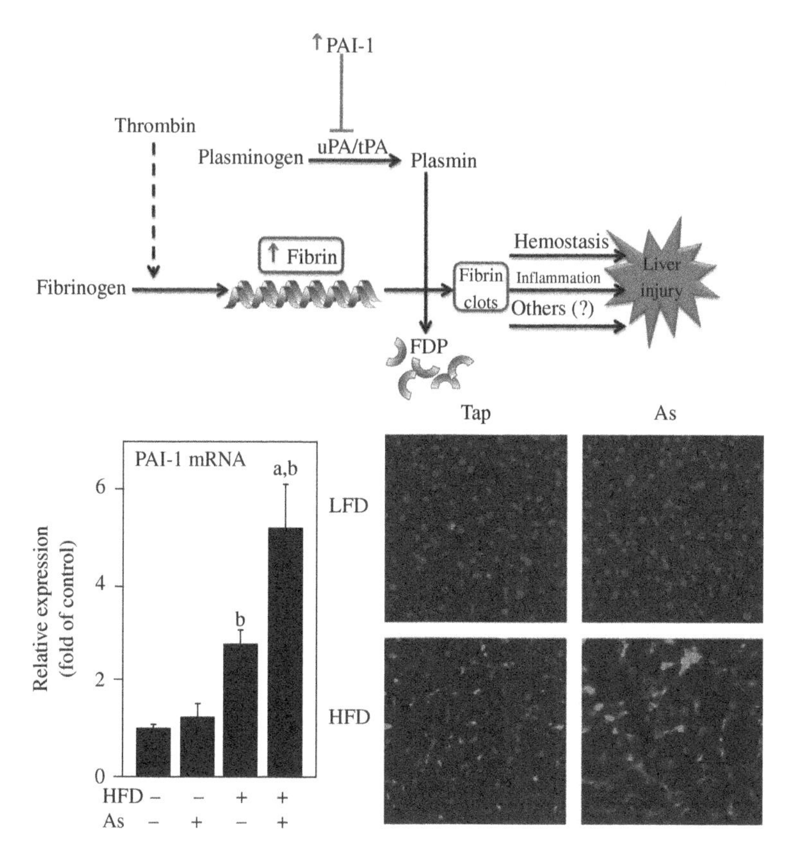

FIGURE 11.4 Effect of high-fat diet and arsenic on hepatic inflammation. Upper panel: Crosslinked fibrin deposition is initiated by activation of the coagulation cascade via thrombin. PAI-1 inhibits the activity of the plasminogen activators (uPA and tPA), blocking the activation of plasminogen to plasmin and, thereby, blunting degradation of fibrin matrices (fibrinolysis). The balance between fibrin deposition and degradation determines whether fibrin accumulates in vivo. This effect could contribute to liver damage via slowing blood flow in the liver (hemostasis), enhancing inflammation, as well as other potential mechanisms. Lower left panel: Effect of high fat diet and arsenic on hepatic mRNA expression plasminogen activator inhibitor-1 (PAI-1) is shown. Real-time RT-PCR data are means ± SE (n=6–10) and are expressed as fold of control. a, p<0.05 compared to low fat diet. b, p<0.05 compared to tap. Lower right: Representative photomicrographs (200×) depicting fibrin immunofluorescence (green color) are shown. From Tan et al. [74]. © Elsevier. (*See insert for color representation of the figure.*)

caused by arsenic exposure in the HFD group could contribute to hepatic inflammation by enhancing fibrin ECM deposition. Previous studies have shown that arsenic exposure induces PAI-1 expression in cultured cells (e.g., [40]), and in vivo in animal models [71], and is associated with arsenic exposure in humans [85]. Interestingly, in light of the work in experimental NAFLD, the latter also showed a significant interaction between arsenic exposure and higher body mass index, such that the increased levels of PAI-1 associated with arsenic exposure were stronger among people with higher body mass index [85]. The induction of PAI-1 may not only have implications for cardiovascular effects of arsenic but may also contribute to hepatic injury during arsenic exposure.

11.4 SUMMARY AND CONCLUSIONS

The ubiquity of arsenic exposure, coupled with its toxicity, makes it a significant health concern for the world population. Given its distribution in the body, concentrations of arsenic in the liver can be much higher than found in the periphery. This factor coupled with the fact that the liver is a major site of metabolism of arsenic (see Chapter 4) may make it unsurprising that liver toxicity is common with high arsenic exposure. Major mechanisms by which arsenic is proposed to be overtly hepatotoxic include altered lipid metabolism, oxidative stress, and enhanced inflammation (see Figs. 11.1, 11.2, and 11.3). However, the concentrations of arsenic exposure required to cause direct hepatotoxicity are high and are therefore likely only concerns in areas of high (ppm) arsenic contamination. However, some studies indicate that arsenic at concentrations relevant to other water supplies may cause more subtle hepatic changes. Furthermore, some work has suggested that arsenic at these concentrations may sensitize the liver to damage from a second hepatotoxic insult. These latter findings are alarming, based on the increasing rate of liver diseases (e.g., NAFLD) in the developed world, wherein arsenic contamination is generally below levels required to cause direct hepatotoxicity.

REFERENCES

[1] S. Ahmad, K.T. Kitchin, W.R. Cullen, Arsenic species that cause release of iron from ferritin and generation of activated oxygen. Arch. Biochem. Biophys. 382 (2000) 195–202.

[2] R.G. Allen, M. Tresini, Oxidative stress and gene regulation. Free Radic. Biol. Med. 28 (2000) 463–499.

[3] G.E. Arteel, L. Guo, T. Schlierf, J.I. Beier, J.P. Kaiser, T.S. Chen, M. Liu, D.J. Conklin, H.L. Miller, C. von Montfort, J.C. States, Subhepatotoxic exposure to arsenic enhances lipopolysaccharide-induced liver injury in mice. Toxicol. Appl. Pharmacol. 226 (2008) 128–139.

[4] J.I. Beier, G.E. Arteel, C.J. McClain, Advances in alcoholic liver disease. Curr. Gastroenterol. Rep. 13 (2011) 56–64.

[5] J.I. Beier, J.P. Luyendyk, L. Guo, C. von Montfort, D.E. Staunton, G.E. Arteel, Fibrin accumulation plays a critical role in the sensitization to lipopolysaccharide-induced liver injury caused by ethanol in mice. Hepatology 49 (2009) 1545–1553.

[6] I. Bergheim, L. Guo, M.A. Davis, J.C. Lambert, J.I. Beier, I. Duveau, J.P. Luyendyk, R.A. Roth, G.E. Arteel, Metformin prevents alcohol-induced liver injury in the mouse: critical role of plasminogen activator inhibitor-1. Gastroenterology 130 (2006) 2099–2112.

[7] I. Bergheim, J.P. Luyendyk, C. Steele, G.K. Russell, L. Guo, R.A. Roth, G.E. Arteel, Metformin prevents endotoxin-induced liver injury after partial hepatectomy. J. Pharmacol. Exp. Ther. 316 (2005) 1053–1061.

[8] I. Bergheim, C.J. McClain, G.E. Arteel, Treatment of alcoholic liver disease. Dig. Dis. 23 (2005) 275–284.

[9] A. Boveris, B. Chance, The mitochondrial generation of hydrogen peroxide. General properties and effect of hyperbaric oxygen. Biochem. J. 134 (1973) 707–716.

[10] E. Cadenas, K.J. Davies, Mitochondrial free radical generation, oxidative stress, and aging. Free Radic. Biol. Med. 29 (2000) 222–230.

[11] M. Cave, K.C. Falkner, M. Ray, S. Joshi-Barve, G. Brock, R. Khan, M. Bon Homme, C.J. McClain, Toxicant-associated steatohepatitis in vinyl chloride workers. Hepatology 51 (2010) 474–481.

[12] Centers for Disease Control and Prevention. Behavioral Risk Factor Surveillance System: Turning Information into Action. http://www.cdc.gov/brfss/. 2007. Accessed February 5, 2007.

[13] C.Y. Chang, C.K. Argo, A.M. Al Osaimi, S.H. Caldwell, Therapy of NAFLD: antioxidants and cytoprotective agents. J. Clin. Gastroenterol. 40 (2006) S51–S60.

[14] S. Choi, A.M. Diehl, Role of inflammation in nonalcoholic steatohepatitis. Curr. Opin. Gastroenterol. 21 (2005) 702–707.

[15] J.M. Clark, The epidemiology of nonalcoholic fatty liver disease in adults. J. Clin. Gastroenterol. 40 (2006) S5–S10.

[16] C.T. D'Angio, J.N. Finkelstein, Oxygen regulation of gene expression: a study in opposites. Mol. Genet. Metab. 71 (2000) 371–380.

[17] S. Das, A. Santra, S. Lahiri, D.N. Guha Mazumder, Implications of oxidative stress and hepatic cytokine (TNF-alpha and IL-6) response in the pathogenesis of hepatic collagenesis in chronic arsenic toxicity. Toxicol. Appl. Pharmacol. 204 (2005) 18–26.

[18] C.P. Day, Non-alcoholic fatty liver disease: current concepts and management strategies. Clin. Med. 6 (2006) 19–25.

[19] C.P. Day, O.F. James, Steatohepatitis: a tale of two "hits"? Gastroenterology 114 (1998) 842–845.

[20] A. De Vizcaya-Ruiz, O. Barbier, R. Ruiz-Ramos, M.E. Cebrian, Biomarkers of oxidative stress and damage in human populations exposed to arsenic. Mutat. Res. 674 (2009) 85–92.

[21] A.M. Diehl, Z.P. Li, H.Z. Lin, S.Q. Yang, Cytokines and the pathogenesis of non-alcoholic steatohepatitis. Gut 54 (2005) 303–306.

[22] W. Droge, Free radicals in the physiological control of cell function. Physiol. Rev. 82 (2002) 47–95.

[23] R.R. Engel, A.H. Smith, Arsenic in drinking water and mortality from vascular disease: an ecologic analysis in 30 counties in the United States. Arch. Environ. Health 49 (1994) 418–427.

[24] G. Ermak, K.J. Davies, Calcium and oxidative stress: from cell signaling to cell death. Mol. Immunol. 38 (2002) 713–721.

[25] G.C. Farrell, Drugs and steatohepatitis. Semin. Liver Dis. 22 (2002) 185–194.

[26] K.R. Feingold, C. Grunfeld, Tumor necrosis factor-alpha stimulates hepatic lipogenesis in the rat in vivo. J. Clin. Invest. 80 (1987) 184–190.

[27] K.M. Flegal, M.D. Carroll, R.J. Kuczmarski, C.L. Johnson, Overweight and obesity in the United States: prevalence and trends, 1960–1994. Int. J. Obes. Relat. Metab. Disord. 22 (1998) 39–47.

[28] S.J. Flora, Arsenic-induced oxidative stress and its reversibility. Free Radic. Biol. Med. 51 (2011) 257–281.

[29] H.J. Forman, M. Torres, Redox signaling in macrophages. Mol. Aspects Med. 22 (2001) 189–216.

[30] M. Franklin, W.B. Bean, R.C. Hardin, Fowler's solution as an etiologic agent in cirrhosis. Am. J. Med. Sci. 219 (1950) 589–596.

[31] F.J. Frost, T. Muller, H.V. Petersen, B. Thomson, K. Tollestrup, Identifying US populations for the study of health effects related to drinking water arsenic. J. Expo. Anal. Environ. Epidemiol. 13 (2003) 231–239.

[32] P.E. Ganey, J.P. Luyendyk, J.F. Maddox, R.A. Roth, Adverse hepatic drug reactions: inflammatory episodes as consequence and contributor. Chem. Biol. Interact. 150 (2004) 35–51.

[33] D.N. Guha Mazumder, A.K. Chakraborty, A. Ghose, J.D. Gupta, D.P. Chakraborty, S.B. Dey, N. Chattopadhyay, Chronic arsenic toxicity from drinking tubewell water in rural West Bengal. Bull. World Health Organ. 66 (1988) 499–506.

[34] I. Hardardottir, W. Doerrler, K.R. Feingold, C. Grunfeld, Cytokines stimulate lipolysis and decrease lipoprotein lipase activity in cultured fat cells by a prostaglandin independent mechanism. Biochem. Biophys. Res. Commun. 186 (1992) 237–243.

[35] J.A. Hewett, R.A. Roth, The coagulation system, but not circulating fibrinogen, contributes to liver injury in rats exposed to lipopolysaccharide from gram-negative bacteria. J. Pharmacol. Exp. Ther. 272 (1995) 53–62.

[36] J.B. Hoek, J.G. Pastorino, Ethanol, oxidative stress, and cytokine-induced liver cell injury. Alcohol 27 (2002) 63–68.

[37] S.R. Holdsworth, N.M. Thomson, E.F. Glasgow, R.C. Atkins, The effect of defibrination on macrophage participation in rabbit nephrotoxic nephritis: studies using glomerular culture and electronmicroscopy. Clin. Exp. Immunol. 37 (1979) 38–43.

[38] K.G. Ishak, H.J. Zimmerman, M.B. Ray, Alcoholic liver disease: pathologic, pathogenetic and clinical aspects. Alcohol. Clin. Exp. Res. 15 (1991) 45–66.

[39] H. Jaeschke, C.V. Smith, J.R. Mitchell, Reactive oxygen species during ischemia-reflow injury in isolated perfused rat liver. J. Clin. Invest. 81 (1988) 1240–1246.

[40] S.J. Jiang, T.M. Lin, H.L. Wu, H.S. Han, G.Y. Shi, Decrease of fibrinolytic activity in human endothelial cells by arsenite. Thromb. Res. 105 (2002) 55–62.

[41] D.M. Jolliffe, A history of the use of arsenicals in man. J. R. Soc. Med. 86 (1993) 287–289.

[42] H. Kamata, H. Hirata, Redox regulation of cellular signalling. Cell. Signal. 11 (1999) 1–14.

[43] K.T. Kitchin, Recent advances in arsenic carcinogenesis: modes of action, animal model systems, and methylated arsenic metabolites. Toxicol. Appl. Pharmacol. 172 (2001) 249–261.

[44] E.K. Kruithof, Plasminogen activator inhibitors: a review. Enzyme 40 (1988) 113–121.

[45] Z. Li, A.M. Diehl, Innate immunity in the liver. Curr. Opin. Gastroenterol. 19 (2003) 565–571.

[46] J.D. Loike, J. el Khoury, L. Cao, C.P. Richards, H. Rascoff, J.T. Mandeville, F.R. Maxfield, S.C. Silverstein, Fibrin regulates neutrophil migration in response to interleukin 8, leukotriene B4, tumor necrosis factor, and formyl-methionyl-leucyl-phenylalanine. J. Exp. Med. 181 (1995) 1763–1772.

[47] M.P. Longnecker, J.L. Daniels, Environmental contaminants as etiologic factors for diabetes. Environ. Health Perspect. 109 Suppl 6 (2001) 871–876.

[48] J. Ludwig, T.R. Viggiano, D.B. McGill, B.J. Oh, Nonalcoholic steatohepatitis: Mayo Clinic experiences with a hitherto unnamed disease. Mayo Clin. Proc. 55 (1980) 434–438.

[49] D.N. Mazumder, Effect of chronic intake of arsenic-contaminated water on liver. Toxicol. Appl. Pharmacol. 206 (2005) 169–175.

[50] A.J. McCullough, Pathophysiology of nonalcoholic steatohepatitis. J. Clin. Gastroenterol. 40 (2006) S17–S29.

[51] A.H. Mokdad, E.S. Ford, B.A. Bowman, W.H. Dietz, F. Vinicor, V.S. Bales, J.S. Marks, Prevalence of obesity, diabetes, and obesity-related health risk factors, 2001. JAMA 289 (2003) 76–79.

[52] J.S. Morris, M. Schmid, S. Newman, P.J. Scheuer, S. Sherlock, Arsenic and noncirrhotic portal hypertension. Gastroenterology 66 (1974) 86–94.

[53] V. Nachiappan, D. Curtiss, B.E. Corkey, L. Kilpatrick, Cytokines inhibit fatty acid oxidation in isolated rat hepatocytes: synergy among TNF, IL-6, and IL-1. Shock 1 (1994) 123–129.

[54] National Center for Health Statistics (NCHS). Health, United States, 2001, with Urban and Rural Health Chart Book, National Center for Health Statistics, Hyattsville, MD, 2001.

[55] M. Navasa, D.A. Gordon, N. Hariharan, H. Jamil, J.K. Shigenaga, A. Moser, W. Fiers, A. Pollock, C. Grunfeld, K.R. Feingold, Regulation of microsomal triglyceride transfer protein mRNA expression by endotoxin and cytokines. J. Lipid Res. 39 (1998) 1220–1230.

[56] J.V. Neel, Diabetes mellitus: a "thrifty" genotype rendered detrimental by "progress"? Am. J. Hum. Genet. 14 (1962) 353–362.

[57] C.L. Ogden, M.D. Carroll, L.R. Curtin, M.A. McDowell, C.J. Tabak, K.M. Flegal, Prevalence of overweight and obesity in the United States, 1999–2004. JAMA 295 (2006) 1549–1555.

[58] L. Patrick, Toxic metals and antioxidants: Part II. The role of antioxidants in arsenic and cadmium toxicity. Altern. Med. Rev. 8 (2003) 106–128.

[59] D.S. Paul, F.S. Walton, R.J. Saunders, M. Styblo, Characterization of the impaired glucose homeostasis produced in C57BL/6 mice by chronic exposure to arsenic and high-fat diet. Environ. Health Perspect. 119 (2011) 1104–1109.

[60] J.M. Pearson, A.E. Schultze, K.A. Schwartz, M.A. Scott, J.M. Davis, R.A. Roth, The thrombin inhibitor, hirudin, attenuates lipopolysaccharide-induced liver injury in the rat. J. Pharmacol. Exp. Ther. 278 (1996) 378–383.

[61] D. Pessayre, A. Mansouri, B. Fromenty, Nonalcoholic steatosis and steatohepatitis. V. Mitochondrial dysfunction in steatohepatitis. Am. J. Physiol. Gastrointest. Liver Physiol. 282 (2002) G193–G199.

[62] S. Pooya, S. Blaise, G.M. Moreno, J. Giudicelli, J.M. Alberto, R.M. Gueant-Rodriguez, E. Jeannesson, N. Gueguen, A. Bressenot, B. Nicolas, Y. Malthiery, J.L. Daval, L. Peyrin-Biroulet, J.P. Bronowicki, J.L. Gueant, Methyl donor deficiency impairs fatty acid oxidation through PGC-1alpha hypomethylation and decreased ER-alpha, ERR-alpha, and HNF-4alpha in the rat liver. J. Hepatol. 57 (2012) 344–351.

[63] P. Roingeard, Hepatitis C virus diversity and hepatic steatosis. J. Viral Hepat. 20 (2013) 77–84.

[64] A. Santra, G.J. Das, B.K. De, B. Roy, D.N. Guha Mazumder, Hepatic manifestations in chronic arsenic toxicity. Indian J. Gastroenterol. 18 (1999) 152–155.

[65] A. Santra, A. Maiti, S. Das, S. Lahiri, S.K. Charkaborty, D.N. Mazumder, Hepatic damage caused by chronic arsenic toxicity in experimental animals. J. Toxicol. Clin. Toxicol. 38 (2000) 395–405.

[66] R.F. Schwabe, E. Seki, D.A. Brenner, Toll-like receptor signaling in the liver. Gastroenterology 130 (2006) 1886–1900.

[67] C.K. Sen, Cellular thiols and redox-regulated signal transduction. Curr. Top. Cell Regul. 36 (2000) 1–30.

[68] H. Sies, Oxidative stress: introductory remarks. in: H. Sies (Ed.), Oxidative Stress, Academic Press, London, 1985, pp. 1–8.

[69] H. Sies, Biochemistry of oxidative stress. Angew. Chem. Int. Ed. Engl. 25 (1986) 1058–1071.

[70] A.H. Smith, C. Hopenhayn-Rich, M.N. Bates, H.M. Goeden, I. Hertz-Picciotto, H.M. Duggan, R. Wood, M.J. Kosnett, M.T. Smith, Cancer risks from arsenic in drinking water. Environ. Health Perspect. 97 (1992) 259–267.

[71] N.V. Soucy, D. Mayka, L.R. Klei, A.A. Nemec, J.A. Bauer, A. Barchowsky, Neovascularization and angiogenic gene expression following chronic arsenic exposure in mice. Cardiovasc. Toxicol. 5 (2005) 29–41.

[72] A.C. Straub, K.A. Clark, M.A. Ross, A.G. Chandra, S. Li, X. Gao, P.J. Pagano, D.B. Stolz, A. Barchowsky, Arsenic-stimulated liver sinusoidal capillarization in mice requires NADPH oxidase-generated superoxide. J. Clin. Invest. 118 (2008) 3980–3989.

[73] A.C. Straub, D.B. Stolz, M.A. Ross, A. Hernandez-Zavala, N.V. Soucy, L.R. Klei, A. Barchowsky, Arsenic stimulates sinusoidal endothelial cell capillarization and vessel remodeling in mouse liver. Hepatology 45 (2007) 205–212.

[74] M. Tan, R.H. Schmidt, J.I. Beier, W.H. Watson, H. Zhong, J.C. States, G.E. Arteel, Chronic subhepatotoxic exposure to arsenic enhances hepatic injury caused by high fat diet in mice. Toxicol. Appl. Pharmacol. 257 (2011) 356–364.

[75] R. Taub, Liver regeneration: from myth to mechanism. Nat. Rev. Mol. Cell Biol. 5 (2004) 836–847.

[76] H. Tilg, The role of cytokines in non-alcoholic fatty liver disease. Dig. Dis. 28 (2010) 179–185.

[77] United States Geological Survey. Arsenic in Ground Water of the United States. http://water.usgs.gov/nawqa/trace/arsenic/. 2007. Accessed February 7, 2007.

[78] V.J. van Ginneken, Liver fattening during feast and famine: an evolutionary paradox. Med. Hypotheses 70 (2008) 924–928.

[79] D.E. Vaughan, PAI-1 and atherothrombosis. J. Thromb. Haemost. 3 (2005) 1879–1883.

[80] M.P. Waalkes, J. Liu, H. Chen, Y. Xie, W.E. Achanzar, Y.S. Zhou, M.L. Cheng, B.A. Diwan, Estrogen signaling in livers of male mice with hepatocellular carcinoma induced by exposure to arsenic in utero. J. Natl. Cancer Inst. 96 (2004) 466–474.

[81] M.P. Waalkes, J. Liu, J.M. Ward, B.A. Diwan, Enhanced urinary bladder and liver carcinogenesis in male CD1 mice exposed to transplacental inorganic arsenic and postnatal diethylstilbestrol or tamoxifen. Toxicol. Appl. Pharmacol. 215 (2006) 295–305.

[82] I.R. Wanless, F. Wong, L.M. Blendis, P. Greig, E.J. Heathcote, G. Levy, Hepatic and portal vein thrombosis in cirrhosis: possible role in development of parenchymal extinction and portal hypertension. Hepatology 21 (1995) 1238–1247.

[83] A.H. Welch, S.A. Watkins, D.R. Helsel, M.J. Focazio, Arsenic in ground-water resources of the United States. U.S. Geological Survey Fact Sheet FS-063-00 (2000). http://co.water.usgs.gov/trace/pubs/fs-063-00/. Accessed April 13, 2015.

[84] K. Wouters, P.J. van Gorp, V. Bieghs, M.J. Gijbels, H. Duimel, D. Lutjohann, A. Kerksiek, R. van Kruchten, N. Maeda, B. Staels, M. van Bilsen, R. Shiri-Sverdlov, M.H. Hofker, Dietary cholesterol, rather than liver steatosis, leads to hepatic inflammation in hyperlipidemic mouse models of nonalcoholic steatohepatitis. Hepatology 48 (2008) 474–486.

[85] F. Wu, F. Jasmine, M.G. Kibriya, M. Liu, O. Wojcik, F. Parvez, R. Rahaman, s. Roy, R. Paul-Brutus, S. Segers, V. Slavkovich, T. Islam, D. Levy, J.L. Mey, G.A. van, J.H. Graziano, H. Ahsan, Y. Chen, Association between arsenic exposure from drinking water and plasma levels of cardiovascular markers. Am. J. Epidemiol. 175 (2012) 1252–1261.

[86] K. Yamanaka, M. Hoshino, M. Okamoto, R. Sawamura, A. Hasegawa, S. Okada, Induction of DNA damage by dimethylarsine, a metabolite of inorganic arsenics, is for the major part likely due to its peroxyl radical. Biochem. Biophys. Res. Commun. 168 (1990) 58–64.

[87] S.Q. Yang, H.Z. Lin, M.D. Lane, M. Clemens, A.M. Diehl, Obesity increases sensitivity to endotoxin liver injury: implications for the pathogenesis of steatohepatitis. Proc. Natl. Acad. Sci. U. S. A. 94 (1997) 2557–2562.

12

GENETIC EPIDEMIOLOGY OF SUSCEPTIBILITY TO ARSENIC-INDUCED DISEASES

Mayukh Banerjee[1] and Ashok K. Giri[2]

[1] *Department of Physiology, University of Alberta, Edmonton, AB, Canada*
[2] *Molecular and Human Genetics Division, Indian Institute of Chemical Biology, Kolkata, India*

12.1 INTRODUCTION

Chronic toxicity ensuing from long periods of exposure to low doses of arsenic is a pandemic issue. As per the latest estimates, this menace plagues around 150 million human beings spread over 70 different countries worldwide [67, 74]. It is thus of little surprise that arsenic poisoning has been dubbed as "the worst mass poisoning in the history of mankind." To add to this furor, arsenic also happens to be number 1 in the Agency for Toxic Substances and Disease Registry (ATSDR) list of carcinogenic materials. There is unequivocal experimental evidence demonstrating that chronic exposure to arsenic leads to the heightened development of multiorgan cancers including those of skin, lungs, liver, and urinary bladder specifically in human beings. In addition, it can also give rise to precancerous dermatological lesions like hyperkeratosis (palmar and plantar) and other noncancerous lesions including hyper- and hypopigmentation and raindrop pigmentation. Although, skin symptoms are considered to be hallmarks of arsenic toxicity, a plethora of epidemiological studies prove beyond any conceivable doubt that chronic arsenic toxicity can also manifest itself in the form of nondermatological disorders such as peripheral neuropathy,

Arsenic: Exposure Sources, Health Risks, and Mechanisms of Toxicity, First Edition.
Edited by J. Christopher States.
© 2016 John Wiley & Sons, Inc. Published 2016 by John Wiley & Sons, Inc.

respiratory distress, and ocular problems like conjunctivitis [34]. Arsenic, thus not only kills people but is also responsible for diminished life expectancy and impaired life quality. The major problem that the arsenic-exposed populations face is lack of any medication that is effective in combating this toxic insult. Consequently, identification of highly susceptible individuals is required to prevent them from getting exposed preferentially.

12.2 ARSENIC SUSCEPTIBILITY

It is long known that not all individuals show unequivocal response to arsenic toxicity. The latency periods and the intensity of the ailments are known to vary widely from person to person. While first dermatological symptoms can appear as early as 6 months after the onset of exposure, some individuals may not develop these symptoms until after an exposure period of 10 years or even longer. More importantly, epidemiological data shows that only 10–15% of the exposed individuals will eventually develop the dermatological effects at all, even after prolonged exposure. Majority of the individuals would not develop the characteristic skin lesions even after being exposed for over 10 years (although, they may develop non-dermatological symptoms). This observation clearly points out that some individuals are far more prone to the devastating effects of arsenic exposure than others, and hence constitute the "high risk" group.

What makes certain individuals especially susceptible to toxic effects of arsenic exposure has been studied for quite some time. While the obvious factor seems to be the extent of exposure (both with regard to the period of exposure and the amount of intake), there are well-documented cases where a group of individuals having similar level of exposure time and concentration have differential disease outcomes. This fact certainly points to some inherent property of each individual that determines his/her degree of susceptibility and the physiological response to chronic arsenic exposure. In this regard, the genetic composition of each individual seems to be indicated. The human genome roughly consists of three billion bases, and minor variations in the genome are sufficient to generate interindividual differences. Such variations are generated in the genome via mutations and genetic polymorphisms. Genetic polymorphisms are particularly important in this respect, as their physiological effects are usually much more subtle than that of mutations and result in subtle interindividual differences and are responsible for making every individual genetically unique. Consequently, in theory, if such polymorphisms are present in the genes whose products are involved in metabolism/transport of arsenic in the body, they might alter the processing of arsenic and thus confer susceptibility/resistance to the individuals harboring them. Hence, many studies attempted to assess the role of genetic polymorphisms in conferring susceptibility/resistance to chronic arsenic toxicity. In the following sections, the data so far gleaned in course of those studies and the areas that still could be investigated to get a clearer picture as to how individuals can have unique responses to arsenic exposure at the physiological level due to their unique genetic composition are discussed.

12.3 CANDIDATE GENES

The first issue to address before embarking on a genetic susceptibility study is to determine the genes (and polymorphisms thereof) which might have a bearing on arsenic toxicity in the physiological milieu. Theoretically, any genes, whose product is involved in metabolism or transport of arsenic (and metabolites thereof), are likely candidates. In addition, the genes that play key roles in mediating the effects of arsenic would also be important in this regard. In order to streamline the putative genes that might be implicated in the process, it is mandatory to take a cursory look at the metabolism of arsenic inside the human body and its impacts at the molecular level. The next section will briefly recount the different theories on arsenic metabolism.

In the classical pathway (see Chapter 4 for detailed discussion of arsenic metabolism), proposed as early as 1945 by Challenger (and subsequently corroborated by Cullen and Reimer) [18, 26], inorganic arsenic is converted in the liver into dimethyl arsenic species through alternating steps of reduction and oxidative methylation. According to this pathway, inorganic pentavalent arsenic is first converted to inorganic trivalent arsenic, which then undergoes first round of oxidative methylation in the presence of the methyl donor *S*-adenosyl-L-methionine (SAM) to form monomethylarsinic acid (MMA^{V}) which is further reduced to the corresponding trivalent form (monomethylarsonous acid MMA^{III}). This MMA^{III} then undergoes a second round of oxidative methylation (again in the presence of SAM) to form dimethylarsinic acid (DMA^{V}), which is then reduced to form dimethylarsinous acid (DMA^{III}), which is believed to be the final product of arsenic metabolism in humans and is excreted through urine. Thus, the important genes involved in this pathway include those encoding for arsenate reductases GSTO, *AS3MT* and *PNP*. In this regard, other members of the GST family might also be important due to the broad overlapping substrate specificity between the family members.

In 2005, Hayakawa *et al.*, proposed an alternative scheme for arsenic metabolism [39]. This new scheme suggested the involvement of glutathione in several steps of arsenic metabolism. According to this hypothesis, inorganic arsenite is first conjugated to glutathione to form arsenic triglutathione (ATG), in the presence of cellular glutathione, and subsequently methylated to monomethyarsonic diglutathione (MADG). This MADG can then either be oxidized to MMA^{V} via the intermediate formation of MMA^{III}, or it can be further methylated and subsequently oxidized to yield DMA^{V} as another end product. Hence, unlike the classical pathway, in the alternative pathway, both MMA^{V} and DMA^{V} are end products, which are then excreted via urine. Very recently, Rehman and Naranmandura [76] have proposed a third mechanism of arsenic metabolism, in which, inorganic arsenic first binds to protein moieties in the liver and then undergoes consecutive rounds of reductive methylation in presence of *S*-aednosyl-methionine (by conjugating with thiol groups of cysteinyl residues on proteins) and produces MMA^{V} and DMA^{V} as the end products, as in the case of alternative pathway.

The other susceptibility factors would reside in the mechanisms via which arsenic exerts its toxic effects on the body, and the pathways that promote such mechanisms and also the ones that combat them. How inorganic arsenic and metabolites thereof cause toxicity has been much debated, and several mechanisms have been proposed. In all probability, several concomitant mechanisms are involved and are perhaps interrelated, rather than being isolated pathways bringing about disease outcomes. The mechanisms have been dealt with in detail in several chapters of the current volume. Thus, in the following section, we will provide only a cursory glance at the different theories regarding arsenic-induced toxicity and try to reconcile them in the physiological perspective.

Arsenic classically has been known as a nonmutagenic carcinogen. Although, it does not induce point mutations, arsenic is, however, a well-documented clastogen, giving rise to chromosomal damage. Although, arsenic is not known to interact directly with the genetic material, nonetheless, it achieves its genotoxic effect by modulating several physiological processes, notably the generation reactive oxygen species (ROS) and reactive nitrogen species (RNS) in the system. The primary ROS likely to be induced by arsenic is the superoxide radical (O_2^-), which is subsequently converted to hydrogen peroxide (H_2O_2) in the presence of superoxide dismutase. H_2O_2 can be converted further to hypochlorous acid (HOCl) and chloride anion (Cl^-) by myeloperoxidase. The hydrogen peroxide formed also can convert spontaneously to the highly reactive hydroxyl radical ($OH^\bullet$) in the presence of cellular ferrous iron (Fe^{2+}) by a mechanism known as the Fenton reaction. On the other hand, the cells have enzymes like catalase and glutathione peroxidase which tend to keep the ROS in check by converting them to relatively harmless products like water and molecular oxygen, utilizing the ROS molecules as their substrates. In addition, inorganic arsenic exposure also induces generation of RNS, by inducing production of nitric oxide (NO) *in vitro* and *in vivo*. So, all the genes encoding for such enzymes which play important roles in mediating the level of ROS and RNS, and polymorphisms thereof which can affect their functioning, are likely to have effects on individual susceptibility to arsenic-induced toxic effects.

This enhanced level of ROS, in turn, can interact with the genetic material in the cells and induce considerable amount of genetic damage. In fact, epidemiological studies on chronically arsenic-exposed populations have unambiguously shown the presence of heightened oxidative DNA adducts compared to unexposed controls [33, 43]. Other studies, using Fpg-mediated single-cell gel electrophoresis also show the presence of oxidative DNA damage in arsenic-exposed individuals [14, 71]. Chronic arsenic exposure leads to DNA damage, which can give rise to chromosomal abnormalities that are considered harbingers of carcinogenesis. In order to combat such DNA damage, cells are equipped with a battery of enzymes specialized to identify and repair the DNA lesions. Depending on the nature of the induced damage, several distinct yet overlapping DNA repair systems have evolved, each specialized for repairing specific kinds of damage. Consequently, it stands to reason that genetic variation in any of these enzymes, which can alter the functioning or even their level of expression, will have significant impact on the efficacy of the repair system and will thus contribute to the susceptibility of the individuals facing arsenic exposure.

Besides affecting the genetic material, another potent mechanism by which arsenic is known to bring about its deleterious effects is epigenomic alteration. Studies have shown that chronic exposure to arsenic can significantly alter the methylation status of chromatin, thereby adversely affecting the normal transcription rate of several physiologically important genes [10, 72, 81, 84]. Both aberrant promoter hyper and hypomethylation primarily have been reported. Hence, the enzymes mediating the process of DNA methylation and also those determining the availability of the methyl donor in the system are expected to have an impact on susceptibility to arsenic-induced toxicity and disease. An additional factor that can also make an individual prone to any kind of toxic insult is the immune status of the individual concerned. So, genes encoding for mediators of immune response (i.e., interleukins or chemokines) will also be important in defining how susceptible a person is to the toxic effect of arsenic exposure.

12.4 GENETIC SUSCEPTIBILITY TO ARSENIC-INDUCED DISEASES

Having outlined the different classes of genes whose variations might affect the extent to which an individual will be prone to the toxic effects of chronic arsenic exposure in the preceding sections, we will now have a look at the actual studies that have been conducted in exposed populations. Although several association studies have been conducted with respect to arsenic toxicity, most of them have used one or more well-defined arsenic-induced cancer(s) as the endpoint. Studies regarding noncancerous effects of arsenic are meagre in the extreme. In the upcoming section, we will document the various association studies from different exposed populations, and attempt to highlight their functional significance from wherever such studies (*in vivo* or *in vitro*) are available.

12.4.1 Role of Genetic Susceptibility in Arsenic Metabolism

As stated earlier, the nature of metabolism and the level of body retention of arsenic to a large extent determine the susceptibility to arsenic-induced diseases. In this section, the role of genetic polymorphisms on these aspects will be documented. Glutathione *S*-transferases (GSTs) comprise a large superfamily of enzymes that play key roles in phase II cellular detoxification. The members have broad overlapping substrate specificity and are usually involved in converting a wide variety of chemicals into the corresponding water-soluble glutathione conjugates that can subsequently be excreted via urine. This group thus forms an ideal set of candidates to probe for possible association with arsenic metabolism/detoxification and ensuing disease outcomes resulting from the imbalance in arsenic retention and excretion due to an alteration in their functional capacities in different variant forms. It is therefore, not surprising that several studies have looked at the putative association of different GST family members with arsenic toxicity.

GSTT1 and *GSTM1* are the most well-known members of this group. Each has a widely studied null variant, in which the entire gene is deleted, thereby resulting in a complete absence of the functional protein. As early as 2005, Kile *et al.* [52] demonstrated that *GSTT1* null individuals had significantly higher body retention of arsenic compared to wild-type individuals as measured by toe-nail arsenic concentration, an index for long-term arsenic deposition in the body. It could be argued that individuals with higher body retention of arsenic would be more likely to suffer from toxic effects than those who can efficiently excrete arsenic more efficiently. However, in a subsequent study, Marcos *et al.* [61] could find no such association of *GSTT1* genotypes with urinary arsenic excretion profiles. Negative results were also obtained for association of *GSTT1* variants with urinary and hair arsenic content from a study in China [58]. In another study on arsenic-exposed population from Argentina, no association was found between *GSTT1* variants and arsenic methylation pattern [82]. Similar negative results for *GSTT1* null variants with respect to urinary profile of arsenic metabolites were also obtained from studies on exposed populations from China [83], Vietnam [3], and Chile [17]. In stark contrast, in another study on arsenic-exposed females in Argentina, *GSTT1* null genotype was seen to significantly affect the urinary excretion of MMA and DMA [78]. Thus, the role of *GSTT1* polymorphisms with respect to arsenic metabolism is highly debatable with more negative results than positive ones across different populations.

Initial association studies regarding the association of *GSTM1* null variant with arsenic metabolism, as measured by urinary profiles of arsenic metabolites, did not show any statistically significant correlation [58, 61]. However, in the study by Lin *et al.* [10], which comprised of two different clans from China, *GSTM1* null males and *GSTM1* null members of only one clan were shown to have significantly elevated level of hair arsenic content. Positive association of higher urinary MMA excretion with *GSTM1* null genotype was however reported in exposed Argentinian female population by Steinmaus *et al.* [82]. Similar evidence was provided by another study by Schläwicke *et al.* also in an Argentine population [78]. Higher urinary excretion of DMA^V in *GSTM1* wild-type individuals was reported by Asuga *et al.* [3] studying a Vietnamese population. However, further negative results [17, 83] failing to show any association between *GSTM1* variants and arsenic metabolism has raised questions regarding the absolute importance of *GSTM1* in arsenic metabolism and susceptibility.

GSTO1 is another member of the GST superfamily that has attracted considerable attention among the workers in the field of arsenic susceptibility because of its ability to catalyse the reduction of MMA^V. The first insight regarding the importance of polymorphic variants of this gene in arsenic metabolism came in 2003, when 2 out of 75 subjects in a study by Marnell *et al.* [62] showed unique patterns of urinary arsenic metabolite profile (had high levels of urinary inorganic arsenic compared to methylated forms), and they both were heterozygous for E155del and Glu208Lys variants. In a subsequent study, *GSTO1* Glu155del heterozygous genotype was demonstrated to have higher urinary concentration of As^V than homozygous wild-type individuals [2]. Further corroboration of the role of *GSTO1* polymorphism in arsenic metabolism came when Chung *et al.* showed individuals carrying wild-type *GSTO1* 140 Ala/Ala genotype excreted significantly higher MMA in urine compared to those who carried the *GSTO1* 140 Ala/Asp or Asp/Asp genotype [25], and similar results were also obtained recently by Rodrigues *et al.* [77] in a Bangladeshi population. However, a plethora of studies also showed negative association of arsenic metabolism pattern with *GSTO1* genotypes. Ala236Val variation was shown by Paiva *et al.* [70] not to be associated in a Chilean population [70], while functional studies showed that *GSTO1* Glu155del did not alter sensitivity to arsenic trioxide [80]. Also, there was no difference in the urinary arsenic profiles among the different variants of *GSTO1* Ala140Asp polymorphism in a Chinese population exposed to arsenic [91], nor in a Central European population [59]. Also, no difference in the ratio of di- and monomethylated arsenic species in urine was found to be associated with any *GSTO1* genotype in a study from Mexico [66]. Similarly, studies on *GSTO2* polymorphisms do not show any unequivocal association with arsenic metabolism patterns. Chung *et al.* [22, 25] studied the effect of Asn142Asp polymorphism on urinary arsenic metabolites and showed that individuals heterozygous for the polymorphic locus had significantly lower baseline value of urinary inorganic arsenic and MMA than individuals homozygous for Asp variant; while, in a very small study, Paiva *et al.*, showed that for the same polymorphism, individuals carrying the 142Asp allele presented almost 4% more DMA excreted in urine, compared with those individuals carrying the normal genotype [69]. In a large population-based study from Bangladesh, these findings were corroborated as the individuals with at least one Asn allele were found to have significantly lower urinary excretion of As^{III}, MMA, and DMA compared to the wildtype [77]. However, negative results from several studies put a question on the role played by this gene variant in arsenic metabolism [3, 91].

A few studies have probed the role of *GSTP1* polymorphisms with respect to arsenic metabolism, The first study, conducted in 2006, showed a mild trend (albeit statistically nonsignificant), where individuals homozygous for the Val variant of the Ile105Val polymorphism excreted higher amount of DMA in their urine compared to the other genotypes [61]. A more detailed study on the Vietnamese population demonstrated that individuals heterozygous for this particular genotype had significantly lower inorganic arsenic but higher DMA^V excretion through urine, and also, higher retention of inorganic arsenic and MMA^V in the hairs compared to the Ile/Ile genotype [3] and further corroborated by another study by the same group [4].

From a cursory glance at the arsenic metabolism pathways (both classical and alternative), it is readily apparent that *AS3MT* is the most important enzyme that catalyses the transfer of methyl groups to arsenic and regulates its metabolism. It is, therefore, not surprising that several groups have attempted to correlate the polymorphic variations in this gene with arsenic metabolism patterns. The earliest study came from Mexico, where three polymorphic sites in this gene were found to be strongly associated with higher ratio of di- and monomethylated arsenic species in urine of children (7–11 years of age), but not in adults [66]. In a subsequent study, three intronic polymorphisms in *AS3MT* (G12390C, C14215T, and G35991A) were again demonstrated to be associated with a lower percentage of MMA and a higher percentage of DMA in urine, which corroborated the earlier results [78, 79]. Further supportive evidence for the role of this gene came from a study on Central European population, where, a nonsynonymous polymorphism (M287T) was shown to be associated with higher urinary excretion of MMA in the heterozygote state, compared to the wild-type homozygote [59]. Similar results were also obtained by Hernandez *et al.* [41, 42]. Several *AS3MT* polymorphisms were also found to be associated with altered urinary arsenic metabolite profile from a Vietnamese population [2], a Taiwanese population [24], and populations from Bangladesh and Mexico [32, 77]. Thus, there is unequivocal evidence of involvement of polymorphic variants of this gene with interindividual variances in methylation capacity of arsenic, and hence its body retention and excretion across several exposed populations worldwide. Although, so many polymorphisms of this gene (both intronic and exonic) have been shown to be involved, the demonstration of the existence of a large genomic region sharing strong linkage disequilibrium with polymorphisms associated with arsenic metabolism might well indicate that the observed phenotype cannot be unequivocally assigned to a unique single-nucleotide polymorphism [36].

Six studies, till date, have examined the effect of 5,10-methylenetetrahydrofolate reductase (*MTHFR*) variants on arsenic metabolism. Lindberg *et al.* [59] showed that individuals homozygous for the variant form of A222V of these gene have significantly elevated urinary levels of MMA but lower DMA compared to those bearing the wild-type homozygous genotype, while carriers of variant allele of *MTHFR* A222V polymorphism were found to affect the percentage of all metabolites, as well as the DMA/MMA ratio in pregnant women in a study from Argentina [78]. In a very small study of a Chinese population, Deng *et al.* [30] showed that individuals with genotype of CT/TT of *MTHFR* C677T polymorphism had the increased percentage of urinary As^{3+} and decreased percentage of urinary DMA^V compared to the wild-type homozygotes and was corroborated by another study from Taipei [24]. Also, subjects with the TT/AA variant of *MTHFR* 677/1298 excreted a significantly higher proportion of ingested arsenic as inorganic arsenic and a lower proportion as DMA^V [82]. However, Porter *et al.* [75] could not find any association between *MTHFR* variants and arsenic metabolism.

Few other genes have been studied for their putative association with differential capacity of arsenic metabolism, but the number of studies for them is far too few to enable the drawing of any concrete conclusion as to the role played by them. In one study, purine nucleoside phosphorylase (*PNP*) variants were found to increase the excretion of DMA in urine [22], while another solitary study reported the significant association of variant genotypes of cystathionine-β-synthase gene (*CBS*), but not with methionine synthase (*MTR*), thymidylate synthase (*TYMS*), dihydrofolate reductase (*DHFR*), serine hydroxymethyltransferase 1 (*SHMT1*) [75]. However, no association of *CBS* variants were found in a study by Chung *et al.* [24] in a Taiwanese population. Similarly, inconclusive results were obtained for variations in the genes involved in one-carbon metabolism such as choline dehydrogenase (*CHDH*) (rs9001, rs7626693) and 5-methyltetrahydrofolate-homocysteine methyltransferase reductase (*MTRR*) (rs1801394) and genes involved in reduction reactions, glutaredoxin (*GLRX*) (rs3822751), and peroxiredoxin 2 (*PRDX2*) (rs10427027, rs12151144) from a solitary study [79]. The same is true for variants of myeloperoxidase (*MPO*) and sulfotransferase (*SULT*) genes about their effect on arsenic methylation capability [49, 79]. AA variants in A2756G polymorphism of methionine synthase (*MS*) gene were shown to have significantly higher homocysteine levels than those with the AG or GG genotype and this elevated level of homocysteine might facilitate better methylation of arsenic, but that hypothesis still needs further validation [24].

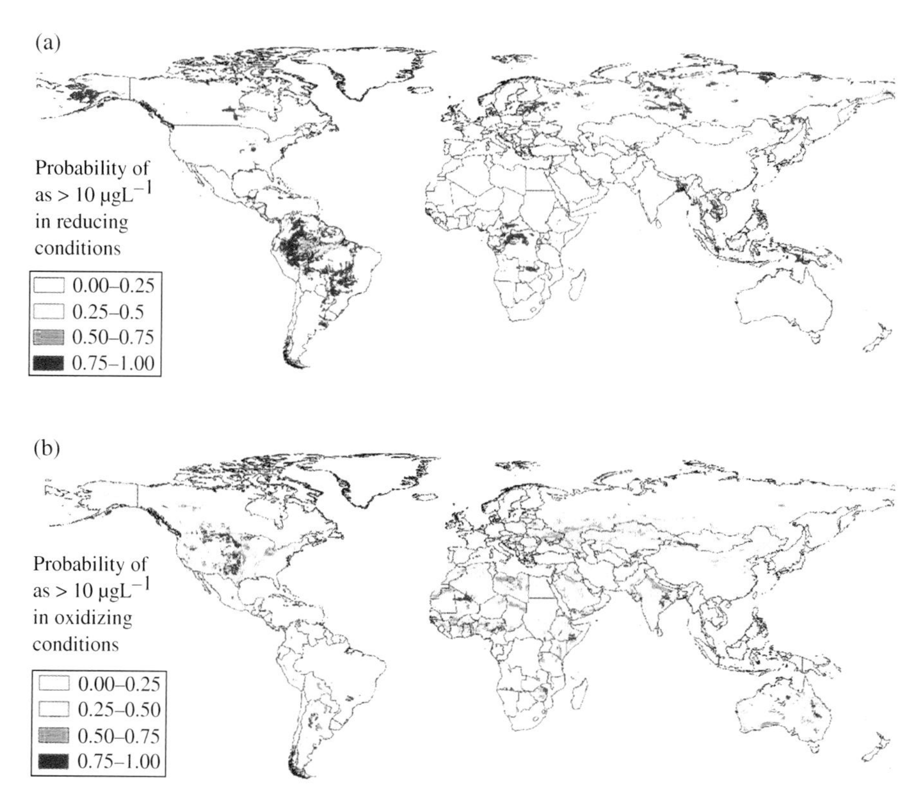

FIGURE 2.2 Modeled probability of geogenic arsenic in groundwater exceeding 10 μg/L in (a) reducing and (b) high pH/oxidizing conditions. These figures give a guide as to the likely widespread extent of the two major types geogenic arsenic in groundwaters, although (i) significant differences exist between modeled high hazard areas and actual high hazards, for example, notably in the eastern United States and (ii) geothermal high geogenic arsenic groundwaters and anthropogenically polluted and mineral deposit impacted groundwaters are not considered in the model (Reprinted with permission from Amini et al. [9]. © American Chemical Society).

Arsenic: Exposure Sources, Health Risks, and Mechanisms of Toxicity, First Edition.
Edited by J. Christopher States.
© 2016 John Wiley & Sons, Inc. Published 2016 by John Wiley & Sons, Inc.

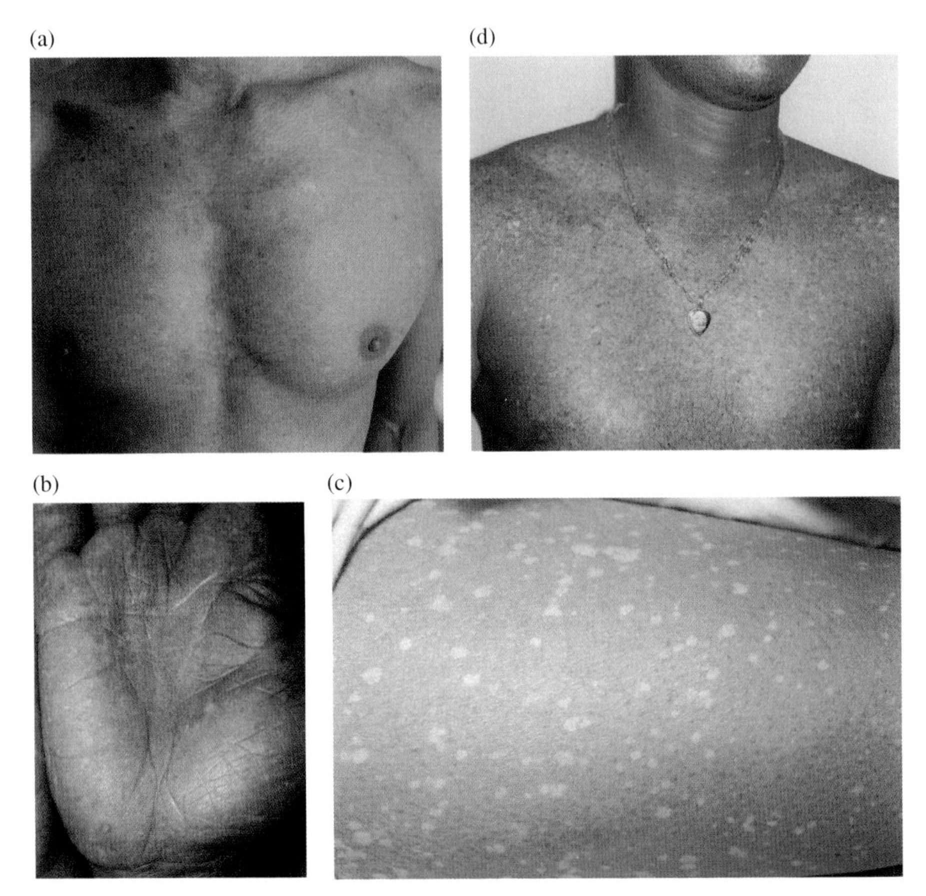

FIGURE 6.1 Pigmentary changes induced by arsenicosis. Small, distinct hyperpigmented macules on chest due to arsenicosis (a). Diffuse pigmentation on palm in arsenicosis (b). Well defined hypopigmented macules on thigh due to arsenicosis (c). Extensive xerodermoid dyschromia on chest in arsenicosis (d).

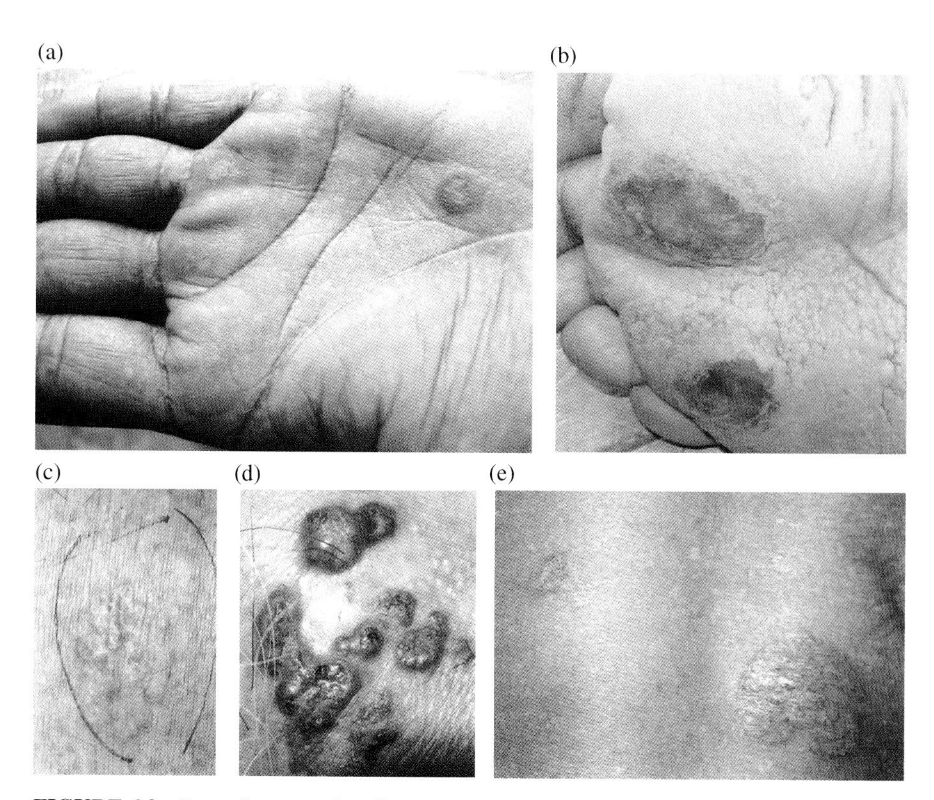

FIGURE 6.2 Premalignant and malignant lesions induced by arsenicosis. Arsenical keratosis on palm (a). Large arsenical keratosis on sole mimicking callosity (b). Classical Bowen's disease in arsenicosis (c). Pigmented nodular basal cell carcinoma in arsenicosis (d). Multiple superficial basal cell carcinoma on back with typical arsenical pigmentary changes in the background (e).

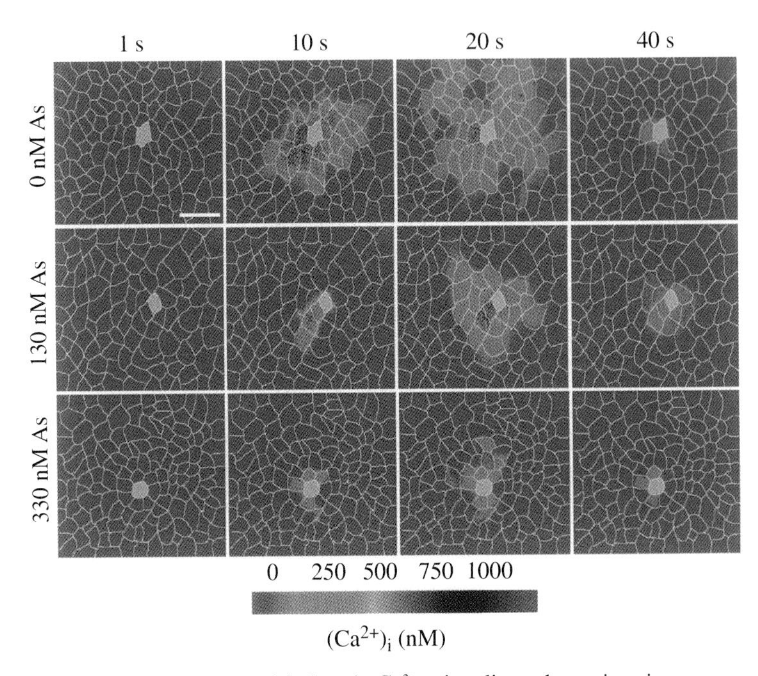

FIGURE 7.3 Localized wound-induced Ca^{2+} signaling dynamics is suppressed in 16HBE14o- cells by chronic low-dose arsenic exposure. Representative images of 16HBE14o- cell monolayers subjected to a localized wound (1–2 cells; gray area) at approximately 3 s and monitored for $[Ca^{2+}]_i$ over 60 s. Color scale at bottom indicates approximate $[Ca^{2+}]_i$ (nM); cells are outlined with white borders; white bar represents 50 μm (From Sherwood et al. [115]. © Oxford University Press).

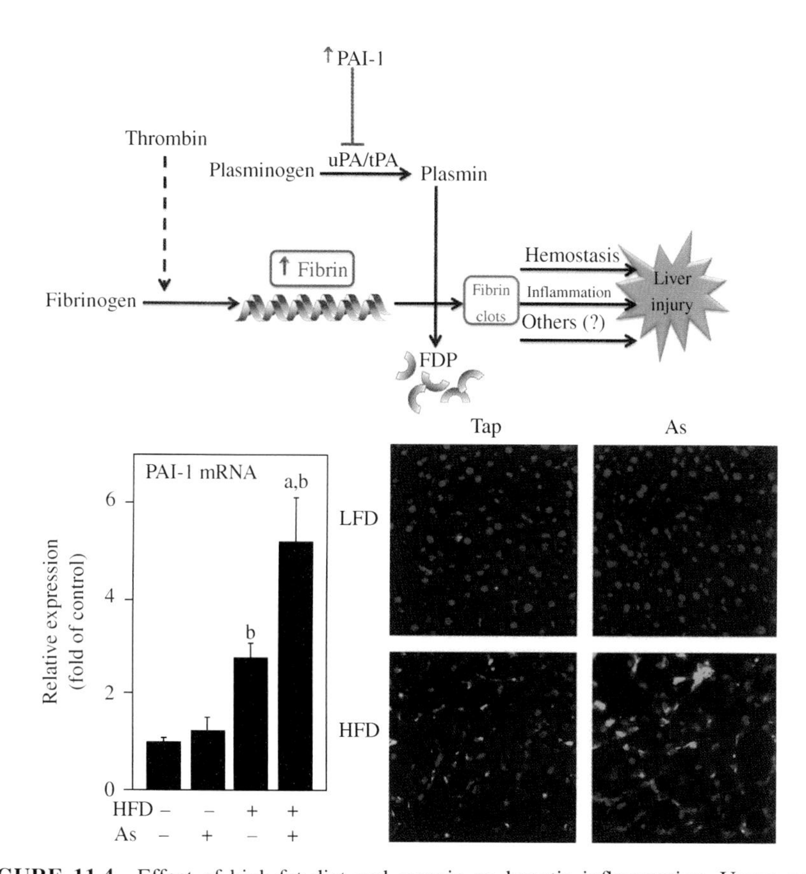

FIGURE 11.4 Effect of high-fat diet and arsenic on hepatic inflammation. Upper panel: Crosslinked fibrin deposition is initiated by activation of the coagulation cascade via thrombin. PAI-1 inhibits the activity of the plasminogen activators (uPA and tPA), blocking the activation of plasminogen to plasmin and, thereby, blunting degradation of fibrin matrices (fibrinolysis). The balance between fibrin deposition and degradation determines whether fibrin accumulates in vivo. This effect could contribute to liver damage via slowing blood flow in the liver (hemostasis), enhancing inflammation, as well as other potential mechanisms. Lower left panel: Effect of high fat diet and arsenic on hepatic mRNA expression plasminogen activator inhibitor-1 (PAI-1) is shown. Real-time RT-PCR data are means ± SE ($n=6$–10) and are expressed as fold of control. a, $p<0.05$ compared to low fat diet. b, $p<0.05$ compared to tap. Lower right: Representative photomicrographs (200×) depicting fibrin immunofluorescence (green color) are shown. From Tan et al. [74]. © Elsevier.

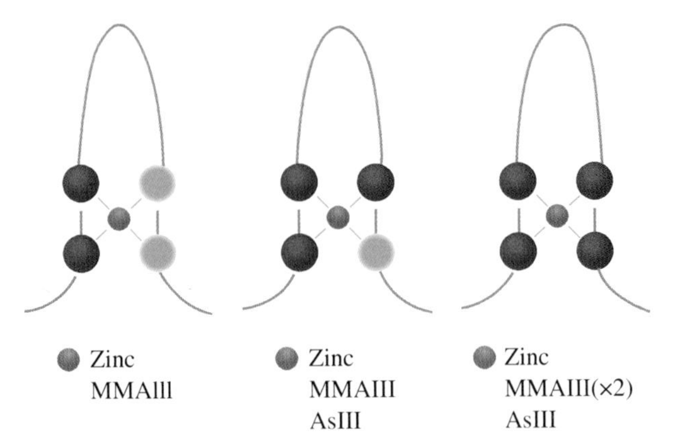

FIGURE 13.1 Schematic representation of a classical zinc finger. Zinc finger motifs are domains of typically 30–40 residues that coordinate zinc ions through cysteine and/or histidine residues. The most common C2H2 configuration is depicted in the schematic, but other configurations representing different placement of cysteine and histidine residues and number of cysteine residues (0–4) have been identified and are depicted in Krishna, Majumdar et al. [65]. Zinc finger motifs may occur singly (e.g., XPA) or in multiple motifs (e.g., TFIIIA) within proteins. Zinc forms bonds with cysteine (red circle) or histidine (yellow circle), but arsenic is only able to bind to cysteine residues. The differences in arsenite (AsIII) and monomethylarsonous (MMAIII) acid selectivity based on current findings is indicated in the figure.

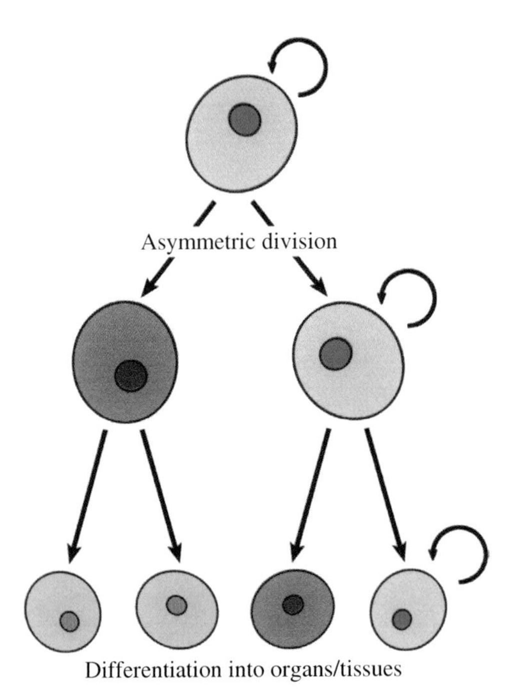

FIGURE 17.2 Stem cells (blue) can undergo the process of asymmetric division, where the cell divides into one identical daughter stem cell and one partially differentiated progenitor cell (red). The progenitor cells can undergo further differentiation to form the terminally differentiated cells of organs and tissues (tan). Curved arrow indicates self-renewal.

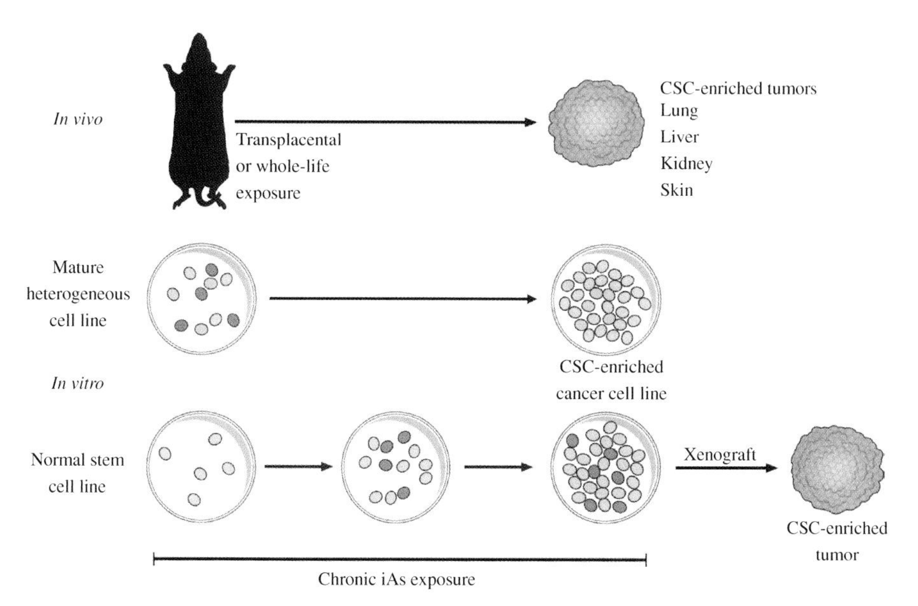

FIGURE 17.3 Arsenic-induced formation of cancer stem cell-enriched cell lines and tumors. Top: Transplacental or "whole life" exposure to arsenicals can result in the formation of cancer stem cell-enriched tumors when the offspring mice reach adulthood. Target tissues include the lung, liver, kidney, and skin. Middle and bottom: chronic exposure to low levels of arsenic can induce malignant transformation of nontumorigenic mature heterogeneous cell lines and of normal stem cell lines. This transformation can lead to cell lines highly enriched in cancer stem-like cells. Xenograft studies using these cell lines can lead to highly aggressive tumors enriched in cancer stem-like cells. Stem cell = blue, progenitor cell = red, and terminally differentiated cell = tan.

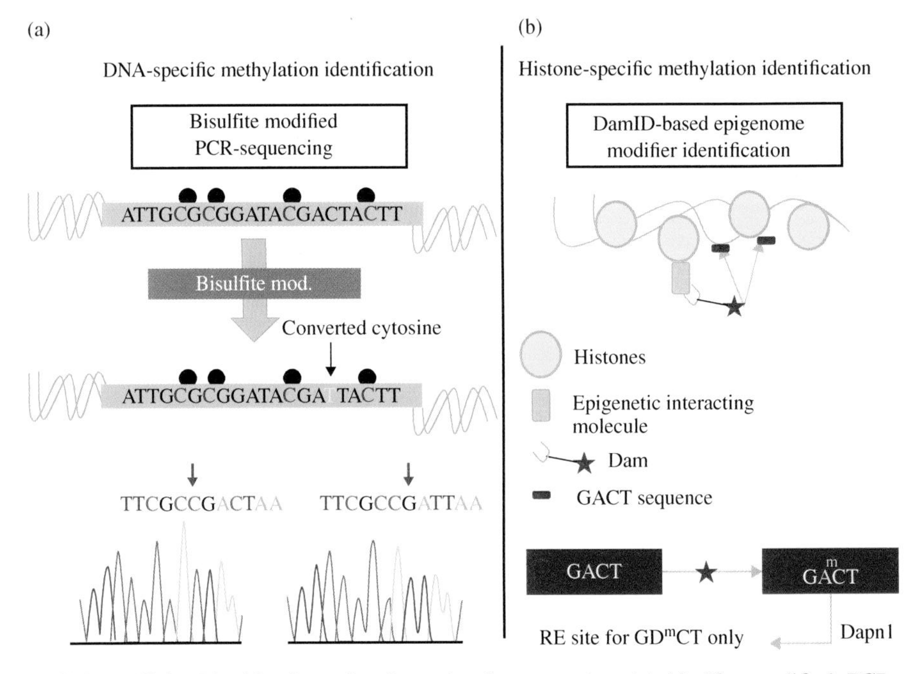

FIGURE 18.2 Identification of epigenetic changes using (a) bisulfite-modified PCR-sequencing and (b) DamID method.

12.4.2 Genetic Susceptibility to Arsenic-Induced Skin Lesions and Skin Cancers

While variation in the arsenic methylation patterns and metabolism can have subtle effects on an individual's susceptibility to the toxic effects of arsenic, it is however, the extent of skin lesions and skin cancers that ultimately define quantitatively and qualitatively how much a person is prone to that toxic insult. Furthermore, in absence of reliable data depicting the relationship between urinary arsenic profiles and the type/extent of skin lesions induced, the more direct approach to measure an individual's proneness to arsenic-induced physiological damage is to find out the associations of the genetic variants directly with the skin lesions (including cancers). Hence, in the existing literature, we find numerous studies that have attempted to determine the putative associations of candidate gene variations with different kinds of arsenic-induced skin lesions. In this section, we will review the evidence for genetic susceptibility in giving rise to arsenic-induced skin lesions, the hallmarks of arsenicosis.

Arsenic is known to give rise to three different kinds of skin cancers as well as precancerous and noncancerous skin lesions. One of the genes and its variants thereof, which have recurrently been associated with several types of cancers, is *p53*. Hence, this gene and its variations seem to be an ideal candidate to be studied with respect to arsenic-induced skin cancers/precancerous skin lesions. In one of the earliest studies regarding the association of *p53* Pro72Arg variation, it was shown that in individuals homozygous for Pro genotype are at 2.18 times higher risk of developing arsenic-induced skin cancers in a Taiwanese population [20]. Interestingly, in a subsequent study on a highly exposed Indian population, De Chaudhuri *et al.* [29] showed that this variation is also associated with higher risk of developing arsenic-induced precancerous keratosis; but, in this case, the Arg homozygous individuals were shown to be at risk, unlike the previous study. This study also showed the association of other *p53* variants with increased risk of induction of arsenic-induced premalignant keratosis, viz., homozygous genotype of no duplication polymorphism at intron 3 (16 bp duplication polymorphism at intron 3) [29]. In a subsequent study from the same group, it was shown that the individuals with the Arg homozygous genotype also had heightened chromosomal aberrations, thereby suggesting a mechanistic role for this polymorphism in bringing about the cancerous outcome [28].

Several groups have also looked at the effects of GST family member polymorphisms directly on the induction of skin lesions. In studies on exposed population from India, *GSTM1* null variant was found to have a protective effect but, *GSTT1*, *GSTP1*, *GSTO1*, and *GSTO2* polymorphisms were not associated with the development of arsenic-induced skin lesions [27, 35]. On the other hand, *GSTT1* wildtype and *GSTP1* GG were shown to be associated with increased risk of skin lesions in a Bangladesh population, but not *GSTM1* and the effect of *GSTT1* was probably mediated by modulating the secondary methylation ratio [63, 64]. However, neither *GSTT1* nor M1 homozygous deletions were found to be associated with an increased susceptibility to arsenic-induced skin lesions in a large cohort-based study on Chinese population, although, *GSTP1* Ile105Val polymorphism showed positive association [57, 58]. On the other hand, Ahsan *et al.* [6] examined the role of three *GSTO1* polymorphisms in a Bangladesh population and found them all to be associated. The *GSTO1* rs4925AA genotype had elevated risk of skin lesions, while the variant alleles of rs11509438 and rs11509437 had reduced risk of skin lesions. Additionally, individuals with the AAG/AAG diplotype (i.e., the diplotype containing the high-risk alleles for all three SNPs) also had higher risk of development of arsenic-induced skin lesions [6].

Other arsenic metabolism genes that have been probed for their associations with arsenic-induced skin lesions are *PNP*, *AS3MT*, and *MTHFR*. While no association was found for *AS3MT* polymorphism in an Indian population, individuals having at least one Thr allele for Met387Thr were found to be at risk of developing premalignant skin lesions in another study from Mexico [27, 86]. Three polymorphisms in *PNP* were found to be associated with the risk of developing arsenic-induced premalignant hyperkeratosis in a solitary study [27], while another single study that examined the role of *MTHFR* polymorphisms reported no association between 677 C>T polymorphism and arsenic-induced skin lesions [19].

In addition to the metabolism genes, DNA repair pathway genes also happen to be good candidates, since arsenic is known to give rise to significant genetic damage and chromosomal damage, which in turn are the forerunners of carcinogenic outcome. Several studies have assessed this possibility in a wide range of DNA repair pathways, genes and polymorphisms thereof. In a comprehensive study on a Caucasian population, Thirumaran *et al.* [85] demonstrated that the variant allele for T241M (C>T) polymorphism in the *XRCC3* gene had a protective effect on the development of arsenic-induced basal cell carcinoma of the skin, while, C-allele for E185Q (G>C) polymorphism in the *NBS1* gene conferred heightened risk for the same only in males [85]. In a study on Bangladesh population, Breton *et al.* [16] looked at associations between four common base excision repair (BER) genetic polymorphisms X-ray repair cross-complementing group 1 (*XRCC1*) Arg399Gln, *XRCC1* Arg194Trp, human 8-oxoguanine DNA glycosylase (*hOGG1*) Ser326Cys and apurinic/apyrimidinic endonuclease (*APE1*) Asp148Glu and arsenic-induced skin lesions including melanosis and keratosis, and could find that *XRCC1* 194Arg/Arg polymorphism conferred a threefold larger odds ratio for skin lesions compared with *XRCC1* 194Trp/Trp, but not for any other polymorphism examined [16]. Nucleotide excision repair pathway was also found to be involved as studies from Indian population showed a risk for development of premalignant hyperkeratosis in individuals with *ERCC2* codon 751 homozygous for Lys genotype, compared to those who had at least one variant Gln allele [11]. Ahsan *et al.* [7] also showed a similar trend in a Bangladesh population, but the association was statistically nonsignificant perhaps due to a small sample size. Further evidence of involvement for this polymorphism came from a study in a Chinese population, where the Lys/Lys variant of codon 751 was found to be associated with higher risk of developing arsenic-induced skin lesions [56], thereby corroborating the study by Banerjee *et al.* [11]. While another study on the Bangladesh population failed to show any association between Lys751Gln and Asp312Asn polymorphisms of *ERCC2* with arsenic-induced skin lesions, it nevertheless showed that both these variations could modify the relationship between tendency to sunburn and skin lesions [65]. The homologous recombination repair pathway gene was also found to be playing a role, with individuals with at least one Met allele for the Thr241Met polymorphism being at a reduced risk of developing arsenic-induced precancerous lesions [54].

Other genes whose polymorphisms have been examined with respect to the susceptibility to develop arsenic-induced skin lesions include the promoter polymorphisms of *CAT, MPO, IL-10*, and *TNF-α*. While no associations were observed for *CAT* and *MPO* in the total study population from Bangladesh (perhaps owing to a small sample size), individuals with GA/AA (–308 *TNF-α)* and TA/AA (–3575 *IL-10*) genotypes were at higher risk of developing arsenic-induced skin lesions in an Indian population [5, 13], probably by regulating the expression of the corresponding cytokine.

12.4.3 Genetic Susceptibility to Arsenic-Induced Nondermatological Cancers

In addition to skin cancers, arsenic also gives rise to a plethora of nondermatological cancers. A few studies have probed the genetic susceptibility to such arsenic-induced nondermatological cancers, although, the studies are much fewer those on skin cancers. In a study on Taiwanese population of arsenic-exposed urinary transitional cell carcinoma subjects, Chen *et al.* [21] could not find any association with *GSTT1* null variant or *p53* codon 72 polymorphism, while Hsu *et al.* [46] demonstrated that *GSTT1* null genotype was marginally associated with the risk of urothelial carcinoma in Taiwanese population, but not the polymorphic variants of *GSTM1*, *GSTP1*, *GSTO1*, or *GSTO2*. Moore *et al.* [68], however, showed that the polymorphisms *GSTT1*, *GSTM1*, *NQO1* (NAD(P)H dehydrogenase, quinone 1), and *MTHFR* could modulate the risk of arsenic-induced bladder cancer, but primarily by interaction with smoking. In another study, it was shown that *GSTP1* Ile105Val homozygous individuals had a higher risk of developing arsenic-induced bladder cancer, but not the polymorphisms in *GSTO2*, *GSTZ1*, *AQP3*, *AS3MT*, *GSTM1*, or *GSTT1* [55]. Also, *MTHFR* polymorphisms did not appear to modulate the risk of developing arsenic-induced bladder cancer [51].Variation in *AS3MT* and *MTHFR* was found to be associated with bladder cancer among those exposed to relatively low concentrations of inorganic arsenic in a study by Beebe-Dimmer *et al.* [15]. On the contrary, *TP53* (tumor protein P53) or *CCND1* (cyclin D1) polymorphisms did not show any association with arsenic-induced urothelial carcinoma, but subjects carrying the *p21*Arg/Arg genotype (p21 = CDKN1A, cyclin-dependent kinase inhibitor 1A) had an increased risk for the same [23]. *MPO* (myeloperoxidase) and *SULT1A1* (sulfotransferase family, cytosolic, 1A, phenol-preferring, member 1) polymorphisms did not show any direct association with arsenic-induced urothelial carcinoma, but might still modulate the development of such cancerous outcome via their differential effects on arsenic methylation pattern [49]. The Met allele of *XRCC3* (X-ray repair complementing defective repair in Chinese hamster cells 3) Thr241Met polymorphism was found to be positively associated with arsenic-induced bladder cancer [8]. Negative results were also obtained for association of polymorphisms in the *MTHFR* and methionine synthase (*sic*; currently 5-methyltetrahydrofolate-homocysteine methyltransferase, *MTR*) genes with arsenic-induced urothelial carcinoma [24], while homozygous variant genotype of *GSTO2* 142 Asp/Asp was inversely associated with urothelial carcinoma risk [25]. Furthermore, SNPs in the fibrous sheath interacting protein 1 gene (*FSIP1*) and the solute carrier family 39, member 2 gene (*SLC39A2*) in the ZIP gene family of metal transporters were also found to modulate the risk of development of chronic arsenic ingestion-induced bladder cancer in a US population [50]. Again, Adonis *et al.* [1] could not find any association between *GSTM1* or *CYP1A1* polymorphisms with arsenic-induced lung cancer in a Chilean population, but they showed that the relative lung cancer risk for the total sample with the *CYP1A1**2A/null *GSTM1* genotype was significantly higher and increased further among the smokers [1]. This result suggests an interaction between genetic variants and smoking habit toward arsenic-induced lung cancer. In a solitary study, the risk of developing arsenic-induced renal cell carcinoma was also found to be influenced by variant genotypes of *p53* Arg72Pro or *MDM2* T309G [48].

12.4.4 Genetic Susceptibility to Other Arsenic-Induced Diseases

The effect of genetic polymorphisms on nondermatological and noncancerous health outcomes is sparsely studied. It was shown that NADPH oxidase (*NOX1*), manganese superoxide dismutase (*SOD2*; Superoxide Dismutase 2, Mitochondrial), and endothelial nitric oxide synthase (*NOS3I*; nitric oxide synthase 3 (endothelial cell)) polymorphisms, but not catalase (*CAT*), might play a role in the development of arsenic-related hypertension, especially in subjects with high triglyceride levels [47], but this has not been validated by any further study either from the same group or on a different population. Risk of atherosclerosis was also found to be modulated by interplay of arsenic exposure with polymorphic variants of *p53* and *GSTP1* [87]. A marked elevated risk of carotid atherosclerosis was also observed in subjects with arsenic exposure of greater than 50 µg/L in drinking water and those who carried the purine nucleoside phosphorylase (*PNP*) A-T haplotype and at least either of the *AS3MT* risk polymorphism or GSTO risk haplotypes [45]. Further reports on the role of genetic variants in arsenic-induced atherosclerosis were found from studies that showed the involvement of apolipoprotein E (*APOE*), chemokine (C—C Motif) ligand 2 (*CCL2* aka *MCP1*), heme oxygenase (decycling) 1 (*HMOX1* aka *HO-1*) [44, 90]. In fact, *HMOX1* promoter polymorphism was found to confer protection against arsenic-induced cardiovascular mortality [44, 88]. Risk of coronary heart disease was also found to be modulated by *AS3MT* polymorphisms in an arsenic-exposed rural population from Texas, USA. [37].The presence of at least one Met allele (Met/Met or Thr/Met) was protective towards development of peripheral neuropathy and conjunctivitis, while individuals with GA/AA (−308 *TNF-α*) and TA/AA (−3575 *IL-10*) genotypes were at higher risk of developing arsenic-induced ocular and respiratory diseases [13, 54]. In a very recent study, individuals with C allele in M287T or G4956C of *AS3MT* gene were shown to have elevated risk of being diabetic, although, the results did not reach statistical significance, probably due to low sample size in a Mexican population [31, 31], but the results warrant further experimental validation in larger cohorts.

12.5 TRANSCRIPTOMIC AND PROTEOMIC APPROACHES TO IDENTIFY ARSENIC SUSCEPTIBILITY

Although, the state-of-the-art "omics" technologies offer a real-time view of the intricate changes that are taking place in an organism concurrent with exposure, this approach has still not been widely used in human subjects with respect to chronic arsenic exposure. In this section, we will be focusing on studies applying transcriptomic and proteomic technologies directly in samples obtained from exposed human populations. As of this writing, only four studies have examined the effects of chronic arsenic exposure on gene expression in chronically exposed human populations, and the results have not been unequivocal. This is also due to the differences in the tissues used for the experiments. While Lu *et al.* [60] showed that around 60 genes involved in the important biological responses like DNA damage, apoptosis and cell cycle regulation were differentially expressed in liver tissues from arsenic-exposed individuals compared to controls, the three other studies looked at the effect on peripheral blood lymphocytes. In a study on 64 Taiwanese individuals, 62 transcripts were shown to have significantly differing expression profiles in intermediate and high arsenic-exposed groups compared to the low-exposure groups, of which, several belonged to the cytokine and growth factor families [89]. In another study comprising of 40 Bangladeshi individuals 312 genes were found to be differentially expressed (including several members of the chemokine family), although, no difference in expression profiles could be found in exposed individuals with and without arsenic-induced skin lesions [9]. Several modulators of immunogenic responses were also found to be differentially expressed in another study on peripheral blood lymphocytes of arsenic-exposed population from New Hampshire [8]. While it seems from the ongoing discussion that molecules involved in immune responses are involved, the results need to be interpreted with utmost caution. It should be borne in mind that all of these studies used unexposed individuals as controls, and thus, the differences observed might actually be responses to arsenic toxicity and not necessarily biomarkers for susceptibility.

Similarly, the number of proteomic studies is also few and far between, and the differences in the methods applied and the tissues used often make drawing a consistent conclusion difficult. Using surface-enhanced laser desorption/ionization time-of-flight (SELDI-TOF) mass spectrometry (MS) method on plasma samples from exposed Bangladeshi population, Harezlak *et al.* [38] showed that 24 protein peaks had a significantly differing expression profile in exposed individuals (14 upregulated and 10 downregulated), although, functional validation was not carried out for any of them. In another study, also involving SELDI-TOF-MS on urine samples from exposed Nevada population, underexpression of urinary beta-defensin 1 has been proposed to be a biomarker for arsenic exposure, although, further studies in other populations will be required to confirm the findings [40]. There are two more studies on human populations that employed proteomic techniques, but they looked at exposure to a mixture of metals including arsenic in an occupational scenario, thereby making it difficult to draw any conclusion on the effect of arsenic alone [53, 92].

12.6 MULTIPRONGED APPROACH FOR EARLY DETECTION OF INDIVIDUALS SUSCEPTIBLE TO ARSENIC TOXITY

While all the aforementioned studies have interesting scientific outputs as deliverables, clearly, the issue to address at the population and public health level is to devise effective strategies to identify highly susceptible individuals in potentially exposed populations as early as possible. This will be beneficial from the preventive perspective and would also enable palliative measures to be adopted at the earliest possible time point in an exposure scenario. However, keeping in mind the subtle effects of several low-penetrance genes in determining the ultimate susceptibility of any given individual toward arsenic toxicity, and their interaction with innumerable environmental factors, it seems highly unlikely that a simple PCR based genotyping from blood samples would be enough to identify individuals at risk. Consequently, it seems indicated that a multipronged approach, such as combining state-of-the-art genomic technology with classical cytogenetic assays (e.g., chromosomal aberration, micronucleus assay), has to be adopted for this purpose. An example is the study in which individuals with *ERCC2* codon 751 Lys/Lys genotype were predicted to be at a higher risk of developing arsenic-induced premalignant hyperkeratotic lesions [11]. In order to validate the causal effect of this variant, a classical cytogenetic approach was employed and showed that individuals with the risk genotype actually do have heightened DNA damage (as measured by chromosomal aberration assay) compared to individuals with the nonrisk genotypes (with equal exposure in terms of arsenic intake and time of exposure). Results from several other studies also showed that a relationship that exists between the genetic damage incurred and certain risk genotypes might actually reflect a causal relationship [12, 28, 35, 54]. In addition, apart from the total exposure status, the relative proportion of different arsenic metabolites in the urine could also provide useful information regarding the individual susceptibility of a person in a given exposure scenario. Therefore, a combination of blood-based genomic and genetic assays, coupled to analytical techniques could be a successful approach to identify individuals at high risk of arsenic toxicity.

12.7 CONCLUSIONS

Thus, while numerous studies on associations between genetic polymorphisms disease outcomes of chronic arsenic exposure have been conducted in different populations all over the world, a clear picture is yet to emerge. For most of the variants and the endpoints involved, contradictory results have been reported, thereby making the task of drawing unequivocal conclusions extremely difficult. From the ongoing discussion, it appears that *AS3MT* is the only gene whose variants seem to be universally implicated in imparting arsenic metabolism efficiency and thereby determining the arsenic load in the body across several populations. In addition, it also seems reasonable to conclude that several of the DNA repair pathway gene polymorphisms are implicated in giving rise to disease outcomes resulting from arsenic exposure. While it is clear that genetic polymorphisms do play an important role in defining the susceptibility of an individual to the toxic effects of chronic arsenic exposure, it is still impossible to say with certainty which combination of genetic variants will impart a high susceptibility to develop a specific disease. It would, however, be fair to say that susceptibility to arsenic toxicity is perhaps an outcome of several low penetrance genetic variants and the particular genetic composition of an individual coupled to his/her exposure load, metabolic efficiency, dietary and life style factors. In a recent paper, Pierce *et al.* [73] carried out genome-wide association studies (GWAS) in a large arsenic-exposed cohort in Bangladesh and showed that the genomic variations at chromosome 10q24.32 are associated both with arsenic metabolism patterns as well as arsenic induced-skin lesions and might actually reflect variations of the *AS3MT* gene. However, these results need to be validated in other populations before a consensus can be reached. Hence, further large-scale cohort-based studies involving whole genome sequencing are required to shed more light on this issue.

REFERENCES

[1] M. Adonis, V. Martinez, P. Marin, D. Berrios, L. Gil, Smoking habit and genetic factors associated with lung cancer in a population highly exposed to arsenic. Toxicol. Lett. 159 (2005) 32–37.

[2] T. Agusa, H. Iwata, J. Fujihara, T. Kunito, H. Takeshita, T.B. Minh, P.T. Trang, P.H. Viet, S. Tanabe, Genetic polymorphisms in AS3MT and arsenic metabolism in residents of the Red River Delta, Vietnam. Toxicol. Appl. Pharmacol. 236 (2009) 131–141.

[3] T. Agusa, H. Iwata, J. Fujihara, T. Kunito, H. Takeshita, T.B. Minh, P.T. Trang, P.H. Viet, S. Tanabe, Genetic polymorphisms in glutathione S-transferase (GST) superfamily and arsenic metabolism in residents of the Red River Delta, Vietnam. Toxicol. Appl. Pharmacol. 242 (2010) 352–362.

[4] T. Agusa, T. Kunito, N.M. Tue, V.T. Lan, J. Fujihara, H. Takeshita, T.B. Minh, P.T. Trang, S. Takahashi, P.H. Viet, S. Tanabe, H. Iwata, Individual variations in arsenic metabolism in Vietnamese: the association with arsenic exposure and GSTP1 genetic polymorphism. Metallomics 4 (2012) 91–100.

[5] H. Ahsan, Y. Chen, M.G. Kibriya, M.N. Islam, V.N. Slavkovich, J.H. Graziano, R.M. Santella, Susceptibility to arsenic-induced hyperkeratosis and oxidative stress genes myeloperoxidase and catalase. Cancer Lett. 201 (2003) 57–65.

[6] H. Ahsan, Y. Chen, M.G. Kibriya, V. Slavkovich, F. Parvez, F. Jasmine, M.V. Gamble, J.H. Graziano, Arsenic metabolism, genetic susceptibility, and risk of premalignant skin lesions in Bangladesh. Cancer Epidemiol. Biomarkers Prev. 16 (2007) 1270–1278.

[7] H. Ahsan, Y. Chen, Q. Wang, V. Slavkovich, J.H. Graziano, R.M. Santella, DNA repair gene XPD and susceptibility to arsenic-induced hyperkeratosis. Toxicol. Lett. 143 (2003) 123–131.

[8] A.S. Andrew, R.A. Mason, K.T. Kelsey, A.R. Schned, C.J. Marsit, H.H. Nelson, M.R. Karagas, DNA repair genotype interacts with arsenic exposure to increase bladder cancer risk. Toxicol. Lett. 187 (2009) 10–14.

[9] M. Argos, M.G. Kibriya, F. Parvez, F. Jasmine, M. Rakibuz-Zaman, H. Ahsan, Gene expression profiles in peripheral lymphocytes by arsenic exposure and skin lesion status in a Bangladeshi population. Cancer Epidemiol. Biomarkers Prev. 15 (2006) 1367–1375.

[10] K.A. Bailey, R.C. Fry, Arsenic-associated changes to the epigenome: what are the functional consequences? Curr. Environ. Health Rep. 1 (2014) 22–34.

[11] M. Banerjee, J. Sarkar, J.K. Das, A. Mukherjee, A.K. Sarkar, L. Mondal, A.K. Giri, Polymorphism in the ERCC2 codon 751 is associated with arsenic-induced premalignant hyperkeratosis and significant chromosome aberrations. Carcinogenesis 28 (2007) 672–676.

[12] M. Banerjee, N. Sarma, R. Biswas, J. Roy, A. Mukherjee, A.K. Giri, DNA repair deficiency leads to susceptibility to develop arsenic-induced premalignant skin lesions. Int. J. Cancer 123 (2008) 283–287.

[13] N. Banerjee, S. Nandy, J.K. Kearns, A.K. Bandyopadhyay, J.K. Das, P. Majumder, S. Basu, S. Banerjee, T.J. Sau, J.C. States, A.K. Giri, Polymorphisms in the TNF-alpha and IL10 gene promoters and risk of arsenic-induced skin lesions and other nondermatological health effects. Toxicol. Sci. 121 (2011) 132–139.

[14] A. Basu, A. Som, S. Ghoshal, L. Mondal, R.C. Chaubey, H.N. Bhilwade, M.M. Rahman, A.K. Giri, Assessment of DNA damage in peripheral blood lymphocytes of individuals susceptible to arsenic induced toxicity in West Bengal, India. Toxicol. Lett. 159 (2005) 100–112.

[15] J.L. Beebe-Dimmer, P.T. Iyer, J.O. Nriagu, G.R. Keele, S. Mehta, J.R. Meliker, E.M. Lange, A.G. Schwartz, K.A. Zuhlke, D. Schottenfeld, K.A. Cooney, Genetic variation in glutathione S-transferase omega-1, arsenic methyltransferase and methylene-tetrahydrofolate reductase, arsenic exposure and bladder cancer: a case-control study. Environ. Health 11 (2012) 43.

[16] C.V. Breton, W. Zhou, M.L. Kile, E.A. Houseman, Q. Quamruzzaman, M. Rahman, G. Mahiuddin, D.C. Christiani, Susceptibility to arsenic-induced skin lesions from polymorphisms in base excision repair genes. Carcinogenesis 28 (2007) 1520–1525.

[17] D.D. Caceres, F. Werlinger, M. Orellana, M. Jara, R. Rocha, S.A. Alvarado, Q. Luis, Polymorphism of glutathione S-transferase (GST) variants and its effect on distribution of urinary arsenic species in people exposed to low inorganic arsenic in tap water: an exploratory study. Arch. Environ. Occup. Health 65 (2010) 140–147.

[18] F. Challenger, Biological methylation. Chem. Rev. 36 (1945) 315–361.

[19] L. Chen, X.B. Guo, F.R. Deng, H. Liu, Y. Jin, Z. Wang, K. Zhang, Study on the relationship between methylene tetrahydrofolate reductase gene (677C-->T) mutation and skin lesions in endemic arsenic poisoning. Wei Sheng Yan Jiu 34 (2005) 146–148.

[20] Y.C. Chen, L. Xu, Y.L. Guo, H.J. Su, Y.M. Hsueh, T.J. Smith, L.M. Ryan, M.S. Lee, S.C. Chaor, J.Y. Lee, D.C. Christiani, Genetic polymorphism in p53 codon 72 and skin cancer in southwestern Taiwan. J. Environ. Sci. Health A Tox. Hazard. Subst. Environ. Eng. 38 (2003) 201–211.

[21] Y.C. Chen, L. Xu, Y.L. Guo, H.J. Su, T.J. Smith, L.M. Ryan, M.S. Lee, D.C. Christiani, Polymorphisms in GSTT1 and p53 and urinary transitional cell carcinoma in southwestern Taiwan: a preliminary study. Biomarkers 9 (2004) 386–394.

[22] C.J. Chung, Y.M. Hsueh, C.H. Bai, Y.K. Huang, Y.L. Huang, M.H. Yang, C.J. Chen, Polymorphisms in arsenic metabolism genes, urinary arsenic methylation profile and cancer. Cancer Causes Control 20 (2009) 1653–1661.

[23] C.J. Chung, C.J. Huang, Y.S. Pu, C.T. Su, Y.K. Huang, Y.T. Chen, Y.M. Hsueh, Polymorphisms in cell cycle regulatory genes, urinary arsenic profile and urothelial carcinoma. Toxicol. Appl. Pharmacol. 232 (2008) 203–209.

[24] C.J. Chung, Y.S. Pu, C.T. Su, H.W. Chen, Y.K. Huang, H.S. Shiue, Y.M. Hsueh, Polymorphisms in one-carbon metabolism pathway genes, urinary arsenic profile, and urothelial carcinoma. Cancer Causes Control 21 (2010) 1605–1613.

[25] C.J. Chung, Y.S. Pu, C.T. Su, C.Y. Huang, Y.M. Hsueh, Gene polymorphisms of glutathione S-transferase omega 1 and 2, urinary arsenic methylation profile and urothelial carcinoma. Sci. Total Environ. 409 (2011) 465–470.

[26] W.R. Cullen, K.J. Reimer, Arsenic speciation in the environment. Chem. Rev. 89 (1989) 713–764.

[27] S. De Chaudhuri, P. Ghosh, N. Sarma, P. Majumdar, T.J. Sau, S. Basu, S. Roychoudhury, K. Ray, A.K. Giri, Genetic variants associated with arsenic susceptibility: study of purine nucleoside phosphorylase, arsenic (+3) methyltransferase, and glutathione S-transferase omega genes. Environ. Health Perspect. 116 (2008) 501–505.

[28] S. De Chaudhuri, M. Kundu, M. Banerjee, J.K. Das, P. Majumdar, S. Basu, S. Roychoudhury, K.K. Singh, A.K. Giri, Arsenic-induced health effects and genetic damage in keratotic individuals: involvement of p53 arginine variant and chromosomal aberrations in arsenic susceptibility. Mutat. Res. 659 (2008) 118–125.

[29] S. De Chaudhuri, J. Mahata, J.K. Das, A. Mukherjee, P. Ghosh, T.J. Sau, L. Mondal, S. Basu, A.K. Giri, S. Roychoudhury, Association of specific p53 polymorphisms with keratosis in individuals exposed to arsenic through drinking water in West Bengal, India. Mutat. Res. 601 (2006) 102–112.

[30] F.R. Deng, X.B. Guo, L. Chen, Z.Q. Wang, K. Zhang, Relationship between 5,10-methylenetetrahydrofolate reductase genetic polymorphism and arsenic metabolism. Beijing Da Xue Xue Bao 39 (2007) 149–152.

[31] Z. Drobna, L.M. Del Razo, G.G. Garcia-Vargas, L.C. Sanchez-Pena, A. Barrera-Hernandez, M. Styblo, D. Loomis, Environmental exposure to arsenic, AS3MT polymorphism and prevalence of diabetes in Mexico. J. Expo. Sci. Environ. Epidemiol. 23 (2013) 151–155.

[32] K. Engstrom, M. Vahter, S.J. Mlakar, G. Concha, B. Nermell, R. Raqib, A. Cardozo, K. Broberg, Polymorphisms in arsenic(+III oxidation state) methyltransferase (AS3MT) predict gene expression of AS3MT as well as arsenic metabolism. Environ. Health Perspect. 119 (2011) 182–188.

[33] J. Fujihara, M. Soejima, T. Yasuda, Y. Koda, T. Kunito, H. Iwata, S. Tanabe, H. Takeshita, Polymorphic trial in oxidative damage of arsenic exposed Vietnamese. Toxicol. Appl. Pharmacol. 256 (2011) 174–178.

[34] P. Ghosh, M. Banerjee, S. De Chaudhuri, R. Chowdhury, J.K. Das, A. Mukherjee, A.K. Sarkar, L. Mondal, K. Baidya, T.J. Sau, A. Banerjee, A. Basu, K. Chaudhuri, K. Ray, A.K. Giri, Comparison of health effects between individuals with and without skin lesions in the population exposed to arsenic through drinking water in West Bengal, India. J. Expo. Sci. Environ. Epidemiol. 17 (2007) 215–223.

[35] P. Ghosh, A. Basu, J. Mahata, S. Basu, M. Sengupta, J.K. Das, A. Mukherjee, A.K. Sarkar, L. Mondal, K. Ray, A.K. Giri, Cytogenetic damage and genetic variants in the individuals susceptible to arsenic-induced cancer through drinking water. Int. J. Cancer 118 (2006) 2470–2478.

[36] P. Gomez-Rubio, M.M. Meza-Montenegro, E. Cantu-Soto, W.T. Klimecki, Genetic association between intronic variants in AS3MT and arsenic methylation efficiency is focused on a large linkage disequilibrium cluster in chromosome 10. J. Appl. Toxicol. 30 (2010) 260–270.

[37] G. Gong, S.E. O'Bryant, Low-level arsenic exposure, AS3MT gene polymorphism and cardiovascular diseases in rural Texas counties. Environ. Res. 113 (2012) 52–57.

[38] J. Harezlak, M.C. Wu, M. Wang, A. Schwartzman, D.C. Christiani, X. Lin, Biomarker discovery for arsenic exposure using functional data. Analysis and feature learning of mass spectrometry proteomic data. J. Proteome Res. 7 (2008) 217–224.

[39] T. Hayakawa, Y. Kobayashi, X. Cui, S. Hirano, A new metabolic pathway of arsenite: arsenic-glutathione complexes are substrates for human arsenic methyltransferase Cyt19. Arch. Toxicol. 79 (2005) 183–191.

[40] C.M. Hegedus, C.F. Skibola, M. Warner, D.R. Skibola, D. Alexander, S. Lim, N.L. Dangleben, L. Zhang, M. Clark, R.M. Pfeiffer, C. Steinmaus, A.H. Smith, M.T. Smith, L.E. Moore, Decreased urinary beta-defensin-1 expression as a biomarker of response to arsenic. Toxicol. Sci. 106 (2008) 74–82.

[41] A. Hernandez, N. Xamena, C. Sekaran, H. Tokunaga, A. Sampayo-Reyes, D. Quinteros, A. Creus, R. Marcos, High arsenic metabolic efficiency in AS3MT287Thr allele carriers. Pharmacogenet. Genomics 18 (2008) 349–355.

[42] A. Hernandez, N. Xamena, J. Surralles, C. Sekaran, H. Tokunaga, D. Quinteros, A. Creus, R. Marcos, Role of the Met(287)Thr polymorphism in the AS3MT gene on the metabolic arsenic profile. Mutat. Res. 637 (2008) 80–92.

[43] P. Hinhumpatch, P. Navasumrit, K. Chaisatra, J. Promvijit, C. Mahidol, M. Ruchirawat, Oxidative DNA damage and repair in children exposed to low levels of arsenic in utero and during early childhood: application of salivary and urinary biomarkers. Toxicol. Appl. Pharmacol. 273 (2013) 569–579.

[44] Y.C. Hsieh, F.I. Hsieh, L.M. Lien, Y.L. Chou, H.Y. Chiou, C.J. Chen, Risk of carotid atherosclerosis associated with genetic polymorphisms of apolipoprotein E and inflammatory genes among arsenic exposed residents in Taiwan. Toxicol. Appl. Pharmacol. 227 (2008) 1–7.

[45] Y.C. Hsieh, L.M. Lien, W.T. Chung, F.I. Hsieh, P.F. Hsieh, M.M. Wu, H.P. Tseng, H.Y. Chiou, C.J. Chen, Significantly increased risk of carotid atherosclerosis with arsenic exposure and polymorphisms in arsenic metabolism genes. Environ. Res. 111 (2011) 804–810.

[46] L.I. Hsu, W.P. Chen, T.Y. Yang, Y.H. Chen, W.C. Lo, Y.H. Wang, Y.T. Liao, Y.M. Hsueh, H.Y. Chiou, M.M. Wu, C.J. Chen, Genetic polymorphisms in glutathione S-transferase (GST) superfamily and risk of arsenic-induced urothelial carcinoma in residents of southwestern Taiwan. J. Biomed. Sci. 18 (2011) 51.

[47] Y.M. Hsueh, P. Lin, H.W. Chen, H.S. Shiue, C.J. Chung, C.T. Tsai, Y.K. Huang, H.Y. Chiou, C.J. Chen, Genetic polymorphisms of oxidative and antioxidant enzymes and arsenic-related hypertension. J. Toxicol. Environ. Health A 68 (2005) 1471–1484.

[48] C.Y. Huang, C.T. Su, J.S. Chu, S.P. Huang, Y.S. Pu, H.Y. Yang, C.J. Chung, C.C. Wu, Y.M. Hsueh, The polymorphisms of P53 codon 72 and MDM2 SNP309 and renal cell carcinoma risk in a low arsenic exposure area. Toxicol. Appl. Pharmacol. 257 (2011) 349–355.

[49] S.K. Huang, A.W. Chiu, Y.S. Pu, Y.K. Huang, C.J. Chung, H.J. Tsai, M.H. Yang, C.J. Chen, Y.M. Hsueh, Arsenic methylation capability, myeloperoxidase and sulfotransferase genetic polymorphisms, and the stage and grade of urothelial carcinoma. Urol. Int. 82 (2009) 227–234.

[50] M.R. Karagas, A.S. Andrew, H.H. Nelson, Z. Li, T. Punshon, A. Schned, C.J. Marsit, J.S. Morris, J.H. Moore, A.L. Tyler, D. Gilbert-Diamond, M.L. Guerinot, K.T. Kelsey,

SLC39A2 and FSIP1 polymorphisms as potential modifiers of arsenic-related bladder cancer. Hum. Genet. 131 (2012) 453–461.

[51] M.R. Karagas, S. Park, H.H. Nelson, A.S. Andrew, L. Mott, A. Schned, K.T. Kelsey, Methylenetetrahydrofolate reductase (MTHFR) variants and bladder cancer: a population-based case-control study. Int. J. Hyg. Environ. Health 208 (2005) 321–327.

[52] M.L. Kile, E.A. Houseman, E. Rodrigues, T.J. Smith, Q. Quamruzzaman, M. Rahman, G. Mahiuddin, L. Su, D.C. Christiani, Toenail arsenic concentrations, GSTT1 gene polymorphisms, and arsenic exposure from drinking water. Cancer Epidemiol. Biomarkers Prev. 14 (2005) 2419–2426.

[53] B. Kossowska, I. Dudka, G. Bugla-Ploskonska, A. Szymanska-Chabowska, W. Doroszkiewicz, R. Gancarz, R. Andrzejak, J. Antonowicz-Juchniewicz, Proteomic analysis of serum of workers occupationally exposed to arsenic, cadmium, and lead for biomarker research: a preliminary study. Sci. Total Environ. 408 (2010) 5317–5324.

[54] M. Kundu, P. Ghosh, S. Mitra, J.K. Das, T.J. Sau, S. Banerjee, J.C. States, A.K. Giri, Precancerous and non-cancer disease endpoints of chronic arsenic exposure: the level of chromosomal damage and XRCC3 T241M polymorphism. Mutat. Res. 706 (2011) 7–12.

[55] C. Lesseur, D. Gilbert-Diamond, A.S. Andrew, R.M. Ekstrom, Z. Li, K.T. Kelsey, C.J. Marsit, M.R. Karagas, A case-control study of polymorphisms in xenobiotic and arsenic metabolism genes and arsenic-related bladder cancer in New Hampshire. Toxicol. Lett. 210 (2012) 100–106.

[56] G.F. Lin, H. Du, J.G. Chen, H.C. Lu, W.C. Guo, K. Golka, J.H. Shen, Association of XPD/ERCC2 G23591A and A35931C polymorphisms with skin lesion prevalence in a multiethnic, arseniasis-hyperendemic village exposed to indoor combustion of high arsenic coal. Arch. Toxicol. 84 (2010) 17–24.

[57] G.F. Lin, H. Du, J.G. Chen, H.C. Lu, W.C. Guo, H. Meng, T.B. Zhang, X.J. Zhang, D.R. Lu, K. Golka, J.H. Shen, Arsenic-related skin lesions and glutathione S-transferase P1 A1578G (Ile105Val) polymorphism in two ethnic clans exposed to indoor combustion of high arsenic coal in one village. Pharmacogenet. Genomics 16 (2006) 863–871.

[58] G.F. Lin, H. Du, J.G. Chen, H.C. Lu, J.X. Kai, Y.S. Zhou, W.C. Guo, X.J. Zhang, D.R. Lu, K. Golka, J.H. Shen, Glutathione S-transferases M1 and T1 polymorphisms and arsenic content in hair and urine in two ethnic clans exposed to indoor combustion of high arsenic coal in Southwest Guizhou, China. Arch. Toxicol. 81 (2007) 545–551.

[59] A.L. Lindberg, R. Kumar, W. Goessler, R. Thirumaran, E. Gurzau, K. Koppova, P. Rudnai, G. Leonardi, T. Fletcher, M. Vahter, Metabolism of low-dose inorganic arsenic in a central European population: influence of sex and genetic polymorphisms. Environ. Health Perspect. 115 (2007) 1081–1086.

[60] T. Lu, J. Liu, E.L. LeCluyse, Y.S. Zhou, M.L. Cheng, M.P. Waalkes, Application of cDNA microarray to the study of arsenic-induced liver diseases in the population of Guizhou, China. Toxicol. Sci. 59 (2001) 185–192.

[61] R. Marcos, V. Martinez, A. Hernandez, A. Creus, C. Sekaran, H. Tokunaga, D. Quinteros, Metabolic profile in workers occupationally exposed to arsenic: role of GST polymorphisms. J. Occup. Environ. Med. 48 (2006) 334–341.

[62] L.L. Marnell, G.G. Garcia-Vargas, U.K. Chowdhury, R.A. Zakharyan, B. Walsh, M.D. Avram, M.J. Kopplin, M.E. Cebrian, E.K. Silbergeld, H.V. Aposhian, Polymorphisms in the human monomethylarsonic acid (MMA V) reductase/hGSTO1 gene and changes in urinary arsenic profiles. Chem. Res. Toxicol. 16 (2003) 1507–1513.

[63] K.M. McCarty, Y.C. Chen, Q. Quamruzzaman, M. Rahman, G. Mahiuddin, Y.M. Hsueh, L. Su, T. Smith, L. Ryan, D.C. Christiani, Arsenic methylation, GSTT1, GSTM1, GSTP1 polymorphisms, and skin lesions. Environ. Health Perspect. 115 (2007) 341–345.

[64] K.M. McCarty, L. Ryan, E.A. Houseman, P.L. Williams, D.P. Miller, Q. Quamruzzaman, M. Rahman, G. Mahiuddin, T. Smith, E. Gonzalez, L. Su, D.C. Christiani, A case-control study of GST polymorphisms and arsenic related skin lesions. Environ. Health 6 (2007) 5.

[65] K.M. McCarty, T.J. Smith, W. Zhou, E. Gonzalez, Q. Quamruzzaman, M. Rahman, G. Mahiuddin, L. Ryan, L. Su, D.C. Christiani, Polymorphisms in XPD (Asp312Asn and Lys751Gln) genes, sunburn and arsenic-related skin lesions. Carcinogenesis 28 (2007) 1697–1702.

[66] M.M. Meza, L. Yu, Y.Y. Rodriguez, M. Guild, D. Thompson, A.J. Gandolfi, W.T. Klimecki, Developmentally restricted genetic determinants of human arsenic metabolism: association between urinary methylated arsenic and CYT19 polymorphisms in children. Environ. Health Perspect. 113 (2005) 775–781.

[67] D. Mondal, M. Banerjee, M. Kundu, N. Banerjee, U. Bhattacharya, A.K. Giri, B. Ganguli, R.S. Sen, D.A. Polya, Comparison of drinking water, raw rice and cooking of rice as arsenic exposure routes in three contrasting areas of West Bengal, India. Environ. Geochem. Health 32 (2010) 463–477.

[68] L.E. Moore, J.K. Wiencke, M.N. Bates, S. Zheng, O.A. Rey, A.H. Smith, Investigation of genetic polymorphisms and smoking in a bladder cancer case-control study in Argentina. Cancer Lett. 211 (2004) 199–207.

[69] L. Paiva, A. Hernandez, V. Martinez, A. Creus, D. Quinteros, R. Marcos, Association between GSTO2 polymorphism and the urinary arsenic profile in copper industry workers. Environ. Res. 110 (2010) 463–468.

[70] L. Paiva, R. Marcos, A. Creus, M. Coggan, A.J. Oakley, P.G. Board, Polymorphism of glutathione transferase Omega 1 in a population exposed to a high environmental arsenic burden. Pharmacogenet. Genomics 18 (2008) 1–10.

[71] J. Palus, D. Lewinska, E. Dziubaltowska, M. Stepnik, J. Beck, K. Rydzynski, R. Nilsson, DNA damage in leukocytes of workers occupationally exposed to arsenic in copper smelters. Environ. Mol. Mutagen. 46 (2005) 81–87.

[72] S. Paul, N. Banerjee, A. Chatterjee, T.J. Sau, J.K. Das, P.K. Mishra, P. Chakrabarti, A. Bandyopadhyay, A.K. Giri, Arsenic-induced promoter hypomethylation and over-expression of ERCC2 reduces DNA repair capacity in humans by non-disjunction of the ERCC2-Cdk7 complex. Metallomics 6 (2014) 864–873.

[73] B.L. Pierce, M.G. Kibriya, L. Tong, F. Jasmine, M. Argos, S. Roy, R. Paul-Brutus, R. Rahaman, M. Rakibuz-Zaman, F. Parvez, A. Ahmed, I. Quasem, S.K. Hore, S. Alam, T. Islam, V. Slavkovich, M.V. Gamble, M. Yunus, M. Rahman, J.A. Baron, J.H. Graziano, H. Ahsan, Genome-wide association study identifies chromosome 10q24.32 variants associated with arsenic metabolism and toxicity phenotypes in Bangladesh. PLoS Genet. 8 (2012) e1002522.

[74] D. Polya, L. Charlet, Rising arsenic risk? Nat. Geosci. 2 (2009) 383–384.

[75] K.E. Porter, A. Basu, A.E. Hubbard, M.N. Bates, D. Kalman, O. Rey, A. Smith, M.T. Smith, C. Steinmaus, C.F. Skibola, Association of genetic variation in cystathionine-beta-synthase and arsenic metabolism. Environ. Res. 110 (2010) 580–587.

[76] K. Rehman, H. Naranmandura, Arsenic metabolism and thioarsenicals. Metallomics 4 (2012) 881–892.

[77] E.G. Rodrigues, M. Kile, E. Hoffman, Q. Quamruzzaman, M. Rahman, G. Mahiuddin, Y. Hsueh, D.C. Christiani, GSTO and AS3MT genetic polymorphisms and differences in urinary arsenic concentrations among residents in Bangladesh. Biomarkers 17 (2012) 240–247.

[78] E.K. Schlawicke, K. Broberg, G. Concha, B. Nermell, M. Warholm, M. Vahter, Genetic polymorphisms influencing arsenic metabolism: evidence from Argentina. Environ. Health Perspect. 115 (2007) 599–605.

[79] E.K. Schlawicke, B. Nermell, G. Concha, U. Stromberg, M. Vahter, K. Broberg, Arsenic metabolism is influenced by polymorphisms in genes involved in one-carbon metabolism and reduction reactions. Mutat. Res. 667 (2009) 4–14.

[80] E. Schmuck, J. Cappello, M. Coggan, J. Brew, J.A. Cavanaugh, A.C. Blackburn, R.T. Baker, H.J. Eyre, G.R. Sutherland, P.G. Board, Deletion of Glu155 causes a deficiency of glutathione transferase Omega 1-1 but does not alter sensitivity to arsenic trioxide and other cytotoxic drugs. Int. J. Biochem. Cell Biol. 40 (2008) 2553–2559.

[81] W.J. Seow, M.L. Kile, A.A. Baccarelli, W.C. Pan, H.M. Byun, G. Mostofa, Q. Quamruzzaman, M. Rahman, X. Lin, D.C. Christiani, Epigenome-wide DNA methylation changes with development of arsenic-induced skin lesions in Bangladesh: a case-control follow-up study. Environ. Mol. Mutagen. 55 (2014) 449–456.

[82] C. Steinmaus, L.E. Moore, M. Shipp, D. Kalman, O.A. Rey, M.L. Biggs, C. Hopenhayn, M.N. Bates, S. Zheng, J.K. Wiencke, A.H. Smith, Genetic polymorphisms in MTHFR 677 and 1298, GSTM1 and T1, and metabolism of arsenic. J. Toxicol. Environ. Health A 70 (2007) 159–170.

[83] L. Su, Y. Jin, Y. Cheng, S. Lin, Study on the relationship between GSTM1, GSTT1 gene polymorphisms and arsenic methylation level. Wei Sheng Yan Jiu 37 (2008) 432–434.

[84] M. Tellez-Plaza, W.Y. Tang, Y. Shang, J.G. Umans, K.A. Francesconi, W. Goessler, M. Ledesma, M. Leon, M. Laclaustra, J. Pollak, E. Guallar, S.A. Cole, M.D. Fallin, A. Navas-Acien, Association of global DNA methylation and global DNA hydroxymethylation with metals and other exposures in human blood DNA samples. Environ. Health Perspect. 122 (2014) 946–954.

[85] R.K. Thirumaran, J.L. Bermejo, P. Rudnai, E. Gurzau, K. Koppova, W. Goessler, M. Vahter, G.S. Leonardi, F. Clemens, T. Fletcher, K. Hemminki, R. Kumar, Single nucleotide polymorphisms in DNA repair genes and basal cell carcinoma of skin. Carcinogenesis 27 (2006) 1676–1681.

[86] O.L. Valenzuela, Z. Drobna, E. Hernandez-Castellanos, L.C. Sanchez-Pena, G.G. Garcia-Vargas, V.H. Borja-Aburto, M. Styblo, L.M. Del Razo, Association of AS3MT polymorphisms and the risk of premalignant arsenic skin lesions. Toxicol. Appl. Pharmacol. 239 (2009) 200–207.

[87] Y.H. Wang, M.M. Wu, C.T. Hong, L.M. Lien, Y.C. Hsieh, H.P. Tseng, S.F. Chang, C.L. Su, H.Y. Chiou, C.J. Chen, Effects of arsenic exposure and genetic polymorphisms of p53, glutathione S-transferase M1, T1, and P1 on the risk of carotid atherosclerosis in Taiwan. Atherosclerosis 192 (2007) 305–312.

[88] M.M. Wu, H.Y. Chiou, C.L. Chen, Y.H. Wang, Y.C. Hsieh, L.M. Lien, T.C. Lee, C.J. Chen, GT-repeat polymorphism in the heme oxygenase-1 gene promoter is associated with cardiovascular mortality risk in an arsenic-exposed population in northeastern Taiwan. Toxicol. Appl. Pharmacol. 248 (2010) 226–233.

[89] M.M. Wu, H.Y. Chiou, I.C. Ho, C.J. Chen, T.C. Lee, Gene expression of inflammatory molecules in circulating lymphocytes from arsenic-exposed human subjects. Environ. Health Perspect. 111 (2003) 1429–1438.

[90] M.M. Wu, H.Y. Chiou, T.C. Lee, C.L. Chen, L.I. Hsu, Y.H. Wang, W.L. Huang, Y.C. Hsieh, T.Y. Yang, C.Y. Lee, P.K. Yip, C.H. Wang, Y.M. Hsueh, C.J. Chen, GT-repeat polymorphism in the heme oxygenase-1 gene promoter and the risk of carotid atherosclerosis related to arsenic exposure. J. Biomed. Sci. 17 (2010) 70.

[91] Y. Xu, X. Li, Q. Zheng, H. Wang, Y. Wang, G. Sun, Lack of association of glutathione-S-transferase omega 1(A140D) and omega 2 (N142D) gene polymorphisms with urinary arsenic profile and oxidative stress status in arsenic-exposed population. Mutat. Res. 679 (2009) 44–49.

[92] R. Zhai, S. Su, X. Lu, R. Liao, X. Ge, M. He, Y. Huang, S. Mai, X. Lu, D. Christiani, Proteomic profiling in the sera of workers occupationally exposed to arsenic and lead: identification of potential biomarkers. Biometals 18 (2005) 603–613.

PART III

MECHANISMS OF TOXICITY

13

ARSENIC INTERACTION WITH ZINC FINGER MOTIFS

LAURIE G. HUDSON[1], KAREN L. COOPER[1], SUSAN R. ATLAS[2], BRENEE S. KING[3] AND KE JIAN LIU[1]

[1] *Pharmaceutical Sciences, University of New Mexico, Albuquerque, NM, USA*
[2] *Physics and Astronomy Department, University of New Mexico, Albuquerque, NM, USA*
[3] *Human Nutrition, Kansas State University, Manhattan, KS, USA*

13.1 INTRODUCTION

Humans come into contact with arsenic through ingestion and inhalation of contaminated water, air, and soil, and additional sources may include occupational and industrial exposures. Drinking water levels in excess of the US Environmental Protection Agency and World Health Organization maximum contaminant levels of 10 ppb are believed to affect more than 140 million people worldwide making arsenic a significant public health concern. Numerous health consequences are attributed to arsenic exposure including neurological, cardiovascular, pulmonary, and metabolic diseases; endocrine disruption; poor reproductive outcomes; immunological dysfunction; cancer; and other toxicities [14, 55, 89, 98, 117, 121]. The diversity of adverse human response to arsenic likely reflects multiple modes of action.

Understanding the mechanisms that underlie actions of arsenic within living systems is complicated by the existence of various forms of arsenic, each with different toxic potential. The major inorganic arsenic species include trivalent arsenite and pentavalent arsenate. By most measures, arsenite is more toxic than arsenate. Furthermore, inorganic arsenic is subject to biomethylation; inorganic arsenic is metabolized through successive oxidative methylation and reduction steps to its

Arsenic: Exposure Sources, Health Risks, and Mechanisms of Toxicity, First Edition.
Edited by J. Christopher States.
© 2016 John Wiley & Sons, Inc. Published 2016 by John Wiley & Sons, Inc.

trivalent and pentavalent mono- and dimethylated metabolites [4, 32, 33, 46, 121]. These methylated molecules are rapidly excreted, and in the past methylation had been thought only to represent a detoxification pathway. More recent *in vitro* and *in vivo* experimental studies find that the trivalent methylated metabolites monomethylarsonous (MMAIII) and dimethylarsinous (DMAIII) acid display greater toxicity and/or carcinogenic potential than inorganic arsenite [18, 36, 63, 118]. However, because most direct human exposure is to inorganic arsenic, all forms of arsenic are considered to be a concern.

Numerous mechanisms account for the adverse effects of arsenic exposure, including alteration of signal transduction cascades, oxidative stress, genotoxicity, and epigenetic modifications, among others which have been reviewed elsewhere [1, 34, 62, 98, 107, 122]. Because of shared physiochemical properties, arsenic can compete for phosphorus in a number of uptake mechanisms and biochemical pathways leading to perturbation of essential cellular processes and modification of phosphorylation-dependent signal transduction cascades [23]. Oxidative stress is another well-studied mechanism of arsenic toxicity. Arsenic exposure may stimulate the production of free radicals in cells [39, 55, 112, 113]. Major arsenic-induced reactive oxygen species include superoxide anion, hydroxyl radical, hydrogen peroxide, singlet oxygen, and peroxyl radicals. These radicals can cause damage to lipids, proteins, and DNA. For example, DNA lesions related to oxidative stress include strand breaks and formation of 8-hydroxydeoxyguanosine [39, 55, 112, 113]. In addition to DNA damage produced by reactive oxygen species, low levels of these species can serve as mitogenic signals and modulate redox-sensitive transcription factors, leading to changes in cell growth and gene expression [8, 40, 47, 66, 123]. Broad changes in cellular properties and behavior may also occur through epigenetic modifications of histones and DNA and regulation of micro RNAs due to arsenic exposure [14, 102, 103, 105]. Although arsenic exerts its actions through numerous and diverse mechanisms, it is the observation that arsenic inhibits DNA repair that led to identification of direct zinc finger protein targets of arsenic, and these targets will be the focus of the following sections.

13.1.1 Arsenic Inhibition of DNA Repair

There are many reported observations that arsenic exposure enhances the mutagenic potential of other DNA damaging and genotoxic stimuli. Inorganic arsenic increases mutation rates when combined with ultraviolet radiation (UVR), benzo[*a*]pyrene, X-rays, alkylating agents, and DNA cross-linkers in bacterial models and cultured mammalian cells [43, 80, 104, 130]. This increase in mutagenesis is evident at low, non-cytotoxic concentrations of arsenite and is associated with inhibition of base excision repair (BER) and nucleotide excision repair (NER) [30, 31, 42, 43, 73, 80, 93, 100, 104, 125, 130]. In addition, there is evidence that DNA repair inhibition is not restricted to inorganic arsenic. Trivalent methylated arsenicals inhibited repair of benzo[*a*]pyrene diol epoxide (BPDE)-induced lesions at lower concentrations when compared to inorganic arsenite [109]. Arsenic appears to disrupt DNA repair through multiple mechanisms including modulation of DNA repair protein expression and direct targeting of specific DNA repair proteins containing zinc finger DNA-binding domains. Arsenic disruption of zinc finger function is presumed to decrease DNA repair activity of the target proteins [8, 14, 35, 105, 127]. The potential for zinc finger DNA repair proteins to be direct arsenic targets is plausible based on the established affinities of arsenic toward closely spaced sulfhydryl (SH) groups that are present in zinc finger structures.

13.2 ZINC FINGER PROTEIN TARGETS OF ARSENIC

Zinc is essential in biological systems due to the critical cellular processes regulated by zinc-binding proteins. It is estimated that up to 10% of the human proteome consists of zinc-binding proteins. Proteins containing zinc finger structures are the most abundant class of zinc metalloproteins representing 2–3% of human genes [82]. Zinc finger proteins perform numerous functions including, but not limited to, DNA and RNA recognition, transcriptional activation, regulation of apoptosis, and protein binding [68].The zinc finger motif was first identified in the DNA binding transcription factor TFIIIA from Xenopus laevis [88] and found to be critical for the DNA binding function of this protein. Broadly speaking, the term zinc finger refers to a sequence of amino acids consisting of histidine and/or cysteine residues that coordinate zinc and stabilize a small, discrete structure [64, 85] (Fig. 13.1). Release or substitution of the coordinated zinc ion typically disrupts protein function. There is little consensus on what constitutes a true "zinc finger" protein. The term was originally applied to structures most closely resembling the approximately 30 aa residue repeats identified in TFIIIA, but this designation has been expanded to apply to many classes of structural (rather than catalytic) zinc-binding motifs [64, 85]. Depending on the classification scheme applied, up to 20 or more zinc finger domains have been proposed [64, 65, 85], but based on an analysis of zinc finger structures identified in the Protein Data Bank (PDB) database, Krishna et al. proposed a structure-based classification consisting of eight groups, of which the two cysteines plus two histidines (C2H2) zinc finger group is the most common [65]. Although proteins may share similar C2H2 motifs, some proteins have multiple and adjacent zinc fingers (TFIIIA) or may have separated, paired or triple-C2H2 fingers [52]. Different zinc finger proteins may perform similar functions despite differences in structure. For example, four cysteines (C4) zinc fingers of the typical DNA-binding domain for nuclear receptors such as the glucocorticoid receptor, and C2H2 zinc fingers common to many transcription factors, interact with DNA [3].

The ability of zinc fingers to interact with inorganic ions other than zinc is particularly relevant to metal toxicology. Substitution of zinc with another metal is believed to disrupt the coordination sphere in the finger environment and consequently the zinc finger function [101]. Analysis of metal interaction with various zinc finger domains revealed that K_d values for zinc or other metals vary greatly between specific zinc fingers. For example, Cd(II) and Cu(II) inhibited bacterial formamidopyrimidine [fapy]-DNA glycosylase (Fpg), Xeroderma pigmentosum group A (XPA), and poly [ADP-ribose] polymerase (PARP)-1, whereas Co(II) and

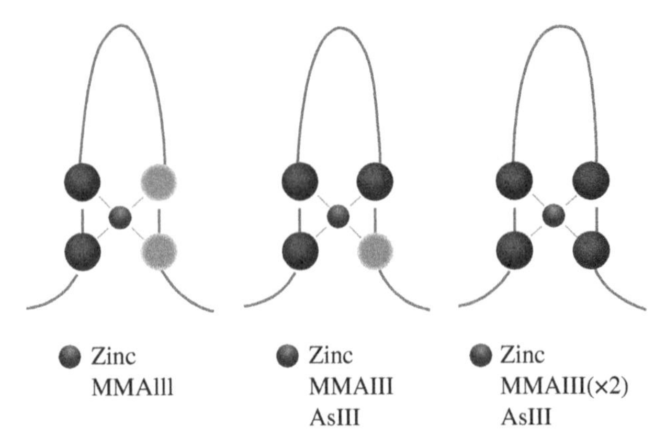

FIGURE 13.1 Schematic representation of a classical zinc finger. Zinc finger motifs are domains of typically 30–40 residues that coordinate zinc ions through cysteine and/or histidine residues. The most common C2H2 configuration is depicted in the schematic, but other configurations representing different placement of cysteine and histidine residues and number of cysteine residues (0–4) have been identified and are depicted in Krishna, Majumdar et al. [65]. Zinc finger motifs may occur singly (e.g., XPA) or in multiple motifs (e.g., TFIIIA) within proteins. Zinc forms bonds with cysteine (red/dark circle) or histidine (yellow/light circle), but arsenic is only able to bind to cysteine residues. The differences in arsenite (AsIII) and monomethylarsonous (MMAIII) acid selectivity based on current findings is indicated in the figure. (*See insert for color representation of the figure.*)

Ni(II) only inhibited XPA and PARP1 and arsenite inhibition was restricted to PARP1 [44]. This work indicates that not all zinc finger structures are equivalent targets for a specific metal ion, although the basis for the observed differences in selectivity is unknown at this time. Trivalent arsenicals have been reported to interact with and inactivate certain zinc finger proteins. The following sections will focus on the human zinc finger protein targets of arsenic that are most strongly validated by experimental evidence.

13.2.1 Xeroderma Pigmentosum Complementation Group A Protein (XPA)

Large, bulky, helix-distorting lesions caused by chemical agents (i.e., benzo(*a*) pyrene), cross-linking agents (i.e., cisplatin), or exposure to UVR are repaired by the NER pathway. This is a versatile repair pathway that removes chemically and structurally distinct lesions from the mammalian genome [13, 41, 129]. There are two sub pathways within NER, transcription-on the template strand of active RNA polymerase II transcribed genes, while GG-NER repairs lesions over the entire genome. XPA is an essential DNA damage recognition protein required for efficient NER. Although XPA has no enzymatic activity, it functions as a core component in both arms of NER reactions by interacting with damaged DNA. XPA also plays a role in subsequent steps of NER through interaction with other proteins in the repair complex (e.g., replication protein A) [94, 111]. Mutations in XPA compromise DNA repair and render individuals highly susceptible to UVR-induced skin cancer. Within its DNA-binding domain, XPA contains a single zinc finger motif where zinc is coordinated by four cysteine residues [119]. Substitution of any of these cysteines severely decreases NER activity [90]. XPA was investigated as a potential arsenic target based on the evidence that arsenic inhibited NER, centrality of XPA in NER, and the vicinal sulfhydryl groups conferred by the C4 zinc finger.

The majority of studies to investigate arsenic interaction with XPA have taken *in vitro* biochemical approaches using peptide representing the zinc finger domain of XPA or purified protein [5, 6, 91, 97, 109]. Investigations on the ability of different arsenicals to release zinc from an XPA zinc finger peptide revealed that all trivalent forms of arsenic promoted zinc release with little effect evident for the pentavalent metabolites MMA(V) and DMA(V) at concentrations up to 1 mM [109]. In contrast, zinc release was noted at MMA(III) and DMA(III) concentrations greater than 10 μM, while arsenite concentrations in excess of 100 μM were required for robust zinc release. In a more direct analysis, electrospray-induced (ESI) mass spectrometry (MS) revealed that the reaction of the XPA zinc finger peptide with MMAIII yielded stable complexes under ESI-MS conditions. However, the ESI-MS approach did not detect arsenite adducts, suggesting that these complexes were less stable [97]. In contrast, Matrix-assisted laser desorption/ionization time-of-flight (MALDI-TOF) MS analysis demonstrated that arsenite, as well as MMAIII, bound XPA zinc finger peptide suggesting that analytical methods may influence the ability to detect arsenite binding (134). It is interesting to note that only a single arsenite molecule was bound to the XPA zinc finger; however, two MMAIII molecules could be detected in a complex with the C4 XPA zinc finger peptide [97, 134]. This suggests that each pair of cysteine residues is capable of interacting with MMAIII and provides evidence for potential differences in protein target interaction based on the form of arsenic.

Most studies on purified protein have focused on potential perturbation of the zinc finger-mediated DNA-binding capacity of XPA. Direct incubation of XPA with up to 1 mM arsenite had no effect on DNA binding to UVC damaged oligonucleotide [5]. However, arsenite at 100:1 molar excess did decrease zinc-bound XPA interaction with mitomycin C-damaged DNA indicating that arsenite is capable of interacting with XPA protein, albeit at high concentrations [91]. These two *in vitro* studies suggested that XPA was not a high-affinity direct molecular target for arsenite. More recent studies suggest that within the *in vivo* cellular context arsenite causes loss of zinc from XPA. XPA immunoprecipitated from cells exposed to low micromolar levels of arsenite was substantially zinc depleted [133]. Furthermore, incubation of cells with the same concentration of MMAIII also decreased zinc content in isolated XPA [134]. Collectively, these findings illustrate differences in the outcome between analysis of arsenite interaction with zinc finger peptides versus protein, and analysis of protein *in vitro* versus cell-based studies. Because XPA binding to DNA requires an intact zinc finger, the loss of zinc from XPA in cells predicts that arsenic disrupts the zinc finger-mediated functions of XPA in cells.

13.2.2 PARP1

Poly(ADP-ribose)polymerases (PARPs) are an ancient family of enzymes consisting of 18 members. PARP1 is the founding family member and the most abundant PARP in cells [77]. PARP1 is a 113 kDa chromatin-associated enzyme that is involved in numerous cellular processes including DNA repair, cell cycle control, apoptotic signaling, and transcriptional regulation [22, 108]. The PARP1 protein has three major domains: the DNA-binding domain (residues 1–374), the automodification domain (residues 375–525), and the catalytic domain (residues 526–1014) [70]. The DNA-binding domain is located in the N terminus of the PARP1 protein and contains two zinc fingers each containing three cysteines plus one histidine (C3H1) responsible for recognition of damaged DNA. These zinc fingers are unique in that they recognize altered DNA structures rather than specific sequences [17, 38]. The zinc finger 1 and zinc finger 2 domains of the PARP1 DNA-binding domain perform different functions [2, 38, 72]. The zinc finger 2 domain displays high DNA-binding affinity, whereas the zinc finger 1 domain is required for DNA-dependent PARP1 activity. A recent low-resolution structure of the N-terminal half of PARP1 revealed structural flexibility in the form of hinges connecting the two zinc fingers with a third zinc finger [74]. The third zinc finger of PARP1 is a C4 motif and believed to couple the DNA binding and catalytic activities of PARP1 leading to the well-documented elevation of PARP1 catalytic activity upon binding to damaged DNA [71, 74, 120].

Because arsenic inhibits repair of oxidative DNA damage and PARP1 has an established role in repair of these lesions through the BER DNA repair pathway [26, 53, 114], PARP1 was investigated as a candidate arsenic target. Mass spectrometry analysis demonstrates that arsenite and MMA(III) bind a synthetic peptide representing the zinc finger 1 domain of PARP1, and displace zinc from the peptide in a dose-dependent manner [31, 128, 132, 133, 134]. Both DNA-binding zinc fingers are vulnerable to arsenite binding [31]. The evidence indicates that arsenite binding displaces three hydrogens, suggesting that a single arsenite molecule binds to the C3H1 zinc finger peptide with trivalent coordination [133]. In the case of MMAIII, two hydrogens are displaced from the peptide, indicating differences between MMAIII and arsenite interaction with the PARP1 zinc finger peptide [128].

Activation of PARP enzymes leads to the production of poly(ADP-ribose) (PAR) making it possible to evaluate PARP activity *in situ* using immunochemical detection of PAR [31, 45, 124]. Measurement of PAR revealed that DNA damage-stimulated PARP1 activity is highly sensitive to arsenic inhibition (detectable at 100–200 nM) [31, 45]. Arsenite inhibited PARP1 activity in HelaS3 cells at submicromolar concentrations; a significant decrease in poly(ADP-ribosyl)ation was detected at 10 nM arsenite with approximately 40% of PARP activity remaining at 0.5 μM arsenite. Similarly, in human keratinocytes, inhibition of PARP activity (up to 90% inhibition) is readily evident at submicromolar concentrations of arsenite [31, 133]. Continuous exposure of urothelial cells to 50 nM MMAIII led to a small but measurable decrease of PARP activity and exposure to 1 μM MMAIII caused approximately 30% PARP inhibition [128]. These findings might suggest that arsenite causes a greater magnitude of PARP inhibition than MMAIII, or may reflect cell-type differences. Further studies will be necessary to resolve this question.

The levels of PAR in cells are maintained by PARP-dependent production and removal of PAR moieties from proteins by poly(ADP-ribose) glycohydrolase (PARG). PARG contains both endo- and exoglycosidase activity [11]. Evidence that the inhibition of PAR production by arsenic is due to PARP activity and intrinsic to this putative target rather than an increase in PARG activity is provided by analysis of protein isolated from arsenic-exposed cells. A dose-dependent decrease in PARP1 activity is evident in immunoprecipitated PARP1 protein, and decreased activity corresponds with zinc loss from PARP1 [31, 100]. PARP1 activity is largely retained when cells are co-treated with arsenite and supplemental zinc [31, 100]. Similarly, continuous 4-week zinc supplementation restored PARP1 activity and reduced the genotoxicity associated with MMAIII exposure of urothelial cells [128]. These findings suggest that the exquisite sensitivity of PARP1 to trivalent arsenic (arsenite or MMAIII) [45, 100, 124] is due to the interaction of arsenic with the PARP1 zinc fingers within the cellular environment leading to zinc depletion and impairment of DNA binding and catalytic activities governed by these domains. The observation that supplemental zinc offsets the effects of arsenic on PARP1 activity provides further support for the importance of the zinc finger motif as a target for arsenic.

13.2.3 PML-RARα Fusion Protein

The PML/RARα fusion protein arising from a chromosomal translocation is the driving mutation in acute promyelocytic leukemia (APL). Both retinoic acid receptor (RAR)α and promyelocytic leukemia protein PML are zinc finger proteins representing two different classes of zinc finger motifs. RARα is a member of the steroid hormone receptor superfamily, and the zinc finger motif confers DNA binding. The functions of the zinc finger domain of PML are not well understood. The aberrant fusion protein heterodimerizes with RXRα, thereby disrupting transcription and differentiation and promoting proliferation of leukemia-initiating cells [25, 83, 87, 99]. Arsenic trioxide (ATO) is used therapeutically to treat APL and causes degradation of PML/RARα leading to the observed clinical benefit [57, 69, 75]. ATO treatment causes a shift in PML and PML/RARα, but not RARα, from a soluble to a detergent-insoluble form [131], suggesting that the PML zinc finger is the ATO target.

Studies using organic arsenical probes demonstrated binding of these probes to cysteine residues in zinc fingers located within the RBCC (N-terminal RING finger/B-box/coiled coil) domain of PML-RARα and PML [131]. Binding of the arsenic probes induced oligomerization, leading to increased interaction with the small ubiquitin-like protein modifier (SUMO)—conjugating enzyme UBC9. This increased interaction resulted in enhanced SUMOylation of PML-RARα and its subsequent degradation. Another study [54] demonstrated that an arsenic-biotin derivative bound to PML in Chinese hamster ovary cells and ATO bound to bacterially produced, purified proteins at a stoichiometry of 0.5 arsenic atoms per polypeptide. Biotin-As bound to PML and PML-RARα, but not to RARα, further supporting the conclusion that the PML zinc finger is the salient arsenic target [131]. Further investigation identified a dicysteine motif (C212/C213) of PML as a binding site and PMLC212A, C213A, or C212/213A mutants lost labeling by the fluorescent biarsenical fluorescein derivative labeling reagent termed "FlAsH" [54]. MALDI-TOF analysis demonstrated that apo PML bound one or two arsenic atoms and further analysis using extended x-ray absorption fine structure (EXAFS) and x-ray absorption near-edge structure (XANES) spectroscopy indicated that arsenic coordinated with three sulfur atoms in the three conserved cysteines in both zinc finger 1 and zinc finger 2 [131].

Although these reports provide support that PML-RARα is a direct target for ATO, there are some caveats to the studies. Both studies used arsenic conjugates as probes which may differ in binding characteristics from the therapeutically administered ATO. In the study by Zhang et al., two organic arsenicals, p-aminophenylarsine oxide (PAPAO) conjugated to biotin and the biarsenical resorufin derivative labeling reagent termed "ReAsH" were used [131]. ReAsH is an organic arsenic derivative which fluoresces red when its arsenic moieties bind to vicinal thiols in target proteins [84]. The other study used FlAsH which is a diarsenical that binds two CC motifs [78] in addition to dithiasorolan-biotin [54]. By using these probes, the trivalent properties of the arsenic were modified, and could lead to altered target protein interaction. Nonetheless, this observation of arsenic targeting PML-RARα leading to a therapeutic outcome deserves more detailed investigation.

13.2.4 Steroid Hormone Receptors

Members of the steroid hormone receptor superfamily are ligand-regulated transcription factors that control a wide range of physiological processes [67, 86, 116]. These proteins are characterized by two C4 zinc finger domains that are responsible for DNA recognition and binding, and cysteine-rich regions in the ligand-binding domain. Epidemiological findings of human health effects and experimental studies indicate that arsenic acts as an endocrine disruptor [50, 58, 106], but evidence that arsenic binding to zinc finger motifs represents the predominant mechanism for disruption of endocrine functions is inconclusive.

Findings supporting arsenic interaction with zinc finger motifs within steroid hormone receptors are provided by peptide-binding studies. Kitchin and Wallace conducted detailed and quantitative studies using radioactive arsenite (^{73}As) [59]. A peptide representing one of the two C4 zinc finger regions in the estrogen receptor-α DNA-binding domain bound arsenite with a K_d of 2.2 μM. The result indicates that this motif is capable of binding arsenic and suggests that DNA-binding zinc finger motifs of steroid hormone receptors may be direct targets [59]. In addition, the same study found that a peptide containing three cysteine residues derived from the ligand-binding region of estrogen receptor-α bound arsenite with high affinity (K_d 1.32 μM). Because arsenite was shown to interact with two distinct and functionally critical cysteine-rich domains of the receptor, one cannot definitively assign the interaction which is ultimately responsible for the observed biological effects.

In cells, the impact of arsenic on steroid hormone receptor-regulated gene expression has established that arsenic modulates receptor function. The response to arsenic is complex; expression of reporter genes regulated by steroid hormone receptor-response elements is enhanced at low concentrations and inhibited at higher concentrations for the glucocorticoid, mineralocorticoid, progesterone, and androgen receptors [9, 10, 24, 56]. Similarly, arsenic disrupts estrogen receptor-mediated gene expression [15, 16, 24, 126], suggesting that there is a shared mechanism of action for these closely related steroid hormone receptor family members. However, differences emerge in other assays. Arsenite interferes with ligand binding to the glucorticoid receptor *in vitro* [115], but this property was not shared by other members of the receptor family [24, 76]. Mutational analysis of the glucocorticoid receptor indicated that the DNA-binding domain was responsible for the arsenic-mediated effects [9], and mutation of the two free cysteine residues (not the zinc coordinating residues) failed to block the arsenic effect [9]. These studies did not include mutation of cysteine residues within the zinc finger motifs, and there are no reports on zinc content of steroid hormone receptors isolated from arsenic-exposed cells, so it is unclear whether zinc finger domains represent the critical arsenic target. Systematic studies will need to be performed to establish whether arsenic interferes with all or most members of the steroid hormone receptor superfamily or if there is selectivity for certain family members.

13.3 ZINC FINGER SELECTIVITY

It is becoming increasing clear that although zinc finger motifs may serve as arsenic targets, not all zinc fingers are equally vulnerable to attack by arsenic. Although it has been long recognized that arsenic is capable of binding to vicinal cysteine residues, there is mounting evidence that vicinal cysteines may not be sufficient for arsenic to target zinc finger proteins.

13.3.1 Peptide Studies

Analysis of binding affinities of ^{73}As for peptides display distinct K_d values based on the number of thiol sites [59]. Arsenite did not bind to peptides lacking cysteine residues even when they contained histidines. Monothiol sites or two sulfhydryls spaced 17 amino acids apart had measured K_d values in excess of 100 μM [59]. Another report demonstrated that 10 μM arsenite strongly disrupted the oxidative folding of riboflavin-binding protein (a protein with 18 cysteines) even in the presence of 5 mM reduced glutathione (a monothiol compound) further indicating poor interaction of arsenite with single thiol groups [61]. In contrast, peptides modeled on the estrogen receptor with two or more nearby cysteines (two to five intervening amino acids) had high affinity with K_d values in the 1–4 μM range.

Mass spectrometry analysis provides evidence that arsenite interacts with peptides and zinc finger proteins containing C4 and C3H1, but not C2H2 zinc finger motifs. Arsenite addition to a native C3H1 PARP zinc finger peptide leads to an arsenite-bound PARP zinc finger signal as detected by MALDI-TOF [31]. Inductively coupled plasma atomic emission spectroscopy (ICP-AES) demonstrates that arsenite forms a covalent bond to the PARP1 zinc finger peptide thus confirming the affinity of arsenite toward the zinc finger (Fig. 13.2a). In contrast, arsenite did not bind to peptides representing the PARP1 zinc finger 1 with a single cysteine to histidine substitution to generate a vicinal C2H2 motif [133]. Arsenite interacted readily with a mutationally generated PARP1 C4 zinc finger motif [133], thereby demonstrating a distinct preference for C3H1 and C4 zinc finger structures. Similarly, arsenite did not interact with the native C2H2 peptide derived from the DNA repair protein aprataxin, but bound to a C3H1 peptide produced by a single histidine to cysteine substitution within the aprataxin zinc finger [133]. This result is consistent with the reports by Kitchin and Wallace on the binding of trivalent arsenite to zinc finger synthetic peptides [59, 60] and indicates preferential arsenite binding to zinc finger motifs with three or four cysteine residues.

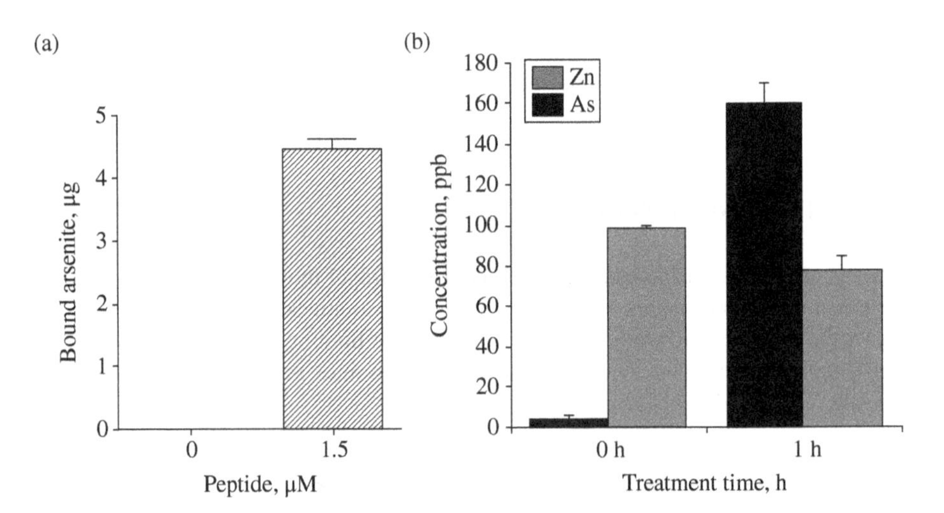

FIGURE 13.2 Arsenite binding to PARP1 zinc finger peptide and PARP1 DNA-binding domain amino acids 1–374. (a) Synthesized PARP1 zinc finger peptide (1.5 µM) was incubated with arsenite (3 µM) for 30 min. The free arsenic was removed by polyacrylamide desalting columns, and the fractions containing peptide were collected. The concentration of arsenic bound to peptide was detected by ICP/AES. (b) A plasmid containing PARP1 DNA-binding domain (DBD) tagged with streptavidin was transfected into in normal human keratinocytes and incubated for 72 h. Cells were subsequently treated with 2 µM arsenite and total protein collected at 0 and 1 h post treatment. For arsenic-binding assays (black bars), the DBD was immunoprecipitated from 1 mg of total cellular protein with biotin-coated polystyrene beads and resulting pellet analyzed for total arsenic by ICP-MS. For determination of zinc release (gray bars), PARP1 protein was immunoprecipitated from cells treated with arsenite as above, eluted from the pellet and denatured to release zinc. Concentration was determined via the absorbance spectrum shift of 4,(2-pyridylazo)-resorcinol in the presence of zinc compared to a standard curve. The data are presented as the means ± SD from three experiments.

Another study conducted a detailed analysis of arsenic interaction with zinc finger motifs using multiple analytical approaches [132]. The authors compared arsenic interaction with zinc finger motifs including transcription factor specificity protein (Sp) 1 (three C2H2 domains), HIV-1 nucleocapsid protein p7(NCp7) (two C3H domains), and estrogen receptor-α (two C4 domains). Single-domain motifs displayed marked differences in the K_d values with greater affinity of arsenic for C4>C3H≫C2H2. The binding affinity of the C2H2 motif was approximately two orders of magnitude lower than either the C4 or C3H motifs. Interestingly, the C2H2 multidomain peptide of Sp1 demonstrated a higher binding affinity for arsenite relative to the single zinc finger domain Sp1 C2H2 peptide [132]. The K_d values of arsenite for the dual C3H1 and C4 motifs did not differ significantly from the single-motif structures. ESI-MS of the arsenite-bound single zinc finger peptides with differing number of cysteine residues (C2H2, C3H, and C4 motifs) indicated that both C3H and C4 motifs bind to one arsenic atom via three deprotonated cysteine residues [132] as reported previously [133]. The dual C3H1 and C4 motif peptides bound two arsenic atoms with the deprotonation of six cysteine residues indicating that each zinc finger motif was capable of binding arsenic. In contrast, the predominant product of the single C2H2 zinc finger motif following incubation with arsenic was apopeptide with negligible detection of arsenic-complexed product. Interestingly, the multidomain C2H2 Sp1 zinc finger peptide generated mono- and di-arsenic-bound products, suggesting arsenic coordination across zinc finger motifs [132]. Circular dichroism (CD) and nuclear magnetic resonance (NMR) spectra indicate that arsenic binding to a C3H1 or C4 zinc finger motif results in an unfolded structure [132].

Somewhat different findings were obtained in an analysis of arsenite interaction with a C3H1 zinc finger model peptide from NCp7 [28]. Fluorescence spectroscopy, CD, and ESI-MS revealed that arsenic interaction differs between apopeptide and zinc-bound peptide. Arsenic binding to apopeptide led to a disordered structure as seen by Zhao et al. [132], but arsenic interaction with zinc-bound peptide led to the formation of a mixed As–Zn-peptide complex that retained partial zinc finger conformation [28]. This observation suggests that a mixed complex containing both arsenic and zinc can occur, but it has yet to be determined if this is unique to the NCp7 zinc finger or is a general property of arsenic interaction with zinc-bound zinc fingers. Similarly, detailed studies have yet to be conducted with trivalent methylated metabolites to assess preferential binding and relative affinities for different zinc finger motifs. The two studies by Demicheli et al. and Zhao et al. support the conclusion that arsenic binding does not recapitulate the structure conferred by zinc coordination and provide additional evidence that arsenic interaction with zinc finger proteins is likely to disrupt protein function [28, 132].

13.3.2 Cell-Based Studies

Zinc coordination within a zinc finger domain is critical to the function of this motif. Peptide studies demonstrating arsenite selectivity for zinc finger configurations containing three or more cysteine residues may or may not reflect what occurs within a cellular environment. Studies of the zinc content of putative target proteins isolated from arsenite-exposed cells support the conclusion reached through peptide analysis [133]. Arsenite was present in ectopically expressed PARP1 DNA-binding domain isolated from treated cells by immunoprecipitation and analyzed by ICP-AES (Fig. 13.2b) providing direct evidence for arsenic binding to a native zinc finger structure in cells. With regard to zinc finger selectivity, cell treatment with arsenite led to reduced zinc content of immunoprecipitated C3H1 (PARP1) and C4 (XPA) zinc finger proteins. In contrast, the C2H2 zinc finger proteins aprataxin and Sp-1 retained zinc following cell exposure to arsenite as did an expressed PARP1 harboring zinc fingers mutated to C2H2 configurations [133, 134]. It is interesting that endogenous Sp1 protein did not appear to interact with arsenite based on zinc release from the protein, despite evidence that the triple zinc finger peptide derived from Sp1-bound arsenite [132]. This finding suggests that additional protein structural considerations beyond the isolated zinc finger sequences influence arsenite interaction with zinc finger motifs. The observation of preferential arsenite interaction based on the number of cysteine residues within a zinc finger is significant because the majority of zinc finger proteins in the human genome are of the C2H2 variety, suggesting that arsenite may have distinct and restricted number of zinc finger targets in cells.

The ability of zinc to offset the actions of arsenite in target protein function provides further evidence that arsenite interaction with zinc finger motifs is functionally significant. When cells were co-exposed to arsenite and increasing concentrations of zinc, followed by PARP activation by UVR exposure, inclusion of zinc restored PARP1 activity in a concentration-dependent manner [31]. Similarly, zinc reversed arsenite-dependent augmentation of UVR-induced oxidative DNA damage and DNA strand breaks [31, 100]. The increased accumulation in DNA damage upon co-exposure of arsenite and UVR is meaningful because it leads to genotoxicity (i.e., increased mutation frequency). Ultraviolet radiation alone is mutagenic, but exposure to 2 μM arsenite increased UVR-induced mutations by nearly twofold and co-exposure with arsenite and zinc eliminated mutations due to UVR + arsenite (Fig. 13.3). Furthermore, the ability of zinc to offset the arsenite augmentation of DNA damage is retained *in vivo* in UVR-exposed mouse skin [20]. Collectively, these findings indicate that arsenite may compete with zinc in the zinc finger domains of PARP1, thereby interfering with zinc-dependent functions of PARP1, including enzyme activity and repair of DNA damage. It is likely that the activities of other zinc finger protein targets of arsenic are similarly impaired in cells and *in vivo*.

13.4 PERSPECTIVES AND FUTURE DIRECTIONS

To date, there is convincing evidence that one of the mechanisms of arsenic toxicity is related to interaction with, and functional disruption of, zinc finger protein targets. A limited number of targets have been identified, which have been discussed in this chapter. There is accumulating evidence that not all zinc finger proteins are high-affinity targets for arsenic, and there are many remaining questions that need to be resolved in order to identify relevant targets responsible for specific adverse responses to arsenic exposure.

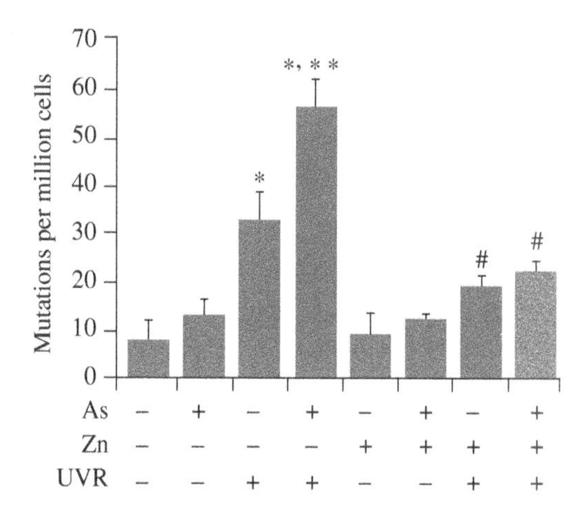

FIGURE 13.3 Arsenite enhances UVR-stimulated mutation frequency. Immortalized human keratinocytes (HaCaT cells) were cultured and treated with arsenite (As, 2 μM), zinc (Zn, 5 μM), both or left untreated (NT) for 24 h before exposure to UV (3 kJ/m^2). Cells were then plated in triplicate and placed in media containing 6-thioguanine (5 μg/mL) for mutant selection per traditional HPRT mutation protocols. Bars represent mutants/million clonable cells ± SEM for each treatment group. *, significantly different from untreated control (NT); **, significantly different from UV alone; and #, significantly different from As+UVR; $p<0.05$; $n \geq 3$.

13.4.1 Arsenic Metabolites

Arsenic disruption of zinc finger proteins is an emerging area of research and to date, the effects of arsenite have received more attention than the methylated metabolites. MMAIII binds to a PARP1 zinc finger peptide [36], and three trivalent arsenicals (arsenite, MMA(III), and DMA(III)) all released zinc from a synthetic peptide based on the zinc finger region of XPA [109]. However, relative K_d values for these zinc finger targets are unavailable for each of these various arsenic species for the same peptide.

Kitchin and Wallace noted that unidentate binding of trivalent arsenicals (arsenite) to a peptide containing a single cysteine residue has a lifetime estimated to be 0.01–1 s [60]. Bidentate binding between arsenite and two cysteines at a dithiol site has an estimated lifetime of approximately 1–2 min, whereas the coordination complexes to tridentate binding of arsenite to three cysteines can persist for 1–2 h. This provides a plausible basis for selectivity for C3H1 and C4 zinc finger targets. These observations also raise the question of whether similar differences in complex stability occur when arsenic is capable of tridentate interaction (arsenite) or bidentate or unidentate binding (MMA or DMA, respectively) to a C3H1 or C4 zinc finger motif. If tridentate binding is required for stable interaction of arsenic with target proteins, it would suggest that the methylated arsenic species may not be able to bind for sufficient duration to elicit robust responses through the specific target in biological systems. However, MMAIII and DMAIII are bioactive and may interact with a wider range of potential targets based on bidentate binding. On the other hand, little is known regarding the concentration of the trivalent arsenic metabolites across different tissues and whether it is possible to achieve the concentrations required to elicit responses in experimental systems throughout the human body. These issues will only be resolved with further investigation.

13.4.2 Zinc Finger Structures

In all zinc finger proteins, zinc is complexed through cysteine and/or histidine residues [79], yet the actual structures of zinc fingers can differ substantially (Fig. 13.4). The generic term "zinc finger" does not encompass structural considerations that might have bearing on the ability of arsenic to interact with the zinc coordination site. Krishna et al. proposed a classification of zinc finger based on spatial structures and fold groups [65]. Using this methodology, they found that each available zinc finger structure could be placed into one of eight fold groups based on the main chain conformation and secondary structure around the zinc-binding site. The majority of zinc fingers fall into three fold groups: the C2H2-like zinc finger, the treble clef zinc finger, and the zinc ribbon. C2H2-like zinc finger is characterized by a β-hairpin followed by an α-helix that forms a left-handed ββα unit. The treble clef motif consists of a β-hairpin at the N-terminus and an α-helix at the C-terminus and the zinc ribbon has two zinc-knuckles with a core structure comprising two β-hairpins. The less-common structures include Gag knuckle composed of two short β-strands connected by a turn (zinc knuckle) followed by a short helix or a loop, Zn_2/Cys_6-like finger which consists of zinc-binding domains in which two coordinating residues are from a helix and two are from a loop, transcriptional adaptor zinc-binding TAZ2

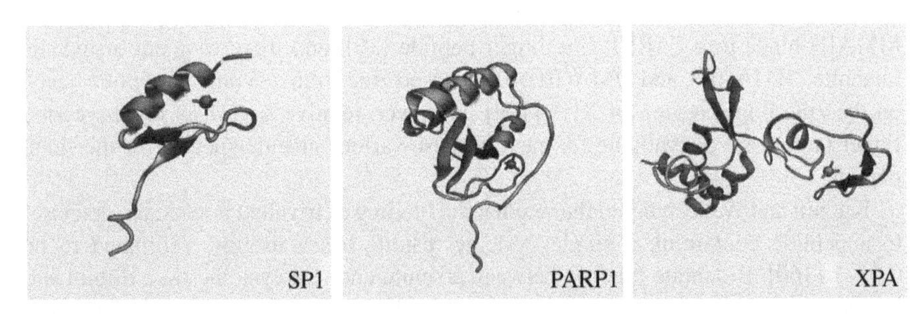

FIGURE 13.4 Representative zinc finger structures. Zinc finger protein ribbon structures were created and rendered using the VMD code of Humphrey et al. [49], starting from PDB structures (www.rcsb.org/pdb). SP1 (C2H2 zinc finger) solution NMR structure [92], showing tetrahedral coordinating residues within 2.4 A. PARP1 (C3H1 zinc finger) solution NMR structure [38], showing coordinating residues within 2.5 A. XPA (C4 zinc finger)-binding domain from solution NMR showing coordinating residues within 2.3 A [51]. Zinc atoms are depicted as small dark gray circles with bonds to coordinating residues (larger dark and light gray circles). α-Helices are shown as coiled ribbons or thick tubes, β-strands as arrows, and nonrepetitive coils or loops as lines or thin tubes. The direction of the polypeptide chain is shown locally by the arrows.

domain-like proteins characterized by zinc-coordinating residues located at the termini of α-helices, short zinc-binding loops that do not have regular secondary structural elements, and metallothioneins that also lack clearly defined regular secondary structural elements [65].

Currently, it is unknown which of these spatial features and folds in zinc finger structures may be necessary for high-affinity arsenic binding and may account for the observed differences in metal binding by zinc finger proteins. For example, both XPA and FPG contain single C4 zinc finger motifs yet showed distinct differences in metal sensitivity [5]. Gaining a greater understanding of the structural considerations for arsenic binding is an important challenge for the field. Collectively, zinc finger motifs are involved in protein:protein, protein:RNA, and protein:DNA interactions, suggesting that arsenic disruption of zinc finger function may play a wider role in arsenic toxicity than can be attributed to the limited number of currently identified targets.

13.4.3 Influences of the Cellular Environment on Arsenic Species

It is evident that arsenic interaction with purified zinc finger proteins *in vitro* does not fully reflect the outcome when target proteins are analyzed *in situ* (e.g., PARP activity by detection of poly (ADP) ribose product) or when target protein is isolated from arsenic-treated cells [31, 133]. The majority of studies conducted using purified protein *in vitro* find little impact of arsenic on protein activity or zinc content at environmentally meaningful concentrations [5, 6, 91, 97, 109]. In contrast, protein immunoprecipitated from arsenic exposed cells or *in situ* analysis reveals significant impact at low arsenic concentrations [5, 31, 43, 45, 133], as illustrated in Figure 13.5.

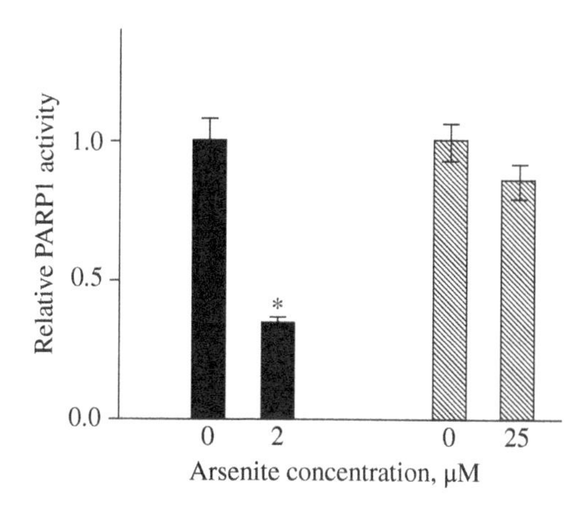

FIGURE 13.5 Effects of arsenite on PARP1 activity. HaCat cells were treated with indicated concentrations of arsenite for 48 h (solid black bars). PARP1 was isolated by immunoprecipitation. Alternatively, PARP1 was first isolated by immunoprecipitation from untreated HaCat cells and then the purified PARP1 protein was incubated with the indicated concentration of arsenite for 48 h (hashed bars). PARP1 DNA-binding activity was determined by electrophoretic mobility shift assay. Data are presented as means ±S.D. *, significantly different from 0 arsenite group, $p<0.05$, $n=4$.

The reasons for this difference are currently unknown, but may be related to properties within living cells that affect arsenic itself.

An important aspect of biological response to arsenic is valency. Trivalent forms of arsenic (inorganic or methylated) are reportedly more toxic than the corresponding pentavalent forms. For example, the zinc finger of XPA released zinc in the presence of inorganic and methylated trivalent, but not pentavalent arsenic species [109]. The greater toxicity of arsenite when compared to arsenate is likely due to its strong covalent reactivity toward SH-groups and disruption of essential proteins [29, 61]. Pentavalent arsenic forms are reduced by thiols, such as glutathione (GSH) to trivalent species [7, 27, 97, 110]. The reduction of arsenate to arsenite is intimately linked to distribution, metabolism, and fate of arsenic in organisms. GSH plays a key role in arsenate metabolism, as demonstrated by retention of arsenate in GSH-depleted rats [21]. Thus, GSH levels are a variable in arsenic response in living systems.

Although inorganic arsenic represents the majority of human exposure, methylated metabolites are also implicated in arsenic toxicity and cells differ in expression and activity of the methyltransferases responsible for arsenic metabolism. There currently exist paradoxical findings regarding the role of methylation in arsenic carcinogenesis. On the one hand, methylated arsenic metabolites are reportedly more carcinogenic to the bladder where the metabolites are concentrated in urine [12, 18, 19, 36, 37, 81]. On the other hand, arsenite is highly enriched in epidermal appendages [81]. Although the skin accumulates arsenic [48], the ratios of organic and inorganic forms are unknown for the epidermis (target tissue for arsenic-associated keratinocytic tumors). Keratinocytes are notably deficient in methylating inorganic arsenic [95, 118], and arsenite (100 nM, 28 weeks) effectively caused malignant transformation of keratinocytes at concentrations similar to those required for urothelial cell transformation by MMAIII (50 nM, 52 weeks) [12, 96]. These findings suggest that arsenic metabolism to methylated forms may be a variable in the response of different tissues to arsenic, and the ultimate toxic arsenic species may be based on local tissue concentration of a specific form of arsenic.

13.5 SUMMARY

Because certain proteins, such as PARP1, are exquisitely sensitive to arsenic exposure it is critical to identify specific targets for arsenic to provide insights into mechanisms responsible for observed human toxicities, and to inform potential mechanism-based interventions. There is now sufficient evidence to support select zinc finger proteins as direct arsenic targets based on arsenic interaction with certain zinc finger motifs. The limited number of proteins identified thus far is unlikely to represent the full range of relevant targets, so more research will be required to identify the determinants of high-affinity interaction, and whether inorganic arsenite or trivalent methylated metabolites have same or different affinities for various zinc finger configurations and structures leading to overlapping or distinct targets. A systematic analysis to compare affinities of different arsenic species for binding to the same zinc finger structures should shed light on this question. Further understanding of how arsenic interacts with the zinc finger motif and how the intracellular milieu contributes to arsenic disruption of zinc fingers and disruption of protein function are other key topics that will require additional consideration. Although interaction with zinc finger motifs is, but one of many mechanisms by which arsenic exerts toxicity, identification of additional targets may provide important links between arsenic exposure and specific adverse human health effects.

REFERENCES

[1] Agency for Toxic Substances and Disease Registry. EPA Report Toxicological Profile for Arsenic. US Department of Health and Human Services, Public Health Service Agency for Toxic Substances and Disease Registry, Atlanta, GA, 2007.

[2] A.A. Ali, G. Timinszky, R. Arribas-Bosacoma, M. Kozlowski, P.O. Hassa, M. Hassler, A.G. Ladurner, L.H. Pearl, A.W. Oliver, The zinc-finger domains of PARP1 cooperate to recognize DNA strand breaks, Nat Struct Mol Biol 19 (2012) 685–692.

[3] A.I. Anzellotti, N.P. Farrell, Zinc metalloproteins as medicinal targets, Chem Soc Rev 37 (2008) 1629–1651.

[4] H.V. Aposhian, R.A. Zakharyan, M.D. Avram, A. Sampayo-Reyes, M.L. Wollenberg, A review of the enzymology of arsenic metabolism and a new potential role of hydrogen peroxide in the detoxication of the trivalent arsenic species, Toxicol Appl Pharmacol 198 (2004) 327–335.

[5] M. Asmuss, L.H. Mullenders, A. Eker, A. Hartwig, Differential effects of toxic metal compounds on the activities of Fpg and XPA, two zinc finger proteins involved in DNA repair, Carcinogenesis 21 (2000) 2097–2104.

[6] M. Asmuss, L.H. Mullenders, A. Hartwig, Interference by toxic metal compounds with isolated zinc finger DNA repair proteins, Toxicol Lett 112–113 (2000) 227–231.

[7] F. Bertolero, G. Pozzi, E. Sabbioni, U. Saffiotti, Cellular uptake and metabolic reduction of pentavalent to trivalent arsenic as determinants of cytotoxicity and morphological transformation, Carcinogenesis 8 (1987) 803–808.

[8] D. Beyersmann, A. Hartwig, Carcinogenic metal compounds: recent insight into molecular and cellular mechanisms, Arch Toxicol 82 (2008) 493–512.

[9] J.E. Bodwell, L.A. Kingsley, J.W. Hamilton, Arsenic at very low concentrations alters glucocorticoid receptor (GR)-mediated gene activation but not GR-mediated gene repression: complex dose-response effects are closely correlated with levels of activated GR and require a functional GR DNA binding domain, Chem Res Toxicol 17 (2004) 1064–1076.

[10] J.E. Bodwell, J.A. Gosse, A.P. Nomikos, J.W. Hamilton, Arsenic disruption of steroid receptor gene activation: complex dose-response effects are shared by several steroid receptors, Chem Res Toxicol 19 (2006) 1619–1629.

[11] M.E. Bonicalzi, J.F. Haince, A. Droit, G.G. Poirier, Regulation of poly(ADP-ribose) metabolism by poly(ADP-ribose) glycohydrolase: where and when? Cell Mol Life Sci 62 (2005) 739–750.

[12] T.G. Bredfeldt, B. Jagadish, K.E. Eblin, E.A. Mash, A.J. Gandolfi, Monomethylarsonous acid induces transformation of human bladder cells, Toxicol Appl Pharmacol 216 (2006) 69–79.

[13] U. Camenisch, H. Nageli, XPA gene, its product and biological roles, Adv Exp Med Biol 637 (2008) 28–38.

[14] Y. Chervona, A. Arita, M. Costa, Carcinogenic metals and the epigenome: understanding the effect of nickel, arsenic, and chromium, Metallomics 4 (2012) 619–627.

[15] S.K. Chow, J.Y. Chan, K.P. Fung, Suppression of cell proliferation and regulation of estrogen receptor alpha signaling pathway by arsenic trioxide on human breast cancer MCF-7 cells, J Endocrinol 182 (2004) 325–337.

[16] S.K. Chow, J.Y. Chan, K.P. Fung, Inhibition of cell proliferation and the action mechanisms of arsenic trioxide (As_2O_3) on human breast cancer cells, J Cell Biochem 93 (2004) 173–187.

[17] N.J. Clark, M. Kramer, U.M. Muthurajan, K. Luger, Alternative modes of binding of poly(ADP-ribose) polymerase 1 to free DNA and nucleosomes, J Biol Chem 287 (2012) 32430–32439.

[18] S.M. Cohen, L.L. Arnold, M. Eldan, A.S. Lewis, B.D. Beck, Methylated arsenicals: the implications of metabolism and carcinogenicity studies in rodents to human risk assessment, Crit Rev Toxicol 36 (2006) 99–133.

[19] S.M. Cohen, T. Ohnishi, L.L. Arnold, X.C. Le, Arsenic-induced bladder cancer in an animal model, Toxicol Appl Pharmacol 222 (2007) 258–263.

[20] K.L. Cooper, B.S. King, M.M. Sandoval, K.J. Liu, L.G. Hudson, Reduction of arsenite-enhanced ultraviolet radiation DNA damage by supplemental zinc, Toxicol Appl Pharmacol 269 (2013) 81–88.

[21] I. Csanaky, Z. Gregus, Role of glutathione in reduction of arsenate and of gamma-glutamyltranspeptidase in disposition of arsenite in rats, Toxicology 207 (2005) 91–104.

[22] D. D'Amours, S. Desnoyers, I. D'Silva, G.G. Poirier, Poly(ADP-ribosyl)ation reactions in the regulation of nuclear functions, Biochem J 342 (Pt 2) (1999) 249–268.

[23] S.U. Dani, The arsenic for phosphorus swap is accidental, rather than a facultative one, and the question whether arsenic is nonessential or toxic is quantitative, not a qualitative one, Sci Total Environ 409 (2011) 4889–4890.

[24] J.C. Davey, J.E. Bodwell, J.A. Gosse, J.W. Hamilton, Arsenic as an endocrine disruptor: effects of arsenic on estrogen receptor-mediated gene expression in vivo and in cell culture, Toxicol Sci 98 (2007) 75–86.

[25] H. de The, M. Le Bras, V. Lallemand-Breitenbach, The cell biology of disease: acute promyelocytic leukemia, arsenic, and PML bodies, J Cell Biol 198 (2012) 11–21.

[26] M. De Vos, V. Schreiber, F. Dantzer, The diverse roles and clinical relevance of PARPs in DNA damage repair: current state of the art, Biochem Pharmacol 84 (2012) 137–146.

[27] M. Delnomdedieu, M.M. Basti, J.D. Otvos, D.J. Thomas, Reduction and binding of arsenate and dimethylarsinate by glutathione: a magnetic resonance study, Chem Biol Interact 90 (1994) 139–155.

[28] C. Demicheli, F. Frezard, F.A. Pereira, D.M. Santos, J.B. Mangrum, N.P. Farrell, Interaction of arsenite with a zinc finger CCHC peptide: evidence for formation of an As-Zn-peptide mixed complex, J Inorg Biochem 105 (2011) 1753–1758.

[29] P.J. Dilda, P.J. Hogg, Arsenical-based cancer drugs, Cancer Treat Rev 33 (2007) 542–564.

[30] W. Ding, L.G. Hudson, X. Sun, C. Feng, K.J. Liu, As(III) inhibits ultraviolet radiation-induced cyclobutane pyrimidine dimer repair via generation of nitric oxide in human keratinocytes, Free Radic Biol Med 45 (2008) 1065–1072.

[31] W. Ding, W. Liu, K.L. Cooper, X.J. Qin, P.L. de Souza Bergo, L.G. Hudson, K.J. Liu, Inhibition of poly(ADP-ribose) polymerase-1 by arsenite interferes with repair of oxidative DNA damage, J Biol Chem 284 (2009) 6809–6817.

[32] E. Dopp, A.D. Kligerman, R.A. Diaz-Bone, Organoarsenicals. Uptake, metabolism, and toxicity, Met Ions Life Sci 7 (2010) 231–265.

[33] Z. Drobna, F.S. Walton, D.S. Paul, W. Xing, D.J. Thomas, M. Styblo, Metabolism of arsenic in human liver: the role of membrane transporters, Arch Toxicol 84 (2010) 3–16.

[34] I.L. Druwe, R.R. Vaillancourt, Influence of arsenate and arsenite on signal transduction pathways: an update, Arch Toxicol 84 (2010) 585–596.

[35] T.R. Durham, E.T. Snow, Metal ions and carcinogenesis, EXS (2006) 97–130.

[36] K.E. Eblin, T.G. Bredfeldt, A.J. Gandolfi, Immortalized human urothelial cells as a model of arsenic-induced bladder cancer, Toxicology 248 (2008) 67–76.

[37] C. Escudero-Lourdes, M.K. Medeiros, M.C. Cardenas-Gonzalez, S.M. Wnek, J.A. Gandolfi, Low level exposure to monomethyl arsonous acid-induced the over-production of inflammation-related cytokines and the activation of cell signals associated with tumor progression in a urothelial cell model, Toxicol Appl Pharmacol 244 (2010) 162–173.

[38] S. Eustermann, H. Videler, J.C. Yang, P.T. Cole, D. Gruszka, D. Veprintsev, D. Neuhaus, The DNA-binding domain of human PARP-1 interacts with DNA single-strand breaks as a monomer through its second zinc finger, J Mol Biol 407 (2011) 149–170.

[39] S.J. Flora, Arsenic-induced oxidative stress and its reversibility, Free Radic Biol Med 51 (2011) 257–281.

[40] M. Genestra, Oxyl radicals, redox-sensitive signalling cascades and antioxidants, Cell Signal 19 (2007) 1807–1819.

[41] N. Goosen, Scanning the DNA for damage by the nucleotide excision repair machinery, DNA Repair (Amst) 9 (2010) 593–596.

[42] A. Hartmann, G. Speit, Effect of arsenic and cadmium on the persistence of mutagen-induced DNA lesions in human cells, Environ Mol Mutagen 27 (1996) 98–104.

[43] A. Hartwig, U.D. Groblinghoff, D. Beyersmann, A.T. Natarajan, R. Filon, L.H. Mullenders, Interaction of arsenic(III) with nucleotide excision repair in UV-irradiated human fibroblasts, Carcinogenesis 18 (1997) 399–405.

[44] A. Hartwig, M. Asmuss, I. Ehleben, U. Herzer, D. Kostelac, A. Pelzer, T. Schwerdtle, A. Burkle, Interference by toxic metal ions with DNA repair processes and cell cycle control: molecular mechanisms, Environ Health Perspect 110 (Suppl 5) (2002) 797–799.

[45] A. Hartwig, A. Pelzer, M. Asmuss, A. Burkle, Very low concentrations of arsenite suppress poly(ADP-ribosyl)ation in mammalian cells, Int J Cancer 104 (2003) 1–6.

[46] A.V. Hirner, A.W. Rettenmeier, Methylated metal(loid) species in humans, Met Ions Life Sci 7 (2010) 465–521.

[47] C. Huang, Q. Ke, M. Costa, X. Shi, Molecular mechanisms of arsenic carcinogenesis, Mol Cell Biochem 255 (2004) 57–66.

[48] M.F. Hughes, E.M. Kenyon, B.C. Edwards, C.T. Mitchell, L.M. Razo, D.J. Thomas, Accumulation and metabolism of arsenic in mice after repeated oral administration of arsenate, Toxicol Appl Pharmacol 191 (2003) 202–210.

[49] W. Humphrey, A. Dalke, K. Schulten, VMD: visual molecular dynamics, J Mol Graph 14 (1996) 33–38, 27–38.

[50] I. Iavicoli, L. Fontana, A. Bergamaschi, The effects of metals as endocrine disruptors, J Toxicol Environ Health B Crit Rev 12 (2009) 206–223.

[51] T. Ikegami, I. Kuraoka, M. Saijo, N. Kodo, Y. Kyogoku, K. Morikawa, K. Tanaka, M. Shirakawa, Solution structure of the DNA- and RPA-binding domain of the human repair factor XPA, Nat Struct Biol 5 (1998) 701–706.

[52] S. Iuchi, Three classes of C2H2 zinc finger proteins, Cell Mol Life Sci 58 (2001) 625–635.

[53] M. Javle, N.J. Curtin, The role of PARP in DNA repair and its therapeutic exploitation, Br J Cancer 105 (2011) 1114–1122.

[54] M. Jeanne, V. Lallemand-Breitenbach, O. Ferhi, M. Koken, M. Le Bras, S. Duffort, L. Peres, C. Berthier, H. Soilihi, B. Raught, H. de The, PML/RARA oxidation and arsenic binding initiate the antileukemia response of As2O3, Cancer Cell 18 (2010) 88–98.

[55] K. Jomova, Z. Jenisova, M. Feszterova, S. Baros, J. Liska, D. Hudecova, C.J. Rhodes, M. Valko, Arsenic: toxicity, oxidative stress and human disease, J Appl Toxicol 31 (2011) 95–107.

[56] R.C. Kaltreider, A.M. Davis, J.P. Lariviere, J.W. Hamilton, Arsenic alters the function of the glucocorticoid receptor as a transcription factor, Environ Health Perspect 109 (2001) 245–251.

[57] T. Kamimura, T. Miyamoto, M. Harada, K. Akashi, Advances in therapies for acute promyelocytic leukemia, Cancer Sci 102 (2011) 1929–1937.

[58] S. Kapaj, H. Peterson, K. Liber, P. Bhattacharya, Human health effects from chronic arsenic poisoning—a review, J Environ Sci Health A Tox Hazard Subst Environ Eng 41 (2006) 2399–2428.

[59] K.T. Kitchin, K. Wallace, Arsenite binding to synthetic peptides based on the Zn finger region and the estrogen binding region of the human estrogen receptor-alpha, Toxicol Appl Pharmacol 206 (2005) 66–72.

[60] K.T. Kitchin, K. Wallace, Dissociation of arsenite-peptide complexes: triphasic nature, rate constants, half-lives, and biological importance, J Biochem Mol Toxicol 20 (2006) 48–56.

[61] K.T. Kitchin, K. Wallace, The role of protein binding of trivalent arsenicals in arsenic carcinogenesis and toxicity, J Inorg Biochem 102 (2008) 532–539.

[62] K.T. Kitchin, R. Conolly, Arsenic-induced carcinogenesis—oxidative stress as a possible mode of action and future research needs for more biologically based risk assessment, Chem Res Toxicol 23 (2010) 327–335.

[63] A.D. Kligerman, C.L. Doerr, A.H. Tennant, K. Harrington-Brock, J.W. Allen, E. Winkfield, P. Poorman-Allen, B. Kundu, K. Funasaka, B.C. Roop, M.J. Mass, D.M. DeMarini, Methylated trivalent arsenicals as candidate ultimate genotoxic forms of arsenic: induction of chromosomal mutations but not gene mutations, Environ Mol Mutagen 42 (2003) 192–205.

[64] A. Klug, The discovery of zinc fingers and their applications in gene regulation and genome manipulation, Annu Rev Biochem 79 (2010) 213–231.

[65] S.S. Krishna, I. Majumdar, N.V. Grishin, Structural classification of zinc fingers: survey and summary, Nucleic Acids Res 31 (2003) 532–550.

[66] Y. Kumagai, D. Sumi, Arsenic: signal transduction, transcription factor, and biotransformation involved in cellular response and toxicity, Annu Rev Pharmacol Toxicol 47 (2007) 243–262.

[67] R. Kumar, I.J. McEwan, Allosteric modulators of steroid hormone receptors: structural dynamics and gene regulation, Endocr Rev 33 (2012) 271–299.

[68] J.H. Laity, B.M. Lee, P.E. Wright, Zinc finger proteins: new insights into structural and functional diversity, Curr Opin Struct Biol 11 (2001) 39–46.

[69] V. Lallemand-Breitenbach, J. Zhu, Z. Chen, H. de The, Curing APL through PML/RARA degradation by As_2O_3, Trends Mol Med 18 (2012) 36–42.

[70] M.F. Langelier, K.M. Servent, E.E. Rogers, J.M. Pascal, A third zinc-binding domain of human poly(ADP-ribose) polymerase-1 coordinates DNA-dependent enzyme activation, J Biol Chem 283 (2008) 4105–4114.

[71] M.F. Langelier, D.D. Ruhl, J.L. Planck, W.L. Kraus, J.M. Pascal, The Zn3 domain of human poly(ADP-ribose) polymerase-1 (PARP-1) functions in both DNA-dependent poly(ADP-ribose) synthesis activity and chromatin compaction, J Biol Chem 285 (2010) 18877–18887.

[72] M.F. Langelier, J.L. Planck, S. Roy, J.M. Pascal, Crystal structures of poly(ADP-ribose) polymerase-1 (PARP-1) zinc fingers bound to DNA: structural and functional insights into DNA-dependent PARP-1 activity, J Biol Chem 286 (2011) 10690–10701.

[73] S.F. Lee-Chen, C.T. Yu, K.Y. Jan, Effect of arsenite on the DNA repair of UV-irradiated Chinese hamster ovary cells, Mutagenesis 7 (1992) 51–55.

[74] W. Lilyestrom, M.J. van der Woerd, N. Clark, K. Luger, Structural and biophysical studies of human PARP-1 in complex with damaged DNA, J Mol Biol 395 (2010) 983–994.

[75] J.X. Liu, G.B. Zhou, S.J. Chen, Z. Chen, Arsenic compounds: revived ancient remedies in the fight against human malignancies, Curr Opin Chem Biol 16 (2012) 92–98.

[76] S. Lopez, Y. Miyashita, S.S. Simons, Jr., Structurally based, selective interaction of arsenite with steroid receptors, J Biol Chem 265 (1990) 16039–16042.

[77] A. Ludwig, B. Behnke, J. Holtlund, H. Hilz, Immunoquantitation and size determination of intrinsic poly(ADP-ribose) polymerase from acid precipitates. An analysis of the in vivo status in mammalian species and in lower eukaryotes, J Biol Chem 263 (1988) 6993–6999.

[78] N.W. Luedtke, R.J. Dexter, D.B. Fried, A. Schepartz, Surveying polypeptide and protein domain conformation and association with FlAsH and ReAsH, Nat Chem Biol 3 (2007) 779–784.

[79] J.P. Mackay, M. Crossley, Zinc fingers are sticking together, Trends Biochem Sci 23 (1998) 1–4.

[80] A. Maier, B.L. Schumann, X. Chang, G. Talaska, A. Puga, Arsenic co-exposure potentiates benzo[a]pyrene genotoxicity, Mutat Res 517 (2002) 101–111.

[81] B.K. Mandal, Y. Ogra, K. Anzai, K.T. Suzuki, Speciation of arsenic in biological samples, Toxicol Appl Pharmacol 198 (2004) 307–318.

[82] W. Maret, Exploring the zinc proteome, J Anal At Spectrom 19 (2004) 15–19.

[83] J.H. Martens, A.B. Brinkman, F. Simmer, K.J. Francoijs, A. Nebbioso, F. Ferrara, L. Altucci, H.G. Stunnenberg, PML-RARalpha/RXR alters the epigenetic landscape in acute promyelocytic leukemia, Cancer Cell 17 (2010) 173–185.

[84] B.R. Martin, B.N. Giepmans, S.R. Adams, R.Y. Tsien, Mammalian cell-based optimization of the biarsenical-binding tetracysteine motif for improved fluorescence and affinity, Nat Biotechnol 23 (2005) 1308–1314.

[85] J.M. Matthews, M. Sunde, Zinc fingers—folds for many occasions, IUBMB Life 54 (2002) 351–355.

[86] I.J. McEwan, Nuclear receptors: one big family, Methods Mol Biol 505 (2009) 3–18.

[87] J.H. Mikesch, H. Gronemeyer, C.W. So, Discovery of novel transcriptional and epigenetic targets in APL by global ChIP analyses: emerging opportunity and challenge, Cancer Cell 17 (2010) 112–114.

[88] J. Miller, A.D. McLachlan, A. Klug, Repetitive zinc-binding domains in the protein transcription factor IIIA from Xenopus oocytes, EMBO J 4 (1985) 1609–1614.

[89] E. Mitchell, S. Frisbie, B. Sarkar, Exposure to multiple metals from groundwater-a global crisis: geology, climate change, health effects, testing, and mitigation, Metallomics 3 (2011) 874–908.

[90] I. Miyamoto, N. Miura, H. Niwa, J. Miyazaki, K. Tanaka, Mutational analysis of the structure and function of the xeroderma pigmentosum group A complementing protein. Identification of essential domains for nuclear localization and DNA excision repair, J Biol Chem 267 (1992) 12182–12187.

[91] D.J. Mustra, A.J. Warren, D.E. Wilcox, J.W. Hamilton, Preferential binding of human XPA to the mitomycin C-DNA interstrand crosslink and modulation by arsenic and cadmium, Chem Biol Interact 168 (2007) 159–168.

[92] S. Oka, Y. Shiraishi, T. Yoshida, T. Ohkubo, Y. Sugiura, Y. Kobayashi, NMR structure of transcription factor Sp1 DNA binding domain, Biochemistry 43 (2004) 16027–16035.

[93] T. Okui, Y. Fujiwara, Inhibition of human excision DNA repair by inorganic arsenic and the co-mutagenic effect in V79 Chinese hamster cells, Mutat Res 172 (1986) 69–76.

[94] C.J. Park, B.S. Choi, The protein shuffle. Sequential interactions among components of the human nucleotide excision repair pathway, FEBS J 273 (2006) 1600–1608.

[95] T.J. Patterson, M. Ngo, P.A. Aronov, T.V. Reznikova, P.G. Green, R.H. Rice, Biological activity of inorganic arsenic and antimony reflects oxidation state in cultured human keratinocytes, Chem Res Toxicol 16 (2003) 1624–1631.

[96] J. Pi, B.A. Diwan, Y. Sun, J. Liu, W. Qu, Y. He, M. Styblo, M.P. Waalkes, Arsenic-induced malignant transformation of human keratinocytes: involvement of Nrf2, Free Radic Biol Med 45 (2008) 651–658.

[97] K. Piatek, T. Schwerdtle, A. Hartwig, W. Bal, Monomethylarsonous acid destroys a tetrathiolate zinc finger much more efficiently than inorganic arsenite: mechanistic considerations and consequences for DNA repair inhibition, Chem Res Toxicol 21 (2008) 600–606.

[98] L.C. Platanias, Biological responses to arsenic compounds, J Biol Chem 284 (2009) 18583–18587.

[99] E. Puccetti, M. Ruthardt, Acute promyelocytic leukemia: PML/RARalpha and the leukemic stem cell, Leukemia 18 (2004) 1169–1175.

[100] X.J. Qin, L.G. Hudson, W. Liu, G.S. Timmins, K.J. Liu, Low concentration of arsenite exacerbates UVR-induced DNA strand breaks by inhibiting PARP-1 activity, Toxicol Appl Pharmacol 232 (2008) 41–50.

[101] S.M. Quintal, Q.A. dePaula, N.P. Farrell, Zinc finger proteins as templates for metal ion exchange and ligand reactivity. Chemical and biological consequences, Metallomics 3 (2011) 121–139.

[102] J.F. Reichard, A. Puga, Effects of arsenic exposure on DNA methylation and epigenetic gene regulation, Epigenomics 2 (2010) 87–104.

[103] X. Ren, C.M. McHale, C.F. Skibola, A.H. Smith, M.T. Smith, L. Zhang, An emerging role for epigenetic dysregulation in arsenic toxicity and carcinogenesis, Environ Health Perspect 119 (2011) 11–19.

[104] T.G. Rossman, Enhancement of UV-mutagenesis by low concentrations of arsenite in *E. coli*, Mutat Res 91 (1981) 207–211.

[105] T.G. Rossman, C.B. Klein, Genetic and epigenetic effects of environmental arsenicals, Metallomics 3 (2011) 1135–1141.

[106] T. Sakurai, S. Himeno, Endocrine disruptive effects of inorganic arsenicals, Environ Sci 13 (2006) 101–106.

[107] K. Salnikow, A. Zhitkovich, Genetic and epigenetic mechanisms in metal carcinogenesis and cocarcinogenesis: nickel, arsenic, and chromium, Chem Res Toxicol 21 (2008) 28–44.

[108] V. Schreiber, F. Dantzer, J.C. Ame, G. de Murcia, Poly(ADP-ribose): novel functions for an old molecule, Nat Rev Mol Cell Biol 7 (2006) 517–528.

[109] T. Schwerdtle, I. Walter, A. Hartwig, Arsenite and its biomethylated metabolites interfere with the formation and repair of stable BPDE-induced DNA adducts in human cells and impair XPAzf and Fpg, DNA Repair (Amst) 2 (2003) 1449–1463.

[110] N. Scott, K.M. Hatlelid, N.E. MacKenzie, D.E. Carter, Reactions of arsenic(III) and arsenic(V) species with glutathione, Chem Res Toxicol 6 (1993) 102–106.

[111] S.M. Shell, Y. Zou, Other proteins interacting with XP proteins, Adv Exp Med Biol 637 (2008) 103–112.

[112] H. Shi, L.G. Hudson, K.J. Liu, Oxidative stress and apoptosis in metal ion-induced carcinogenesis, Free Radic Biol Med 37 (2004) 582–593.

[113] H. Shi, X. Shi, K.J. Liu, Oxidative mechanism of arsenic toxicity and carcinogenesis, Mol Cell Biochem 255 (2004) 67–78.

[114] M. Shrivastav, L.P. De Haro, J.A. Nickoloff, Regulation of DNA double-strand break repair pathway choice, Cell Res 18 (2008) 134–147.

[115] S.S. Simons, Jr., P.K. Chakraborti, A.H. Cavanaugh, Arsenite and cadmium(II) as probes of glucocorticoid receptor structure and function, J Biol Chem 265 (1990) 1938–1945.

[116] V. Stanisic, D.M. Lonard, B.W. O'Malley, Modulation of steroid hormone receptor activity, Prog Brain Res 181 (2010) 153–176.

[117] J.C. States, A. Barchowsky, I.L. Cartwright, J.F. Reichard, B.W. Futscher, R.C. Lantz, Arsenic toxicology: translating between experimental models and human pathology, Environ Health Perspect 119 (2011) 1356–1363.

[118] M. Styblo, L.M. Del Razo, L. Vega, D.R. Germolec, E.L. LeCluyse, G.A. Hamilton, W. Reed, C. Wang, W.R. Cullen, D.J. Thomas, Comparative toxicity of trivalent and pentavalent inorganic and methylated arsenicals in rat and human cells, Arch Toxicol 74 (2000) 289–299.

[119] K. Tanaka, N. Miura, I. Satokata, I. Miyamoto, M.C. Yoshida, Y. Satoh, S. Kondo, A. Yasui, H. Okayama, Y. Okada, Analysis of a human DNA excision repair gene involved in group A xeroderma pigmentosum and containing a zinc-finger domain, Nature 348 (1990) 73–76.

[120] Z. Tao, P. Gao, D.W. Hoffman, H.W. Liu, Domain C of human poly(ADP-ribose) polymerase-1 is important for enzyme activity and contains a novel zinc-ribbon motif, Biochemistry 47 (2008) 5804–5813.

[121] E.J. Tokar, L. Benbrahim-Tallaa, J.M. Ward, R. Lunn, R.L. Sams, 2nd, M.P. Waalkes, Cancer in experimental animals exposed to arsenic and arsenic compounds, Crit Rev Toxicol 40 (2010) 912–927.

[122] E.J. Tokar, W. Qu, M.P. Waalkes, Arsenic, stem cells, and the developmental basis of adult cancer, Toxicol Sci 120 (Suppl 1) (2011) S192–S203.

[123] M. Valko, C.J. Rhodes, J. Moncol, M. Izakovic, M. Mazur, Free radicals, metals and antioxidants in oxidative stress-induced cancer, Chem Biol Interact 160 (2006) 1–40.

[124] I. Walter, T. Schwerdtle, C. Thuy, J.L. Parsons, G.L. Dianov, A. Hartwig, Impact of arsenite and its methylated metabolites on PARP-1 activity, PARP-1 gene expression and poly(ADP-ribosyl)ation in cultured human cells, DNA Repair (Amst) 6 (2007) 61–70.

[125] T.C. Wang, J.S. Huang, V.C. Yang, H.J. Lan, C.J. Lin, K.Y. Jan, Delay of the excision of UV light-induced DNA adducts is involved in the coclastogenicity of UV light plus arsenite, Int J Radiat Biol 66 (1994) 367–372.

[126] W.H. Watson, J.D. Yager, Arsenic: extension of its endocrine disruption potential to interference with estrogen receptor-mediated signaling, Toxicol Sci 98 (2007) 1–4.

[127] A. Witkiewicz-Kucharczyk, W. Bal, Damage of zinc fingers in DNA repair proteins, a novel molecular mechanism in carcinogenesis, Toxicol Lett 162 (2006) 29–42.

[128] S.M. Wnek, C.L. Kuhlman, J.M. Camarillo, M.K. Medeiros, K.J. Liu, S.S. Lau, A.J. Gandolfi, Interdependent genotoxic mechanisms of monomethylarsonous acid: role of ROS-induced DNA damage and poly(ADP-ribose) polymerase-1 inhibition in the malignant transformation of urothelial cells, Toxicol Appl Pharmacol 257 (2011) 1–13.

[129] R.D. Wood, Mammalian nucleotide excision repair proteins and interstrand crosslink repair, Environ Mol Mutagen 51 (2010) 520–526.

[130] J.W. Yager, J.K. Wiencke, Enhancement of chromosomal damage by arsenic: implications for mechanism, Environ Health Perspect 101 (Suppl 3) (1993) 79–82.

[131] X.W. Zhang, X.J. Yan, Z.R. Zhou, F.F. Yang, Z.Y. Wu, H.B. Sun, W.X. Liang, A.X. Song, V. Lallemand-Breitenbach, M. Jeanne, Q.Y. Zhang, H.Y. Yang, Q.H. Huang, G.B. Zhou, J.H. Tong, Y. Zhang, J.H. Wu, H.Y. Hu, H. de The, S.J. Chen, Z. Chen, Arsenic trioxide controls the fate of the PML-RARalpha oncoprotein by directly binding PML, Science 328 (2010) 240–243.

[132] L. Zhao, S. Chen, L. Jia, S. Shu, P. Zhu, Y. Liu, Selectivity of arsenite interaction with zinc finger proteins, Metallomics 4 (2012) 988–994.

[133] X. Zhou, X. Sun, K.L. Cooper, F. Wang, K.J. Liu, L.G. Hudson, Arsenite interacts selectively with zinc finger proteins containing C3H1 or C4 motifs, J Biol Chem 286 (2011) 22855–22863.

[134] X. Zhou, X. Sun, C. Mobarak, A.J. Gandolfi, S.W. Burchiel, L.G. Hudson, K.J. Liu. Differential binding of monomethylarsonous acid compared to arsenite and arsenic trioxide with zinc finger peptides and proteins. Chem Res Toxicol 27(4) (2014) 690–698.

14

ROLE IN CHEMOTHERAPY

Koren K. Mann and Maryse Lemaire
Department of Oncology, Lady Davis Institute for Medical Research, McGill University, Montreal, QC, Canada

14.1 INTRODUCTION

Although arsenic compounds are well-known carcinogens, arsenicals have been evaluated for the treatment of various diseases for more than 2000 years [69]. In Chinese traditional medicine, arsenic was administered in various formats against several cancers: as a topical form against breast and skin cancers, as well as oral or intravenous formulations for leukemia. Paracelsus (1493–1541 A.D.) may have used realgar (As_4S_4) to treat cancer patients [64]. In the eighteenth century, Fowler's solution (1% potassium arsenite) was commonly used orally to treat a number of diseases, including Hodgkin's lymphoma and leukemia [107]. From 1700s to 1900s, arsenic was indeed the main treatment for all forms of leukemia. In the 1930s, arsenic was the treatment of choice for chronic myeloid leukemia (CML), until alkylating antineoplastic agents and radiotherapy were introduced in the 1950s [144]. Because of concerns about its toxicity, arsenical use was then discontinued. Arsenic use was again renewed in the 1970s when several studies described impressive results in acute promyelocytic leukemia (APL) patients, a population of patients who, at the time, had a dismal prognosis [43, 139, 142]. Ultimately, this led to the approval of arsenic trioxide for the treatment of relapsed and refractory APL by several governmental agencies, including the United States Food and Drug Administration in 2000. Even though the mention of arsenic tends to generate apprehension, its use has significantly improved patient survival outcomes. However, it is not completely

Arsenic: Exposure Sources, Health Risks, and Mechanisms of Toxicity, First Edition.
Edited by J. Christopher States.
© 2016 John Wiley & Sons, Inc. Published 2016 by John Wiley & Sons, Inc.

understood how arsenic achieves its clinical effectiveness. In addition, the use of arsenic has been restricted to APL, although a significant number of investigations have the potential to broaden the clinical setting for arsenic trioxide and develop new arsenicals with enhanced efficacy.

14.2 INORGANIC ARSENICALS COMPOUNDS AS CHEMOTHERAPEUTIC DRUGS

Arsenic exists in several states in nature, but the majority of its applications in cancer are as arsenic oxides. Arsenic trioxide (As_2O_3, ATO, diarsenic oxide, arsenolite, Trisenox®) and tetraarsenic oxide (As_4O_6, TAO) are inorganic arsenic compounds that have both shown antitumor properties, although the mechanisms have yet to be fully elucidated.

14.2.1 Arsenic Trioxide

ATO has extensive antitumor activity both *in vitro* and *in vivo*. This compound has been used in Chinese traditional medicine for centuries, and it is still the most widely used arsenical against cancer, particularly against leukemia. ATO is approved for use in acute promyelocytic leukemia (APL), but may have potential in other malignancies as well.

14.2.1.1 Antitumor Properties of Arsenic Trioxide in APL APL is the M3 subtype of acute myeloid leukemia (AML), a cancer of the myeloid lineage of white blood cells. This rare malignant disease is characterized by the accumulation of abnormal promyelocytes both in the bone marrow and in the blood. It was first described in the 1950s [54] and had a 0% 5-year survival rate until the elucidation of the key pathogenic events in the 1970s. APL is characterized by specific chromosomal translocations, all of which involve the retinoic acid receptor α (RARα) gene, resulting in the creation of several fusion proteins [122].

RARs are important ligand-inducible transcription factors that control the expression of target genes involved in normal myeloid differentiation. In the vast majority of APL cases, at (15;17) translocation is involved, which results in a fusion between the promyelocytic leukemia (PML) gene and the *RARα* gene [32, 132]. The resulting PML–RARα fusion protein binds to DNA in the promoters of RAR target genes with high affinity, and recruits repressors of transcription, such as nuclear co-repressor molecule (NCoR) and histone deacetylases (HDAC) [193]. As a result, it blocks normal myelocyte differentiation, and it is consequently a key factor in APL leukemogenesis [13, 32] (Fig. 14.1). In addition to disrupting RAR-signaling, PML is disrupted. In normal differentiated cells, PML homodimers form via disulfide bonds and localize within nuclear bodies [61]. Studies have shown that PML-RARα disrupts the normal PML nuclear bodies, resulting in sparse smaller fragments [19, 42], which may also have a driver role in APL. PML can prevent initiation and progression of cancers [166]. In fact, PML normally interacts with p53, a tumor-suppressor

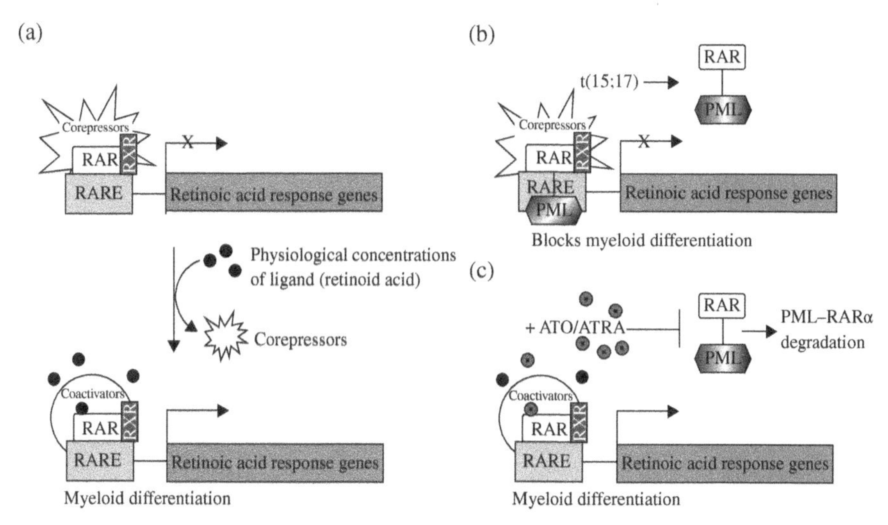

FIGURE 14.1 Consequences of promyelocytic leukemia (PML)–retinoic acid receptor α (RARα) fusion protein on RARα-mediated granulocytic differentiation. (a) In the absence of the endogenous ligand all-*trans* retinoic acid (ATRA), co-repressors bind the RARα/retinoid X receptor (RXR) heterodimers and inhibit gene transcription. Upon binding of ATRA, co-repressors are released and co-activators are recruited, resulting in transcription of target genes and subsequent granulocytic differentiation. (b) The t(15;17) translocation results in a fusion between the PML gene and the RARα gene. The PML–RARα fusion protein binds to RARα target gene promoters and recruits co-repressors, inhibiting transcription and subsequent myeloid differentiation. This occurs even in the presence of physiologic concentrations of retinoic acid. (c) The addition of pharmacologic concentrations of ATRA and/or arsenic trioxide (ATO) triggers PML–RARα degradation. ATRA binds the RARα moiety, while ATO induces PML–RARα degradation via the PML moiety. Moreover, following PML/RARα degradation, wild-type RARα can initiate transcription of the target genes and myeloid differentiation.

known to induce apoptosis upon its accumulation. Cells that do not express PML are unable to induce p53 acetylation and phosphorylation with subsequent apoptosis [51, 55, 119]. Consequently, PML-RARα negatively regulates the function of both partners [105] (Fig. 14.2).

In 70% of patients, conventional APL chemotherapy with daunorubicin and cytarabine achieves complete remission [66, 147], which is characterized by the disappearance of leukemic cells in bone marrow and peripheral blood [112]. However, these complete remissions are rarely sustained. The introduction of all-*trans* retinoic acid (ATRA) in APL treatment dramatically changed the prognosis, increasing the remission rate to 90%. ATRA is arguably the first example of differentiation therapy, because it causes the accumulated promyelocytes to differentiate into granulocytes [44, 135, 147, 167]. At pharmacologic concentrations, ATRA binds to PML-RARα, dissociates the NCoR–HDAC-repressive complex and allows transcription of the target genes, resulting in blast differentiation and apoptosis of the APL cells. Also, by degrading PML-RARα, ATRA allows the PML proteins produced from the normal allele to reform the nuclear bodies [194]. However, patients can become

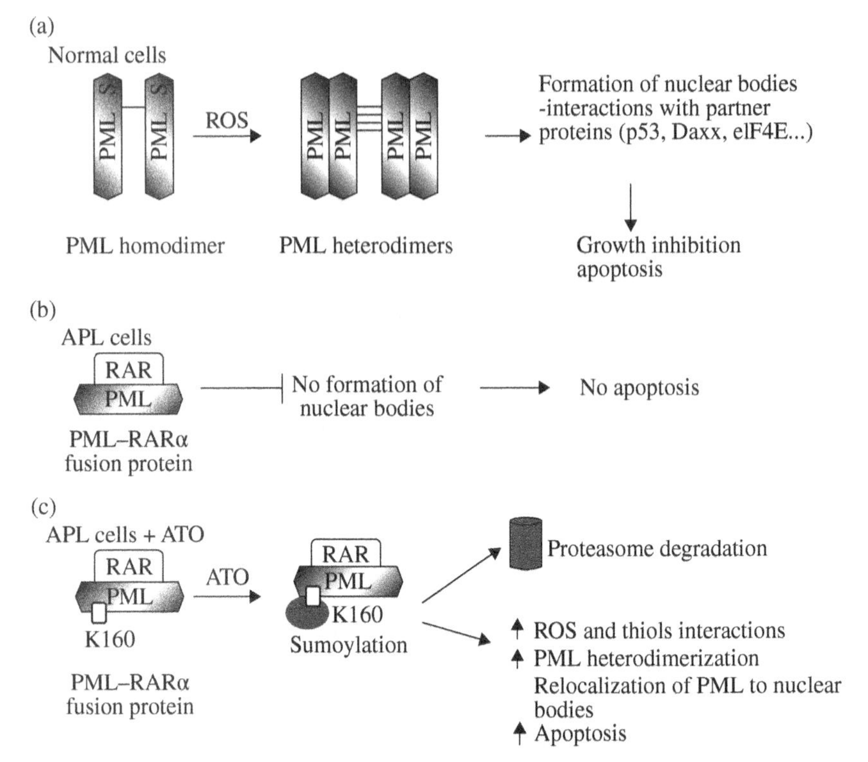

FIGURE 14.2 Consequences of promyelocytic leukemia (PML)–retinoic acid receptor α (RARα) fusion protein expression on PML function. (a) In normal cells, the PML protein interacts with partners to inhibit cell proliferation and to induce apoptosis. Under oxidative stimuli, PML heterodimerizes and forms the core of nuclear bodies, where it generates its effects. (b) In APL cells, PML–RARα oncoprotein disrupts PML localization and abrogates normal PML functions. (c) Arsenic trioxide (ATO) induces sumoylation of PML on K160, which triggers the oncoprotein degradation through the proteasome. ATO also produces reactive oxygen species (ROS), which favors homodimerization of the normal PML proteins produced from the nontranslocated allele by cross-linking PML disulfide bonds. ATO may also physically interact with the PML through the thiol group and directly trigger its dimerization. As a result, nuclear bodies will re-form and PML functions will be restored.

resistant to ATRA and relapse after the therapy [96]. Fortunately, these patients are likely to be sensitive to ATO.

14.2.1.2 Clinical Studies As a single agent, ATO prolongs the survival of refractory/relapsed APL patients [47]. In clinical trials [18], 70–75% of APL patients achieved a complete remission. Moreover, ATO treatment induces complete remission in 82% of patients previously treated with ATRA, who relapsed less than 1 year after their treatment, and in 55% of patients who had relapsed more than 1 year after ATRA therapy. The approved standard regimen consists of one daily ATO intravenous administration for up to a maximum of 60 days [18], followed by a 2-year consolidation therapy protocol. Consolidation treatment is given after disappearance of APL cells following the initial therapy to kill any remaining cells, including leukemia-initiating cells. ATO treatment indeed extends the survival of refractory/relapsed APL patients, but is remarkably more efficient in combination with ATRA, reaching a 91% 5-year overall survival [75, 140, 168]. It is also now well established that ATO can also achieve complete remission in newly diagnosed APL patients. Clinical studies suggest that patients from all risk groups could benefit from ATO as induction therapy instead of reserving it for use at relapse [140].

The major ATO-associated side effects are hepatotoxicity and cardiovascular toxicities, which can limit its clinical use [57, 84]. For instance, ATO can alter APL patient cardiac function by delaying cardiac repolarization, represented by a prolongation of the QT interval on the electrocardiogram and of the QT heart rate corrected (QTc) interval [8]. This latter observation may be due to ATO-induced TGFβ1 secretion from cardiac fibroblasts, which has been reported to accentuate long QT syndrome [26]. As a result, it can cause fatal torsade de pointes [116]. Another limiting side effect of ATO is the potential development of a life-threatening syndrome called the "APL differentiation syndrome." During differentiation therapy with either ATRA or ATO, a massive release of inflammatory molecules, or "cytokine storm," is associated with promyelocyte differentiation causing fever, edema, renal and hepatic dysfunction, and pleural and pericardial effusions [128]. Differentiation syndrome occurs in 25% of APL patients and can be aggressively managed.

14.2.1.3 ATO Import and Export Arsenic efficacy is dependent on its capacity to enter the target cells. Cellular uptake of arsenic is known to involve aquaporins 3, 7, and 9 [89, 110] and the glucose/fructose transporters 1 and 5 (GLUT1 and GLUT5). Inhibition of aquaporin 9 significantly reduces cellular accumulation of arsenic [184]. Interestingly, inhibitors of GLUT5 abrograte arsenic uptake and cellular accumulation [14], while specific GLUT1 inhibitors that are known to block glucose uptake do not modify intracellular concentrations of arsenic, suggesting the use of GLUT1 by arsenic through a different pathway from glucose [63]. This suggestion is supported by data showing that mutation on specific GLUT1 residues that decrease glucose uptake leads to increased uptake of arsenic [63].

Once inside the cell, arsenic is modified through a series of oxidation/reduction and methylation reactions. All the resulting arsenicals can react with thiol groups of cysteine-containing molecules, such as glutathione (GSH), to generate thiol-arsenical complexes [136]. Arsenic treatment results in several glutathionylated-arsenic species that are targeted for export, such as arsenic triglutathione, methylarsonous diglutathione, and dimethylarsinous GSH [121]. Cellular export takes place through drug exporter proteins, namely ATP-binding cassette, subfamily C (CFTR/MRP), member 1 (ABCC1). ABCC1 exports exclusively glutathionylated substrates from the cell and efficiently transports glutathionylated arsenicals. Overexpressing ABCC1 prevents ATO antitumor activity, while inhibition of ABCC1 enhances ATO-induced cell apoptosis [37, 100].

14.2.1.4 Mechanism of Action: Differentiation Versus Apoptosis In contrast to ATRA that primarily targets the PML-RARα fusion protein (Fig. 14.1), ATO-induced antitumor properties likely involve several mechanisms (Fig. 14.3). Like ATRA, ATO induces both APL cell differentiation through degradation of the PML-RARα oncoprotein [20]. In addition, ATO activates intracellular signal transduction pathways resulting in growth inhibition, cell death, and anti-angiogenesis. ATO treatment is also known to induce reactive oxygen species (ROS) [120] and to arrest growth of APL cells, although the stage of cell cycle is controversial [81]. In our experience, at clinically achievable ATO concentrations, APL cells are arrested in the G0/G1 transition of cell cycle (Koren Mann, Wilson Miller, unpublished data).

Differentiation While ATRA targets the RARα moiety of the PML-RARα to induce differentiation, ATO targets the PML-RARα protein for degradation through the PML moiety [20, 195] (Fig. 14.2). ATO-induced PML-RARα degradation is

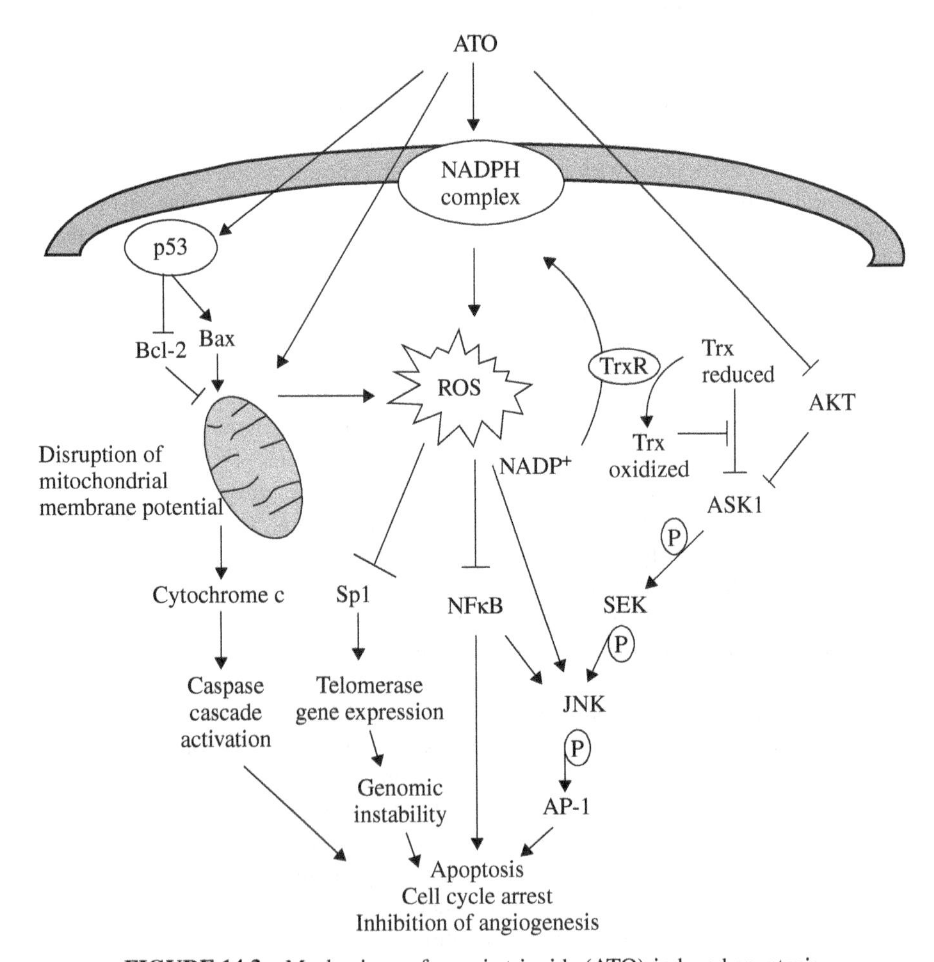

FIGURE 14.3 Mechanisms of arsenic trioxide (ATO)-induced apoptosis.

dependent on sumoylation and protein degradation. Sumoylation of the PML lysine 160 (K160) is required for recruitment to the nuclear bodies of PML partner proteins [109]. Sumoylation of K160 in the PML, both wild-type and PML/RARα, is necessary for 11S proteasome regulatory component recruitment, which will eventually trigger PML-RARα degradation [76, 149]. ATO exposure increases PML sumoylation on K160 [33], which leads to PML-RARα degradation. Additionally, ATO-induced oxidative stress promotes homodimerization of the normal PML proteins that are produced from the nontranslocated allele by cross-linking PML disulfide bonds. Moreover, arsenic may physically interact with the PML through the thiol group on cysteine residues and directly trigger its dimerization. As a result, nuclear bodies will reform, and PML will associate to the nuclear matrix [33, 61]. The efficacy of ATO in APL is linked to the degradation of the PML-RARα, because in the 1–2% of APL patients with the t(11;17) chromosomal translocation generating the PLZF-RARα protein fusion, ATO has no effect [71].

Apoptosis In addition to inducing differentiation in APL cells, ATO triggers apoptosis in APL cells, as well as in other cell types. ATO-induced apoptosis likely involves signaling via several pathways and an exhaustive list has yet to be generated. There is evidence that ATO activates extrinsic Fas-dependent, Daxx-mediated apoptosis [153], but also that it can directly target the mitochondrial membrane [20, 73, 78, 107] and inhibit telomerase activity by the loss of Sp1 transcriptional regulation [23, 24]. Here, we will discuss the major pathways involved: p53, redox/GSH, stress kinases (MAPK), and NFκB (Fig. 14.3).

P53 As discussed earlier, PML interacts with the nuclear matrix and localizes within nuclear bodies, where it modulates p53 signaling to induce apoptosis under particular conditions. The presence of PML-RARα suppresses p53 signaling, which consequently has a role in APL biogenesis by preventing cell death [59, 119]. ATO treatment of cancer cells mediates p53 accumulation, which results in caspase activation and apoptosis. However, there are contradictory data for whether p53 is required for ATO-induced death. Cells expressing wild-type or mutant p53 are sensitive to ATO-mediated cytotoxicity, but may utilize different death pathways [86].

REDOX HOMEOSTASIS In APL, PML/RARα degradation is an important component of ATO efficacy, but it is not sufficient for its antitumor activity. For instance, arsenic treatment still degrades the PML/RARα oncoprotein in arsenic-resistant APL cells [30, 56, 74, 107]. Thus, ATO sensitivity seems to be dependent on other intracellular factors, such as ROS production, GSH levels, and GSH-dependent antioxidant enzymes [106]. Arsenicals, including ATO, are known to generate ROS. ATO activates the membrane nicotinamide adenine dinucleotide phosphate (NADPH) oxidase complex *in vitro* [11, 156], resulting in ROS generation, mitochondrial membrane potential disruption, and subsequent caspase pathway activation [21, 120, 156]. Moreover, ROS production results in mitochondrial DNA damage and ATO-induced apoptosis [80]. Intracellular antioxidants, such as GSH, protect the cells against excess ROS production. ROS trigger GSH oxidation, which will be reduced by GSH-reductase using NAPDH as electron donor [114]. Thus, GSH-regulating proteins are involved in scavenging ROS and maintaining the cellular redox homeostasis [25]. Therefore, by preventing ROS-induced damage, high intracellular GSH levels may impede ATO efficacy [28], in addition to their role in enhancing cellular export of arsenic, as discussed earlier. Indeed arsenic-resistant APL cells have high GSH levels [28, 30]. Addition of *N*-acetylcysteine, a compound that increases intracellular GSH levels, blocks the ATO-induced leukemic cells toxicity, in part, through inhibition of ROS production [25]. In addition to GSH, several other enzymes regulate the cellular redox status including thioredoxin (Trx) reductase, which control redox reactions and also protect cells from oxidative damage [56, 74, 107]. ATO inhibits Trx reductase activity, resulting in cellular growth inhibition, which can contribute to ATO cytotoxicity [92].

STRESS KINASE PATHWAYS As discussed before, ATO has many cellular effects that may interact with one another. For example, ATO-induced ROS leads to activation of mitogen-activated protein kinase (MAPK) pathways in a dose- and cell-specific manner [145]. c-Jun N-terminal kinase (JNK) is a member of the MAPK family, which is activated by ATO. When activated, JNK phosphorylates members of the activating protein 1 (AP1) transcription factor complex, as well as other proteins known to be involved in regulating cell death [31, 145]. ATO activates JNK as a result of its phosphorylation by the upstream SEK (MAP2K4 *aka* MKK4) signaling cascade [31] and through the inhibition of JNK phosphatases [17]. Importantly, ROS production and subsequent JNK activation are required for ATO-induced death in APL cells [31]. In contrast, p38 MAPK may provide survival signals and inhibition of p38 indeed increases ATO-induced JNK apoptosis and differentiation of APL cells [39, 48, 65, 93, 156]. Thus, utilization of a mitogen-activated protein kinase kinase 1 (MAP2K1; *aka* MEK1) inhibitor in combination with ATO could increase its antileukemic activity. This suggests that p38 and JNK are opposing kinases, and only when JNK activation prevails does apoptosis ensue.

There is also a link between glutathione pathways and MAPK pathways. The GSH transferase P1-1 (GSTP1-1) is involved in ATO-induced cell death by regulating JNK activation. Upon increased intracellular ROS, GSTP1-1 oligomerizes and dissociates from JNK, allowing its activation [138]. Accordingly, JNK activation also can be limited when JNK is complexed with GSH and GSTP1-1 [9, 138]. ATO exposure of APL cells results in concomitant GSTP1-1 polymerization and JNK activation, both leading to cell death [1, 10].

Less is known about the upstream molecules that activate JNK. Mitogen-activated protein kinase kinase kinase 5 (MAP3K5 aka apoptosis stimulating kinase 1, ASK1) is a MAP3K upstream of mitogen-activated protein kinase kinase (MAP2K4; aka SEK1, MKK4) and JNK. In APL cells, ASK1 inhibits ATO-induced apoptosis [176], while ASK1$^{-/-}$ fibroblasts are protected from ATO-induced death (Koren Mann, unpublished data). It has been suggested that ATO may activate ASK1 through oxidation of thioredoxin (Trx), an inhibitor of ASK1 [134]. The oxidized form of Trx cannot bind to ASK1 to inhibit its kinase activity; so as a result, ASK1 is activated [134]. Further studies are required to determine definitively the role of ASK1 in ATO-induced activation and then to evaluate the potential of ASK1 inhibitors in combination with ATO.

Activation of phosphatidylinositol 3-kinase (PI3K) leads to the production of phosphatidylinositol (3,4,5)-triphosphate (PIP3)and to subsequent activation of AKT and mTOR, which are important players in cell survival [157]. Importantly, AKT inhibits MAPK and JNK activation [12]. ATO not only triggers cell death but also inhibits cell survival signals. ATO decreases AKT activity, and sometimes AKT expression, while inducing, JNK-dependent apoptosis [95]. Leukemic cells expressing constitutive PI3K/AKT are resistant to ATO-induced apoptosis [146]. Moreover, in APL cells, inhibition of this pathway increases ATO-induced apoptosis [68].

NUCLEAR FACTOR-κB The transcription factor nuclear factor-kappa B (NFκB) is repressed by the inhibitory-κB protein (IκB), which can be activated by environmental stressors. IκB protein is phosphorylated by IκB kinase (IKK), causing its dissociation from NFκB, which can then translocate to the nucleus and regulate multiple inflammatory target genes that play a role in pathogenesis of hematologic malignancies [68]. *In vitro* arsenic exposure decreases NFκB activation and blunts the inflammatory response. Indeed, high arsenic concentrations inhibit IKK by binding to cysteine 179 located in the catalytic activation loop, which consequently will inhibit NFκB release [67]. At clinically relevant concentrations, there is evidence that ATO represses NFκB. For instance, ATO inhibits tumor necrosis factor alpha (TNFα), PMA and GM-CSF-induced NFκB activation in human cancer cells lines [67, 130]. As a result of NFκB inhibition, proinflammatory responses are reduced and cells may become more sensitive to apoptosis [67]. Additionally, ATO-mediated NFκB inhibition diminished survival protein expression [79]. Moreover, NFκB activation is linked with JNK repression. ATO treatment is able to activate JNK in leukemic cell lines through NFκB inhibition, which leads to apoptosis [101].

14.2.1.5 Effects of ATO on Leukemia-Initiating Cells Leukemia-initiating cells (LICs) resemble stem cells; they are progenitor cells, are able to renew themselves, and are in a quiescent state. This latter point allows them to escape chemotherapy targeting proliferating cells. Importantly, LIC clearance is required to achieve long-term remission. In APL, the PML-RARα oncoprotein is required for LIC self-renewal [111]. PML modulates p53 signaling, which in turn, controls LIC quiescence and self-renewal capacities [60, 88, 189]. Thus, PML/RARα inhibits wild-type PML regulation of p53 and its control of LIC. ATO treatment, and its combination with ATRA, induces LIC eradication in APL mice [111, 191]. While relieving the transcriptional block imposed by PML-RARα is enough to differentiate APL blasts, PML-RARα degradation is required for LIC clearance and disease eradication [111]. ATO may also be important for LIC clearance from other types of malignancies. For instance, in the CML mouse model, the PML moiety is required for LIC maintenance, and ATO has been shown to decrease PML expression [60].

14.2.1.6 Potential of ATO in Other Malignancies Promising *in vitro* and *in vivo* observations suggest that ATO, because it mediates effects through several pathways, could be used in other cancer types. Many studies have been performed in a variety of solid tumors and hematological malignancies that show some efficacy. ATO treatment is also reported to be effective in a rat liver cancer model [148], but ATO has failed to show clinical benefit in a trial for hepatocellular carcinoma [82]. Similar negative results are reported with phase II trial with pancreatic cancer [3] and with renal cell carcinoma patients [158]. However, patients with advanced hormone-refractory prostate cancer markedly reduced their prostate-specific antigen levels under ATO treatment [45]. In gliomas and cervical tumor cells, ATO was able to reduce the cancer stem cell population [190]. Current phase II clinical trials are testing ATO in the settings of gastric, ovarian, bladder, neuroblastoma, lung and breast cancer, gliomas, and cervical cancer and results are pending (*Clinicaltrial.gov*). However, as of yet, no results indicate that ATO will have significant antitumor effects in solid tumors as a single agent.

In contrast to solid tumors, hematologic malignancies may be more sensitive to the effects of ATO. ATO treatment in myelodysplastic syndrome, AML or CML patients expressing the AML1/MDS1/EVI1 fusion protein has been shown to be effective. Sensitivity may occur through ATO-induced degradation of the fusion oncoprotein, and thus ATO could be a valuable pharmacologic compound to use in this setting [137].

Historically, ATO therapy was commonly used in CML patients, but inhibitors of tyrosine kinases are today's first line of treatment. Imatinib (Gleevec®), an inhibitor of tyrosine kinase receptors [182, 187], targets the constitutively active tyrosine kinase fusion protein, BCR-ABL, which is characteristic of CML [129]. However, a significant proportion of patients in the accelerated blast crisis will fail treatment. ATO has been shown to reduce BCR-ABL levels and can induce apoptosis through a BCR-ABL-independent pathway [123, 124]. Thus, combination therapy with Gleevec was evaluated. Hematologic response was seen in a few patients with blast or chronic phase CML [127].

Combination Therapy The chemotherapeutic use of ATO is limited in malignancies other than APL, mainly because of the cardiovascular toxicity seen at the higher effective doses required for antitumor effects in these settings. However, combination therapies with ATO are currently being evaluated and identification of an effective combination could allow the use of lower doses of ATO to achieve antitumor effects without excessive toxicities.

BUTHIONINE SULFOXIMINE As discussed earlier, GSH levels are key modulators of ATO-induced cytotoxicity and increased GSH levels protect against ATO-induced ROS production and death. Therefore, approaches to decrease GSH levels were studied in an effort to enhance sensitivity to ATO-induced apoptosis. Administration of the antioxidant buthionine sulfoximine (BSO), an inhibitor of GSH synthesis, sensitizes APL cells to ATO-induced growth inhibition and apoptosis [30, 196]. Interestingly, BSO can also increase sensitivity in ATO-resistant cell lines [30, 46].

However, the combination of ATO and BSO has not been developed successfully for clinical use, because there were manufacturing problems and BSO enhanced myelotoxicity (Gerald Batist, personal communication).

ASCORBIC ACID Although known as an antioxidant, ascorbic acid can also be a pro-oxidant. Like BSO, ascorbic acid decreases GSH levels and consequently, influences the cellular redox system. Moreover, ascorbic acid has oxidative properties when combined with ATO, which potentiate ROS production and ATO-mediated apoptosis [28]. *In vitro* and *in vivo* co-treatment of ATO with ascorbic acid significantly increased cytotoxicity in APL and multiple myeloma [5, 6, 28, 50]. Ascorbic acid inhibits the activation of IKK *in vitro*, which restrains NFκB functions, in a manner similar to ATO [5, 28, 50]. Thus, ascorbic acid and ATO work cooperatively to decrease NFκB-mediated survival signals. One clinical study has evaluated the efficacy of ATO administration in conjunction with ascorbic acid in heavily pretreated multiple myeloma patients, and modest responses were achieved (two partial response and four stable disease) [6]. Additional trials are underway in AML, non-APL, and results are pending (*clinicaltrials.gov*).

TROLOX Trolox (6-hydroxy-2,5,7,8-tetramethylchroman-2-carboxylic acid) is a hydrophilic vitamin E derivative that can act as an antioxidant [29] but, like ascorbic acid, has pro-oxidant properties when used in conjunction with ATO [35]. Interestingly, this compound increases ATO-induced apoptosis *in vitro*, while it confers additional protection in normal cells [35, 36]. Indeed, it reduces mitochondrial membrane potential and induces release of the cytochrome c in lymphoma cells; while *in vivo*, it decreases ATO-induced toxicity and markers of oxidative stress in normal liver [36]. Moreover, trolox increases ATO sensitivity in ATO-resistant cell lines [35]. The addition of trolox to ATO treatment in mice transplanted with lymphoma, increases their lifespan and decreases tumor-metastasis significantly more than ATO alone [36]. Taken together, these observations suggest that trolox might be a valuable compound to potentiate ATO effects, while protecting normal tissue, but further experiments are required to evaluate the clinical efficacy of this combination.

HYPERTHERMIA Ovarian cancer patients typically have a bad prognosis, and acquired resistance to the standard cisplatin treatment is common [62, 117]. A newer investigational therapy in ovarian cancer is that of hyperthermia [53, 155]. Both hyperthermia and arsenic sensitize tumor cells to cisplatin by preventing resistance-associated mechanisms. The combination of sodium arsenite and hyperthermia has demonstrated potential antitumor activity against epithelial ovarian cancer cells. Arsenic [113] and arsenic with hyperthermia [108] inhibit nucleotide excision DNA repair protein XPC, induce expression of the mismatch DNA repair protein MSH2 (associated with cisplatin sensitivity), and increase ovarian cell uptake of cisplatin in a xenograft mouse model of ovarian cancer. This latter observation has been proposed to result from competition between arsenic and cisplatin for the same drug exporter and detoxification pathway, leading to intracellular cisplatin accumulation.

In vitro studies show that the inhibition of XPC expression by arsenic co-treatment is dependent on presence of p53 [124].

MITOTIC ARREST Many investigators have used sodium arsenite for *in vitro* studies rather than ATO. Chemically, these agents are both converted to arsenous acid and exist as $As(OH)_3$ in neutral solution [126]. Arsenite has long been known to disrupt mitosis; but in normal diploid cells, this disruption results only in a prolonged delay from which the cells can recover albeit with an increase in aneuploidy [178]. A wide variety of cancer cell lines including melanoma; HeLa; glioma; and lung, ovarian, and breast cancer cells undergo an apoptotic death after mitotic arrest induced by ATO or sodium arsenite exposure [83, 103, 175, 179, 181]. This apoptotic death follows centrosome fragmentation [151, 180] and can be prevented by induction of the cyclin dependent kinase inhibitor CDKN1A (*aka* p21) in p53 expressing cells [150]. It is likely that induction of CDKN1A serves to inhibit cyclin-dependent kinase 1 (CDK1), thus releasing the cells from the CDK1 dependent mitotic arrest [104].The centrosome fragmentation can be augmented by inhibition of the heat shock system [151, 174]. Thus, the ability to induce mitotic arrest and apoptosis also may play a role in chemotherapy of solid tumors.

14.2.2 Tetra-Arsenic Oxide

Tetra-arsenic oxide (As_4S_6, TAO) is an arsenic oxide compound that is used in Korean traditional medicine, and has promising antitumor activity. *In vitro*, TAO inhibits prostate, gastric, head and neck, and cervical cancer cell proliferation and produces more ROS and apoptosis in myeloid leukemia cells than ATO [2, 27]. TAO-induced apoptosis requires cytochrome-c release and caspase-3 activation [118], and is p53-independent [27]. TAO has demonstrated antiangiogenic capacities by arresting tumor cells and endothelial cells in either G1 or G2/M [118, 169]. In support of this, TAO reduces pulmonary metastatic nodules in a murine melanoma model. TAO can be administered orally [118, 183]. Interestingly, despite differences described here, TAO and ATO may yield the same active compound in aqueous solution [171]. Further experiments are required to evaluate the potential antitumor activity of TAO in humans.

14.2.3 Arsenic Sulfides

One disadvantage of ATO is that it must be administered by intravenous infusion to avoid severe liver damage. Thus, any arsenic compound with similar chemotherapeutic effects that could be given orally will favor patient compliance to treatment and improve the long-term consolidation therapy. Realgar (As_4S_4, *Xiong Huang*, As_2S_2, AsS) and orpiment (As_2S_3) have been used in traditional Chinese medicines for centuries against a wide variety of diseases, including leukemia [171]. These arsenic sulfide minerals can be administered orally and have limited toxicity. As of today, there are 22 registered formulas containing realgar recorded in the Chinese Pharmacopoeia [85]. As a result, researchers have found As_4S_4 and As_2S_3 attractive to explore as chemotherapeutic agents.

14.2.3.1 Clinical Efficacy Medicinal herbs that contain realgar, such as *Indigo naturalis*, *Salvia miltiorrhiza*, or *Radix astragali*, have been tested since the early 1980s against CML, AML, and APL. Realgar was shown to be the active ingredient in the medicinal herbs. In one of the trials using a realgar-*Indigo naturalis* formula (composed of realgar, *Indigo naturalis* and *Salvia miltiorrhiza*), 72% of 86 CML patients achieved a complete remission, and increasing realgar concentrations improved its efficacy [192]. The rate of responses achieved in APL patients was comparable to ATO, although less significant toxicity was observed [58, 125, 160]. Realgar was studied in combination with the tyrosine kinase receptor inhibitor, imatinib (Gleevec) [182, 187]. Similar to ATO, realgar showed promising combinatorial potential with imatinib both *in vitro* and in a murine model of CML by targeting the BCR-ABL oncoprotein [182].

Realgar also has been tested as a single agent in one pilot study against APL [91]. In newly diagnosed patients, as well as in relapsed patients, realgar treatment resulted in hematologic complete remission with an estimated 1-year overall survival rate of 86%. The dose was well-tolerated with subjects experiencing only transient adverse effects, including asymptomatic QT prolongation and signs of liver toxicity (although incidence of fatty liver was high). Unlike ATO, no myelosuppression was reported, but mild temporary leukocytosis was observed in newly diagnosed patients only [91]. Moreover, realgar administration has been evaluated for maintenance therapy in 103 patients that had previously achieved complete remission after ATRA and chemotherapy treatment. Estimated 1-year survival rate under realgar maintenance treatment is 96%. At the molecular level, 79% of PML-RARα positive patients became negative after realgar treatment. Morphologic changes in APL promyelocytes were observed and pointed toward apoptosis or degeneration more than differentiation [91]. Also, it should be noted that from this trial, one newly diagnosed APL patient displayed complete remission after treatment with chemically pure orpiment (As_2S_3) treatment [91, 172].

Realgar is insoluble in water, which results in a poor bioavailability. Recent studies demonstrated that reducing the particle size of realgar improved its bioavailability and activity [172]. In fact, realgar nanoparticles demonstrated size-dependent cytotoxicity when tested on a human umbilical vein endothelial cell line, such that smaller nanoparticles inhibited cell growth through apoptosis better than the bigger particles. Because realgar nanoparticles are effective inhibitors of umbilical vein endothelial cell growth, it is suggested that they could suppress angiogenesis [34]. Realgar nanoparticles (100–150 nm) also inhibit proliferation and induce differentiation of human myelocytic leukemia cell lines [7, 94, 164, 165], even more than realgar powder [177]. Realgar nanoparticles also inhibited growth and triggered apoptosis of a human cervical cancer cell line [85, 87]. In addition, realgar nanoparticles given to mice bearing melanoma by transdermal administration were effective against the tumors with modest side effects [188]. Recently, realgar fluorescent quantum dots (<20 nm) were engineered, and they were found to inhibit the proliferation of several human ovarian cancer cell lines without being cytotoxic to fibroblasts [159, 160, 165]. However, as of this writing, no clinical trials have assessed efficacy or pharmacokinetics or chemotherapeutic potential of realgar nanoparticles.

14.2.3.2 Mechanisms of Action In Chinese traditional medicine, realgar is thought to purge "toxic heat" in the body, which are the cancer products that accumulate in the human body and are thought to be the major cause for cancer progression [141]. However, the exact mechanism of action of realgar is complicated, unclear, and may vary according to the other components of the mixtures studied. These other components may act as synergists or adjuvants. For instance, in the realgar-*Indigo naturalis* formula used in clinical trials, three active ingredients were identified that potentiate each other's effects. *In vitro* experiments showed that tanshinone IIA and indirubin, the two other active components in the formula, increased realgar uptake by the APL cells, through up-regulation of the transmembrane aquaglyceroporin-9 [160].

As a single molecule, realgar was perceived to function similarly to ATO. Indeed, both ATO and realgar produce ROS *in vitro* and *in vivo*, and both compounds induce apoptosis in APL and CML cells [99, 124, 139, 196]. Moreover, similar to ATO, realgar exposure of APL cells causes reorganization of PML bodies [197], inhibition of the PI3-K/Akt pathway and down-regulation of Bcl-x(L). Likewise, *in vitro* realgar exposure induces G2/M growth arrest in cancer cells [4] similar to ATO [174]. Also, in CML cells and mice, realgar induced a ligase involved in ubiquitination, c-CBL. Realgar bound to c-CBL and increased ubiquitination and proteolysis of its BCR-ABL oncoprotein substrate to favor apoptosis [99, 186]. ATO is also efficient at targeting BCR-ABL oncoprotein, but targeting is achieved through Rho-JNK pathway [123]. Also, realgar does not change superoxide or GSH levels [161]. Furthermore, multiple myeloma cells treated with realgar exhibit a different gene expression pattern than those treated with ATO [162, 163].

To some extent, realgar nanoparticles also have different mechanisms of action from raw realgar. Although realgar nanoparticles produce ROS *in vivo* like powder realgar [173], dissimilarities are also observed; realgar nanoparticles induce lipid peroxidation, increase lactate dehydrogenase, and reduce membrane fluidity, whereas realgar powder has only small effects on cell membrane integrity [94, 185]. Moreover, although realgar nanoparticles have demonstrated capability of inducing APL cell differentiation like ATO, differentiation pathways are different, as realgar-induced differentiation involved increased activity of serine/threonine phosphatases type 1 (PP1) and type 2A (PP2A). Selective inhibition of PP1 and PP2A suppressed realgar-induced differentiation, while it did not have any effect on ATO-induced differentiation in leukemic cells [101, 185]. Additionally, while ATO recruits the 11S proteasome regulatory components, realgar particles induce the 26S proteasome subunits PSMC2 and PSMD1, which contribute to the degradation of PML/RARα and subsequent differentiation [77, 94].

Part of the reason for the yet undefined mode of action of realgar is the lack of information regarding metabolites. Inorganic arsenic is known to be metabolized through a series of reduction, oxidation, and methylation reactions, and it is thought that arsenic toxicity is dependent upon its methylation status. Realgar exposure generates urine arsenic metabolites, including monomethylarsonate (MMA^{V}) and dimethyl arsenate (DMA^{V}). However, it is unknown if there is different metabolism pattern between ATO and realgar. When contrasted with ATO, realgar exposure leads to As–S, As–O, the presence of unknown metabolites, and a lack of methylated trivalent forms in urine [52, 70].

14.3 ORGANIC ARSENICALS AS CHEMOTHERAPEUTIC DRUGS

Numerous organic arsenicals have been explored, and some have shown promising effects *in vitro* against both APL and non-APL cell lines. These include phenylarsonous acid [115, 133] and *S*-dimethylarsino-thiosuccinic acid (MER1) [49]. However, few of them have entered clinical trials. Melarsoprol (Arsobal®) is an arsenical that was tested in clinical trials, but the results were disappointing [72, 131, 143]. Other arsenicals with improved therapeutic index are in development, such as darinaparsin, GMZ27, and GSAO [22, 40, 97]. These have lower toxicity and some have demonstrated promising results in clinical trials.

14.3.1 Darinaparsin

S-dimethylarsino-GSH (darinaparsin, SGLU-1, ZIO-101-C, Zinapar™) is a metabolite of inorganic arsenic biotransformation, which consists of dimethylarsenic conjugated to a GSH molecule. Darinaparsin has distinct, while partially overlapping, effects on several signaling pathways from ATO. To induce apoptosis, both arsenicals act on JNK- and caspase-dependant pathways; whereas, unlike ATO, darinaparsin activity does not require decreased Bcl-2 protein expression to trigger apoptosis [102]. In multiple studies, darinaparsin showed greater antitumor activity than ATO against a wide variety of tumor cell lines [37, 97, 102]. Its promising antitumor activity in mice against leukemia, multiple myeloma, and tumors has made darinaparsin an attractive drug for clinical development [15, 37, 152].

14.3.1.1 Clinical Efficacy Exploratory phase I and II clinical trials on heavily pretreated patients were conducted with darinaparsin both for the treatment of hematologic cancers and solid tumors. Some activity was observed for the intravenous formulation of darinaparsin in several hematological cancers, including acute myelogenous leukemia, myelodysplastic syndromes, peripheral T-cell lymphoma, myeloproliferative disorder, refractory Hodgkin and non-Hodgkin lymphoma, multiple myeloma, and also in solid tumors [57, 154]. Notably, intravenous darinaparsin was well tolerated in patients, without generating liver toxicity even in the hepatocellular carcinoma subjects [170]. However, among all studies, three patients developed a neutropenic fever [57, 90] and six patients (21%) had a slight increase in QT interval (grade 1–2). However, there was no correlation between prolonged compensated QT and increased exposure, suggesting that the observations were not related to darinaparsin plasma concentrations [57].

The overall response rate from hematological malignancies was 17%. However, in lymphoma patients, one complete remission, one unconfirmed complete remission, three partial response, and four stable disease responses were observed [59]. Interestingly, in AML, two complete remissions and four partial responses were observed [7]. Moreover, out of seven patients with peripheral T-cell lymphoma, one had a complete remission, one had an unconfirmed complete remission, and two had prolonged stable disease [59]. As result, the United States Food and Drug Administration granted darinaparsin *orphan drug status* in 2010 for the treatment of peripheral T-cell lymphoma. At this time, multiple ongoing/recruiting/completing clinical trials are evaluating safety and efficacy of darinaparsin as a single agent in subjects with hematological malignancies (*clinicaltrials.gov*) and solid tumors (*clinicaltrials.gov*: NCT01139346; NCT00591422; NCT00591396; NCT00423306; NCT00592163). Trials are ongoing to evaluate the combination of darinaparsin with cyclophosphamide, doxorubicin, vincristine, and prednisone (CHOP) in the lymphoma setting (*clinicaltrials.gov*: NCT01139359). In contrast to hematological malignancies, none of the patients bearing solid tumors that received intravenous darinaparsin as single agent have had objective responses, but minor responses and prolonged stable disease were observed [154, 170].

The oral bioavailability of darinaparsin also was evaluated in phase I clinical trials conducted in patients with relapsed and refractory advanced solid tumors. Even if the aim of these studies was not to evaluate drug's efficacy, one subject experienced a partial response (head and neck cancer), and 15 had prolonged stable disease (head and neck, lymphoma, colon, and pancreatic cancers) [96]. Although atrial fibrillation and congestive heart failure occurred in one patient [96], oral darinaparsin was well tolerated and the maximum tolerated dose has not been reached. As of this writing, three clinical trials are evaluating oral darinaparsin administration in patients with solid tumors (*clinicaltrials.gov*: NCT01139346; NCT00591422; NCT00592163).

14.3.1.2 Mechanisms of Action The exact mechanism of action of darinaparsin is complex and may vary according to cell types [104]. Previous reports showed that both ATO and darinaparsin induce ROS and require JNK activation for apoptosis. However, a series of studies on darinaparsin suggest that it utilizes a partially different mechanism of action compared to ATO. For instance, ATO-resistant cell lines are sensitive to darinaparsin [35, 107]. Also, darinaparsin does not induce APL cell differentiation [109], PML-RARα degradation, or PML nuclear body restoration as seen with ATO [20, 37]. While both ATO and darinaparsin are both known to produce ROS, darinaparsin induces ROS to a greater extent [35, 157]. Moreover, darinaparsin has been shown to inhibit microtubule polymerization [107], while reports with ATO seem to diverge [16, 41, 83, 97].

ROS Production and Involvement of Mitochondria Darinaparsin activates the NADPH oxidase complex, generating ROS, which disrupts the mitochondrial membrane potential and induces pro-apoptotic BH3-only proteins (Noxa, Bim, and Bmf) [109]. Also, darinaparsin inhibits the mitochondrial transport chain [35, 105], which results in cytochrome c release and tumor cell apoptosis. Importantly, darinaparsin produced more superoxide that ATO, generating more ROS and inducing more apoptosis than ATO [35]. Moreover, darinaparsin may inhibit tumor cell growth by generating mitochondrial swelling, which consequently will deplete ATP production.

ROS production can induce adaptive cellular stress responses. Under stress conditions, nonsurvival mRNAs will be co-sequestered with translational proteins to form stress granules, coordinated by microtubules. Addition of darinaparsin initiates the cell's stress response and produces a large amount of smaller and sparser stress granules compared to ATO [107]. These observations suggest incomplete stress granule formation under darinaparsin exposure *in vitro*. Incomplete formation will prevent subsequent disassembly [107], resulting in sequestration of important mRNAs that are normally required to restore cellular functions [71] and eventually apoptosis [155]. In fact, darinaparsin and ATO inhibit microtubule polymerization, which is required for proper stress granule formation [84, 107].

GSH Involvement Cellular import and export of darinaparsin may be dependent upon its GSH moiety. Indeed, at equimolar concentrations of ATO and darinaparsin, the latter results in higher intracellular arsenic levels [37, 98, 101]. Exogenous addition of GSH or cysteine to myeloma cell lines has no effect on ATO uptake, while it prevents darinaparsin import [109]. On the other hand, export of ATO is facilitated partially through drug exporter ABCC1 and blocking ABCC1 increases intracellular arsenic levels and ATO-induced death. Inhibition of ABCC1 has no effect on intracellular arsenic levels or death induced by darinaparsin [35]. Moreover, expression of ABCC1 prevents ATO activity by exporting it out of the cells, while ABCC1-expressing cells are sensitive to darinaparsin [107]. This observation suggests that tumors resistant to chemotherapy because of multidrug transporters might be sensitive to darinaparsin.

14.3.2 GMZ27

Dipropil-*S*-glycerol arsenic (GMZ27) is a novel arsenic-lipid derivative that has been reported to inhibit the proliferation of AML cells better than ATO. However, unlike ATO, GMZ27 only slightly induces AML cell differentiation. Like darinaparsin, GMZ27 activity is not related to PML-RARα degradation. Also, GMZ27 induces ROS production and simultaneously decreases intracellular GSH. Moreover, it disrupts the mitochondrial membrane potential, inducing the caspase cascade, and consequently inducing apoptosis [19]. *In vivo* mouse experiments showed that GMZ27 has a LD50 10 times higher than ATO [20]. Thus far, this compound has not been evaluated in patients.

14.3.3 GSAO

4-(*N*-(*S*-glutathionylacetyl)amino) phenylarsenous acid (GSAO) was synthesized by combining phenylarsonous acid and a GSH molecule [37]. This compound has been identified recently as a selective inhibitor of angiogenic endothelial cells [36, 40]. This interesting property may lead to GSAO pharmaceutical development, because it is known that angiogenesis is required for the maintenance of the tumor and its metastatic competence [14]. In fact, GSAO has demonstrated antitumor activity against murine prostate, lung, and pancreas tumors as well as fibrosarcomas [40] with minimal toxicity.

14.3.3.1 Mechanisms of Action GSAO is first cleaved at the cell surface by glutamyltranspeptidase, which is required for its entry into the cytosol [38]. Once inside the cell, it is cleaved again by a dipeptidase. The resulting by-product perturbs the mitochondrial inner membrane transporter, adenine nucleotide translocase, by binding in a calcium-dependent manner to cysteine thiols [40, 41]. This perturbation results in reduction of ATP production, disruption of mitochondrial membrane potential, and induction of apoptosis [38].

Interestingly, GSAO inhibits proliferation of endothelial cells more efficiently than of tumor cells. Unlike tumor cells, endothelial cells minimally express multidrug transporters, and consequently do not efflux GSAO efficiently [36]. Moreover, it should be noted that an alternative compound, namely 4-(*N*-(*S*-penicillaminylacetyl) amino) phenylarsenous acid or PENAO has been synthesized to overcome the limiting step of cell-surface cleavage. This compound accumulates more quickly in cells, and has better antitumor efficacy both *in vitro* and *in vivo* [38]. A phase I trial is currently recruiting (*australiancancertrials.gov.au*; Trial ID 362898).

14.4 CONCLUSIONS

The therapeutic use of ATO over the past centuries has evolved to its standard use in modern medicine in APL patients. ATO, along with ATRA, provided insight into the molecular pathology of APL and realized the possibility of efficiently targeting oncogenic fusion proteins. As we understand more about the pathways involved in ATO-induced death, we may be able to extend the therapeutic spectrum of ATO. In particular, combination therapies may prove effective in malignancies where toxicity limits usefulness. Further development of therapy with organoarsenicals and improvement of treatment with the utilization of arsenic nanoparticles seem promising approaches that could benefit cancer patients.

REFERENCES

[1] V. Adler, Z. Yin, S.Y. Fuchs, M. Benezra, L. Rosario, K.D. Tew, M.R. Pincus, M. Sardana, C.J. Henderson, C.R. Wolf, R.J. Davis, Z. Ronai, Regulation of JNK signaling by GSTp. EMBO J. 18 (1999) 1321–1334.

[2] W.S. Ahn, S.M. Bae, K.H. Lee, Y.W. Kim, J.M. Lee, S.E. Namkoong, I.P. Lee, C.K. Kim, J.S. Seo, J.I. Sin, Y.W. Kim, Comparison of effects of As2O3 and As4O6 on cell growth inhibition and gene expression profiles by cDNA microarray analysis in SiHa cells. Oncol. Rep. 12 (2004) 573–580.

[3] M. Aklilu, H.L. Kindler, S. Nattam, A. Brich, E.E. Vokes, A multi-center phase II study of arsenic trioxide (AT) in patients (Pts) with advanced pancreatic cancer (PC) refractory to gemcitabine. J. Clin. Oncol. 22 (2004) 4114.

[4] Y.L. An, F. Nie, Z.Y. Wang, D.S. Zhang, Preparation and characterization of realgar nanoparticles and their inhibitory effect on rat glioma cells. Int. J. Nanomedicine 6 (2011) 3187–3194.

[5] T. Bachleitner-Hofmann, B. Gisslinger, E. Grumbeck, H. Gisslinger, Arsenic trioxide and ascorbic acid: synergy with potential implications for the treatment of acute myeloid leukaemia? Br. J. Haematol. 112 (2001) 783–786.

[6] N.J. Bahlis, J. McCafferty-Grad, I. Jordan-McMurry, J. Neil, I. Reis, M. Kharfan-Dabaja, J. Eckman, M. Goodman, H.F. Fernandez, L.H. Boise, K.P. Lee, Feasibility and correlates of arsenic trioxide combined with ascorbic acid-mediated depletion of intracellular glutathione for the treatment of relapsed/refractory multiple myeloma. Clin. Cancer Res. 8 (2002) 3658–3668.

[7] P. Balaz, M. fabian, M. Pastorek, D. Cholujova, J. Sedlak, Mechanochemical preparation and anticancer effect of realgar As4S4 nanoparticles. Mater. Lett. 63 (2009) 1542–1544.

[8] J.T. Barbey, J.C. Pezzullo, S.L. Soignet, Effect of arsenic trioxide on QT interval in patients with advanced malignancies. J. Clin. Oncol. 21 (2003) 3609–3615.

[9] S. Bernardini, L. Bellincampi, S. Ballerini, M. Ranalli, A. Pastore, C. Cortese, G. Federici, Role of GST P1-1 in mediating the effect of etoposide on human neuroblastoma cell line Sh-Sy5y. J. Cell. Biochem. 86 (2002) 340–347.

[10] S. Bernardini, M. Nuccetelli, N.I. Noguera, L. Bellincampi, P. Lunghi, A. Bonati, K. Mann, W.H. Miller, Jr., G. Federici, C.F. Lo, Role of GSTP1-1 in mediating the effect of As2O3 in the Acute Promyelocytic Leukemia cell line NB4. Ann. Hematol. 85 (2006) 681–687.

[11] L. Bernstam, J. Nriagu, Molecular aspects of arsenic stress. J. Toxicol. Environ. Health B Crit. Rev. 3 (2000) 293–322.

[12] M.A. Bjornsti, P.J. Houghton, The TOR pathway: a target for cancer therapy. Nat. Rev. Cancer 4 (2004) 335–348.

[13] J. Borrow, A.D. Goddard, D. Sheer, E. Solomon, Molecular analysis of acute promyelocytic leukemia breakpoint cluster region on chromosome 17. Science 249 (1990) 1577–1580.

[14] M. Calatayud, J.A. Barrios, D. Velez, V. Devesa, In vitro study of transporters involved in intestinal absorption of inorganic arsenic. Chem. Res. Toxicol. 25 (2012) 446–453.

[15] R.A. Campbell, E. Sanchez, H. Chen, L. Turker, M. Li, O. Trac, D. Shalitin, M.S. Gordon, S. Pang, B. Bonavida, J. Said, B. Wallner, R.P. Gale, J.R. Berenson, ZIO-101, a novel organic arsenic, inhibits human myeloma cell growth in a SCID-hu model. Blood 108 (2006) 3462.

[16] M. Carre, G. Carles, N. Andre, S. Douillard, J. Ciccolini, C. Briand, D. Braguer, Involvement of microtubules and mitochondria in the antagonism of arsenic trioxide on paclitaxel-induced apoptosis. Biochem. Pharmacol. 63 (2002) 1831–1842.

[17] M. Cavigelli, W.W. Li, A. Lin, B. Su, K. Yoshioka, M. Karin, The tumor promoter arsenite stimulates AP-1 activity by inhibiting a JNK phosphatase. EMBO J. 15 (1996) 6269–6279.

[18] Cell Therapeutics. Trisenox-TM. 2000. Food and Drug Administration (USFDA).

[19] K.S. Chang, Y.H. Fan, M. Andreeff, J. Liu, Z.M. Mu, The PML gene encodes a phosphoprotein associated with the nuclear matrix. Blood 85 (1995) 3646–3653.

[20] G.Q. Chen, J. Zhu, X.G. Shi, J.H. Ni, H.J. Zhong, G.Y. Si, X.L. Jin, W. Tang, X.S. Li, S.M. Xong, Z.X. Shen, G.L. Sun, J. Ma, P. Zhang, T.D. Zhang, C. Gazin, T. Naoe, S.J. Chen, Z.Y. Wang, Z. Chen, In vitro studies on cellular and molecular mechanisms of arsenic trioxide (As2O3) in the treatment of acute promyelocytic leukemia: As2O3 induces NB4 cell apoptosis with downregulation of Bcl-2 expression and modulation of PML-RAR alpha/PML proteins. Blood 88 (1996) 1052–1061.

[21] Y.C. Chen, S.Y. Lin-Shiau, J.K. Lin, Involvement of reactive oxygen species and caspase 3 activation in arsenite-induced apoptosis. J. Cell. Physiol. 177 (1998) 324–333.

[22] X. Cheng, M. Golemovic, F. Giles, R. Zingaro, M.Z. Gao, E.J. Freireich, M. Andreeff, H.M. Kantarjian, S. Verstovsek. Organic arsenic lipid derivatives are more potent and less toxic than inorganic arsenic trioxide in preclinical testing. Blood 104 11 (2004), 1803.

[23] W.C. Chou, H.Y. Chen, S.L. Yu, L. Cheng, P.C. Yang, C.V. Dang, Arsenic suppresses gene expression in promyelocytic leukemia cells partly through Sp1 oxidation. Blood 106 (2005) 304–310.

[24] W.C. Chou, A.L. Hawkins, J.F. Barrett, C.A. Griffin, C.V. Dang, Arsenic inhibition of telomerase transcription leads to genetic instability. J. Clin. Invest. 108 (2001) 1541–1547.

[25] W.C. Chou, C. Jie, A.A. Kenedy, R.J. Jones, M.A. Trush, C.V. Dang, Role of NADPH oxidase in arsenic-induced reactive oxygen species formation and cytotoxicity in myeloid leukemia cells. Proc. Natl. Acad. Sci. U. S. A. 101 (2004) 4578–4583.

[26] W. Chu, C. Li, X. Qu, D. Zhao, X. Wang, X. Yu, F. Cai, H. Liang, Y. Zhang, X. Zhao, B. Li, G. Qiao, D. Dong, Y. Lu, Z. Du, B. Yang, Arsenic-induced interstitial myocardial fibrosis reveals a new insight into drug-induced long QT syndrome. Cardiovasc. Res. 96 (2012) 90–98.

[27] W.H. Chung, B.H. Sung, S.S. Kim, H. Rhim, H.J. Kuh, Synergistic interaction between tetra-arsenic oxide and paclitaxel in human cancer cells in vitro. Int. J. Oncol. 34 (2009) 1669–1679.

[28] J. Dai, R.S. Weinberg, S. Waxman, Y. Jing, Malignant cells can be sensitized to undergo growth inhibition and apoptosis by arsenic trioxide through modulation of the glutathione redox system. Blood 93 (1999) 268–277.

[29] M.J. Davies, L.G. Forni, R.L. Willson, Vitamin E analogue Trolox C. E.s.r. and pulse-radiolysis studies of free-radical reactions. Biochem. J. 255 (1988) 513–522.

[30] K. Davison, S. Cote, S. Mader, W.H. Miller, Glutathione depletion overcomes resistance to arsenic trioxide in arsenic-resistant cell lines. Leukemia 17 (2003) 931–940.

[31] K. Davison, K.K. Mann, S. Waxman, W.H. Miller, Jr., JNK activation is a mediator of arsenic trioxide-induced apoptosis in acute promyelocytic leukemia cells. Blood 103 (2004) 3496–3502.

[32] H. de The, C. Chomienne, M. Lanotte, L. Degos, A. Dejean, The t(15;17) translocation of acute promyelocytic leukaemia fuses the retinoic acid receptor alpha gene to a novel transcribed locus. Nature 347 (1990) 558–561.

[33] H. de The, B.M. Le, V. Lallemand-Breitenbach, The cell biology of disease: acute promyelocytic leukemia, arsenic, and PML bodies. J. Cell Biol. 198 (2012) 11–21.

[34] Y. Deng, H. Xu, K. Huang, X. Yang, C. Xie, J. Wu, Size effects of realgar particles on apoptosis in a human umbilical vein endothelial cell line: ECV-304. Pharmacol. Res. 44 (2001) 513–518.

[35] Z. Diaz, M. Colombo, K.K. Mann, H. Su, K.N. Smith, D.S. Bohle, H.M. Schipper, W.H. Miller, Jr., Trolox selectively enhances arsenic-mediated oxidative stress and apoptosis in APL and other malignant cell lines. Blood 105 (2005) 1237–1245.

[36] Z. Diaz, A. Laurenzana, K.K. Mann, T.A. Bismar, H.M. Schipper, W.H. Miller, Jr., Trolox enhances the anti-lymphoma effects of arsenic trioxide, while protecting against liver toxicity. Leukemia 21 (2007) 2117–2127.

[37] Z. Diaz, K.K. Mann, S. Marcoux, M. Kourelis, M. Colombo, P.B. Komarnitsky, W.H. Miller, Jr., A novel arsenical has antitumor activity toward As2O3-resistant and MRP1/ABCC1-overexpressing cell lines. Leukemia 22 (2008) 1853–1863.

[38] P.J. Dilda, P.J. Hogg, Arsenical-based cancer drugs. Cancer Treat. Rev. 33 (2007) 542–564.

[39] B. Dolniak, E. Katsoulidis, N. Carayol, J.K. Altman, A.J. Redig, M.S. Tallman, T. Ueda, R. Watanabe-Fukunaga, R. Fukunaga, L.C. Platanias, Regulation of arsenic trioxide-induced cellular responses by Mnk1 and Mnk2. J. Biol. Chem. 283 (2008) 12034–12042.

[40] A.S. Don, O. Kisker, P. Dilda, N. Donoghue, X. Zhao, S. Decollogne, B. Creighton, E. Flynn, J. Folkman, P.J. Hogg, A peptide trivalent arsenical inhibits tumor angiogenesis by perturbing mitochondrial function in angiogenic endothelial cells. Cancer Cell 3 (2003) 497–509.

[41] X.F. Duan, Y.L. Wu, H.Z. Xu, M. Zhao, H.Y. Zhuang, X.D. Wang, H. Yan, G.Q. Chen, Synergistic mitosis-arresting effects of arsenic trioxide and paclitaxel on human malignant lymphocytes. Chem. Biol. Interact. 183 (2010) 222–230.

[42] J.A. Dyck, G.G. Maul, W.H. Miller, Jr., J.D. Chen, A. Kakizuka, R.M. Evans, A novel macromolecular structure is a target of the promyelocyte-retinoic acid receptor oncoprotein. Cell 76 (1994) 333–343.

[43] A.M. Evens, M.S. Tallman, R.B. Gartenhaus, The potential of arsenic trioxide in the treatment of malignant disease: past, present, and future. Leuk. Res. 28 (2004) 891–900.

[44] P. Fenaux, C. Chastang, S. Chevret, M. Sanz, H. Dombret, E. Archimbaud, M. Fey, C. Rayon, F. Huguet, J.J. Sotto, C. Gardin, P.C. Makhoul, P. Travade, E. Solary, N. Fegueux, D. Bordessoule, J.S. Miguel, H. Link, B. Desablens, A. Stamatoullas, E. Deconinck, F. Maloisel, S. Castaigne, C. Preudhomme, L. Degos, A randomized comparison of all transretinoic acid (ATRA) followed by chemotherapy and ATRA plus chemotherapy and the role of maintenance therapy in newly diagnosed acute promyelocytic leukemia. The European APL Group. Blood 94 (1999) 1192–1200.

[45] R.E. Gallagher, A. Ferrari, A. Kaubisch, D. Makower, C. Stein, L. Rajdev, R. Gucalp, S. Wadler, J. Mandeli, C. Sarta. Arsenic trioxide (ATO) in metastatic hormone-refractory prostate cancer (HRPC): results of phase II trial T99-0077. J. Clin. Oncol. 22 14 Suppl, 4638.

[46] R.B. Gartenhaus, S.N. Prachand, M. Paniaqua, Y. Li, L.I. Gordon, Arsenic trioxide cytotoxicity in steroid and chemotherapy-resistant myeloma cell lines: enhancement of apoptosis by manipulation of cellular redox state. Clin. Cancer Res. 8 (2002) 566–572.

[47] A. Ghavamzadeh, K. Alimoghaddam, S. Rostami, S.H. Ghaffari, M. Jahani, M. Iravani, S.A. Mousavi, B. Bahar, M. Jalili, Phase II study of single-agent arsenic trioxide for the front-line therapy of acute promyelocytic leukemia. J. Clin. Oncol. 29 (2011) 2753–2757.

[48] N. Giafis, E. Katsoulidis, A. Sassano, M.S. Tallman, L.S. Higgins, A.R. Nebreda, R.J. Davis, L.C. Platanias, Role of the p38 mitogen-activated protein kinase pathway in the generation of arsenic trioxide-dependent cellular responses. Cancer Res. 66 (2006) 6763–6771.

[49] M. Golemovic, A. Quintas-Cardama, T. Manshouri, N. Orsolic, H. Duzkale, M. Johansen, E.J. Freireich, H. Kantarjian, R.A. Zingaro, S. Verstovsek, MER1, a novel organic arsenic derivative, has potent PML-RARalpha-independent cytotoxic activity against leukemia cells. Invest. New Drugs 28 (2010) 402–412.

[50] J.M. Grad, N.J. Bahlis, I. Reis, M.M. Oshiro, W.S. Dalton, L.H. Boise, Ascorbic acid enhances arsenic trioxide-induced cytotoxicity in multiple myeloma cells. Blood 98 (2001) 805–813.

[51] A. Guo, P. Salomoni, J. Luo, A. Shih, S. Zhong, W. Gu, P.P. Pandolfi, The function of PML in p53-dependent apoptosis. Nat. Cell Biol. 2 (2000) 730–736.

[52] H.R. Hansen, A. Raab, M. Jaspars, B.F. Milne, J. Feldmann, Sulfur-containing arsenical mistaken for dimethylarsinous acid [DMA(III)] and identified as a natural metabolite in urine: major implications for studies on arsenic metabolism and toxicity. Chem. Res. Toxicol. 17 (2004) 1086–1091.

[53] C.W. Helm, The role of hyperthermic intraperitoneal chemotherapy (HIPEC) in ovarian cancer. Oncologist 14 (2009) 683–694.

[54] L.K. Hillestad, Acute promyelocytic leukemia. Acta Med. Scand. 159 (1957) 189–194.

[55] T.G. Hofmann, A. Moller, H. Sirma, H. Zentgraf, Y. Taya, W. Droge, H. Will, M.L. Schmitz, Regulation of p53 activity by its interaction with homeodomain-interacting protein kinase-2. Nat. Cell Biol. 4 (2002) 1–10.

[56] A. Holmgren, M. Bjornstedt, Thioredoxin and thioredoxin reductase. Methods Enzymol. 252 (1995) 199–208.

[57] P.J. Hosein, M.D. Craig, M.S. Tallman, R.V. Boccia, B.L. Hamilton, J.J. Lewis, I.S. Lossos, A multicenter phase II study of darinaparsin in relapsed or refractory Hodgkin's and non-Hodgkin's lymphoma. Am. J. Hematol. 87 (2012) 111–114.

[58] S. Huang, A. Guo, Y. Xiang, X. Wang, H. Lin, L. Fu, Clinical study on the treatment of acute promyelocytic leukemia with Composite Indigo Naturalis tablets. Chin. J. Hematol. 16 (1995) 26–28.

[59] A. Insinga, S. Monestiroli, S. Ronzoni, R. Carbone, M. Pearson, G. Pruneri, G. Viale, E. Appella, P. Pelicci, S. Minucci, Impairment of p53 acetylation, stability and function by an oncogenic transcription factor. EMBO J. 23 (2004) 1144–1154.

[60] K. Ito, R. Bernardi, A. Morotti, S. Matsuoka, G. Saglio, Y. Ikeda, J. Rosenblatt, D.E. Avigan, J. Teruya-Feldstein, P.P. Pandolfi, PML targeting eradicates quiescent leukaemia-initiating cells. Nature 453 (2008) 1072–1078.

[61] M. Jeanne, V. Lallemand-Breitenbach, O. Ferhi, M. Koken, B.M. Le, S. Duffort, L. Peres, C. Berthier, H. Soilihi, B. Raught, H. de Thé, PML/RARA oxidation and arsenic binding initiate the antileukemia response of As2O3. Cancer Cell 18 (2010) 88–98.

[62] A. Jemal, R. Siegel, E. Ward, Y. Hao, J. Xu, M.J. Thun, Cancer statistics, 2009. CA Cancer J. Clin. 59 (2009) 225–249.

[63] X. Jiang, J.R. McDermott, A.A. Ajees, B.P. Rosen, Z. Liu, Trivalent arsenicals and glucose use different translocation pathways in mammalian GLUT1. Metallomics 2 (2010) 211–219.

[64] D.M. Jolliffe, A history of the use of arsenicals in man. J. R. Soc. Med. 86 (1993) 287–289.

[65] P. Kannan-Thulasiraman, E. Katsoulidis, M.S. Tallman, J.S. Arthur, L.C. Platanias, Activation of the mitogen- and stress-activated kinase 1 by arsenic trioxide. J. Biol. Chem. 281 (2006) 22446–22452.

[66] H.M. Kantarjian, M.J. Keating, R.S. Walters, E.H. Estey, K.B. McCredie, T.L. Smith, W.T. Dalton, Jr., A. Cork, J.M. Trujillo, E.J. Freireich, Acute promyelocytic leukemia. M.D. Anderson Hospital experience. Am. J. Med. 80 (1986) 789–797.

[67] P. Kapahi, T. Takahashi, G. Natoli, S.R. Adams, Y. Chen, R.Y. Tsien, M. Karin, Inhibition of NF-kappa B activation by arsenite through reaction with a critical cysteine in the activation loop of Ikappa B kinase. J. Biol. Chem. 275 (2000) 36062–36066.

[68] M. Karin, Nuclear factor-kappaB in cancer development and progression. Nature 441 (2006) 431–436.

[69] C.D. Klaassen Heavy metals and heavy-metal antagonists. in: J. G. Hardman, L. E. Limbird, P. B. Molinoff, R. W. Ruddon, and A. G. Gilman (Eds.), *The Pharmacological Basis of Therapeutics*, McGraw-Hill, New York, 1996.

[70] I. Koch, S. Sylvester, V.W. Lai, A. Owen, K.J. Reimer, W.R. Cullen, Bioaccessibility and excretion of arsenic in Niu Huang Jie Du Pian pills. Toxicol. Appl. Pharmacol. 222 (2007) 357–364.

[71] M.H. Koken, M.T. Daniel, M. Gianni, A. Zelent, J. Licht, A. Buzyn, P. Minard, L. Degos, B. Varet, H. de Thé, Retinoic acid, but not arsenic trioxide, degrades the PLZF/RARalpha fusion protein, without inducing terminal differentiation or apoptosis, in a RA-therapy resistant t(11;17)(q23;q21) APL patient. Oncogene 18 (1999) 1113–1118.

[72] A. Konig, L. Wrazel, R.P. Warrell, Jr., R. Rivi, P.P. Pandolfi, A. Jakubowski, J.L. Gabrilove, Comparative activity of melarsoprol and arsenic trioxide in chronic B-cell leukemia lines. Blood 90 (1997) 562–570.

[73] G. Kroemer, H. de Thé, Arsenic trioxide, a novel mitochondriotoxic anticancer agent? J. Natl. Cancer Inst. 91 (1999) 743–745.

[74] Y. Kumagai, D. Sumi, Arsenic: signal transduction, transcription factor, and biotransformation involved in cellular response and toxicity. Annu. Rev. Pharmacol. Toxicol. 47 (2007) 243–262.

[75] V. Lallemand-Breitenbach, M.C. Guillemin, A. Janin, M.T. Daniel, L. Degos, S.C. Kogan, J.M. Bishop, H. de Thé, Retinoic acid and arsenic synergize to eradicate leukemic cells in a mouse model of acute promyelocytic leukemia. J. Exp. Med. 189 (1999) 1043–1052.

[76] V. Lallemand-Breitenbach, M. Jeanne, S. Benhenda, R. Nasr, M. Lei, L. Peres, J. Zhou, J. Zhu, B. Raught, H. de Thé, Arsenic degrades PML or PML-RARalpha through a SUMO-triggered RNF4/ubiquitin-mediated pathway. Nat. Cell Biol. 10 (2008) 547–555.

[77] V. Lallemand-Breitenbach, J. Zhu, F. Puvion, M. Koken, N. Honore, A. Doubeikovsky, E. Duprez, P.P. Pandolfi, E. Puvion, P. Freemont, T.H. de, Role of promyelocytic leukemia (PML) sumolation in nuclear body formation, 11S proteasome recruitment, and As2O3-induced PML or PML/retinoic acid receptor alpha degradation. J. Exp. Med. 193 (2001) 1361–1371.

[78] N. Larochette, D. Decaudin, E. Jacotot, C. Brenner, I. Marzo, S.A. Susin, N. Zamzami, Z. Xie, J. Reed, G. Kroemer, Arsenite induces apoptosis via a direct effect on the mitochondrial permeability transition pore. Exp. Cell Res. 249 (1999) 413–421.

[79] A. Lemarie, C. Morzadec, D. Merino, O. Micheau, O. Fardel, L. Vernhet, Arsenic trioxide induces apoptosis of human monocytes during macrophagic differentiation through nuclear factor-kappaB-related survival pathway down-regulation. J. Pharmacol. Exp. Ther. 316 (2006) 304–314.

[80] L. Li, J. Wang, R.D. Ye, G. Shi, H. Jin, X. Tang, J. Yi, PML/RARalpha fusion protein mediates the unique sensitivity to arsenic cytotoxicity in acute promyelocytic leukemia cells: mechanisms involve the impairment of cAMP signaling and the aberrant regulation of NADPH oxidase. J. Cell. Physiol. 217 (2008) 486–493.

[81] Y. Li, X. Qu, J. Qu, Y. Zhang, J. Liu, Y. Teng, X. Hu, K. Hou, Y. Liu, Arsenic trioxide induces apoptosis and G2/M phase arrest by inducing Cbl to inhibit PI3K/Akt signaling and thereby regulate p53 activation. Cancer Lett. 284 (2009) 208–215.

[82] C.C. Lin, C. Hsu, C.H. Hsu, W.L. Hsu, A.L. Cheng, C.H. Yang, Arsenic trioxide in patients with hepatocellular carcinoma: a phase II trial. Invest. New Drugs 25 (2007) 77–84.

[83] Y.H. Ling, J.D. Jiang, J.F. Holland, R. Perez-Soler, Arsenic trioxide produces polymerization of microtubules and mitotic arrest before apoptosis in human tumor cell lines. Mol. Pharmacol. 62 (2002) 529–538.

[84] A. List, M. Beran, J. DiPersio, J. Slack, N. Vey, C.S. Rosenfeld, P. Greenberg, Opportunities for Trisenox (arsenic trioxide) in the treatment of myelodysplastic syndromes. Leukemia 17 (2003) 1499–1507.

[85] J. Liu, Y. Lu, Q. Wu, R.A. Goyer, M.P. Waalkes, Mineral arsenicals in traditional medicines: orpiment, realgar, and arsenolite. J. Pharmacol. Exp. Ther. 326 (2008) 363–368.

[86] Q. Liu, S. Hilsenbeck, Y. Gazitt, Arsenic trioxide-induced apoptosis in myeloma cells: p53-dependent G1 or G2/M cell cycle arrest, activation of caspase-8 or caspase-9, and synergy with APO2/TRAIL. Blood 101 (2003) 4078–4087.

[87] R. Liu, D. Pu, Y. Liu, Y. Cheng, L. Yin, T. Li, L. Zhao, Induction of SiHa cells apoptosis by nanometer realgar suspension and its mechanism. J. Huazhong Univ. Sci. Technolog. Med. Sci. 28 (2008) 317–321.

[88] Y. Liu, S.E. Elf, Y. Miyata, G. Sashida, Y. Liu, G. Huang, G.S. Di, J.M. Lee, A. DeBlasio, S. Menendez, J. Antipin, B. Reva, A. Koff, S.D. Nimer, p53 regulates hematopoietic stem cell quiescence. Cell Stem Cell 4 (2009) 37–48.

[89] Z. Liu, Roles of vertebrate aquaglyceroporins in arsenic transport and detoxification. Adv. Exp. Med. Biol. 679 (2010) 71–81.

[90] I. Lossos, M.D. Craig, M.S. Tallman, R.V. Boccia, P.R. Conkling, C. Becerra, P.B. Komarnitsky, B.L. Hamilton, J. Lewis, W.H. Miller. Novel organic arsenic molecule darinaparsin: development of IV and oral forms. J. Clin. Oncol. 27 15S, 8501. 2009.

Ref Type: Abstract

[91] D.P. Lu, J.Y. Qiu, B. Jiang, Q. Wang, K.Y. Liu, Y.R. Liu, S.S. Chen, Tetra-arsenic tetrasulfide for the treatment of acute promyelocytic leukemia: a pilot report. Blood 99 (2002) 3136–3143.

[92] J. Lu, E.H. Chew, A. Holmgren, Targeting thioredoxin reductase is a basis for cancer therapy by arsenic trioxide. Proc. Natl. Acad. Sci. U. S. A. 104 (2007) 12288–12293.

[93] P. Lunghi, A. Tabilio, F. Lo-Coco, P.G. Pelicci, A. Bonati, Arsenic trioxide (ATO) and MEK1 inhibition synergize to induce apoptosis in acute promyelocytic leukemia cells. Leukemia 19 (2005) 234–244.

[94] L.Y. Luo, J. Huang, B.D. Gou, T.L. Zhang, K. Wang, Induction of human promyelocytic leukemia HL-60 cell differentiation into monocytes by arsenic sulphide: involvement of serine/threonine protein phosphatases. Leuk. Res. 30 (2006) 1399–1405.

[95] K.K. Mann, M. Colombo, W.H. Miller, Jr., Arsenic trioxide decreases AKT protein in a caspase-dependent manner. Mol. Cancer Ther. 7 (2008) 1680–1687.

[96] K.K. Mann, W. Shao, W.H. Miller, Jr., The biology of acute promyelocytic leukemia. Curr. Oncol. Rep. 3 (2001) 209–216.

[97] K.K. Mann, B. Wallner, I.S. Lossos, W.H. Miller, Jr., Darinaparsin: a novel organic arsenical with promising anticancer activity. Expert Opin. Investig. Drugs 18 (2009) 1727–1734.

[98] T. Manshouri, S.V. Kala, F. Ashoori, R. Zingaro, E.J. Freireich, M. Andreeff, H.M. Kantarjian, S. Verstovsek, Comparison of uptake and intracellular induced structural changes of arsenic trioxide, an inorganic compound, and organic arsenic derivative S-dimethylarsino-glutathione (SGLU; ZIO-101) in NB4 acute promyelocytic leukemia (APL) cells. Blood 106 (2005) 4446.

[99] J.H. Mao, X.Y. Sun, J.X. Liu, Q.Y. Zhang, P. Liu, Q.H. Huang, K.K. Li, Q. Chen, Z. Chen, S.J. Chen, As4S4 targets RING-type E3 ligase c-CBL to induce degradation of BCR-ABL in chronic myelogenous leukemia. Proc. Natl. Acad. Sci. U. S. A. 107 (2010) 21683–21688.

[100] T.A. Mason, E. Kolobova, J. Liu, J.T. Roland, C. Chiang, J.R. Goldenring, Darinaparsin is a multivalent chemotherapeutic which induces incomplete stress response with disruption of microtubules and Shh signaling. PLoS One 6 (2011) e27699.

[101] J. Mathieu, F. Besancon, Arsenic trioxide represses NF-kappaB activation and increases apoptosis in ATRA-treated APL cells. Ann. N. Y. Acad. Sci. 1090 (2006) 203–208.

[102] S.M. Matulis, A.A. Morales, L. Yehiayan, C. Croutch, D. Gutman, Y. Cai, K.P. Lee, L.H. Boise, Darinaparsin induces a unique cellular response and is active in an arsenic trioxide-resistant myeloma cell line. Mol. Cancer Ther. 8 (2009) 1197–1206.

[103] S.C. McNeely, A.C. Belshoff, B.F. Taylor, T.W. Fan, M.J. McCabe, Jr., A.R. Pinhas, J.C. States, Sensitivity to sodium arsenite in human melanoma cells depends upon susceptibility to arsenite-induced mitotic arrest. Toxicol. Appl. Pharmacol. 229 (2008) 252–261.

[104] S.C. McNeely, B.F. Taylor, J.C. States, Mitotic arrest-associated apoptosis induced by sodium arsenite in A375 melanoma cells is BUBR1-dependent. Toxicol. Appl. Pharmacol. 231 (2008) 61–67.

[105] A. Melnick, J.D. Licht, Deconstructing a disease: RARalpha, its fusion partners, and their roles in the pathogenesis of acute promyelocytic leukemia. Blood 93 (1999) 3167–3215.

[106] W.H. Miller, Jr., Molecular targets of arsenic trioxide in malignant cells. Oncologist 7 Suppl 1 (2002) 14–19.

[107] W.H. Miller, Jr., H.M. Schipper, J.S. Lee, J. Singer, S. Waxman, Mechanisms of action of arsenic trioxide. Cancer Res. 62 (2002) 3893–3903.

[108] C.S. Muenyi, V.A. States, J.H. Masters, T.W. Fan, C.W. Helm, J.C. States, Sodium arsenite and hyperthermia modulate cisplatin-DNA damage responses and enhance platinum accumulation in murine metastatic ovarian cancer xenograft after hyperthermic intraperitoneal chemotherapy (HIPEC). J. Ovarian Res. 4 (2011) 9.

[109] S. Muller, M.J. Matunis, A. Dejean, Conjugation with the ubiquitin-related modifier SUMO-1 regulates the partitioning of PML within the nucleus. EMBO J. 17 (1998) 61–70.

[110] H. Naranmandura, Y. Ogra, K. Iwata, J. Lee, K.T. Suzuki, M. Weinfeld, X.C. Le, Evidence for toxicity differences between inorganic arsenite and thioarsenicals in human bladder cancer cells. Toxicol. Appl. Pharmacol. 238 (2009) 133–140.

[111] R. Nasr, M.C. Guillemin, O. Ferhi, H. Soilihi, L. Peres, C. Berthier, P. Rousselot, M. Robledo-Sarmiento, V. Lallemand-Breitenbach, B. Gourmel, D. Vitoux, P.P. Pandolfi, C. Rochette-Egly, J. Zhu, H. de Thé, Eradication of acute promyelocytic leukemia-initiating cells through PML-RARA degradation. Nat. Med. 14 (2008) 1333–1342.

[112] National Study Commission on Cytotoxic Exposure. Recommendations for handling cytotoxic agents. 2014. Massachusetts College of Pharmacy and Allied Health Sciences, Available from Louis P. Jeffrey, ScD, Chairman, National Study Commission on Cytotoxic Exposure.

[113] M. Nollen, F. Ebert, J. Moser, L.H. Mullenders, A. Hartwig, T. Schwerdtle, Impact of arsenic on nucleotide excision repair: XPC function, protein level, and gene expression. Mol. Nutr. Food Res. 53 (2009) 572–582.

[114] J. Nordberg, E.S. Arner, Reactive oxygen species, antioxidants, and the mammalian thioredoxin system. Free Radic. Biol. Med. 31 (2001) 1287–1312.

[115] C. Oetken, W.M. von, M. Autero, T. Ruutu, L.C. Andersson, T. Mustelin, Phenylarsine oxide augments tyrosine phosphorylation in hematopoietic cells. Eur. J. Haematol. 49 (1992) 208–214.

[116] K. Ohnishi, H. Yoshida, K. Shigeno, S. Nakamura, S. Fujisawa, K. Naito, K. Shinjo, Y. Fujita, H. Matsui, A. Takeshita, S. Sugiyama, H. Satoh, H. Terada, R. Ohno, Prolongation of the QT interval and ventricular tachycardia in patients treated with arsenic trioxide for acute promyelocytic leukemia. Ann. Intern. Med. 133 (2000) 881–885.

[117] R.F. Ozols, Treatment goals in ovarian cancer. Int. J. Gynecol. Cancer 15 Suppl 1 (2005) 3–11.

[118] M.J. Park, I.C. Park, I.J. Bae, K.M. Seo, S.H. Lee, S.I. Hong, C.K. Eun, W. Zhang, C.H. Rhee, Tetraarsenic oxide, a novel orally administrable angiogenesis inhibitor. Int. J. Oncol. 22 (2003) 1271–1276.

[119] M. Pearson, R. Carbone, C. Sebastiani, M. Cioce, M. Fagioli, S. Saito, Y. Higashimoto, E. Appella, S. Minucci, P.P. Pandolfi, P.G. Pelicci, PML regulates p53 acetylation and premature senescence induced by oncogenic Ras. Nature 406 (2000) 207–210.

[120] H. Pelicano, L. Feng, Y. Zhou, J.S. Carew, E.O. Hileman, W. Plunkett, M.J. Keating, P. Huang, Inhibition of mitochondrial respiration: a novel strategy to enhance drug-induced apoptosis in human leukemia cells by a reactive oxygen species-mediated mechanism. J. Biol. Chem. 278 (2003) 37832–37839.

[121] A.J. Percy, J. Gailer, Methylated trivalent arsenic-glutathione complexes are more stable than their arsenite analog. Bioinorg. Chem. Appl. 2008 (2008) 539082.

[122] F. Piazza, C. Gurrieri, P.P. Pandolfi, The theory of APL. Oncogene 20 (2001) 7216–7222.

[123] S. Potin, J. Bertoglio, J. Breard, Involvement of a Rho-ROCK-JNK pathway in arsenic trioxide-induced apoptosis in chronic myelogenous leukemia cells. FEBS Lett. 581 (2007) 118–124.

[124] E. Puccetti, S. Guller, A. Orleth, N. Bruggenolte, D. Hoelzer, O.G. Ottmann, M. Ruthardt, BCR-ABL mediates arsenic trioxide-induced apoptosis independently of its aberrant kinase activity. Cancer Res. 60 (2000) 3409–3413.

[125] L. Qian, Y. Zhao. The cooperation group of phase II clinical trial of compound Huangdai Tablet. Chin. J. Hematol. 27, 801–804. 2006.

[126] A. Ramirez-Solis, R. Mukopadhyay, B.P. Rosen, T.L. Stemmler, Experimental and theoretical characterization of arsenite in water: insights into the coordination environment of As-O. Inorg. Chem. 43 (2004) 2954–2959.

[127] F. Ravandi-Kashani, J. Ridgeway, S. Nishimura, Pilot study of combination of imatinib mesylate and Trisenox (As2O3) in patients with accelerated and blast phase CML. Blood 102 (2003) 314b.

[128] E.M. Rego, G.C. De Santis, Differentiation syndrome in promyelocytic leukemia: clinical presentation, pathogenesis and treatment. Mediterr. J. Hematol. Infect. Dis. 3 (2011) e2011048.

[129] R. Ren, Mechanisms of BCR-ABL in the pathogenesis of chronic myelogenous leukaemia. Nat. Rev. Cancer 5 (2005) 172–183.

[130] R.R. Roussel, A. Barchowsky, Arsenic inhibits NF-kappaB-mediated gene transcription by blocking IkappaB kinase activity and IkappaBalpha phosphorylation and degradation. Arch. Biochem. Biophys. 377 (2000) 204–212.

[131] P. Rousselot, J. Larghero, B. Arnulf, J. Poupon, B. Royer, A. Tibi, I. Madelaine-Chambrin, P. Cimerman, S. Chevret, O. Hermine, H. Dombret, B.J. Claude, F.J. Paul, A clinical and pharmacological study of arsenic trioxide in advanced multiple myeloma patients. Leukemia 18 (2004) 1518–1521.

[132] J.D. Rowley, H.M. Golomb, C. Dougherty, 15/17 translocation, a consistent chromosomal change in acute promyelocytic leukaemia. Lancet 1 (1977) 549–550.

[133] N. Sahara, A. Takeshita, M. Kobayashi, K. Shigeno, S. Nakamura, K. Shinjo, K. Naito, M. Maekawa, T. Horii, K. Ohnishi, K. Kitamura, T. Naoe, H. Hayash, R. Ohno, Phenylarsine oxide (PAO) more intensely induces apoptosis in acute promyelocytic leukemia and As2O3-resistant APL cell lines than As2O3 by activating the mitochondrial pathway. Leuk. Lymphoma 45 (2004) 987–995.

[134] M. Saitoh, H. Nishitoh, M. Fujii, K. Takeda, K. Tobiume, Y. Sawada, M. Kawabata, K. Miyazono, H. Ichijo, Mammalian thioredoxin is a direct inhibitor of apoptosis signal-regulating kinase (ASK) 1. EMBO J. 17 (1998) 2596–2606.

[135] M.A. Sanz, G. Martin, M. Gonzalez, A. Leon, C. Rayon, C. Rivas, D. Colomer, E. Amutio, F.J. Capote, G.A. Milone, J. De La Serna, J. Roman, E. Barragan, J. Bergua, L. Escoda, R. Parody, S. Negri, M.J. Calasanz, P. Bolufer, Risk-adapted treatment of acute promyelocytic leukemia with all-trans-retinoic acid and anthracycline monochemotherapy: a multicenter study by the PETHEMA group. Blood 103 (2004) 1237–1243.

[136] N. Scott, K.M. Hatlelid, N.E. MacKenzie, D.E. Carter, Reactions of arsenic(III) and arsenic(V) species with glutathione. Chem. Res. Toxicol. 6 (1993) 102–106.

[137] D. Shackelford, C. Kenific, A. Blusztajn, S. Waxman, R. Ren, Targeted degradation of the AML1/MDS1/EVI1 oncoprotein by arsenic trioxide. Cancer Res. 66 (2006) 11360–11369.

[138] H. Shen, S. Tsuchida, K. Tamai, K. Sato, Identification of cysteine residues involved in disulfide formation in the inactivation of glutathione transferase P-form by hydrogen peroxide. Arch. Biochem. Biophys. 300 (1993) 137–141.

[139] Z.X. Shen, G.Q. Chen, J.H. Ni, X.S. Li, S.M. Xiong, Q.Y. Qiu, J. Zhu, W. Tang, G.L. Sun, K.Q. Yang, Y. Chen, L. Zhou, Z.W. Fang, Y.T. Wang, J. Ma, P. Zhang, T.D. Zhang, S.J. Chen, Z. Chen, Z.Y. Wang, Use of arsenic trioxide (As2O3) in the treatment of acute promyelocytic leukemia (APL): II. Clinical efficacy and pharmacokinetics in relapsed patients. Blood 89 (1997) 3354–3360.

[140] Z.X. Shen, Z.Z. Shi, J. Fang, B.W. Gu, J.M. Li, Y.M. Zhu, J.Y. Shi, P.Z. Zheng, H. Yan, Y.F. Liu, Y. Chen, Y. Shen, W. Wu, W. Tang, S. Waxman, T.H. de, Z.Y. Wang, S.J. Chen, Z. Chen, All-trans retinoic acid/As2O3 combination yields a high quality remission and survival in newly diagnosed acute promyelocytic leukemia. Proc. Natl. Acad. Sci. U. S. A. 101 (2004) 5328–5335.

[141] Shen-Nong. TCM Approaches to Cancer Treatment. 2006. http://www.shen-nong.com/eng/exam/specialties_cancertreatment.html. Accessed April 11, 2015.

[142] S.L. Soignet, P. Maslak, Z.G. Wang, S. Jhanwar, E. Calleja, L.J. Dardashti, D. Corso, A. DeBlasio, J. Gabrilove, D.A. Scheinberg, P.P. Pandolfi, R.P. Warrell, Jr., Complete remission after treatment of acute promyelocytic leukemia with arsenic trioxide. N. Engl. J. Med. 339 (1998) 1341–1348.

[143] S.L. Soignet, W.P. Tong, S. Hirschfeld, R.P. Warrell, Jr., Clinical study of an organic arsenical, melarsoprol, in patients with advanced leukemia. Cancer Chemother. Pharmacol. 44 (1999) 417–421.

[144] D.J. Stephens, The therapeutic effect of solution of potassium arsenite in chronic myelogenous leukemia. Ann. Intern. Med. 1936 (1936) 1488–1502.

[145] D. Sumi, Y. Shinkai, Y. Kumagai, Signal transduction pathways and transcription factors triggered by arsenic trioxide in leukemia cells. Toxicol. Appl. Pharmacol. 244 (2010) 385–392.

[146] G. Tabellini, A. Cappellini, P.L. Tazzari, F. Fala, A.M. Billi, L. Manzoli, L. Cocco, A.M. Martelli, Phosphoinositide 3-kinase/Akt involvement in arsenic trioxide resistance of human leukemia cells. J. Cell. Physiol. 202 (2005) 623–634.

[147] M.S. Tallman, J.W. Andersen, C.A. Schiffer, F.R. Appelbaum, J.H. Feusner, A. Ogden, L. Shepherd, C. Willman, C.D. Bloomfield, J.M. Rowe, P.H. Wiernik, All-trans-retinoic acid in acute promyelocytic leukemia. N. Engl. J. Med. 337 (1997) 1021–1028.

[148] B. Tan, J.F. Huang, Q. Wei, H. Zhang, R.Z. Ni, Anti-hepatoma effect of arsenic trioxide on experimental liver cancer induced by 2-acetamidofluorene in rats. World J. Gastroenterol. 11 (2005) 5938–5943.

[149] M.H. Tatham, M.C. Geoffroy, L. Shen, A. Plechanovova, N. Hattersley, E.G. Jaffray, J.J. Palvimo, R.T. Hay, RNF4 is a poly-SUMO-specific E3 ubiquitin ligase required for arsenic-induced PML degradation. Nat. Cell Biol. 10 (2008) 538–546.

[150] B.F. Taylor, S.C. McNeely, H.L. Miller, G.M. Lehmann, M.J. McCabe, Jr., J.C. States, p53 suppression of arsenite-induced mitotic catastrophe is mediated by p21CIP1/WAF1. J. Pharmacol. Exp. Ther. 318 (2006) 142–151.

[151] B.F. Taylor, S.C. McNeely, H.L. Miller, J.C. States, Arsenite-induced mitotic death involves stress response and is independent of tubulin polymerization. Toxicol. Appl. Pharmacol. 230 (2008) 235–246.

[152] J. Tian, H. Zhao, R. Nolley, S.W. Reese, S.R. Young, X. Li, D.M. Peehl, S.J. Knox, Darinaparsin: solid tumor hypoxic cytotoxin and radiosensitizer. Clin. Cancer Res. 18 (2012) 3366–3376.

[153] S. Torii, D.A. Egan, R.A. Evans, J.C. Reed, Human Daxx regulates Fas-induced apoptosis from nuclear PML oncogenic domains (PODs). EMBO J. 18 (1999) 6037–6049.

[154] A.M. Tsimberidou, L.H. Camacho, S. Verstovsek, C. Ng, D.S. Hong, C.K. Uehara, C. Gutierrez, S. Daring, J. Stevens, P.B. Komarnitsky, B. Schwartz, R. Kurzrock, A phase I clinical trial of darinaparsin in patients with refractory solid tumors. Clin. Cancer Res. 15 (2009) 4769–4776.

[155] J. van der Zee, Heating the patient: a promising approach? Ann. Oncol. 13 (2002) 1173–1184.

[156] A. Verma, M. Mohindru, D.K. Deb, A. Sassano, S. Kambhampati, F. Ravandi, S. Minucci, D.V. Kalvakolanu, L.C. Platanias, Activation of Rac1 and the p38 mitogen-activated protein kinase pathway in response to arsenic trioxide. J. Biol. Chem. 277 (2002) 44988–44995.

[157] I. Vivanco, C.L. Sawyers, The phosphatidylinositol 3-Kinase AKT pathway in human cancer. Nat. Rev. Cancer 2 (2002) 489–501.

[158] J. Vuky, R. Yu, L. Schwartz, R.J. Motzer, Phase II trial of arsenic trioxide in patients with metastatic renal cell carcinoma. Invest. New Drugs 20 (2002) 327–330.

[159] J. Wang, M. Lin, T. Zhang, Y. Yan, P.C. Ho, Q.H. Xu, K.P. Loh, Arsenic(II) sulfide quantum dots prepared by a wet process from its bulk. J. Am. Chem. Soc. 130 (2008) 11596–11597.

[160] L. Wang, G.B. Zhou, P. Liu, J.H. Song, Y. Liang, X.J. Yan, F. Xu, B.S. Wang, J.H. Mao, Z.X. Shen, S.J. Chen, Z. Chen, Dissection of mechanisms of Chinese medicinal formula Realgar-Indigo naturalis as an effective treatment for promyelocytic leukemia. Proc. Natl. Acad. Sci. U. S. A. 105 (2008) 4826–4831.

[161] L.W. Wang, Y.L. Shi, N. Wang, B.D. Gou, T.L. Zhang, K. Wang, Association of oxidative stress with realgar-induced differentiation in human leukemia HL-60 cells. Chemotherapy 55 (2009) 460–467.

[162] M. Wang, S. Liu, P. Liu, Gene expression profile of multiple myeloma cell line treated by arsenic trioxide. J. Huazhong Univ. Sci. Technolog. Med. Sci. 27 (2007) 646–649.

[163] M.C. Wang, S.X. Liu, P.B. Liu, Gene expression profile of multiple myeloma cell line treated by realgar. J. Exp. Clin. Cancer Res. 25 (2006) 243–249.

[164] N. Wang, L.W. Wang, B. Gou, T.L. Zhang, Preparation of realgar nanoparticle suspension and its inhibition effect on the proliferation of human myelocytic leukaemia HL-60 cells. J. Disper. Sci. Technol. 30 (2005) 237–240.

[165] N. Wang, L.W. Wang, B.D. Gou, T.L. Zhang, K. Wang, Realgar-induced differentiation is associated with MAPK pathways in HL-60 cells. Cell Biol. Int. 32 (2008) 1497–1505.

[166] Z.G. Wang, D. Ruggero, S. Ronchetti, S. Zhong, M. Gaboli, R. Rivi, P.P. Pandolfi, PML is essential for multiple apoptotic pathways. Nat. Genet. 20 (1998) 266–272.

[167] Z.Y. Wang, Ham-Wasserman lecture: treatment of acute leukemia by inducing differentiation and apoptosis. Hematology. Am. Soc. Hematol. Educ. Program 2003) 1–13.

[168] Z.Y. Wang, Z. Chen, Acute promyelocytic leukemia: from highly fatal to highly curable. Blood 111 (2008) 2505–2515.

[169] S.H. Woo, M.J. Park, S. An, H.C. Lee, H.O. Jin, S.J. Lee, H.S. Gwak, I.C. Park, S.I. Hong, C.H. Rhee, Diarsenic and tetraarsenic oxide inhibit cell cycle progression and bFGF- and VEGF-induced proliferation of human endothelial cells. J. Cell. Biochem. 95 (2005) 120–130.

[170] J. Wu, C. Henderson, L. Feun, V.P. Van, P. Gold, H. Zheng, T. Ryan, L.S. Blaszkowsky, H. Chen, M. Costa, B. Rosenzweig, M. Nierodzik, H. Hochster, F. Muggia, G. Abbadessa, J. Lewis, A.X. Zhu, Phase II study of darinaparsin in patients with advanced hepatocellular carcinoma. Invest. New Drugs 28 (2010) 670–676.

[171] J. Wu, Y. Shao, J. Liu, G. Chen, P.C. Ho, The medicinal use of realgar (As(4)S(4)) and its recent development as an anticancer agent. J. Ethnopharmacol. 135 (2011) 595–602.

[172] J.Z. Wu, P.C. Ho, Evaluation of the in vitro activity and in vivo bioavailability of realgar nanoparticles prepared by cryo-grinding. Eur. J. Pharm. Sci. 29 (2006) 35–44.

[173] J.Z. Wu, P.C. Ho, Comparing the relative oxidative DNA damage caused by various arsenic species by quantifying urinary levels of 8-hydroxy-2′-deoxyguanosine with isotope-dilution liquid chromatography/mass spectrometry. Pharm. Res. 26 (2009) 1525–1533.

[174] Y.C. Wu, W.Y. Yen, T.C. Lee, L.H. Yih, Heat shock protein inhibitors, 17-DMAG and KNK437, enhance arsenic trioxide-induced mitotic apoptosis. Toxicol. Appl. Pharmacol. 236 (2009) 231–238.

[175] Y.C. Wu, W.Y. Yen, L.H. Yih, Requirement of a functional spindle checkpoint for arsenite-induced apoptosis. J. Cell. Biochem. 105 (2008) 678–687.

[176] W. Yan, A. Arai, M. Aoki, H. Ichijo, O. Miura, ASK1 is activated by arsenic trioxide in leukemic cells through accumulation of reactive oxygen species and may play a negative role in induction of apoptosis. Biochem. Biophys. Res. Commun. 355 (2007) 1038–1044.

[177] H.Q. Ye, L. Gan, X.L. Yang, H.B. Xu, Membrane toxicity accounts for apoptosis induced by realgar nanoparticles in promyelocytic leukemia HL-60 cells. Biol. Trace Elem. Res. 103 (2005) 117–132.

[178] L.H. Yih, I.C. Ho, T.C. Lee, Sodium arsenite disturbs mitosis and induces chromosome loss in human fibroblasts. Cancer Res. 57 (1997) 5051–5059.

[179] L.H. Yih, S.W. Hsueh, W.S. Luu, T.H. Chiu, T.C. Lee, Arsenite induces prominent mitotic arrest via inhibition of G2 checkpoint activation in CGL-2 cells. Carcinogenesis 26 (2005) 53–63.

[180] L.H. Yih, Y.Y. Tseng, Y.C. Wu, T.C. Lee, Induction of centrosome amplification during arsenite-induced mitotic arrest in CGL-2 cells. Cancer Res. 66 (2006) 2098–2106.

[181] L.H. Yih, Y.C. Wu, N.C. Hsu, H.H. Kuo, Arsenic trioxide induces abnormal mitotic spindles through a PIP4KIIgamma/Rho pathway. Toxicol. Sci. 128 (2012) 115–125.

[182] T. Yin, Y.L. Wu, H.P. Sun, G.L. Sun, Y.Z. Du, K.K. Wang, J. Zhang, G.Q. Chen, S.J. Chen, Z. Chen, Combined effects of As4S4 and imatinib on chronic myeloid leukemia cells and BCR-ABL oncoprotein. Blood 104 (2004) 4219–4225.

[183] M.H. Yoo, J.T. Kim, C.H. Rhee, M.J. Park, I.J. Bae, N.Y. Yi, M.B. Jeong, S.M. Jeong, T.C. Nam, K.M. Seo, Reverse effects of tetraarsenic oxide on the angiogenesis induced by nerve growth factor in the rat cornea. J. Vet. Med. Sci. 66 (2004) 1091–1095.

[184] Y. Yoshino, B. Yuan, T. Kaise, M. Takeichi, S. Tanaka, T. Hirano, D.L. Kroetz, H. Toyoda, Contribution of aquaporin 9 and multidrug resistance-associated protein 2 to differential sensitivity to arsenite between primary cultured chorion and amnion cells prepared from human fetal membranes. Toxicol. Appl. Pharmacol. 257 (2011) 198–208.

[185] S. Yuksel, G. Saydam, R. Uslu, U.A. Sanli, E. Terzioglu, F. Buyukececi, S.B. Omay, Arsenic trioxide and methylprednisolone use different signal transduction pathways in leukemic differentiation. Leuk. Res. 26 (2002) 391–398.

[186] J. Zhang, J.C. Wang, Y.H. Han, L.F. Wang, S.P. Ji, S.X. Liu, X.P. Liu, L.B. Yao, High expression of bcl-x(L) in K562 cells and its role in the low sensitivity of K562 to realgar-induced apoptosis. Acta Haematol. 113 (2005) 247–254.

[187] Q.Y. Zhang, J.H. Mao, P. Liu, Q.H. Huang, J. Lu, Y.Y. Xie, L. Weng, Y. Zhang, Q. Chen, S.J. Chen, Z. Chen, A systems biology understanding of the synergistic effects of arsenic sulfide and Imatinib in BCR/ABL-associated leukemia. Proc. Natl. Acad. Sci. U. S. A. 106 (2009) 3378–3383.

[188] Q.H. Zhao, Y. Zhang, Y. Liu, H.L. Wang, Y.Y. Shen, W.J. Yang, L.P. Wen, Anticancer effect of realgar nanoparticles on mouse melanoma skin cancer in vivo via transdermal drug delivery. Med. Oncol. 27 (2010) 203–212.

[189] Z. Zhao, J. Zuber, E. Diaz-Flores, L. Lintault, S.C. Kogan, K. Shannon, S.W. Lowe, p53 loss promotes acute myeloid leukemia by enabling aberrant self-renewal. Genes Dev. 24 (2010) 1389–1402.

[190] Y. Zhen, S. Zhao, Q. Li, Y. Li, K. Kawamoto, Arsenic trioxide-mediated Notch pathway inhibition depletes the cancer stem-like cell population in gliomas. Cancer Lett. 292 (2010) 64–72.

[191] X. Zheng, A. Seshire, B. Ruster, G. Bug, T. Beissert, E. Puccetti, D. Hoelzer, R. Henschler, M. Ruthardt, Arsenic but not all-trans retinoic acid overcomes the aberrant

stem cell capacity of PML/RARalpha-positive leukemic stem cells. Haematologica 92 (2007) 323–331.

[192] A.X. Zhou, Z.W. Chen, J.M. Yang, T.E. Wang, L. Yang, Q. Wang, R. Ma, F. Liu, J.F. Zheng, C.S. Deng, B.S. Yao, Z.F. Wang, N.P. Hu, Z.X. Wang. Clinical Study on Chronic Myelogenous Leukemia Treated with QingHuang San in 86 Cases. World Integrative Medicine Congress, Beijing, China, September 20–25, 1997.

[193] G.B. Zhou, W.L. Zhao, Z.Y. Wang, S.J. Chen, Z. Chen, Retinoic acid and arsenic for treating acute promyelocytic leukemia. PLoS Med. 2 (2005) e12.

[194] J. Zhu, M. Gianni, E. Kopf, N. Honore, M. Chelbi-Alix, M. Koken, F. Quignon, C. Rochette-Egly, T.H. de, Retinoic acid induces proteasome-dependent degradation of retinoic acid receptor alpha (RARalpha) and oncogenic RARalpha fusion proteins. Proc. Natl. Acad. Sci. U. S. A. 96 (1999) 14807–14812.

[195] J. Zhu, M.H. Koken, F. Quignon, M.K. Chelbi-Alix, L. Degos, Z.Y. Wang, Z. Chen, T.H. de, Arsenic-induced PML targeting onto nuclear bodies: implications for the treatment of acute promyelocytic leukemia. Proc. Natl. Acad. Sci. U. S. A. 94 (1997) 3978–3983.

[196] X.H. Zhu, Y.L. Shen, Y.K. Jing, X. Cai, P.M. Jia, Y. Huang, W. Tang, G.Y. Shi, Y.P. Sun, J. Dai, Z.Y. Wang, S.J. Chen, T.D. Zhang, S. Waxman, Z. Chen, G.Q. Chen, Apoptosis and growth inhibition in malignant lymphocytes after treatment with arsenic trioxide at clinically achievable concentrations. J. Natl. Cancer Inst. 91 (1999) 772–778.

[197] A. Zimber, Q.D. Nguyen, C. Gespach, Nuclear bodies and compartments: functional roles and cellular signalling in health and disease. Cell. Signal. 16 (2004) 1085–1104.

15

GENOTOXICITY

Ana María Salazar and Patricia Ostrosky-Wegman

Instituto de Investigaciones Biomédicas, Universidad Nacional Autónoma de México (UNAM), Mexico City, Mexico

15.1 INTRODUCTION

A compound is considered genotoxic if it induces DNA damage at subcytotoxic concentrations. Arsenic (As) has been shown to be genotoxic in a wide variety of experimental designs and biological endpoints. The induction by arsenic of chromosomal aberrations, micronuclei, DNA strand breakage, sister chromatid exchanges, and oxidative DNA damage is well documented. The genotoxicity of arsenic has been observed at comparatively low concentrations in cultured human peripheral lymphocytes and in several mammalian cell lines. In addition, arsenic induces chromosomal aberrations and micronuclei in vivo, as has been found in experiments with mice as well as in numerous epidemiological studies. Few studies have been performed in especially vulnerable populations, such as pregnant women, newborns, and children. DNA damage has clear implications for human health. Arsenic exposure has been associated with cancer, diabetes, and cardiovascular and neurodegenerative diseases [101].

Inorganic and organic arsenic species appear in the environment as trivalent, pentavalent, and organoarsenic forms. The trivalent inorganic arsenic (As^{III}) is the major inorganic form and is responsible for the genotoxic potential of arsenic. The As^{III}, which, at a physiological pH, is dissolved in water-forming arsenite, is transported into hepatocytes by simple and facilitated diffusion using transmembrane proteins, such as aquaglyceroporins and hexose permease transporters [70, 96, 118]. While inside the hepatic cells, the arsenite is biotransformed, leading to the formation of

Arsenic: Exposure Sources, Health Risks, and Mechanisms of Toxicity, First Edition.
Edited by J. Christopher States.
© 2016 John Wiley & Sons, Inc. Published 2016 by John Wiley & Sons, Inc.

trivalent and pentavalent mono-, di-, and trimethylated species [7, 51, 112, 116]. The pentavalent inorganic arsenate (As^V) is frequently present in surface water and, in the organism, is reduced to arsenite (for review, see Chapter 4).

The organic forms are mainly the methylated metabolites monomethylarsonic acid (MMA) and dimethylarsinic acid (DMA). The genotoxicity of the organoarsenicals MMA^V and DMA^V is restricted to higher concentrations (in the millimolar dose range), while the methylated species MMA^{III} and DMA^{III} are genotoxic at micromolar doses [21, 54, 108]. Although methylation has been considered as a detoxification process, the trivalent methylated metabolites have been shown in several test systems to be at least as or even more genotoxic than inorganic arsenic. The genotoxic potential of other organoarsenic compounds continues to be reported, such as roxarsone (4-hydroxy-3-nitrobenzenearsonic acid), which is widely used as a chicken feed additive and was recently shown to induce micronuclei and DNA damage by the in vitro comet assay in mammalian cells [137].

Oxidative stress is the main mechanism postulated to explain arsenic clastogenicity [129]. The data available indicate that arsenic may act indirectly, altering the regulation of DNA repair or integrity as a result of the production of reactive oxygen species (ROS) [68], which might also induce oxidative damage to proteins, inducing the impairment of proliferative signaling [32]. Aneuploidy, polyploidy, and mitotic disruption produced by arsenic exposure result from sulfhydryl binding [53]. A key role is played by p53 and $p21^{CIP1/WAF1}$ (CDKN1A; cyclin-dependent kinase 1A) [104].

This chapter provides an overview of the genotoxic effects of in vitro and in vivo arsenic exposure in cellular and animal models, as well as the results of human occupational and environmental arsenic exposure. Arsenic genotoxicity mechanisms are discussed.

15.2 MUTAGENICITY, CO-MUTAGENICITY, AND CO-GENOTOXICITY

Arsenicals have been shown to be nonmutagenic in bacterial and standard mammalian cell mutation assays which measure mutation at a single locus [40, 68]; thus, it has been concluded that arsenic does not cause point mutations. Arsenite does not induce point mutations in bacterial systems: arsenite was not significantly mutagenic in *Escherichia coli* selecting for tryptophan revertants or in the induction of ouabain- and thioguanine-resistant mutants in mammalian cells [98]. It has been shown that arsenic does not induce gene mutations in two genetic loci in Syrian hamster embryo cells [59]. Evaluation of the mutagenicity of methylated species in mouse lymphoma cells, *Salmonella* and by prophage induction in *E. coli*, indicates that the trivalent methylated arsenicals were not point mutagens [54]. The ability of arsenic compounds to induce gene mutations in mammalian cells has come under discussion as certain authors [125] have reported that sodium arsenite produces gene mutations (large-scale rearrangements, frameshifts, and base-pair substitutions) in the *supF* gene using the pZ189 shuttle vector system in DNA repair proficient GM637 human fibroblast. More recently, several arsenic compounds were evaluated using the mouse lymphoma assay, finding they were mutagenic at 10 and 20 μM [111].

What has become evident is that arsenic enhances the frequency of chemical mutations in a synergistic manner under the combined exposure with DNA damage agents in vitro. Thus, arsenite increases the mutagenicity of ultraviolet radiation (UV) in *E. coli* [97]. Furthermore, sodium arsenite at relatively low concentrations is co-mutagenic with *N*-methyl-*N*-nitrosourea (MNU) at the *hprt* locus in Chinese hamster V79 cells. With a nick translation assay, which measures DNA strand breakage by incorporating radioactive deoxyribonucleoside monophosphate at the 3′OH ends of the strands, it was demonstrated that the inhibition of MNU-induced DNA repair by arsenite occurs after the incision step [61], suggesting that the inhibition of DNA ligase activity by arsenite could be a possible mechanism of its co-mutagenesis [62]. Concerning arsenic co-genotoxicity, arsenite potentiates the x-ray- and UV-induced chromosomal damage in human lymphocytes and primary human fibroblast cultures [49]. In addition, chromosomal aberrations in lymphocytes exposed simultaneously to both arsenite and 1,3-butadiene diepoxide were increased [128], while co-incubation experiments with arsenic and antimony produces an increase in the micronuclei (MN) frequency in human lymphocytes [107]. Recent data indicating that arsenite and monomethylarsonous acid inhibit poly(ADP-ribose) polymerase 1 (PARP1) and discussed in Chapter 13 suggest a potential mechanism for the clastogenic effect of arsenic exposure.

15.3 *IN VITRO* GENOTOXICITY

Arsenic has been shown to be genotoxic in a wide variety of in vitro endpoints. The induction of gene amplification, chromosomal aberrations (CA), micronuclei (MN), DNA strand breaks (DSB), sister chromatid exchanges (SCE), and oxidative DNA damage is well documented. Several cell types have been used routinely, including human peripheral lymphocytes as well as various established cell lines.

Arsenite-induced gene amplification (a sign of genomic instability) at the dihydrofolate reductase *(dhfr)* locus in human and rodent cells but failed to cause the amplification in SV40-transformed human keratinocytes [10, 85, 100].

Significantly elevated frequencies of SCE, CA, and cell transformation were observed after arsenite treatment at nontoxic doses in Syrian hamster embryo cells [59]. The induction of SCE and DSB were detected at 2.5 and 5 μM doses in a human lung cell line. However, a reduction in the migration of DNA at a higher dose used in this study (10 μM) was observed, suggesting that another mode of action for arsenic is the formation of adducts [84]. DNA adducts and DNA-protein cross-links have been confirmed in both human leukemia cells and Chinese hamster ovary cells and also in human hepatic cells [93, 124]. The effects of arsenic have also been widely documented in primary cultures. Using the comet assay, arsenic did not produce DSB in leukocytes treated for 2 and 24 h with 20–150 μM arsenic [38]. Nevertheless, in cultured lymphocytes treated with MMA and DMA metabolites for 24 h, an increase in DSB was observed, suggesting that arsenical compounds produce more damage in cycling cells [110]. It is worth mentioning that MMA^{III} and DMA^{III} were more potent inducers of DSB than arsenite [76]. Because inorganic arsenic induces neurotoxic effects and neurological defects, other putative target cells were explored, finding a significant dose-dependent increase of DSB in the primary cultures of rat astrocytes treated with 2.5, 5, and 10 μM sodium arsenite [17].

Rasmussen and Menzel [95] reported a different sensitivity to arsenic-induced SCE in human lymphocytes and lymphoblastoid cell lines established by transformation with the Epstein–Barr virus. Notably, the lymphoblastoid cell lines retained their arsenic sensitivity after cryopreservation and subsequent revival [95]. Additionally, an individual susceptibility to the aneugenic effect of arsenite was observed in human lymphocytes treated in vitro [120]. Furthermore, the effect of sodium arsenite on cultured peripheral lymphocytes obtained from individuals with and without arsenic-related skin lesions showed an individual susceptibility among a population exposed to arsenic through drinking water [72]. Thus, several observations suggest that the metabolism of arsenic compounds varies among different individuals; therefore, biotransformation could be a crucial factor in the susceptibility to the toxic effects of arsenic.

Formation of CA and MN is evident in in vitro studies. Trivalent but not pentavalent arsenic at comparable doses significantly induced chromosomal aberrations (chromosomal and chromatid breakage) in cultured peripheral lymphocytes and mammalian cell lines at relatively low concentration ranges [18, 87, 95]. It is known that endoreduplication frequently leads to the formation of hyperploidy; arsenite also induces chromosomal endoreduplication in cultured human fibroblasts and Chinese hamster ovary cells [45, 57], but no endoreduplication was observed with arsenate treatment [57]. It is important to mention that these studies were performed using elevated concentrations of arsenic (50–200 μM ranges). In addition, hyperploidy and aneuploidy were observed in lymphocyte cultures treated with sodium arsenite at low concentrations [94, 102, 120].

Arsenite (0.5 and 2 μM) was reported to significantly induce MN in isolated human peripheral lymphocytes using the cytokinesis-block with cytochalasin B [10]. Previously, the induction of MN-containing chromosomal fragments was proposed to be the prominent effect of arsenic by Eastmond and Tucker [25]. The evaluation of MN-positive (MN+) or -negative (MN−) for centrosomes was conducted using an antikinetochore antibody; MN+ were less common than MN− in cultured human lymphocytes at 3 and 6 μM arsenite. In contrast, the prevalence of MN+ was greater after higher arsenite exposure (15 and 20 μM), indicating that arsenite-induced aneuploidy in the telomerase-immortalized diploid human tGM24 fibroblasts [104]. However, comparing the trivalent inorganic (arsenite, iAs^{III}) and organic (monomethylarsonous acid, MMA^{III}) species, an increase in the MN+ frequency was found using the fluorescence *in situ* hybridization technique with a centromere probe in peripheral human lymphocytes treated with 2 μM MMA^{III} [21]. Thus, this evidence shows a clear aneuploidogenic capacity of MMA^{III}.

In conclusion, increases in the MN frequency and chromosomal damage are well documented. In general, in vitro data suggest that arsenite acts as an aneugen. The aneugenic effect may be explained by interference with the spindle apparatus, disrupting the mitotic spindle of the cell [46, 63, 64, 94, 135].

15.4 *IN VIVO* GENOTOXICITY

15.4.1 Animal Studies

Chromosomal aberrations and micronuclei have been observed in mice after oral exposure to comparatively low concentrations of arsenite. Subsequent to 24 h of treatment with sodium and potassium arsenite at 10 mg/kg by intraperitoneal injection, micronuclei in polychromatic erythrocytes are induced in the bone marrow of BALB/c and C57BL mice, while another arsenical compound, such as As_2S_3, did not produce MN [117]. In a study performed to evaluate the genotoxic effects of exposure over long period of time, Swiss albino mice were exposed to sodium arsenite in drinking water at concentrations of 10, 50, 100, and 200 mg/L for a period of 3 months. In this study, arsenic induced an increase in DNA strand breakage in bone marrow and testicular cells; the magnitude of the DNA damage was greater in the bone marrow cells than in the testicular cells [13].

The generation of DNA damage by the administration of methylarsenic species has been demonstrated in rodents. DNA strand breakage was high in the lungs after at 12 h of DMAA administration but not in other tissues [129]. As a part of experiments carried out to elucidate the networks involved in hepatic arsenic-related genotoxicity, an elevation of MN in polychromatic erythrocytes in the peripheral blood was observed in Golden Syrian hamsters which received drinking water with 15 mg/L of sodium arsenite for 18 weeks [41]. The induction of dose-dependent DNA damage measured by comet assay was detected in the ovarian tissue of rats exposed to different concentrations (50, 100, and 200 ppm) of sodium arsenite in the drinking water for 28 days [1].

15.4.2 Human Studies

Numerous genotoxicity studies in relation to arsenic exposure have included exposed individuals from several populations, mainly after long-term exposure via the drinking water and/or ambient air. However, studies in relation to occupational exposure are few. Several cell types have been used routinely for these studies, including peripheral lymphocytes, buccal, and bladder cells.

15.4.2.1 Environmental Exposure to Arsenic After long-term exposure to arsenic-contaminated drinking water, the great majority of biomonitoring studies found elevated frequencies of DNA lesions, such as micronuclei or chromosomal aberrations [1, 8, 12, 22, 24, 33, 65, 71, 73, 89]. An elevation of chromosomal aberration frequency in lymphocytes, as well as high micronuclei frequency in buccal and urothelial cells, were detected in individuals from a Mexican group chronically exposed to high arsenic (408 μg/L) compared to a group exposed to low arsenic (30 μg/L) in drinking water [33]. The MN frequency was elevated in a study performed in exfoliated bladder cells obtained from a Chilean population chronically exposed to high and low arsenic levels (600 and 15 μg/L, respectively). Moreover, when the group exposed to high levels was assessed by urinary arsenic levels, the high-exposure group (urinary arsenic 84–1893 μg/L) had more MN+ than MN– [83]. In another study of a Chilean population exposed to arsenic levels in the drinking water as high 750 μg/L, a significant increase in the frequency of MN in peripheral blood lymphocytes also was observed [74]. The same research group carried out a study to evaluate the frequency of MN in buccal cells, but no significant differences were observed in this study [75]. In contrast, a study of an Indian population exposed to arsenic through the drinking water (214.7 ± 9.03 μg/L) showed that MN frequencies in the exposed group were elevated 5.33-fold over unexposed levels in lymphocytes, 4.63-fold in the oral mucosa cells, and 4.71-fold in urothelial cells, indicating that the chronic ingestion of arsenic in drinking water by the exposed subjects is linked to the enhanced frequency of micronuclei in all the three cell types [11]. Increased rates of CA were reported previously in the same population [71]. Recently, an association between arsenic exposure, dermatological lesions, hypertension, and chromosomal abnormalities was evaluated among Iranian people. When comparing two Iranian populations, one exposed to 1031 μg/L arsenic in drinking water, and the other (control group) to nondetectable levels; it was found that the incidence of hyperkeratosis was 34 times higher in the exposed group, together with an increase in diastolic blood pressure and chromosomal abnormalities [22].

The MN frequency was also elevated in peripheral blood lymphocytes from women and children from northwestern Argentina exposed to 205 μg/L arsenic in drinking water. The data obtained indicate that the proportion of MN appears to originate from whole chromosome loss, as was supported by the fluorescence in situ hybridization technique. This study is a one of few studies regarding genotoxicity induced by arsenic in a population of children [24].

Two other studies evaluated DNA damage in children exposed to arsenic and lead in Mexico. One evaluated children exposed simultaneously to both metals and did not find a significant association between basal DNA damage and urinary arsenic nor with lead levels in the blood. However, lymphocytes from 58% of the children did not respond to the peroxide challenge and had a more severe basal DNA damage, suggesting that the DNA repair capacity is diminished [80]. The other study evaluated three groups of children living in a mining site in Mexico [48, 134]. The DNA damage was higher in the blood cells of the children exposed to arsenic, whereas the capacity for DNA repair in response to peroxide was found to be diminished in both exposed groups [48].

Other markers of DNA damage evaluated in individuals exposed to arsenic have shown positive correlations between the marker and the exposure: DNA fragmentation in buccal cells [27]; P53 expression in lymphocytes from nonmelanoma skin cancer patients [103]; and oxidative DNA adducts, such as 8-hydroxy-2-deoxyguanosine (8-OHdG), in arsenic-related skin neoplasms or in the urine of patients from an accidental oral intake of arsenic [78, 133].

It is worth mentioning that some studies have shown that genetic polymorphism might play a role in the susceptibility to arsenic-induced DNA damage. For example, in a case control study performed in India, in 206 cases with skin lesions and 215 controls without skin lesions having similar arsenic exposure, the presence of at least one Met allele in the XRCC3 T241M polymorphism was found to be a protective genotype, decreasing the frequency of chromosomal aberrations, skin lesions, peripheral neuropathy, and conjunctivitis [58]. In another study performed in children and adults exposed to elevated levels of drinking water arsenic in northern Mexico, a dose-dependent increase in DNA damage was observed and the arsenic 3 methyltransferase (AS3MT) Met287Thr polymorphism was found to correlate with the highest DNA damage observed with the comet assay [106].

15.4.2.2 Occupational Exposure to Arsenic Several studies have detected elevated DNA damage in human peripheral lymphocytes of workers occupationally exposed to arsenic in copper smelters [60, 86, 91, 123]. More recently, a significant increase in the MN frequency in peripheral blood lymphocytes and in buccal epithelial cells of smelter workers was observed. While the fluorescence in situ hybridization (FISH) technique revealed the presence of clastogenic and aneugenic effects in peripheral blood lymphocytes in both groups, the clastogenic effect was slightly more pronounced in the smelter workers; however, the difference was not statistically significant [60]. The level of DNA damage, expressed as the median tail moment, was significantly higher in the leukocytes of workers than in the controls. Using a variant of the comet assay conducted with formamidopyrimidine glycosylase (FPG) digestion to detect oxidative DNA damage indicated that the lesions were present in leukocytes from both the exposed and control groups, but the levels of damage were significantly higher among the workers. Incubation of the cells in culture resulted in a significant reduction in the levels of DNA damage, especially among leukocytes from the workers, suggesting that the DNA damage could be repaired [91]. In contrast, no significant differences in micronuclei frequencies were detected in lymphocytes from smelting plant workers exposed to arsenic [90]. Occupational exposure to arsenic among workers in a glass plant in India showed DNA damage in leukocytes from the group with higher levels of arsenic in the blood [123].

15.5 POSSIBLE MECHANISMS

15.5.1 Role of Oxidative Stress in Mediating the Genotoxic Response

Oxidative stress has been proposed as a plausible general mode of action for arsenic toxicity and carcinogenesis [40, 56, 68]. Data on covalent binding between arsenic and DNA structures have led to the proposal that the DNA damage observed during arsenic exposure is indirect [52], occurring mainly as a result of ROS induction, which generates chromosomal aberrations, DNA strand breakage, DNA adducts, and DNA-protein cross-links [124]. Figure 15.1 illustrates the major proposed mechanisms of arsenic-induced genotoxicity; it is evident that oxidative stress plays a main role. Oxidative stress is characterized by the generation of several reactive oxygen species (ROS), such as superoxide anion (O_2^-), hydroxyl radical (·OH), hydrogen peroxide (H_2O_2), singlet oxygen ($^1O_2$), and peroxyl radical (LOO), among others [119].

The primary species formed in arsenic-induced oxidative stress is O_2^-, followed by a cascade of secondary ROS, such as H_2O_2 and ·OH [109]. The O_2^- and H_2O_2

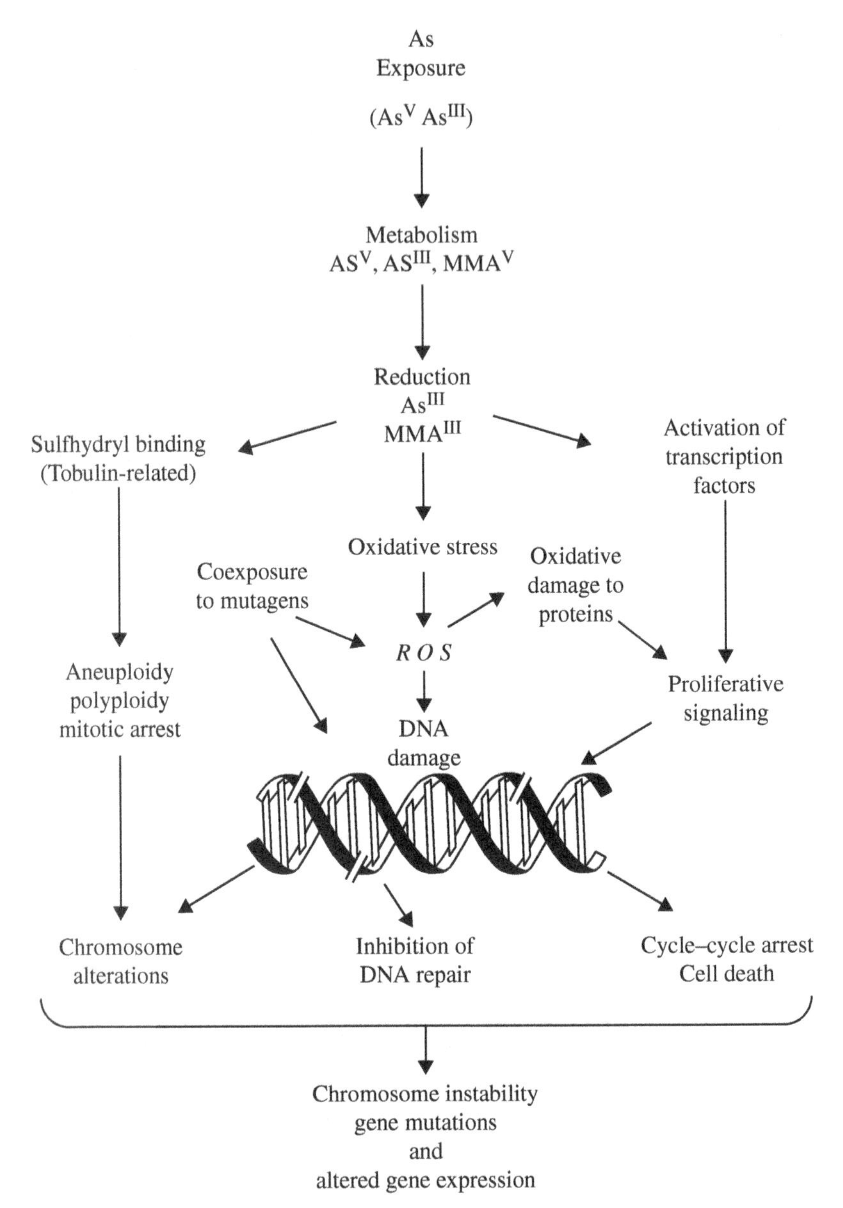

FIGURE 15.1 Proposed mechanisms of genotoxicity induced by arsenic.

species are the predominant reactive species produced by endothelial cells after arsenite exposure and an increase in oxygen cell consumption contributing to increased ROS production, stimulating cell signaling and activating transcription factors [9]. It has also been proposed that arsenic can affect the expression of stress-response genes, DNA repair and damage-responsive genes, and the activation of transcription factors such as the AP-1 complex. Arsenite can also increase proinflammatory cytokines, which could influence the response to acute arsenic toxicity [67]. Alternatively, the dimethylarsenic peroxyl radical and H_2O_2 anion radical produced from the reaction between molecular oxygen and dimethylarsine suggest that these two radicals cause DNA single-strand breakage [132]. In contrast, ROS scavengers can suppress arsenic-induced oxidative stress and its mutagenic and cytotoxic effects in cells [29, 40, 68].

Alternatively, DNA modifications due to arsenic-induced ROS can produce oxidative damage, with the hydroxyl radical (·OH) attack on DNA, generating multiple purine, pyrimidine, and deoxyribose oxidation products, such as the 8-hydroxy-2′-deoxyguanosine (8OHdG), being the most studied [35]. Markers of oxidative stress, including 8-oxodG were detected after DMA^V administration to bronchiolar Clara cells from mice [4]. It has also been demonstrated that the presence of 8-OHdG is associated with arsenic-related human keratoses, squamous cell carcinoma and basal cell epithelioma [3]. As^{III} can induce 8-OHdG and promote genomic instability by damaging DNA, inducing oncogene expression (including several factors regulating cell cycle progression) in human breast cancer MCF-7 adenocarcinoma epithelial cells [101].

Arsenic-induced single-strand DNA breakage is likely caused through ROS, either directly by free-radical attack on the DNA bases or indirectly during the course of base excision repair (BER) mechanisms [55]. BER is the predominant repair pathway for DNA lesions caused by ROS and the first candidate in arsenic-related DNA repair [26]. Arsenic-induced ROS have been shown to promote breakage in murine lungs [132]. Arsenic can induce DNA damage by interfering with the DNA repair processes [39, 61]. In addition, DMA^V affects DNA repair and replication mechanisms in human alveolar cells, leading to the persistence of DNA damage (mainly apurinic/apyrimidinic sites) and the generation of DNA breakage as a consequence [130, 131]. Arsenic was shown to alter BER mechanisms in both lung fibroblasts and keratinocytes, increasing the levels of BER-related enzymes and repair capacity [113]. DNA enzymes, such as polymerase β and DNA ligase I, participate in the BER mechanism and are known to be modulated by arsenic exposure [26].

It has been proposed that As^{III} acts at the transcriptional level to repress a group of genes encoding for DNA repair enzymes participating in BER and NER mechanisms mainly through its down-regulation [36]. In human exposed populations, exposure to arsenic in the drinking water was correlated to the decreased expression of ERCC1, XPB, and XPF in lymphocytes from exposed individuals [5, 6]. OGG1 expression was strongly associated with arsenic concentration, revealing the involvement of mechanisms related to the effects of arsenic-mediated ROS on DNA [82]. In a recent study, hypertension was shown to be correlated with high urinary arsenic levels and with the polymorphisms OGG1 Cys-Cys and MnSOD Val-Ala/Ala-Ala [19].

Because mitochondria are a major source of intracellular ROS, arsenic-mediated disruption of their function can lead to an increase in intracellular ROS levels and, subsequently, to an increased mutagenic potential, either directly or by decreasing DNA repair capacity [92]. These results show arsenic-associated mitochondrial dysfunction, mitochondrial DNA (mtDNA) depletion, and induction of mtDNA deletions in mammalian cells [92]. Moreover, the suppression of arsenic-induced apoptosis in HeLa cells by the antioxidant action of *N*-acetyl-cysteine prevents mitochondrial membrane depolarization [126]. More evidence has shown that arsenic opens the permeability transition pores by condensing the mitochondrial matrix [16].

Arsenic is also able to alter the generation of reactive nitrogen species (RNS) [119]. Nitric oxide is a reactive radical generated in biological tissues by specific nitric oxide synthases (NOS), which metabolize arginine to citrulline and are important in oxidative biological signaling [2]. It has been shown that nitric oxide possesses toxic effects, such as pro-oxidant activity and genotoxicity [66]. In fact, mitochondrial damage can mediate the genotoxicity of arsenic in mammalian cells, and this damage can lead to the release of superoxide anions, which then react with nitric oxide to produce the highly reactive peroxynitrites [69].

15.5.2 DNA Damage: Oxidative Stress and P53

The p53 tumor suppressor gene (*TP53*) plays a key role in safeguarding the integrity of the human genome; P53 is a regulatory transcriptional factor activated in response to several stress signals, including DNA damage, resulting in the transactivation of target genes (for review see Ref. [122]). The P53 transactivation of downstream genes such as *CDKN1A* helps to block cell cycle progression, allowing time for DNA repair before replication [23, 50], or causes apoptosis (via induction of Bax and down-regulation of Bcl-2) in heavily damaged cells [81]. The activation of P53 involves an increase in protein levels as well as structural changes in the protein. Post-translational modifications generally result in P53 stabilization and accumulation in the nucleus, where P53 interacts with sequence-specific sites on its target genes [42, 127]. The role of P53 as a mediator of the DNA damage response is demonstrated by the genome instability and aneuploidy observed in many P53-deficient cancer cells [14, 30, 37]. Interestingly, the P53 pathway is extremely sensitive to DNA strand breakage [44]. Moreover, a mutation in *TP53* has been found in different tumors [34].

Therefore, if *TP53* is an important component affected during the carcinogenesis process, this gene could be a potential target of arsenic. Evidence of the possible interaction of arsenic with *TP53* is demonstrated in several studies. Exposure of human cells to increasing concentrations (0.1–50 μM) of arsenite for 24 h resulted in a dose-dependent increase in the level of P53 protein expression, especially in cells with wild-type *TP53* [105]. Several arsenic compounds were tested for their ability to increase the cellular level of P53; As^{III} and dimethylated compounds (DMA^{III} and DMA^{V}) activate P53, whereas As^{V} and monomethylated compounds (MMA^{III} and MMA^{V}) had no effect [28]. DNA damage by arsenic results in an accumulation of the P53 protein, mainly via post-translational stabilization [47]. In addition, the potential role of *TP53* in arsenic-induced cell cycle arrest has been suggested via *CDKN1A* ("*p21*") [121, 136]. Exposure to 5 μM arsenite induced DSB in human fibroblasts, accompanied by phosphorylation and accumulation of the P53 and increases in P53 target genes, including $p21^{CIP1/WAF1}$ and MDM2 [136]. Arsenite caused a modest increase in P53 protein levels in normal human fibroblasts exposed to a long-term (14 days) low dose (0.1 μM), while toxic concentrations (50 μM) increased P53 levels after short-term exposure (18 h) [121]. Arsenic was found to interact indirectly by upregulating the expression of its regulators, such as the MDM2 protein [47], or targets of P53, such as GADD45 [15]. Remarkably, arsenite failed to increase P53 accumulation in the presence of Wortmannin or in Ataxia telangiectasia fibroblasts, implicating the activation via the ATM kinase pathway. It has been proposed that the absence of normal P53 functioning along with increased positive growth signaling in the presence of DNA damage may contribute to defective DNA repair and account for the co-mutagenic effects of arsenite [99].

Furthermore, the participation of oxidative stress is also implicated in the biochemical pathway of P53. In fact, ROS production might contribute to the cytostatic and pro-apoptotic effect of P53. Thus, the antiproliferative effects of ROS involve the activation of P53, not only indirectly (by producing DNA damage), but also by promoting P53 phosphorylation [43]. P53 promotes mitochondrial respiration by maintaining cytochrome oxidase activity in cell lines [77]. Nevertheless, if high levels of ROS are produced, inhibition of the P53 activity occurs by the oxidation of cysteine residues [20]. Thus, P53 might be a key piece in the genotoxic damage induced by arsenic. Amazingly, arsenic exposure produces ROS, causing apoptosis directly or via the activation of the P53 pathway (reviewed in Ref. [109]) by binding to the sulfhydryls (reviewed in Ref. [53]).

15.5.3 The Role of P53 and CDKN1A (p21$^{CIP1/WAF1}$) in Arsenic-Induced Genetic Damage

Arsenic in the trivalent form (arsenite) causes both aneuploidy and clastogenesis. Both effects have been proposed as mechanisms for arsenic carcinogenesis [88].

The loss of the mitotic checkpoint, abnormal centrosome amplification and defects in the kinetochore-microtubule attachment are events that disrupt chromosome segregation and contribute to aneuploidogenesis [31]. In fact, the spindle assembly checkpoint, activated immediately upon entry into mitosis, is the main mitotic checkpoint controlling mitotic progression, ensuring the proper alignment of the chromosomes at the metaphase plate before segregation. The major regulator of the spindle assembly checkpoint is the mitotic checkpoint complex composed of MAD2, BUB1, and MAD3/BUBR1 [114]. However, it is likely that regulation may occur by other pathways, such as ATM and p21$^{CIP1/WAF1}$, which suppress aneuploidy in a genome instability background; P53 may also play a role in suppressing aneuploidy by regulating mitotic progression either directly or indirectly through p21$^{CIP1/WAF1}$ [115].

In this context, arsenite induces the accumulation of mitotic cells with nonfunctional P53 but not in P53-expressing cells. In addition, the P53-deficient cells undergo apoptosis, whereas most of P53-expressing cells exited from the G_2/M phase and are arrested in the G_1 phase [79]. Furthermore, P53-dependent p21$^{CIP1/WAF1}$ induction releases cells from mitotic arrest-associated apoptosis [115]. Thus, the effect of the suppression of P53 and p21$^{CIP1/WAF1}$ by siRNA on chemical-induced aneuploidy and clastogenesis in telomerase-immortalized diploid human fibroblasts was investigated, using a FISH technique with a centromeric DNA probe. MN+ in cells transfected with scrambled control siRNA or mock-transfected increased after arsenite exposure,

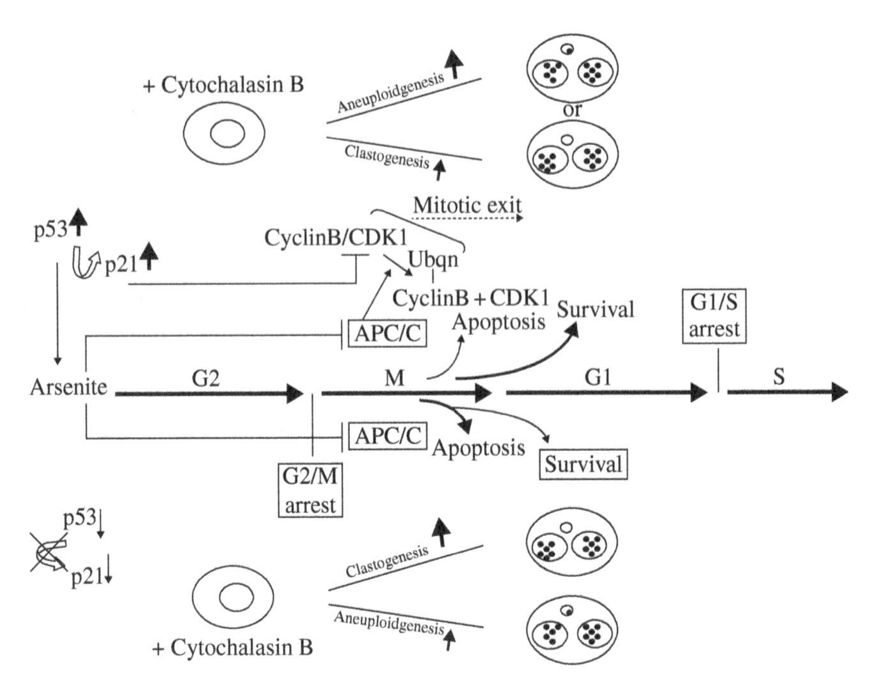

FIGURE 15.2 Schematic of proposed role of p21$^{CIP1/WAF}$ induction in exit from arsenite-induced mitotic delay and the effect on aneuploidy. Reprinted with permission from Salazar et al. [104].

indicating that arsenite-induced aneuploidy, while the suppression of p21$^{CIP1/WAF1}$, like P53 suppression, decreased the fraction of MN with centromeric signals (MN+) [104]. These results suggest that cells lacking normal P53 function cannot become aneuploid because they die by mitotic arrest-associated apoptosis (mitotic catastrophe), while cells with a normal P53 function are able to exit from mitotic arrest and can become aneuploid (Fig. 15.2) [104]. Elucidation of this recovery process might provide information relevant to the mechanisms of both arsenic carcinogenesis and chemotherapy by mitotic disruption.

15.6 CONCLUSIONS

There is evidence supporting the induction of transformation by arsenic, but arsenic has also been shown to have a low mutagenic activity. Arsenic is also proposed to act as a co-mutagen and a co-genotoxicant, exacerbating damage induced by other agents. Thus, arsenic increases UV-mediated DNA breakage by interfering with the repair process. Arsenic is a known inducer of chromosomal and chromatid aberrations as well as DNA strand breakage even at low concentrations. It has been proposed that much of the DNA damage observed during arsenic exposure is indirect, occurring mainly as a result of ROS induction, which generates DNA adducts, DNA strand breakage, cross-links, and chromosomal aberrations. Human cell lines exposed to arsenic exhibit DNA-protein adducts, as well as sister chromatid exchanges. DNA modifications due to arsenic-induced ROS can produce oxidative damage, which can be measured through the presence of the products of guanine oxidation in position 8. Methylated trivalent species, such as MMA^{III} and DMA^{III}, were found to be potent genotoxins. Cytogenetic aberrations and micronuclei have been observed in lymphocytes as well in oral and bladder exfoliated cells of arsenic-exposed populations.

Arsenic has been shown to increase p53 expression, and some data suggest that the suppression of p53 and $p21^{CIP/WAF1}$ is key in arsenite-induced aneuploidy.

Because oxidative stress induction by arsenic is well documented, it has been proposed that free oxygen radicals are the main mechanism of DNA damage, producing chromosomal aberrations, DNA strand breakage, and adducts. Oxidative stress is known to accelerate the development of diseases such as cancer, diabetes, Alzheimer's, Parkinson's, and cardiovascular diseases.

REFERENCES

[1] Z. Akram, S. Jalali, S.A. Shami, L. Ahmad, S. Batool, O. Kalsoom, Genotoxicity of sodium arsenite and DNA fragmentation in ovarian cells of rat. Toxicol. Lett. 190 (2009) 81–85.

[2] W.K. Alderton, C.E. Cooper, R.G. Knowles, Nitric oxide synthases: structure, function and inhibition. Biochem. J. 357 (2001) 593–615.

[3] Y. An, Z. Gao, Z. Wang, S. Yang, J. Liang, Y. Feng, K. Kato, M. Nakano, S. Okada, K. Yamanaka, Immunohistochemical analysis of oxidative DNA damage in arsenic-related human skin samples from arsenic-contaminated area of China. Cancer Lett. 214 (2004) 11–18.

[4] Y. An, K. Kato, M. Nakano, H. Otsu, S. Okada, K. Yamanaka, Specific induction of oxidative stress in terminal bronchiolar Clara cells during dimethylarsenic-induced lung tumor promoting process in mice. Cancer Lett. 230 (2005) 57–64.

[5] A.S. Andrew, J.L. Burgess, M.M. Meza, E. Demidenko, M.G. Waugh, J.W. Hamilton, M.R. Karagas, Arsenic exposure is associated with decreased DNA repair in vitro and in individuals exposed to drinking water arsenic. Environ. Health Perspect. 114 (2006) 1193–1198.

[6] A.S. Andrew, M.R. Karagas, J.W. Hamilton, Decreased DNA repair gene expression among individuals exposed to arsenic in United States drinking water. Int. J. Cancer 104 (2003) 263–268.

[7] H.V. Aposhian, M.M. Aposhian, Arsenic toxicology: five questions. Chem. Res. Toxicol. 19 (2006) 1–15.

[8] M. Banerjee, N. Banerjee, P. Ghosh, J.K. Das, S. Basu, A.K. Sarkar, J.C. States, A.K. Giri, Evaluation of the serum catalase and myeloperoxidase activities in chronic arsenic-exposed individuals and concomitant cytogenetic damage. Toxicol. Appl. Pharmacol. 249 (2010) 47–54.

[9] A. Barchowsky, L.R. Klei, E.J. Dudek, H.M. Swartz, P.E. James, Stimulation of reactive oxygen, but not reactive nitrogen species, in vascular endothelial cells exposed to low levels of arsenite. Free Radic. Biol. Med. 27 (1999) 1405–1412.

[10] J.C. Barrett, P.W. Lamb, T.C. Wang, T.C. Lee, Mechanisms of arsenic-induced cell transformation. Biol. Trace Elem. Res. 21 (1989) 421–429.

[11] A. Basu, P. Ghosh, J.K. Das, A. Banerjee, K. Ray, A.K. Giri, Micronuclei as biomarkers of carcinogen exposure in populations exposed to arsenic through drinking water in West Bengal, India: a comparative study in three cell types. Cancer Epidemiol. Biomarkers Prev. 13 (2004) 820–827.

[12] A. Basu, J. Mahata, A.K. Roy, J.N. Sarkar, G. Poddar, A.K. Nandy, P.K. Sarkar, P.K. Dutta, A. Banerjee, M. Das, K. Ray, S. Roychaudhury, A.T. Natarajan, R. Nilsson, A.K. Giri, Enhanced frequency of micronuclei in individuals exposed to arsenic through drinking water in West Bengal, India. Mutat. Res. 516 (2002) 29–40.

[13] R. Biswas, S. Poddar, A. Mukherjee, Investigation on the genotoxic effects of long-term administration of sodium arsenite in bone marrow and testicular cells in vivo using the comet assay. J. Environ. Pathol. Toxicol. Oncol. 26 (2007) 29–37.

[14] F. Borel, O.D. Lohez, F.B. Lacroix, R.L. Margolis, Multiple centrosomes arise from tetraploidy checkpoint failure and mitotic centrosome clusters in p53 and RB pocket protein-compromised cells. Proc. Natl. Acad. Sci. U. S. A. 99 (2002) 9819–9824.

[15] J.J. Bower, S.S. Leonard, F. Chen, X. Shi, As(III) transcriptionally activates the gadd45a gene via the formation of H2O2. Free Radic. Biol. Med. 41 (2006) 285–294.

[16] J. Bustamante, L. Nutt, S. Orrenius, V. Gogvadze, Arsenic stimulates release of cytochrome c from isolated mitochondria via induction of mitochondrial permeability transition 6. Toxicol. Appl. Pharmacol. 207 (2005) 110–116.

[17] I. Catanzaro, G. Schiera, G. Sciandrello, G. Barbata, F. Caradonna, P. Proia, I. Di Liegro, Biological effects of inorganic arsenic on primary cultures of rat astrocytes. Int. J. Mol. Med. 26 (2010) 457–462.

[18] T. Chakraborty, M. De, Clastogenic effects of inorganic arsenic salts on human chromosomes in vitro. Drug Chem. Toxicol. 32 (2009) 169–173.

[19] S.C. Chen, C.C. Chen, C.Y. Kuo, C.H. Huang, C.H. Lin, Z.Y. Lu, Y.Y. Chen, H.S. Lee, R.H. Wong, Elevated risk of hypertension induced by arsenic exposure in Taiwanese rural residents: possible effects of manganese superoxide dismutase (MnSOD) and 8-oxoguanine DNA glycosylase (OGG1) genes. Arch. Toxicol. 86 (2012) 869–878.

[20] C.S. Cobbs, M. Samanta, L.E. Harkins, G.Y. Gillespie, B.A. Merrick, L.A. MacMillan-Crow, Evidence for peroxynitrite-mediated modifications to p53 in human gliomas: possible functional consequences. Arch. Biochem. Biophys. 394 (2001) 167–172.

[21] R. Colognato, F. Coppede, J. Ponti, E. Sabbioni, L. Migliore, Genotoxicity induced by arsenic compounds in peripheral human lymphocytes analysed by cytokinesis-block micronucleus assay. Mutagenesis 22 (2007) 255–261.

[22] S. Dastgiri, M. Mosaferi, M.A. Fizi, N. Olfati, S. Zolali, N. Pouladi, P. Azarfam, Arsenic exposure, dermatological lesions, hypertension, and chromosomal abnormalities among people in a rural community of northwest Iran. J. Health Popul. Nutr. 28 (2010) 14–22.

[23] F.A. Derheimer, M.B. Kastan, Multiple roles of ATM in monitoring and maintaining DNA integrity. FEBS Lett. 584 (2010) 3675–3681.

[24] F.N. Dulout, C.A. Grillo, A.I. Seoane, C.R. Maderna, R. Nilsson, M. Vahter, F. Darroudi, A.T. Natarajan, Chromosomal aberrations in peripheral blood lymphocytes from native Andean women and children from northwestern Argentina exposed to arsenic in drinking water. Mutat. Res. 370 (1996) 151–158.

[25] D.A. Eastmond, J.D. Tucker, Identification of aneuploidy-inducing agents using cytokinesis-blocked human lymphocytes and an antikinetochore antibody. Environ. Mol. Mutagen. 13 (1989) 34–43.

[26] F. Ebert, A. Weiss, M. Bultemeyer, I. Hamann, A. Hartwig, T. Schwerdtle, Arsenicals affect base excision repair by several mechanisms. Mutat. Res. 715 (2011) 32–41.

[27] Z. Feng, Y. Xia, D. Tian, K. Wu, M. Schmitt, R.K. Kwok, J.L. Mumford, DNA damage in buccal epithelial cells from individuals chronically exposed to arsenic via drinking water in Inner Mongolia, China. Anticancer Res 21 (2001) 51–57.

[28] M. Filippova, P.J. Duerksen-Hughes, Inorganic and dimethylated arsenic species induce cellular p53. Chem. Res. Toxicol. 16 (2003) 423–431.

[29] S.J. Flora, Arsenic-induced oxidative stress and its reversibility. Free Radic. Biol. Med. 51 (2011) 257–281.

[30] T. Fujiwara, M. Bandi, M. Nitta, E.V. Ivanova, R.T. Bronson, D. Pellman, Cytokinesis failure generating tetraploids promotes tumorigenesis in p53-null cells. Nature 437 (2005) 1043–1047.

[31] K. Fukasawa, Centrosome amplification, chromosome instability and cancer development. Cancer Lett. 230 (2005) 6–19.

[32] P.R. Gentry, T.B. McDonald, D.E. Sullivan, A.M. Shipp, J.W. Yager, H.J. Clewell, III, Analysis of genomic dose-response information on arsenic to inform key events in a mode of action for carcinogenicity. Environ. Mol. Mutagen. 51 (2010) 1–14.

[33] M.E. Gonsebatt, L. Vega, A.M. Salazar, R. Montero, P. Guzman, J. Blas, L.M. Del Razo, G. Garcia-Vargas, A. Albores, M.E. Cebrian, M. Kelsh, P. Ostrosky-Wegman, Cytogenetic effects in human exposure to arsenic. Mutat. Res. 386 (1997) 219–228.

[34] M.S. Greenblatt, W.P. Bennett, M. Hollstein, C.C. Harris, Mutations in the p53 tumor suppressor gene: clues to cancer etiology and molecular pathogenesis. Cancer Res. 54 (1994) 4855–4878.

[35] B. Halliwell, Oxidative stress and cancer: have we moved forward? Biochem. J. 401 (2007) 1–11.

[36] H.K. Hamadeh, K.J. Trouba, R.P. Amin, C.A. Afshari, D. Germolec, Coordination of altered DNA repair and damage pathways in arsenite-exposed keratinocytes. Toxicol. Sci. 69 (2002) 306–316.

[37] W. Hanel, U.M. Moll, Links between mutant p53 and genomic instability. J. Cell. Biochem. 113 (2012) 433–439.

[38] A. Hartmann, G. Speit, Comparative investigations of the genotoxic effects of metals in the single cells gel (SCG) assay and the sister chromatid exchange (SCE) test. Environ. Mol. Mutagen. 23 (1994) 299–305.

[39] A. Hartwig, T. Schwerdtle, Interactions by carcinogenic metal compounds with DNA repair processes: toxicological implications. Toxicol. Lett. 127 (2002) 47–54.

[40] T.K. Hei, S.X. Liu, C. Waldren, Mutagenicity of arsenic in mammalian cells: role of reactive oxygen species. Proc. Natl. Acad. Sci. U. S. A. 95 (1998) 8103–8107.

[41] A. Hernandez, A. Sampayo-Reyes, R. Marcos, Identification of differentially expressed genes in the livers of chronically i-As-treated hamsters. Mutat. Res. 713 (2011) 48–55.

[42] R. Hess, B. Plaumann, A.S. Lutum, C. Haessler, B. Heinz, M. Fritsche, G. Brandner, Nuclear accumulation of p53 in response to treatment with DNA-damaging agents. Toxicol. Lett. 72 (1994) 43–52.

[43] L.J. Hofseth, S. Saito, S.P. Hussain, M.G. Espey, K.M. Miranda, Y. Araki, C. Jhappan, Y. Higashimoto, P. He, S.P. Linke, M.M. Quezado, I. Zurer, V. Rotter, D.A. Wink, E. Appella, C.C. Harris, Nitric oxide-induced cellular stress and p53 activation in chronic inflammation. Proc. Natl. Acad. Sci. U. S. A. 100 (2003) 143–148.

[44] L.C. Huang, K.C. Clarkin, G.M. Wahl, Sensitivity and selectivity of the DNA damage sensor responsible for activating p53-dependent G1 arrest. Proc. Natl. Acad. Sci. U. S. A. 93 (1996) 4827–4832.

[45] R.N. Huang, I.C. Ho, L.H. Yih, T.C. Lee, Sodium arsenite induces chromosome endoreduplication and inhibits protein phosphatase activity in human fibroblasts. Environ. Mol. Mutagen. 25 (1995) 188–196.

[46] S.C. Huang, T.C. Lee, Arsenite inhibits mitotic division and perturbs spindle dynamics in HeLa S3 cells. Carcinogenesis 19 (1998) 889–896.

[47] Y. Huang, J. Zhang, K.T. McHenry, M.M. Kim, W. Zeng, V. Lopez-Pajares, C.C. Dibble, J.P. Mizgerd, Z.M. Yuan, Induction of cytoplasmic accumulation of p53: a mechanism for low levels of arsenic exposure to predispose cells for malignant transformation. Cancer Res. 68 (2008) 9131–9136.

[48] Y. Jasso-Pineda, F. Diaz-Barriga, J. Calderon, L. Yanez, L. Carrizales, I.N. Perez-Maldonado, DNA damage and decreased DNA repair in peripheral blood mononuclear cells in individuals exposed to arsenic and lead in a mining site. Biol. Trace Elem. Res. 146 (2012) 141–149.

[49] A.N. Jha, M. Noditi, R. Nilsson, A.T. Natarajan, Genotoxic effects of sodium arsenite on human cells. Mutat. Res. 284 (1992) 215–221.

[50] M.B. Kastan, O. Onyekwere, D. Sidransky, B. Vogelstein, R.W. Craig, Participation of p53 protein in the cellular response to DNA damage. Cancer Res. 51 (1991) 6304–6311.

[51] K.T. Kitchin, Recent advances in arsenic carcinogenesis: modes of action, animal model systems, and methylated arsenic metabolites. Toxicol. Appl. Pharmacol. 172 (2001) 249–261.

[52] K.T. Kitchin, K. Wallace, Evidence against the nuclear in situ binding of arsenicals: oxidative stress theory of arsenic carcinogenesis. Toxicol. Appl. Pharmacol. 232 (2008) 252–257.

[53] K.T. Kitchin, K. Wallace, The role of protein binding of trivalent arsenicals in arsenic carcinogenesis and toxicity. J. Inorg. Biochem. 102 (2008) 532–539.

[54] A.D. Kligerman, C.L. Doerr, A.H. Tennant, K. Harrington-Brock, J.W. Allen, E. Winkfield, P. Poorman-Allen, B. Kundu, K. Funasaka, B.C. Roop, M.J. Mass, D.M. DeMarini, Methylated trivalent arsenicals as candidate ultimate genotoxic forms of arsenic: induction of chromosomal mutations but not gene mutations. Environ. Mol. Mutagen. 42 (2003) 192–205.

[55] A.D. Kligerman, S.I. Malik, J.A. Campbell, Cytogenetic insights into DNA damage and repair of lesions induced by a monomethylated trivalent arsenical. Mutat. Res. 695 (2010) 2–8.

[56] A.D. Kligerman, A.H. Tennant, Insights into the carcinogenic mode of action of arsenic. Toxicol. Appl. Pharmacol. 222 (2007) 281–288.

[57] T.S. Kochhar, W. Howard, S. Hoffman, L. Brammer-Carleton, Effect of trivalent and pentavalent arsenic in causing chromosome alterations in cultured Chinese hamster ovary (CHO) cells. Toxicol. Lett. 84 (1996) 37–42.

[58] M. Kundu, P. Ghosh, S. Mitra, J.K. Das, T.J. Sau, S. Banerjee, J.C. States, A.K. Giri, Precancerous and non-cancer disease endpoints of chronic arsenic exposure: the level of chromosomal damage and XRCC3 T241M polymorphism. Mutat. Res.. 706 (2011) 7–12.

[59] T.C. Lee, M. Oshimura, J.C. Barrett, Comparison of arsenic-induced cell transformation, cytotoxicity, mutation and cytogenetic effects in Syrian hamster embryo cells in culture. Carcinogenesis 6 (1985) 1421–1426.

[60] D. Lewinska, J. Palus, M. Stepnik, E. Dziubaltowska, J. Beck, K. Rydzynski, A.T. Natarajan, R. Nilsson, Micronucleus frequency in peripheral blood lymphocytes and buccal mucosa cells of copper smelter workers, with special regard to arsenic exposure. Int. Arch. Occup. Environ. Health 80 (2007) 371–380.

[61] J.H. Li, T.G. Rossman, Inhibition of DNA ligase activity by arsenite: a possible mechanism of its comutagenesis. Mol. Toxicol. 2 (1989) 1–9.

[62] J.H. Li, T.G. Rossman, Mechanism of comutagenesis of sodium arsenite with N-methyl-N-nitrosourea. Biol. Trace Elem. Res. 21 (1989) 373–381.

[63] Y.M. Li, J.D. Broome, Arsenic targets tubulins to induce apoptosis in myeloid leukemia cells. Cancer Res. 59 (1999) 776–780.

[64] Y.H. Ling, J.D. Jiang, J.F. Holland, R. Perez-Soler, Arsenic trioxide produces polymerization of microtubules and mitotic arrest before apoptosis in human tumor cell lines. Mol. Pharmacol. 62 (2002) 529–538.

[65] S.H. Liou, J.C. Lung, Y.H. Chen, T. Yang, L.L. Hsieh, C.J. Chen, T.N. Wu, Increased chromosome-type chromosome aberration frequencies as biomarkers of cancer risk in a blackfoot endemic area. Cancer Res. 59 (1999) 1481–1484.

[66] F. Liu, K.Y. Jan, DNA damage in arsenite- and cadmium-treated bovine aortic endothelial cells. Free Radic. Biol. Med. 28 (2000) 55–63.

[67] J. Liu, M.B. Kadiiska, Y. Liu, T. Lu, W. Qu, M.P. Waalkes, Stress-related gene expression in mice treated with inorganic arsenicals. Toxicol. Sci. 61 (2001) 314–320.

[68] S.X. Liu, M. Athar, I. Lippai, C. Waldren, T.K. Hei, Induction of oxyradicals by arsenic: implication for mechanism of genotoxicity. Proc. Natl. Acad. Sci. U. S. A. 98 (2001) 1643–1648.

[69] S.X. Liu, M.M. Davidson, X. Tang, W.F. Walker, M. Athar, V. Ivanov, T.K. Hei, Mitochondrial damage mediates genotoxicity of arsenic in mammalian cells. Cancer Res. 65 (2005) 3236–3242.

[70] Z. Liu, M. Styblo, B.P. Rosen, Methylarsonous acid transport by aquaglyceroporins. Environ. Health Perspect. 114 (2006) 527–531.

[71] J. Mahata, A. Basu, S. Ghoshal, J.N. Sarkar, A.K. Roy, G. Poddar, A.K. Nandy, A. Banerjee, K. Ray, A.T. Natarajan, R. Nilsson, A.K. Giri, Chromosomal aberrations and sister chromatid exchanges in individuals exposed to arsenic through drinking water in West Bengal, India. Mutat. Res. 534 (2003) 133–143.

[72] J. Mahata, P. Ghosh, J.N. Sarkar, K. Ray, A.T. Natarajan, A.K. Giri, Effect of sodium arsenite on peripheral lymphocytes in vitro: individual susceptibility among a population exposed to arsenic through the drinking water. Mutagenesis 19 (2004) 223–229.

[73] J. Maki-Paakkanen, P. Kurttio, A. Paldy, J. Pekkanen, Association between the clastogenic effect in peripheral lymphocytes and human exposure to arsenic through drinking water. Environ. Mol. Mutagen. 32 (1998) 301–313.

[74] V. Martinez, A. Creus, W. Venegas, A. Arroyo, J.P. Beck, T.W. Gebel, J. Surralles, R. Marcos, Evaluation of micronucleus induction in a Chilean population environmentally exposed to arsenic. Mutat. Res. 564 (2004) 65–74.

[75] V. Martinez, A. Creus, W. Venegas, A. Arroyo, J.P. Beck, T.W. Gebel, J. Surralles, R. Marcos, Micronuclei assessment in buccal cells of people environmentally exposed to arsenic in northern Chile. Toxicol. Lett. 155 (2005) 319–327.

[76] M.J. Mass, A. Tennant, B.C. Roop, W.R. Cullen, M. Styblo, D.J. Thomas, A.D. Kligerman, Methylated trivalent arsenic species are genotoxic. Chem. Res. Toxicol. 14 (2001) 355–361.

[77] S. Matoba, J.G. Kang, W.D. Patino, A. Wragg, M. Boehm, O. Gavrilova, P.J. Hurley, F. Bunz, P.M. Hwang, p53 regulates mitochondrial respiration. Science 312 (2006) 1650–1653.

[78] M. Matsui, C. Nishigori, S. Toyokuni, J. Takada, M. Akaboshi, M. Ishikawa, S. Imamura, Y. Miyachi, The role of oxidative DNA damage in human arsenic carcinogenesis: detection of 8-hydroxy-2′-deoxyguanosine in arsenic-related Bowen's disease. J. Invest. Dermatol. 113 (1999) 26–31.

[79] S.C. McNeely, X. Xu, B.F. Taylor, W. Zacharias, M.J. McCabe, Jr., J.C. States, Exit from arsenite-induced mitotic arrest is p53 dependent. Environ. Health Perspect. 114 (2006) 1401–1406.

[80] J. Mendez-Gomez, G.G. Garcia-Vargas, L. Lopez-Carrillo, E.S. Calderon-Aranda, A. Gomez, E. Vera, M. Valverde, M.E. Cebrian, E. Rojas, Genotoxic effects of environmental exposure to arsenic and lead on children in region Lagunera, Mexico. Ann. N. Y. Acad. Sci. 1140 (2008) 358–367.

[81] T. Miyashita, S. Krajewski, M. Krajewska, H.G. Wang, H.K. Lin, D.A. Liebermann, B. Hoffman, J.C. Reed, Tumor suppressor p53 is a regulator of bcl-2 and bax gene expression in vitro and in vivo. Oncogene 9 (1994) 1799–1805.

[82] J. Mo, Y. Xia, T.J. Wade, M. Schmitt, X.C. Le, R. Dang, J.L. Mumford, Chronic arsenic exposure and oxidative stress: OGG1 expression and arsenic exposure, nail selenium, and skin hyperkeratosis in Inner Mongolia. Environ. Health Perspect. 114 (2006) 835–841.

[83] L.E. Moore, A.H. Smith, C. Hopenhayn-Rich, M.L. Biggs, D.A. Kalman, M.T. Smith, Micronuclei in exfoliated bladder cells among individuals chronically exposed to arsenic in drinking water. Cancer Epidemiol. Biomarkers Prev. 6 (1997) 31–36.

[84] S.A. Mouron, C.A. Grillo, F.N. Dulout, C.D. Golijow, Induction of DNA strand breaks, DNA-protein crosslinks and sister chromatid exchanges by arsenite in a human lung cell line. Toxicol. In Vitro 20 (2006) 279–285.

[85] K. Mure, A.N. Uddin, L.C. Lopez, M. Styblo, T.G. Rossman, Arsenite induces delayed mutagenesis and transformation in human osteosarcoma cells at extremely low concentrations. Environ. Mol. Mutagen. 41 (2003) 322–331.

[86] I. Nordenson, G. Beckman, L. Beckman, S. Nordstrom, Occupational and environmental risks in and around a smelter in northern Sweden. II. Chromosomal aberrations in workers exposed to arsenic. Hereditas 88 (1978) 47–50.

[87] I. Nordenson, A. Sweins, L. Beckman, Chromosome aberrations in cultured human lymphocytes exposed to trivalent and pentavalent arsenic. Scand. J. Work Environ. Health 7 (1981) 277–281.

[88] NRC Arsenic in Drinking Water, National Academy Press, Washington, DC, 1999.

[89] P. Ostrosky-Wegman, M.E. Gonsebatt, R. Montero, L. Vega, H. Barba, J. Espinosa, A. Palao, C. Cortinas, G. Garcia-Vargas, L.M. Del Razo, Lymphocyte proliferation kinetics and genotoxic findings in a pilot study on individuals chronically exposed to arsenic in Mexico. Mutat. Res. 250 (1991) 477–482.

[90] L. Paiva, V. Martinez, A. Creus, D. Quinteros, R. Marcos, Evaluation of micronucleus frequencies in blood lymphocytes from smelting plant workers exposed to arsenic. Environ. Mol. Mutagen. 49 (2008) 200–205.

[91] J. Palus, D. Lewinska, E. Dziubaltowska, M. Stepnik, J. Beck, K. Rydzynski, R. Nilsson, DNA damage in leukocytes of workers occupationally exposed to arsenic in copper smelters. Environ. Mol. Mutagen. 46 (2005) 81–87.

[92] M.A. Partridge, S.X. Huang, E. Hernandez-Rosa, M.M. Davidson, T.K. Hei, Arsenic induced mitochondrial DNA damage and altered mitochondrial oxidative function: implications for genotoxic mechanisms in mammalian cells. Cancer Res. 67 (2007) 5239–5247.

[93] P. Ramirez, L.M. Del Razo, M.C. Gutierrez-Ruiz, M.E. Gonsebatt, Arsenite induces DNA-protein crosslinks and cytokeratin expression in the WRL-68 human hepatic cell line. Carcinogenesis 21 (2000) 701–706.

[94] P. Ramirez, D.A. Eastmond, J.P. Laclette, P. Ostrosky-Wegman, Disruption of microtubule assembly and spindle formation as a mechanism for the induction of aneuploid cells by sodium arsenite and vanadium pentoxide. Mutat. Res. 386 (1997) 291–298.

[95] R.E. Rasmussen, D.B. Menzel, Variation in arsenic-induced sister chromatid exchange in human lymphocytes and lymphoblastoid cell lines. Mutat. Res. 386 (1997) 299–306.

[96] B.P. Rosen, Z. Liu, Transport pathways for arsenic and selenium: a minireview. Environ. Int. 35 (2009) 512–515.

[97] T.G. Rossman, Enhancement of UV-mutagenesis by low concentrations of arsenite in E. coli. Mutat. Res. 91 (1981) 207–211.

[98] T.G. Rossman, D. Stone, M. Molina, W. Troll, Absence of arsenite mutagenicity in E. coli and Chinese hamster cells. Environ. Mutagen. 2 (1980) 371–379.

[99] T.G. Rossman, A.N. Uddin, F.J. Burns, M.C. Bosland, Arsenite is a cocarcinogen with solar ultraviolet radiation for mouse skin: an animal model for arsenic carcinogenesis. Toxicol. Appl. Pharmacol. 176 (2001) 64–71.

[100] T.G. Rossman, D. Wolosin, Differential susceptibility to carcinogen-induced amplification of SV40 and dhfr sequences in SV40-transformed human keratinocytes. Mol. Carcinog. 6 (1992) 203–213.

[101] R. Ruiz-Ramos, L. Lopez-Carrillo, A.D. Rios-Perez, A. De Vizcaya-Ruiz, M.E. Cebrian, Sodium arsenite induces ROS generation, DNA oxidative damage, HO-1 and c-Myc proteins, NF-kappaB activation and cell proliferation in human breast cancer MCF-7 cells. Mutat. Res. 674 (2009) 109–115.

[102] D.S. Rupa, D.A. Eastmond, Chromosomal alterations affecting the 1cen-1q12 region in buccal mucosal cells of betel quid chewers detected using multicolor fluorescence in situ hybridization. Carcinogenesis 18 (1997) 2347–2351.

[103] A.M. Salazar, E. Calderon-Aranda, M.E. Cebrian, M. Sordo, A. Bendesky, A. Gomez-Munoz, L. Acosta-Saavedra, P. Ostrosky-Wegman, p53 expression in circulating lymphocytes of non-melanoma skin cancer patients from an arsenic contaminated region in Mexico. A pilot study. Mol. Cell. Biochem. 255 (2004) 25–31.

[104] A.M. Salazar, H.L. Miller, S.C. McNeely, M. Sordo, P. Ostrosky-Wegman, J.C. States, Suppression of p53 and p21CIP1/WAF1 reduces arsenite-induced aneuploidy. Chem. Res. Toxicol. 23 (2010) 357–364.

[105] A.M. Salazar, P. Ostrosky-Wegman, D. Menendez, E. Miranda, A. Garcia-Carranca, E. Rojas, Induction of p53 protein expression by sodium arsenite. Mutat. Res. 381 (1997) 259–265.

[106] A. Sampayo-Reyes, A. Hernandez, N. El-Yamani, C. Lopez-Campos, E. Mayet-Machado, C.B. Rincon-Castaneda, M.L. Limones-Aguilar, J.E. Lopez-Campos, M.B. de Leon, S. Gonzalez-Hernandez, D. Hinojosa-Garza, R. Marcos, Arsenic induces DNA damage in environmentally exposed Mexican children and adults. Influence of GSTO1 and AS3MT polymorphisms. Toxicol. Sci. 117 (2010) 63–71.

[107] N. Schaumloffel, T. Gebel, Heterogeneity of the DNA damage provoked by antimony and arsenic. Mutagenesis 13 (1998) 281–286.

[108] T. Schwerdtle, I. Walter, I. Mackiw, A. Hartwig, Induction of oxidative DNA damage by arsenite and its trivalent and pentavalent methylated metabolites in cultured human cells and isolated DNA. Carcinogenesis 24 (2003) 967–974.

[109] H. Shi, X. Shi, K.J. Liu, Oxidative mechanism of arsenic toxicity and carcinogenesis. Mol. Cell. Biochem. 255 (2004) 67–78.

[110] M. Sordo, L.A. Herrera, P. Ostrosky-Wegman, E. Rojas, Cytotoxic and genotoxic effects of As, MMA, and DMA on leukocytes and stimulated human lymphocytes. Teratog. Carcinog. Mutagen. 21 (2001) 249–260.

[111] C. Soriano, A. Creus, R. Marcos, Gene-mutation induction by arsenic compounds in the mouse lymphoma assay. Mutat. Res. 634 (2007) 40–50.

[112] M. Styblo, Z. Drobna, I. Jaspers, S. Lin, D.J. Thomas, The role of biomethylation in toxicity and carcinogenicity of arsenic: a research update. Environ. Health Perspect. 110 Suppl 5 (2002) 767–771.

[113] P. Sykora, E.T. Snow, Modulation of DNA polymerase beta-dependent base excision repair in cultured human cells after low dose exposure to arsenite. Toxicol. Appl. Pharmacol. 228 (2008) 385–394.

[114] A.L. Tan, P.C. Rida, U. Surana, Essential tension and constructive destruction: the spindle checkpoint and its regulatory links with mitotic exit. Biochem. J. 386 (2005) 1–13.

[115] B.F. Taylor, S.C. McNeely, H.L. Miller, G.M. Lehmann, M.J. McCabe, Jr., J.C. States, p53 suppression of arsenite-induced mitotic catastrophe is mediated by p21CIP1/WAF1. J. Pharmacol. Exp. Ther. 318 (2006) 142–151.

[116] D.J. Thompson, A chemical hypothesis for arsenic methylation in mammals. Chem. Biol. Interact. 88 (1993) 89–114.

[117] H. Tinwell, S.C. Stephens, J. Ashby, Arsenite as the probable active species in the human carcinogenicity of arsenic: mouse micronucleus assays on Na and K arsenite, orpiment, and Fowler's solution. Environ. Health Perspect. 95 (1991) 205–210.

[118] M. Torres-Avila, P. Leal-Galicia, L.C. Sanchez-Pena, L.M. Del Razo, M.E. Gonsebatt, Arsenite induces aquaglyceroporin 9 expression in murine livers. Environ. Res. 110 (2010) 443–447.

[119] M. Valko, C.J. Rhodes, J. Moncol, M. Izakovic, M. Mazur, Free radicals, metals and antioxidants in oxidative stress-induced cancer. Chem. Biol. Interact. 160 (2006) 1–40.

[120] L. Vega, M.E. Gonsebatt, P. Ostrosky-Wegman, Aneugenic effect of sodium arsenite on human lymphocytes in vitro: an individual susceptibility effect detected. Mutat. Res. 334 (1995) 365–373.

[121] B.L. Vogt, T.G. Rossman, Effects of arsenite on p53, p21 and cyclin D expression in normal human fibroblasts: a possible mechanism for arsenite's comutagenicity. Mutat. Res. 478 (2001) 159–168.

[122] K.H. Vousden, D.P. Lane, p53 in health and disease. Nat. Rev. Mol. Cell Biol. 8 (2007) 275–283.

[123] S.B. Vuyyuri, M. Ishaq, D. Kuppala, P. Grover, Y.R. Ahuja, Evaluation of micronucleus frequencies and DNA damage in glass workers exposed to arsenic. Environ. Mol. Mutagen. 47 (2006) 562–570.

[124] T.S. Wang, T.Y. Hsu, C.H. Chung, A.S. Wang, D.T. Bau, K.Y. Jan, Arsenite induces oxidative DNA adducts and DNA-protein cross-links in mammalian cells. Free Radic. Biol. Med. 31 (2001) 321–330.

[125] J.K. Wiencke, J.W. Yager, A. Varkonyi, M. Hultner, L.H. Lutze, Study of arsenic mutagenesis using the plasmid shuttle vector pZ189 propagated in DNA repair proficient human cells. Mutat. Res. 386 (1997) 335–344.

[126] S.H. Woo, I.C. Park, M.J. Park, H.C. Lee, S.J. Lee, Y.J. Chun, S.H. Lee, S.I. Hong, C.H. Rhee, Arsenic trioxide induces apoptosis through a reactive oxygen species-dependent pathway and loss of mitochondrial membrane potential in HeLa cells. Int. J. Oncol. 21 (2002) 57–63.

[127] Y. Xu, Regulation of p53 responses by post-translational modifications. Cell Death Differ. 10 (2003) 400–403.

[128] J.W. Yager, J.K. Wiencke, Enhancement of chromosomal damage by arsenic: implications for mechanism. Environ. Health Perspect. 101 Suppl 3 (1993) 79–82.

[129] K. Yamanaka, A. Hasegawa, R. Sawamura, S. Okada, Dimethylated arsenics induce DNA strand breaks in lung via the production of active oxygen in mice. Biochem. Biophys. Res. Commun. 165 (1989) 43–50.

[130] K. Yamanaka, H. Hayashi, K. Kato, A. Hasegawa, S. Okada, Involvement of preferential formation of apurinic/apyrimidinic sites in dimethylarsenic-induced DNA strand breaks and DNA-protein crosslinks in cultured alveolar epithelial cells. Biochem. Biophys. Res. Commun. 207 (1995) 244–249.

[131] K. Yamanaka, H. Hayashi, M. Tachikawa, K. Kato, A. Hasegawa, N. Oku, S. Okada, Metabolic methylation is a possible genotoxicity-enhancing process of inorganic arsenics. Mutat. Res. 394 (1997) 95–101.

[132] K. Yamanaka, S. Okada, Induction of lung-specific DNA damage by metabolically methylated arsenics via the production of free radicals. Environ. Health Perspect. 102 Suppl 3 (1994) 37–40.

[133] H. Yamauchi, Y. Aminaka, K. Yoshida, G. Sun, J. Pi, M.P. Waalkes, Evaluation of DNA damage in patients with arsenic poisoning: urinary 8-hydroxydeoxyguanine. Toxicol. Appl. Pharmacol. 198 (2004) 291–296.

[134] L. Yanez, E. Garcia-Nieto, E. Rojas, L. Carrizales, J. Mejia, J. Calderon, I. Razo, F. Diaz-Barriga, DNA damage in blood cells from children exposed to arsenic and lead in a mining area. Environ. Res. 93 (2003) 231–240.

[135] L.H. Yih, I.C. Ho, T.C. Lee, Sodium arsenite disturbs mitosis and induces chromosome loss in human fibroblasts. Cancer Res. 57 (1997) 5051–5059.

[136] L.H. Yih, T.C. Lee, Arsenite induces p53 accumulation through an ATM-dependent pathway in human fibroblasts. Cancer Res. 60 (2000) 6346–6352.

[137] Y. Zhang, J. Ying, J. Chen, C. Hu, Assessing the genotoxic potentials of roxarsone in V79 cells using the alkaline Comet assay and micronucleus test. Mutat. Res. 741 (2012) 65–69.

16

ARSENIC AND SIGNAL TRANSDUCTION

INGRID L. DRUWE AND RICHARD R. VAILLANCOURT
Department of Pharmacology and Toxicology, University of Arizona College of Pharmacy, Tucson, AZ, USA

16.1 PRODUCTION OF REACTIVE OXYGEN SPECIES

Arsenite is known to chemically modify cysteine residues in proteins, with a preference for dithiols [20, 44]. In bacteria, the *arsRDABC* operon confers resistance to arsenite. The *arsD* and *arsR* genes encode the proteins ArsD and ArsR, respectively. The function of ArsD and ArsR is to repress *arsRDABC* transcription. When arsenite concentrations increase in the bacterial environment, the ArsD and ArsR proteins bind arsenite which de-represses transcription of the *arsRDABC* operon because ArsD and ArsR are released from the arsRDABC promoter DNA [52]. For ArsR, it was shown by mutagenesis that dithiols from cysteines 32 and 34 were needed for activity [87], while vicinal dithiols from cysteines 12–13 and 112–113 were needed for activity of ArsD [52]. These data suggest that arsenite affects protein activity through what appears to be covalent modification of cysteine residues in proteins. The preference for modification of cysteine also suggests that arsenite can affect the redox status within the cell.

It has been demonstrated that arsenite has the ability to induce the formation of ROS in a wide variety of cells including human vascular smooth muscle cells, human epithelial bladder cells [25], human-hamster hybrid cells, vascular endothelial cells, and Chinese hamster ovary (CHO) cells to name a few [51, 78, 86]. In the presence of iron (Fe^{2+}), arsenite undergoes a Fenton reaction and produces ROS resulting in

Arsenic: Exposure Sources, Health Risks, and Mechanisms of Toxicity, First Edition.
Edited by J. Christopher States.
© 2016 John Wiley & Sons, Inc. Published 2016 by John Wiley & Sons, Inc.

cellular damage [28, 51, 86]. DNA damage is one of the most prominent toxic effects produced by ROS in cells exposed to arsenite [17, 28, 30], which frequently results in apoptosis [17, 19, 28, 41, 81]. Chattopadhyay *et al.* showed that chronic arsenite exposure of 0.3 mg/L in rats resulted in depletion of glutathione and increased oxidized glutathione and lipid peroxidation in the brain [12, 13]. Additionally, it has been shown that hydrogen peroxide-resistant CHO cells are also resistant to arsenite toxicity [10] and conversely that antioxidant-deficient cells are hypersensitive to arsenite [104]. The Bing-Hua Jiang group at the University of Pittsburg used both the human lung epithelial cell line, BEAS-2B and the adenocarcinoma cell line A549 to study the effects of arsenic on angiogenesis. In their study, they showed that arsenic increased phosphorylated ERK and AKT activity in a dose-dependent manner starting at 5 μM in the BEAS-2B cell line and at 2.5 μM in the A549 cell line after 6 h of arsenite exposure. Their studies have shown that acute arsenite exposure induced both hydrogen peroxide as well as superoxide ROS [57, 104]. In addition, these studies showed that ROS production by arsenic was both cell line and dose dependent, as ROS were not produced by the BEAS-2B cell line after 4 h of treatment with exposures lower than 2.5 μM arsenite. Likewise, Druwe *et al.* found that in HepG2 cells arsenite exposures between 0.13 and 2.0 μM did not result in ROS production but did result in the C-reactive protein (CRP) production [23].

Nuclear factor erythroid 2-related factor 2 (Nrf2) is a transcription factor of interest in arsenic-dependent signal transduction pathways due to the regulation of genes containing antioxidant-response elements (ARE). Nrf2 is a central regulator of a variety of antioxidant enzymes such as NAD(P)H dehydrogenase (NQO-1) which is responsible for reducing quinones into hydroquinones, thereby preventing these from resulting in radical species; heme oxygenase (HO-1) which catalyzes the degradation of heme and is induced by oxidative stress; glutamate-cysteine ligase catalytic subunit (GCLC), which is the rate-limiting enzyme produced in the synthesis of glutathione; catalase (CAT) which is responsible for the decomposition of hydrogen peroxide into hydrogen and water, while superoxide dismutase (SOD) is responsible for the decomposition of superoxide, respectively, into hydrogen peroxide and oxygen. The transcription of these enzymes is carefully controlled through antioxidant-response elements located within their promoter regions [75, 105]. ROS have the ability to activate signal transduction pathways as part of normal system physiology. For example, in pancreatic β cells glucose increases the intracellular accumulation of H_2O_2, which then results in the secretion of insulin [75]. β cells express low levels of antioxidant enzymes, thereby rendering them more susceptible to oxidative damage and present an interesting model for the study of ROS as signaling molecules. Pi *et al.* found that the accumulation of H_2O_2 produced in response to increased levels of glucose was H_2O_2 specific and that other ROS-producing molecules, such as arsenite, suppressed the secretion of insulin and induced the nuclear accumulation of Nrf-2 and the induction of its target genes, such as HO-1, NQO-1, and γ-glutamate cysteine ligase catalytic subunit [75, 105]. The activity of Nrf2 itself is tightly regulated by Keap1 at multiple levels: (i) Keap1 senses disturbances in cellular redox conditions and modulates the Nrf2 response accordingly; (ii) Keap1 in a complex with Cul3, functions as a E3 ubiquitin ligase and constantly targets Nrf2 for ubiquitination and degradation; and (iii) when Nrf2 is activated, E3 ubiquitin ligase activity is inhibited leading to an increase in Nrf2 levels and increased translocation of Nrf2 into the nucleus [105]. Wang *et al.* showed that arsenite activates the Nrf2 pathway in UROtsa cells. In addition, they showed that arsenite enhanced the interaction between Keap1 and Cul3, which resulted in impaired dynamic assembly/disassembly of the E3 ubiquitin ligase for Nrf2 and thus decreased Nrf2 degradation. Interestingly, the researchers also found that induction of Nrf2 by arsenite is independent of the Cys151 residue in Keap1 that is required for Nrf2 activation by other environmental stressors such as tert-butylhydroquinone and sulforaphane [105]. This novel finding provides evidence for distinct mechanisms of Nrf2 activation by arsenite and may provide a molecular target for future therapeutics though much more research is needed in this area [105].

Oxidative stress has also been associated with increased cellular proliferation after exposure to low levels of arsenite [110]. In human embryonic lung fibroblasts (HELFs), low levels of arsenite (0.5 μM) stimulate cellular proliferation, while higher concentrations (5–10 μM) actually inhibited proliferation and cell growth. These observations correlated positively with ROS levels and arsenite concentration [110]. In low-level arsenite-treated groups, the activity of the antioxidant enzyme superoxide dismutase was significantly increased over nontreated cells but inhibited in the cells treated with the higher concentration of arsenite, thereby providing evidence that there is a concentration-dependent relationship between arsenite and ROS production [26, 110].

16.2 G-PROTEIN-COUPLED RECEPTORS

G-protein-coupled receptors (GPCR) consist of seven transmembrane domains that bind ligands and transduce signals into the cell [70]. When a ligand binds to a GPCR, it causes the receptor to undergo a conformational change, which allows it to act as a guanine nucleotide exchange factor (GEF). The GPCR can then activate an associated G-protein by exchanging its bound GDP for a GTP. The G-protein's α subunit, together with the bound GTP, dissociates from the β and γ subunits to further affect intracellular signaling proteins or target functional proteins directly depending on the α subunit types, $G_{\alpha s}$, $G_{\alpha i/o}$, $G_{\alpha q/11}$, or $G_{\alpha 12/13}$.

Of the transmembrane class of receptors, GPCRs are the largest single family, and are involved in a broad number of diseases including cardiovascular, metabolic, neurodegenerative, psychiatric, and oncologic diseases. The expression of GPCRs on the plasma membrane makes GPCRs readily accessible by various ligands including hydrophilic hormones, pharmacological drugs and environmental toxicants. GPCRs display a nonuniform pattern of expression in different tissues and cell types [95]. For this reason, GPCRs are also the target of approximately 40% of all modern pharmaceutical agents. For this reason, it is of note that arsenic has the ability to both activate and inhibit GPCRs, some of the most pharmaceutically targeted receptors.

In vascular endothelial or liver sinusoidal endothelial cells, arsenite stimulates the sphingosine-1-phosphate receptor ($S1P_1$), a member of the GPCR family [91]. Given

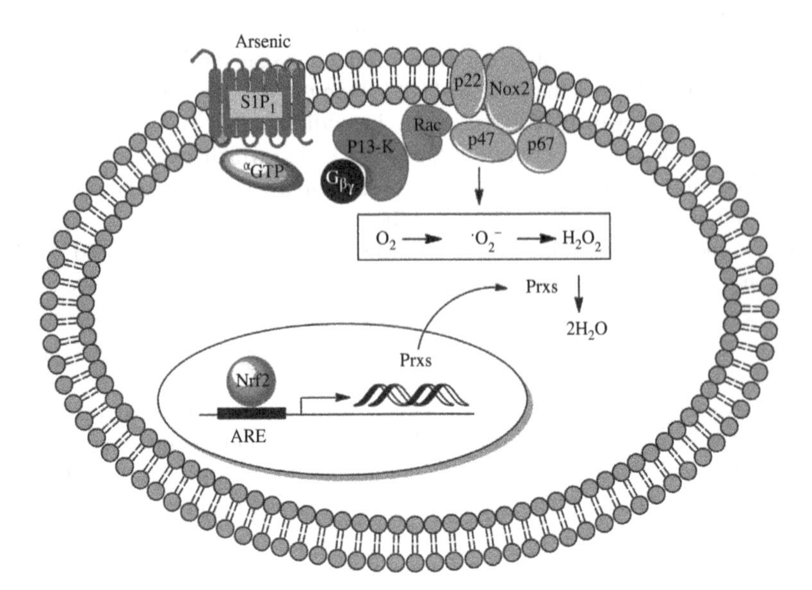

FIGURE 16.1 Regulation of the $S1P_1$ GPCR and the NADPH oxidase signaling pathway by arsenic. Endothelial cells undergo angiogenesis following gene expression and tube formation after exposure to arsenite, and the angiogenic response is inhibited with VPC23019, a competitive inhibitor to the sphingosine-1-phosphate type 1 and type 3 receptors. Arsenite also activates the small GTP-binding protein Rac, presumably through Gβγ regulation of the p110β subunit of PI3-K. Activation of the NADPH oxidase (Nox2) enzyme complex and associated subunits (p22, p47, p67) generates ROS. Nrf2 transcriptional activity increases in response to oxidative stress resulting in peroxiredoxin (Prxs) gene expression. Studies have shown that loss of Nrf2 expression results in greater arsenic-induced toxicity in cell culture and in mice, likely due to loss of expression of antioxidant-response genes.

that arsenic exposure typically occurs through ingestion of contaminated drinking water, it is particularly unsettling to know that arsenic functions as a sphingosine-1-phosphate receptor agonist mimetic. In their studies, Straub *et al.* co-exposed human microvascular endothelial cells (HMVEC) with pertussis toxin, a selective inhibitor of the G_o and G_i family of G proteins, and arsenite [91]. Their results showed that pertussis toxin abrogated arsenite-dependent responses, further re-enforcing the fact that arsenite was acting through an extracellular receptor and capable of generating ROS through Rac and NADPH oxidase (Fig. 16.1). Moreover, incubating human microvascular endothelial cells with the $S1P_1$ inhibitor, VPC23019, or performing small interfering RNA knockdown of $S1P_1$ expression blocked arsenic-stimulated HMVEC angiogenic gene expression and tube formation, but did not affect induction of either HMOX1 or IL-8. These results demonstrate that arsenic is capable of stimulating signaling pathways that elicit physiological responses. Rac signaling is essential in cell survival and cell motility; signaling events that are also known hallmarks of tumorigenic transformation and metastasis [92]. Straub *et al.* noted arsenic cotreatment with the $S1P_1$ inhibitor, VPC23019 led to the subsequent inhibition of Rac indicating that the $S1P_1$ receptor activates Rac [91]. Knowing that arsenite can activate Rac through the sphingosine-1-phosphate receptor may help explain how arsenite contributes to the development of cancer and may identify the sphingosine-1-phosphate receptor as a therapeutic target in cancer prevention.

In liver sinusoidal endothelial cells, extensive fenestrations and weak connections between cells enable liver hepatocytes to pass nutrients and waste molecules for efficient metabolism under normal physiological conditions. In the presence of arsenite, pathological defenestration and capillarization of sinusoidal endothelial cells induce the expression of junctional platelet endothelial cell adhesion molecule (PECAM)-1 [90]. In addition, the NADPH oxidase system is required for capillarization of sinusoidal endothelial cells, which indicates that superoxide and reactive oxygen species (ROS) are also part of the pathological signaling mechanism. In another study investigating the role of the NADPH oxidase pathway, Suzuki *et al.* used apocynin, an inhibitor of NADPH oxidase, and demonstrated decreased cytotoxicity of rat bladder epithelial cells [94]. Since the Rac1 GTPase is activated by arsenite through stimulation of the sphingosine-1-phosphate receptor [90], at least in endothelial cells and possibly epithelial cells, arsenite regulates a signal transduction pathway that begins with a GPCR, G protein (G_i or G_o), and an effector enzyme (i.e., NADPH oxidase) that is capable of increasing PECAM-1 gene transcription just like other canonical GPCR pathways.

Work from Aaron Barchowshy's group added to our understanding of arsenic signaling through GPCR, leading to a better appreciation to the role of arsenic in metabolic syndrome and cardiovascular disease risk. In their studies, Klei *et al.* showed that environmentally relevant levels of arsenic (0.1–1.0 μM) had the ability to inhibit progenitor cell differentiation into adipocytes [47]. Human mesenchymal stem cells (hMSCs) exposed to arsenite showed decreased fat droplet formation, decreased transcriptional promoters of differentiation and decreased expression of differentiated adipocyte markers, such as adiponectin and perilipin. Interestingly, preincubation of hMSC with pertussis toxin (Ptx) resulted in a 90% recovery of adipogenesis. Additionally, Klei *et al.* showed that selective competitive antagonists of G_i-coupled endothelin-1 type A and B receptors were approximately 60% effective in blocking arsenic inhibition and a combination of antagonists to both receptors was 85% effective. However, they found that antagonists to the $S1P_1$ were ineffective in blocking the effects of adipogenesis inhibition caused by arsenic. These data showed that adipocyte differentiation requires the GPCR endothelin-1.

The sphingosine and endothelin-1 receptors are not the only GPCRs that are affected by arsenic. The purinergic GPCR P2Y is inhibited by physiologically relevant doses of arsenite [85]. The P2Y receptors are expressed in almost all human tissues, and their biological function is based on their G-protein coupling. P2Y receptors are typically stimulated by nucleotide ligands such as ATP, ADP, UTP, UDP, and UDP-glucose. For example, ATP stimulation of P2Y receptors located in airway epithelia trigger the production of inositol trisphosphate (IP_3) through activation of G_q, and ultimately lead to the release of Ca^{2+} from intracellular stores. Exposure to as little as 0.8 μM (60 ppb) arsenite leads to a decrease in intracellular Ca^{2+} release in 16HBE14o cells, a bronchial epithelial cell model, indicating that in this model system, unlike the endothelial cell model described by Straub *et al.*, arsenite acts as a GPCR inhibitor rather than an activator [85]. Further adding to the complexity of what we know and understand about how arsenic may be behaving as a signal-transducing molecule. Sherwood *et al.* also discovered that in addition to inhibiting the P2Y receptor, arsenite also inhibits the ionotropic P2X receptor, another member of the purinergic receptor family, but not a member of the GPCR family of receptors. Purinergic signaling is crucial in innate activities such as ciliary beat, which is the mechanism used by pulmonary cells to remove xenobiotics that enter the airway as well as mucus in order to keep the airway clean. Because arsenic was found to inhibit both P2Y- and P2X-mediated calcium signaling typically induced by ATP, it is possible that inhaled arsenic from contaminated air can rapidly reach the airway epithelium where purinergic signaling is important for ciliary beat, water transport, and other lung physiology that compromises airway defense mechanisms thereby contributing to chronic lung disease.

The results of these studies show that the effects of arsenic on GPCR signaling are tissue and environment specific. These studies also confirm what many in the signaling field believed all along; that is, arsenic is not merely a cellular toxicant, but has pleiotropic effects because it is also capable of both activating and inhibiting cellular responses. It has been shown that arsenic has the ability to bind to zinc finger domains on molecules that have vicinal sulfhydryl groups [44, 46] (also see Chapter 13). Interestingly, the S_1P and P2Y receptors both contain a Cys X_6 Cys X_{3-7} Cys motif which could in theory compete for binding of the GPCR with their conventional ligands and then perhaps lead to the observed activation or inhibition of their respective signaling cascades.

16.3 COX-2

Skin cancer was one of the first recognized maladies associated with arsenic exposure. Due to the growth-promoting activity associated with prostanoids, arsenic-dependent expression of cyclooxygenase-2 (Cox-2) was investigated in normal human epidermal keratinocytes [96]. Cox-2 expression increased 1.3-fold after 8 h of continuous exposure to 2.5 μM arsenite. The increased Cox-2 expression was accompanied by an increase in PGE_2 and phosphoERK1/2, while p38 MAP kinase was not activated by arsenite. In a prior study using human umbilical vein endothelial cells, HUVEC, and the immortalized human endothelial cell line, ECV304, Tsai *et al.* demonstrated an arsenite-dependent induction of Cox-2 [97]. The use of inhibitors against nuclear factor-kappaB (NF-κB) [pyrrolidine dithiocarbamate (PDTC)] and p38 MAP kinase (SB203580) demonstrated that Cox-2 induction was dependent on the IKK and NF-κB pathway, and not on the p38 MAP kinase pathway. Although arsenite-dependent ROS was not measured in this study, 20 μM arsenite, which can induce ROS [57], was typically used strongly suggesting that Cox-2 induction may have been induced by ROS. Arsenite-dependent induction of oxidative stress was observed when bovine aortic endothelial (BAE) cells were exposed to arsenite concentrations as low as 0.5 μM. Bunderson *et al.* found Cox-2 induction and peroxynitrite, a strong oxidant produced by coupling nitric oxide and superoxide anion, in BAE cells [8].

Peroxynitrite production often leads to nitration of tyrosines. In BAE cells, nitration of Cox-2 on tyrosine residues was detected using nitrotyrosine antibody and Western blots. However, the consequences of Cox-2 nitration are unknown, although it has been reported that nitration of Cox-2 results in its physical association with inducible nitric oxide synthase (iNOS) in diabetic Wistar rats [53]. In addition, total Cox-2 and iNOS activity were reported to increase, but the activity of the nitrated enzymes was not measured. Thus is it not clear whether nitration activates these enzymes. A more recent study showed that arsenic trioxide inhibited migration of human gastric cancer SGC-7901 cells and that *N*-acetyl-L-cysteine (NAC, a radical scavenger) and celecoxib (Cox-2 inhibitor) inhibited the activity of arsenic trioxide suggesting that ROS and Cox-2 mediated the migratory activity of SGC-7901 cells [58]. In contrast, arsenic trioxide caused the degradation of Cox-2 in human acute myeloid leukemia, HL60 cells [32]. Taken together, these results demonstrate that arsenite affects Cox-2 expression by mechanisms that involve induction, nitration, and degradation of Cox-2, further emphasizing that arsenite has inflammatory and growth promoting activity that involve Cox-2.

16.4 NUCLEAR RECEPTORS AND TRANSCRIPTION

Over 20 years ago, it was recognized that arsenite inhibited the binding of dexamethasone with the glucocorticoid receptor suggesting that arsenite binds to the steroid binding domain of the receptor [60]. Further investigation into the role of arsenite in glucocorticoid receptor activity using reporter gene assays demonstrated that low (up to 1 μM) arsenite exposures stimulated glucocorticoid receptor transcriptional activity while higher exposures inhibited transcription [4]. An analysis of the glucocorticoid receptor domains indicated that arsenite-dependent modulation of transcription targeted the central DNA-binding domain. These results demonstrate the differential effects of arsenite that are mediated by the arsenite exposure. In this case, low arsenite concentrations stimulate transcription while higher concentrations inhibit transcription.

More recently, the retinoid X receptor-α (RXRα) has been shown to be a target of arsenite. RXRα forms a heterodimer with the pregnane X receptor (PXR) to form the PXR/RXRα transcriptional unit and transcribes the CYP3A4 gene. The human HepG2 hepatocellular carcinoma cell line was transfected with a reporter gene containing the rat proximal promoter of CYP3A23, a CYP3A family member regulated by the PXR. Exposing the transfected cells to 5 μM arsenite for 6 h resulted in a 50% decrease in transcriptional activity [66]. On further evaluation, it was found that arsenite increased the proteasomal degradation of RXRα without affecting PXR expression. Taken together, these results demonstrate that arsenite inhibits receptors through different mechanisms. Arsenite appears to act directly on the transcriptional activity of the glucocorticoid receptor by binding to the DNA-binding domain, while the transcriptional activity of the RXRα receptor is inhibited due to proteolysis. These studies highlight the potentially broad impact of arsenite on cellular physiology given that glucocorticoids control so many cellular processes ranging from immune responses to glucose metabolism, while RXRα controls the expression of CYP3A4, an enzyme that controls the metabolism of approximately 60% of all drugs.

Acute promyelocytic leukemia (APL) is associated with chromosomal translocations that fuse DNA and hormone-binding domains of retinoic acid receptor α (RARα) with ectopic DNA reading frames. The promyelocytic leukemia (PML) protein fuses with RARα to produce the PML-RARα fusion protein, which is the most common form of translocation resulting in aberrant cell growth and a leukemic phenotype. Although all-*trans* retinoic acid (ATRA) can induce complete disease remission, using ATRA often results in significant adverse effects [29] and almost all patients develop retinoid resistance [21, 107]. Arsenic trioxide (As_2O_3) has emerged as the standard of therapy for the treatment of APL [84, 88, 93]. Unlike the glucocorticoid receptor where arsenite binds the DNA-binding domain, arsenic trioxide appears to affect RARα activity indirectly by activating the ERK MAP kinase pathway. Arsenic trioxide prevents the transcriptional repressor, silencing mediator of retinoic acid and thyroid hormone receptor (SMRT), from interacting with the PML-RARα oncoprotein [38]. Moreover, ERK-dependent phosphorylation sites on the PML protein have been mapped indicating that the target of ERK regulation is through the PML-RARα fusion protein [33]. Thus, PML-RARα nuclear receptor function through ERK MAP kinase-dependent inhibition of co-repressor, SMRT, appears to be the mechanism by which arsenic trioxide provides efficacious therapeutic benefit for APL patients. It remains to be determined how arsenic trioxide activates the ERK MAP kinase pathway in acute promyelocytic leukemia cells. Arsenic trioxide activates the epidermal growth factor (EGF) receptor in human epidermoid carcinoma A431 cells, which results in ERK activation [59]. If a similar mechanism occurs in leukemic cells, then arsenic trioxide-dependent activation of the EGF receptor could account for ERK activation. However, the EGF receptor is not expressed in HL60 cells, an acute promyelocytic leukemia cell line [16, 109] and likely not expressed in other leukemic cells, so activation of the ERK MAP kinase pathway in leukemic cells would need to occur upstream of ERK in the absence of the EGF receptor.

16.5 NRF2

Nrf2 is a key protein needed for the transcription of a variety of genes involved in phase 2 drug-metabolizing enzymes, antioxidative enzymes, and membrane transport proteins through antioxidant-response elements [39]. Expression of Nrf2 is highly regulated with expression relatively high in cancer cell lines and tissues, while ubiquitously expressed at low basal levels in normal tissues (reviewed in Ref. [49]). Pi *et al.* investigated the role of arsenite in Nrf2 expression using HaCaT cells, an immortalized human keratinocyte cell line [77]. Although in their study chronic arsenic exposure did not exceed 24 h, Nrf2 expression was increased after 2 h of 2.5 μM arsenite exposure and remained elevated throughout the 24 h time course. Under basal conditions, Keap1 functions as an E3 ubiquitin ligase targeting Nrf2 for ubiquitination and degradation, but arsenite appears to allow Nrf2 to escape this degradation pathway.

The Keap1-Cul3 E3 ubiquitin ligase complex regulates the ubiquitination and protein stability of Nrf2 (reviewed in Ref. [111]). Within Keap1, cysteine 151 is recognized as essential for ubiquitin-dependent degradation of Nrf2 as this thiol is directly alkylated by electrophilic agents [27, 61]. By preventing Nrf2 degradation, electrophilic agents thereby induce higher expression of Nrf2 for transcriptional activation of genes with AREs. Wang *et al.* investigated the mechanism by which arsenite and its methylated metabolite, monomethylarsonous acid [MMA(III)], induced the expression of Nrf2 [105]. Arsenite and MMA(III) induced the Nrf2-dependent response in human bladder urothelial (UROtsa) cells by enhancing Nrf2 protein expression by inhibiting Nrf2 ubiquitination and proteasomal degradation. However unlike other known inducers of Nrf2, such as tert-butylhydroquinone or sulforaphane, arsenite- and MMA(III)-dependent induction of Nrf2 did not require cysteine 151 of Keap1 indicating a different mechanism between these groups of Nrf2 inducers.

16.6 DNA MODIFICATION

Defining the mechanism by which arsenic causes cellular malignant transformation has remained elusive. However, recent efforts in this area have led to DNA modification as a common theme. When Zhao *et al.* chronically exposed the rat liver cell line TRL 1215 to arsenite for 18 weeks, they observed highly aggressive, malignant tumor formation after inoculation of cells into nude mice [113]. Concurrent with malignant transformation was global DNA hypomethylation and overexpression of genes such as metallothionein. DNA hypomethylation was attributed to depletion of cellular *S*-adenosyl-methionine (SAM), since metabolism of inorganic arsenic in the liver requires SAM as a cofactor in methylation reactions (reviewed in Ref. [11]). It was suggested that arsenic-induced malignant transformation was due to depletion of SAM because of arsenic metabolism. Chronically arsenic-exposed TRL 1215 cells were later referred to as CAsE cells. Subsequent studies demonstrated that the CAsE cells methylated inorganic arsenic two times more efficiently than the parental TRL 1215 cell line [82]. In addition, CAsE cells were more tolerant than the parental TRL 1215 cell line to re-exposure of arsenite as measured by cytotoxicity assay [82]. These data demonstrated that chronic arsenic exposure to the liver cell line, TRL 1215, can re-program these cells as manifested by increased tumorigenicity and arsenic tolerance. However, inorganic arsenic metabolism by TRL 1215 cells has not been sufficiently characterized to state unequivocally that arsenic methylation with concomitant SAM depletion can account for global DNA hypomethylation and malignant cellular transformation.

Not only were CAsE cells tolerant to cytotoxicity after arsenite re-exposure, but CAsE cells also developed a hyperproliferative phenotype when re-exposed to arsenic and compared to the parental cell line. Results obtained from cDNA microarray studies demonstrated that c-myc was one of approximately 80 genes whose expression increased in malignantly transformed CAsE cells exposed to chronic arsenite [15]. Chen *et al.* demonstrated a correlation between c-MYC over-expression and arsenic-dependent malignant transformation. Since hypomethylation of the c-MYC promoter correlates with c-MYC gene overexpression in hepatocellular carcinomas [99], the implication is that c-MYC overexpression due to arsenic exposure could have similar outcomes. Experiments that use cell lines derived from mice with knockout of *c-MYC* or cell lines with *c-MYC* expression knocked down by siRNA would provide a more definitive answer as to the role of c-MYC in arsenic-mediated malignant cellular transformation.

c-MYC is not the only protein whose expression and activation increased due to chronic arsenic exposure. Zhang *et al.* exposed JB6 Cl 41 mouse epidermal cells with 2 μM arsenite for 16 weeks and observed cellular transformation as evidenced by anchorage-independent cell growth [112]. Phosphorylation of serine 422 on translation initiation factor 4B, eIF4B, increased 20-fold, while the phosphorylation of the eIF4B upstream kinase, p70S6K, increased 2.5-fold indicating that chronic arsenic exposure activates the AKT-mTORC1-p70S6K-eIF4B translation initiation pathway. Subsequent experiments with siRNA knockdown of eIF4B in JB6 Cl 41 cells chronically exposed to arsenic demonstrated a decrease in cell proliferation, anchorage-independent cell growth, and protein translation indicating a requirement for protein translation mediated by eIF4B in arsenite-dependent cellular transformation.

Research regarding the mechanism of arsenic-induced malignant cellular transformation has advanced from specific proteins and the corresponding messenger RNA (mRNA) to microRNA (miRNA) expression. miRNAs are small noncoding RNAs that regulate protein-coding gene expression by interacting with mRNA, resulting in mRNA degradation and suppression of protein translation (reviewed in Refs. [2, 34, 36]). miRNA microarray technology has been used to assess differential miRNA expression in response to chronic arsenic exposure. Wang *et al.* showed by microarray that expression of three members of the miR-200 family (miR-200b, 200c, and 205) were downregulated more than twofold in human bronchial epithelial cells chronically exposed to 2.5 μM arsenite for 16 weeks [106]. Additional evaluation of miR-200 expression by qPCR in human bronchial epithelial cells showed that all five members of miR-200 family were depleted as a result of chronic arsenic exposure. Further mechanistic studies showed increased DNA methylation of the miR-200 promoters which has been reported as important for silencing the expression of the miR-200 family [102]. These data demonstrate that chronic, low-dose arsenic exposure can alter miRNA expression resulting in malignant cellular transformation raising the question as to how is miR-200 expression regulated.

The transcriptional repressor zinc finger E-box-binding homeodomain 1 (ZEB1) protein is associated with cancer cell metastasis [89] and directly suppresses transcription of miR-200 family members [6, 9]. Thus, the role of ZEB1 was investigated in arsenite-induced cellular transformation. During the time course of chronic exposure to 2.5 μM arsenite, human bronchial epithelial cells began to express ZEB1 starting at 8 weeks of exposure [106]. At the same time, arsenite induced expression of vimentin, a mesenchymal cell marker, indicating that human bronchial epithelial cells began the epithelial to mesenchymal transition (EMT) at a time when ZEB1 expression increased. Conversely, when human bronchial epithelial cells stably expressed miR-200b and were exposed to chronic arsenite, the cells maintained their epithelial cell morphology by expressing high levels of E-cadherin and were unable to achieve anchorage-independent cell growth as measured by growth in soft agar [106]. These data demonstrate, once again, that cells chronically maintained in low concentrations of arsenite develop a malignant phenotype. Wang *et al.* took the arsenic signaling field a step further by demonstrating a reciprocal relationship between ZEB1 and miR-200 family member expression. How arsenite affects ZEB1 and miR-200 family member expression will require additional investigation.

16.7 CELL CYCLE

The cell cycle consists of four major phases in which the cell receives external cues to replicate DNA and then divide into two cells. During the G_1 phase, the cell responds to growth factors, mitogenic agents and nutrients, and evaluates the integrity of its genome before moving on to the S phase. If DNA repair is not necessary, then the cell progresses to the S phase and DNA replication occurs. During the G_2 phase, the cell prepares for cell division and divides during the M phase. As the cell proceeds through each phase of the cell cycle, expression of various proteins increases or decreases depending on the required function of the respective proteins. Cyclin-dependent kinases (CDK), such as CDK4/6 and CDK2, are expressed throughout the cell cycle. However, CDKs are not active without their cyclin partner whose expression is typically regulated at unique phases within the cell cycle. For example, D-type cyclin expression is increased in the presence of mitogenic factors during the early G_1 phase of the cell cycle. The CDK-cyclin D heterodimer functions to phosphorylate target proteins required for progression of the cell cycle. Conversely, Cdk inhibitors including INK family members (CDKN2A (aka INK4a, p16), CDKN2B (aka INK4b, p15), and CDKN2C (aka INK4c, p18)) and CIP family members (CDKN1A (aka CIP1, p21), CDKN1B (aka KIP1, p27), and CDKN1C (akaKIP2, p57)) modulate the activity of CDK-cyclin complexes. For example, CDKN1A expression is induced in response to DNA damage and results in cell cycle arrest in the G_1 phase, thus allowing the DNA repair machinery to repair the DNA damage. If cellular DNA damage is beyond repair, then cell death or apoptotic pathways are activated to discard damaged cells so that mutated DNA is not replicated, thus minimizing the onset of malignant cell growth and cancer.

In MC/CAR multiple myeloma cells, arsenic trioxide-induced expression of the Cdk inhibitor, CDKN1A, which associates with cyclin E. The MC/CAR cells arrested in the G_1 and G_2-M phases of the cell cycle. In addition, expression of the cell survival protein, Bcl-2, decreased contributing to the apoptotic cell death of MC/CAR cells [71]. In the non-small cell lung cancer cell line, H460, arsenic trioxide induced up-regulation of cyclin B, M-phase arrest, phosphorylation of Bcl-2 and apoptosis [55]. In addition to cyclins B and D, and the Cdk inhibitor, CDKN1A, arsenite also affects other proteins involved in the cell cycle. In A375 human malignant melanoma cells, arsenite induced mitotic arrest correlated with BUBR1 phosphorylation [62]. BUBR1 is a protein that functions in the mitotic checkpoint protein complex that is required to prevent unequal segregation of chromosomes during cell division. Failure to properly segregate chromosomes can lead to cellular apoptosis.

In normal human fibroblasts exposed to both arsenite and ionizing radiation, cyclin D1 expression increased, while CDKN1A expression decreased resulting in a decreased block in cell cycle progression after DNA damage [101]. It was hypothesized that the DNA damage induced by ionizing radiation and the positive growth signaling associated with increased cyclin D1 expression contributes to the comutagenic effects of arsenite when present with other carcinogens. Using the HaCat keratinocyte cell line, Ouyang *et al.* demonstrated that elevated cyclin D1 expression required NF-κB activation [68]. Dominant-negative mutant IkappaB kinase beta (IKKβ) impaired arsenite-induced cyclin D1 expression and IKKβ knockout mouse embryo fibroblasts failed to express cyclin D1 in response to arsenite. These results demonstrate that arsenite activates intracellular pathways, like the NF-κB pathway, to affect cell cycle responses. In addition, arsenite activates the NF-κB pathway for induction of Cox-2 in mouse epidermal JB6 Cl41 cells [69] and induction of CRP in the human hepatic cell line, HepG2 [23]. Thus in addition to modulating expression of cell cycle proteins such as cyclins B and D1, CDKN1A, and BUBR1, arsenite induces proinflammatory proteins such as Cox-2 and CRP strongly suggesting that arsenite signals, at least in part, through inflammatory signaling mechanisms.

16.8 ARSENIC AND MAP KINASE SIGNALING

When researchers first began investigating the role of mitogen-activated protein (MAP) kinases in arsenic signaling, the studies often used nonphysiologically relevant doses of arsenic, and for this reason those studies will not be considered in this section. The MAP kinases consist of three major families: extracellular signal-regulated kinases (ERK) 1 and 2, c-Jun *N*-terminal kinases (JNKs), and p38. The MAP kinases are proline-directed serine/threonine kinases that phosphorylate transcription factors and other intracellular proteins to either increase or decrease their activity. The ERK MAP kinases are often associated with cellular proliferation. When rat lung epithelial cells (LECs) were exposed to relatively low (2 μM) and high (40 μM) doses of arsenite, the cells had a proliferative and apoptotic response, respectively [50]. The proliferative response correlated with ERK1/2 activation and stimulation of activator protein-1 (AP-1) DNA binding, while the LEC apoptotic response correlated with activation of JNK and expression of the pro-apoptotic protein, Bax. AP-1 DNA binding activity was also observed in the SV-40 immortalized human urothelial cell line, UROtsa, using 1 μM arsenite [22]. In contrast, low doses of arsenite (1–5 μM) stimulated proliferation of porcine aortic endothelial cells, but those doses of arsenite did not correlate with ERK1/2 activation [1]. In fact, porcine aortic endothelial cell proliferation correlated with superoxide and H_2O_2 production. Together, these data suggest that arsenite activates ERK1/2 in some cell types, but there is no absolute correlation between ERK1/2 activation and cell proliferation. However, how arsenite activates ERK1/2 in cells like rat lung epithelial cells remains an unanswered question, but as Liu *et al.* demonstrated with arsenic trioxide and the EGF receptor [59], if these cells express the EGF receptor, then this receptor is worthy of investigation as the cause of ERK activation in rat lung epithelial cells.

Generally, measurements of MAP kinase activity in response to arsenic occur after acute exposure to cell lines. Those acute exposures typically require nonphysiological doses of arsenite (>20 μM), thus the relevance of those studies are highly debatable. However, chronic exposure to low doses of arsenite for a period of days was sufficient to activate the JNK MAP kinase signaling pathway. For example, exposing human embryonic lung fibroblasts (HELF) to 1 μM arsenite continuously for 10–30 passages was sufficient to activate JNK1, but not JNK2 [54]. JNK1 activation also correlated with anchorage-independent growth of arsenite-transformed HELF and knock down of JNK1 expression by siRNA decreased the number of cell colonies in soft agar. In contrast, Liu *et al.* demonstrated that the JNK2 MAPK pathway was required for arsenite-dependent induction of heat shock protein (HSP) 27 and 70 expression [56]. These experiments demonstrate that arsenite exposure can activate MAP kinase signaling proteins, although there are subtle differences in response among cell lines.

A chronic, low-dose arsenite approach was used to measure Ras activation. Ras is a small GTP-binding protein that functions upstream of the Raf-MEK-ERK pathway. Stimulating spontaneously immortalized human epidermal cells with 2 μM arsenite for 3 days was sufficient to activate Ras [72]. The activation of Ras also correlated with colony-forming activity. Moreover, incubating spontaneously immortalized human epidermal cells with AG1478, a tyrosine kinase inhibitor that targets the EGF receptor, inhibited Ras activity and colony forming activity suggesting that the proliferative response to chronic arsenite exposure is due to activation of the EGF receptor and Ras (Fig. 16.2). Exactly how arsenite regulates these proteins remains to be determined.

From the types of studies cited earlier, it is clear that physiologically relevant doses of arsenic activate MAP kinase signaling pathways. Exactly how arsenite activates JNK1 versus JNK2 in the same cell or how arsenite activates the EGF receptor remain to be determined. Undoubtedly, there are likely subtle differences in how arsenite and EGF activate the EGF receptor. More specific reagents or more sensitive assays may be required to dissect those differences in EGF receptor signaling.

In the last few years, arsenite exposure has been associated with cellular autophagy. Autophagy is a cellular response that is associated with stress and cell death whereby the cellular machinery degrades damaged or superfluous organelles. Degradation of these cellular constituents occurs within a vesicular structure called the autophagosome. Human lymphoblastoid cells undergo autophagy when exposed to 6 μM arsenite for 96 h [5]. Similarly, arsenic trioxide induced autophagy in the human glioblastoma cell line, U118-MG [18]. The autophagic activity in U188-MG cells correlated with the MAP kinase activity, including ERK, JNK, and p38. Additional studies will be needed to further dissect the mechanism by which arsenite and arsenic trioxide activate the various MAP kinases.

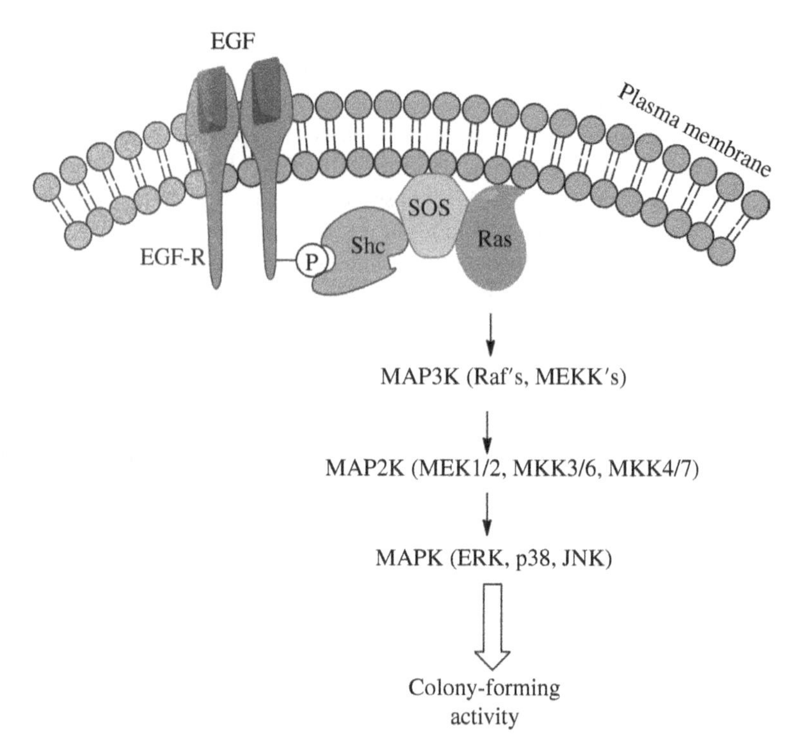

FIGURE 16.2 Regulation of the EGF signaling pathway by arsenic. EGF activates the EGF receptor resulting in tyrosine phosphorylation of the receptor, and then recruitment of Shc and other proteins (not shown). The SH2 domain of Shc mediates the interaction between Shc and the EGF receptor. The son-of-sevenless (SOS) guanine nucleotide-exchange protein is recruited and exchanges GDP for GTP on the small G protein, Ras. GTP-bound Ras activates Raf that functions upstream of MEK1/2 and ERK in the ERK MAPK pathway.

16.9 ARSENIC AND THE PHOSPHATIDYLINOSITOL 3-KINASE SIGNALING PATHWAY

Epidemiological studies have linked arsenic exposure through the drinking water supply with diabetes in humans [64, 65]. It has been shown that exposing 3T3-L1 adipocytes with 5 μM arsenite for 4 h inhibited insulin-stimulated glucose uptake by approximately 25%, while a dose of 100 μM resulted in 50% inhibition although the physiological relevance of such a dose is debatable [73]. These results demonstrated that arsenite perturbs the insulin signaling pathway and supports the hypothesis that arsenic is associated with insulin resistance. Researchers have been working to investigate the mechanism by which arsenite affects the insulin signaling pathway.

Phosphatidylinositol 3-kinase (PI3-K) is a critical enzyme linking the insulin receptor with protein kinase B (PKB/Akt) and the translocation of intracellular vesicles containing the glucose transporter-4 (GLUT4) to the plasma membrane to facilitate cellular glucose uptake. PI3-K phosphorylates phosphatidylinositol-4,5-bisphosphate (PIP_2) at the 3-position to produce phosphatidylinositol-3,4,5-bisphosphate (PIP_3).

PIP_3 is required for activation of phosphoinositide-dependent kinase (PDK), which phosphorylates Akt at threonine 308 in the activation loop. Phosphatase and tensin homolog on chromosome 10 (PTEN) is the lipid phosphatase that dephosphorylates PIP_3. Inhibition of PTEN effectively increases the activity of the Akt signaling pathway downstream of PI3-K.

As stated earlier, exposing 3T3-L1 adipocytes to arsenite prevented insulin-stimulated glucose uptake implying an inhibition of PI3-K activity [73]. However, Paul *et al.* detected no changes in PI3-K activity in response to arsenite when measuring formation of radioactively labeled PIP_3. This result suggested that arsenite might be increasing PTEN activity, which would decrease PIP_3 levels and inhibit PDK and Akt signaling downstream of PI3-K. However, Paul *et al.* detected no changes in PTEN activity in response to arsenite exposure. Thus, the mechanism by which arsenite inhibits insulin-stimulated glucose uptake has not been fully elucidated.

The use of dominant-negative mutant proteins has helped elucidate the role of signaling proteins in arsenic responses. For example, expression of a mutant form of Akt which has the two critical phosphorylation sites (e.g., T308 and S473) mutated to alanine prevented arsenite-dependent activation of IKKβ in mouse epidermal JB6 Cl41 cells [67]. In addition, an acute exposure to 5 μM arsenite for 20 min activated PI3-K as measured by production of radioactively labeled PIP_3 [67]. These results helped define an arsenite activated signaling axis consisting of PI3K, Akt, IKKβ, and NFκB (Fig. 16.3). Activation of this pathway increases cyclin D1 expression and regulates the cell cycle in mouse epidermal JB6 and likely plays an important role in regulating insulin-stimulated glucose uptake in insulin responsive cells. However, one should not overlook that the PI3K signaling cascade can be regulated through inhibition of PTEN. Wan *et al.* recently demonstrated that PTEN was inactivated by arsenic trioxide-induced oxidative stress, thus effectively shifting the equilibrium for more PIP3 than PIP2 enabling activation of downstream enzymes of PI3-K [103]. PTEN is a member of the cysteine-based protein tyrosine phosphatase family, and recent advancements in dissecting the role of arsenic in the binding to sulfhydryls within zinc finger domains will undoubtedly help elucidate the inhibitory actions of arsenic on PTEN activity.

The PI3-K pathway also plays an important role in arsenic-dependent regulation of p53 DNA- binding activity. In normal human keratinocyte (NHK) cells, low doses of arsenite increased the DNA-binding activity of p53 [83]. Since p53 plays a key role in regulating stress and DNA damage responses, leading to either growth arrest or apoptosis, the status of p53 in cell lines used to study arsenic signal transduction is an important consideration when evaluating arsenic-dependent physiological responses. The status of p53 may need special consideration in studies designed to evaluate long-term arsenite or MMA(III) exposure since many cell lines used in such studies have mutated and inactive p53. As a result, the interpretation of studies with inactive p53 may not accurately reflect human arsenic exposures.

Studies of human populations chronically exposed to arsenic in drinking water indicated that nitric oxide metabolites were low [76]. These results began to reveal the possible causes of vascular disease associated with chronic arsenic exposure.

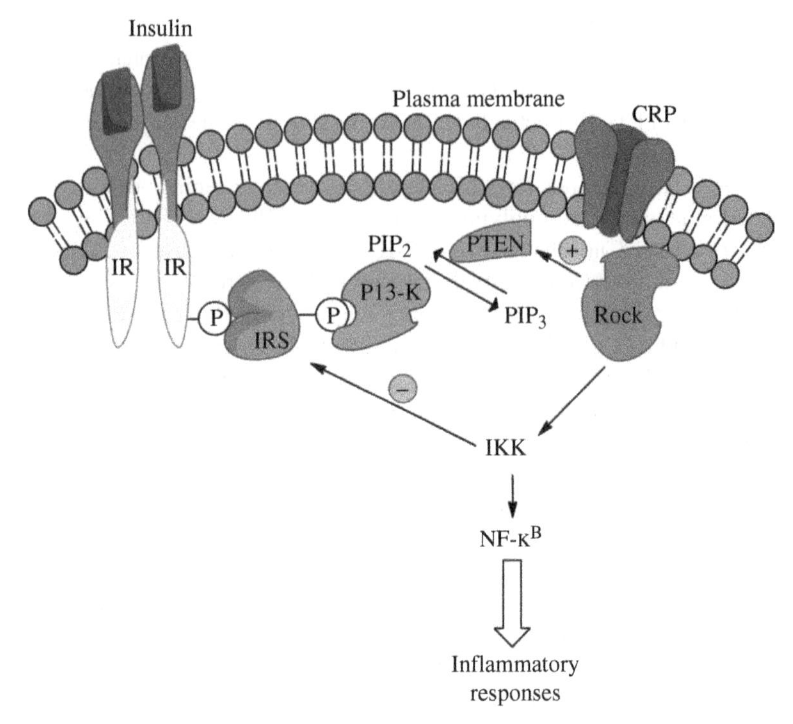

FIGURE 16.3 Regulation of the insulin signaling pathway by inflammation. CRP secretion from HepG2 cells is induced by arsenite. It is proposed that CRP-responsive cells bind circulating CRP which results in the activation of RhoA-associated kinase (ROCK) and subsequent activation of the NF-κB pathway. IKK phosphorylates IRS at serine 307 which diminishes the ability of the insulin receptor to activate PI3-K and leads to insulin resistance. In addition, ROCK activates PTEN, thus decreasing the PIP_3 pool for downstream signaling for phosphoinositide-dependent kinase (PDK) and Akt.

Research into the mechanism for arsenic-induced vascular disease using porcine aortic endothelial cells showed that endothelial nitric oxide synthase (eNOS) activity decreased in response to arsenite exposure after 24 h [98]. The activity of eNOS was also correlated with Akt activity, thus linking arsenite-dependent regulation of eNOS with the PI3-K pathway.

16.10 ARSENIC METABOLITES AND SIGNAL TRANSDUCTION

Historically, methylation of inorganic arsenic had been viewed as a detoxification pathway with different animal species having variable metabolic capabilities owing to their susceptibility to arsenic toxicity [100]. Methylated arsenic species were not viewed as having signal transduction activity. However, the chemical synthesis of monomethylarsonous acid (MMA(III)) allowed researchers to investigate both its cytotoxic as well as its signaling properties. In 2000, Petrick *et al.* demonstrated that methylated arsenic species displayed a rank order with regard to cytotoxicity with MMA(III) being more cytotoxic than arsenite, arsenate, monomethylarsonic acid (MMA(V)), or dimethylarsenic acid (DMA(V)) [74]. Later, Bredfelt *et al.* demonstrated that the human bladder cell line, UROtsa, chronically exposed to 50 nM MMA(III) exhibited a transformed phenotype with anchorage-independent growth in soft agar after 24 weeks and formed tumors in SCID mice after 52 weeks [7]. Further investigation into the signaling properties of MMA(III) demonstrated an increase in ROS and metallothionein expression in URotsa cells after acute exposures [24, 26], while chronic exposure to MMA(III) induced Cox-2 expression after 12 weeks [25]. Moreover, chronic MMA(III) exposure resulted in EGF receptor over-expression demonstrating that like arsenite, MMA(III) can modulate signaling pathways linked to cell proliferation. These results provided compelling evidence to demonstrate that MMA(III) is not only a by-product of inorganic arsenic metabolism, but is also capable of activating signal transduction pathways involved in cellular proliferation and carcinogenesis.

Malignant transformation of cells chronically exposed to low doses of arsenite or MMA(III) has provided evidence for re-programming of the epigenetic landscape for selected genes whose activity favors cell proliferation. Overall DNA hypomethylation has been linked with chronic arsenite exposure [14, 113], suggesting that cells adapted to long-term arsenite exposure through epigenetic mechanisms. Researchers began investigating histone modifications for clues as to how arsenite was altering cellular phenotype. Histone phosphorylation/dephosphorylation and acetylation/deacetylation of various amino acids are common post-translational modifications associated with changes in gene expression. Using the JB6 mouse epidermal cell line Cl 41, He *et al.* demonstrated that low doses of arsenite could induce phosphorylation of histone H3 on serine 10, which is generally associated with chromosomal condensation [35]. p90 ribosomal S6 kinase (RSK) is regulated by the ERK MAP kinases. In cells deficient in RSK, serine 10 of histone H3 was not phosphorylated in response to 10 μM arsenite, indicating a connection between histone H3 phosphorylation and ERK MAP kinase activation.

Arsenic is classified as a human carcinogen, but generally demonstrates weak mutagenic activity in most assays. However, exposing primary human foreskin fibroblasts to 5 μM arsenite resulted in micronuclei formation, which could be reversed with *S*-adenosyl-methionine, suggesting that methylated arsenic species are also involved in micronuclei formation [80]. Micronuclei are fragments of the cell nucleus that either contains whole chromosomes or chromosomal fragments. Phosphorylation of the histone variant, H2A.X, is indicative of DNA double strand breaks and 5 μM arsenite induced phosphorylation of H2A.X in human HepG2 cells [80]. Although SAM did not affect H2A.X phosphorylation in these experiments, these results strongly suggest that DNA repair mechanisms are activated in response to arsenite. Presumably, cells have a threshold in which DNA damage and repair rates are capable of negating the deleterious effects of arsenic. However, chronic arsenic exposures as low as 1 μM arsenite or 50 nM MMA(III) are sufficient to induce cellular malignant transformation demonstrating that DNA damage and repair mechanisms can be exceeded.

Chromatin immunoprecipitation (ChIP) assays using antibodies directed against specifically modified histones (e.g., acetylated histone H3) has been coupled with microarray hybridization of gene promoters to further interrogate the histone modification state of DNA after chronic arsenite or MMA(III) exposure. Using chronically exposed UROtsa cells as a model, Jensen *et al.* demonstrated that arsenite and MMA(III) alter the acetylation state of histone H3 in various gene promoters [40]. In selected gene promoters, histone H3 hypoacetylation correlated with DNA hypermethylation and concomitant repression of gene expression. Although arsenite and MMA(III) altered the acetylation of different promoter regions of the genome, there were 17 promoter regions whose changes were common to both arsenic species suggesting an overlapping mechanism of action.

Promoter microarray and DNA microarray experiments are generally viewed as hypothesis generating experiments since such experiments can generate volumes of data. The results of such studies require validation through alternative approaches and techniques. For example, the SPARC (secreted protein acidic and rich in cysteine) gene promoter was identified in a DNA promoter microarray study as hypoacetylated [40], while Larson *et al.* independently validated the down-regulated expression of SPARC in arsenic transformed UROtsa cells and different stages of urothelial carcinoma [48]. Although promoter microarray combined with ChIP is a powerful approach to dissect mechanisms of arsenic toxicity and carcinogenesis, these technologies do not provide all the answers. Models organisms like yeast still provide important information when studying arsenic. Jo *et al.* demonstrated that the yeast gene, *SAS2* (something about silencing), is required for resistance to arsenite in yeast [42]. The human ortholog for Sas2p is MYST1, a histone 4 lysine 16 (H4K16) acetyltransferase. When MYST1 expression was knocked down in UROtsa cells, cell viability decreased in response to acute exposure to 5 μM arsenite or 1 μM MMA(III) suggesting that acetylation of H4K16 is required for resistance to arsenic exposure [42]. Presumably, the study of genes upregulated due to acetylated H4K16 will be informative in understanding resistance to arsenic exposure.

16.11 ARSENIC BINDING TO CYSTEINE RESIDUES

Arsenic binding to proteins and affecting cell signaling events has long been studied as has arsenic's affinity for sulfhydryls. The principle of arsenic binding to sulfhydryl-rich areas such as are found in hair has long been used to analyze the hair of potential arsenic poisoning victims. A number of different groups have tried to gain a deeper insight into this phenomenon. Hoffman and Lane looked into direct methods to study arsenical interactions with proteins. In the methods they developed, Hoffman and Lane used the arsenical phenylarsine oxide (PAO) and constructed a PAO affinity resin in order to bind proteins from 3T3-L1 adipocyte cell lysates for analysis. Among the proteins that they found to bind to the arsenical resin were the insulin responsive transporter GLUT4, and both the α and β subunits of tubulin. Based on this work, the researchers suggested that the arsenic-binding motif was cysteine-glycine-cysteine (CGC) and was approximated to have a distance of 40 Å between thiols [37]. Further studies by Menzel *et al.* using a similar approach to Hoffman and Lane but utilizing human lymphocytes showed that in addition to tubulin, actin also bound to their arsenic resin column [63]. Menzel *et al.* noted that analysis of both the actin and tubulin amino acid sequences showed no indication of a CGC arsenic-binding motif, and suggested that a secondary or tertiary structure of the protein is likely involved in an arsenic binding motif. They further suggested that this motif would likely be characterized by either two or three thiol groups [63]. Bhattacharjee and Rosen also tried to gain further insight by studying an arsenic stimulated *Escherichia coli* ArsA-ATPase. The complex is an ATP-dependent arsenic transporter that removes arsenic from the interior of the *Escherichia coli* bacteria and allows for arsenic resistance. Bhattacharjee and Rosen further cloned point mutations into the ArsA-ATPase that replaced cysteines at Cys^{113}, Cys^{172}, and Cys^{422} with serines. They concluded that all three cysteines were needed to confer arsenic activation to the ArsA-ATPase. Further study of the crystallographic structure of ArsA-ATPase revealed that the arsenic binding motif is due to the tertiary structure of the protein and that a critical interthiol distance of 3–6 Å was required for arsenic binding [3]. Shi *et al.* sought to address the problem of the arsenic-binding motif by using the repressor protein ArsR that binds to the arsenic regulatory element of the *arsa* gene [87]. Shi *et al.* used directed point mutations on the ArsR protein and substituted cysteines with glycines. Interestingly, Shi *et al.* found that two cysteines were necessary for arsenic binding and subsequent release of ArsR from the regulatory region of the *ars* operon, but that three cysteines could form a soft metal complex similar to that found by Bhattacharjee and Rosen with the ArsA-ATPase allosteric site. Because Shi *et al.* controlled the position of the cysteine residues via point mutations, they arrived at a more direct estimate of the interthiol distance between vicinal dithiols than was originally estimated. Based on their calculations, they proposed a triangular pyramidal structure with a distance of 3.2 Å on each side [87].

In 2008, Kitchin and Wallace set out to look at the mechanisms underlying arsenic carcinogenesis [46]. They hypothesized that arsenic binding to various proteins was a key mechanism involved in arsenic carcinogenesis, and they employed radioactive [^{73}As] to investigate this hypothesis. Their data suggested that arsenic had the ability to bind to a number of *in vivo* protein targets including tubulin, poly(ADP-ribose) polymerase (PARP-1), thioredoxin reductase, estrogen receptor α, arsenic (+3) methyltransferase (ASMT) and Keap-1. Many of these proteins contained not only vicinal cysteines but specifically C2H2, C3H, and C4 zinc finger domains [43, 45]. Similarly, Qin *et al.* demonstrated that low-dose arsenite combined with ultraviolet radiation (UVR) greatly enhanced DNA strand breaks in HaCat keratinocytes [79]. Pretreatment of HaCat cells with zinc diminished the exacerbating effects of arsenite on UVR-dependent DNA strand breaks. Moreover, PARP-1 activity was restored with zinc. These data were consistent with a prior study of normal epidermal keratinocytes in which genes involved in DNA repair were repressed in response to low dose arsenite [31]. Although PARP-1 was not included in the list of DNA repair genes whose expression was repressed by arsenite, specific inhibition of PARP-1 enzyme activity by arsenite effectively served the same function in HaCat keratinocytes. It was later shown by mass spectrometry that arsenite binds to the zinc finger of PARP-1, with a preference for zinc fingers with three or four cysteine residues further demonstrating the specificity of arsenite binding to sulfhydryls [114] (see also Chapter 13). Based on the data cited earlier, Wnek *et al.* set out to determine whether MMA(III) bound to zinc fingers as well. They found that MMA(III) had the ability to bind to the zinc finger motif in PARP-1 and to inhibit PARP-1 activity and DNA repair in a dose dependent manner. Additionally, they found that the presence of zinc supplementation restored PARP-1 activity [108]. In summary, there are clearly chemical differences between arsenite and MMA(III), but the binding to sulfhydryls within zinc finger domains appears to be an emerging theme that is common to both arsenic species and may account for the signal transduction and toxicology associated with arsenic exposure.

16.12 CONCLUDING REMARKS

Despite all that we have learned about arsenic over the last decade or so, many of the signaling properties associated with arsenite and its metabolites still remain an enigma. We have learned that arsenic trioxide has become a standard of care in the treatment of acute promyelocytic leukemia due to its ability to induce apoptosis. The activity of arsenite can be inhibited by a receptor antagonist to the G_i-coupled endothelin-1 type A and B receptors. Does that mean the arsenite really binds to the endothelin-1 receptor? If so, how is that happening? It is safe to say that when it comes to arsenite, there are many observations, but few answers as to how a given response occurs. At the molecular level, the fact that arsenite can bind to cysteine and histidine residues in zinc finger domains may start defining molecular recognition motifs that attract arsenic in biological processes. The notion that arsenic binds to vicinal sulfhydryls has been bandied about for many years, but perhaps our current analytical methods will provide investigators with unequivocal data to support such a claim.

Over the years, the study of arsenic signal transduction has likely suffered from assays that lack sensitivity and uniformity or consistency in exposure. The lack of assay sensitivity has forced investigators to use nonphysiologically relevant arsenic doses in order to detect a response. For example, in the MAP kinase field, environmentally relevant doses of arsenite hardly affect kinase phosphorylation, while the use of dominant-negative mutant forms of kinases often provide the “expected” inhibitory effect in a signaling pathway. However in such experiments, one is always left with the difficult task of interpreting the results. Does overexpression of a dominant-negative protein only inhibit the pathway of interest? It seems unlikely given the concentration of protein expressed. The use of siRNA where knock down of protein expression has become a standard approach to evaluate the role of a given protein in a signaling pathway will undoubtedly play a big role in future research in arsenic signal transduction. Certainly, there are potential pitfalls associated with any technology, but the use of siRNA has already helped make advancements in the arsenic signal transduction field and holds promise for the future. In addition, the arsenic signal transduction field could benefit from the use of animal models engineered to express reporter genes. Arsenic affects proteins that regulate transcription directly or indirectly via epigenetic modifications. Animals expressing reporter genes would facilitate *in vivo* studies of specific tissues and organs that are susceptible to arsenic exposure.

The issue of chronic versus acute dosing studies is an issue of concern in studying arsenic signal transduction. Environmentally relevant human exposures typically occur over long periods of time. How are such exposures replicated in cell lines in which most signal transduction studies are performed? This question is a challenge because properly addressing it deals directly with the issue of biological relevance. As scientists, we strive to answer important research questions. In this chapter, we have identified some of the roles that arsenic plays in signal transduction pathways. We all recognize that there is much more to know and learn about arsenic. Observations made from the many epidemiological studies have helped guide us in our signal transduction studies. But to truly make an impact in this field of arsenic signal transduction, we must strive to move from descriptive studies to address questions that tell us how arsenic affects biological systems. Addressing the questions that tell us how arsenic behaves in biological systems will enable us to develop rationally appropriate strategies and therapies that can improve human health. Not only will this research improve human health, but it will better guide risk assessment and inform public policies.

REFERENCES

[1] A. Barchowsky, R.R. Roussel, L.R. Klei, P.E. James, N. Ganju, K.R. Smith, E.J. Dudek, Low levels of arsenic trioxide stimulate proliferative signals in primary vascular cells without activating stress effector pathways. Toxicol. Appl. Pharmacol. 159 (1999) 65–75.

[2] D.P. Bartel, MicroRNAs: genomics, biogenesis, mechanism, and function. Cell 116 (2004) 281–297.

[3] H. Bhattacharjee, B.P. Rosen, Spatial proximity of Cys113, Cys172, and Cys422 in the metalloactivation domain of the ArsA ATPase. J. Biol. Chem. 271 (1996) 24465–24470.

[4] J.E. Bodwell, J.A. Gosse, A.P. Nomikos, J.W. Hamilton, Arsenic disruption of steroid receptor gene activation: complex dose-response effects are shared by several steroid receptors. Chem. Res. Toxicol. 19 (2006) 1619–1629.

[5] A.M. Bolt, R.M. Byrd, W.T. Klimecki, Autophagy is the predominant process induced by arsenite in human lymphoblastoid cell lines. Toxicol. Appl. Pharmacol. 244 (2010) 366–373.

[6] C.P. Bracken, P.A. Gregory, N. Kolesnikoff, A.G. Bert, J. Wang, M.F. Shannon, G.J. Goodall, A double-negative feedback loop between ZEB1-SIP1 and the microRNA-200 family regulates epithelial-mesenchymal transition. Cancer Res. 68 (2008) 7846–7854.

[7] T.G. Bredfeldt, B. Jagadish, K.E. Eblin, E.A. Mash, A.J. Gandolfi, Monomethylarsonous acid induces transformation of human bladder cells. Toxicol. Appl. Pharmacol. 216 (2006) 69–79.

[8] M. Bunderson, J.D. Coffin, H.D. Beall, Arsenic induces peroxynitrite generation and cyclooxygenase-2 protein expression in aortic endothelial cells: possible role in atherosclerosis. Toxicol. Appl. Pharmacol. 184 (2002) 11–18.

[9] U. Burk, J. Schubert, U. Wellner, O. Schmalhofer, E. Vincan, S. Spaderna, T. Brabletz, A reciprocal repression between ZEB1 and members of the miR-200 family promotes EMT and invasion in cancer cells. EMBO Rep. 9 (2008) 582–589.

[10] O. Cantoni, A. Guidarelli, P. Sestili, F. Mannello, G. Gazzanelli, F. Cattabeni, Hydrogen peroxide cytotoxicity under conditions of normal or reduced catalase activity in H2O2-sensitive and -resistant Chinese hamster ovary (CHO) cell variants. Toxicol. Lett. 73 (1994) 193–199.

[11] D.E. Carter, H.V. Aposhian, A.J. Gandolfi, The metabolism of inorganic arsenic oxides, gallium arsenide, and arsine: a toxicochemical review. Toxicol. Appl. Pharmacol. 193 (2003) 309–334.

[12] S. Chattopadhyay, S. Bhaumik, M. Purkayastha, S. Basu, C.A. Nag, G.S. Das, Apoptosis and necrosis in developing brain cells due to arsenic toxicity and protection with antioxidants. Toxicol. Lett. 136 (2002) 65–76.

[13] A.N. Chaudhuri, S. Basu, S. Chattopadhyay, G.S. Das, Effect of high arsenic content in drinking water on rat brain. Indian J. Biochem. Biophys. 36 (1999) 51–54.

[14] H. Chen, S. Li, J. Liu, B.A. Diwan, J.C. Barrett, M.P. Waalkes, Chronic inorganic arsenic exposure induces hepatic global and individual gene hypomethylation: implications for arsenic hepatocarcinogenesis. Carcinogenesis 25 (2004) 1779–1786.

[15] H. Chen, J. Liu, B.A. Merrick, M.P. Waalkes, Genetic events associated with arsenic-induced malignant transformation: applications of cDNA microarray technology. Mol. Carcinog. 30 (2001) 79–87.

[16] L.L. Chen, M.L. Clawson, S. Bilgrami, G. Carmichael, A sequence-specific single-stranded DNA-binding protein that is responsive to epidermal growth factor recognizes an S1 nuclease-sensitive region in the epidermal growth factor receptor promoter. Cell Growth Differ. 4 (1993) 975–983.

[17] Y.C. Chen, S.Y. Lin-Shiau, J.K. Lin, Involvement of reactive oxygen species and caspase 3 activation in arsenite-induced apoptosis. J. Cell Physiol. 177 (1998) 324–333.

[18] H.W. Chiu, Y.S. Ho, Y.J. Wang, Arsenic trioxide induces autophagy and apoptosis in human glioma cells in vitro and in vivo through downregulation of survivin. J. Mol. Med. (Berl) 89 (2011) 927–941.

[19] H. de la Fuente, D. Portales-Perez, L. Baranda, F. Diaz-Barriga, V. Saavedra-Alanis, E. Layseca, R. Gonzalez-Amaro, Effect of arsenic, cadmium and lead on the induction of apoptosis of normal human mononuclear cells. Clin. Exp. Immunol. 129 (2002) 69–77.

[20] M. Delnomdedieu, M.M. Basti, J.D. Otvos, D.J. Thomas, Transfer of arsenite from glutathione to dithiols: a model of interaction. Chem. Res. Toxicol. 6 (1993) 598–602.

[21] L. Delva, M. Cornic, N. Balitrand, F. Guidez, J.M. Miclea, A. Delmer, F. Teillet, P. Fenaux, S. Castaigne, L. Degos, Resistance to all-trans retinoic acid (ATRA) therapy in relapsing acute promyelocytic leukemia: study of in vitro ATRA sensitivity and cellular retinoic acid binding protein levels in leukemic cells. Blood 82 (1993) 2175–2181.

[22] Z. Drobna, I. Jaspers, D.J. Thomas, M. Styblo, Differential activation of AP-1 in human bladder epithelial cells by inorganic and methylated arsenicals. FASEB J 17 (2003) 67–69.

[23] I.L. Druwe, J.J. Sollome, P. Sanchez-Soria, R.N. Hardwick, T.D. Camenisch, R.R. Vaillancourt, Arsenite activates NFkappaB through induction of C-reactive protein. Toxicol. Appl. Pharmacol. 261 (2012) 263–270.

[24] K.E. Eblin, M.E. Bowen, D.W. Cromey, T.G. Bredfeldt, E.A. Mash, S.S. Lau, A.J. Gandolfi, Arsenite and monomethylarsonous acid generate oxidative stress response in human bladder cell culture. Toxicol. Appl. Pharmacol. 217 (2006) 7–14.

[25] K.E. Eblin, T.G. Bredfeldt, S. Buffington, A.J. Gandolfi, Mitogenic signal transduction caused by monomethylarsonous acid in human bladder cells: role in arsenic-induced carcinogenesis. Toxicol. Sci. 95 (2007) 321–330.

[26] K.E. Eblin, A.M. Hau, T.J. Jensen, B.W. Futscher, A.J. Gandolfi, The role of reactive oxygen species in arsenite and monomethylarsonous acid-induced signal transduction in human bladder cells: acute studies. Toxicology 250 (2008) 47–54.

[27] A.L. Eggler, Y. Luo, R.B. van Breemen, A.D. Mesecar, Identification of the highly reactive cysteine 151 in the chemopreventive agent-sensor Keap1 protein is method-dependent. Chem. Res. Toxicol. 20 (2007) 1878–1884.

[28] K. Felix, S.K. Manna, K. Wise, J. Barr, G.T. Ramesh, Low levels of arsenite activates nuclear factor-kappaB and activator protein-1 in immortalized mesencephalic cells. J. Biochem. Mol. Toxicol. 19 (2005) 67–77.

[29] S.R. Frankel, A. Eardley, G. Lauwers, M. Weiss, R.P. Warrell, Jr., The "retinoic acid syndrome" in acute promyelocytic leukemia. Ann. Intern. Med. 117 (1992) 292–296.

[30] J.R. Gurr, F. Liu, S. Lynn, K.Y. Jan, Calcium-dependent nitric oxide production is involved in arsenite-induced micronuclei. Mutat. Res. 416 (1998) 137–148.

[31] H.K. Hamadeh, K.J. Trouba, R.P. Amin, C.A. Afshari, D. Germolec, Coordination of altered DNA repair and damage pathways in arsenite-exposed keratinocytes. Toxicol. Sci. 69 (2002) 306–316.

[32] S.S. Han, K. Kim, E.R. Hahm, C.H. Park, B.F. Kimler, S.J. Lee, S.H. Lee, W.S. Kim, C.W. Jung, K. Park, J. Kim, S.S. Yoon, J.H. Lee, S. Park, Arsenic trioxide represses constitutive activation of NF-kappaB and COX-2 expression in human acute myeloid leukemia, HL-60. J. Cell Biochem. 94 (2005) 695–707.

[33] F. Hayakawa, M.L. Privalsky, Phosphorylation of PML by mitogen-activated protein kinases plays a key role in arsenic trioxide-mediated apoptosis. Cancer Cell 5 (2004) 389–401.

[34] L. He, G.J. Hannon, MicroRNAs: small RNAs with a big role in gene regulation. Nat. Rev. Genet. 5 (2004) 522–531.

[35] Z. He, W.Y. Ma, G. Liu, Y. Zhang, A.M. Bode, Z. Dong, Arsenite-induced phosphorylation of histone H3 at serine 10 is mediated by Akt1, extracellular signal-regulated kinase 2, and p90 ribosomal S6 kinase 2 but not mitogen- and stress-activated protein kinase 1. J. Biol. Chem. 278 (2003) 10588–10593.

[36] O. Hobert, Common logic of transcription factor and microRNA action. Trends Biochem. Sci. 29 (2004) 462–468.

[37] R.D. Hoffman, M.D. Lane, Iodophenylarsine oxide and arsenical affinity chromatography: new probes for dithiol proteins. Application to tubulins and to components of the insulin receptor-glucose transporter signal transduction pathway. J. Biol. Chem. 267 (1992) 14005–14011.

[38] S.H. Hong, Z. Yang, M.L. Privalsky, Arsenic trioxide is a potent inhibitor of the interaction of SMRT corepressor with its transcription factor partners, including the PML-retinoic acid receptor alpha oncoprotein found in human acute promyelocytic leukemia. Mol. Cell Biol. 21 (2001) 7172–7182.

[39] T. Ishii, K. Itoh, M. Yamamoto, Roles of Nrf2 in activation of antioxidant enzyme genes via antioxidant responsive elements. Methods Enzymol. 348 (2002) 182–190.

[40] T.J. Jensen, P. Novak, K.E. Eblin, A.J. Gandolfi, B.W. Futscher, Epigenetic remodeling during arsenical-induced malignant transformation. Carcinogenesis 29 (2008) 1500–1508.

[41] X.H. Jiang, B.C. Wong, S.T. Yuen, S.H. Jiang, C.H. Cho, K.C. Lai, M.C. Lin, H.F. Kung, S.K. Lam, Arsenic trioxide induces apoptosis in human gastric cancer cells through up-regulation of p53 and activation of caspase-3. Int. J. Cancer 91 (2001) 173–179.

[42] W.J. Jo, X. Ren, F. Chu, M. Aleshin, H. Wintz, A. Burlingame, M.T. Smith, C.D. Vulpe, L. Zhang, Acetylated H4K16 by MYST1 protects UROtsa cells from arsenic toxicity and is decreased following chronic arsenic exposure. Toxicol. Appl. Pharmacol. 241 (2009) 294–302.

[43] K.T. Kitchin, K. Wallace, Arsenite binding to synthetic peptides based on the Zn finger region and the estrogen binding region of the human estrogen receptor-alpha, Toxicol. Appl. Pharmacol. 206 (2005) 66–72.

[44] K.T. Kitchin, K. Wallace, Arsenite binding to synthetic peptides: the effect of increasing length between two cysteines. J. Biochem. Mol. Toxicol. 20 (2006) 35–38.

[45] K.T. Kitchin, K. Wallace, Dissociation of arsenite-peptide complexes: triphasic nature, rate constants, half-lives, and biological importance, J. Biochem. Mol. Toxicol. 20 (2006) 48–56.

[46] K.T. Kitchin, K. Wallace, Evidence against the nuclear in situ binding of arsenicals—oxidative stress theory of arsenic carcinogenesis. Toxicol. Appl. Pharmacol. 232 (2008) 252–257.

[47] L.R. Klei, D.Y. Garciafigueroa, A. Barchowsky, Arsenic activates endothelin-1 Gi protein-coupled receptor signaling to inhibit stem cell differentiation in adipogenesis. Toxicol. Sci. 131 (2013) 512–520.

[48] J. Larson, T. Yasmin, D.A. Sens, X.D. Zhou, M.A. Sens, S.H. Garrett, J.R. Dunlevy, L. Cao, S. Somji, SPARC gene expression is repressed in human urothelial cells (UROtsa) exposed to or malignantly transformed by cadmium or arsenite. Toxicol. Lett. 199 (2010) 166–172.

[49] A. Lau, N.F. Villeneuve, Z. Sun, P.K. Wong, D.D. Zhang, Dual roles of Nrf2 in cancer. Pharmacol. Res. 58 (2008) 262–270.

[50] A.T. Lau, M. Li, R. Xie, Q.Y. He, J.F. Chiu, Opposed arsenite-induced signaling pathways promote cell proliferation or apoptosis in cultured lung cells. Carcinogenesis 25 (2004) 21–28.

[51] S.S. Leonard, J.J. Bower, X. Shi, Metal-induced toxicity, carcinogenesis, mechanisms and cellular responses. Mol. Cell Biochem. 255 (2004) 3–10.

[52] S. Li, Y. Chen, B.P. Rosen, Role of vicinal cysteine pairs in metalloid sensing by the ArsD As(III)-responsive repressor. Mol. Microbiol. 41 (2001) 687–696.

[53] Y. Li, J. Qi, K. Liu, B. Li, H. Wang, J. Jia, Peroxynitrite-induced nitration of cyclooxygenase-2 and inducible nitric oxide synthase promotes their binding in diabetic angiopathy. Mol. Med. 16 (2010) 335–342.

[54] Y. Li, L. Shen, H. Xu, Y. Pang, Y. Xu, M. Ling, J. Zhou, X. Wang, Q. Liu, Up-regulation of cyclin D1 by JNK1/c-Jun is involved in tumorigenesis of human embryo lung fibroblast cells induced by a low concentration of arsenite 1. Toxicol. Lett. 206 (2011) 113–120.

[55] Y.H. Ling, J.D. Jiang, J.F. Holland, R. Perez-Soler, Arsenic trioxide produces polymerization of microtubules and mitotic arrest before apoptosis in human tumor cell lines. Mol. Pharmacol. 62 (2002) 529–538.

[56] J. Liu, D. Zhang, X. Mi, Q. Xia, Y. Yu, Z. Zuo, W. Guo, X. Zhao, J. Cao, Q. Yang, A. Zhu, W. Yang, X. Shi, J. Li, C. Huang, p27 suppresses arsenite-induced Hsp27/Hsp70 expression through inhibiting JNK2/c-Jun- and HSF-1-dependent pathways. J. Biol. Chem. 285 (2010) 26058–26065.

[57] L.Z. Liu, Y. Jiang, R.L. Carpenter, Y. Jing, S.C. Peiper, B.H. Jiang, Role and mechanism of arsenic in regulating angiogenesis. PLoS One 6 (2011) e20858.

[58] Y. Liu, W. Zhang, X. Zhang, Y. Qi, D. Huang, Y. Zhang, Arsenic trioxide inhibits invasion/migration in SGC-7901 cells by activating the reactive oxygen species-dependent cyclooxygenase-2/matrix metalloproteinase-2 pathway. Exp. Biol. Med. (Maywood.) 236 (2011) 592–597.

[59] Z.M. Liu, H.S. Huang, As2O3-induced c-Src/EGFR/ERK signaling is via Sp1 binding sites to stimulate p21WAF1/CIP1 expression in human epidermoid carcinoma A431 cells. Cell Signal. 18 (2006) 244–255.

[60] S. Lopez, Y. Miyashita, S.S. Simons, Jr., Structurally based, selective interaction of arsenite with steroid receptors. J. Biol. Chem. 265 (1990) 16039–16042.

[61] Y. Luo, A.L. Eggler, D. Liu, G. Liu, A.D. Mesecar, R.B. van Breemen, Sites of alkylation of human Keap1 by natural chemoprevention agents. J. Am. Soc. Mass Spectrom. 18 (2007) 2226–2232.

[62] S.C. McNeely, B.F. Taylor, J.C. States, Mitotic arrest-associated apoptosis induced by sodium arsenite in A375 melanoma cells is BUBR1-dependent. Toxicol. Appl. Pharmacol. 231 (2008) 61–67.

[63] D.B. Menzel, H.K. Hamadeh, E. Lee, D.M. Meacher, V. Said, R.E. Rasmussen, H. Greene, R.N. Roth, Arsenic binding proteins from human lymphoblastoid cells. Toxicol. Lett. 105 (1999) 89–101.

[64] A. Navas-Acien, E.K. Silbergeld, R. Pastor-Barriuso, E. Guallar, Arsenic exposure and prevalence of type 2 diabetes in US adults. JAMA 300 (2008) 814–822.

[65] A. Navas-Acien, E.K. Silbergeld, R.A. Streeter, J.M. Clark, T.A. Burke, E. Guallar, Arsenic exposure and type 2 diabetes: a systematic review of the experimental and epidemiological evidence. Environ. Health Perspect. 114 (2006) 641–648.

[66] T.L. Noreault-Conti, A. Fellows, J.M. Jacobs, H.W. Trask, S.C. Strom, R.M. Evans, S.A. Wrighton, P.R. Sinclair, J.F. Sinclair, R.C. Nichols, Arsenic decreases RXRalpha-dependent transcription of CYP3A and suppresses immune regulators in hepatocytes. Int. Immunopharmacol. 12 (2012) 651–656.

[67] W. Ouyang, J. Li, Q. Ma, C. Huang, Essential roles of PI-3K/Akt/IKKbeta/NFkappaB pathway in cyclin D1 induction by arsenite in JB6 Cl41 cells. Carcinogenesis 27 (2006) 864–873.

[68] W. Ouyang, Q. Ma, J. Li, D. Zhang, Z.G. Liu, A.K. Rustgi, C. Huang, Cyclin D1 induction through IkappaB kinase beta/nuclear factor-kappaB pathway is responsible for arsenite-induced increased cell cycle G1-S phase transition in human keratinocytes. Cancer Res. 65 (2005) 9287–9293.

[69] W. Ouyang, D. Zhang, Q. Ma, J. Li, C. Huang, Cyclooxygenase-2 induction by arsenite through the IKKbeta/NFkappaB pathway exerts an antiapoptotic effect in mouse epidermal Cl41 cells. Environ. Health Perspect. 115 (2007) 513–518.

[70] P.S. Park, Ensemble of G protein-coupled receptor active states. Curr. Med. Chem. 19 (2012) 1146–1154.

[71] W.H. Park, J.G. Seol, E.S. Kim, J.M. Hyun, C.W. Jung, C.C. Lee, B.K. Kim, Y.Y. Lee, Arsenic trioxide-mediated growth inhibition in MC/CAR myeloma cells via cell cycle arrest in association with induction of cyclin-dependent kinase inhibitor, p21, and apoptosis. Cancer Res. 60 (2000) 3065–3071.

[72] T.J. Patterson, R.H. Rice, Arsenite and insulin exhibit opposing effects on epidermal growth factor receptor and keratinocyte proliferative potential. Toxicol. Appl. Pharmacol. 221 (2007) 119–128.

[73] D.S. Paul, A.W. Harmon, V. Devesa, D.J. Thomas, M. Styblo, Molecular mechanisms of the diabetogenic effects of arsenic: inhibition of insulin signaling by arsenite and methylarsonous acid. Environ. Health Perspect. 115 (2007) 734–742.

[74] J.S. Petrick, F. Ayala-Fierro, W.R. Cullen, D.E. Carter, A.H. Vasken, Monomethylarsonous acid (MMA(III)) is more toxic than arsenite in Chang human hepatocytes 1. Toxicol. Appl. Pharmacol. 163 (2000) 203–207.

[75] J. Pi, Y. Bai, J.M. Reece, J. Williams, D. Liu, M.L. Freeman, W.E. Fahl, D. Shugar, J. Liu, W. Qu, S. Collins, M.P. Waalkes, Molecular mechanism of human Nrf2 activation and degradation: role of sequential phosphorylation by protein kinase CK2. Free Radic. Biol. Med. 42 (2007) 1797–1806.

[76] J. Pi, Y. Kumagai, G. Sun, H. Yamauchi, T. Yoshida, H. Iso, A. Endo, L. Yu, K. Yuki, T. Miyauchi, N. Shimojo, Decreased serum concentrations of nitric oxide metabolites among Chinese in an endemic area of chronic arsenic poisoning in inner Mongolia. Free Radic. Biol. Med. 28 (2000) 1137–1142.

[77] J. Pi, W. Qu, J.M. Reece, Y. Kumagai, M.P. Waalkes, Transcription factor Nrf2 activation by inorganic arsenic in cultured keratinocytes: involvement of hydrogen peroxide. Exp. Cell Res. 290 (2003) 234–245.

[78] Y. Qian, V. Castranova, X. Shi, New perspectives in arsenic-induced cell signal transduction. J. Inorg. Biochem. 96 (2003) 271–278.

[79] X.J. Qin, L.G. Hudson, W. Liu, G.S. Timmins, K.J. Liu, Low concentration of arsenite exacerbates UVR-induced DNA strand breaks by inhibiting PARP-1 activity. Toxicol. Appl. Pharmacol. 232 (2008) 41–50.

[80] T. Ramirez, H. Stopper, R. Hock, L.A. Herrera, Prevention of aneuploidy by S-adenosylmethionine in human cells treated with sodium arsenite. Mutat. Res. 617 (2007) 16–22.

[81] G.J. Roboz, S. Dias, G. Lam, W.J. Lane, S.L. Soignet, R.P. Warrell, Jr., S. Rafii, Arsenic trioxide induces dose- and time-dependent apoptosis of endothelium and may exert an antileukemic effect via inhibition of angiogenesis. Blood 96 (2000) 1525–1530.

[82] E.H. Romach, C.Q. Zhao, L.M. Del Razo, M.E. Cebrian, M.P. Waalkes, Studies on the mechanisms of arsenic-induced self tolerance developed in liver epithelial cells through continuous low-level arsenite exposure. Toxicol. Sci. 54 (2000) 500–508.

[83] M. Sandoval, M. Morales, R. Tapia, A.L. del Carmen, M. Sordo, P. Ostrosky-Wegman, A. Ortega, E. Lopez-Bayghen, p53 response to arsenic exposure in epithelial cells: protein kinase B/Akt involvement. Toxicol. Sci. 99 (2007) 126–140.

[84] Z.X. Shen, G.Q. Chen, J.H. Ni, X.S. Li, S.M. Xiong, Q.Y. Qiu, J. Zhu, W. Tang, G.L. Sun, K.Q. Yang, Y. Chen, L. Zhou, Z.W. Fang, Y.T. Wang, J. Ma, P. Zhang, T.D. Zhang, S.J. Chen, Z. Chen, Z.Y. Wang, Use of arsenic trioxide (As2O3) in the treatment of acute promyelocytic leukemia (APL): II. Clinical efficacy and pharmacokinetics in relapsed patients. Blood 89 (1997) 3354–3360.

[85] C.L. Sherwood, R.C. Lantz, J.L. Burgess, S. Boitano, Arsenic alters ATP-dependent Ca(2)+ signaling in human airway epithelial cell wound response. Toxicol. Sci. 121 (2011) 191–206.

[86] H. Shi, L.G. Hudson, W. Ding, S. Wang, K.L. Cooper, S. Liu, Y. Chen, X. Shi, K.J. Liu, Arsenite causes DNA damage in keratinocytes via generation of hydroxyl radicals. Chem. Res. Toxicol. 17 (2004) 871–878.

[87] W. Shi, J. Dong, R.A. Scott, M.Y. Ksenzenko, B.P. Rosen, The role of arsenic-thiol interactions in metalloregulation of the ars operon. J. Biol. Chem. 271 (1996) 9291–9297.

[88] S.L. Soignet, P. Maslak, Z.G. Wang, S. Jhanwar, E. Calleja, L.J. Dardashti, D. Corso, A. DeBlasio, J. Gabrilove, D.A. Scheinberg, P.P. Pandolfi, R.P. Warrell, Jr., Complete remission after treatment of acute promyelocytic leukemia with arsenic trioxide. N. Engl. J. Med. 339 (1998) 1341–1348.

[89] S. Spaderna, O. Schmalhofer, M. Wahlbuhl, A. Dimmler, K. Bauer, A. Sultan, F. Hlubek, A. Jung, D. Strand, A. Eger, T. Kirchner, J. Behrens, T. Brabletz, The transcriptional repressor ZEB1 promotes metastasis and loss of cell polarity in cancer. Cancer Res. 68 (2008) 537–544.

[90] A.C. Straub, K.A. Clark, M.A. Ross, A.G. Chandra, S. Li, X. Gao, P.J. Pagano, D.B. Stolz, A. Barchowsky, Arsenic-stimulated liver sinusoidal capillarization in mice requires NADPH oxidase-generated superoxide. J. Clin. Invest. 118 (2008) 3980–3989.

[91] A.C. Straub, L.R. Klei, D.B. Stolz, A. Barchowsky, Arsenic requires sphingosine-1-phosphate type 1 receptors to induce angiogenic genes and endothelial cell remodeling. Am. J. Pathol. 174 (2009) 1949–1958.

[92] D. Sun, D. Xu, B. Zhang, Rac signaling in tumorigenesis and as target for anticancer drug development. Drug Resist. Updat. 9 (2006) 274–287.

[93] H.D. Sun, D. Xu, B. Zhang, Ai-Lin I treated 32 cases of acute promyelocytic leukemia. Chin. J. Integrat. Chin. West Med. 12 (1992) 170–171.

[94] S. Suzuki, L.L. Arnold, K.L. Pennington, S. Kakiuchi-Kiyota, S.M. Cohen, Effects of co-administration of dietary sodium arsenite and an NADPH oxidase inhibitor on the rat bladder epithelium. Toxicology 261 (2009) 41–46.

[95] S.L. Tressel, G. Koukos, B. Tchernychev, S.L. Jacques, L. Covic, A. Kuliopulos, Pharmacology, biodistribution, and efficacy of GPCR-based pepducins in disease models. Methods Mol. Biol. 683 (2011) 259–275.

[96] K.J. Trouba, D.R. Germolec, Micromolar concentrations of sodium arsenite induce cyclooxygenase-2 expression and stimulate p42/44 mitogen-activated protein kinase phosphorylation in normal human epidermal keratinocytes. Toxicol. Sci. 79 (2004) 248–257.

[97] S.H. Tsai, Y.C. Liang, L. Chen, F.M. Ho, M.S. Hsieh, J.K. Lin, Arsenite stimulates cyclooxygenase-2 expression through activating IkappaB kinase and nuclear factor kappaB in primary and ECV304 endothelial cells. J. Cell Biochem. 84 (2002) 750–758.

[98] T.C. Tsou, F.Y. Tsai, Y.W. Hsieh, L.A. Li, S.C. Yeh, L.W. Chang, Arsenite induces endothelial cytotoxicity by down-regulation of vascular endothelial nitric oxide synthase. Toxicol. Appl. Pharmacol. 208 (2005) 277–284.

[99] T. Tsujiuchi, M. Tsutsumi, Y. Sasaki, M. Takahama, Y. Konishi, Hypomethylation of CpG sites and c-myc gene overexpression in hepatocellular carcinomas, but not hyperplastic nodules, induced by a choline-deficient L-amino acid-defined diet in rats. Jpn. J. Cancer Res. 90 (1999) 909–913.

[100] M. Vahter, Methylation of inorganic arsenic in different mammalian species and population groups. Sci. Prog. 82 (Pt 1) (1999) 69–88.

[101] B.L. Vogt, T.G. Rossman, Effects of arsenite on p53, p21 and cyclin D expression in normal human fibroblasts—a possible mechanism for arsenite's comutagenicity. Mutat. Res. 478 (2001) 159–168.

[102] L. Vrba, T.J. Jensen, J.C. Garbe, R.L. Heimark, A.E. Cress, S. Dickinson, M.R. Stampfer, B.W. Futscher, Role for DNA methylation in the regulation of miR-200c and miR-141 expression in normal and cancer cells. PLoS One 5 (2010) e8697.

[103] X. Wan, A.T. Dennis, C. Obejero-Paz, J.L. Overholt, J. Heredia-Moya, K.L. Kirk, E. Ficker, Oxidative inactivation of the lipid phosphatase phosphatase and tensin homolog on chromosome ten (PTEN) as a novel mechanism of acquired long QT syndrome. J. Biol. Chem. 286 (2011) 2843–2852.

[104] T.S. Wang, Y.F. Shu, Y.C. Liu, K.Y. Jan, H. Huang, Glutathione peroxidase and catalase modulate the genotoxicity of arsenite. Toxicology 121 (1997) 229–237.

[105] X.J. Wang, Z. Sun, W. Chen, Y. Li, N.F. Villeneuve, D.D. Zhang, Activation of Nrf2 by arsenite and monomethylarsonous acid is independent of Keap1-C151: enhanced Keap1-Cul3 interaction. Toxicol. Appl. Pharmacol. 230 (2008) 383–389.

[106] Z. Wang, Y. Zhao, E. Smith, G.J. Goodall, P.A. Drew, T. Brabletz, C. Yang, Reversal and prevention of arsenic-induced human bronchial epithelial cell malignant transformation by microRNA-200b. Toxicol. Sci. 121 (2011) 110–122.

[107] R.P. Warrell, Jr., Retinoid resistance in acute promyelocytic leukemia: new mechanisms, strategies, and implications. Blood 82 (1993) 1949–1953.

[108] S.M. Wnek, C.L. Kuhlman, J.M. Camarillo, M.K. Medeiros, K.J. Liu, S.S. Lau, A.J. Gandolfi, Interdependent genotoxic mechanisms of monomethylarsonous acid: role of ROS-induced DNA damage and poly(ADP-ribose) polymerase-1 inhibition in the malignant transformation of urothelial cells. Toxicol. Appl. Pharmacol. 257 (2011) 1–13.

[109] Y.H. Xu, N. Richert, S. Ito, G.T. Merlino, I. Pastan, Characterization of epidermal growth factor receptor gene expression in malignant and normal human cell lines. Proc. Natl. Acad. Sci. U. S. A. 81 (1984) 7308–7312.

[110] P. Yang, X.Q. He, L. Peng, A.P. Li, X.R. Wang, J.W. Zhou, Q.Z. Liu, The role of oxidative stress in hormesis induced by sodium arsenite in human embryo lung fibroblast (HELF) cellular proliferation model. J, Toxicol. Environ. Health A 70 (2007) 976–983.

[111] D.D. Zhang, Mechanistic studies of the Nrf2-Keap1 signaling pathway. Drug Metab. Rev. 38 (2006) 769–789.

[112] Y. Zhang, Q. Wang, X. Guo, R. Miller, Y. Guo, H.S. Yang, Activation and up-regulation of translation initiation factor 4B contribute to arsenic-induced transformation. Mol. Carcinog. 50 (2011) 528–538.

[113] C.Q. Zhao, M.R. Young, B.A. Diwan, T.P. Coogan, M.P. Waalkes, Association of arsenic-induced malignant transformation with DNA hypomethylation and aberrant gene expression. Proc. Natl. Acad. Sci. U. S. A 94 (1997) 10907–10912.

[114] X. Zhou, X. Sun, K.L. Cooper, F. Wang, K.J. Liu, L.G. Hudson, Arsenite interacts selectively with zinc finger proteins containing C3H1 or C4 motifs. J. Biol. Chem. 286 (2011) 22855–22863.

17

STEM CELL TARGETING AND ALTERATION BY ARSENIC

YUANYUAN XU, ERIK J. TOKAR AND MICHAEL P. WAALKES

Inorganic Toxicology Group, National Toxicology Program Laboratory, National Institute of Environmental Health Sciences, Research Triangle Park, NC, USA

17.1 INTRODUCTION

Arsenic is a naturally occurring element that is ubiquitous in the environment. Exposure to arsenic induces a variety of disorders, but its exact mechanisms are not well-defined. Stem cells are capable of self-renewal, are conditionally "immortal," and can differentiate into any type of cell or tissue in the body. Dysregulation of these characteristics likely results in abnormal growth and disease development. Recent studies provide compelling evidence that arsenic targets stem cells for the development of various disorders including cancer and developmental abnormalities. Paradoxically, arsenic can also act as an effective chemotherapeutic by targeting and eradicating stem cells/cancer stem cells in various malignancies. This chapter discusses the multiple effects and mechanisms of arsenic on stem cells and the resultant or potential disease manifestations or cures. This discussion will focus only on those studies that specifically and directly look at stem cells in their work.

17.2 STEM CELLS

The widely used definition of a stem cell (SC) originated in the hematopoietic field, in which SCs are defined as single cells that are clonal precursors of both identical daughter SCs, as well as a defined set of differentiated progeny [28, 97]. SCs play

Arsenic: Exposure Sources, Health Risks, and Mechanisms of Toxicity, First Edition.
Edited by J. Christopher States.
© 2016 John Wiley & Sons, Inc. Published 2016 by John Wiley & Sons, Inc.

a central role in an organism during all stages of growth and development and particularly in tissue repair response to toxic insult. Embryonic SCs are pluripotent and able to differentiate into all types of cells in the body to form normal tissues and generate organs. Adult SCs, also sometimes called organ-specific SCs or progenitor cells, are generally rare, quiescent cells that reside in a hierarchical structure within tissues. Compared with embryonic SCs, adult SCs are more restricted in their potential, functioning as the body's natural source of cells for specific tissue homeostasis and repair. In a healthy organism, adult SCs are capable of both self-renewal, to maintain the SC pool, and differentiation, to support life-long production of all mature cells within an organ [11, 66], and the delicate balance between these two processes is tightly regulated by a complex loop of genetic determinants and signaling factors throughout the entire lifetime of an organism. Recently developed are induced pluripotent SCs, where fully differentiated adult cells have been genetically "de-differentiated" to take on embryonic SC-like gene expression patterns and characteristics via the introduction of various combinations of genes and factors important for maintaining pluripotency [8, 80]. Although similar to embryonic SCs in many ways, some important differences between the cell types remain, and the use of induced pluripotent SCs for clinical applications is still somewhat controversial and requires additional research [8, 44]. The most recognized and perhaps most important property of SCs is self-renewal, by which SCs give rise to at least one multipotent daughter SC [67, 70, 92, 98]. In recent years, the putative organ-specific SCs with high expression of SC maintenance-related genes (i.e., Oct-4, p63, Bmi-1, Hedgehog, Wnt, and Notch) have been identified from various human tissues with cell surface markers, such as CD133, CD34, CD44, CD24, and CD117. [54, 67, 71, 92].

17.3 CANCER AS A DISEASE OF STEM CELLS

In recent decades, the cancer stem cell (CSC) hypothesis has become a hot topic in cancer research. CSCs alone appear capable of generating or repopulating tumors [12, 67]. Though CSCs are a small subpopulation of tumor cells, they are thought to be the driving force of tumor initiation, progression, metastasis, and relapse [67, 70, 71]. The theory that tumorigenesis is driven by a subpopulation of pluripotent cells has developed over the last century and a half. The idea originated in the 1870s when it was first hypothesized that tumors may develop from a small population of tumor cells which are "misplaced" during embryonic development [16, 22]. In 1937, Furth and Kahn posited that a single leukemic cell could induce hematopoietic malignancies after injection into a mouse [27], which is remarkably insightful as it was the first suggestion that certain cells within a cancer population possess a high capacity to repopulate cancers and form new tumors. In the 1960s, the existence of self-renewing SCs was shown by Till and McCulloch and colleagues following studies using spleen colonies formed from transplanted mouse bone marrow cells [7, 60, 81]. The existence of CSCs was established in the 1990s, when Dick and colleagues successfully identified and purified human leukemia-initiating cells possessing distinct

SC characteristics with specific cell surface markers [13, 51]. The concept of CSCs suggests that many tumors are hierarchically organized with the putative CSCs being at the top of the hierarchical structure [14, 20, 71]. Since then, the existence of CSCs has been shown in various human solid tumors including brain, breast, pancreas, lung, liver, colon, and prostate as well as hematological cancers using cell-surface markers combined with phenotypic traits such as high tumorigenic capacity in xenograft models or various *in vitro* assays [65, 67, 92]. Markers for CSCs can vary considerably based on tumor type and stage, and even different mutations within individual tumors. Thus, it is difficult to identify general markers for all CSCs. Defining reliable markers for identifying CSCs is an active area of research.

The notion that cancer is a disease of SCs is appealing for several reasons. First, carcinogenesis is a process that requires multiple mutations. Most tissues are composed of restricted progenitors or terminally differentiated cells with limited life span, which makes the accumulation of multiple mutations a low probability. Furthermore, it is difficult for terminally differentiated cells to form heterogeneous tumor structures containing cells at different stages and degrees of differentiation and transformation, although dedifferentiation has been suggested to occur in some cancers [26]. In addition, the efficiency of tumorigenesis by cells from established cancer cell lines and even cells directly from human cancer tissues is very low [59], suggesting that the majority of cancer cells lack the ability to regenerate tumors. CSCs provide explanations for all the aforementioned issues. Further evidence that cancer is a SC-based disease comes from the observation that the malignant CSCs share many parallels with SCs including the (i) ability to self-renew although distorted; (ii) capacity for differentiation although into deranged final products; (iii) capacity of "organogenictiy" to the extent that a tumor is organ-like; (iv) resistance to drugs and induction of apoptosis; (v) active telomerase expression; (vi) high membrane transporter activity; and (vii) the ability to migrate locally and if required (or if a tumor unfortunately systemically disseminates) to distant parts of the body [67, 71].

Just as normal SCs function through self-renewal and differentiation to generate and replenish normal tissues, CSCs have the parallel characteristics of tumor initiation, maintenance, and repopulation only with disrupted signaling pathways [12, 54, 67, 71, 92]. CSCs likely account for relapses by survival after chemotherapy. Thus, cancer can be considered an abnormal and highly distorted "organ" that originated from, and is driven by CSCs.

The origination of CSCs is still incompletely defined. CSCs are likely to be derived from normal SCs based on the vast similarities between the two cell types, of which the capacity for self-renewal is the most important one involved in neoplastic proliferation. CSCs have enhanced expression of many cell surface and/or self-renewal markers also seen in SCs [12, 54, 65, 67, 92]. CSCs also exhibit differentiation capacity which likely leads to the commonly observed heterogenic structure of tumors. In some cases, CSC properties are also seen in progenitor cells or differentiated cells that have acquired mutations or malignancies [67, 70]. Various signaling pathways (i.e., WNT, Hedgehog, Notch, and BMI) that are closely related to normal SC maintenance, lineage determination, and differentiation are commonly disrupted during malignant transformation, particularly in the CSC subpopulation of tumor cells [54, 67, 71].

17.4 STEM CELL STIMULATION BY ARSENIC EXPOSURE

Accumulating evidence indicates that SCs may be the target cell phenotype during arsenic-induced carcinogenesis [79, 84, 94]. Arsenic exposure alters SC dynamics, blocking the exit of SCs into differentiation, and induces an overabundance of CSCs in various *in vivo* and *in vitro* model systems [68, 69, 79, 82, 83, 94].

17.4.1 *In Vitro* Studies

Arsenic is a well-known human skin carcinogen [38]. The keratinocyte SC is probably a key target in arsenic-induced skin cancers. As a self-renewing epithelium, the epidermis contains SCs, transit amplifying cells, and cells in various stages of maturation. Arsenic suppressed the differentiation of cultured keratinocytes, which indicates that the metalloid probably acts on SCs or transit amplifying cells by preventing them from leaving the mitotic pool [42]. In this regard, a low micromolar level of sodium arsenite was found to alter SC dynamics in human keratinocytes [68]. In this study, cultured normal human epidermal cells and spontaneously immortalized human epidermal keratinocytes were treated with 2 μM of sodium arsenite. Keratinocytes usually lose colony formation capacity when they have reduced proliferative potential and are committed to the differentiation pathway. Arsenic treatment attenuated the decrease of colony formation capacity induced by confluence or suspension in keratinocytes. These data indicate that arsenic maintains the cell proliferative potential and probably blocks the exit of putative target SCs from the germinative compartment into differentiation. This action of arsenic may be based on inhibiting the decline of β1-integrin and activating transcriptional β-catenin activity, both of which are potential epidermal SC markers, as well as the preventing of loss of epidermal growth factor receptor protein [68, 69].

Two additional *in vitro* studies provide further evidence that arsenic targets the SC population as it induces an overabundance of CSCs during malignant transformation of a human keratinocyte cell line, specifically the HaCaT cell line [43, 79]. In one study, the HaCaT skin keratinocyte cell line was transformed by chronic exposure to 100 nM of sodium arsenite exposure for 30 weeks [79]. SCs were isolated from arsenic-transformed and control keratinocytes using magnetic beads labeled with the cell surface marker CD34, a common marker for skin SCs and CSCs [89]. The number of isolated CSCs from arsenic-transformed keratinocytes was 2.5 times higher than isolated control SCs.

A holoclone is a colony of cells enriched in SCs or CSCs derived from a single cell and appears to indicate the self-renewal capability typical of SCs or CSCs [52]. CSCs isolated from arsenic-transformed keratinocytes generated 3.5 times more holoclones than isolated control SCs [79]. In contrast, HaCaT cells malignantly transformed by UV [31] did not show overproduction of CD34-positive putative CSCs or holoclones.

CSCs from arsenic-transformed keratinocytes showed increased transcription of common SC/CSC markers (p63) and keratinocyte SC/CSC markers (CD34, K5, K14, K15, and K19) and dramatically stronger immunofluorescence for Rac1 [79], a key gene in skin SC population dynamics [10]. In addition, when compared with control SCs and the total population of arsenic-transformed cells, CD34-positive CSCs showed elevated matrix metalloproteinase (MMP)-9 secretion and enhanced colony formation, both indicative of malignant phenotype, indicating that the arsenic-induced malignant phenotype is particularly pronounced in the isolated CSCs [79]. In a similar study, Jiang et al. exposed HaCaT cells to inorganic arsenite (1.0 μM) for approximately 15 weeks to induce malignant transformation of the cells. During this transformation, the arsenic-treated cells underwent epithelial-to-mesenchymal transition (EMT) as demonstrated by a concomitant change in morphology, a decrease in the epithelial cell marker, E-cadherin, and an increase in mesenchymal cell markers, *N*-cadherin and vimentin. Arsenic-treated cells showed an increase in non-adherent sphere formation and an increase in *K5* and *CD34* mRNA levels compared with control cells [43], suggesting that arsenic induces an acquisition of a CSC-like phenotype, a process which may be linked to the induction of EMT [57]. The increases in transcription factor, Snail, and in the NF-κB pathway likely play a role in the observed malignant transformation, induction of EMT, and/or acquisition of CSC-like phenotype seen in the arsenic-treated cells [57, 108]. Together, these studies [43, 79] suggest that arsenic can target human skin SCs during transformation and induce a CSC overabundance possibly via distortion of SC self-renewal and the NF-κB and Snail pathways.

The prostate is a probable carcinogenic human target of arsenic [38]. A human prostate epithelial stem/progenitor cell line (WPE-stem) was endowed with a CSC-like phenotype following 18 weeks of exposure to an environmentally relevant level of sodium arsenite (5 μM), becoming highly invasive, showing increases in colony forming efficiency (anchorage-independent growth) in soft agar, formation of free-floating prostaspheres, and activity of secreted MMP-9 [83]. The malignantly transformed CSCs formed highly pleomorphic, aggressive tumors with immature epithelial- and mesenchymal-like cells, suggesting a highly pluripotent cell of origin [83]. During the transformation, dysregulated expression (an early depletion and subsequent reactivation) of SC self-renewal genes (*p63*, *ABCG2*, *BMI-1*, *sonic hedgehog*, *OCT-4*, *NOTCH-1*) occurred. This reactivation of an SC-like expression program does not appear to be specific to arsenic, but also occurs during the progression of normal hematopoietic and epithelial SCs to leukemic SCs and epithelial CSCs [49, 99] and reflects the dysregulation of SC self-renewal that is likely to be a key phenomenon and mechanism during acquisition of malignancy by normal SCs [49, 67, 71, 92, 94]. Phosphatase and tensin homolog (PTEN) tumor suppressor gene is frequently inactivated in cancers [19]. Loss of this tumor suppressor gene also plays a key role in SC differentiation and enhances self-renewal ability without altering pluripotency of SCs [19]. More specifically in prostate SCs, depletion of PTEN leads to the expansion of CSC-like cells, tumor initiation, and metastasis [95, 96], and PTEN knockout increases CSC-like sphere formation, clonogenicity, and tumorigenicity [21]. A rapid and progressive depletion of *PTEN* expression occurred during acquisition of CSC phenotype and reactivation of SC self-renewal genes in human prostate SCs [82, 83] and *PTEN* was also depleted in arsenic-transformed human skin SCs [79]. Together, these studies suggest that *PTEN* depletion may play an important role in arsenic-induced malignancy and, more specifically, CSC formation.

Arsenic is also recognized as a potential kidney carcinogen in humans [38] and can cause renal cancers in rodents [86]. A rodent kidney SC/PC cell line, RIMM-18, was exposed to chronic low-level sodium arsenite (500 nM) for 28 weeks [87]. Compared with control cells, arsenite-exposed cells showed increases in several characteristics common to cancer cells, including MMP-2 and MMP-9 hypersecretion, colony formation in soft agar, invasive ability, and cyclooxygenase-2 expression. Arsenite-exposed cells also showed an increase in the number and overall size of free-floating spheres; and when plated in Matrigel, these arsenite-induced spheres demonstrated much more aggressive properties indicative of CSC-like cells. Arsenic exposure aberrantly altered the expression of several genes (*Wnt3, Wt-1, β-catenin, Bmp-7*) critical for normal kidney development and kidney SC maintenance and self-renewal. Two of these genes, *Wnt3* and *Wt-1*, showed a "U-shaped" expression trend similar to that seen during the malignant transformation and recruitment of human prostate SCs [82, 101]. Additionally, the markedly increased expression of renal SC/CSC markers (*Ncam, Osr1, Cd133, Cd24*) suggests that arsenic exposure causes an expansion of these cells during transformation. This study [87], combined with the recent development of a transplacental arsenic-induced kidney cancer model in mice [86], suggests that arsenic targets a SC/PC phenotype for kidney carcinogenesis and that exposure during stages of high SC numbers and activity (i.e., *in utero*) may be necessary for arsenic-induced kidney cancer in animals.

EMT is an important process driving differentiation and de-differentiation in tumor initiation and metastasis. The induction of EMT endowed mammary epithelial cells with expression of SC markers ($CD44^{high}/CD24^{low}$) and the self-renewal traits associated with normal SCs/CSCs (i.e., increased mammosphere formation and tumor-initiating capacity) [57, 62]. A normal human bronchial epithelial cell line (HBE cells) was treated with 1.0 μM sodium arsenite for 15 weeks [100], and similar to the study described earlier [43], the arsenic-treated cells showed hallmarks of EMT. These hallmarks included alteration from an epithelial to a mesenchymal morphology, increased expression of vimentin and *N*-cadherin, and decreased expression of E-cadherin. During the arsenic-induced EMT, the cells also showed increased spheroid formation capacity, increased transcript levels of lung SC/CSC surface markers, *CD133* and *CD44*, and an enrichment of side population cells, all indicative a overproduction of SCs/CSCs [100]. At the end of arsenic treatment, transcript levels of *OCT4*, *BMI-1*, and *ALDH1* were all pronounced [100], which is consistent with observed arsenic-induced dysregulation of self-renewal genes [82]. The observed EMT and overproduction of CSCs caused by arsenic in HBE cells may be regulated both by HIF-2α through TWIST1 and BMI-1 signaling pathways [100].

Arsenic has also been shown to disrupt the differentiation of mouse embryonal carcinoma cells [32], rodent myoblasts [33, 76, 103], and human and mouse mesenchymal SCs [35, 48, 102]. Arsenic exposure (0, 0.1, 0.5, 1.0 μM sodium arsenite, for up to 9 days) suppresses the differentiation of P19 embryonal carcinoma cells into muscle cells and neurons, as evidenced by changes in cell morphology and decreased expression of myosin heavy chain and Tuj1 [32]. The repression of SC differentiation caused by arsenic was related to decreased expression of muscle- and neuron-specific transcription factors (i.e., Pax3, Myf5, MyoD, myogenin, neurogenin 1, eurogenin 2, and NeuroD) regulated through repressed Wnt/β-catenin signaling pathways in the early stages of embryogenesis. Moreover, arsenic appeared to inhibit differentiation of these mouse embryonic SCs without altering their proliferation, based on the observance of increased transcript levels of the SC pluripotency gene, *Nanog* [32]. It should be noted here that, although they are undifferentiated cells, the P19 cell line was derived from an experimental mouse teratocarcinoma [90]. While normal SCs and CSCs share many fundamental characteristics, including self-renewal and multipotency, many of these characteristics are dramatically altered in CSCs. Therefore, it requires further investigation to determine if arsenic would have similar effects on normal embryonic SCs as it had on the P19 embryonal carcinoma cells [32]. Myoblasts are embryonic progenitor cells that differentiate into muscle cells. Exposure of mouse myoblasts to 20 nM sodium arsenite caused a delayed differentiation of these cells via reduced expression of Igf-1, a potent inducer of muscle differentiation, and increased expression of Ezh2, which epigenetically represses terminal muscle differentiation [33, 76]. Transcription factors that play a key role in muscle cell differentiation, myogenin and Mef2c, showed decreased expression in arsenic exposed cells, likely due to histone modifications and alterations in methylation patterns. Rat myoblasts treated with sodium arsenite (1–10 μM; 48 h) showed a decrease in migration and adhesion rates, numbers of focal adhesions, and total F-actin content, all factors possibly mediated by the reduction of tyrosine phosphorylation of focal adhesion kinase [103]. Mesenchymal SCs can differentiate into adipocytes, connective tissue cells important for the synthesis and storage of fat. Several similar studies recently showed that inorganic arsenic, and its methylated metabolites inhibit the differentiation of mesenchymal SCs into adipocytes through the decreased expression of peroxisome proliferator-activated receptor gamma CCAAT enhancer binding proteins, two transcriptional programs essential for differentiation [35, 48, 102]. The arsenic inhibition of differentiation could be prevented, at least partially, by Ptx or endothelin-1 receptors [48] and possibly the up-regulation of the Wnt signaling pathway [102]. Together, these studies indicate that arsenic exposure can alter normal differentiation of SCs via several pathways which may, in turn, lead to developmental abnormalities, metabolic and cardiovascular disease, or other disorders, at least in muscle and adipogenic cells.

Several studies have examined the toxicity of acute *in vivo* exposure of arsenicals on various SC/PCs. Sodium arsenite was shown to increase the proliferation of female human and murine hematopoietic progenitor cells, but had no effect on male cells of either species while the methylated metabolite, monomethylarsonous acid (MMA^{III}), was toxic to cells of both genders in both species [24]. ATO decreased the viability and differentiation of CD34+ human hematopoietic SCs, a process likely related to induce mitochondrial damage [91]. These studies suggest that hematopoietic SC/PCs may be cellular targets of arsenicals, which could help explain the therapeutic efficacy of arsenicals in treating some leukemia (see section later). ATO also induces apoptosis in adult rat hepatic SCs [2] which is seemingly in contrast to the increase in liver CSCs in tumors in mice exposed to whole life inorganic arsenic [85]. Trivalent arsenate metabolites (arsenite, MMA^{III}, and dimethylarsinous acid), but not pentavalent metabolites, were toxic at low levels (<1 mg/L) to mouse neural progenitor cells through generation of reactive oxygen species and activation of the apoptotic pathways [72]. Similarly, sodium arsenite (1–10 μM) inhibited differentiation and self-renewal, induced the mitochondrial apoptotic pathway, and suppressed the PI3K-AKT pathway in human neural SCs and mouse embryonic SCs [40, 41]. In this latter study, an immortalized mouse neural SC line was highly resistant to sodium arsenite, which, the authors suggest, may be due to the substantial changes in gene regulation in the immortalized cell line compared with primary neural SCs, including markedly increased production of interleukin-6 [41]. Treatment of human embryonic SC-derived embryoid bodies with sodium arsenite induced significant developmental toxicity via down-regulation of gene expression of all three germ layers. These effects were correlated with high mortality and developmental defects in rat pups born to dams that received 0.1 mg/kg sodium arsenite in their drinking water, *ad libitum*, for up to day 20 of gestation. Monoisoamyl dimercaptosuccinic acid prevented and/or reversed these arsenic-induced effects *in vitro* and *in vivo* [25] providing a possible therapeutic approach to the effects of early life arsenic exposure.

17.4.2 *In Vivo* Studies and Models of Arsenic Specificity

Both human and animal studies suggest that fetal or early life exposure to arsenic increases the risk of cancers later in life [58, 74, 93, 94, 105]. With a short biological half-life (~4–5 days), fetal or early life arsenic exposure does not appear to accumulate in the body or directly participate in the formation of cancer during later life stages. Fetal or early life exposure-induced adulthood cancers would require a long-lived target cell population that retains the capacity for self-renewal even after the initial lesion, such as a SC population. Fetal SCs may be very sensitive to transplacental carcinogenesis based on their abundance and high activity. During developmental stages, normal SCs show strictly controlled self-renewal, which generally allows for repopulation of the SC compartment when required and stops upon replenishment of the SC pool. Dysregulation of this self-renewal process could lead to uncontrolled SC expansion, and possibly CSC initiation, and may be a key early event in carcinogenesis.

Fetal SCs are highly active in global proliferative growth, differentiation for organogenesis and abundant in number, which makes them a probable key target for transplacental carcinogenesis [4, 93], and particularly in arsenic carcinogenesis [82, 93, 94]. Indeed, Waalkes and colleagues have found that fetal exposure to arsenic altered SC dynamics during skin carcinogenesis in adult mice, resulting in markedly increased CD34-positive skin CSCs in the arsenic-induced skin squamous cell carcinomas [94]. In this study, Tg.AC mice, a strain sensitive to skin carcinogenesis via activation of the *v-Ha-ras* transgene likely in keratinocyte SCs, were given drinking water with sodium arsenite at 0 (control), 42.5, or 85 ppm from gestation days 8 to 18. After birth and weaning, offspring were treated with 12-*O*-tetradecanoyl phorbol-13-acetate (TPA) (2 μg/0.1 ml acetone, twice a week to a shaved area of the dorsal skin) through adulthood to 40 weeks of age. Although TPA treatment alone induced squamous cell carcinoma (SCC), fetal arsenic exposure plus subsequent TPA treatment dramatically increased the incidence and multiplicity of SCC in the offspring mice as adults compared with control mice and mice receiving only TPA treatment. This increase showed a highly significant arsenic dose relationship. Malignant tumors after fetal arsenic plus subsequent TPA were also more aggressive with more frequent local muscle invasion, infiltration of dermis, and mitotic figures, as well as molecular indicators of aggressiveness, such as an increase of the cell proliferation gene, *cyclin D1*, and a decrease of the tumor suppressor gene, *p16*. Chemically induced skin tumors arise from carcinogenic "hits" in the epidermal SC compartment of the hair follicle [30], which is also a key locale for initial *v-Ha-ras* transgene expression in Tg.AC skin carcinogenesis [37, 89]. Thus, the increase of SCC in adult mice following fetal arsenic plus subsequent TPA treatment indicates an SC origin for the malignancies. Tumor expression of *v-Ha-ras* was three times higher with fetal arsenic plus adult TPA compared with TPA alone, which indicates arsenic plus TPA treatment increased tumor response. Moreover, the expression of the keratinocyte SC/CSC marker, CD34 [88], was greatly increased in tumors induced by fetal arsenic plus subsequent TPA treatment compared with TPA treatment alone, suggesting that the keratinocyte SC is a crucial target in skin carcinogenesis and, in particular, arsenic-induced skin carcinogenesis. Immunohistological analysis also found that the number of CD34-positive cells was dramatically increased in tumors induced by fetal arsenic plus subsequent TPA treatment compared with TPA treatment alone. CD34-positive tumor cells are probable CSCs in murine epidermal skin tumors as evidenced by their capacity to initiate secondary tumors with SC properties [56]. Thus, fetal exposure to arsenic appears to induce a remarkable increase in the number of keratinocyte CSCs. In addition, Rac1, an important protein in the skin SC replication and self-renewal signaling pathway, was significantly increased in tumors induced by fetal arsenic plus subsequent TPA treatment compared with TPA treatment alone. Altogether, these data indicate that fetal arsenic exposure modifies the SC response to carcinogen exposure in adulthood, potentially leading to an overabundance of CSCs (i.e., CD34-positive skin tumor cells) and distorts the dynamics of the SC population (i.e., increased Rac1 expression).

In most cases, humans are exposed to chemical carcinogens throughout their whole life and not just during individual life stages. However, most rodent bioassays begin exposure to chemicals several weeks after birth for 2 years, which may lead to false-negative results [36]. Thus Waalkes et al. developed a whole-life arsenic exposure model in CD1 mice [85]. In this study, breeder mice received drinking water containing 0 (control), 6, 12, or 24 ppm, arsenic starting 2 weeks before mating, then pregnant female mice received the same treatments through birth and weaning, and the pups continued to receive these treatments through adulthood for 2 years until euthanasia. Arsenic exposure caused increases of lung adenocarcinoma and liver hepatocellular carcinoma, two target sites and tumor types in humans exposed to arsenic, and increases in gallbladder tumors, uterine carcinoma, and ovarian tumors. Moreover, ALDH1A and CD133, two common markers for liver and lung SCs/CSCs, were strongly and widely present in liver hepatocellular carcinoma and lung adenocarcinoma resulting from whole-life arsenic-exposed mice compared with spontaneous tumors in control mice and ENU-induced tumors in CD1 mice. These data are consistent with prior work suggesting that arsenic induces overabundance of CSCs in resultant cancers.

Interestingly, the repeated observance of CSC overabundance in various model systems during arsenic transformation may be specific to, or at least particularly pronounced with, this metalloid [83]. The nontumorigenic human prostate epithelial cell line, RWPE-1, has been malignantly transformed by arsenic, cadmium, and *N*-methyl-*N*-nitrosourea (MNU) to form several isogenic malignant cell lines. Compared with the parental control cell line (RWPE-1), the arsenic-transformed cell line forms free-floating spheroids at a significantly higher rate (8.7-fold of control), compared with the cadmium (3.1-fold) and MNU (2.3-fold) transformants [84]. Spheres formed by cancer/tumor cell lines have been shown to be enriched with CSC-like cells [29]. Dissociation of these spheres into single cells to measure their clonogenic capacity in soft agar, an assay commonly used to quantify CSC-like cells in cancer or tumor cell lines [77], showed that the arsenic-transformant formed a markedly increased number of colonies compared with control (59 vs 0.67 colonies), whereas the cadmium and MNU transformants showed no increase in colony formation (1.7 and 1.0 vs 0.67 colonies, respectively). Moreover, even though cells from all the malignant transformants could form holoclones, the arsenic transformant formed them at a much higher rate and produced the only holoclones that could be subcloned and propagated on a long-term basis and showed high expression levels of several SC/CSC markers [84]. These data indicate that the arsenic transformants produced holoclones with a significantly higher self-renewal capacity and an overabundance of putative CSCs.

Similar results have been seen in skin cells, both *in vitro* and *in vivo*. An *in vitro* study using human epidermal cell lines suggests that arsenic, but not cadmium, targets SCs and alters SC dynamics [68]. When HaCaT cells, a human keratinocyte cell line, were exposed to arsenic or UV radiation both groups of cells acquired a malignant phenotype, but only the arsenic transformants showed an overabundance of putative CSCs (CD34-positive cells) and enhanced holoclone formation rate [79].

In Tg.AC mice, *in utero* arsenic exposure enhanced the formation of skin tumors and showed a unique CSC overabundance in tumors compared with TPA-alone treatment [94]. This does not mean that production of CSCs is not involved in carcinogenic process induced by other carcinogenic agents. For example, chronic exposure to a low level of cadmium produces CSCs during malignant transformation of human breast epithelial cells, MCF-10A [9]. However, with arsenic the relative amount of CSCs to total cells appears to consistently increase with malignant transformation or tumor initiation [79, 84, 85, 94]. Further study is required to see if this overabundance of CSCs is true for other target tissues of arsenic carcinogenesis.

17.4.3 Arsenic Targets SCs as a Cancer Therapeutic Drug

Just as arsenic can act as a carcinogen, some forms, mainly arsenic trioxide (ATO), have been used as an effective chemotherapeutic, especially for various hematological malignancies [63, 64, 106, 107]. CSCs are believed to be the main challenge of cancer chemotherapeutics and tumor relapse and, thus, targeting these cells may be a critical mechanism for arsenic as a chemotherapeutic. CSCs stay in a quiescent state with rare division which makes them resistant to radio- and chemotherapies, as they are not only noncycling but also have a low-rate of metabolism thus making them insensitive to toxicants that affect energy metabolism and signaling pathway inhibitors. ATO targets leukemia-initiating cells and sensitizes them to chemotherapy by breaking their dormancy. Degradation of the fusion oncogene promyelocytic leukemia protein (PML)/retinoic acid receptor-α (RARA) plays an important role in apoptotic resistance and self-renewal of hematopoietic and leukemic SCs [17, 18, 39, 63]. Loss of PML induces an exit from quiescence of SCs, which was more profound in leukemic SCs compared with hematopoietic SCs, suggesting a possible therapeutic window for leukemia patients. ATO targeted the PML protein and caused an increase in general RNA in cultured hematopoietic SCs, indicative of hematopoietic SCs activation [39]. Similarly, ATO disrupts or abolishes the self-renewal/maintenance and also inhibits the growth of leukemic SCs through inhibition or degradation of PML/RARA [39, 107]. When combined with other chemotherapeutics, ATO increases the efficiency of the antileukemic treatment or toxicity to SCs. Targeting and degradation of PML/RARA oncoprotein effectively cures 70% of acute promyelocytic leukemia cases with ATO alone and 90% of cases when ATO is combined with retinoic acid [50]. ATO can be combined with interferon-α to cure murine adult T-cell leukemia via loss of leukemia-initiating activity with similar results seen in human ATL [23]. Another form of arsenic, arsenic disulfide, has been shown to induce apoptosis and induce differentiation of acute promyelocytic leukemia cells by acting through the PI3K pathway, effects that were synergized with a PI3K inhibitor [34]. Moreover, both human and mouse leukemic SCs were more sensitive to pro-apoptotic stimuli after ATO treatment [39].

ATO also appears to act as a chemotherapeutic in some solid malignancies through a similar targeting of CSCs. CD133 is a common marker of CSCs in various tumors, such as brain, pancreas, prostate, lung, liver, and colon tumors. CD133+ tumor cells are indicated to be responsible for tumor initiation, propagation and relapse [61] and are highly resistant to traditional radio- and chemotherapy [5, 55]. ATO effectively induced apoptosis of CD133+ gallbladder carcinoma cells, downregulated CD133 expression and decreased the number of CD133+ cells [3]. In addition, CD133 overexpression prevented gallbladder carcinoma cells from arsenic-induced apoptosis through AKT signaling pathways [3]. The antagonizing or inhibition of pathways crucial for SC maintenance and self-renewal may be another mechanism by which ATO targets and inhibits growth or induces apoptosis of SCs. Beachy and colleagues demonstrated that ATO inhibits the growth of medulloblastoma and basal cell carcinoma via antagonism of the hedgehog pathway [46, 47], while Zhang and colleagues showed that apoptosis of glioma SCs is regulated via an ATO-induced down-regulation of Sox2 [78].

The exact reasons why ATO appears to act mainly as an anticancer agent, while other inorganic arsenicals act as carcinogens are not well-understood. One possible explanation may be related to differences in exposures or treatments, where ATO treatments are generally at higher doses for shorter periods of time, and inorganic arsenical exposures are lower doses for longer periods. A more plausible explanation, however, may be the seemingly opposing effects these agents have on SCs/CSCs. For example, while both forms of arsenic appear to target SCs/CSCs, ATO induces apoptosis and/or differentiation, breaks SC dormancy/quiescence, downregulates SC/CSC markers, and eradicates SCs/CSCs, while other inorganic arsenicals in most cases seem to have the opposite effects which results in the maintenance and transformation of SCs/CSCs. Given the importance, and perhaps necessity, of SCs/CSCs in driving cancer initiation and tumorigenesis [67, 71, 92], these opposing effects would help explain the anticancer effects of ATO and the carcinogenic effects of other inorganic arsenicals.

17.4.4 Mechanisms Involved in Stem Cell-Induced Dysfunction by Arsenic

17.4.4.1 Antiapoptosis and Hyper-Resistance to Arsenic Generally, SCs are more resistant to toxic insults. The observation that arsenic targets SCs and induces an overabundance of CSCs may relate, at least in part, to the toxicant-resistant and antiapoptotic properties of SCs. To test this hypothesis, another study was done to compare the resistance and hyper-adaptability of human prostate epithelial SCs (WPE-stem) and their parental cells (RWPE-1) [84]. Both innate and acquired resistance to acute and chronic arsenic-induced cytolethality was higher for WPE-stem cells compared with parental RWPE-1 cells. The SCs also showed significantly higher expression of antiapoptotic (i.e., BCL2 and MT), stress related (i.e., NFE2L2, SOD1, and PRODH), and arsenic adaptation (i.e., ABCC1 and GSTP1) factors and significantly lower expression of pro-apoptotic factors (i.e., BAX and caspases 3, 7, 8, and 9) compared with their parental RWPE-1 cells. The production of CSCs, assessed by free-floating sphere formation, colony formation in soft agar, holoclone serial passage, was pronounced in arsenic-transformed RWPE-1 cells. Interestingly, the overabundance of CSCs during acquisition of malignant phenotype related to resistance and hyper-adaptability of SCs may be specific to arsenic, because similar results were not seen in isogenic cadmium- or *N*-methyl-*N*-nitrosourea-induced malignant RWPE-1 transformants [84]. Similarly, the *in vitro* study conducted in keratinocytes suggests that only arsenic targets in SCs alter SC dynamics compared with cadmium [68]. When HaCaT cells were exposed to arsenic or ultra violet light, both groups of cells showed a malignant phenotype, but only arsenic transformants showed overabundance of putative CSCs (a high level of CD34-positive cells) and enhanced holoclone formation rate [79]. Furthermore, *in vivo* study conducted with CD1 mice showed that exposure to arsenic in drinking water throughout the "whole life" induced the lung and liver carcinomas which were rich in CSCs compared with spontaneous tumors or *N*-nitrosomethylurea-induced lung adenocarcinoma [85].

17.4.4.2 Recruitment of Normal SC into CSCs by Neighboring Malignant Cells Like normal SCs, CSCs reside in a niche where their fate is subject to many microenvironmental factors. Neighboring cells can influence the differentiation and homeostasis of nearby SCs by releasing various growth factors, chemokines, and cytokines into the SC microenvironment [1]. Malignant cells often release extracellular factors modifying tumor behavior. Recently, our group found that normal human SCs were induced into an oncogenic phenotype by noncontact co-culture with arsenic-transformed malignant epithelial cells. In this co-culture study, normal human prostate epithelial SCs (WPE-stem) were co-cultured using transwell with noncontiguous arsenic-transformed isogenic malignant prostate epithelial cells (CAsE-PE) [101]. The transwell has 0.4 μm pores that prevent cell migration through the membrane but allow soluble factors to pass. The distance between WPE-stem and malignant epithelial cells in this co-culture system is 1 mm, approximately the width of 50 to 100 prostate epithelial cells (Fig. 17.1). WPE-stem co-cultured with normal prostate epithelial cells (RWPE-1) was used as the control. After a few weeks of noncontact malignant epithelial co-culture, MMP-9 and MMP-2 were hyper-secreted by WPE-stem cells. Markedly increased production of free-floating spheroids and a rapid persistent suppression of tumor suppression gene, *PTEN*, occurred in WPE-stem cells after malignant epithelial co-culture. Secreted metalloproteinase activity and colony formation of free-floating spheroid cells formed after malignant epithelial co-culture also dramatically increased. Colonies with highly branched ductal-like structure were produced in Matrigel by single-cell suspension from free-floating spheroid SCs formed by malignant epithelial co-culture. These findings were very consistent with what had been seen in direct arsenic transformation of WPE-stem cells. After malignant epithelial co-culture, WPE-stem cells also exhibited depolarized spindle-like morphology, increased expression of mesenchymal cell marker *VIMENTIN*, and decreased expression of epithelial cell maker *E-CAD*, all indicative of EMT, which often occurs when early stage tumors are converted into invasive cancer [45]. All these data indicate that normal WPE-stem cells acquire an oncogenic phenotype and high aggressiveness after malignant epithelial co-culture. Meanwhile, WPE-stem cells still kept self-renewal ability based on the observance that colonies

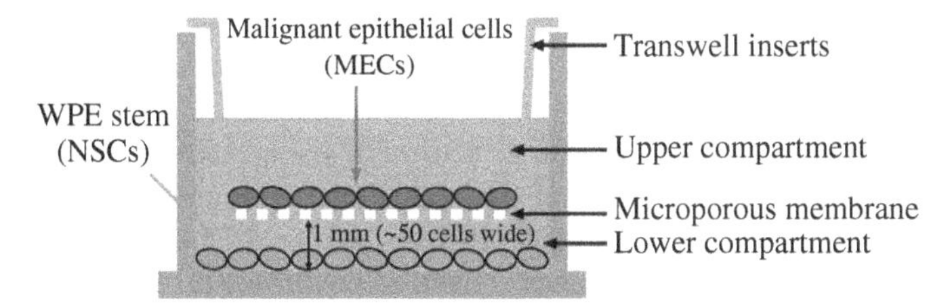

FIGURE 17.1 Diagrammatic representation of the transwell insert co-culture system used in Xu et al. [101]. Malignant epithelial cells (MECs) are plated in the upper compartment and normal stem cells are plated in the bottom compartment. The microporous membrane allows for sharing of culture medium and secreted factors between the two cell types while preventing direct cell–cell contact. The distance between the two cell types is approximately 1 mm, or the equivalent width of approximately 50 epithelial cells.

from co-cultured spheroid SCs could be serial passaged more than three times. In addition, a time-related loss (in first 2 weeks of co-culture) and regain (by 3 weeks) "U"-shaped gene transcription of prostate SC-associated genes *NOTCH-1*, *OCT-4*, *BMI-1*, *WNT-3*, and *p63* was observed in WPE-stem cells co-cultured with malignant epithelial cells. This phenomenon again indicates that the dysregulation of self-renewal program may act as an important mechanism during malignant transformation of SCs. The transforming effects on normal SCs by nearby malignant epithelial cells may or may not be specific to arsenic, and further work is required in this area.

Since there was no physical cell contact or detectable arsenic in the co-culture system, the transformation of normal SCs into cancer-stem-cell-like phenotype is probably due to the secreted factors from arsenic-transformed malignant epithelia. Interleukin-6 (IL-6), a cytokine involved in tumor microenvironment control, was hyper-secreted by arsenic-transformed prostate epithelial (CAsE-PE) cells. Exposure to IL-6 duplicated response of WPE-stem cells to arsenic-transformed malignant cell co-culture, such as hyper-secretion of metalloproteinase-9 activity, increase in expression of *P63* and *K5*, and decrease in expression of *PTEN* and *E-CAD*. Thus, arsenic may not only directly target SCs but also "recruit" normal SCs indirectly that are near an arsenic-induced malignancy via secretion of soluble factors. This CSC "recruitment" phenomenon potentially constitutes a new phenomenon in cancer growth and extension and helps explain the overabundance of CSCs, which has been repeatedly observed in various *in vitro* and *in vivo* model systems of arsenic carcinogenesis.

17.4.4.3 Dysregulation of Self-Renewal Cancer is considered to be a SC disease when mutations arrest the process for SC progression from SC to transit-amplifying progenitor cells to differentiated cells [73]. The widely observed link between cancer and tissue repair suggests that cancer may arise from alteration of self-renewal pathways that locks SCs in an abnormal active self-renewing state [6, 71]. Normal SCs are accepted to have high self-renewing capacity but produce clones able to eventually make millions of cells. However, CSCs can remain dormant in a niche or exit into the cell cycle when required [6]. The self-renewing capacity of SCs is mainly based on asymmetrical division which gives rise to a sister SC that preserves SC identity and a progenitor with capacity to divide several times before it differentiates into mature cells (Fig. 17.2). This asymmetric division does not expand the SC number under normal conditions. During tissue regeneration, after toxic insults or with disease states, quiescent SCs can be activated, symmetrically dividing into two daughter SCs and enlarge the SC pool. Thus, the long-term quiescent SC pool likely serves as a reservoir of highly potent cells available to respond to life-threatening conditions [6, 71]. Once the emergency is over, SCs will stop proliferating and return to a quiescent status. Thus, normal SCs have a limited period of symmetric division. Both normal SCs and CSCs have quiescent and proliferative subpopulations; however, CSCs are expected to have a longer period of proliferation [6]. Compared with their normal counterparts, mammary CSCs show more symmetric division and possess an unlimited replicative potential [15]. Though there is no direct evidence showing that arsenic increases symmetric division of SCs, accumulating studies suggest

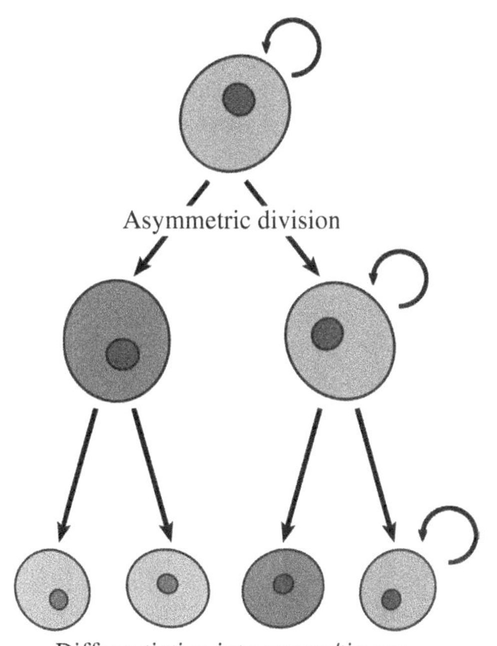

FIGURE 17.2 Stem cells (blue) can undergo the process of asymmetric division, where the cell divides into one identical daughter stem cell and one partially differentiated progenitor cell (red). The progenitor cells can undergo further differentiation to form the terminally differentiated cells of organs and tissues (tan). Curved arrow indicates self-renewal. (*See insert for color representation of the figure.*)

that arsenic has the capacity of disturbing SC dynamics and distorting the self-renewal pathways of SCs during malignant transformation [68, 83, 94]. In human keratinocytes exposed to arsenic, the SCs were blocked from differentiation and retained in a germinative status [68]. Also in prostate SCs, arsenic exhibited a survival selection and overproduction of CSCs during malignant transformation [83, 84]. Additionally, depletion and reactivation of self-renewal genes occurred with concurrent depletion of tumor suppressor PTEN [83], which would contribute to retaining SCs in an activated self-renewal state instead of a return to quiescence. This dysregulated self-renewal in SCs is potentially a crucial mechanism to increase the pool size of target cells for arsenic, thereby resulting in a CSC overabundance and greatly enhancing the probability of tumor initiation and progression.

17.5 SUMMARY AND CONCLUSION

Millions of people worldwide are exposed to potentially unhealthy levels of arsenicals, primarily from their drinking water [38]. The effects of this exposure on disease outcome are well-documented, but the exact mechanisms and/or target cell are not as well-known. Emerging evidence from *in vitro* and *in vivo* studies shows that arsenicals can target SCs to both cause and cure various diseases. For disease induction, inorganic arsenicals disrupt SC differentiation, aberrantly alter pathways and/or factors crucial for normal SC self-renewal and maintenance, and alter SC population dynamics. These effects lead to a hyper-adaptability to the metalloid and, in turn, a CSC overabundance in resultant arsenic-induced tumors or cancer cell lines (Fig. 17.3). In this regard, a recent study shows that epithelial cells malignantly transformed by arsenic can "recruit" normal SCs to acquire a cancer phenotype, which may be an important mechanism helping explain multifocal cancers commonly seen in individuals exposed to the metalloid. As a cancer treatment, arsenic trioxide targets and eradicates leukemic SCs which results in long-term remission or cure of hematological malignancies. Transplacental and whole-life arsenic exposure studies in mice have shown an impressive ability to produce CSC-enriched tumors in the offspring mice as adults. Additionally, accumulating evidence in humans and mice suggests that fetal and/or developmental arsenic exposure is strongly associated with various diseases later in life, including cancer and metabolic disorders [53, 58, 74, 75, 104, 105]. It should be

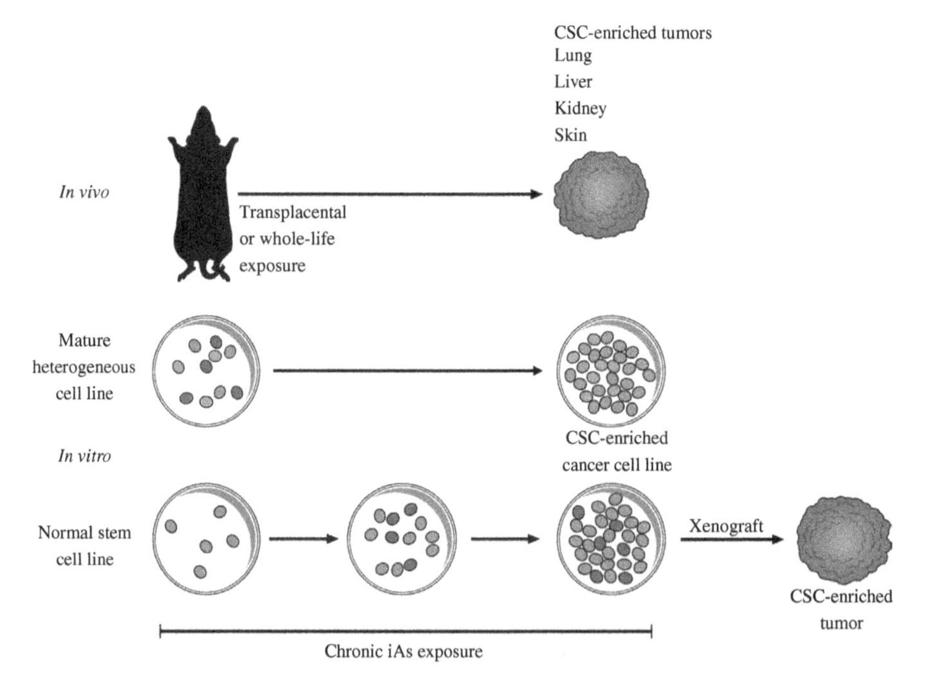

FIGURE 17.3 Arsenic-induced formation of cancer stem cell-enriched cell lines and tumors. Top: Transplacental or "whole life" exposure to arsenicals can result in the formation of cancer stem cell-enriched tumors when the offspring mice reach adulthood. Target tissues include the lung, liver, kidney, and skin. Middle and bottom: chronic exposure to low levels of arsenic can induce malignant transformation of nontumorigenic mature heterogeneous cell lines and of normal stem cell lines. This transformation can lead to cell lines highly enriched in cancer stem-like cells. Xenograft studies using these cell lines can lead to highly aggressive tumors enriched in cancer stem-like cells. Stem cell = blue, progenitor cell = red, and terminally differentiated cell = tan. (*See insert for color representation of the figure.*)

noted that these latter studies did not look directly at SCs. However, given the effects of arsenic on SCs in the mouse transplacental studies and the *in vitro* studies, it is reasonable to suggest that the apparent arsenic-induced fetal/early life basis of these other diseases could be associated with an effect on SCs. This hypothesis requires further study. Together, the studies discussed here provide strong evidence that SCs play a key role in this arsenic-based disease outcome. Additional studies are warranted to help further define the role that SCs play in arsenic-induced disease initiation and manifestation, particularly when fetal and/or early life exposure is a factor.

REFERENCES

[1] D.E. Abbott, C.M. Bailey, L.M. Postovit, E.A. Seftor, N. Margaryan, R.E. Seftor, M.J. Hendrix, The epigenetic influence of tumor and embryonic microenvironments: how different are they? Cancer Microenvironment: Official Journal of the International Cancer Microenvironment Society 1 (2008) 13–21.

[2] S. Agarwal, S. Roy, A. Ray, S. Mazumder, S. Bhattacharya, Arsenic trioxide and lead acetate induce apoptosis in adult rat hepatic stem cells, Cell Biology and Toxicology 25 (2009) 403–413.

[3] Z. Ai, H. Pan, T. Suo, C. Lv, Y. Wang, S. Tong, H. Liu, Arsenic oxide targets stem cell marker CD133/prominin-1 in gallbladder carcinoma, Cancer Letters 310 (2011) 181–187.

[4] L.M. Anderson, B.A. Diwan, N.T. Fear, E. Roman, Critical windows of exposure for children's health: cancer in human epidemiological studies and neoplasms in experimental animal models, Environmental Health Perspectives 108 Suppl 3 (2000) 573–594.

[5] S. Bao, Q. Wu, R.E. McLendon, Y. Hao, Q. Shi, A.B. Hjelmeland, M.W. Dewhirst, D.D. Bigner, J.N. Rich, Glioma stem cells promote radioresistance by preferential activation of the DNA damage response, Nature 444 (2006) 756–760.

[6] P.A. Beachy, S.S. Karhadkar, D.M. Berman, Tissue repair and stem cell renewal in carcinogenesis, Nature 432 (2004) 324–331.

[7] A.J. Becker, E.A. McCulloch, J.E. Till, Cytological demonstration of the clonal nature of spleen colonies derived from transplanted mouse marrow cells, Nature 197 (1963) 452–454.

[8] M. Bellin, M.C. Marchetto, F.H. Gage, C.L. Mummery, Induced pluripotent stem cells: the new patient? Nature Reviews. Molecular Cell Biology 13 (2012) 713–726.

[9] L. Benbrahim-Tallaa, E.J. Tokar, B.A. Diwan, A.L. Dill, J.F. Coppin, M.P. Waalkes, Cadmium malignantly transforms normal human breast epithelial cells into a basal-like phenotype, Environmental Health Perspectives 117 (2009) 1847–1852.

[10] S.A. Benitah, M. Frye, M. Glogauer, F.M. Watt, Stem cell depletion through epidermal deletion of Rac1, Science 309 (2005) 933–935.

[11] C. Blanpain, V. Horsley, E. Fuchs, Epithelial stem cells: turning over new leaves, Cell 128 (2007) 445–458.

[12] S. Bomken, K. Fiser, O. Heidenreich, J. Vormoor, Understanding the cancer stem cell, British Journal of Cancer 103 (2010) 439–445.

[13] D. Bonnet, J.E. Dick, Human acute myeloid leukemia is organized as a hierarchy that originates from a primitive hematopoietic cell, Nature Medicine 3 (1997) 730–737.

[14] R.W. Cho, M.F. Clarke, Recent advances in cancer stem cells, Current Opinion in Genetics & Development 18 (2008) 48–53.

[15] A. Cicalese, G. Bonizzi, C.E. Pasi, M. Faretta, S. Ronzoni, B. Giulini, C. Brisken, S. Minucci, P.P. Di Fiore, P.G. Pelicci, The tumor suppressor p53 regulates polarity of self-renewing divisions in mammary stem cells, Cell 138 (2009) 1083–1095.

[16] V. Conheim, Congenitales, quergestreiftes muskelsarkom der nieren, Virchows Archiv für Pathologische Anatomie und Physiologie und für Klinische Medizin 65 (1875) 64–69.

[17] H. de The, Z. Chen, Acute promyelocytic leukaemia: novel insights into the mechanisms of cure, Nature Reviews 10 (2010) 775–783.

[18] H. de The, Z. Chen, Acute promyelocytic leukaemia: novel insights into the mechanisms of cure, Nature Reviews Cancer 10 (2012) 775–783.

[19] A. Di Cristofano, P.P. Pandolfi, The multiple roles of PTEN in tumor suppression, Cell 100 (2000) 387–390.

[20] J.E. Dick, Stem cell concepts renew cancer research, Blood 112 (2008) 4793–4807.

[21] A. Dubrovska, S. Kim, R.J. Salamone, J.R. Walker, S.M. Maira, C. Garcia-Echeverria, P.G. Schultz, V.A. Reddy, The role of PTEN/Akt/PI3K signaling in the maintenance and viability of prostate cancer stem-like cell populations, Proceedings of the National Academy of Sciences of the United States of America 106 (2009) 268–273.

[22] F. Durante, Nesso fisio-pathologico tra la struttura dei nei materni e la genesi di alcuni tumori maligni, Arch Memor Observ Chir Pract 11 (1874) 217–226.

[23] H. El Hajj, M. El-Sabban, H. Hasegawa, G. Zaatari, J. Ablain, S.T. Saab, A. Janin, R. Mahfouz, R. Nasr, Y. Kfoury, C. Nicot, O. Hermine, W. Hall, H. de The, A. Bazarbachi, Therapy-induced selective loss of leukemia-initiating activity in murine adult T cell leukemia, The Journal of Experimental Medicine 207 (2010) 2785–2792.

[24] D. Ferrario, C. Croera, R. Brustio, A. Collotta, G. Bowe, M. Vahter, L. Gribaldo, Toxicity of inorganic arsenic and its metabolites on haematopoietic progenitors "in vitro": comparison between species and sexes, Toxicology 249 (2008) 102–108.

[25] S.J. Flora, A. Mehta, Monoisoamyl dimercaptosuccinic acid abrogates arsenic-induced developmental toxicity in human embryonic stem cell-derived embryoid bodies: comparison with in vivo studies, Biochemical Pharmacology 78 (2009) 1340–1349.

[26] D. Friedmann-Morvinski, E.A. Bushong, E. Ke, Y. Soda, T. Marumoto, O. Singer, M.H. Ellisman, I.M. Verma, Dedifferentiation of neurons and astrocytes by oncogenes can induce gliomas in mice, Science 338 (2012) 1080–1084.

[27] J. Furth, M.C. Kahn, The transmission of leukaemia of mice with a single cell, American Journal of Cancer 31 (1937) 276–282.

[28] F.H. Gage, Mammalian neural stem cells, Science 287 (2000) 1433–1438.

[29] A.J. Ghods, D. Irvin, G. Liu, X. Yuan, I.R. Abdulkadir, P. Tunici, B. Konda, S. Wachsmann-Hogiu, K.L. Black, J.S. Yu, Spheres isolated from 9L gliosarcoma rat cell line possess chemoresistant and aggressive cancer stem-like cells, Stem Cells 25 (2007) 1645–1653.

[30] A. Glick, A. Ryscavage, R. Perez-Lorenzo, H. Hennings, S. Yuspa, N. Darwiche, The high-risk benign tumor: evidence from the two-stage skin cancer model and relevance for human cancer, Molecular Carcinogenesis 46 (2007) 605–610.

[31] Y.Y. He, J. Pi, J.L. Huang, B.A. Diwan, M.P. Waalkes, C.F. Chignell, Chronic UVA irradiation of human HaCaT keratinocytes induces malignant transformation associated with acquired apoptotic resistance, Oncogene 25 (2006) 3680–3688.

[32] G.M. Hong, L.J. Bain, Arsenic exposure inhibits myogenesis and neurogenesis in P19 stem cells through repression of the beta-catenin signaling pathway, Toxicological Sciences: An Official Journal of the Society of Toxicology 129 (2012) 146–156.

[33] G.M. Hong, L.J. Bain, Sodium arsenite represses the expression of myogenin in C2C12 mouse myoblast cells through histone modifications and altered expression of Ezh2, Glp, and Igf-1, Toxicology and Applied Pharmacology 260 (2012) 250–259.

[34] Z. Hong, M. Xiao, Y. Yang, Z. Han, Y. Cao, C. Li, Y. Wu, Q. Gong, X. Zhou, D. Xu, L. Meng, D. Ma, J. Zhou, Arsenic disulfide synergizes with the phosphoinositide 3-kinase inhibitor PI-103 to eradicate acute myeloid leukemia stem cells by inducing differentiation, Carcinogenesis 32 (2011) 1550–1558.

[35] Y. Hou, P. Xue, C.G. Woods, X. Wang, J. Fu, K. Yarborough, W. Qu, Q. Zhang, M.E. Andersen, J. Pi, Association between arsenic suppression of adipogenesis and induction of CHOP10 via the endoplasmic reticulum stress response, Environmental Health Perspectives 121 (2013) 237–243.

[36] J. Huff, M.F. Jacobson, D.L. Davis, The limits of two-year bioassay exposure regimens for identifying chemical carcinogens, Environmental Health Perspectives 116 (2008) 1439–1442.

[37] M.C. Humble, C.S. Trempus, J.W. Spalding, R.E. Cannon, R.W. Tennant, Biological, cellular, and molecular characteristics of an inducible transgenic skin tumor model: a review, Oncogene 24 (2005) 8217–8228.

[38] IARC, (Ed.), *IARC Monographs on the Evaluation of Carcinogenic Risks to Humans*, IARC Press, Leon, France, 2012.

[39] K. Ito, R. Bernardi, A. Morotti, S. Matsuoka, G. Saglio, Y. Ikeda, J. Rosenblatt, D.E. Avigan, J. Teruya-Feldstein, P.P. Pandolfi, PML targeting eradicates quiescent leukaemia-initiating cells, Nature 453 (2008) 1072–1078.

[40] V.N. Ivanov, T.K. Hei, Induction of apoptotic death and retardation of neuronal differentiation of human neural stem cells by sodium arsenite treatment, Experimental Cell Research 319 (2013) 875–887.

[41] V.N. Ivanov, G. Wen, T.K. Hei, Sodium arsenite exposure inhibits AKT and Stat3 activation, suppresses self-renewal and induces apoptotic death of embryonic stem cells, Apoptosis 18 (2013) 188–200.

[42] B.A. Jessen, Q. Qin, M.A. Phillips, D.L. Phillips, R.H. Rice, Keratinocyte differentiation marker suppression by arsenic: mediation by AP1 response elements and antagonism by tetradecanoylphorbol acetate, Toxicology and Applied Pharmacology 174 (2001) 302–311.

[43] R. Jiang, Y. Li, Y. Xu, Y. Zhou, Y. Pang, L. Shen, Y. Zhao, J. Zhang, J. Zhou, X. Wang, Q. Liu, EMT and CSC-like properties mediated by the IKKbeta/IkappaBalpha/RelA signal pathway via the transcriptional regulator, Snail, are involved in the arsenite-induced neoplastic transformation of human keratinocytes, Archives of Toxicology 87 (2012) 991–1000.

[44] Y. Jung, G. Bauer, J.A. Nolta, Concise review: induced pluripotent stem cell-derived mesenchymal stem cells: progress toward safe clinical products, Stem Cells 30 (2012) 42–47.

[45] Y. Kang, J. Massague, Epithelial-mesenchymal transitions: twist in development and metastasis, Cell 118 (2004) 277–279.

[46] J. Kim, J.J. Lee, D. Gardner, P.A. Beachy, Arsenic antagonizes the Hedgehog pathway by preventing ciliary accumulation and reducing stability of the Gli2 transcriptional effector,

Proceedings of the National Academy of Sciences of the United States of America 107 (2010) 13432–13437.

[47] J. Kim, B.T. Aftab, J.Y. Tang, D. Kim, A.H. Lee, M. Rezaee, J. Kim, B. Chen, E.M. King, A. Borodovsky, G.J. Riggins, E.H. Epstein, Jr., P.A. Beachy, C.M. Rudin, Itraconazole and arsenic trioxide inhibit hedgehog pathway activation and tumor growth associated with acquired resistance to smoothened antagonists, Cancer Cell 23 (2013) 23–34.

[48] L.R. Klei, D.Y. Garciafigueroa, A. Barchowsky, Arsenic activates endothelin-1 gi protein-coupled receptor signaling to inhibit stem cell differentiation in adipogenesis, Toxicological Sciences 131 (2013) 512–520.

[49] A.V. Krivtsov, D. Twomey, Z. Feng, M.C. Stubbs, Y. Wang, J. Faber, J.E. Levine, J. Wang, W.C. Hahn, D.G. Gilliland, T.R. Golub, S.A. Armstrong, Transformation from committed progenitor to leukaemia stem cell initiated by MLL-AF9, Nature 442 (2006) 818–822.

[50] V. Lallemand-Breitenbach, J. Zhu, Z. Chen, H. de The, Curing APL through PML/RARA degradation by As2O3, Trends in Molecular Medicine 18 (2012) 36–42.

[51] T. Lapidot, C. Sirard, J. Vormoor, B. Murdoch, T. Hoang, J. Caceres-Cortes, M. Minden, B. Paterson, M.A. Caligiuri, J.E. Dick, A cell initiating human acute myeloid leukaemia after transplantation into SCID mice, Nature 367 (1994) 645–648.

[52] H. Li, X. Chen, T. Calhoun-Davis, K. Claypool, D.G. Tang, PC3 human prostate carcinoma cell holoclones contain self-renewing tumor-initiating cells, Cancer Research 68 (2008) 1820–1825.

[53] J. Liaw, G. Marshall, Y. Yuan, C. Ferreccio, C. Steinmaus, A.H. Smith, Increased childhood liver cancer mortality and arsenic in drinking water in northern Chile, Cancer Epidemiology, Biomarkers and Prevention 17 (2008) 1982–1987.

[54] N.A. Lobo, Y. Shimono, D. Qian, M.F. Clarke, The biology of cancer stem cells, Annual Review of Cell and Developmental Biology 23 (2007) 675–699.

[55] S. Ma, T.K. Lee, B.J. Zheng, K.W. Chan, X.Y. Guan, CD133+ HCC cancer stem cells confer chemoresistance by preferential expression of the Akt/PKB survival pathway, Oncogene 27 (2008) 1749–1758.

[56] I. Malanchi, H. Peinado, D. Kassen, T. Hussenet, D. Metzger, P. Chambon, M. Huber, D. Hohl, A. Cano, W. Birchmeier, J. Huelsken, Cutaneous cancer stem cell maintenance is dependent on beta-catenin signalling, Nature 452 (2008) 650–653.

[57] S.A. Mani, W. Guo, M.J. Liao, E.N. Eaton, A. Ayyanan, A.Y. Zhou, M. Brooks, F. Reinhard, C.C. Zhang, M. Shipitsin, L.L. Campbell, K. Polyak, C. Brisken, J. Yang, R.A. Weinberg, The epithelial-mesenchymal transition generates cells with properties of stem cells, Cell 133 (2008) 704–715.

[58] G. Marshall, C. Ferreccio, Y. Yuan, M.N. Bates, C. Steinmaus, S. Selvin, J. Liaw, A.H. Smith, Fifty-year study of lung and bladder cancer mortality in Chile related to arsenic in drinking water, Journal of the National Cancer Institute 99 (2007) 920–928.

[59] J.R. Masters, Human cancer cell lines: fact and fantasy, Nature Reviews Molecular Cell Biology 1 (2000) 233–236.

[60] E.A. McCulloch, J.E. Till, The radiation sensitivity of normal mouse bone marrow cells, determined by quantitative marrow transplantation into irradiated mice, Radiation Research 13 (1960) 115–125.

[61] D. Mizrak, M. Brittan, M.R. Alison, CD133: molecule of the moment, The Journal of Pathology 214 (2008) 3–9.

[62] A.P. Morel, M. Lievre, C. Thomas, G. Hinkal, S. Ansieau, A. Puisieux, Generation of breast cancer stem cells through epithelial-mesenchymal transition, PLoS One 3 (2008) e2888.

[63] R. Nasr, M.C. Guillemin, O. Ferhi, H. Soilihi, L. Peres, C. Berthier, P. Rousselot, M. Robledo-Sarmiento, V. Lallemand-Breitenbach, B. Gourmel, D. Vitoux, P.P. Pandolfi, C. Rochette-Egly, J. Zhu, H. de The, Eradication of acute promyelocytic leukemia-initiating cells through PML-RARA degradation, Nature Medicine 14 (2008) 1333–1342.

[64] R. Nasr, V. Lallemand-Breitenbach, J. Zhu, M.C. Guillemin, H. de The, Therapy-induced PML/RARA proteolysis and acute promyelocytic leukemia cure, Clinical Cancer Research: An Official Journal of the American Association for Cancer Research 15 (2009) 6321–6326.

[65] C.A. O'Brien, A. Kreso, C.H. Jamieson, Cancer stem cells and self-renewal, Clinical Cancer Research: An Official Journal of the American Association for Cancer Research 16 (2010) 3113–3120.

[66] S.H. Orkin, L.I. Zon, Hematopoiesis: an evolving paradigm for stem cell biology, Cell 132 (2008) 631–644.

[67] R. Pardal, M.F. Clarke, S.J. Morrison, Applying the principles of stem-cell biology to cancer, Nature Reviews. Cancer 3 (2003) 895–902.

[68] T.J. Patterson, T.V. Reznikova, M.A. Phillips, R.H. Rice, Arsenite maintains germinative state in cultured human epidermal cells, Toxicology and Applied Pharmacology 207 (2005) 69–77.

[69] T.J. Patterson, R.H. Rice, Arsenite and insulin exhibit opposing effects on epidermal growth factor receptor and keratinocyte proliferative potential, Toxicology and Applied Pharmacology 221 (2007) 119–128.

[70] K. Polyak, W.C. Hahn, Roots and stems: stem cells in cancer, Nature Medicine 12 (2006) 296–300.

[71] T. Reya, S.J. Morrison, M.F. Clarke, I.L. Weissman, Stem cells, cancer, and cancer stem cells, Nature 414 (2001) 105–111.

[72] R.A. Rocha, J.V. Gimeno-Alcaniz, R. Martin-Ibanez, J.M. Canals, D. Velez, V. Devesa, Arsenic and fluoride induce neural progenitor cell apoptosis, Toxicology Letters 203 (2011) 237–244.

[73] S. Sell, Cancer stem cells and differentiation therapy, Tumour Biology: The Journal of the International Society for Oncodevelopmental Biology and Medicine 27 (2006) 59–70.

[74] A.H. Smith, G. Marshall, Y. Yuan, C. Ferreccio, J. Liaw, O. von Ehrenstein, C. Steinmaus, M.N. Bates, S. Selvin, Increased mortality from lung cancer and bronchiectasis in young adults after exposure to arsenic in utero and in early childhood, Environmental Health Perspectives 114 (2006) 1293–1296.

[75] J.C. States, A.V. Singh, T.B. Knudsen, E.C. Rouchka, N.O. Ngalame, G.E. Arteel, Y. Piao, M.S. Ko, Prenatal arsenic exposure alters gene expression in the adult liver to a proinflammatory state contributing to accelerated atherosclerosis, PLoS One 7 (2012) e38713.

[76] A.A. Steffens, G.M. Hong, L.J. Bain, Sodium arsenite delays the differentiation of C2C12 mouse myoblast cells and alters methylation patterns on the transcription factor myogenin, Toxicology and Applied Pharmacology 250 (2011) 154–161.

[77] J. Stingl, P. Eirew, I. Ricketson, M. Shackleton, F. Vaillant, D. Choi, H.I. Li, C.J. Eaves, Purification and unique properties of mammary epithelial stem cells, Nature 439 (2006) 993–997.

[78] H. Sun, S. Zhang, Arsenic trioxide regulates the apoptosis of glioma cell and glioma stem cell via down-regulation of stem cell marker Sox2, Biochemical and Biophysical Research Communications 410 (2011) 692–697.

[79] Y. Sun, E.J. Tokar, M.P. Waalkes, Overabundance of putative cancer stem cells in human skin keratinocyte cells malignantly transformed by arsenic, Toxicological Sciences: An Official Journal of the Society of Toxicology 125 (2012) 20–29.

[80] K. Takahashi, S. Yamanaka, Induction of pluripotent stem cells from mouse embryonic and adult fibroblast cultures by defined factors, Cell 126 (2006) 663–676.

[81] J.E. Till, E.A. McCulloch, L. Siminovitch, A stochastic model of stem cell proliferation, based on the growth of spleen colony-forming cells, Proceedings of the National Academy of Sciences 51 (1964) 29–36.

[82] E.J. Tokar, L. Benbrahim-Tallaa, J.M. Ward, R. Lunn, R.L. Sams, 2nd, M.P. Waalkes, Cancer in experimental animals exposed to arsenic and arsenic compounds, Critical Reviews in Toxicology 40 (2010) 912–927.

[83] E.J. Tokar, B.A. Diwan, M.P. Waalkes, Arsenic exposure transforms human epithelial stem/progenitor cells into a cancer stem-like phenotype, Environmental Health Perspectives 118 (2010) 108–115.

[84] E.J. Tokar, W. Qu, J. Liu, W. Liu, M.M. Webber, J.M. Phang, M.P. Waalkes, Arsenic-specific stem cell selection during malignant transformation, Journal of the National Cancer Institute 102 (2010) 638–649.

[85] E.J. Tokar, B.A. Diwan, J.M. Ward, D.A. Delker, M.P. Waalkes, Carcinogenic effects of "whole-life" exposure to inorganic arsenic in CD1 mice, Toxicological Sciences: An Official Journal of the Society of Toxicology 119 (2011) 73–83.

[86] E.J. Tokar, B.A. Diwan, M.P. Waalkes, Renal, hepatic, pulmonary and adrenal tumors induced by prenatal inorganic arsenic followed by dimethylarsinic acid in adulthood in CD1 mice, Toxicology Letters 209 (2012) 179–185.

[87] E.J. Tokar, R.J. Person, Y. Sun, A.O. Perantoni, M.P. Waalkes, Chronic exposure of renal stem cells to inorganic arsenic induces a cancer phenotype, Chemical Research in Toxicology 26 (2013) 96–105.

[88] C.S. Trempus, R.J. Morris, C.D. Bortner, G. Cotsarelis, R.S. Faircloth, J.M. Reece, R.W. Tennant, Enrichment for living murine keratinocytes from the hair follicle bulge with the cell surface marker CD34, The Journal of Investigative Dermatology 120 (2003) 501–511.

[89] C.S. Trempus, R.J. Morris, M. Ehinger, A. Elmore, C.D. Bortner, M. Ito, G. Cotsarelis, J.G. Nijhof, J. Peckham, N. Flagler, G. Kissling, M.M. Humble, L.C. King, L.D. Adams, D. Desai, S. Amin, R.W. Tennant, CD34 expression by hair follicle stem cells is required for skin tumor development in mice, Cancer Research 67 (2007) 4173–4181.

[90] M.A. van der Heyden, L.H. Defize, Twenty one years of P19 cells: what an embryonal carcinoma cell line taught us about cardiomyocyte differentiation, Cardiovascular Research 58 (2003) 292–302.

[91] L. Vernhet, C. Morzadec, J. van Grevenynghe, B. Bareau, M. Corolleur, T. Fest, O. Fardel, Inorganic arsenic induces necrosis of human CD34-positive haematopoietic stem cells, Environmental Toxicology 23 (2008) 263–268.

[92] J.E. Visvader, G.J. Lindeman, Cancer stem cells in solid tumours: accumulating evidence and unresolved questions, Nature Reviews. Cancer 8 (2008) 755–768.

[93] M.P. Waalkes, J. Liu, B.A. Diwan, Transplacental arsenic carcinogenesis in mice, Toxicology and Applied Pharmacology 222 (2007) 271–280.

[94] M.P. Waalkes, J. Liu, D.R. Germolec, C.S. Trempus, R.E. Cannon, E.J. Tokar, R.W. Tennant, J.M. Ward, B.A. Diwan, Arsenic exposure in utero exacerbates skin cancer response in adulthood with contemporaneous distortion of tumor stem cell dynamics, Cancer Research 68 (2008) 8278–8285.

[95] S. Wang, J. Gao, Q. Lei, N. Rozengurt, C. Pritchard, J. Jiao, G.V. Thomas, G. Li, P. Roy-Burman, P.S. Nelson, X. Liu, H. Wu, Prostate-specific deletion of the murine Pten tumor suppressor gene leads to metastatic prostate cancer, Cancer Cell 4 (2003) 209–221.

[96] S. Wang, A.J. Garcia, M. Wu, D.A. Lawson, O.N. Witte, H. Wu, Pten deletion leads to the expansion of a prostatic stem/progenitor cell subpopulation and tumor initiation, Proceedings of the National Academy of Sciences of the United States of America 103 (2006) 1480–1485.

[97] I.L. Weissman, Stem cells: units of development, units of regeneration, and units in evolution, Cell 100 (2000) 157–168.

[98] M.S. Wicha, S. Liu, G. Dontu, Cancer stem cells: an old idea—a paradigm shift, Cancer Research 66 (2006) 1883–1890; discussion 1895–1886.

[99] D.J. Wong, H. Liu, T.W. Ridky, D. Cassarino, E. Segal, H.Y. Chang, Module map of stem cell genes guides creation of epithelial cancer stem cells, Cell Stem Cell 2 (2008) 333–344.

[100] Y. Xu, Y. Li, Y. Pang, M. Ling, L. Shen, X. Yang, J. Zhang, J. Zhou, X. Wang, Q. Liu, EMT and stem cell-like properties associated with HIF-2alpha are involved in arsenite-induced transformation of human bronchial epithelial cells, PLoS One 7 (2012) e37765.

[101] Y. Xu, E.J. Tokar, Y. Sun, M.P. Waalkes, Arsenic-transformed malignant prostate epithelia can convert noncontiguous normal stem cells into an oncogenic phenotype, Environmental Health Perspectives 120 (2012) 865–871.

[102] S. Yadav, M. Anbalagan, Y. Shi, F. Wang, H. Wang, Arsenic inhibits the adipogenic differentiation of mesenchymal stem cells by down-regulating peroxisome proliferator-activated receptor gamma and CCAAT enhancer-binding proteins, Toxicology In Vitro 27 (2013) 211–219.

[103] S.L. Yancy, E.A. Shelden, R.R. Gilmont, M.J. Welsh, Sodium arsenite exposure alters cell migration, focal adhesion localization and decreases tyrosine phosphorylation of focal adhesion kinase in H9C2 myoblasts, Toxicological Sciences 84 (2005) 278–286.

[104] T. Yorifuji, T. Tsuda, H. Doi, P. Grandjean, Cancer excess after arsenic exposure from contaminated milk powder, Environmental Health and Preventive Medicine 16 (2012) 164–170.

[105] Y. Yuan, G. Marshall, C. Ferreccio, C. Steinmaus, J. Liaw, M. Bates, A.H. Smith, Kidney cancer mortality: fifty-year latency patterns related to arsenic exposure, Epidemiology 21 (2010) 103–108.

[106] X.W. Zhang, X.J. Yan, Z.R. Zhou, F.F. Yang, Z.Y. Wu, H.B. Sun, W.X. Liang, A.X. Song, V. Lallemand-Breitenbach, M. Jeanne, Q.Y. Zhang, H.Y. Yang, Q.H. Huang, G.B. Zhou, J.H. Tong, Y. Zhang, J.H. Wu, H.Y. Hu, H. de The, S.J. Chen, Z. Chen, Arsenic trioxide controls the fate of the PML-RARalpha oncoprotein by directly binding PML, Science 328 (2010) 240–243.

[107] X. Zheng, A. Seshire, B. Ruster, G. Bug, T. Beissert, E. Puccetti, D. Hoelzer, R. Henschler, M. Ruthardt, Arsenic but not all-trans retinoic acid overcomes the aberrant stem cell capacity of PML/RARalpha-positive leukemic stem cells, Haematologica 92 (2007) 323–331.

[108] L.F. Zhu, Y. Hu, C.C. Yang, X.H. Xu, T.Y. Ning, Z.L. Wang, J.H. Ye, L.K. Liu, Snail overexpression induces an epithelial to mesenchymal transition and cancer stem cell-like properties in SCC9 cells, Laboratory Investigation 92 (2012) 744–752.

18

EPIGENETICS AND ARSENIC TOXICITY

Somnath Paul[1] and Pritha Bhattacharjee[2]

[1]*Molecular Biology and Human Genetics Division, Council for Scientific and Industrial Research – Indian Institute of Chemical Biology, Kolkata, India*
[2]*Department of Environmental Science, University of Calcutta, Kolkata, India*

18.1 INTRODUCTION

In 1942, Conrad Waddington coined the term "epigenetics" as "the branch of biology which studies the causal interactions between genes and their products which bring the phenotype into being." In the modern context, it can be defined as all transgenerationally transmitted heritable changes in gene expression that do not involve any changes in the coding sequence [23] and as a form of potentially reversible DNA modification. In contrast, the heritable changes coded within the nucleotide sequence are known as genetics. Years of genetic research elucidated the role of the alteration in nuclear sequence leading to mutations and single-nucleotide polymorphisms in disease outcomes. New-generation findings point out the role of epigenetics as a causal force in such cases. Epigenetic mechanisms alter the accessibility of chromatin for transcriptional regulation, and thus epigenetic alterations serve as an important regulator of gene expression. Chromatin structure within nucleosomes plays a key role, where open frame of euchromatin region is suitable for gene expression as it is more accessible by transcriptional machinery; on the other hand, closed and compact heterochromatin region is often associated with transcriptional repression. Moreover,

Arsenic: Exposure Sources, Health Risks, and Mechanisms of Toxicity, First Edition.
Edited by J. Christopher States.
© 2016 John Wiley & Sons, Inc. Published 2016 by John Wiley & Sons, Inc.

epigenetic signatures in the nucleosome lead to histone modifications, a potential player in variable transcriptional states of the chromatin.

Thus far, three major epigenetic mechanisms, that is, (i) DNA methylation, (ii) histone modifications, and (iii) noncoding RNA-mediated gene silencing have been identified. Thus, it is imperative to include the role of "epigenomics" with studies of other "omics" (genomics, transcriptomics, proteomics, and metabolomics) in modern molecular biology. About 6000 publications came to light from its inception in the 1960s, but the major breakthroughs being made only after the 1990s. A variety of environmental factors can cause disease manifestation through epigenetic deregulation. DNA methylation alteration, especially at promoter regions, can alter gene expression. In general, gene expression and methylation are inversely correlated, but in an *in vitro* study by Boellmann et al. [5] showed that only 57% genes follow this general principle. Molecular cross-talk across the pathways is thought to be the major guiding force behind the gene expression regulation. The noncoding RNA (ncRNA), which is generated after post-transcriptional modifications, also regulates gene expression [63]. A recent study has shown that loss of miR-886 by its promoter hypermethylation suppresses the target genes *PLK1* and *TGFβ1*, and thus is used as a new therapeutic target for small-cell lung cancer [9]. Here, we will illustrate the epigenetic regulations in general, role of arsenic as an epimutagen and how epigenetic modifications can lead to different diseases including cancer (Fig. 18.1). We will briefly discuss the technologies to identify the epigenetic deregulation and scope for future studies.

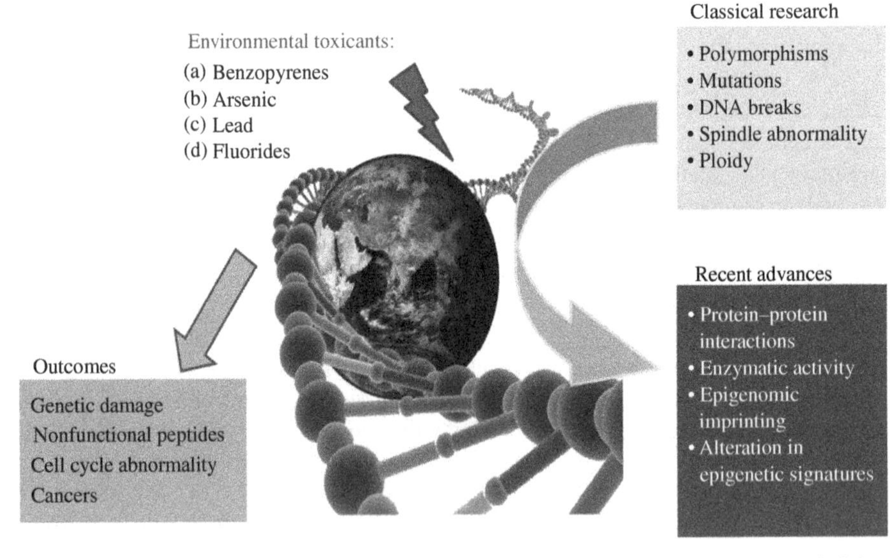

FIGURE 18.1 Environmental toxicant and disease outcome: unraveling the underlying molecular mechanisms. Identifications of mutations or polymorphisms in genome are now replaced with high-throughput techniques exploring epigenome.

18.2 MAJOR EPIGENETIC REGULATING MECHANISMS

18.2.1 Nucleic Acid Methylation

Covalent modification and addition of a methyl moiety from *S*-adenosylmethionine onto the fifth carbon of cytosine (C) forms 5-methylcytosine (5-mC) within the DNA double helix is called "transmethylation." This modification is predominantly found in the cytosine–guanine-rich region of the genome called "C_pG islands" and are highly associated with the promoter regions of genes. This transmethylation is catalyzed by DNA methyltransferases (DNMTs) using *S*-adenosylmethionine [2]. DNMT1 maintains the methylation status, which is necessary to preserve after every replication cycle, and this methylation activity is dependent upon a folate-rich diet [66]. DNMT3a and DNMT3b, on the other hand, are *de novo* methyl transferases that set up DNA methylation pattern early in developmental stages. The fidelity of the various epigenetic regulatory enzymes is enumerated in Table 18.1. Methylation of DNA, RNA, histones, and other proteins dictates replicational, transcriptional, and translational fidelity; mismatch repair; and chromatin remodeling, all being classified as epigenetic modifications of the genome, and thus plays important roles in cancer and aging [42]. Along with 5-methylcytosine (5mC) modifications of DNA, another modification, 5-hydroxymethylcytosine (5hmC), has been recently reported in Purkinjee neurons, embryonic stem cells, and several human cell lines [40, 60, 64].

RNA editing, whereby certain residues within RNAs, especially t-RNAs are modified by transmethylation, occurs to maintain the structural and functional integrity of the RNA [29]. Adenosine to inosine transformation is another type of epigenetic-mediated RNA editing, which has been understood recently [50]. The deamination of adenine to inosine is enzymatically controlled by certain specialized superfamily of enzymes called "adenosine deaminase" acting on RNA (ADAR; Table 18.1) [37]. If this process is hampered, alteration of the translation machinery can yield aberrant proteins that may lead to cancer [22].

18.2.2 Histone Modifications

Histone modifications provide another important mechanism of epigenetic regulation. The nucleosome core particle consists of stretches of DNA (~146 bp) wrapped in left-handed superhelical turns around a histone octamer consisting of two copies of each core histone (H2A, H2B, H3, and H4) [43]. Although histone H1 is not part of the nucleosome "bead," histone H1 stabilizes the DNA strand within the "bead," forming the "pearl necklace" [76]. Nucleosomal histones are subject to a wide variety of covalent modifications including acetylation, methylation, phosphorylation, citrullination, ubiquitination, sumoylation, ADP ribosylation, deamination, and proline isomerization takes place [39]. Specific histone modifications can act sequentially or in combination to form a recognizable "code" playing an important regulatory role in transcriptional activation or repression (Table 18.2), as well as DNA repair and replication. Histone acetylation, methylation, and phosphorylation are the most commonly studied epigenetic histone modifications.

TABLE 18.1 A Comprehensive List of Epigenetic Regulatory Enzymes Found in Humans

Enzyme	Functions
Nucleic acids	
DNMT1	Methylates the C_pG residues and are normally active in maintaining the methylation profile of the DNA, especially when DNA is hemimethylated. It associates with the chromatin during G2/M transition phase and maintains the methylation status of the genome, postreplication, and mitosis. It is also responsible for maintaining methylation patterns established in development.
DNMT2	Presently called "TRDMT1," it is primarily associated with methylation of the cytosine-38 of the aspartic acid t-RNA. It is highly conserved and the functional aspect of the methylation of cytosine-38 enables the tRNA to maintain its structural integrity
DNMT3A	This is a DNA methyltransferase that functions in *de novo* methylation, rather than maintenance methylation.
DNMT3B	It preferentially methylates nucleosomal DNA within the core "pearl" of the nucleosome.
ADAR	Catalyzes hydrolytic de-amination of the adenosine and converts it to inosine, within the mRNA specially, referred as A to I RNA editing. This is important to maintain the structural and functional integrity of the RNA and any alteration may destabilize the RNA conformation.
Histones	
GCN5	General control nonderepressible-5 (GCN5), first identified in yeasts, is a type A HAT within the GNAT (Gcn5-related *N*-acetyltransferases) subfamily. It is also found in humans and its prime target within the nucleosomes are H2B, H3, and H4. It contains a bromodomain, like all Type A HATs
MYST	Abbreviated from co-ordinated association of four HAT enzymes, namely, MOZ, Ybf2 (Sas3), Sas2, and Tip60. They are typically characterized by zinc finger domains that help in acetylation of the lysine residues of H2A, H3, and H4, being closely intercalated with the DNA. They are Type A HATs, found within the nucleus.
HAT1	This is one of the lesser-known type B HATs, found within the cytosol and rapidly acetylates the newly synthesized histones by the cellular translational machinery. This is important for incorporation of the histones with the nascently replicated DNA by *de novo* deposition. HAT1 specifically targets the H4 histone moiety.

Histone acetyltransferases (HATs) and histone deacetylases (HDACs) are the key enzymes involved in histone acetylation, which is a dynamic and reversible process [28]. HATs are generally of two types (Table 18.1); Type A HATs are present within the nucleus and contain a bromodomain that regulates acetylation of the histones within the nucleosomes. Type B HATs are found within the cytoplasm and regulate acetylation of the synthesized histones within the cytosol, before they are transported to the nucleus and incorporated within the nucleosomal complex.

TABLE 18.2 Epigenetic Modifications of Histones and Their Functional Significance

Histone Mark	Sites of Modifications	Functional Significance
Acetylated lysine (K)	H3 (9,14, 18) H4 (5, 8, 13, 16) H2A H2B	Activation
Phosphorylated serine (S)/threonine (T)	H3 (3, 10, 28) H2A H2B	Activation
Methylated arginine (R)	H3 (17, 23) H4 (3)	Activation
Methylated lysine (K)	H3 (4, 36, 79) H3 (9, 27) & H4 (20)	Activation Repression
Ubiquitinylated lysine (K)	H2B (120) H2A (119)	Activation Repression

TABLE 18.3 Comprehensive List of the Major Enzymes Associated with Histone Modifications

Histone	Acetylation	Deacetylation	Methylation	Demethylation	Phos-phorylation	Ubiq-uitination
H2A	CBP/p300					Bmi/Ring1A
H2B	CBP/p300				MST1	RNF20/ RNF40
H3	PCAF/Sc RTT109		MLL/SET/ ASH1/NSD	JHDM1/ GASC1/JMUD	AuroraB/ HASPIN/ MSK1	
H4	TIP60/HOB1	SIRT2	PRMT/SUV4		CKII	

While acetylation of lysine residues maintains the transcriptionally open chromatin conformation, deacetylation of these residues favors chromatin condensation [17, 62]. Like acetylation, histone methylation is also a reversible process. However, unlike acetylation, which occurs only on lysine residues at the histone tail, histone methylation occurs on both lysine and arginine residues [45, 73]. A list of modifications and associated enzymes that bring about such changes within the histones is provided in Table 18.3.

18.2.3 Noncoding RNA-Mediated Pathway

In addition to regulating gene expression at the transcriptional and post-transcriptional level, noncoding RNAs (ncRNAs) play a major role in control of epigenetic patterns. ncRNAs including miRNA (micro RNA), piRNA (piwi-interacting RNA), siRNA (small interfering RNA), and lncRNA (long noncoding RNA) play a major role in RNA-directed gene silencing, chromatin modification, structural changes to chromatin, and gene promoter regulation [24]. These ncRNAs are vital in transcription progression, and interact with mRNA templates and stall the translation mechanism inhibiting protein synthesis [3, 16]. This inhibition alters protein expression patterns within cells and regulates multiple intracellular processes.

The association of overexpression of certain miRNAs like miR17-92 with carcinogenic progression as it negatively regulates tumor suppressor genes or genes that control cell differentiation has been observed [21, 79]. The small noncoding RNA gene products, such as miRNAs, are highly conserved, and are transcribed mainly from the nonprotein coding regions of the genome. Each of these miRNAs is thought to deactivate several hundred genes by binding to the 3′UTR of mRNAs with incomplete base pairing. Because of this suppressive effect of miRNAs on gene expression, elimination or reduction of miRNAs that target either oncogene or tumor suppressor genes could have immense potential in early diagnosis or therapeutics. Thus miRNA profiling has been used in search of biomarkers for many cancers and also nonmalignant diseases [12, 26, 27, 47, 52]; an added advantage is the minimum invasiveness of sample collection and ease of storage.

18.3 SIGNIFICANCE OF EPIGENETIC STUDIES IN ARSENIC TOXICITY

Consumption of drinking water above the maximum contaminant level of 10 μg/L is associated with a multitude of adverse health effects including cancer [33]. Millions of people in more than 70 nations are exposed to arsenic via either environmental or occupational exposure (see Chapters 1 and 2). It is interesting that arsenic is not a classical mutagen, but a potent human carcinogen. Although epigenetic alterations, with special emphasis in DNA methylation have been studied in different types of cancers (Table 18.4), very few studies have been done so far for arsenic-induced cancers. To understand the underlying molecular basis of such toxicity, intensive research (*in vitro, in vivo,* population based) focusing mainly on genetic aspects identified a number of mechanisms including chromosomal damage, inefficient DNA repair, alteration in gene expression, and protein profiling that may lead to arsenic toxicity and carcinogenicity [4]. The epigenetic machineries influence a number of biological processes and act as an interface between environmental toxicant and disease endpoints, which needs to be studied in depth for arsenic toxicity, preferably in humans, since suitable animal models are quite limited (see Chapter 19).

Chronic exposure to arsenic depletes *S*-adenosylmethionine within TRL1215 murine hepatocytes and leads to global hypomethylation [82]. Since biotransformation of inorganic arsenic leads to the formation of monomethylasonic acid (MMA) and dimethylarsenic acid (DMA) within the body, it is believed that arsenic exposure may lead to *S*-adenosylmethionine depletion, hence contributing to global hypomethylation. In a study by Xie et al. [74], newborn male mice born to dams exposed to a known carcinogenic dose of sodium arsenite (85 ppm) in the drinking water lead to significant reduction in methylation in GC-rich regions of hepatic DNA. The authors hypothesized that the reduction of DNA methylation in GC-rich regions could be due to the inhibition of DNMT. Liver is the main organ for arsenic metabolism and thus a

TABLE 18.4 List of Some DNA-Methylation Related Cancers Compiled Using DiseaseMeth

Cancer Type	Gene	Chromosome Number
Brain cancer	ANGEL2 (NM_144567)	1
	CASP8 (NM_033355)	2
	PLAC8L1 (NM_001029869)	5
	RAET1G (NM_001001788)	6
	NRTN (NM_004558)	19
	NM_001006117 (NM_001006117.2)	hChr Y
Esophageal cancer	NFYC (NM_014223)	1
	CENPC-1 (NM_001812)	4
	MTSS1 (NM_014751)	8
	TEX2 (NM_018469)	17
	PHF16 (NM_014735)	hChr X
Breast cancer	UGT1A1 (NM_000463)	2
	SH3BP2 (NM_003023)	4
Gastric cancer	RASSF1 (NM_170714)	3
	IRS2 (NM_003749)	13
Wilm's tumor	DAPK1 (NM_004938)	9

Source: Database for the above compilation: http://202.97.205.78/diseasemeth/.

target for arsenic toxicity as well. Chronic arsenic exposure and its associated hepatic damage and cardiovascular risk were observed in a population in West Bengal, India [19]. A microarray-based study revealed aberrant expression in 80 genes including significant upregulation of oncogenes, like *Myc*, *Ras*, and *Met* [11]. Surprisingly, the expression of the tumor-suppressor gene *TP53* did not change. On the contrary, Chanda et al. [10] have demonstrated that promoter hypermethylation of *TP53* and *CDKN2A* (aka *p16-INK4A*) takes place in people chronically exposed to arsenic. Since methylation pattern is very dynamic, such studies on larger cohorts along with high throughput techniques will be helpful to delineate the molecular mechanism. Several studies have implied that the metabolic rate of biotransformation of *i*As to DMA is associated with higher incidence of skin cancers [13, 77], bladder cancer [14, 32], and peripheral vascular disease [67]. In a separate study, it was shown that indeed, concentrations as low as 0.2 mM arsenite could inhibit both DNMT1 and DNMT3a along with *S*-adenosylmethionine depletion and global DNA hypomethylation in human keratinocytes [53]. Methylation profile in arsenic exposed Mexican population ($N=16$) also indicated that global DNA hypomethylation with DNA hypermethylation of specific genes that includes tumor suppressor genes may lead to carcinogenesis in humans [57].

Arsenic disrupts epigenetic machineries by targeting DNMTs, as one of the mechanisms, which results in loss of methylation specifically to the methylated-silenced genes [18]. This may lead to re-expression of oncogenic genes. While, decrease in DNMT activity may be a reason toward arsenic-induced global hypomethylation, growth arrest and DNA damage 45 (*GADD45*) might play an important role in arsenic-induced demethylation. GADD45 has been reported to be expressed due to oxidative stress within the cells. It has a demethylating enzyme domain that can demethylate the genome upon stress response [68]. Recent studies have found a strong association of GADD45A overexpression with arsenic exposure, and this expression has been proposed to be a prime contender for DNA hypomethylation, in view of reports of GADD45-mediated DNA demethylation [48, 80], although there is some disagreement about the role of GADD45 in DNA methylation [35].

Arsenic regulates histone modifications which include methylation and acetylation. Histone methylations are again tightly controlled by both DNA methylation and demethylation processes. Arsenic trioxide (As_2O_3) treatment also affected the levels of another histone H3lysine9 (H3K9)-specific methyltransferase, Setdb1/Eset. Specifically, arsenic treatment, which induces the degradation of promyelocytic leukemia (PML) protein, resulted in the disappearance of Sedb1 signals from the nuclei of mouse embryos [15, 80]. The global level of H4K16 acetylation in human bladder epithelial cells was reduced in a dose- and time-dependent manner by both As(III) and MMA(III) treatment. Moreover, knockdown of *MYST1*, the gene responsible for H4K16 acetylation, resulted in increased cytotoxicity from arsenical exposure in human bladder epithelial cells, suggesting that H4K16 acetylation may be important for resistance to arsenic-induced toxicity [36, 80]. The di- and trimethylated forms of the lysine (K) residues within the histones are important to impart chromatin conformation and determine the supercoiled state. *In vitro* studies have shown that arsenic alters the methylation of these residue types and have been associated with development of several types of cellular dysfunctions and even cancers. For example, altered histone methylation patterns were observed in arsenic-transformed human urothelial cells. The transformed cells experienced a loss of H3K27me3 and H3K9me2, both of which are marks of transcriptional repression, and an increase in H3K4me2 in the promoter region of *WNT5A* gene, which is important in cellular development and differentiation and is often improperly regulated in numerous cancers [34, 80]. Meng et al. [46, 80] observed that miRNA-29a showed a positive therapeutic implication in HepG2 cells, when treated with arsenic trioxide. Alternatively, miRNA-21 has been implicated in cancer progression via activation of ERK/NFκB pathway [41, 80]. Upon treatment with arsenic, miRNA-21 expression increased; increased miRNA-21 expression also was induced by reactive oxygen species, which can be formed within the cellular microenvironment after arsenic exposure.

The multilayered affects induced by arsenic makes it difficult to understand *prima facie* the molecular toxicity of arsenic within the cellular system. In a comparative outlook, alteration of the epigenetic signatures within the DNA and histones, along with changes in miRNA profiles, might lead to an overall imbalance within the cellular system that tips toward arsenic-induced toxicity and carcinogenesis. Efforts at genome-wide study of arsenic-induced DNA methylation and histone modification changes associated with cellular transformation have shown dramatic effects of chronic incubation with inorganic arsenic or its trivalent monomethylated metabolite MMA3 [34, 55, 56, 72]. Chronic exposure of immortalized human urothelial cells to 50 nM MMA3 induced DNA methylation changes that did not revert after removal of the arsenical [34, 72]. Agglomerative DNA methylation changes were observed in prostate epithelial cells undergoing chronic arsenic exposure-induced malignant transformation *in vitro* and the DNA methylation changes were more often associated with histone H3 lysine-27 trimethylation stem cell domains [55]. Later work in this system reported a suggested linkage between arsenical exposure-induced histone H3 lysine 9 methylation and DNA methylation changes in these cells [56].

18.4 EPIGENOMIC IMPLICATIONS

Epigenetic interactions are vital in development and in cell cycle regulation. ncRNA is a regulator of Hox loci, and as such plays an important role in cellular differentiation [8, 80]. Recently, Thompson et al. [65] related post-translational modification of histones with chromatin structure organization and carcinogenesis. Benzo[*a*]pyrene, chromium, and cadmium, in addition to arsenic, are epimutagens; they modify gene expression patterns in key developmental stages in carcinogenesis [54]. Nucleic acid adducts from benzo[*a*]pyrene-like toxicants might not always alter the nucleic acid methylation pattern, but may lead to nucleic acid degradation and damage [1, 31]. These adducts (benzopyrene isoforms) in DNA, RNA, and proteins pose a significant threat of cancer genesis hampering normal cell survival [51]. In this context, Maier et al. [44] analyzed toxicity in co-exposure experiments and concluded arsenic enhances benzo[*a*]pyrene-DNA adduct formation in murine hepatoma cells [44]. These results show the capability of arsenic as epimutagen and enhancer for other toxic substances.

Chronic arsenicosis has been of pandemic concern. The geogenic incidence and exposure of arsenic through ground water and diet to humans have caused several arsenic-exposure-induced symptoms like dermatological and nondermatological cancers, peripheral neuropathy, and respiratory problems. Epigenetic implications are fundamental to understanding arsenic toxicity because *S*-adenosylmethionine is consumed in arsenic metabolism, and it is the main component of the methylation pool within the cellular microenvironment. Hence, it is very important to understand the mechanism of rejuvenating the epigenetic marks, postarsenic toxicity.

18.5 TECHNIQUES TO IDENTIFY EPIGENETIC MODIFICATIONS

The epigenetic modifications of DNA, histone, and RNA can be identified by a number of techniques ranging from standard molecular biological methods for aiming single target, to high-throughput technologies aiming 10^5 genes at a go. Most commonly used methods to identify the epigenetic modifications are enlisted in the following text.

18.5.1 For DNA Methylation

18.5.1.1 Bisufite Modification Bisulfite treatment can distinguish between cytosine and 5mC, where treatment converts unmethylated cytosines to uracils, but 5mCs remain unaltered. Therefore, sequencing after bisulfite treatment will read unmethylated cytosines (now uracils) as thymines, while 5mC is read as cytosine. A representative figure of the process is given in Figure 18.2. Thus comparison with the reference sequence (i.e., sequence before bisulfite treatment) will reveal the extent

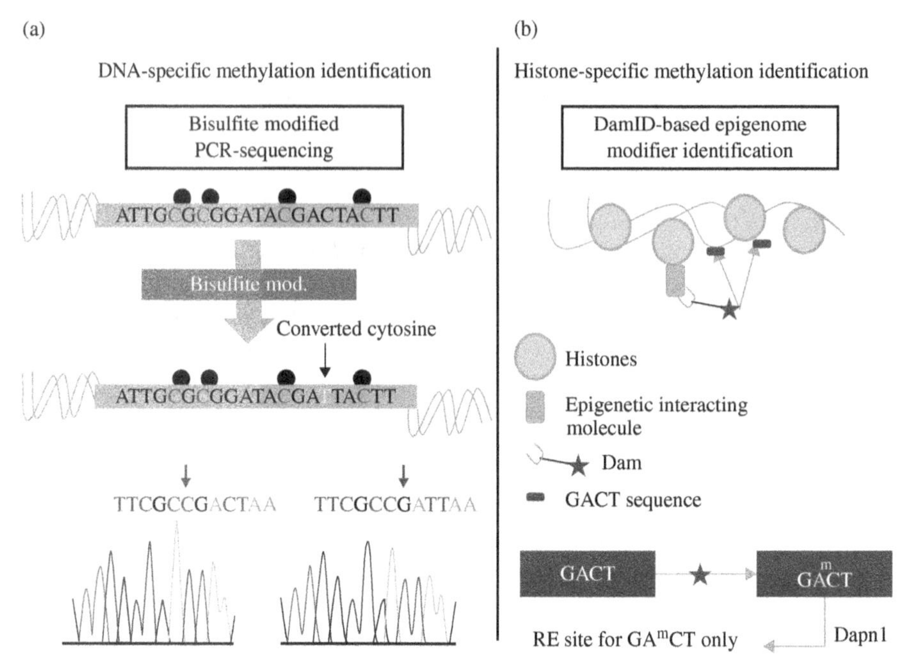

FIGURE 18.2 Identification of epigenetic changes using (a) bisulfite-modified PCR-sequencing and (b) DamID method. (*See insert for color representation of the figure.*)

of methylation. A number of techniques available for identifying DNA methylation status after bisulfite treatment, including restriction enzyme, Sanger sequencing, pyrosequencing, microarray, and PCR (bisulfite-specific PCR and COBRA, combined bisulfite restriction analysis). However, this method is unable to distinguish between 5mC and 5hmC. Recent advancements utilize separate antibodies for 5mC and 5hmC for detection using dot blots [20, 59, 70]. It is important to mention here that the DNA isolation method can be a potential source of global DNA methylation [61]; so, researchers should be careful during experimental design and data interpretation.

18.5.1.2 Next-Generation Sequencing (NGS) NGS is a high-throughput approach to sequence analysis which provides better resolution compared to the traditional Sanger-sequencing process [81]. Since 80% of cytosines present within human CpG islands are methylated, NGS is mostly preferred with its availability of versatile platforms like Roche 454 sequencer [38, 78], Illumina Solexa sequencing [30], and SoLiD™ by Applied Biosystems [6].

18.5.1.3 Enrichment-Based Technologies Enrichment-based technologies include methylated DNA immunoprecipitated sequencing (MeDIP-seq) and methylated DNA-binding chromatin sequencing (MBD-Seq). MeDIP utilizes antibodies that specifically recognize methylated DNA. The immunoprecipitated methylated DNA can be identified by a number of downstream methods including PCR, microarray hybridization array, and next-generation sequencing. In the MBD-sequencing approach, methylated DNA fragments are captured in a process that enriches the mCpG density [7]. MBD-Seq gives an overall view of the methylated genome, and the profile is altered when cells are exposed to certain epimutagens.

18.5.2 For Histone Modification

18.5.2.1 ChIP-Based Technologies, ChIP-chip and Chip-Seq Chromatin immunoprecipitation (ChIP) is a powerful technique for analyzing histone modifications as well as binding sites for proteins that bind either directly or indirectly to DNA. ChIP experiments have facilitated the high-throughput genome-wide mapping of the epigenome. Briefly, chromatin extracts are prepared following fixation with formaldehyde. This bulk chromatin is then sheared into fragments of useful size. Aliquots are then incubated with an antibody specific for the particular type of modification and purified DNA is analyzed by genome-wide hybridization to a microarray (ChIP-chip) or by sequencing (ChIP-Seq) via an NGS platform. For an example, NimbleGen promoter array covers 19,222 promoters in the human genome that overlap with 17,289 genes on the Illumina Human-6 v2 BeadChip expression array. Comparative ChIP-based epigenetic mapping analysis can reveal the significant epigenetic modifications in target cells and correlation of these changes with key physiological alterations might explain the underlying mechanism of disease manifestation.

18.5.2.2 DNA Adenosine Methylation Identification (DamID) DamID has emerged as a powerful tool to decipher transcriptional networks, to study chromatin-associated proteins, and to monitor higher-order chromatin organization on a genome-wide scale. This method enables us to have the molecular picture of the interacting proteins or substrates onto the histones that might have a causal effect on the epigenetic signature of the genome. A representative strategy for DamID process is shown in Figure 18.2.

As cross-talk among the networks is responsible for setting an epigenetic signature for a particular disease; genome-wide epigenetic profiling is much preferred over any particular modification. Moreover, high-throughput technologies like NGS and MeDip are now available at more affordable cost, which provide quick identification of epigenetic changes, that is, DNA methylation, histone modification, and protein–chromatin interactions. There also are some other simple techniques available to identify histone modifications, for example, nucleoenzymatic assays, which identify and characterize histone modifying activities like methylation, acetylation, and phosphorylation [25], and pull-down assays that allow for the identification of novel histone effector proteins and utilizes biotinylated histone peptides modified at various residues [73].

18.6 FUTURE PERSPECTIVES

Long-term exposure to xenobiotics (like arsenic) either through occupational and environmental sources result in multitude of health problems including cancer. All along, epigenetic studies have become turning points in modern science. Identification of epigenetic biomarkers, by comparing the epigenetic changes in tumors versus normal surrogate tissue and/or overall DNA methylation status in exposed versus unexposed individuals, may serve as a useful tool in assessing environmental exposure and cancer risk. Molecular mechanisms in diseases are better explained when epigenomics are considered in totality with genomics, transcriptomics, and proteomics; modern technological advancements help this fusion to grow. Moreover, reversibility of epigenetic events unravels new targets for chemotherapeutic intervention [69]. Reversion of epigenetic marks that silence tumor-suppressor genes will surely lead to the development of alternative chemotherapies and so is the case with, histone deacetylase inhibitors, and engineered zinc finger protein transcription factors [33, 58].

Recently, in 2013, Yan et al. published a report showing that arsenic trioxide upregulates Pirh2-complex leading to enhanced proteosomal degradation of an oncogenic protein ΔNp63 [75]. Liver toxicity and altered ratios of arsenic metabolites MMA and DMA were observed in promyelocytic leukemia patients treated with arsenic trioxide suggesting that high arsenic doses can modify the methylation of arsenic [71]. A later study indicated that although removal of arsenic from drinking water efficiently reduced dermatological lesions, simultaneously nondermatological health effects were drastically worsened [49]. This variability opens a newer outlook on understanding arsenic toxicity and amelioration. Maybe, a better understanding of epigenetic regulation will elaborate the mechanistic significance of altered epigenetic marks and reactivation of silenced genes or suppression of activated genes induced by arsenic exposure can prove promising as an alternative therapeutic tool in arsenic toxicity and carcinogenesis.

REFERENCES

[1] W.M. Baird, L. Diamond, The nature of benzo(*a*)pyrene-DNA adducts formed in hamster embryo cells depends on the length of time of exposure to benzo(*a*)pyrene. Biochem. Biophys. Res. Commun. 77 (1977) 162–167.

[2] F.F. Becker, P. Holton, M. Ruchirawat, J.N. Lapeyre, Perturbation of maintenance and de novo DNA methylation in vitro by UVB (280-340 nm)-induced pyrimidine photodimers. Proc. Natl. Acad. Sci. U. S. A. 82 (1985) 6055–6059.

[3] I. Behm-Ansmant, J. Rehwinkel, E. Izaurralde, MicroRNAs silence gene expression by repressing protein expression and/or by promoting mRNA decay. Cold Spring Harb. Symp. Quant. Biol. 71 (2006) 523–530.

[4] P. Bhattacharjee, M. Banerjee, A.K. Giri, Role of genomic instability in arsenic-induced carcinogenicity. A review. Environ. Int. 53 (2013) 29–40.

[5] F. Boellmann, L. Zhang, H.J. Clewell, G.P. Schroth, E.M. Kenyon, M.E. Andersen, R.S. Thomas, Genome-wide analysis of DNA methylation and gene expression changes in the mouse lung following subchronic arsenate exposure. Toxicol. Sci. 117 (2010) 404–417.

[6] C.A. Bormann Chung, V.L. Boyd, K.J. McKernan, Y. Fu, C. Monighetti, H.E. Peckham, M. Barker, Whole methylome analysis by ultra-deep sequencing using two-base encoding. PLoS One 5 (2010) e9320.

[7] A.B. Brinkman, F. Simmer, K. Ma, A. Kaan, J. Zhu, H.G. Stunnenberg, Whole-genome DNA methylation profiling using MethylCap-seq. Methods 52 (2010) 232–236.

[8] H.W. Brock, J.W. Hodgson, S. Petruk, A. Mazo, Regulatory noncoding RNAs at Hox loci. Biochem. Cell Biol. 87 (2009) 27–34.

[9] J. Cao, Y. Song, N. Bi, J. Shen, W. Liu, J. Fan, G. Sun, T. Tong, J. He, Y. Shi, X. Zhang, N. Lu, Y. He, H. Zhang, K. Ma, X. Luo, L. Lv, H. Deng, J. Cheng, J. Zhu, L. Wang, Q. Zhan, DNA methylation-mediated repression of miR-886-3p predicts poor outcome of human small cell lung cancer. Cancer Res. 73 (2013) 3326–3335.

[10] S. Chanda, U.B. Dasgupta, D. Guhamazumder, M. Gupta, U. Chaudhuri, S. Lahiri, S. Das, N. Ghosh, D. Chatterjee, DNA hypermethylation of promoter of gene p53 and p16 in arsenic-exposed people with and without malignancy. Toxicol. Sci. 89 (2006) 431–437.

[11] H. Chen, J. Liu, B.A. Merrick, M.P. Waalkes, Genetic events associated with arsenic-induced malignant transformation: applications of cDNA microarray technology. Mol. Carcinog. 30 (2001) 79–87.

[12] X. Chen, Y. Ba, L. Ma, X. Cai, Y. Yin, K. Wang, J. Guo, Y. Zhang, J. Chen, X. Guo, Q. Li, X. Li, W. Wang, Y. Zhang, J. Wang, X. Jiang, Y. Xiang, C. Xu, P. Zheng, J. Zhang, R. Li, H. Zhang, X. Shang, T. Gong, G. Ning, J. Wang, K. Zen, J. Zhang, C.Y. Zhang, Characterization of microRNAs in serum: a novel class of biomarkers for diagnosis of cancer and other diseases. Cell Res. 18 (2008) 997–1006.

[13] Y.C. Chen, Y.L. Guo, H.J. Su, Y.M. Hsueh, T.J. Smith, L.M. Ryan, M.S. Lee, S.C. Chao, J.Y. Lee, D.C. Christiani, Arsenic methylation and skin cancer risk in southwestern Taiwan. J. Occup. Environ. Med. 45 (2003) 241–248.

[14] Y.C. Chen, H.J. Su, Y.L. Guo, Y.M. Hsueh, T.J. Smith, L.M. Ryan, M.S. Lee, D.C. Christiani, Arsenic methylation and bladder cancer risk in Taiwan. Cancer Causes Control 14 (2003) 303–310.

[15] S. Cho, J.S. Park, Y.K. Kang, Dual functions of histone-lysine N-methyltransferase Setdb1 protein at promyelocytic leukemia-nuclear body (PML-NB): maintaining PML-NB structure and regulating the expression of its associated genes. J. Biol. Chem. 286 (2011) 41115–41124.

[16] C.Y. Chu, T.M. Rana, Translation repression in human cells by microRNA-induced gene silencing requires RCK/p54. PLoS Biol. 4 (2006) e210.

[17] W.D. Cress, E. Seto, Histone deacetylases, transcriptional control, and cancer. J. Cell. Physiol. 184 (2000) 1–16.

[18] X. Cui, T. Wakai, Y. Shirai, N. Yokoyama, K. Hatakeyama, S. Hirano, Arsenic trioxide inhibits DNA methyltransferase and restores methylation-silenced genes in human liver cancer cells. Hum. Pathol. 37 (2006) 298–311.

[19] N. Das, S. Paul, D. Chatterjee, N. Banerjee, N.S. Majumder, N. Sarma, T.J. Sau, S. Basu, S. Banerjee, P. Majumder, A.K. Bandyopadhyay, J.C. States, A.K. Giri, Arsenic exposure through drinking water increases the risk of liver and cardiovascular diseases in the population of West Bengal, India. BMC Public Health 12 (2012) 639.

[20] T. Davis, R. Vaisvila, High sensitivity 5-hydroxymethylcytosine detection in Balb/C brain tissue. J. Vis. Exp. pii (2011) 2661.

[21] B. Diosdado, M.A. van de Wiel, J.S. Terhaar Sive Droste, S. Mongera, C. Postma, W.J. Meijerink, B. Carvalho, G.A. Meijer, MiR-17-92 cluster is associated with 13q gain and c-myc expression during colorectal adenoma to adenocarcinoma progression. Br. J. Cancer 101 (2009) 707–714.

[22] D. Dominissini, S. Moshitch-Moshkovitz, N. Amariglio, G. Rechavi, Adenosine-to-inosine RNA editing meets cancer. Carcinogenesis 32 (2011) 1569–1577.

[23] A.P. Feinberg, B. Tycko, The history of cancer epigenetics. Nat. Rev. Cancer 4 (2004) 143–153.

[24] E.J. Finnegan, M. Wang, P. Waterhouse, Gene silencing: fleshing out the bones. Curr. Biol. 11 (2001) R99–R102.

[25] W. Fischle, In nucleo enzymatic assays for the identification and characterization of histone modifying activities. Methods 36 (2005) 362–367.

[26] Y. Fu, Z. Yi, X. Wu, J. Li, F. Xu, Circulating microRNAs in patients with active pulmonary tuberculosis. J. Clin. Microbiol. 49 (2011) 4246–4251.

[27] J. Gao, Q.G. Liu, The role of miR-26 in tumors and normal tissues (review). Oncol. Lett. 2 (2011) 1019–1023.

[28] M.A. Glozak, E. Seto, Histone deacetylases and cancer. Oncogene 26 (2007) 5420–5432.

[29] M.G. Goll, F. Kirpekar, K.A. Maggert, J.A. Yoder, C.L. Hsieh, X. Zhang, K.G. Golic, S.E. Jacobsen, T.H. Bestor, Methylation of tRNAAsp by the DNA methyltransferase homolog Dnmt2. Science 311 (2006) 395–398.

[30] H. Gu, C. Bock, T.S. Mikkelsen, N. Jager, Z.D. Smith, E. Tomazou, A. Gnirke, E.S. Lander, A. Meissner, Genome-scale DNA methylation mapping of clinical samples at single-nucleotide resolution. Nat. Methods 7 (2010) 133–136.

[31] G. Huang, H. Guo, T. Wu, Genetic variations of CYP2B6 gene were associated with plasma BPDE-Alb adducts and DNA damage levels in coke oven workers. Toxicol. Lett. 211 (2012) 232–238.

[32] S.K. Huang, A.W. Chiu, Y.S. Pu, Y.K. Huang, C.J. Chung, H.J. Tsai, M.H. Yang, C.J. Chen, Y.M. Hsueh, Arsenic methylation capability, heme oxygenase-1 and NADPH quinone oxidoreductase-1 genetic polymorphisms and the stage and grade of urothelial carcinomas. Urol. Int. 80 (2008) 405–412.

[33] A.C. Jamieson, J.C. Miller, C.O. Pabo, Drug discovery with engineered zinc-finger proteins. Nat. Rev. Drug Discov. 2 (2003) 361–368.

[34] T.J. Jensen, R.J. Wozniak, K.E. Eblin, S.M. Wnek, A.J. Gandolfi, B.W. Futscher, Epigenetic mediated transcriptional activation of WNT5A participates in arsenical-associated malignant transformation. Toxicol. Appl. Pharmacol. 235 (2009) 39–46.

[35] S.G. Jin, C. Guo, G.P. Pfeifer, GADD45A does not promote DNA demethylation. PLoS Genet. 4 (2008) e1000013.

[36] W.J. Jo, X. Ren, F. Chu, M. Aleshin, H. Wintz, A. Burlingame, M.T. Smith, C.D. Vulpe, L. Zhang, Acetylated H4K16 by MYST1 protects UROtsa cells from arsenic toxicity and is decreased following chronic arsenic exposure. Toxicol. Appl. Pharmacol. 241 (2009) 294–302.

[37] L.P. Keegan, A. Leroy, D. Sproul, M.A. O'Connell, Adenosine deaminases acting on RNA (ADARs): RNA-editing enzymes. Genome Biol. 5 (2004) 209.

[38] Y. Korshunova, R.K. Maloney, N. Lakey, R.W. Citek, B. Bacher, A. Budiman, J.M. Ordway, W.R. McCombie, J. Leon, J.A. Jeddeloh, J.D. McPherson, Massively parallel bisulfite pyrosequencing reveals the molecular complexity of breast cancer-associated cytosine-methylation patterns obtained from tissue and serum DNA. Genome Res. 18 (2008) 19–29.

[39] T. Kouzarides, SnapShot: histone-modifying enzymes. Cell 128 (2007) 802.

[40] S. Kriaucionis, N. Heintz, The nuclear DNA base 5-hydroxymethylcytosine is present in Purkinje neurons and the brain. Science 324 (2009) 929–930.

[41] M. Ling, Y. Li, Y. Xu, Y. Pang, L. Shen, R. Jiang, Y. Zhao, X. Yang, J. Zhang, J. Zhou, X. Wang, Q. Liu, Regulation of miRNA-21 by reactive oxygen species-activated ERK/NF-kappaB in arsenite-induced cell transformation. Free Radic. Biol. Med. 52 (2012) 1508–1518.

[42] W.A. Loenen, S-adenosylmethionine: jack of all trades and master of everything? Biochem. Soc. Trans. 34 (2006) 330–333.

[43] K. Luger, T.J. Rechsteiner, A.J. Flaus, M.M. Waye, T.J. Richmond, Characterization of nucleosome core particles containing histone proteins made in bacteria. J. Mol. Biol. 272 (1997) 301–311.

[44] A. Maier, B.L. Schumann, X. Chang, G. Talaska, A. Puga, Arsenic co-exposure potentiates benzo[a]pyrene genotoxicity. Mutat. Res. 517 (2002) 101–111.

[45] C. Martin, Y. Zhang, The diverse functions of histone lysine methylation. Nat. Rev. Mol. Cell Biol. 6 (2005) 838–849.

[46] X.Z. Meng, T.S. Zheng, X. Chen, J.B. Wang, W.H. Zhang, S.H. Pan, H.C. Jiang, L.X. Liu, microRNA expression alteration after arsenic trioxide treatment in HepG-2 cells. J. Gastroenterol. Hepatol. 26 (2011) 186–193.

[47] P.S. Mitchell, R.K. Parkin, E.M. Kroh, B.R. Fritz, S.K. Wyman, E.L. Pogosova-Agadjanyan, A. Peterson, J. Noteboom, K.C. O'Briant, A. Allen, D.W. Lin, N. Urban, C.W. Drescher, B.S. Knudsen, D.L. Stirewalt, R. Gentleman, R.L. Vessella, P.S. Nelson, D.B. Martin, M. Tewari, Circulating microRNAs as stable blood-based markers for cancer detection. Proc. Natl. Acad. Sci. U. S. A. 105 (2008) 10513–10518.

[48] C. Niehrs, A. Schafer, Active DNA demethylation by Gadd45 and DNA repair. Trends Cell Biol. 22 (2012) 220–227.

[49] S. Paul, N. Das, P. Bhattacharjee, M. Banerjee, J.K. Das, N. Sarma, A. Sarkar, A.K. Bandyopadhyay, T.J. Sau, S. Basu, S. Banerjee, P. Majumder, A.K. Giri, Arsenic-induced toxicity and carcinogenicity: a two-wave cross-sectional study in arsenicosis individuals in West Bengal, India. J. Expo. Sci. Environ. Epidemiol. 23 (2013) 156–162.

[50] N. Paz, E.Y. Levanon, N. Amariglio, A.B. Heimberger, Z. Ram, S. Constantini, Z.S. Barbash, K. Adamsky, M. Safran, A. Hirschberg, M. Krupsky, I. Ben-Dov, S. Cazacu, T. Mikkelsen, C. Brodie, E. Eisenberg, G. Rechavi, Altered adenosine-to-inosine RNA editing in human cancer. Genome Res. 17 (2007) 1586–1595.

[51] J.C. Pelling, T.J. Slaga, J. DiGiovanni, Formation and persistence of DNA, RNA, and protein adducts in mouse skin exposed to pure optical enantiomers of 7 beta,8 alpha-dihydroxy-9 alpha,10 alpha-epoxy-7,8,9,10-tetrahydrobenzo(a)pyre ne in vivo. Cancer Res. 44 (1984) 1081–1086.

[52] G. Rabinowits, C. Gercel-Taylor, J.M. Day, D.D. Taylor, G.H. Kloecker, Exosomal microRNA: a diagnostic marker for lung cancer. Clin. Lung Cancer 10 (2009) 42–46.

[53] J.F. Reichard, M. Schnekenburger, A. Puga, Long term low-dose arsenic exposure induces loss of DNA methylation. Biochem. Biophys. Res. Commun. 352 (2007) 188–192.

[54] M. Schnekenburger, G. Talaska, A. Puga, Chromium cross-links histone deacetylase 1-DNA methyltransferase 1 complexes to chromatin, inhibiting histone-remodeling marks critical for transcriptional activation. Mol. Cell. Biol. 27 (2007) 7089–7101.

[55] P.L. Severson, E.J. Tokar, L. Vrba, M.P. Waalkes, B.W. Futscher, Agglomerates of aberrant DNA methylation are associated with toxicant-induced malignant transformation 2. Epigenetics 7 (2012) 1238–1248.

[56] P.L. Severson, E.J. Tokar, L. Vrba, M.P. Waalkes, B.W. Futscher, Coordinate H3K9 and DNA methylation silencing of ZNFs in toxicant-induced malignant transformation 1. Epigenetics 8 (2013) 1080–1088.

[57] L. Smeester, J.E. Rager, K.A. Bailey, X. Guan, N. Smith, G. Garcia-Vargas, L.M. Del Razo, Z. Drobna, H. Kelkar, M. Styblo, R.C. Fry, Epigenetic changes in individuals with arsenicosis. Chem. Res. Toxicol. 24 (2011) 165–167.

[58] A.W. Snowden, L. Zhang, F. Urnov, C. Dent, Y. Jouvenot, X. Zhong, E.J. Rebar, A.C. Jamieson, H.S. Zhang, S. Tan, C.C. Case, C.O. Pabo, A.P. Wolffe, P.D. Gregory, Repression of vascular endothelial growth factor A in glioblastoma cells using engineered zinc finger transcription factors. Cancer Res. 63 (2003) 8968–8976.

[59] C.X. Song, C. He, Bioorthogonal labeling of 5-hydroxymethylcytosine in genomic DNA and diazirine-based DNA photo-cross-linking probes. Acc. Chem. Res. 44 (2011) 709–717.

[60] C.X. Song, K.E. Szulwach, Y. Fu, Q. Dai, C. Yi, X. Li, Y. Li, C.H. Chen, W. Zhang, X. Jian, J. Wang, L. Zhang, T.J. Looney, B. Zhang, L.A. Godley, L.M. Hicks, B.T. Lahn, P. Jin, C. He, Selective chemical labeling reveals the genome-wide distribution of 5-hydroxymethylcytosine. Nat. Biotechnol. 29 (2011) 68–72.

[61] C. Soriano-Tarraga, J. Jimenez-Conde, E. Giralt-Steinhauer, A. Ois, A. Rodriguez-Campello, E. Cuadrado-Godia, I. Fernandez-Cadenas, J. Montaner, G. Lucas, R. Elosua, J. Roquer, DNA isolation method is a source of global DNA methylation variability measured with LUMA. Experimental analysis and a systematic review. PLoS One 8 (2013) e60750.

[62] D.E. Sterner, S.L. Berger, Acetylation of histones and transcription-related factors. Microbiol. Mol. Biol. Rev. 64 (2000) 435–459.

[63] M. Szymanski, J. Barciszewski, Noncoding RNAs in biology and disease. eLS. 2009. http://onlinelibrary.wiley.com/doi/10.1002/9780470015902.a0021437/full. Accessed April 14, 2015.

[64] M. Tahiliani, K.P. Koh, Y. Shen, W.A. Pastor, H. Bandukwala, Y. Brudno, S. Agarwal, L.M. Iyer, D.R. Liu, L. Aravind, A. Rao, Conversion of 5-methylcytosine to 5-hydroxymethylcytosine in mammalian DNA by MLL partner TET1. Science 324 (2009) 930–935.

[65] P.M. Thompson, T. Gotoh, M. Kok, P.S. White, G.M. Brodeur, CHD5, a new member of the chromodomain gene family, is preferentially expressed in the nervous system. Oncogene 22 (2003) 1002–1011.

[66] J. Trasler, L. Deng, S. Melnyk, I. Pogribny, F. Hiou-Tim, S. Sibani, C. Oakes, E. Li, S.J. James, R. Rozen, Impact of Dnmt1 deficiency, with and without low folate diets, on tumor numbers and DNA methylation in Min mice. Carcinogenesis 24 (2003) 39–45.

[67] C.H. Tseng, Y.K. Huang, Y.L. Huang, C.J. Chung, M.H. Yang, C.J. Chen, Y.M. Hsueh, Arsenic exposure, urinary arsenic speciation, and peripheral vascular disease in blackfoot disease-hyperendemic villages in Taiwan. Toxicol. Appl. Pharmacol. 206 (2005) 299–308.

[68] M. Vairapandi, Characterization of DNA demethylation in normal and cancerous cell lines and the regulatory role of cell cycle proteins in human DNA demethylase activity. J. Cell. Biochem. 91 (2004) 572–583.

[69] A. Villar-Garea, M. Esteller, DNA demethylating agents and chromatin-remodelling drugs: which, how and why? Curr. Drug Metab. 4 (2003) 11–31.

[70] H. Wang, S. Guan, A. Quimby, D. Cohen-Karni, S. Pradhan, G. Wilson, R.J. Roberts, Z. Zhu, Y. Zheng, Comparative characterization of the PvuRts1I family of restriction enzymes and their application in mapping genomic 5-hydroxymethylcytosine. Nucleic Acids Res. 39 (2011) 9294–9305.

[71] H. Wang, S. Xi, Z. Liu, Y. Yang, Q. Zheng, F. Wang, Y. Xu, Y. Wang, Y. Zheng, G. Sun, Arsenic methylation metabolism and liver injury of acute promyelocytic leukemia patients undergoing arsenic trioxide treatment. Environ. Toxicol. 28 (2013) 267–275.

[72] S.M. Wnek, T.J. Jensen, P.L. Severson, B.W. Futscher, A.J. Gandolfi, Monomethylarsonous acid produces irreversible events resulting in malignant transformation of a human bladder cell line following 12 weeks of low-level exposure 3. Toxicol. Sci. 116 (2010) 44–57.

[73] J. Wysocka, Identifying novel proteins recognizing histone modifications using peptide pull-down assay. Methods 40 (2006) 339–343.

[74] Y. Xie, J. Liu, L. Benbrahim-Tallaa, J.M. Ward, D. Logsdon, B.A. Diwan, M.P. Waalkes, Aberrant DNA methylation and gene expression in livers of newborn mice transplacentally exposed to a hepatocarcinogenic dose of inorganic arsenic. Toxicology 236 (2007) 7–15.

[75] W. Yan, X. Chen, Y. Zhang, J. Zhang, Y.S. Jung, X. Chen, Arsenic suppresses cell survival via Pirh2-mediated proteasomal degradation of DeltaNp63 protein. J. Biol. Chem. 288 (2013) 2907–2913.

[76] Y. Yoshikawa, Y.S. Velichko, Y. Ichiba, K. Yoshikawa, Self-assembled pearling structure of long duplex DNA with histone H1. Eur. J. Biochem. 268 (2001) 2593–2599.

[77] R.C. Yu, K.H. Hsu, C.J. Chen, J.R. Froines, Arsenic methylation capacity and skin cancer. Cancer Epidemiol. Biomarkers Prev. 9 (2000) 1259–1262.

[78] M. Zeschnigk, M. Martin, G. Betzl, A. Kalbe, C. Sirsch, K. Buiting, S. Gross, E. Fritzilas, B. Frey, S. Rahmann, B. Horsthemke, Massive parallel bisulfite sequencing of CG-rich DNA fragments reveals that methylation of many X-chromosomal CpG islands in female blood DNA is incomplete. Hum. Mol. Genet. 18 (2009) 1439–1448.

[79] L. Zhang, L. Ding, T.H. Cheung, M.Q. Dong, J. Chen, A.K. Sewell, X. Liu, J.R. Yates, III, M. Han, Systematic identification of C. elegans miRISC proteins, miRNAs, and mRNA targets by their interactions with GW182 proteins AIN-1 and AIN-2. Mol. Cell 28 (2007) 598–613.

[80] W.Y. Zhang, F.Q. Xu, C.L. Shan, R. Xiang, L.H. Ye, X.D. Zhang, Gene expression profiles of human liver cells mediated by hepatitis B virus X protein. Acta Pharmacol. Sin. 30 (2009) 424–434.

[81] Y. Zhang, A. Jeltsch, The application of next generation sequencing in DNA methylation analysis. Genes (Basel) 1 (2010) 85–101.

[82] C.Q. Zhao, M.R. Young, B.A. Diwan, T.P. Coogan, M.P. Waalkes, Association of arsenic-induced malignant transformation with DNA hypomethylation and aberrant gene expression. Proc. Natl. Acad. Sci. U. S. A. 94 (1997) 10907–10912.

PART IV

MODELS FOR ARSENIC TOXICOLOGY AND RISK ASSESSMENT

19

CANCER INDUCED BY EXPOSURE TO ARSENICALS IN ANIMALS

Erik J. Tokar, Yuanyuan Xu and Michael P. Waalkes
Inorganic Toxicology Group, National Toxicology Program Laboratory, National Institute of Environmental Health Sciences, Research Triangle Park, NC, USA

19.1 INTRODUCTION

Evidence that arsenicals were carcinogenic to humans was observed as early as the late 1800s as secondary skin cancers in people given medicinal arsenicals [8]. Indeed, the International Agency for Research on Cancer (IARC) linked arsenic exposure to human cancer in one of its first published monographs [9] and considered there to be *compelling evidence* consistently linking inorganic arsenic exposure to human skin cancer. Arsenic and arsenic compounds have now been the subject of another two monographs [10, 12], appeared in the 1987 updating summaries [11] and were covered in the monograph that was a review of the known human carcinogens [13]. Overall, IARC now considers there to be *sufficient evidence* for arsenic and inorganic arsenic compounds as human lung, skin, and urinary bladder carcinogens, and some evidence for activity as kidney, liver, and prostate carcinogens [13]. However, the story is quite different in rodents. In the earliest evaluation, there was a lack of solid evidence for arsenic as an animal carcinogen [9]. Even in the 1980 assessment, the evidence for carcinogenicity of arsenic compounds in animals was considered inadequate [10]. Similarly, in the 1987 review [11], there was *limited evidence* for inorganic arsenicals and no adequate data available on organic arsenicals in animals. The subsequent 2004 monograph, which reviewed arsenic as a drinking water contaminant, considered various additional studies to provide *sufficient evidence* for rodent carcinogenicity of dimethylarsinic

Arsenic: Exposure Sources, Health Risks, and Mechanisms of Toxicity, First Edition.
Edited by J. Christopher States.
© 2016 John Wiley & Sons, Inc. Published 2016 by John Wiley & Sons, Inc.

acid (DMA^V), a biomethylation product of inorganic arsenic, as this organoarsenical produced urinary bladder tumors in rats and lung tumors in mice [12]. However, even in 2004, the available studies on inorganic arsenic in animals, as a whole, were still considered to provide *limited evidence* of carcinogenicity in animals [12].

IARC recently updated all the known human inorganic carcinogens, including arsenic and arsenic compounds, which is now available as a monograph [13]. For arsenicals, this evaluation contained various additional positive animal carcinogenesis studies published after the 2004 evaluation [12]. These additional studies with arsenic in animals include human-relevant routes such as oral exposure in drinking water. Various attempts were made to expose animals during potentially high sensitivity life stages, such as with *in utero* exposure studies. Some studies have appeared after the 2012 IARC evaluation was compiled. Generally, inorganic salts of arsenite or arsenate and DMA^V have been the most used arsenicals in rodent carcinogenesis testing [12, 13], although other methylated arsenicals and inorganics have received some attention.

Arsenicals have been used in many studies before, after, or during application of known carcinogenic agents to increase end-point tumor response (see [13, 26] for a review). The exact meaning of these studies, specifically to the carcinogenicity of arsenicals as a single agent, is incompletely defined and complex. Arsenicals are also effective cancer chemotherapeutics, at least against certain tumors. So the timing of "initiation/promotion" may become very complicated. Because of such issues, this chapter focuses on the data where arsenicals, when given alone, cause cancer. For studies where arsenicals are given with other carcinogens during a rodent's lifetime to produce or enhance tumor outcome, the interested reader is referred to other reviews [13, 26].

Various studies with serious issues are not considered in this chapter. Some studies were not included because of serious design flaws. This exclusion criterion includes any rodent study that did not have tumors defined by histopathological analysis. Similarly, studies without appropriate control groups also were not included. Also not included were those studies that used complex mixtures, such as "Bordeaux mixture," an arsenical-based fungicide that contains not only calcium arsenate but also high levels of copper and did not have appropriate control groups (i.e., fungicide minus arsenic [12]). Work that provided insufficient information to interpret relevant data (e.g., lack of descriptive statistics, lack of specific group sizes, and statistical analysis technique undefined) were considered impossible to independently reevaluate and were, therefore, not included here. If a study only showed a significant tumor increase by combining tumors from different organs in the absence of a discrete target site, it was not included. A strict $p<0.05$ limit was used for statistical significance. These parameters are generally consistent with the IARC review process [12, 13]. Indeed, overall, the interpretations of the IARC 2012 Monograph on arsenic and arsenic compounds [13] were considered the most relevant resource on these sorts of studies.

19.2 ARSENICAL CARCINOGENESIS AFTER ORAL EXPOSURE

Drinking water is the most common route of inorganic arsenic exposure in humans [12, 13], making carcinogenic potential after oral exposure perhaps most the important route to investigate in rodent models. There have been several studies that have investigated chronic oral exposure to arsenicals in rodents. These studies include oral exposure to inorganic arsenic or DMA^V in mice or rats.

In naturally cancer-sensitive or genetically altered mice, three separate studies provide evidence for a carcinogenic potential of arsenicals [6, 7, 14]. Oral sodium arsenate added to the drinking water for 18 months increases lung tumor multiplicity and size in male A/J mice [6], which have a predilection to chemically induced lung cancer. Likewise, chronic (≥50 weeks) exposure to the methylated arsenical metabolite, DMA^V, in the drinking water increases lung cancer incidence and multiplicity in A/J mice [7]. Similarly, chronic oral DMA^V increases lung tumors in mutant *Ogg−/−* mice, a strain unable to repair certain types of oxidative DNA damage, but not in the repair-sufficient *Ogg+/+* mice [14], indicating oxidative DNA damage may lead to lung tumor formation with DMA^V, although direct evidence of this was never shown [14]. Arsenic and inorganic arsenic compounds are considered human lung carcinogens by oral exposure [12, 13], making the findings of oral carcinogenic activity relevant even if in "hypersensitive" animals [6, 7, 14]. Although the methylated arsenical DMA^V might be rare in drinking water, as a biomethylation product of inorganic arsenic, it could be found in the circulation or perhaps locally produced [12, 13]. However, available data indicate that rodent lung cells do not methylate inorganic arsenic [24], which would make local production seem unlikely.

In male rats, oral DMA^V exposure in the drinking water for up to 2 years produced a dose-related induction of urinary bladder carcinoma and combined papilloma or carcinoma (first reported as a short communication in [40]; full report in [39]). DMA^V exposure induced transitional cell carcinoma. Thus, this rodent model [39, 40] produced tumors concordant in site (urinary bladder) and type (transitional cell carcinoma) with a major tumor found in humans exposed to arsenic [12, 13]. These are key data that show the urinary bladder as a target site of arsenical alone in rodents and duplicate a specific, and relatively uncommon, tumor found in humans [39, 40], greatly supporting their impact.

Later work showed that DMA^V added to rat feed induces dose-related increases in urinary bladder tumors over 2 years in female rats but not in males [2]. The malignant urinary bladder tumors were again transitional cell carcinoma, concordant with human bladder tumors and arsenic. Although tumors primarily occurred in female rats, preneoplasia (urothelial cell hyperplasia) increased with dose in the arsenical-treated males [2]. The reason for this gender-based sensitivity is unknown. Tumors were not induced by DMA^V at other sites in rats [2]. Similar groups of male and female mice were treated with DMA^V but no urinary bladder tumors or tumors at other sites were induced [2]. Shen et al. [22] found oral exposure to the organoarsenical biomethylation product, trimethylarsine, over 2 years in the drinking water induces liver adenoma in rats. Oral monomethylarsonic acid (MMA^V) given to rats for up to 2 years did not induce tumors in a dose–response study that included male and female rats and mice [3]. Some researchers think there are issues with the rat in general as a model for arsenic toxicity and carcinogenesis studies, as this species has dramatically different arsenic biokinetics from the human or mouse due to high-affinity arsenical blood-binding proteins [16]. This should not be taken to detract from the remarkable rodent/human tissue concordance with the DMA^V-induced bladder cancers in rats [2, 39].

Overall, oral exposure to inorganic arsenate induced lung tumors in mice [6], while oral exposure to DMA^{V} induced lung tumors in mice in two studies [7, 14] and urinary bladder tumors in rats in two studies [2, 39]. Oral exposure to trimethylarsine induced benign rat liver tumors [22]. The positive oral exposure studies with arsenicals in mice were in strains known to be susceptible to chemical carcinogenesis (e.g., A/J mice). The reason why the adult rat is sensitive to methylated arsenicals or arsenical-induced bladder cancers is not clear, although the high-affinity arsenical blood-binding protein may play a role [16].

19.3 INHALATION EXPOSURE

There are no available inhalation studies for the inorganic arsenicals in rodents [12, 13, 26]. The National Toxicology Program provided clear evidence of gallium arsenide carcinogenicity after inhalation in rodents [17]. In female rats, inhalation of particulate gallium arsenide for about 2 years caused dose-related increases in lung tumors and adrenal medulla pheochromocytomas. In male rats, lung tumors were not seen but dose-related increases in atypical hyperplasia of alveolar epithelium occurred with the inhalation of gallium arsenide. In the female rats, increases also occurred in leukemia at the highest gallium arsenide dose. In male and female mice, inhalation of several doses of gallium arsenide particulate for 2 years did not cause tumors, but induced dose-related increases in lung-epithelial alveolar hyperplasia.

19.4 INTRATRACHEAL INSTILLATIONS

Early on, it was appreciated that inhaled arsenic was carcinogenic in the human lung [9]. Intratracheal instillations of arsenicals in rodents were used in an attempt to duplicate inhalation exposure. Repeated weekly intratracheal instillations (15 times) of calcium arsenate, at a level that caused 15% early mortality, induced lung adenomas in hamsters when assessed over their lifespan [19]. Similarly, hamsters given multiple weekly intratracheal instillations of calcium arsenate at the beginning of the experiment developed increased lung adenoma or combined lung adenoma and carcinoma over their lifetime [41]. Intratracheal instillations of compounds are used to try and mimic inhalation, but, in general, the utility of this type of dosing as a model for inhalation carcinogenesis has several important drawbacks, including the bolus nature of the exposure, and the chance for "overload" of the normal mechanisms for dealing with respired particulates.

19.5 EARLY LIFE ARSENICAL EXPOSURES AND ADULTHOOD CANCER

The early life period in rodents can, in general, be a time of high sensitivity to carcinogens. Exposure in early life has been used to study the carcinogenic potential of arsenicals. In one early study, pregnant mice received a single dose of arsenic trioxide subcutaneously (s.c.) on gestation days 14, 15, 16, or 17, and after birth, the offspring received arsenic trioxide s.c. on *postpartum* days 1, 2, and 3 [20, 21] (same study re-reported). Adult offspring from mothers treated on gestation day 15 and treated postpartum developed more lung tumors (63%) when compared to controls (18%) while the mice exposed *in utero* on gestation days 14, 16, or 17 and treated postpartum did not develop lung tumors in excess of the spontaneous rate. Key parts of the experimental design of this study are not reported, like the number of pregnant females used per group, and if male, female, or pooled offspring were used [20, 21]. Nevertheless, this study indicates that rodents may be sensitive to arsenic-induced lung cancer in early life after perinatal exposure [20, 21].

Several *in utero* exposure studies have been performed using sodium arsenite in mice [26, 34–37]. Many used the inorganic arsenical for transplacental exposure followed by some nonarsenical given after birth to the offspring potentially to promote tumors initiated *in utero*. One study used *in utero* sodium arsenite exposure followed by oral DMA^{V} in the drinking water of the offspring as adults [29] and is discussed here because the exposure is entirely arsenic-based compounds. Another study used methylarsonous acid (MMA^{III}) exposure transplacentally [27]. There are several studies where inorganic arsenic is given to the pregnant mother followed by another agent given to the offspring some time after birth to stimulate carcinogenesis at a particular target site with human relevance [26]. For example, in Tg.AC mice, the combination of *in utero* inorganic arsenic followed by phorbol ester exposure on the skin of the offspring mice produces many more skin squamous cell carcinoma than either treatment alone [33]. The skin is a target of arsenic in humans and squamous cell carcinoma is a common tumor associated with arsenic exposure [12, 13]. Likewise, *in utero* inorganic arsenic followed by diethylstilbestrol after birth induces urinary bladder tumors in mice [34, 35], again duplicating a human target site of arsenic [12, 13]. For a full discussion of use of arsenic in initiation/promotion studies, including transplacental initiation and postnatal promotion, the reader is directed to pertinent reviews [13, 26].

In an early study, pregnant C3H mice were exposed to multiple doses of sodium arsenite (0, 42.5, and 85 ppm) in the drinking water from gestation days 8 to 18, allowed to give birth, then after weaning, tumor formation was assessed in groups of male or female offspring that had been formed based on maternal dose [37]. The levels of maternal arsenic exposure used in this and other transplacental studies [34–37] did not adversely impact dam weights, drinking water consumption, litter size, newborn weights, or weanling weights. During the next 90 weeks postpartum, female offspring exposed to arsenic *in utero* developed dose-related increases in lung adenocarcinoma, benign ovarian tumors, and combined benign or malignant ovarian tumors. Female offspring also developed arsenic dose-related uterine and oviduct preneoplasias after fetal arsenic exposure. After *in utero* arsenic exposure, male offspring developed dose-related increases in liver adenoma, hepatocellular carcinoma, combined liver adenoma or carcinoma, and adrenal cortical adenoma [37].

A following study looked at the carcinogenic effects in C3H mice of various exposures to sodium arsenite (0, 42.5, and 85 ppm) in the drinking water of pregnant dams from gestation days 8 to 18, and subsequent exposure to the tumor-promoting phorbol ester, TPA, applied to the skin in order to potentially promote skin cancers initiated by prenatal arsenic exposure [36]. Over the 2 years after birth, arsenic exposure was not associated with skin tumors regardless of additional treatments. However, arsenic-exposed female offspring developed ovarian tumors and dose-related uterine and oviduct preneoplasia in excess. Arsenic-exposed male offspring showed dose-related increases in liver adenoma, hepatocellular carcinoma, combined adenoma or carcinoma, and adrenal cortical adenoma [36].

By following the *in utero* exposure model first reported by Waalkes et al. [36, 37], a similar recent study examined the effects of gestational exposure to sodium arsenite (0 or 85 ppm) on hepatic tumor formation in C3H mice after reaching 74–84 weeks of age [18]. Largely confirming results from earlier studies [36, 37], Nohara et al. [18] found that gestational exposure to sodium arsenite led to an increase in hepatic tumor incidence, multiplicity, and size in male mice. Furthermore, *in utero* arsenic-induced hepatic tumors often harbored a mutation in the Ha-*ras* oncogene [18], which is consistent with the increase in this oncogene seen in skin tumors induced by gestational arsenic exposure plus later-life phorbal ester exposure in Tg.AC mice [33] and in cell lines transformed by chronic arsenic exposure [4]. Together, these studies suggest that arsenic potentially targets the Ras pathway as a key mechanism during carcinogenesis.

In a study using CD1 mice, pregnant animals received sodium arsenite (0 or 85 ppm) in the drinking water from gestation days 8 to 18, were allowed to give birth, and either female [35] or male [34] offspring were treated with estrogenic compounds after birth to potentially stimulate some tumors sites that could be estrogen sensitive (ovary, uterus, liver, etc.). Female CD1 offspring receiving arsenic alone developed ovarian tumors, combined uterine adenoma or carcinoma, and adrenal cortical adenoma over 90 weeks after birth. Arsenic also increased oviduct hyperplasia as with C3H mice. Male CD1 offspring developed liver adenoma, hepatocellular carcinoma, combined liver adenoma or carcinoma, lung adenocarcinoma, and adrenal cortical adenoma [34]. In arsenic-exposed males, prenatal arsenic also induced renal cystic tubular hyperplasia, considered as a preneoplastic lesion.

Pregnant Tg.AC mice were given drinking water with 0, 42.5, or 85 ppm arsenite from gestation days 8 to 18 and male and female offspring were assessed for tumor burden 36 weeks after birth [31]. *In utero* arsenic exposure increased adrenal cortical adenomas compared with control in all male and female groups. Thus, arsenic *in utero* rapidly induces adrenal tumors in Tg.AC mice.

Various studies had suggested that MMA^{III}, a methylation product of inorganic arsenic, may be a key carcinogenic species (e.g., [25, 38]), although actual tumor endpoint data were absent. Thus, pregnant CD1 mice were given drinking water containing MMA^{III} at 0, 12.5, or 25 ppm from gestation days 8 to 18 and tumors were assessed in male or female offspring over 2 years [27]. Female offspring had MMA^{III}-related increases in epithelial uterine tumors, adrenal cortical adenoma, and total epithelial ovarian tumors. Males had MMA^{III} dose-related increases in hepatocellular carcinoma, adrenal adenoma, and lung adenocarcinoma. Therefore, some similar tumors were produced in the offspring by maternal MMA^{III} exposure and maternal inorganic arsenic exposure during pregnancy [27].

Because human inorganic arsenic exposure occurs during the entire life span, the effects of "whole-life" exposure to inorganic arsenic on carcinogenesis in mice have been studied [29]. In this study, CD1 mice were exposed to 0, 6, 12, or 24 ppm sodium arsenite in the drinking water for 2 weeks before breeding (males and female), during pregnancy and lactation, and after weaning (males and females) through adulthood. Tumors were assessed in groups of male and female offspring until 2 years of age. Exposure-related increases occurred with arsenic in lung adenocarcinoma (both sexes), hepatocellular carcinoma (both sexes), gallbladder tumors (males), uterine carcinomas (females), and ovarian tumors (including carcinomas). Adrenal tumors increased at all exposures (both sexes). Arsenic target sites were quite similar to prior *in utero* exposure only studies. This whole-life arsenic exposure caused more aggressive and frequent tumors at lower exposures and at remarkably similar sites compared to *in utero* only arsenic exposure. It seems possible that the *in utero* arsenic exposure period dictates target site while the exposure during other parts of "whole life" enhanced tumor development [30].

The biomethylation product of arsenic, DMA^V, is a multisite tumor promoter of various nonarsenical carcinogens [13, 26] and alone induces urogenital tumors in rodents [2, 39] but has not been studied combined with inorganic arsenic. To assess the possibility that this arsenical metabolism product would enhance *in utero* stimulated inorganic arsenic carcinogenesis, pregnant CD1 mice were provided drinking water (0 or 85 ppm inorganic arsenic) from gestation days 8 to 18 [29]. After weaning, male offspring received DMA^V (0 or 200 ppm; drinking water) for up to 2 years [29]. While renal tumors did not occur in unexposed controls or in mice treated with DMA alone, gestational arsenic exposure plus later DMA did induce renal tumors (primarily renal cell carcinoma) in the offspring. *In utero* arsenic alone increased hepatocellular carcinomas by threefold over spontaneous rate, and *in utero* arsenic plus post-natal DMA increased this rate further, while DMA alone had no effect. Arsenic alone, DMA alone, and arsenic plus DMA increased lung adenocarcinoma and adrenal adenoma compared to control rates. Overall, DMA during adulthood enhanced lesions initiated by *in utero* arsenic in the liver, and DMA was carcinogenic alone in the lung and adrenal [29].

19.6 OVERVIEW AND DISCUSSION

Evidence that arsenicals are human carcinogens dates back to over 120 years [8]. However, for a very long time, there was great difficulty in producing cancers with arsenic in animals [5, 9, 10]. With relatively recent efforts, multiple studies using various arsenicals and multiple routes of exposure now show arsenicals can act alone as carcinogens in rodents, producing tumors at various sites. Some of the target sites that arsenic produces tumors in rodents are clearly concordant with human target sites [13]. Target sites in humans for which there is considered to be sufficient evidence include the lung, skin, and urinary bladder [12, 13]. Rodent studies have now repeatedly produced lung cancer (oral, *in utero*, whole life, intratracheal, etc.) and urinary bladder cancer (oral) with arsenicals alone. This includes inorganic arsenic and certain methylated arsenicals [13]. Available studies on arsenicals in skin cancer indicate that a second nonarsenical agent is necessary for cancer development in rodents (for a review, see [13, 29]). Human target sites, for which there is considered some evidence of activity for arsenicals (kidney, liver, and prostate [13]), can also be largely duplicated with arsenicals alone in rodents as models for cancers of the liver (oral, *in utero*, whole life, etc.) and kidney (prenatal inorganic arsenic followed by postnatal DMA^{V}) have emerged. No models for arsenic-induced prostate cancer have been developed in rodents. Nonetheless, the finding of lung, urinary bladder, liver, and kidney is a remarkable target-site concordance between humans and rodents, particularly for a human carcinogen that was not considered carcinogenic in rodents just a few years ago [5].

Arsenic remains an important environmental human carcinogen. Millions of people worldwide are exposed to arsenicals, primarily from the drinking water [12], at levels that are very likely to lead to adverse effects. Mechanisms can be difficult to define in humans for a variety of reasons. Rodent carcinogenesis models can provide key insight into the potential mechanisms of an agent. There now should be no question as to the carcinogenic effects of arsenicals in rodents, as the evidence in its totality is strong and compelling with clear target-site concordance with most of the human primary (lung and urinary bladder) and potential (liver and kidney) human targets of arsenic [13]. Not only this, but arsenicals act as complete carcinogens in numerous studies [12, 13, 26]. Furthermore, although not covered in detail in this chapter, arsenicals can act together with other agents to exacerbate carcinogenic outcome in various sites in rodents [13, 26], including the skin. These studies all indicate that arsenicals impact various tumor sites in rodents that have human relevance. Thus, there is little evidence to support a perpetuation of the concept that arsenicals do not cause cancer in rodents.

Based on the number of positive tumor endpoint studies in mice that include a perinatal arsenical exposure component [20, 21, 27–30, 34–37], early life in rodents appears to be a period of high sensitivity to arsenic carcinogenesis. Data are now emerging which indicate that humans show similar sensitivity to arsenical exposure during early life [15, 23, 42, 43]. Several human studies now show that *in utero* or early life inorganic arsenic exposure via drinking water is linked to cancer in adulthood, including cancers of the lung [23], liver [15], and kidney [43]. A tragic human inorganic arsenic poisoning event due to arsenic-contaminated baby food during infancy, resulting in brief but high early life arsenical exposures in the 1950s, has now been associated with several cancers in adulthood, including cancer of the liver [42]. It is noteworthy that multiple mouse studies with an *in utero* arsenical exposure component are also associated with lung and liver cancers [27–30, 34–37] and one study associated *in utero* arsenical exposure with renal cancer [29]. These results show a remarkable tumor concordance between human and rodents when a prenatal exposure component is involved. This remarkable concordance must be considered in light of the fact that it was even questioned if inorganic arsenic was carcinogenic in rodents just a few years ago [5]. In this regard, it would seem prudent for all environmental agents that might be of concern for carcinogenic effects to be tested for transplacental/early life carcinogenic potential. Unfortunately, exposure to potential carcinogens during this life period, which would occur with environmental carcinogens, was long simply ignored in most testing, as was pointed out decades ago [32]. Early life is definitely a time of high sensitivity to chemical carcinogenesis for a variety of reasons [1, 31], so testing environmental agents as if they were occupational carcinogens only in adult rodents because this is a "common" carcinogen bioassay model may miss a key life stage that contributes significantly to lifelong accumulated cancer risk from environmental exposures.

Relatively recent work has greatly added to our understanding of arsenicals as rodent carcinogens. However, much additional work should be performed. For instance, there is only one *in vivo* tumor endpoint study [29] on the carcinogenic effects of the trivalent methylated arsenicals, such as MMA^{III}. Recent *in vitro* data have led to the hypothesis that the trivalent arsenicals, particularly MMA^{III}, which are produced by metabolism of inorganic arsenicals, could be the ultimate carcinogenic species of arsenic because they have high reactivity and toxicity compared with inorganic forms (e.g., [38]). This hypothesis should be tested further in whole animal studies that assess the carcinogenic potential of MMA^{III} as directly compared with inorganic arsenic. Furthermore, humans are exposed to arsenic at all stages of their lifetime, and not just adulthood or *in utero*. Therefore, more studies with various arsenicals, including inorganics and methylated metabolites, should be performed where animals are exposed through their entire life to more accurately mimic environmental arsenic exposure in humans. Such rodent studies are the important next steps in defining mechanisms and potential hazards of arsenic carcinogenesis.

ACKNOWLEDGMENTS

This article may be the work product of an employee or a group of employees of the NIEHS, NIH; however, the statements contained herein do not necessarily represent the statements, opinions, or conclusions of the NIEHS, NIH, or the US government. The content of this publication does not necessarily reflect the views or the policies of the Department of Health and Human Services, nor does mention of trade names, commercial products, or organizations imply endorsement by the US government. The authors report no declarations of interest.

REFERENCES

[1] L.M. Anderson, B.A. Diwan, N.T. Fear, E. Roman, Critical windows of exposure for children's health: cancer in human epidemiological studies and neoplasms in experimental animal models. Environ. Health Perspect. 108 Suppl 3 (2000) 573–594.

[2] L.L. Arnold, M. Eldan, A. Nyska, M. van Gemert, S.M. Cohen, Dimethylarsinic acid: results of chronic toxicity/oncogenicity studies in F344 rats and in B6C3F1 mice. Toxicology 223 (2006) 82–100.

[3] L.L. Arnold, M. Eldan, M. van Gemert, C.C. Capen, S.M. Cohen, Chronic studies evaluating the carcinogenicity of monomethylarsonic acid in rats and mice. Toxicology 190 (2003) 197–219.

[4] L. Benbrahim-Tallaa, R.A. Waterland, M. Styblo, W.E. Achanzar, M.M. Webber, M.P. Waalkes, Molecular events associated with arsenic-induced malignant transformation of human prostatic epithelial cells: aberrant genomic DNA methylation and K-ras oncogene activation. Toxicol. Appl. Pharmacol. 206 (2005) 288–298.

[5] Committee on How Toxicogenomics Could Inform Critical Issues in Carcinogenic Risk Assessment of Environmental Chemicals Toxicogenomic Technologies and Risk Assessment of Environmental Carcinogens: A Workshop Summary, National Academies Press, Washington, DC, 2005.

[6] X. Cui, T. Wakai, Y. Shirai, K. Hatakeyama, S. Hirano, Chronic oral exposure to inorganic arsenate interferes with methylation status of p16INK4a and RASSF1A and induces lung cancer in A/J mice. Toxicol. Sci. 91 (2006) 372–381.

[7] H. Hayashi, M. Kanisawa, K. Yamanaka, T. Ito, N. Udaka, H. Ohji, K. Okudela, S. Okada, H. Kitamura, Dimethylarsinic acid, a main metabolite of inorganic arsenics, has tumorigenicity and progression effects in the pulmonary tumors of A/J mice. Cancer Lett. 125 (1998) 83–88.

[8] J. Hutchinson, On some examples of arsenic-keratosis of the skin and of arsenic-cancer. Trans. Pathol. Soc. London 39 (1988) 352–363.

[9] IARC IARC Monographs on the Evaluation of the Carcinogenic Risk of Chemicals to Humans, Vol. 2, Some Inorganic and Organometallic Compounds, IARC Press, Lyons, 1973.

[10] IARC IARC Monographs on the Evaluation of the Carcinogenic Risk of Chemicals to Humans, Vol. 23, Some Metals and Metallic Compounds, IARC Press, Lyons, 1980.

[11] IARC IARC Monographs on the Evaluation of Carcinogenic Risks to Humans, Suppl. 7, Overall Evaluations of Carcinogenicity: An Updating of IARC Monographs Volumes 1 to 42, IARC Press, Lyons, 1987.

[12] IARC IARC Monographs on the Evaluation of Carcinogenic Risks to Humans, Vol. 84, Some Drinking-Water Disinfectants and Contaminants, Including Arsenic, IARC Press, Lyons, 2004.

[13] IARC IARC Monographs on the Evaluation of Carcinogenic Risks to Humans, Vol. 100C, Arsenic and Arsenic Compounds, IARC Press, Lyons, 2012.

[14] A. Kinoshita, H. Wanibuchi, K. Morimura, M. Wei, D. Nakae, T. Arai, O. Minowa, T. Noda, S. Nishimura, S. Fukushima, Carcinogenicity of dimethylarsinic acid in Ogg1-deficient mice. Cancer Sci. 98 (2007) 803–814.

[15] J. Liaw, G. Marshall, Y. Yuan, C. Ferreccio, C. Steinmaus, A.H. Smith, Increased childhood liver cancer mortality and arsenic in drinking water in northern Chile. Cancer Epidemiol. Biomarkers Prev. 17 (2008) 1982–1987.

[16] M. Lu, H. Wang, X.F. Li, L.L. Arnold, S.M. Cohen, X.C. Le, Binding of dimethylarsinous acid to cys-13alpha of rat hemoglobin is responsible for the retention of arsenic in rat blood. Chem. Res. Toxicol. 20 (2007) 27–37.

[17] National Toxicology Program NTP toxicology and carcinogenesis studies of gallium arsenide (CAS No. 1303-00-0) in F344/N rats and B6C3F1 mice (inhalation studies). Natl. Toxicol. Program Tech. Rep. Ser. 492 (2000) 1–306.

[18] K. Nohara, Y. Tateishi, T. Suzuki, K. Okamura, H. Murai, S. Takumi, F. Maekawa, N. Nishimura, M. Kobori, T. Ito, Late-onset increases in oxidative stress and other tumorigenic activities and tumors with a Ha-ras mutation in the liver of adult male C3H mice gestationally exposed to arsenic. Toxicol. Sci. 129 (2012) 293–304.

[19] G. Pershagen, N.E. Bjorklund, On the pulmonary tumorigenicity of arsenic trisulfide and calcium arsenate in hamsters. Cancer Lett. 27 (1985) 99–104.

[20] P. Rudnai, M. Börzsönyi, Carcinogenic effect of arsenic trioxide in transplacental and neonatal treated CFLP mice. Nat. Sci. 2 (1980) 11–18.

[21] P. Rudnai, M. Börzsönyi, The tumorigenic effect of treatment with arsenic trioxide. Magy. Onkol. 25 (1981) 73–77.

[22] J. Shen, H. Wanibuchi, E.I. Salim, M. Wei, A. Kinoshita, K. Yoshida, G. Endo, S. Fukushima, Liver tumorigenicity of trimethylarsine oxide in male Fischer 344 rats—association with oxidative DNA damage and enhanced cell proliferation. Carcinogenesis 24 (2003) 1827–1835.

[23] A.H. Smith, G. Marshall, Y. Yuan, C. Ferreccio, J. Liaw, E.O. von Ehrenstein, C. Steinmaus, M.N. Bates, S. Selvin, Increased mortality from lung cancer and bronchiectasis in young adults after exposure to arsenic in utero and in early childhood. Environ. Health Perspect. 114 (2006) 1293–1296.

[24] M. Styblo, Z. Drobna, I. Jaspers, S. Lin, D.J. Thomas, The role of biomethylation in toxicity and carcinogenicity of arsenic: a research update. Environ. Health Perspect. 110 Suppl 5 (2002) 767–771.

[25] A.H. Tennant, A.D. Kligerman, Superoxide dismutase protects cells from DNA damage induced by trivalent methylated arsenicals. Environ. Mol. Mutagen. 52 (2011) 238–243.

[26] E.J. Tokar, L. Benbrahim-Tallaa, J.M. Ward, R. Lunn, R.L. Sams, M.P. Waalkes, Cancer in experimental animals exposed to arsenic and arsenic compounds. Crit. Rev. Toxicol. 40 (2010) 912–927.

[27] E.J. Tokar, B.A. Diwan, D.J. Thomas, M.P. Waalkes, Tumors and proliferative lesions in adult offspring after maternal exposure to methylarsonous acid during gestation in CD1 mice. Arch. Toxicol. 86 (2012) 975–982.

[28] E.J. Tokar, B.A. Diwan, M.P. Waalkes, Arsenic exposure in utero and nonepidermal proliferative response in adulthood in Tg.AC mice. Int. J. Toxicol. 29 (2010) 291–296.

[29] E.J. Tokar, B.A. Diwan, M.P. Waalkes, Renal, hepatic, pulmonary and adrenal tumors induced by prenatal inorganic arsenic followed by dimethylarsinic acid in adulthood in CD1 mice. Toxicol. Lett. 209 (2012) 179–185.

[30] E.J. Tokar, B.A. Diwan, J.M. Ward, D.A. Delker, M.P. Waalkes, Carcinogenic effects of "whole-life" exposure to inorganic arsenic in CD1 mice. Toxicol. Sci. 119 (2011) 73–83.

[31] E.J. Tokar, W. Qu, M.P. Waalkes, Arsenic, stem cells, and the developmental basis of adult cancer. Toxicol. Sci. 120 Suppl 1 (2011) S192–S203.

[32] L. Tomatis, Overview of perinatal and multigeneration carcinogenesis. IARC Sci. Publ. (1989) 1–15.

[33] M.P. Waalkes, J. Liu, D.R. Germolec, C.S. Trempus, R.E. Cannon, E.J. Tokar, R.W. Tennant, J.M. Ward, B.A. Diwan, Arsenic exposure in utero exacerbates skin cancer response in adulthood with contemporaneous distortion of tumor stem cell dynamics. Cancer Res. 68 (2008) 8278–8285.

[34] M.P. Waalkes, J. Liu, J.M. Ward, B.A. Diwan, Enhanced urinary bladder and liver carcinogenesis in male CD1 mice exposed to transplacental inorganic arsenic and postnatal diethylstilbestrol or tamoxifen. Toxicol. Appl. Pharmacol. 215 (2006) 295–305.

[35] M.P. Waalkes, J. Liu, J.M. Ward, D.A. Powell, B.A. Diwan, Urogenital carcinogenesis in female CD1 mice induced by in utero arsenic exposure is exacerbated by postnatal diethylstilbestrol treatment. Cancer Res. 66 (2006) 1337–1345.

[36] M.P. Waalkes, J.M. Ward, B.A. Diwan, Induction of tumors of the liver, lung, ovary and adrenal in adult mice after brief maternal gestational exposure to inorganic arsenic: promotional effects of postnatal phorbol ester exposure on hepatic and pulmonary, but not dermal cancers. Carcinogenesis 25 (2004) 133–141.

[37] M.P. Waalkes, J.M. Ward, J. Liu, B.A. Diwan, Transplacental carcinogenicity of inorganic arsenic in the drinking water: induction of hepatic, ovarian, pulmonary, and adrenal tumors in mice. Toxicol. Appl. Pharmacol. 186 (2003) 7–17.

[38] X.J. Wang, Z. Sun, W. Chen, K.E. Eblin, J.A. Gandolfi, D.D. Zhang, Nrf2 protects human bladder urothelial cells from arsenite and monomethylarsonous acid toxicity. Toxicol. Appl. Pharmacol. 225 (2007) 206–213.

[39] M. Wei, H. Wanibuchi, K. Morimura, S. Iwai, K. Yoshida, G. Endo, D. Nakae, S. Fukushima, Carcinogenicity of dimethylarsinic acid in male F344 rats and genetic alterations in induced urinary bladder tumors. Carcinogenesis 23 (2002) 1387–1397.

[40] M. Wei, H. Wanibuchi, S. Yamamoto, W. Li, S. Fukushima, Urinary bladder carcinogenicity of dimethylarsinic acid in male F344 rats. Carcinogenesis 20 (1999) 1873–1876.

[41] A. Yamamoto, A. Hisanaga, N. Ishinishi, Tumorigenicity of inorganic arsenic compounds following intratracheal instillations to the lungs of hamsters. Int. J. Cancer 40 (1987) 220–223.

[42] T. Yorifuji, T. Tsuda, H. Doi, P. Grandjean, Cancer excess after arsenic exposure from contaminated milk powder. Environ. Health Prev. Med. 16 (2011) 164–170.

[43] Y. Yuan, G. Marshall, C. Ferreccio, C. Steinmaus, J. Liaw, M. Bates, A.H. Smith, Kidney cancer mortality: fifty-year latency patterns related to arsenic exposure. Epidemiology 21 (2010) 103–108.

20

ARSENIC-INDUCED CARDIOVASCULAR DISEASE

AARON BARCHOWSKY[1] AND J. CHRISTOPHER STATES[2]

[1] *Department of Environmental and Occupational Health, University of Pittsburgh, Pittsburgh, PA, USA*
[2] *Department of Pharmacology and Toxicology, University of Louisville, Louisville, KY, USA*

20.1 INTRODUCTION

Epidemiologic studies of arsenic exposures in large populations in several different countries report dose-dependent increased risk of cardiovascular disease (CVD) from low-to-moderate, as well as high, arsenic concentrations [9, 42, 59, 74]. CVD, especially coronary heart disease (CHD) is the main noncancer disease endpoint where there is unequivocal causal linkage to arsenic exposure. In addition, CVD may arguably be the most important arsenic-related cancer or noncancer disease risk, since CVD is highly prevalent worldwide and the portion of disease burden attributable to inorganic arsenic exposure may be highly significant [58]. The pathogenic cardiac effects of arsenic are even more evident as the main limitation of arsenic cancer therapies is cardiac arrhythmia and infarctions that are observed in greater than 30% of patients [15, 46]. In addition, the profound effect of arsenic on vascular remodeling and angiogenesis has been suggested to contribute to both tumorigenesis and metabolic diseases. The human evidence for arsenic-induced CVDs is compelling, and there has been a large amount of success in generating animal models that provide mechanistic insight into the causes of pathogenic cardiovascular effects of arsenic. In contrast to cancer and other arsenic-related diseases, mice have been an excellent experimental model, since they may be as sensitive as

Arsenic: Exposure Sources, Health Risks, and Mechanisms of Toxicity, First Edition.
Edited by J. Christopher States.
© 2016 John Wiley & Sons, Inc. Published 2016 by John Wiley & Sons, Inc.

or more sensitive than humans to vascular pathologies caused by low-to-moderate arsenic exposures. Genetic and wild-type mouse models in a number of laboratories have revealed important information regarding enhanced sensitivity to adult CVD caused by *in utero* or perinatal exposures and the apparent U-shaped dose–response relationships that are seen in humans. In addition to mice, the guinea pig model has been effective in revealing mechanistic insight into the cardiac side effects of arsenic therapies. This chapter will discuss the mechanistic insights learned thus far from these animal models of arsenic-induced vascular disease with attention to future directions for this research.

20.2 ARSENIC-INDUCED ATHEROSCLEROSIS AND ISCHEMIC HEART DISEASE

Atherosclerosis is the primary underlying pathology related to coronary artery disease, myocardial ischemic disease, infarction, and ultimately cardiac disease mortality. CVDs, including hypertension, myocardial infarction, and stroke, have been related to high levels of arsenic exposure (>500 μg/L) in China, southwestern Taiwan, Bangladesh, and Chile [7, 8, 49, 70, 71, 85]. Associations of CVD prevalence, incidence, and mortality with moderate-to-high arsenic exposure (<300 μg/L) were found in cross-sectional studies from northeastern Taiwan and prospective studies from Bangladesh [7, 9, 11, 59]. A series of ecological studies in the United States and Spain have found positive association between increased CVD mortality and low arsenic exposure (<100 μg/L) [16, 32, 35, 39, 40, 86]. A prospective cohort study in Native Americans in the Strong Heart Study associated increased risks of low arsenic exposures (<100 μg/L) with incident fatal and nonfatal CVD, CHD, and stroke [42]. Several large prospective epidemiological studies found that CHD and its primary preclinical sign, prolonged QT interval, may be increased by as much as 33% by arsenic exposures between 12 and 150 μg/L in drinking water [9, 12, 41, 42, 59]. This level of increased risk places arsenic as one of the highest risk factors for CHD and a major contributor to disease burden given that CVD is the leading cause of mortality worldwide [58]. In addition, arsenic may be synergistic with other risk factors, such as smoking, in increasing CHD [9, 41, 42]. Age at exposure may also synergize arsenic-related CHD as the rate of myocardial infarctions was greatly increased in young adults who were exposed perinatally to arsenic [85]. An early series of case studies of high-level exposures in Chile revealed that even exposed infants died of CHD with complete blockage of the coronary arteries before the age of 2 [52, 53].

20.3 ANIMAL MODELS OF ATHEROSCLEROSIS AND VASCULAR DYSFUNCTION

There has been a large amount of research in rat and mouse models of atherosclerosis that has advanced understanding of the etiology of arsenic-induced atherogenesis [6, 13, 18, 31, 57, 63, 64]. These studies were facilitated by the use of the $ApoE^{-/-}$ and

$ApoE^{-/-}$/ $LDLr^{-/-}$ mouse models that develop human-like disease with atherosclerotic plaques and are validated tools for atherosclerosis research. However, since these genetic models are spontaneously atherogenic, the findings are limited to identifying mechanisms through which arsenic accelerates, but may not be the root cause of, atherosclerosis. Nonetheless, Simeonova et al. were the first to use the $ApoE^{-/-}$ mouse to demonstrate that exposure to arsenic in drinking water greatly accelerated aortic plaque formation and suggested a role for arsenic induction of inflammatory cytokines in the vessel wall as a possible pathogenic mechanism for the accelerated lesions [57]. However, these studies were confounded by introduction of high fat diet in the latter part of the study. Srivastava et al. [64] were able to demonstrate that only arsenic exposure without high fat diet was able to accelerate atherogenesis in the $ApoE^{-/-}$ mouse. In addition, this group showed dose–response with exposures from 1 to 50 ppm. Curiously, Lemaire et al. [31] showed increased atherosclerosis with 200 ppb exposure than with 1 ppm suggesting a U-shaped dose–response curve. A U-shaped dose–response curve with increased disease at concentrations within the range of human exposures is significant. These results suggest the possibility that even lower exposures may have serious consequences for millions of exposed people. Both groups observed increased levels of proinflammatory cytokines in the arsenic exposed mice, consistent with a role for inflammation in the progression of atherosclerotic disease.

Srivastava et al. [63] also showed that *in utero* exposure of $ApoE^{-/-}$ mice from gestational days 8 to 18 also was able to accelerate disease. These results are highly significant and support the conclusions that fetal arsenic exposure is atherogenic as suggested by the reports of coronary disease in infants in Antofogasta [52, 53]. The animal experiments were performed at relatively high exposure (49 ppm) and did not address dose–response. Thus, it is difficult to infer how much influence low human exposures *in utero* may have on the development of chronic adult disease. This group did show that liver gene expression patterns were changed to a proinflammatory pattern consistent with susceptibility to atherosclerosis [65]. Furthermore, differential gene expression continued to develop weeks after the exposure [45, 65] suggesting that the arsenic exposure may be the first hit in a two-hit scenario. Thus, future studies examining the potential interaction of other factors such as diet with *in utero* arsenic exposure on disease development are desirable.

The $ApoE^{-/-}$ mice exposed to arsenic *in utero* also exhibited decreased vasorelaxation as measured in aortic rings [63]. This deficit in vasorelaxation can contribute to hypertension. Chen et al. published a study linking hypertension to arsenic exposure in Taiwan [8] and reported a dose–response. Since then, multiple studies have followed and confirmed the association of arsenic exposure to hypertension (see [1, 25] for reviews), although the effect may be small. However, there appears to be great variability in the effect size depending on geographic region or ages of the subjects in which the studies are done. A recent study looking at association of arsenic exposure with increased blood pressure in children found a modest effect of *in utero* arsenic exposure on blood pressure in 4.5-year-old children in rural Bangladesh [23]. Arsenic exposures were monitored in the pregnant women by determining urine arsenic in this study. Rather, large increases in urine arsenic were required to observe an increase in blood pressure in the exposed children in the follow-up studies 4.5 years later. In contrast, only a modest arsenic exposure (20–30 ppb vs. <5 ppb) was required to see increased blood pressure in 30- to 60-year-old adults in Iran [38]. Chronic exposure of FVB female mice (100 ppb, 22 weeks) induced hypertension that was associated with left ventricular hypertrophy [54]. The study provides clear evidence of a causal role of arsenic exposure in the development of hypertension. Thus, more comprehensive studies, both epidemiological and experimental, are needed to sort out the effects of age, gender, and geography/ethnicity on susceptibility to hypertension in arsenic-exposed people.

20.3.1 Cardiac Arrhythmia and Infarction

Acquired long QT syndrome (acLQTS) is a significant dose-limiting side effect of antileukemic arsenic therapies, which occurs when circulating arsenic concentrations approach 5 μM [4, 5, 15, 17, 47, 51]. Prolonged QT intervals have also been associated with chronic environmental exposures to arsenic in drinking water when circulating arsenic levels are often submicromolar [44]. LQTS reflects slowed ventricular repolarization at the cellular level and is characterized by a prolongation of the QT interval on the electrocardiogram. As a direct consequence of abnormal repolarization in LQTS, arsenic-sensitive patients may present with syncope, torsades de pointes, arrhythmias, or sudden cardiac death [5, 15, 51]. The cardiac hERG (human ether-a-go-go related gene) potassium channel is the α-subunit of the rapid delayed rectifier current, IKr, in ventricular myocytes. This channel contributes prominently to terminal repolarization and is the target of many chemicals and therapeutics that cause acLQTS [15, 29]. However, in contrast to most of these agents that bind and block the channel, arsenic is devoid of direct acute effects on cardiac repolarization or inhibition of hERG/IKr [15, 29]. Instead, arsenic reduces trafficking of hERG/IKr to myocyte membranes by affecting molecular chaperone proteins, such as Hsp90 [15, 17].

20.3.2 Angiogenesis and Vascular Remodeling

Arsenic exposure induces both angiogenesis and vascular remodeling. The peripheral artery disease seen in arsenic-exposed people is the consequence of an arteriosclerotic process that includes both intimal smooth muscle cell proliferation and atherosclerotic plaque formation. Blackfoot disease is the culmination of arsenic exposure-associated peripheral artery disease long recognized in Taiwan. Pathological examination showed that thromboangiitis obliterans was present in 30% of the patients with the remainder exhibiting arteriosclerosis obliterans [69]. The loss of circulation results in tissue necrosis manifesting as a dry gangrene and autoamputation of the digits. This condition was thought to be restricted to the arsenic-exposed population in Taiwan. However, more recently with high-level exposure in West Bengal, India, and Bangladesh, apparently identical peripheral artery disease culminating in dry gangrene has been documented in these areas as well [21]. It should be noted that the dry gangrene has been associated with the prevalence of diabetes mellitus in both the Taiwanese and Bangladeshi populations. Indeed, the peripheral artery disease associated with arsenic exposure bears more than a passing resemblance to that seen in diabetics.

Arsenic not only stimulates proliferation of smooth muscle cells in large vessels but also promotes formation of new vessels at low concentrations. However, at high concentrations, arsenic exposure is anti-angiogenic. An early report using human umbilical vein endothelial cells showed that sodium arsenite up to 1 μM stimulated but that elevated concentrations (5 μM) suppressed vascular endothelial growth factor (VEGF) expression and *in vitro* angiogenesis [27]. Soucy et al. [61] showed that low arsenic trioxide (ATO) exposure (≤1 μM) stimulated vessel growth in the chorioallantoic membrane assay, and in Matrigel implants in mice but only when administered with a fibroblast growth factor. Curiously, high ATO exposure decreased vessel growth in the chorioallantoic membrane assay. Furthermore, greater effect was noted with low doses. In addition, these authors tested the effects of ATO on tumor growth using mice injected with green fluorescent protein (GFP)-labeled B16-F10 melanoma cells. Tumor growth and lung metastases were seen in all animals and lower ATO doses (≤1 mg/kg) were associated with larger tumor volumes, whereas high dose (5 mg/kg) was associated with tumor volumes smaller than in saline controls although metastases were greater. Similarly, ATO doses of 2.5 and 5.0 mg/kg severely inhibited tumor growth in a lung cancer model [83] and a gastric cancer model [82]. VEGF expression and microvessel density were lower in tumors in arsenic treated mice injected with SGC-7901 gastric cancer cells [82]. Similar findings were reported for the lung cancer model as well [83]. Thus, it appears that low exposures similar to those experienced by humans drinking arsenic-contaminated drinking water stimulates angiogenesis and promotes tumor growth, consistent with the known carcinogenic risk. In contrast, high arsenic doses, similar to those used to treat patients with acute promyelogenous leukemia, appear to inhibit angiogenesis and thus inhibit tumor growth.

Arsenic also causes vessel growth and remodeling in normal tissues. Chronic low arsenic exposures (5–250 ppb 20 weeks) induce microvessel growth in murine cardiac tissue [62]. This vessel growth is accompanied by increased expression of angiogenic and tissue-remodeling gene expression including VEGF, VEGF receptors, plasminogen activator inhibitor-1 (PAI-1), endothelin-1, and matrix metalloproteinase-9 (MMP-9). The sinusoidal endothelium in the liver undergoes architectural changes in response to low-level arsenic exposure [67, 68]. Normally, the sinusoidal endothelium has many fenestrae allowing free exchange of materials. After 5 weeks exposure of C57Bl6 mice to 250 ppb arsenic in drinking water, the sinusoidal endothelium remodeled with the fenestrae closing rendering the appearance more like a capillary endothelium [67]. This closure of the fenestrae was followed by inflammatory infiltration of the liver [68] suggesting that the vascular changes induced by arsenic exposure are primary pathological effects initiating a cascade of pathological sequelae leading to liver disease. The signaling pathways for this remodeling of the nature of the sinusoidal bed to a more capillary-like phenotype is markedly different from that involved in stimulating angiogenesis discussed in the previous paragraph. This remodeling is dependent upon reactive oxygen signaling generated by NAPDH oxidase (Nox2) [66]. The pathological changes also are accompanied by increased platelet/endothelial cell adhesion molecule (PECAM1) expression, protein nitration, and decreased liver clearance of modified albumin. Nox2 knockouts or mice treated with a Nox2 inhibitor were resistant to sinusoidal capillarization by arsenite exposure.

20.4 HYPERTENSION

Results from numerous studies over the past 20 years suggest a strong link between arsenic exposure and hypertension. An early study in the arseniasis region of Taiwan suggested a dose–response relationship between cumulative arsenic exposure and prevalence of hypertension with an overall increase of 1.5-fold compared with residents in nonendemic areas [8]. Since then, epidemiologic studies in various parts of the world also suggest a link between arsenic exposure and hypertension. A systematic review of these studies [1] found support for the linkage, but the studies were deemed too small to find causal evidence. However, animal studies clearly show that chronic arsenic exposure causes increases in both systolic and diastolic blood pressure [54]. The authors also reported that the mice developed left ventricular hypertrophy, a common outcome of chronic hypertension. Consistent with these findings are the findings mentioned earlier in this chapter that *in utero* arsenic exposure had long-lasting inhibitory effects on arterial relaxation [63]. Consistent with this *ex vivo* experiment, a small study in Romania found an adverse effect of arsenic exposure on blood pressure hyperreactivity [30]. There was no effect on steady-state blood pressure, but only on blood pressure responses to stimuli. The authors suggested that these aberrant responses were early indicators of a later onset of hypertension. Similarly, a study in Bangladesh found an association of arsenic exposure with pulse pressure but not hypertension per se [25]. A recent study in an arseniasis endemic area Nadia district in West Bengal, India, reported an increased odds ratio (OR) for hypertension [adjusted OR 2.87 (95% CI = 1.26–4.83)] in the arsenic-exposed populace as compared to those in the low arsenic exposure Hoogly district [22]. A dose–effect relationship was observed with both cumulative arsenic exposure and arsenic level measured in hair showing that both an estimate of exposure and a direct biomarker measurement of internal dose correlate. A recent epidemiological study of residents in Texas showed that AS3MT gene polymorphism as well as modest arsenic exposure ($\leq$15.3 µg/L) both associated with CHD and that hypertension associated with arsenic exposure [20]. CHD has complex etiology and hypertension is but one risk factor. Thus, arsenic metabolic capability may modulate the effect of arsenic exposure on CVD and an impact on risk of hypertension may become evident in a larger study. Consistent with this idea is the association of hypertension with inorganic arsenic, monomethylarsonic acid (MMA), dimethylarsinic acid (DMA), or total arsenic in urine where all correlated with increased systolic or pulse blood pressure but that an increased association with higher urinary concentrations of the arsenic metabolite MMA or lower percentage of DMA was found in a study in Shanxi province in northwest China [34]. This group reported similar findings in a study in Inner Mongolia where it was found that systolic blood pressure increases correlated with cumulative arsenic exposure and a greater correlation with decreased percent DMA in urine [33]. It appears that the preponderance of evidence supports the association of arsenic exposure with the development of hypertension. However, more studies with better designs including large sample sizes, speciation of urinary arsenic species, and AS3MT polymorphism analysis are needed to substantiate the suggested linkage.

20.5 SUBCLINICAL DISEASE

20.5.1 Subclinical Atherosclerosis

Carotid arterial wall thickening and presence of large irregular deposits on the arterial wall can be determined by high-resolution ultra-sound techniques that measure both carotid arterial wall thickness and the presence of plaque. These markers of disease reflect atherosclerotic processes leading to myocardial infarctions and stroke and have been used in prospective studies as predictors of disease. Using these markers, a dose–response relationship was defined in southeastern Taiwanese villagers exposed to high levels of arsenic (median levels from 700 to 930 μg/L) between well water arsenic concentration and carotid atherosclerosis [77]. Similarly, in villagers in northeastern Taiwan where the arsenic levels are low to moderate (median in villages from undetectable to 140 μg/L), there was a significant increase in the prevalence of carotid atherosclerosis with an adjusted OR of 2.6 (95% CI 1.3–5.0) among those with exposure greater than 100 μg/L compared to those with exposure less than 50 μg/L [81]. Similarly, a cross-sectional study in Mexico found that urinary arsenic in children (3–14 years old) was positively associated with increased carotid wall thickening. The results also indicated a dose–response relationship with higher urinary arsenic (>70 μg/L) showing greater thickening than those with more moderate urinary arsenic (35–70 μg/L) and those with low urinary arsenic (<35 μg/L) showing the least [48]. These results strongly support a role for both high and low arsenic exposures in subclinical atherosclerosis, consistent with clear evidence of arsenic exposure causing CVD as discussed earlier.

20.5.2 Inflammatory Markers and Endothelial Cell Dysfunction

The atherosclerotic process is initiated by invasion of circulating monocytes into the sub-endothelial space where they become macrophages and ingest the excess lipid deposits. This process is mediated first by the binding of monocytes to the endothelium, and the binding is mediated by cellular adhesion molecules [e.g., E-selectin, P-selectin, intercellular adhesion molecule-I (ICAM-1), and vascular adhesion molecule-I (VCAM-1)]. Endothelial cells express adhesion molecules on their surface in response to inflammatory stimuli. The soluble parts of the cellular adhesion molecules serve as circulating markers of inflammation. Cross-sectional studies conducted in Bangladesh have shown associations between arsenic exposure in drinking water and elevated levels of circulating markers of inflammation [10, 78]. Moderately high exposure was associated with both soluble ICAM-1 and VCAM-1 in an earlier study conducted in Araihazar [10], but only with soluble VCAM-1 in a later study embedded in the HEALS study [78]. In a study comparing people from arsenic endemic and nonendemic areas of Bangladesh, there were significantly higher levels of oxidized LDL (Ox-LDL) and C-reactive protein (CRP) in addition to ICAM-1 and VCAM-1 in subjects from arsenic-endemic areas than those from nonendemic areas [28]. Significant correlations with arsenic exposure measured in water, hair, and nails were found for all these circulating markers and the relations were significant both before and after adjusting for covariates. Ox-LDL in plasma has long been positively related to CVD [24] and CRP is a circulating acute-phase reactant reflecting systemic inflammation and an independent predictor of CVD [50]. Other circulating markers of inflammation such as PAI-1, MMP-9, and myeloperoxidase and their associations with arsenic exposure have been assessed in studies in Bangladesh and West Bengal [3, 10]. Elevated PAI-1 is associated with an increased risk of atherothrombotic events [73]. PAI-1 is induced by arsenite exposure in experimental systems. Human microvascular endothelial cells increase PAI-1 expression in response to arsenite [26]. Exposure of mice to greater than 500 µg/L arsenic in drinking water induced circulating PAI-1 levels, although dose- and time-dependent PAI-1 expression in cardiac tissue was observed at lower exposures (50–250 µg/L) [62]. Consistent with experimental findings, PAI-1 levels were elevated in Bangladeshis in the highest quartile of well-water arsenic (73.5–500.6 µg/L), with the greatest correlation in those subjects with elevated body mass index [10]. That MMP-9 plays a role in atherosclerotic progression has been known for a long time [19].

20.5.3 QT Interval Lengthening

Direct effects of arsenic poisoning on heart function were sporadically reported prior to the use of ATO to treat acute promyelocytic leukemia (APL) (see [36] for a review). In 1990, Little et al. [36] reported on two patients with arsenic poisoning who demonstrated prolongation of QT-U interval and torsades de pointes. These same cardiac arrhythmias were observed in early studies of APL patients treated with ATO [47, 60, 72]. Curiously, although QT interval prolongation was observed, tachycardia and torsades de pointes were absent when patients were treated with tetra-arsenic tetrasulfide [37, 56].

Laboratory studies have duplicated the induction of cardiac arrhythmias in animal models, although the guinea pig appears much more sensitive than the rabbit [14, 80]. Later work in the guinea pig model showed that the mechanism of QT interval prolongation was by suppression of ether-a-go-go related gene (ERG, alpha subunit of a potassium ion channel encoded by *KCNH2*) and Kir2.1 (inward-rectifier potassium ion channel encoded by *KCNJ2*), and that this suppression was accompanied by induction of miR-1 and miR-133 [55]. Both ERG and Kir2.1 could be suppressed by forced expression of miR-1 or direct intramuscular injection of miR-133, both of which also increased the mortality rate. Antisense knockdown of these microRNAs eliminated the cardiac arrhythmias induced by ATO [55]. Although QT interval prolongation is common among APL patients treated with ATO, modern management approaches mitigate the effects and clinically significant arrhythmias are rare [51].

The reports of cardiac arrhythmias from acute relatively high-dose arsenic exposures prompted investigation of the direct cardiac effects of chronic relatively low-dose exposures. Several studies in Bangladesh, Taiwan, Mongolia, Turkey, and the United States have reported association of QT interval prolongation with chronic arsenic exposure [2, 12, 43, 44, 75, 76, 79, 84]. These studies included a wide range of arsenic exposures. Thus, in humans, arsenic induces cardiac arrhythmias both with acute relatively high exposure doses used in treating APL patients or in cases of poisoning, and with chronic consumption of arsenic-contaminated water.

20.6 SUMMARY AND CONCLUSIONS

Clearly, arsenic exposure induces CVD. Chronic environmental exposures cause arterial disease manifested as atherosclerosis in major arteries and arteriolosclerosis in peripheral vasculature. In addition, remodeling of capillary beds, particularly the liver sinusoidal endothelium, is observed even with relatively low exposures in the mouse. Cardiac arrhythmias initially associated with acute high doses first in poisoning cases and later in the clinic with ATO therapy for APL, also have been associated with chronic environmental exposures. Recent epidemiological findings indicate that the morbidity and mortality resulting from the CVD effects of arsenic exposure may be greater than those from the carcinogenic effects. Thus, considering the magnitude of the environmental arsenic exposure problem, further study of the mechanisms and development of means to mitigate the CVD effects of arsenic exposure is important for world health.

REFERENCES

[1] L.N. Abhyankar, M.R. Jones, E. Guallar, A. Navas-Acien, Arsenic exposure and hypertension: a systematic review. Environ. Health Perspect. 120 (2012) 494–500.

[2] S.A. Ahmad, F. Khatun, M.H. Sayed, M.H. Khan, R. Aziz, M.Z. Hossain, M.H. Faruquee, Electrocardiographic abnormalities among arsenic-exposed persons through groundwater in Bangladesh. J. Health Popul. Nutr. 24 (2006) 221–227.

[3] M. Banerjee, N. Banerjee, P. Ghosh, J.K. Das, S. Basu, A.K. Sarkar, J.C. States, A.K. Giri, Evaluation of the serum catalase and myeloperoxidase activities in chronic arsenic-exposed individuals and concomitant cytogenetic damage. Toxicol. Appl. Pharmacol. 249 (2010) 47–54.

[4] J.T. Barbey, J.C. Pezzullo, S.L. Soignet, Effect of arsenic trioxide on QT interval in patients with advanced malignancies. J. Clin. Oncol. 21 (2003) 3609–3615.

[5] J.T. Barbey, S. Soignet, Prolongation of the QT interval and ventricular tachycardia in patients treated with arsenic trioxide for acute promyelocytic leukemia. Ann. Intern. Med. 135 (2001) 842–843.

[6] M. Bunderson, D.M. Brooks, D.L. Walker, M.E. Rosenfeld, J.D. Coffin, H.D. Beall, Arsenic exposure exacerbates atherosclerotic plaque formation and increases nitrotyrosine and leukotriene biosynthesis. Toxicol. Appl. Pharmacol. 201 (2004) 32–39.

[7] C.J. Chen, H.Y. Chiou, M.H. Chiang, L.J. Lin, T.Y. Tai, Dose-response relationship between ischemic heart disease mortality and long-term arsenic exposure. Arterioscler. Thromb. Vasc. Biol. 16 (1996) 504–510.

[8] C.J. Chen, Y.M. Hsueh, M.S. Lai, M.P. Shyu, S.Y. Chen, M.M. Wu, T.L. Kuo, T.Y. Tai, Increased prevalence of hypertension and long-term arsenic exposure. Hypertension 25 (1995) 53–60.

[9] Y. Chen, J.H. Graziano, F. Parvez, M. Liu, V. Slavkovich, T. Kalra, M. Argos, T. Islam, A. Ahmed, M. Rakibuz-Zaman, R. Hasan, G. Sarwar, D. Levy, A. van Geen, H. Ahsan, Arsenic exposure from drinking water and mortality from cardiovascular disease in Bangladesh: prospective cohort study. BMJ 342 (2011) d2431.

[10] Y. Chen, R.M. Santella, M.G. Kibriya, Q. Wang, M. Kappil, W.J. Verret, J.H. Graziano, H. Ahsan, Association between arsenic exposure from drinking water and plasma levels of soluble cell adhesion molecules. Environ. Health Perspect. 115 (2007) 1415–1420.

[11] Y. Chen, F. Wu, J.H. Graziano, F. Parvez, M. Liu, R.R. Paul, I. Shaheen, G. Sarwar, A. Ahmed, T. Islam, V. Slavkovich, T. Rundek, R.T. Demmer, M. Desvarieux, H. Ahsan, Arsenic exposure from drinking water, arsenic methylation capacity, and carotid intima-media thickness in Bangladesh. Am. J. Epidemiol. 178 (2013) 372–381.

[12] Y. Chen, F. Wu, F. Parvez, A. Ahmed, M. Eunus, T.R. McClintock, T.I. Patwary, T. Islam, A.K. Ghosal, S. Islam, R. Hasan, D. Levy, G. Sarwar, V. Slavkovich, G.A. van, J.H. Graziano, H. Ahsan, Arsenic exposure from drinking water and QT-interval prolongation: results from the Health Effects of Arsenic Longitudinal Study. Environ. Health Perspect. 121 (2013) 427.

[13] T.J. Cheng, J.J. Chuu, C.Y. Chang, W.C. Tsai, K.J. Chen, H.R. Guo, Atherosclerosis induced by arsenic in drinking water in rats through altering lipid metabolism. Toxicol. Appl. Pharmacol. 256 (2011) 146–153.

[14] C.E. Chiang, H.N. Luk, T.M. Wang, P.Y. Ding, Prolongation of cardiac repolarization by arsenic trioxide. Blood 100 (2002) 2249–2252.

[15] A. Dennis, L. Wang, X. Wan, E. Ficker, hERG channel trafficking: novel targets in drug-induced long QT syndrome. Biochem. Soc. Trans. 35 (2007) 1060–1063.

[16] R.R. Engel, A.H. Smith, Arsenic in drinking water and mortality from vascular disease: an ecologic analysis in 30 counties in the United States. Arch. Environ. Health 49 (1994) 418–427.

[17] E. Ficker, Y.A. Kuryshev, A.T. Dennis, C. Obejero-Paz, L. Wang, P. Hawryluk, B.A. Wible, A.M. Brown, Mechanisms of arsenic-induced prolongation of cardiac repolarization. Mol. Pharmacol. 66 (2004) 33–44.

[18] S. Franco, I. Segura, H.H. Riese, M.A. Blasco, Decreased B16F10 melanoma growth and impaired vascularization in telomerase-deficient mice with critically short telomeres. Cancer Res. 62 (2002) 552–559.

[19] Z.S. Galis, G.K. Sukhova, M.W. Lark, P. Libby, Increased expression of matrix metalloproteinases and matrix degrading activity in vulnerable regions of human atherosclerotic plaques. J. Clin. Invest. 94 (1994) 2493–2503.

[20] G. Gong, S.E. O'Bryant, Low-level arsenic exposure, AS3MT gene polymorphism and cardiovascular diseases in rural Texas counties. Environ. Res. 113 (2012) 52–57.

[21] M.D. Guha, U.B. Dasgupta, Chronic arsenic toxicity: studies in West Bengal, India. Kaohsiung J. Med. Sci. 27 (2011) 360–370.

[22] M.D. Guha, I. Purkayastha, A. Ghose, G. Mistry, C. Saha, A.K. Nandy, A. Das, K.K. Majumdar, Hypertension in chronic arsenic exposure: a case control study in West Bengal. J. Environ. Sci. Health A Tox. Hazard. Subst. Environ. Eng. 47 (2012) 1514–1520.

[23] S. Hawkesworth, Y. Wagatsuma, M. Kippler, A.J. Fulford, S.E. Arifeen, L.A. Persson, S.E. Moore, M. Vahter, Early exposure to toxic metals has a limited effect on blood pressure or kidney function in later childhood, rural Bangladesh. Int. J. Epidemiol. 42 (2013) 176–185.

[24] P. Holvoet, A. Mertens, P. Verhamme, K. Bogaerts, G. Beyens, R. Verhaeghe, D. Collen, E. Muls, F. Van de Werf, Circulating oxidized LDL is a useful marker for identifying patients with coronary artery disease. Arterioscler. Thromb. Vasc. Biol. 21 (2001) 844–848.

[25] A.K. Islam, A.A. Majumder, Hypertension in Bangladesh: a review. Indian Heart J. 64 (2012) 319–323.

[26] S.J. Jiang, T.M. Lin, H.L. Wu, H.S. Han, G.Y. Shi, Decrease of fibrinolytic activity in human endothelial cells by arsenite. Thromb. Res. 105 (2002) 55–62.

[27] Y.H. Kao, C.L. Yu, L.W. Chang, H.S. Yu, Low concentrations of arsenic induce vascular endothelial growth factor and nitric oxide release and stimulate angiogenesis in vitro. Chem. Res. Toxicol. 16 (2003) 460–468.

[28] M.R. Karim, M. Rahman, K. Islam, A.A. Mamun, S. Hossain, E. Hossain, A. Aziz, F. Yeasmin, S. Agarwal, M.I. Hossain, Z.A. Saud, F. Nikkon, M. Hossain, A. Mandal, R.O. Jenkins, P.I. Haris, H. Miyataka, S. Himeno, K. Hossain, Increases in oxidized low-density lipoprotein and other inflammatory and adhesion molecules with a concomitant decrease in high-density lipoprotein in the individuals exposed to arsenic in Bangladesh. Toxicol. Sci. 135 (2013) 17–25.

[29] A.N. Katchman, J. Koerner, T. Tosaka, R.L. Woosley, S.N. Ebert, Comparative evaluation of HERG currents and QT intervals following challenge with suspected torsadogenic and nontorsadogenic drugs. J. Pharmacol. Exp. Ther. 316 (2006) 1098–1106.

[30] J. Kunrath, E. Gurzau, A. Gurzau, W. Goessler, E.R. Gelmann, T.T. Thach, K.M. McCarty, C.W. Yeckel, Blood pressure hyperreactivity: an early cardiovascular risk in normotensive men exposed to low-to-moderate inorganic arsenic in drinking water. J. Hypertens. 31 (2013) 361–369.

[31] M. Lemaire, C.A. Lemarie, M.F. Molina, E.L. Schiffrin, S. Lehoux, K.K. Mann, Exposure to moderate arsenic concentrations increases atherosclerosis in ApoE-/- mouse model. Toxicol. Sci. 122 (2011) 211–221.

[32] D.R. Lewis, J.W. Southwick, R. Ouellet-Hellstrom, J. Rench, R.L. Calderon, Drinking water arsenic in Utah: a cohort mortality study. Environ. Health Perspect. 107 (1999) 359–365.

[33] X. Li, B. Li, S. Xi, Q. Zheng, X. Lv, G. Sun, Prolonged environmental exposure of arsenic through drinking water on the risk of hypertension and type 2 diabetes. Environ. Sci. Pollut. Res. Int. 20 (2013) 8151–8161.

[34] X. Li, B. Li, S. Xi, Q. Zheng, D. Wang, G. Sun, Association of urinary monomethylated arsenic concentration and risk of hypertension: a cross-sectional study from arsenic contaminated areas in northwestern China. Environ. Health 12 (2013) 37.

[35] L.D. Lisabeth, H.J. Ahn, J.J. Chen, S. Sealy-Jefferson, J.F. Burke, J.R. Meliker, Arsenic in drinking water and stroke hospitalizations in Michigan. Stroke 41 (2010) 2499–2504.

[36] R.E. Little, G.N. Kay, J.B. Cavender, A.E. Epstein, V.J. Plumb, Torsade de pointes and T-U wave alternans associated with arsenic poisoning. Pacing Clin. Electrophysiol. 13 (1990) 164–170.

[37] D.P. Lu, J.Y. Qiu, B. Jiang, Q. Wang, K.Y. Liu, Y.R. Liu, S.S. Chen, Tetra-arsenic tetra-sulfide for the treatment of acute promyelocytic leukemia: a pilot report. Blood 99 (2002) 3136–3143.

[38] M. Mahram, D. Shahsavari, S. Oveisi, S. Jalilolghadr, Comparison of hypertension and diabetes mellitus prevalence in areas with and without water arsenic contamination. J. Res. Med. Sci. 18 (2013) 408–412.

[39] M.A. Medrano, R. Boix, R. Pastor-Barriuso, M. Palau, J. Damian, R. Ramis, J.L. Del Barrio, A. Navas-Acien, Arsenic in public water supplies and cardiovascular mortality in Spain. Environ. Res. 110 (2010) 448–454.

[40] J.R. Meliker, R.L. Wahl, L.L. Cameron, J.O. Nriagu, Arsenic in drinking water and cerebrovascular disease, diabetes mellitus, and kidney disease in Michigan: a standardized mortality ratio analysis. Environ. Health 6 (2007) 4.

[41] K. Moon, E. Guallar, A. Navas-Acien, Arsenic exposure and cardiovascular disease: an updated systematic review. Curr. Atheroscler. Rep. 14 (2012) 542–555.

[42] K.A. Moon, E. Guallar, J.G. Umans, R.B. Devereux, L.G. Best, K.A. Francesconi, W. Goessler, J. Pollak, E.K. Silbergeld, B.V. Howard, A. Navas-Acien, Association between exposure to low to moderate arsenic levels and incident cardiovascular disease. A prospective cohort study. Ann. Intern. Med. 159 (2013) 649–659.

[43] I. Mordukhovich, R.O. Wright, C. Amarasiriwardena, E. Baja, A. Baccarelli, H. Suh, D. Sparrow, P. Vokonas, J. Schwartz, Association between low-level environmental arsenic exposure and QT interval duration in a general population study. Am. J. Epidemiol. 170 (2009) 739–746.

[44] J.L. Mumford, K. Wu, Y. Xia, R. Kwok, Z. Yang, J. Foster, W.E. Sanders, Chronic arsenic exposure and cardiac repolarization abnormalities with QT interval prolongation in a population-based study. Environ. Health Perspect. 115 (2007) 690–694.

[45] N.N. Ngalame, A.F. Micciche, M.E. Feil, J.C. States, Delayed temporal increase of hepatic Hsp70 in ApoE knockout mice after prenatal arsenic exposure. Toxicol. Sci. 131 (2013) 225–233.

[46] H. Oh, S.B. Bradfute, T.D. Gallardo, T. Nakamura, V. Gaussin, Y. Mishina, J. Pocius, L.H. Michael, R.R. Behringer, D.J. Garry, M.L. Entman, M.D. Schneider, Cardiac progenitor cells from adult myocardium: homing, differentiation, and fusion after infarction. Proc. Natl. Acad. Sci. U. S. A. 100 (2003) 12313–12318.

[47] K. Ohnishi, H. Yoshida, K. Shigeno, S. Nakamura, S. Fujisawa, K. Naito, K. Shinjo, Y. Fujita, H. Matsui, A. Takeshita, S. Sugiyama, H. Satoh, H. Terada, R. Ohno, Prolongation of the QT interval and ventricular tachycardia in patients treated with arsenic trioxide for acute promyelocytic leukemia. Ann. Intern. Med. 133 (2000) 881–885.

[48] C. Osorio-Yanez, J.C. Ayllon-Vergara, G. Aguilar-Madrid, L. Arreola-Mendoza, E. Hernandez-Castellanos, A. Barrera-Hernandez, A. De Vizcaya-Ruiz, L.M. Del Razo, Carotid intima-media thickness and plasma asymmetric dimethylarginine in Mexican children exposed to inorganic arsenic. Environ. Health Perspect. 121 (2013) 1090–1096.

[49] M. Rahman, M. Tondel, S.A. Ahmad, I.A. Chowdhury, M.H. Faruquee, O. Axelson, Hypertension and arsenic exposure in Bangladesh. Hypertension 33 (1999) 74–78.

[50] P.M. Ridker, Connecting the role of C-reactive protein and statins in cardiovascular disease. Clin. Cardiol. 26 (2003) III39–III44.

[51] G.J. Roboz, E.K. Ritchie, R.F. Carlin, M. Samuel, L. Gale, J.L. Provenzano-Gober, T.J. Curcio, E.J. Feldman, P.D. Kligfield, Prevalence, management, and clinical consequences of QT interval prolongation during treatment with arsenic trioxide. J. Clin. Oncol. 32 (2014) 3723–3728.

[52] H.G. Rosenberg, Systemic arterial disease with myocardial infarction. Report on two infants. Circulation 47 (1973) 270–275.

[53] H.G. Rosenberg, Systemic arterial disease and chronic arsenicism in infants. Arch. Pathol. 97 (1974) 360–365.

[54] P. Sanchez-Soria, D. Broka, S.L. Monks, T.D. Camenisch, Chronic low-level arsenite exposure through drinking water increases blood pressure and promotes concentric left ventricular hypertrophy in female mice. Toxicol. Pathol. 40 (2012) 504–512.

[55] H. Shan, Y. Zhang, B. Cai, X. Chen, Y. Fan, L. Yang, X. Chen, H. Liang, Y. Zhang, X. Song, C. Xu, Y. Lu, B. Yang, Z. Du, Upregulation of microRNA-1 and microRNA-133 contributes to arsenic-induced cardiac electrical remodeling. Int. J. Cardiol. 167 (2013) 2798–2805.

[56] J.C. Shen, K.Y. Liu, B. Jiang, X.J. Lu, D.P. Lu, Effect of the tetra-arsenic tetra-sulfide (As4S4) on the corrected QT interval in the treatment of acute promyelocytic leukemia. Zhonghua Xue Ye Xue Za Zhi. 25 (2004) 359–361.

[57] P.P. Simeonova, T. Hulderman, D. Harki, M.I. Luster, Arsenic exposure accelerates atherogenesis in apolipoprotein E(-/-) mice. Environ. Health Perspect. 111 (2003) 1744–1748.

[58] A.H. Smith, C.M. Steinmaus, Arsenic in drinking water. BMJ 342 (2011) d2248.

[59] N. Sohel, L.A. Persson, M. Rahman, P.K. Streatfield, M. Yunus, E.C. Ekstrom, M. Vahter, Arsenic in drinking water and adult mortality: a population-based cohort study in rural Bangladesh. Epidemiology 20 (2009) 824–830.

[60] S.L. Soignet, S.R. Frankel, D. Douer, M.S. Tallman, H. Kantarjian, E. Calleja, R.M. Stone, M. Kalaycio, D.A. Scheinberg, P. Steinherz, E.L. Sievers, S. Coutre, S. Dahlberg, R. Ellison, R.P. Warrell, Jr., United States multicenter study of arsenic trioxide in relapsed acute promyelocytic leukemia. J. Clin. Oncol. 19 (2001) 3852–3860.

[61] N.V. Soucy, M.A. Ihnat, C.D. Kamat, L. Hess, M.J. Post, L.R. Klei, C. Clark, A. Barchowsky, Arsenic stimulates angiogenesis and tumorigenesis in vivo. Toxicol. Sci. 76 (2003) 271–279.

[62] N.V. Soucy, D. Mayka, L.R. Klei, A.A. Nemec, J.A. Bauer, A. Barchowsky, Neovascularization and angiogenic gene expression following chronic arsenic exposure in mice. Cardiovasc. Toxicol. 5 (2005) 29–41.

[63] S. Srivastava, S.E. D'Souza, U. Sen, J.C. States, In utero arsenic exposure induces early onset of atherosclerosis in ApoE-/- mice. Reprod. Toxicol. 23 (2007) 449–456.

[64] S. Srivastava, E.N. Vladykovskaya, P. Haberzettl, S.D. Sithu, S.E. D'Souza, J.C. States, Arsenic exacerbates atherosclerotic lesion formation and inflammation in ApoE-/- mice. Toxicol. Appl. Pharmacol. 241 (2009) 90–100.

[65] J.C. States, A.V. Singh, T.B. Knudsen, E.C. Rouchka, N.O. Ngalame, G.E. Arteel, Y. Piao, M.S. Ko, Prenatal arsenic exposure alters gene expression in the adult liver to a proinflammatory state contributing to accelerated atherosclerosis. PLoS One 7 (2012) e38713.

[66] A.C. Straub, K.A. Clark, M.A. Ross, A.G. Chandra, S. Li, X. Gao, P.J. Pagano, D.B. Stolz, A. Barchowsky, Arsenic-stimulated liver sinusoidal capillarization in mice requires NADPH oxidase-generated superoxide. J. Clin. Invest. 118 (2008) 3980–3989.

[67] A.C. Straub, D.B. Stolz, M.A. Ross, A. Hernandez-Zavala, N.V. Soucy, L.R. Klei, A. Barchowsky, Arsenic stimulates sinusoidal endothelial cell capillarization and vessel remodeling in mouse liver. Hepatology 45 (2007) 205–212.

[68] A.C. Straub, D.B. Stolz, H. Vin, M.A. Ross, N.V. Soucy, L.R. Klei, A. Barchowsky, Low level arsenic promotes progressive inflammatory angiogenesis and liver blood vessel remodeling in mice. Toxicol. Appl. Pharmacol. 222 (2007) 327–336.

[69] C.H. Tseng, Blackfoot disease and arsenic: a never-ending story. J. Environ. Sci. Health C Environ. Carcinog. Ecotoxicol. Rev. 23 (2005) 55–74.

[70] C.H. Tseng, Cardiovascular disease in arsenic-exposed subjects living in the arseniasis-hyperendemic areas in Taiwan. Atherosclerosis 199 (2008) 12–18.

[71] C.H. Tseng, Y.K. Huang, Y.L. Huang, C.J. Chung, M.H. Yang, C.J. Chen, Y.M. Hsueh, Arsenic exposure, urinary arsenic speciation, and peripheral vascular disease in blackfoot disease-hyperendemic villages in Taiwan. Toxicol. Appl. Pharmacol. 206 (2005) 299–308.

[72] D. Unnikrishnan, J.P. Dutcher, N. Varshneya, R. Lucariello, M. Api, S. Garl, P.H. Wiernik, S. Chiaramida, Torsades de pointes in 3 patients with leukemia treated with arsenic trioxide. Blood 97 (2001) 1514–1516.

[73] D.E. Vaughan, PAI-1 and atherothrombosis. J. Thromb. Haemost. 3 (2005) 1879–1883.

[74] T.J. Wade, Y. Xia, K. Wu, Y. Li, Z. Ning, X.C. Le, X. Lu, Y. Feng, X. He, J.L. Mumford, Increased mortality associated with well-water arsenic exposure in Inner Mongolia, China. Int. J. Environ. Res. Public Health 6 (2009) 1107–1123.

[75] C.H. Wang, C.L. Chen, C.K. Hsiao, F.T. Chiang, L.I. Hsu, H.Y. Chiou, Y.M. Hsueh, M.M. Wu, C.J. Chen, Increased risk of QT prolongation associated with atherosclerotic diseases in arseniasis-endemic area in southwestern coast of Taiwan. Toxicol. Appl. Pharmacol. 239 (2009) 320–324.

[76] C.H. Wang, C.K. Hsiao, C.L. Chen, L.I. Hsu, H.Y. Chiou, S.Y. Chen, Y.M. Hsueh, M.M. Wu, C.J. Chen, A review of the epidemiologic literature on the role of environmental arsenic exposure and cardiovascular diseases. Toxicol. Appl. Pharmacol. 222 (2007) 315–326.

[77] C.H. Wang, J.S. Jeng, P.K. Yip, C.L. Chen, L.I. Hsu, Y.M. Hsueh, H.Y. Chiou, M.M. Wu, C.J. Chen, Biological gradient between long-term arsenic exposure and carotid atherosclerosis. Circulation 105 (2002) 1804–1809.

[78] F. Wu, F. Jasmine, M.G. Kibriya, M. Liu, O. Wojcik, F. Parvez, R. Rahaman, S. Roy, R. Paul-Brutus, S. Segers, V. Slavkovich, T. Islam, D. Levy, J.L. Mey, G.A. van, J.H. Graziano, H. Ahsan, Y. Chen, Association between arsenic exposure from drinking water and plasma levels of cardiovascular markers. Am. J. Epidemiol. 175 (2012) 1252–1261.

[79] F. Wu, P. Molinaro, Y. Chen, Arsenic exposure and subclinical endpoints of cardiovascular diseases. Curr. Environ. Health Rep. 1 (2014) 148–162.

[80] M.H. Wu, C.J. Lin, C.L. Chen, M.J. Su, S.S. Sun, A.L. Cheng, Direct cardiac effects of As_2O_3 in rabbits: evidence of reversible chronic toxicity and tissue accumulation of arsenicals after parenteral administration. Toxicol. Appl. Pharmacol. 189 (2003) 214–220.

[81] M.M. Wu, H.Y. Chiou, Y.M. Hsueh, C.T. Hong, C.L. Su, S.F. Chang, W.L. Huang, H.T. Wang, Y.H. Wang, Y.C. Hsieh, C.J. Chen, Effect of plasma homocysteine level and

urinary monomethylarsonic acid on the risk of arsenic-associated carotid atherosclerosis. Toxicol. Appl. Pharmacol. 216 (2006) 168–175.

[82] Y.F. Xiao, S.X. Liu, D.D. Wu, X. Chen, L.F. Ren, Inhibitory effect of arsenic trioxide on angiogenesis and expression of vascular endothelial growth factor in gastric cancer. World J. Gastroenterol. 12 (2006) 5780–5786.

[83] M.H. Yang, Y.S. Zang, H. Huang, K. Chen, B. Li, G.Y. Sun, X.W. Zhao, Arsenic trioxide exerts anti-lung cancer activity by inhibiting angiogenesis. Curr. Cancer Drug Targets 14 (2014) 557–566.

[84] A. Yildiz, M. Karaca, S. Biceroglu, M.T. Nalbantcilar, U. Coskun, F. Arik, F. Aliyev, O. Yiginer, C. Turkoglu, Effect of chronic arsenic exposure from drinking waters on the QT interval and transmural dispersion of repolarization. J. Int. Med. Res. 36 (2008) 471–478.

[85] Y. Yuan, G. Marshall, C. Ferreccio, C. Steinmaus, S. Selvin, J. Liaw, M.N. Bates, A.H. Smith, Acute myocardial infarction mortality in comparison with lung and bladder cancer mortality in arsenic-exposed region II of Chile from 1950 to 2000. Am. J. Epidemiol. 166 (2007) 1381–1391.

[86] K.M. Zierold, L. Knobeloch, H. Anderson, Prevalence of chronic diseases in adults exposed to arsenic-contaminated drinking water. Am. J. Public Health 94 (2004) 1936–1937.

21

INVERTEBRATE MODELS IN ARSENIC RESEARCH: PAST, PRESENT, AND FUTURE

IAIN L. CARTWRIGHT
Molecular Genetics, Biochemistry & Microbiology, University of Cincinnati College of Medicine, Cincinnati, OH, USA

21.1 INTRODUCTION—WHITHER INVERTEBRATES?

For many professional toxicologists, a question regarding the most appropriate organism in which to pursue their studies might well precipitate a spirited debate, but it would probably be fair to say that invertebrate models would likely play a vanishingly small part in the conversation. There are a number of credible arguments for this point of view, among which the most prominent would likely be that a suitable animal model on which to draw conclusions relevant to human health should be one with very strong parallels in terms of metabolic biochemistry, systemic and cellular physiology, and general anatomy, as well as issues such as mode of reproduction, and program of development from embryo through adult. In other words, a vertebrate mammalian model with close evolutionary connections to *Homo sapiens* would be one obvious choice (e.g., chimpanzee). However, for practical reasons, most fundamental medical (and toxicological) research in animal models has been performed on smaller, more manageable mammals, that is, rodents, owing to the availability of pure-bred strains, relative ease of handling, reliability of measurable outcomes, relatively fast reproduction in controlled environments, and to some degree the economic (as well as ethical) considerations which make larger vertebrates much less attractive. Although there is a

Arsenic: Exposure Sources, Health Risks, and Mechanisms of Toxicity, First Edition.
Edited by J. Christopher States.
© 2016 John Wiley & Sons, Inc. Published 2016 by John Wiley & Sons, Inc.

long and rich history of toxicological studies on rats, based in part on sizeable organs and demonstrable metabolic parallels to humans, in recent times the mouse has become the standard biomedical research organism, at least for those who would seek to find molecular explanations and mechanistic pathways associated with human disease. This has largely come about because of the tremendous advances that have occurred in genetic and molecular manipulation in the mouse over the past few decades. With the increasing pursuit of molecular correlates for phenotypic responses and pathological outcomes arising from xenobiotic exposure, it is no surprise that most toxicological work has also been, and will likely continue to be, performed on murine models.

However, there is an important general question worth asking that is not diminished by such considerations: Are there alternative organism-based approaches that are faster, cheaper, more comprehensive in scale and efficiency, and reasonably likely to produce biological data of significance? This question is clearly somewhat loaded in the context of a chapter on invertebrate models in arsenic research, but the answer really depends on what features of the toxicological response are being investigated. For a toxicant such as arsenic, with such a plethora of documented pathological outcomes attributed to its long-term ingestion, a key issue becomes what are the molecular pathways and signaling programs with which arsenic interacts and presumably interferes? A reasonable assumption based on data gathered to this point is that there are many such "toxicity pathways," and several are likely interacting with one another when considered from a systems biology point of view. The question really devolves to one regarding the molecular mechanism(s) by and through which arsenic acts at the cellular level. It has become increasingly clear that many of the biochemical and physiological features of eukaryotic cells and organisms are shared in common across evolutionary time. Moreover, even when considering the simplest single-celled eukaryotes such as yeasts, the evolutionary conservation of genes (documented through a plethora of genome projects) has strongly bolstered this view. Consequently, it seems advantageous to consider, alongside established vertebrate models, a biological model (or models) where the range of experimental approaches available (particularly those involving genetic analysis and manipulation) could allow a broader inquisition into the myriad possibilities that might be in play in the interaction between arsenic (and, obviously, other environmental insults) and biological pathways.

A call for a paradigm shift in toxicity testing toward the uncovering of these so-called "toxicity pathways" has been made in several quarters by organizations such as the National Toxicology Program of NIEHS, the Environmental Protection Agency, and the Human Genome Research Institute at NIH [15, 20, 69]. Included in the vision has been the implementation of extensive *in vitro* testing in appropriate cell types, using genomics technologies and high-throughput approaches for maximum efficiency, and also including the incorporation of lower organism *in vivo* studies. It remains true that the genetics and genomics revolution of the past decade currently provides its most robust *in vivo* discovery platforms in such lower organisms, encouraging researchers to heed this call to incorporate invertebrates into their research portfolios, and foreshadowing their growing importance to the field.

Given these considerations, the model eukaryotic systems to be discussed here, that is, the single-celled fungal yeasts *Saccharomyces cerevisiae* (budding) and

Schizosaccharomyces pombe (fission), the relatively simple multicellular round worm *Caenorhabditis elegans* (a nematode), and the complex multicellular fruit fly *Drosophila melanogaster* (an arthropod), appear likely to provide significant molecular "leads" regarding insight into the molecular pathways by which common xenobiotics (including arsenic) exert their effects. As will be emphasized frequently later, it is the considerable genetic advantages offered by such organisms in experimental pathway discovery that makes them so attractive for consideration. Of course, when new experimental information derived from such systems comes to hand, the next step has to be validation of its relevance in a more traditional model vertebrate. This seems to be an efficient way forward: the simpler eukaryotic model providing a robust platform on which to perform comprehensive genome-wide testing for identification of interacting candidate genes/pathways, followed by careful examination of the same in a suitable vertebrate exposure model. In this commentary, a rather loose definition of invertebrate will be applied—while these model organisms are certainly representative of eukaryotes, the yeasts are fungi while the other two are animals without backbones and thus are true invertebrates according to taxonomy. However, given the history of yeasts as powerful discovery systems in eukaryotic genetics, they will be accorded honorary status of invertebrate for the present purpose. In the following sections, references to "yeast" will refer to *S. cerevisiae* by default; where *S. pombe* was used experimentally it will be explicitly noted.

It is important to throw in a word or two of caution early in this discussion of invertebrates in toxicological (and specifically arsenic) research. There are clearly some significant biological differences between mammals and the organisms considered here. Uptake, distribution, metabolism, and excretion, key features to be considered in pharmacodynamics, are obviously also essential considerations in any holistic model that attempts to describe the interaction of a toxicant such as arsenic with the human body. All of these organisms seem relatively resistant to concentrations of arsenic that would be highly toxic in mammalian tissue culture, but obviously it is very hard to know the actual intracellular concentrations encountered in a given situation owing to variations in the aforementioned dynamic parameters from one organism to the next. Metabolic transformation of arsenic is also lacking in these models, but the transgenic introduction of human genes known to metabolize arsenic (e.g., [64]) provides a versatile experimental platform in which to test the relative *in vivo* toxicity of different metabolites. Obviously, yeast cannot provide data regarding organ or tissue-specific toxicity, but in each case these models display numerous attributes (see later) that are likely to aid in uncovering molecular aspects of many basic "toxicity pathways" that define the interaction of arsenic with physiological systems.

21.2 HISTORICAL USE OF INVERTEBRATES IN ARSENIC RESEARCH

Each of the organisms considered here has its own relative experimental advantages (Table 21.1 lists some general features), and has a long tradition as a model organism for either genetic studies or developmental biology, or both. However, it is really only since around the turn of the current century that the outstanding genetic advantages

TABLE 21.1 Invertebrate Models: Distinctive Features and Relative Advantages

Yeast	*Caenorhabditis elegans*	*Drosophila*
Single-cell eukaryote; liquid or plate culture growth; rapid replication and cell division every 90 min	Simple multicellular eukaryote; growth on *Escherichia coli* lawn plates from embryo, via four larval stages (L1, L2, L3, L4), to adult in ca. 72 h	Complex multicellular eukaryote; growth from embryo, via three larval stages (L1, L2, L3) and one pupal stage, to adult in ca. 10 days
Typically grown as haploids; a and α mating types provide access to standard genetic/mutational analysis through a/α diploid formation followed by sporulation and meiotic gene segregation	Laboratory strains typically self-fertilizing diploid hermaphrodites (XX homozygotes); males (XO) arise at low frequency but have strong competitive advantage in mating with hermaphrodites, hence access to standard genetic/mutational analysis facilitated	Male (XY) and female (XX) diploid organisms with standard sexual reproduction, allowing facile classical genetic analysis; lack of recombination in males a useful feature exploited in mutant strain creation
Completely sequenced genome; ca. 6000 genes, approximately 31% have human gene homology	Completely sequenced genome; ca. 20,000 genes, approximately 40% of genes show homology to human genes	Completely sequenced genome; ca. 13,000 genes, approximately 50% show homology to human genes, but homology with known human disease alleles rises to 75%
Many intracellular signaling pathways conserved with mammals	Many intra- and intercellular signaling pathways conserved with mammals	Very high level of intra- and intercellular signal pathway conservation with mammals
Exhibits basic cellular processes common to all eukaryotes	Completely defined cell lineage; transparent; neuronal/synaptic, apoptotic, and simple movement/behavioral phenotypes highly accessible	Many strong organ/tissue homologies with mammals; sophisticated behavior testing possible, for example, learning/memory/social interaction; frequent human disease model, for example, neurological, cancer, innate immunity, cardiovascular, and metabolic disease
Genetically accessible—facile gene replacement; complete set of haploid and diploid gene knockouts available; simple recombination-mediated gene replacement, gene knockout, gene overexpression, etc.	Genetically accessible—easy gene knockdown via RNAi by feeding *E. coli*-expressing shRNA; large range of gene mutants available; easy transgenic manipulation for overexpression, etc.	Genetically accessible—complete RNAi transcriptome libraries available; very wide range of extant mutants; precise gene targeting relatively easy; simple transgenic creation; robust systems for inducing stage- or tissue-specific transgene expression and for induction of recombination-mediated gene knockouts (mosaic analysis) or gene replacement

exhibited by these organisms, in terms of mutant analyses, genetic screens, etc., have begun to be exploited to any real degree in toxicological research. Prior to that time, these represented cheap, easily cultured, rapidly reproducing, and readily examined experimental platforms on which to conduct fairly simple toxicity assays (effect on growth rates, longevity, live/dead endpoints, etc.). Thus, for example, it was found that the survival of *Drosophila* upon either arsenite or arsenate exposure was dose-dependent, with arsenite being significantly more toxic at all life stages [27]. However, it was also noted that adult *Drosophila* were much more tolerant as regards survival to a given dose of arsenic than were larvae proceeding through development, suggesting that key developmental processes were quite sensitive to arsenic. In conjunction with early studies showing arsenate led to lower fecundity in *Drosophila* at sublethal doses [73], the data strongly suggested that pathways important for fertility, embryogenesis, and early development represent important areas for consideration regarding the mode of action. More recent mammalian data strongly corroborate this view (e.g., [43, 46, 105]).

Another area of particular interest in the early days of arsenic research was the possibility that it could act as a mutagen. It was apparent that one pathological response to long-term arsenic ingestion was the appearance of a variety of organ or tissue cancers, and so it seemed likely that arsenic could be a direct mutagen at the level of DNA, as seemed to be the case for some other metals, such as chromium. However, studies in yeast and flies, as well as a variety of vertebrate animals and *in vitro* cell culture, produced little evidence for direct mutational events on DNA, although clastogenic events involving chromosomal breakage of one kind or another were frequently documented [33]. Levels of sister chromatid exchange, a marker for mitotic recombination, were quite noticeable, and a so-called SMART assay system (Somatic Mutation and Recombination Test) developed in *Drosophila* was consistent with this, noticeably showing that arsenite was an order of magnitude more potent at inducing recombination events than was arsenate [74]. However, it is fair to say that other studies were not able to reproduce such data at lower, nontoxic concentrations [78, 102]. It turns out that the relatively low sensitivity to inorganic arsenicals in these chromosomal instability tests is almost certainly attributable to the lack of arsenic methylation in wild-type *Drosophila* [79]. As will be further explained , this apparent disadvantage of *Drosophila* as compared to vertebrates (where arsenic methylation is observed as part of cellular metabolism) can likely be utilized to great advantage in probing differential effects of methylated versus inorganic arsenicals.

As molecular biological tools became a part of the discovery arsenal for toxicology, such approaches were applied to these organisms with some success, often corroborating or elaborating on features that were concurrently being described in vertebrate models, predominantly using *in vitro* tissue culture. For example, one of the universal findings from yeast to man is that cells exposed to arsenite exhibit a marked stress response, an easily characterized feature of which is typically the expression of one or more heat-shock transcripts and their encoded proteins [88]. It is likely that chronic exposure to arsenite has a pronounced effect on cellular (particularly protein) homeostasis, leading to continuous cellular stress; the opportunity to investigate this and other stress responses from a genetic standpoint, being highly facilitated in these invertebrate models, is likely to be of particular importance in the future for identifying pathways that may be involved in the progression to (certain) arsenic-associated human diseases. Such considerations will be considered more fully in the following sections.

A very interesting (and largely ignored) observation accompanying arsenite (and heat-shock) induced stress in *Drosophila* was that significant changes in the post-translational modification of histone proteins, specifically methylation and acetylation, occurred in parallel [2, 12, 16, 17]. Given the strong contemporary interest in epigenetic changes correlated with arsenic exposure (see Chapter 18) it seems clear that these experiments were far and away ahead of their time, but perhaps it also serves to emphasize the point that important basic information is quite likely to be derived from data generated in an invertebrate model.

21.3 COMBINING GENETICS AND MOLECULAR BIOLOGY IN CONTEMPORARY RESEARCH

As detailed elsewhere in this volume, a good many of the damaging effects of arsenic within cells have been ascribed to reactive oxidative species (ROS) generation, or to the known proclivity of arsenic (specifically As(III) species) to bind with free thiols, for example, in glutathione or in proteins, particularly avidly to vicinally juxtaposed pairs of thiols. As is the case for most toxic metals, cells have accumulated a series of defenses to minimize or ameliorate the impact that arsenic will have on the biochemical machinery operating within. In broad terms, these can be categorized as effects on transport (i.e., uptake and excretion), bioavailability (*in situ* detoxification and/or metabolism), and damage control via various forms of stress response aimed at restoring homeostasis. With the ability to identify specific cellular–arsenic interaction pathways that are made more realistic by combining the technology and tools of molecular biology with the traditional strength of genetic manipulability of these model invertebrates, discoveries in all three areas have helped inform and, in some cases, guide research in vertebrate systems.

21.3.1 Transport

21.3.1.1 Cellular Uptake Since arsenate is an inorganic phosphate analog, it comes as no particular surprise that the uptake of arsenate into cells is mediated by phosphate transporters. Genetic analysis in yeast has identified several *PHO* genes (e.g., *PHO84* and *PHO87*) that lead to arsenate resistance when mutated [9, 10]. These cells are not, however, resistant to arsenite, which can enter cells through several different pathways. At neutral pH, arsenite (unlike arsenate) exists predominantly as an uncharged entity (i.e., $As(OH)_3$) and can enter cells through a pore defined by the yeast *fps1* gene. Fps1p is a member of the aquaglyceroporin (AQP) family of membrane channels that allow uncharged solutes such as glycerol to pass into and out of cells [109]). Subsequent genetic analysis has uncovered an important pathway of regulation [99]. Immediately upstream of *fps1* is the MAP kinase Hog1p, which can regulate closing of the pore via phosphorylation of Fps1p. Hog1p is itself phosphorylated in response to arsenite, and other members of an important stress response-activated MAP kinase cascade upstream of Hog1p are clearly involved in responding to arsenite [99]. The net effect is that the pore becomes less permeable—cells that overexpress Hog1p are relatively arsenite-resistant, while cells deleted for Hog1p are relatively arsenic-sensitive. Things are a little more complicated than this, however, since overexpression of Fps1p leads to arsenite resistance rather than sensitivity—it turns out that the AQP channel is bidirectional in nature and the flow of arsenite out of the cell is more highly favored in this deregulated condition [58]. Most interestingly, even though AQPs do not allow arsenate passage, these genetic manipulations also affect arsenate sensitivity—cells become arsenate-tolerant when Fps1p is overexpressed and sensitive when the *fps1* gene is inactivated. The most reasonable explanation is that yeast cells harbor an efficient arsenate reductase [5, 63], denoted as Acr2p, such that intracellular levels of arsenite are increased as arsenate is transported in by the PHO transporters mentioned earlier. Arsenite then flows out of the cell via AQP down a concentration gradient. Hence, the AQPs play essential roles in both arsenite and arsenate sensitivity [58]. In congruence with these data, it was found that mammalian cells utilize certain members of a more extensive AQP family both for arsenite passage [54] and for the methylated metabolites produced in mammals [56]. Signaling pathways homologous to those revealed by these yeast genetic analyses are clearly worth investigating for allelic variants in humans, given the well-known inter-individual variation in sensitivity to arsenic exposure.

Research in yeast identified a further means of arsenite uptake, namely via the family of hexose permease proteins [55], although it seems likely that arsenite itself is not being recognized as a form of sugar, given that the transport of trivalent methylated arsenicals was not competitive with that of glucose. The importance of these yeast studies was later confirmed in mammalian cells, where similar observations for the GLUT family of hexose transporters suggested distinct translocation pathways used by them for glucose as compared to trivalent arsenicals [35].

21.3.1.2 Cellular Efflux As mentioned earlier, AQP channels play an important role in modulating internal trivalent arsenic levels via both cellular inflow and efflux. Moreover, the reduction of imported arsenate to arsenite in yeast by Acr2p may also occur in human cells, where the Cdc25B and Cdc25C phosphatases, bearing some structural homology to the Acr2 protein, have been shown to possess adventitious arsenate reductase activity also [3]. Thus, the AQP proteins are likely critical components of both As(III) and As(V) sensitivity in all eukaryotic cells.

In yeast, an additional membrane protein, Acr3p, has been characterized as an As(III)-specific antiporter, which can pump arsenite out of the cell against a concentration gradient in exchange for protons [59]. No homologs of this protein seem to exist in higher eukaryotes. However, both yeast and *C. elegans* have demonstrable homologs to the mammalian membrane-associated ATP-binding cassette (ABC) transporter proteins (i.e., multidrug resistance associated protein (MRP) and multidrug resistant (MDR) families) that, when mutated, display significantly heightened sensitivity to arsenite [8, 25]. Effective chelation of arsenite by glutathione

(GSH) in the form $As(GS)_3$, mediated by glutathione S-transferases, is an important component for export specificity by these transporters, at least for the MRP class [48]. With these observations as a starting point, experiments investigating both overexpression and knockout of homologous ABC genes in vertebrates demonstrated that active transport (as opposed to passive flow) of arsenite and its methylated metabolites out of cells in higher eukaryotes relies on both MDR [52, 53] and MRP [39, 47] members of these ABC transporter families. Furthermore, these studies, first conducted with invertebrate models, highlighted the essential role of glutathione in mediating arsenic toxicity, a subject further considered later.

An interesting story has emerged in regard to the function of the eukaryotic homolog of the bacterial *arsA* gene. In bacteria, this gene is part of an arsenic-resistant operon comprising *arsA, B,* and *C* [14]. ArsA is characterized as a membrane-bound ATPase responsible for arsenite efflux from bacterial cells; ArsB is an accessory protein for ArsA function, while ArsC is an arsenate reductase [81]. All eukaryotes appear to contain a homolog of the ArsA protein, generally designated ASNA-1 [45], although variously known in yeast as Arr4p or Get3p. However, although both yeast and *C. elegans* deficient for ASNA-1 show strong arsenite sensitivity [87, 103], it seems clear that the eukaryotic protein is neither membrane-bound nor likely involved in arsenic extrusion. Although the ATPase activity is conserved [87], and is stimulated somewhat by arsenite in *C. elegans* [103], it has been found that ASNA-1 is part of a lumenal complex in the endoplasmic reticulum (ER) that participates in the insertion of so-called tail-anchored proteins into the ER membrane [92]. Most interestingly, ASNA-1 seems to be intimately involved in the pathway of insulin secretion, both in *C. elegans* intestinal cells and in human pancreatic beta cells [41]. A specific larval stage growth-arrest phenotype exhibited by *C. elegans* deficient in ASNA-1 is fully complemented by the human gene, suggesting very strong conservation of function. A recent RNAi-based genetic screen in *C. elegans* for this particular growth-defective phenotype has revealed numerous additional genes, predicted to interact with ASNA-1, and likely to be involved in pathways of insulin signaling [4]. Given the strongly implied connection between arsenic exposure and type 2 diabetes in human populations [61], this invertebrate model holds out particular promise for more detailed mechanistic studies, integrating both genetics and biochemistry.

21.3.2 Intracellular Metabolism

As a result of many years of study regarding the fate of arsenic inside vertebrate cells, there are two particular features that stand out: first, that the intracellular antioxidant glutathione plays an important role in modulating arsenic toxicity (see [48] for a review); second, that inorganic arsenic is modified by methylation to produce mono- and dimethylated species of both the +3 and +5 valence states, with evidence of trimethylated species in some cell types [94, 97]. As mentioned earlier, the conjugation of glutathione with inorganic As(III) species appears to play an essential first step in pathways of ATP-dependent transmembrane export, from yeast (where it is actually intracellularly vacuolarized) to vertebrate cells, and it seems clear that transportation of the methylated As(III) species is also effected via similar GSH-linked conjugates [40, 94, 97]. Given the typically high concentrations of GSH in cells (0.5–10 mM) and the avidity with which As(III) species bind to thiols, it has been estimated that up to 99% of the arsenic in a cell will be present in such GSH conjugates [42]. The biosynthetic pathway for GSH is conserved across all eukaryotes so that the observation that mutation of the *C. elegans* glutamate–cysteine ligase gene (*GCLC*, which catalyzes the first step in the two-step production of the GSH tripeptide) leads to strongly increased sensitivity to arsenic [50], particularly As(III), is fully in accord with the idea that GSH plays an essential protective role toward arsenic toxicity in all animals.

Given these facts, as well as the central importance of GSH in maintaining redox homeostasis in cells, it is perhaps surprising that there has been little effort to look at possible genetic variation in GSH biosynthetic and recycling pathways in humans exposed to arsenic. Future attention to this oversight may be considered all the more appropriate when one considers the data generated by Ortiz et al. [70] in their examination of the genetic basis of tolerance or sensitivity to arsenite in *Drosophila*. This investigation was predicated on the notion that an examination of natural variation in sensitivity to arsenite across geographically widely distributed strains of *Drosophila* could lead to a completely unbiased identification of a gene (or genes) that might contribute to such variation. In short, let Nature show the way! By using a combination of classical genetic analysis with several of the modern genetic manipulation approaches available in this organism, this work demonstrated that the sensitivity phenotype mapped to a very small region of the X chromosome containing the glutathione synthetase gene (*GSS*). Knocking down expression of this gene in living organisms by RNA interference dramatically sensitized them to arsenite, such that no visible larval development occurred at concentrations of arsenite that were 50-fold lower than those that it took to kill a wild type adult. GSS is the second enzyme in the two-step GSH biosynthetic pathway (following GCLC), and these data strongly suggested that normal operation of this pathway is an essential defense against the depredations that can be wrought by arsenic during critical periods of organismal growth and tissue specification. A consideration of such effects in the context of human fetal development and growth is thus not unreasonable.

Metabolism of ingested inorganic arsenic to create multiple methylated species seems to be a virtually universal response to arsenic ingestion in vertebrates, and accrued evidence has led to the current conception that some of these species, particularly those of the +3 valence state, are considerably more toxic to cells than the inorganic precursor [23, 97]. This association has been extended into epidemiologic studies as well, where correlations between the presence of methylated species (and/or polymorphisms in the enzyme that catalyzes their formation) and the incidence of disease in arsenic-exposed individuals have been drawn [93, 104]. The enzyme concerned, arsenic(III) methyltransferase (AS3MT), seems to be encoded in the deuterostome lineage of animals, as well as in bacterial and archaeal species [98]. However, the AS3MT gene appears not to be present in either the protostome or fungal eukaryotic lineages, which includes the various model organisms discussed here. Apparent experimental verification of this, in *Drosophila* at least, was provided by studies showing no evidence for any methylated arsenic metabolites after both larvae and adult flies were fed with inorganic arsenite [79]. The same outcome has been reported for yeast growing in arsenite-containing medium, although no data were actually shown [76]. Given this, one might ask what role can these organisms play in questions regarding arsenic methylation and its effects? A partial answer to this came from an additional observation that when flies were fed the methylated arsenical, dimethylarsinic acid, a strong positive response was obtained in the SMART assay, a measurement of genomic instability via loss of heterozygosity [79]. No response was seen upon administration of inorganic arsenite, therefore strongly implicating the methylated arsenical as the culprit behind such genomic instability, and suggesting that *Drosophila* might be a good vehicle to study this important aspect of arsenic bioactivity. In response to this idea, a transgenic *Drosophila* model in which the human *AS3MT* (*hAS3MT*) gene can be inducibly expressed has been reported [64]. These investigators showed that in the presence of inorganic arsenite, *hAS3MT* transgenic flies were able to metabolize the ingested metal to both mono- and dimethylated derivatives only upon *hAS3MT* gene induction. Such a controllable model will likely have many uses in the future, but an important initial observation was that genomic instability was strongly enhanced only in induced flies fed sublethal amounts of arsenite throughout larval development. Thus, this model provides an experimentally manipulable platform upon which further genetic analysis may readily be performed. Specifically, utilizing the ability to knock down virtually any gene of interest *in vivo* using an available *Drosophila* RNAi transcriptome library [18], these *hAS3MT* transgenic flies can be monitored for suppression or enhancement of any particular phenotype of choice, thus opening a window into the molecular mechanisms and/or pathways compromised in a significant way by methylated arsenicals as compared to inorganic species. A further result from this report was particularly intriguing. Adult flies induced for hAS3MT expression and fed arsenite at relatively toxic levels survived longer than their nontransgenic (and uninduced transgenic) siblings, even though they were more susceptible as larvae (and at lower arsenic doses) to the genomic instability described earlier. Whole-body elemental analysis showed that these induced transgenic flies contained less arsenic than those not expressing hAS3MT [64]. The data are most consistent with the idea that methylated arsenicals are more readily excreted from cells and thus methylation functions as a detoxification mechanism [97]. However, the data also show that, at lower concentrations, methylated arsenicals, while not necessarily being lethal, are much more likely to create molecular lesions that would have long-term pathological consequences in a long-lived vertebrate, such as a human.

These examples, drawn mostly from *Drosophila*, show how such an organism can be utilized to provide new insights, and suggest the potential for further insight when more comprehensive genetic analyses are undertaken. For example, in pursuing the theme of arsenic methylation, great interest has been aroused by the discovery that the *hAS3MT* gene displays numerous polymorphic variants [108], and that some of these are correlated with varying urinary profiles of methylated species and/or susceptibility to disease in arsenic-exposed populations [1, 85, 93, 104]. Recently, an *in vitro* study of the biochemical properties of both the wild-type and an M287T variant of hAS3MT has been reported [19]. Intriguingly, the two variants are differentially sensitive to the presence of GSH in terms of their catalytic properties in conversion of arsenite to mono- and dimethylated species. These interactions could be very usefully studied *in vivo* by adapting (and combining) the two *Drosophila* systems described earlier, that is, those that are mutant for GSH synthesis and others that are transgenic for *hAS3MT* or its polymorphic variants.

21.3.3 Stress and Homeostasis

Virtually all eukaryotic cells exhibit a set of rather highly conserved stress response pathways that are crucial for the maintenance of cellular homeostasis in the face of environmental insult [88]. Most, if not all, of these pathways have the ability to activate a decision "switch" in response to sustained or high-level stress, whereby cellular recovery becomes impossible and the cell will typically progress toward an apoptotic fate. Through observations in many different kinds of cellular systems, it has been well documented that a number of these stress pathways are induced in response to arsenic exposure, and at concentrations usually much lower than are required to see overt phenotypic responses to toxicity, such as cell death. As mentioned earlier, induction of heat-shock stress proteins in response to arsenic has been observed in all eukaryotic cells and organisms for which it has been tested. Also well documented is the generation of ROS in many eukaryotic cells and organisms upon arsenic exposure, typically leading to an oxidative stress response. Numerous signal transduction pathways have been associated with this particular response, and there is good evidence in the literature for arsenic involvement in activating one or both of the JNK and p38 stress-activated protein kinase cascades, presumably by virtue of ROS generation [44]. However, genetic experiments in *S. pombe* show that cells are not simply responding to arsenic as a progenitor of ROS, since the activation of p38, produced by both arsenite and hydrogen peroxide exposure, is mediated by a particular upstream pathway that is essential for response to one toxicant but not the other [80]. This point is brought home again when considering another common response to environmental stress, namely the phosphorylation of eukaryotic initiation factor 2 alpha (eIF2α), which leads to significant downregulation of translation. There are several different kinase species that perform this phosphorylation and, once again, genetic studies in *S. pombe* show that different kinases are activated by hydrogen peroxide as compared to arsenite-induced stress [111], suggesting that the upstream signal transduction pathways responding to these stresses are nonidentical. These two examples, utilizing the power of genetics to explore stress-mediated signal transduction pathways, serve to reinforce the point that although ROS have been widely seen as likely instigators of arsenic-provoked pathologies, eukaryotic cells also respond to arsenic-induced stress at a molecular level that is separate from, or additional to, that mediated purely by ROS generation. The current data support the idea that there are both overlapping and nonoverlapping pathways in the cellular response to ROS and arsenic.

This important theme of using genetic approaches to unravel further details of signaling pathways induced by toxicants lies behind the creation of a multitransgenic gene reporter-based system in *Drosophila* that responds visually, via pathway-specific fluorescent wavelengths, when particular stress-inducing conditions are encountered [13]. Thus, signaling that leads to activation of either or both of the transcription factors AP-1 (a downstream effector of the JNK stress pathway) and Nrf2 (a downstream effector of ROS-induced stress) can be visualized side by side in whole animals and easily compared in individual tissues and at any developmental stage. In fact, preliminary experiments show both reporters strongly expressed in response to arsenite [13]. This system could be subjected to a genetic screen, using RNAi-based gene knockdown for example [108], to determine the precise nature of the "toxicity pathways" induced by arsenite, and should allow nodes of cross-talk and/or intersection of pathways to be identified. Moreover, the system has been constructed very methodically by using a series of precisely defined integration sites engineered into the *Drosophila* genome [28], allowing other stress pathway-linked reporter genes (representing p38 or NF-κB-associated stress, for example) to be incorporated as desired while eliminating variable output related to position effects, low expression insertion sites, or other artifacts that have often plagued previous reporter gene systems. Such considerations are important in the case of arsenic because its chemical properties, both as a metalloid and as a strong electrophile (particularly in regard to its thiol reactivity), appear to check the boxes of a variety of stress sensors. Moreover, this propensity for arsenic to provoke several different kinds of stress response is potentially related to the heterogeneous range of pathological outcomes that prolonged stress derived from chronic arsenic ingestion produces in genetically diverse individuals.

One particularly important additional pathway is the so-called ER stress response (otherwise known as the unfolded protein response or UPR), which is receiving increasing attention for potential involvement in various types of chronic disease, for example, neurodegenerative conditions such as Alzheimer's, inflammatory disease, diabetes, and cancer [107]. The UPR is seen as a crucial homeostatic mechanism for all cells exposed to conditions that lead to protein unfolding, misfolding, or aggregation in the ER. Through activation of three sensor-dependent pathways, the UPR is able to reduce the protein-synthetic capacity of the ER while at the same time increasing the presence of factors such as chaperones that can help refold or otherwise disaggregate proteins, thereby helping restore a condition of protein homeostasis [106]. Proteins that are irreparably damaged can be targeted for proteosomal degradation via translocation to the cytosol in a process known as ER-associated degradation [89]. The course of correct protein folding in the ER is intimately connected with the oxidation and isomerization of protein thiols, a process that arsenic, with its high reactivity toward such functional groups, seems eminently positioned to corrupt. Indeed, several studies in invertebrate systems have called particular attention to the protein synthesis, folding, and degradation pathways as sites where arsenic-induced stress has significant impact. One productive line of research in this area has found that a strongly arsenite-inducible RNA-binding protein (ZFAND2A, *aka* AIRAP, AIP-1) highly conserved from invertebrates to mammals and conferring arsenite sensitivity when knocked down via RNAi in *C. elegans* [90] is associated with the 19S regulatory subunit of cytoplasmic proteasomes [91]. Furthermore, proteasomes isolated from *C. elegans* (and mammalian cells) knocked down for ZFAND2A expression are impaired in function for peptide degradation, suggesting that ZFAND2A plays an important role in adapting core proteasomes to handle the enhanced proteotoxicity that is presumably occurring in the presence of arsenic [91]. Interestingly, ZFAND2A knockdowns also display shortened lifespans and hypersensitivity to the toxic effects of transgenically expressed misfolding-prone proteins [110]. This link to proteotoxicity gains additional strength with the observation that ZFAND2A confers protection in a *C. elegans* model of Alzheimer's disease expressing the human beta amyloid protein [29]. Together, these data imply that strongly proteotoxic effects are induced by arsenic, presumably because of interference in protein folding, maturation, and/or integrity; as such, it might be anticipated that the ER stress pathway would be induced in response to arsenic in addition to the protein folding-sensitive heat-shock response.

While induction of ER stress pathways has been reported from a few studies of arsenic in mammalian cells [6, 51, 66], and a role for ASNA-1 (mentioned in Section 21.3.1.2) has been proposed to be part of an adaptive response to arsenic-induced ER stress [22], some recent observations regarding proteotoxicity in yeast suggest that the ER stress response and associated proteasome degradation pathway will be a very productive area for more detailed investigation regarding the pathological mechanism of action of arsenic [34, 71]. While overtly global genetic and toxicogenomic studies pertaining to this issue are considered in Section 21.4, a mostly biochemical study in yeast has provided striking observations on the ability of arsenite both to induce protein aggregation and to interfere with chaperones engaged in protein folding/refolding [34]. Furthermore, these arsenite-induced aggregates can now serve as "seeds" for the further aggregation of newly synthesized proteins in the act of folding. These are exceptionally important observations, since they provide striking confirmation of the physical, probably chemical, and generally nonproductive association of arsenic with proteins, and suggest that persistent protein misfolding may well be linked to some of the long-term pathological effects of arsenic, as suggested for other ER stress agents [107].

21.4 GLOBAL APPROACHES—NOW AND INTO THE FUTURE

To conclude this overview regarding the contribution of invertebrate systems to our understanding of arsenic toxicity, it is pertinent to reconsider how several of the special features exhibited by these representative organisms are being exploited in approaches that have a more global, or system-wide, perspective. Obviously, given the fact that these organisms are cheap to maintain, easy to grow in relatively large numbers, and boast a repertoire of sophisticated genetic analyses in their research heritage, it makes sense that large-scale genetic screens (i.e., covering the entire genome), along with transcriptomic, proteomic, and metabolomic studies in wild-type and pertinent mutant backgrounds, are increasingly to be found in the literature. High-throughput technology is easily applied in the case of yeast—its unicellularity and liquid medium culture compare favorably to standard cell culture, while its relatively small gene content allows comprehensive mutant screening. While it is harder to subject multicellular organisms to standard high-throughput approaches, a range of assays in growth, reproduction, and behavior have been developed for both worms and flies in ways compatible with what might be termed medium-throughput approaches. These trends can only be expected to continue, and will likely expand in scope, in the foreseeable future.

21.4.1 Yeast

Extensive summaries of the advantages and disadvantages of using the single-celled yeast as a toxicological model in the context of contemporary genomics technology have appeared [21, 68], and will not be reiterated exhaustively here. Experimental platforms and analytical approaches for quantitative expression profiling in the presence of a toxicant at the gene, protein, and metabolite level are readily accessible, and there have been several reports in recent years regarding "omics" responses of yeast to growth in arsenic. One comprehensive analysis of mRNA expression, protein abundance, and sulfur metabolite status in response to arsenite concluded that the flux into sulfur assimilation was significantly enhanced, with greatly increased glutathione biosynthesis and reduced sulfur incorporation into proteins, a notable observation [100]. Such data are strongly consistent with the genetic results pertaining to glutathione biosynthesis in *Drosophila* exposed to arsenite referred to earlier [70]. Other pathways showing significant modulation were protein biosynthesis, proteolytic processing, oxidative stress, and arsenic detoxification, consistent with much other data painstakingly accumulated over the years (and referred to earlier).

These kinds of "toxicogenomics" approaches have allowed a more comprehensive insight into the cellular pathways affected by arsenic. However, the data are essentially correlative; alterations in levels of any biomolecule are really only half the story since they do not allow more specific questions to be answered, such as where in a pathway, with what target(s), and by what mechanism is arsenic interacting to effect the observed changes. Moreover, there is a requirement for a phenotypic assay to relate pathological output to participation of a given pathway (or set of pathways) in the response to arsenic—this has been termed the requirement for "phenotypic anchors" [72]. Yeast is exceptionally well placed to provide this link between particular genes and the phenotypic output through the application of "functional genomics"—basically the ability to screen each and every gene for its effect on a chosen toxicant-induced phenotype. This is because library collections of knockout strains have been constructed for essentially all 6000 genes in the yeast genome (the "deletome"), and are available as heterozygous diploid strains (containing both essential and nonessential gene knockouts), as homozygous diploid strains (containing the nonessential knockouts), as well as haploid strain sets of the nonessential genes [26]. As applied to the discovery of toxicity pathways, this "functional toxicogenomics" approach is extraordinarily valuable, and is likely to throw up candidates that can be tested in the mammalian context assuming appropriate homologs exist. In a general sense, it is only restricted by the availability of appropriate quantitative phenotypic outputs, so in yeast the approach is obviously more limited than in organisms with additional measurable outputs such as growth and development, reproduction, and behavior.

An early application of this type of study with arsenite compared such "phenotypic profiling" with transcriptional profiling and found that most of the pathways uncovered by the functional genomics approach lay upstream of the transcriptional pathways identified by the transcriptomics approach [30], and thus were likely more closely linked to points of direct arsenic interference. This has proven to be a rather general observation for other toxicants too, the regulation of gene expression showing a rather low correlation with the actual requirement of the gene for growth in the presence of the toxicant [26]. Both the Haugen et al. study [30] and a later one [101] confirmed once again the central role played by sulfur metabolism in general, and glutathione biosynthesis in particular, in the response to arsenite, while, at the same time, by integrating the data into gene ontology and interactome databases, identifying a series of other potentially important "toxicity pathways" for future consideration. Other reports utilizing this yeast "deletome" resource have also appeared, providing a rich source of information for comparative purposes [36, 112].

An arsenic-specific functional genomics study with highly significant ramifications compared wild-type and homozygous deletion strains for the ability of either arsenite or its monomethylated derivative [MMA(III)] to inhibit growth [37]. Loss of genes related to several pathways conferred heightened sensitivity to one or the other arsenical, but importantly the two gene sets were only partially overlapping, implicating differentially sensitive toxicity pathways to these two toxicants. While tubulin folding was sensitive to both arsenite and MMA(III), glutathione biosynthesis seemed particularly crucial in resisting the effects of MMA(III), suggesting the highly reactive (and potentially damaging) nature of this arsenical. Such observations seem especially important in the context of thinking about human metabolic variability in the methylation of arsenic. Other pathways uncovered included DNA damage repair and chromatin modification; the human homolog of a yeast histone acetyltransferase gene so identified was subsequently found to confer protection toward arsenite in human cells, seemingly proving beyond doubt the potential value of this type of approach [38]. Furthermore, the human homolog of a specific adenine methyltransferase, the absence of which conferred resistance to arsenite in yeast [37], was later found to promote methylation of MMA to DMA species in human cells, a previously unsuspected reaction that could well play an important metabolic role in humans [76]. Utilizing a broadly similar "deletome" approach, Pan et al. [71] identified a number of pathways involved in resistance to arsenite, including those already described here, as well as several others including oxidative stress response, arsenic transport, and mitotic cell cycle progression. Importantly, they identified a family of tubulin- and actin-folding chaperones (comprising the TRiC and prefoldin complexes) that are especially sensitive to arsenite, and confirmed that the bovine TRiC homolog was impaired in its ability to fold appropriate substrates when exposed to arsenite. These data are highly consistent with reports that have linked arsenic with ER stress, proteotoxicity, and chaperone interaction, referred to in Section 21.3.3. It seems clear that these types of approaches in yeast are providing essential functional insight, and future investigation in mammalian systems of the candidates thrown up in such studies will likely be exceedingly valuable and enlightening.

21.4.2 *Caenorhabditis elegans*

While some of the obvious developmental and genetic advantages of this organism have been mentioned earlier, and are discussed from a toxicological perspective in appropriate reviews [11, 49], the potential for use of *C. elegans* as a multicellular organism that can be subjected to relatively high-throughput approaches based on a variety of phenotypic outputs has yet to be fully realized as far as arsenic-directed research is concerned. Unlike yeast, but in common with *Drosophila* (discussed in the following text), the opportunity exists to monitor many more sophisticated phenotypes appropriate to multicellular systems that undergo staged developmental growth from embryo to adult, and that harbor digestive, muscular, and nervous systems, for example, movement, feeding, size, reproductive success, lifespan, complex behaviors toward stimuli, and others. For example, the response to mercurials in several of these parameters has been reported [62]. Since this nematode worm is transparent and its cellular development highly orchestrated and fully described [95, 96], it is eminently possible to monitor cellular response to arsenic during growth utilizing appropriate GFP-marked reporter genes specific to the GI tract, the musculature, or the neuronal development pathway. With the ability to knock down expression of virtually any gene utilizing RNAi-based technology (achieved merely by feeding with *Escherichia coli* expressing appropriate shRNAs), to screen through mutant collections that cover a sizable portion of the genome, and to overexpress chosen genes through transgenic technology, the goal of performing "functional toxicogenomics" in the search for arsenic toxicity pathways is well within reach. Furthermore, major efforts have been made to automate the recording of quantitative variation in many of those phenotypes that might arise in response to a toxicant such as arsenic, making the prospect of conducting genome-wide screens feasible. Thus, high-throughput sorting based on size, or on fluorescence output, is achievable using a sophisticated COPAS Biosort instrument [7]. More recently, a much simpler (and cheaper) alternative has been described that utilizes a flatbed scanner and its light stimulus to initiate recordable movement in populations of *C. elegans* cultured on typical petri plates [60]. An associated algorithm, WormScan, allows variations in movement, mortality, fecundity, and size to be recorded. The system is highly appropriate for toxicity testing of arsenic and methylated metabolites. Most interestingly, it was used in the identification of mutant alleles of the dihydrolipoamide dehydrogenase (DLD) enzyme that confers resistance to phosphine gas in both worms and crop insect pests. These mutants are also highly sensitive to arsenite, and it was speculated that this is due to the high concentration of reduced lipoamide moieties present as a result of the DLD mutation. By virtue of their dithiol configuration, these lipoamides are strongly reactive with the electrophilic arsenite [86], thereby causing termination of mitochondrial tricarboxylic cycle activity. The kinds of approaches described or alluded to here represent powerful analytical tools that are just waiting to be exploited more fully to provide deeper insight into arsenic toxicity pathways.

21.4.3 *Drosophila*

Until now, there has been little arsenic research reported in *Drosophila* that encompasses a more global, genetically based pathway discovery approach of the type described for yeast. However, the resources available in this organism make it an extremely promising vehicle for future studies. Thus, comprehensive libraries of inducible RNAi knockdown stocks covering virtually the whole transcriptome have been described [18, 67] and can readily be incorporated into screens aimed at characterizing toxicity pathways related to chosen phenotypic outputs. As with *C. elegans*, the variety of potential phenotypes that might be displayed in response to arsenic remains to be fully explored, but longevity, reproduction, mating behavior, and developmental progression are all easily approached, while chromosomal damage/genomic integrity, stress responsiveness, and stem cell commitment and renewal, can all be readily envisaged with many appropriate transgenic reporter tools already in place. Two reviews on the use of *Drosophila* as a toxicological model, specifically emphasizing the utility of the organism in neurotoxicological research, have appeared and are recommended for detailed coverage of experimental approaches [32, 75]. In particular, the Ruden laboratory has been at the forefront of quantitative genetic approaches to uncover pathways affected by exposure to lead (Pb). These workers have pioneered the use of quantitative trait locus (QTL) discovery in toxicogenomics studies, utilizing a panel of recombinant inbred (RI) fly lines as experimental subjects and quantitatively measuring their response to Pb from both a behavioral [31] and a gene expression [83] perspective. By these means, they have been able to map several loci responsive to Pb that form the basis for more detailed gene and pathway identification. It would seem that the application of these techniques to arsenic represents a natural progression. The so-called expression QTLs (eQTLs) derived from comprehensive transcription profiling by these types of RI approaches can additionally be complemented by protein QTLs (pQTLs), metabolomic QTLs (mQTLs), and epigenomic QTLs (epiQTLs) depending on the particular output that is being monitored in response to the chosen toxicant. Moreover, with cheap, high-throughput direct DNA-sequencing techniques increasingly available, selection of arsenic-responsive traits by continuous breeding from a heterogeneous starting population followed by direct identification of sequence variation represents an additional way to home in on candidate genes that are part of "arsenic toxicity pathways" (see [84]). As was noted in the *Drosophila GSS* study [70] described in Section 21.3.2, the beauty of these types of approaches is the use of natural genetic variation to reveal the secrets of individual variable response to a toxicant such as arsenic.

A final example of a potentially fruitful area for investigation utilizing *Drosophila*-based research relates to the recent observations (see Section 21.3.3) concerning the role of arsenic in proteotoxicity—protein aggregation, degradation, unfolding, etc. Many of the most intractable neurodegenerative disorders afflicting people, for example, Alzheimer's, Parkinson's, Huntington's, and ALS, are part of a large group of diseases characterized as proteinopathies, or protein misfolding disorders. A very active research enterprise has grown up around the use of *Drosophila* for basic modeling and process discovery concerning such diseases, and there are innumerable methodological, behavioral, and neurological tools described to assay the processes and pathways affected by these unnatural protein species [57, 77]. Given that described outcomes of chronic exposure to arsenic include encephalopathy and peripheral neuropathy, and that both ER stress and mitochondrial stress are commonly associated with arsenic exposure [65, 66] and several types of neurodegenerative disease [24, 82], it seems important to explore a potential link. *Drosophila* has an abundance of genetic and analytical tools that could significantly aid such an endeavor.

REFERENCES

[1] T. Agusa, H. Iwata, J. Fujihara, T. Kunito, H. Takeshita, T.B. Minh, P.T. Trang, P.H. Viet, S. Tanabe, Genetic polymorphisms in AS3MT and arsenic metabolism in residents of the Red River Delta, Vietnam, Toxicol Appl Pharmacol 236 (2009) 131–141.

[2] A.P. Arrigo, Acetylation and methylation patterns of core histones are modified after heat or arsenite treatment of *Drosophila* tissue culture cells, Nucleic Acids Res 11 (1983) 1389–1404.

[3] H. Bhattacharjee, J. Sheng, A.A. Ajees, R. Mukhopadhyay, B.P. Rosen, Adventitious arsenate reductase activity of the catalytic domain of the human Cdc25B and Cdc25C phosphatases, Biochemistry 49 (2010) 802–809.

[4] O. Billing, B. Natarajan, A. Mohammed, P. Naredi, G. Kao, A directed RNAi screen based on larval growth arrest reveals new modifiers of *C. elegans* insulin signaling, PLoS One 7 (2012) e34507.

[5] P. Bobrowicz, R. Wysocki, G. Owsianik, A. Goffeau, S. Ulaszewski, Isolation of three contiguous genes, ACR1, ACR2 and ACR3, involved in resistance to arsenic compounds in the yeast *Saccharomyces cerevisiae*, Yeast 13 (1997) 819–828.

[6] A.M. Bolt, F. Zhao, S. Pacheco, W.T. Klimecki, Arsenite-induced autophagy is associated with proteotoxicity in human lymphoblastoid cells, Toxicol Appl Pharmacol 264 (2012) 255–261.

[7] W.A. Boyd, S.J. McBride, J.R. Rice, D.W. Snyder, J.H. Freedman, A high-throughput method for assessing chemical toxicity using a *Caenorhabditis elegans* reproduction assay, Toxicol Appl Pharmacol 245 (2010) 153–159.

[8] A. Broeks, B. Gerrard, R. Allikmets, M. Dean, R.H. Plasterk, Homologues of the human multidrug resistance genes MRP and MDR contribute to heavy metal resistance in the soil nematode *Caenorhabditis elegans*, EMBO J 15 (1996) 6132–6143.

[9] M. Bun-Ya, S. Harashima, Y. Oshima, Putative GTP-binding protein, Gtr1, associated with the function of the Pho84 inorganic phosphate transporter in *Saccharomyces cerevisiae*, Mol Cell Biol 12 (1992) 2958–2966.

[10] M. Bun-Ya, K. Shikata, S. Nakade, C. Yompakdee, S. Harashima, Y. Oshima, Two new genes, PHO86 and PHO87, involved in inorganic phosphate uptake in *Saccharomyces cerevisiae*, Curr Genet 29 (1996) 344–351.

[11] S. Caito, S. Fretham, E. Martinez-Finley, S. Chakraborty, D. Avila, P. Chen, M. Aschner, Genome-wide analyses of metal responsive genes in *Caenorhabditis elegans*, Front Genet 3 (2012) 52.

[12] R. Camato, R.M. Tanguay, Changes in the methylation pattern of core histones during heat-shock in *Drosophila* cells, EMBO J 1 (1982) 1529–1532.

[13] N. Chatterjee, D. Bohmann, A versatile PhiC31 based reporter system for measuring AP-1 and Nrf2 signaling in *Drosophila* and in tissue culture, PLoS One 7 (2012) e34063.

[14] C.M. Chen, T.K. Misra, S. Silver, B.P. Rosen, Nucleotide sequence of the structural genes for an anion pump. The plasmid-encoded arsenical resistance operon, J Biol Chem 261 (1986) 15030–15038.

[15] F.S. Collins, G.M. Gray, J.R. Bucher, Toxicology. Transforming environmental health protection, Science 319 (2008) 906–907.

[16] R. Desrosiers, R.M. Tanguay, Further characterization of the posttranslational modifications of core histones in response to heat and arsenite stress in *Drosophila*, Biochem Cell Biol 64 (1986) 750–757.

[17] R. Desrosiers, R.M. Tanguay, Methylation of *Drosophila* histones at proline, lysine, and arginine residues during heat shock, J Biol Chem 263 (1988) 4686–4692.

[18] G. Dietzl, D. Chen, F. Schnorrer, K.C. Su, Y. Barinova, M. Fellner, B. Gasser, K. Kinsey, S. Oppel, S. Scheiblauer, A. Couto, V. Marra, K. Keleman, B.J. Dickson, A genome-wide transgenic RNAi library for conditional gene inactivation in *Drosophila*, Nature 448 (2007) 151–156.

[19] L. Ding, R.J. Saunders, Z. Drobna, F.S. Walton, P. Xun, D.J. Thomas, M. Styblo, Methylation of arsenic by recombinant human wild-type arsenic (+3 oxidation state) methyltransferase and its methionine 287 threonine (M287T) polymorph: role of glutathione, Toxicol Appl Pharmacol 264 (2012) 121–130.

[20] D.J. Dix, K.A. Houck, M.T. Martin, A.M. Richard, R.W. Setzer, R.J. Kavlock, The ToxCast program for prioritizing toxicity testing of environmental chemicals, Toxicol Sci 95 (2007) 5–12.

[21] S.S.C. Dos, M.C. Teixeira, T.R. Cabrito, I. Sa-Correia, Yeast toxicogenomics: genome-wide responses to chemical stresses with impact in environmental health, pharmacology, and biotechnology, Front Genet 3 (2012) 63.

[22] V. Favaloro, M. Spasic, B. Schwappach, B. Dobberstein, Distinct targeting pathways for the membrane insertion of tail-anchored (TA) proteins, J Cell Sci 121 (2008) 1832–1840.

[23] D. Ferrario, C. Croera, R. Brustio, A. Collotta, G. Bowe, M. Vahter, L. Gribaldo, Toxicity of inorganic arsenic and its metabolites on haematopoietic progenitors "in vitro": comparison between species and sexes, Toxicology 249 (2008) 102–108.

[24] E. Ferreiro, I. Baldeiras, I.L. Ferreira, R.O. Costa, A.C. Rego, C.F. Pereira, C.R. Oliveira, Mitochondrial- and endoplasmic reticulum-associated oxidative stress in Alzheimer's disease: from pathogenesis to biomarkers, Int J Cell Biol 2012 (2012) 735206.

[25] M. Ghosh, J. Shen, B.P. Rosen, Pathways of As(III) detoxification in *Saccharomyces cerevisiae*, Proc Natl Acad Sci U S A 96 (1999) 5001–5006.

[26] G. Giaever, A.M. Chu, L. Ni, C. Connelly, L. Riles, S. Veronneau, S. Dow, A. Lucau-Danila, K. Anderson, B. Andre, A.P. Arkin, A. Astromoff, M. El-Bakkoury, R. Bangham, R. Benito, S. Brachat, S. Campanaro, M. Curtiss, K. Davis, A. Deutschbauer, K.D. Entian, P. Flaherty, F. Foury, D.J. Garfinkel, M. Gerstein, D. Gotte, U. Guldener,

J.H. Hegemann, S. Hempel, Z. Herman, D.F. Jaramillo, D.E. Kelly, S.L. Kelly, P. Kotter, D. LaBonte, D.C. Lamb, N. Lan, H. Liang, H. Liao, L. Liu, C. Luo, M. Lussier, R. Mao, P. Menard, S.L. Ooi, J.L. Revuelta, C.J. Roberts, M. Rose, P. Ross-Macdonald, B. Scherens, G. Schimmack, B. Shafer, D.D. Shoemaker, S. Sookhai-Mahadeo, R.K. Storms, J.N. Strathern, G. Valle, M. Voet, G. Volckaert, C.Y. Wang, T.R. Ward, J. Wilhelmy, E.A. Winzeler, Y. Yang, G. Yen, E. Youngman, K. Yu, H. Bussey, J.D. Boeke, M. Snyder, P. Philippsen, R.W. Davis, M. Johnston, Functional profiling of the *Saccharomyces cerevisiae* genome, Nature 418 (2002) 387–391.

[27] S.H. Goldstein, H. Babich, Differential effects of arsenite and arsenate to *Drosophila melanogaster* in a combined adult/developmental toxicity assay, Bull Environ Contam Toxicol 42 (1989) 276–282.

[28] A.C. Groth, M. Fish, R. Nusse, M.P. Calos, Construction of transgenic *Drosophila* by using the site-specific integrase from phage phiC31, Genetics 166 (2004) 1775–1782.

[29] W.M. Hassan, D.A. Merin, V. Fonte, C.D. Link, AIP-1 ameliorates beta-amyloid peptide toxicity in a *Caenorhabditis elegans* Alzheimer's disease model, Hum Mol Genet 18 (2009) 2739–2747.

[30] A.C. Haugen, R. Kelley, J.B. Collins, C.J. Tucker, C. Deng, C.A. Afshari, J.M. Brown, T. Ideker, H.B. Van, Integrating phenotypic and expression profiles to map arsenic-response networks, Genome Biol 5 (2004) R95.

[31] H.V. Hirsch, D. Possidente, S. Averill, T.P. Despain, J. Buytkins, V. Thomas, W.P. Goebel, A. Shipp-Hilts, D. Wilson, K. Hollocher, B. Possidente, G. Lnenicka, D.M. Ruden, Variations at a quantitative trait locus (QTL) affect development of behavior in lead-exposed *Drosophila melanogaster*, Neurotoxicology 30 (2009) 305–311.

[32] H.V. Hirsch, G. Lnenicka, D. Possidente, B. Possidente, M.D. Garfinkel, L. Wang, X. Lu, D.M. Ruden, *Drosophila melanogaster* as a model for lead neurotoxicology and toxicogenomics research, Front Genet 3 (2012) 68.

[33] D. Jacobson-Kram, D. Montalbano, The reproductive effects assessment group's report on the mutagenicity of inorganic arsenic, Environ Mutagen 7 (1985) 787–804.

[34] T. Jacobson, C. Navarrete, S.K. Sharma, T.C. Sideri, S. Ibstedt, S. Priya, C.M. Grant, P. Christen, P. Goloubinoff, M.J. Tamas, Arsenite interferes with protein folding and triggers formation of protein aggregates in yeast, J Cell Sci 125 (2012) 5073–5083.

[35] X. Jiang, J.R. McDermott, A.A. Ajees, B.P. Rosen, Z. Liu, Trivalent arsenicals and glucose use different translocation pathways in mammalian GLUT1, Metallomics 2 (2010) 211–219.

[36] Y.H. Jin, P.E. Dunlap, S.J. McBride, H. Al-Refai, P.R. Bushel, J.H. Freedman, Global transcriptome and deletome profiles of yeast exposed to transition metals, PLoS Genet 4 (2008) e1000053.

[37] W.J. Jo, A. Loguinov, H. Wintz, M. Chang, A.H. Smith, D. Kalman, L. Zhang, M.T. Smith, C.D. Vulpe, Comparative functional genomic analysis identifies distinct and overlapping sets of genes required for resistance to monomethylarsonous acid (MMAIII) and arsenite (AsIII) in yeast, Toxicol Sci 111 (2009) 424–436.

[38] W.J. Jo, X. Ren, F. Chu, M. Aleshin, H. Wintz, A. Burlingame, M.T. Smith, C.D. Vulpe, L. Zhang, Acetylated H4K16 by MYST1 protects UROtsa cells from arsenic toxicity and is decreased following chronic arsenic exposure, Toxicol Appl Pharmacol 241 (2009) 294–302.

[39] S.V. Kala, M.W. Neely, G. Kala, C.I. Prater, D.W. Atwood, J.S. Rice, M.W. Lieberman, The MRP2/cMOAT transporter and arsenic-glutathione complex formation are required for biliary excretion of arsenic, J Biol Chem 275 (2000) 33404–33408.

[40] S.V. Kala, G. Kala, C.I. Prater, A.C. Sartorelli, M.W. Lieberman, Formation and urinary excretion of arsenic triglutathione and methylarsenic diglutathione, Chem Res Toxicol 17 (2004) 243–249.

[41] G. Kao, C. Nordenson, M. Still, A. Ronnlund, S. Tuck, P. Naredi, ASNA-1 positively regulates insulin secretion in *C. elegans* and mammalian cells, Cell 128 (2007) 577–587.

[42] K.T. Kitchin, K. Wallace, Arsenite binding to synthetic peptides based on the Zn finger region and the estrogen binding region of the human estrogen receptor-alpha, Toxicol Appl Pharmacol 206 (2005) 66–72.

[43] C.D. Kozul-Horvath, F. Zandbergen, B.P. Jackson, R.I. Enelow, J.W. Hamilton, Effects of low-dose drinking water arsenic on mouse fetal and postnatal growth and development, PLoS One 7 (2012) e38249.

[44] Y. Kumagai, D. Sumi, Arsenic: signal transduction, transcription factor, and biotransformation involved in cellular response and toxicity, Annu Rev Pharmacol Toxicol 47 (2007) 243–262.

[45] B. Kurdi-Haidar, D. Heath, P. Naredi, N. Varki, S.B. Howell, Immunohistochemical analysis of the distribution of the human ATPase (hASNA-I) in normal tissues and its overexpression in breast adenomas and carcinomas, J Histochem Cytochem 46 (1998) 1243–1248.

[46] R.C. Lantz, B. Chau, P. Sarihan, M.L. Witten, V.I. Pivniouk, G.J. Chen, In utero and postnatal exposure to arsenic alters pulmonary structure and function, Toxicol Appl Pharmacol 235 (2009) 105–113.

[47] E.M. Leslie, A. Haimeur, M.P. Waalkes, Arsenic transport by the human multidrug resistance protein 1 (MRP1/ABCC1). Evidence that a tri-glutathione conjugate is required, J Biol Chem 279 (2004) 32700–32708.

[48] E.M. Leslie, Arsenic-glutathione conjugate transport by the human multidrug resistance proteins (MRPs/ABCCs), J Inorg Biochem 108 (2012) 141–149.

[49] M.C. Leung, P.L. Williams, A. Benedetto, C. Au, K.J. Helmcke, M. Aschner, J.N. Meyer, *Caenorhabditis elegans*: an emerging model in biomedical and environmental toxicology, Toxicol Sci 106 (2008) 5–28.

[50] V.H. Liao, C.W. Yu, *Caenorhabditis elegans* gcs-1 confers resistance to arsenic-induced oxidative stress, Biometals 18 (2005) 519–528.

[51] A.M. Lin, P.L. Chao, S.F. Fang, C.W. Chi, C.H. Yang, Endoplasmic reticulum stress is involved in arsenite-induced oxidative injury in rat brain, Toxicol Appl Pharmacol 224 (2007) 138–146.

[52] J. Liu, H. Chen, D.S. Miller, J.E. Saavedra, L.K. Keefer, D.R. Johnson, C.D. Klaassen, M.P. Waalkes, Overexpression of glutathione S-transferase II and multidrug resistance transport proteins is associated with acquired tolerance to inorganic arsenic, Mol Pharmacol 60 (2001) 302–309.

[53] J. Liu, Y. Liu, D.A. Powell, M.P. Waalkes, C.D. Klaassen, Multidrug-resistance mdr1a/1b double knockout mice are more sensitive than wild type mice to acute arsenic toxicity, with higher arsenic accumulation in tissues, Toxicology 170 (2002) 55–62.

[54] Z. Liu, J. Shen, J.M. Carbrey, R. Mukhopadhyay, P. Agre, B.P. Rosen, Arsenite transport by mammalian aquaglyceroporins AQP7 and AQP9, Proc Natl Acad Sci U S A 99 (2002) 6053–6058.

[55] Z. Liu, E. Boles, B.P. Rosen, Arsenic trioxide uptake by hexose permeases in *Saccharomyces cerevisiae*, J Biol Chem 279 (2004) 17312–17318.

[56] Z. Liu, M. Styblo, B.P. Rosen, Methylarsonous acid transport by aquaglyceroporins, Environ Health Perspect 114 (2006) 527–531.

[57] B. Lu, H. Vogel, *Drosophila* models of neurodegenerative diseases, Annu Rev Pathol 4 (2009) 315–342.

[58] E. Maciaszczyk-Dziubinska, I. Migdal, M. Migocka, T. Bocer, R. Wysocki, The yeast aquaglyceroporin Fps1p is a bidirectional arsenite channel, FEBS Lett 584 (2010) 726–732.

[59] E. Maciaszczyk-Dziubinska, M. Migocka, R. Wysocki, Acr3p is a plasma membrane antiporter that catalyzes As(III)/H(+) and Sb(III)/H(+) exchange in *Saccharomyces cerevisiae*, Biochim Biophys Acta 1808 (2011) 1855–1859.

[60] M.D. Mathew, N.D. Mathew, P.R. Ebert, WormScan: a technique for high-throughput phenotypic analysis of Caenorhabditis elegans, PLoS One 7 (2012) e33483.

[61] E.A. Maull, H. Ahsan, J. Edwards, M.P. Longnecker, A. Navas-Acien, J. Pi, E.K. Silbergeld, M. Styblo, C.H. Tseng, K.A. Thayer, D. Loomis, Evaluation of the association between arsenic and diabetes: a National Toxicology Program workshop review, Environ Health Perspect 120 (2012) 1658–1670.

[62] M.K. McElwee, J.H. Freedman, Comparative toxicology of mercurials in *Caenorhabditis elegans*, Environ Toxicol Chem 30 (2011) 2135–2141.

[63] R. Mukhopadhyay, B.P. Rosen, *Saccharomyces cerevisiae* ACR2 gene encodes an arsenate reductase, FEMS Microbiol Lett 168 (1998) 127–136.

[64] O.J.G. Muniz, J. Shang, B. Catron, J. Landero, J.A. Caruso, I.L. Cartwright, A transgenic *Drosophila* model for arsenic methylation suggests a metabolic rationale for differential dose-dependent toxicity endpoints, Toxicol Sci 121 (2011) 303–311.

[65] H. Naranmandura, S. Xu, T. Sawata, W.H. Hao, H. Liu, N. Bu, Y. Ogra, Y.J. Lou, N. Suzuki, Mitochondria are the main target organelle for trivalent monomethylarsonous acid (MMA(III))-induced cytotoxicity, Chem Res Toxicol 24 (2011) 1094–1103.

[66] H. Naranmandura, S. Xu, S. Koike, L.Q. Pan, B. Chen, Y.W. Wang, K. Rehman, B. Wu, Z. Chen, N. Suzuki, The endoplasmic reticulum is a target organelle for trivalent dimethylarsinic acid (DMAIII)-induced cytotoxicity, Toxicol Appl Pharmacol 260 (2012) 241–249.

[67] J.Q. Ni, R. Zhou, B. Czech, L.P. Liu, L. Holderbaum, D. Yang-Zhou, H.S. Shim, R. Tao, D. Handler, P. Karpowicz, R. Binari, M. Booker, J. Brennecke, L.A. Perkins, G.J. Hannon, N. Perrimon, A genome-scale shRNA resource for transgenic RNAi in *Drosophila*, Nat Methods 8 (2011) 405–407.

[68] M. North, C.D. Vulpe, Functional toxicogenomics: mechanism-centered toxicology, Int J Mol Sci 11 (2010) 4796–4813.

[69] NRC, Toxicity Testing in the 21st Century: A Vision and a Strategy, The National Academies Press, Washington, DC, 2007.

[70] J.G. Ortiz, R. Opoka, D. Kane, I.L. Cartwright, Investigating arsenic susceptibility from a genetic perspective in *Drosophila* reveals a key role for glutathione synthetase, Toxicol Sci 107 (2009) 416–426.

[71] X. Pan, S. Reissman, N.R. Douglas, Z. Huang, D.S. Yuan, X. Wang, J.M. McCaffery, J. Frydman, J.D. Boeke, Trivalent arsenic inhibits the functions of chaperonin complex, Genetics 186 (2010) 725–734.

[72] R. Paules, Phenotypic anchoring: linking cause and effect, Environ Health Perspect 111 (2003) A338–A339.

[73] A.D. Pickett, N.A. Patterson, Arsenates: effect on fecundity in some diptera, Science 140 (1963) 493–494.

[74] P. Ramos-Morales, R. Rodriguez-Arnaiz, Genotoxicity of two arsenic compounds in germ cells and somatic cells of *Drosophila melanogaster*, Environ Mol Mutagen 25 (1995) 288–299.

[75] M.D. Rand, Drosophotoxicology: the growing potential for *Drosophila* in neurotoxicology, Neurotoxicol Teratol 32 (2010) 74–83.

[76] X. Ren, M. Aleshin, W.J. Jo, R. Dills, D.A. Kalman, C.D. Vulpe, M.T. Smith, L. Zhang, Involvement of N-6 adenine-specific DNA methyltransferase 1 (N6AMT1) in arsenic biomethylation and its role in arsenic-induced toxicity, Environ Health Perspect 119 (2011) 771–777.

[77] D.E. Rincon-Limas, K. Jensen, P. Fernandez-Funez, *Drosophila* models of proteinopathies: the little fly that could, Curr Pharm Des 18 (2012) 1108–1122.

[78] M. Rizki, E. Kossatz, N. Xamena, A. Creus, R. Marcos, Influence of sodium arsenite on the genotoxicity of potassium dichromate and ethyl methanesulfonate: studies with the wing spot test in *Drosophila*, Environ Mol Mutagen 39 (2002) 49–54.

[79] M. Rizki, E. Kossatz, A. Velazquez, A. Creus, M. Farina, S. Fortaner, E. Sabbioni, R. Marcos, Metabolism of arsenic in *Drosophila melanogaster* and the genotoxicity of dimethylarsinic acid in the *Drosophila* wing spot test, Environ Mol Mutagen 47 (2006) 162–168.

[80] M.A. Rodriguez-Gabriel, P. Russell, Distinct signaling pathways respond to arsenite and reactive oxygen species in *Schizosaccharomyces pombe*, Eukaryot Cell 4 (2005) 1396–1402.

[81] B.P. Rosen, Families of arsenic transporters, Trends Microbiol 7 (1999) 207–212.

[82] B.D. Roussel, A.J. Kruppa, E. Miranda, D.C. Crowther, D.A. Lomas, S.J. Marciniak, Endoplasmic reticulum dysfunction in neurological disease, Lancet Neurol 12 (2013) 105–118.

[83] D.M. Ruden, L. Chen, D. Possidente, B. Possidente, P. Rasouli, L. Wang, X. Lu, M.D. Garfinkel, H.V. Hirsch, G.P. Page, Genetical toxicogenomics in *Drosophila* identifies master-modulatory loci that are regulated by developmental exposure to lead, Neurotoxicology 30 (2009) 898–914.

[84] D.M. Ruden, Frontiers in toxicogenomics—the grand challenge: to understand how the genome and epigenome interact with the toxic environment, Front Genet 2 (2011) 12.

[85] K. Schlawicke Engstrom, B. Nermell, G. Concha, U. Stromberg, M. Vahter, K. Broberg, Arsenic metabolism is influenced by polymorphisms in genes involved in one-carbon metabolism and reduction reactions, Mutat Res 667 (2009) 4–14.

[86] D.I. Schlipalius, N. Valmas, A.G. Tuck, R. Jagadeesan, L. Ma, R. Kaur, A. Goldinger, C. Anderson, J. Kuang, S. Zuryn, Y.S. Mau, Q. Cheng, P.J. Collins, M.K. Nayak, H.J. Schirra, M.A. Hilliard, P.R. Ebert, A core metabolic enzyme mediates resistance to phosphine gas, Science 338 (2012) 807–810.

[87] J. Shen, C.M. Hsu, B.K. Kang, B.P. Rosen, H. Bhattacharjee, The *Saccharomyces cerevisiae* Arr4p is involved in metal and heat tolerance, Biometals 16 (2003) 369–378.

[88] S.O. Simmons, C.Y. Fan, R. Ramabhadran, Cellular stress response pathway system as a sentinel ensemble in toxicological screening, Toxicol Sci 111 (2009) 202–225.

[89] M.H. Smith, H.L. Ploegh, J.S. Weissman, Road to ruin: targeting proteins for degradation in the endoplasmic reticulum, Science 334 (2011) 1086–1090.

[90] J. Sok, M. Calfon, J. Lu, P. Lichtlen, S.G. Clark, D. Ron, Arsenite-inducible RNA-associated protein (AIRAP) protects cells from arsenite toxicity, Cell Stress Chaperones 6 (2001) 6–15.

[91] A. Stanhill, C.M. Haynes, Y. Zhang, G. Min, M.C. Steele, J. Kalinina, E. Martinez, C.M. Pickart, X.P. Kong, D. Ron, An arsenite-inducible 19S regulatory particle-associated protein adapts proteasomes to proteotoxicity, Mol Cell 23 (2006) 875–885.

[92] S. Stefanovic, R.S. Hegde, Identification of a targeting factor for posttranslational membrane protein insertion into the ER, Cell 128 (2007) 1147–1159.

[93] C. Steinmaus, Y. Yuan, D. Kalman, O.A. Rey, C.F. Skibola, D. Dauphine, A. Basu, K.E. Porter, A. Hubbard, M.N. Bates, M.T. Smith, A.H. Smith, Individual differences in arsenic metabolism and lung cancer in a case-control study in Cordoba, Argentina, Toxicol Appl Pharmacol 247 (2010) 138–145.

[94] M. Styblo, R.L.M. Del, E.L. LeCluyse, G.A. Hamilton, C. Wang, W.R. Cullen, D.J. Thomas, Metabolism of arsenic in primary cultures of human and rat hepatocytes, Chem Res Toxicol 12 (1999) 560–565.

[95] J.E. Sulston, H.R. Horvitz, Post-embryonic cell lineages of the nematode, *Caenorhabditis elegans*, Dev Biol 56 (1977) 110–156.

[96] J.E. Sulston, E. Schierenberg, J.G. White, J.N. Thomson, The embryonic cell lineage of the nematode *Caenorhabditis elegans*, Dev Biol 100 (1983) 64–119.

[97] D.J. Thomas, M. Styblo, S. Lin, The cellular metabolism and systemic toxicity of arsenic, Toxicol Appl Pharmacol 176 (2001) 127–144.

[98] D.J. Thomas, J. Li, S.B. Waters, W. Xing, B.M. Adair, Z. Drobna, V. Devesa, M. Styblo, Arsenic (+3 oxidation state) methyltransferase and the methylation of arsenicals, Exp Biol Med (Maywood) 232 (2007) 3–13.

[99] M. Thorsen, Y. Di, C. Tangemo, M. Morillas, D. Ahmadpour, C. Van der Does, A. Wagner, E. Johansson, J. Boman, F. Posas, R. Wysocki, M.J. Tamas, The MAPK Hog1p modulates Fps1p-dependent arsenite uptake and tolerance in yeast, Mol Biol Cell 17 (2006) 4400–4410.

[100] M. Thorsen, G. Lagniel, E. Kristiansson, C. Junot, O. Nerman, J. Labarre, M.J. Tamas, Quantitative transcriptome, proteome, and sulfur metabolite profiling of the *Saccharomyces cerevisiae* response to arsenite, Physiol Genomics 30 (2007) 35–43.

[101] M. Thorsen, G.G. Perrone, E. Kristiansson, M. Traini, T. Ye, I.W. Dawes, O. Nerman, M.J. Tamas, Genetic basis of arsenite and cadmium tolerance in *Saccharomyces cerevisiae*, BMC Genomics 10 (2009) 105.

[102] N.K. Tripathy, F.E. Wurgler, H. Frei, Genetic toxicity of six carcinogens and six non-carcinogens in the *Drosophila* wing spot test, Mutat Res 242 (1990) 169–180.

[103] Y.Y. Tseng, C.W. Yu, V.H. Liao, *Caenorhabditis elegans* expresses a functional ArsA, FEBS J 274 (2007) 2566–2572.

[104] O.L. Valenzuela, Z. Drobna, E. Hernandez-Castellanos, L.C. Sanchez-Pena, G.G. Garcia-Vargas, V.H. Borja-Aburto, M. Styblo, L.M. Del Razo, Association of AS3MT polymorphisms and the risk of premalignant arsenic skin lesions, Toxicol Appl Pharmacol 239 (2009) 200–207.

[105] M.P. Waalkes, J.M. Ward, J. Liu, B.A. Diwan, Transplacental carcinogenicity of inorganic arsenic in the drinking water: induction of hepatic, ovarian, pulmonary, and adrenal tumors in mice, Toxicol Appl Pharmacol 186 (2003) 7–17.

[106] P. Walter, D. Ron, The unfolded protein response: from stress pathway to homeostatic regulation, Science 334 (2011) 1081–1086.

[107] S. Wang, R.J. Kaufman, The impact of the unfolded protein response on human disease, J Cell Biol 197 (2012) 857–867.

[108] T.C. Wood, O.E. Salavagionne, B. Mukherjee, L. Wang, A.F. Klumpp, B.A. Thomae, B.W. Eckloff, D.J. Schaid, E.D. Wieben, R.M. Weinshilboum, Human arsenic methyltransferase (AS3MT) pharmacogenetics: gene resequencing and functional genomics studies, J Biol Chem 281 (2006) 7364–7373.

[109] R. Wysocki, C.C. Chery, D. Wawrzycka, H.M. Van, R. Cornelis, J.M. Thevelein, M.J. Tamas, The glycerol channel Fps1p mediates the uptake of arsenite and antimonite in *Saccharomyces cerevisiae*, Mol Microbiol 40 (2001) 1391–1401.

[110] C. Yun, A. Stanhill, Y. Yang, Y. Zhang, C.M. Haynes, C.F. Xu, T.A. Neubert, A. Mor, M.R. Philips, D. Ron, Proteasomal adaptation to environmental stress links resistance to proteotoxicity with longevity in *Caenorhabditis elegans*, Proc Natl Acad Sci U S A 105 (2008) 7094–7099.

[111] K. Zhan, J. Narasimhan, R.C. Wek, Differential activation of eIF2 kinases in response to cellular stresses in *Schizosaccharomyces pombe*, Genetics 168 (2004) 1867–1875.

[112] X. Zhou, A. Arita, T.P. Ellen, X. Liu, J. Bai, J.P. Rooney, A.D. Kurtz, C.B. Klein, W. Dai, T.J. Begley, M. Costa, A genome-wide screen in *Saccharomyces cerevisiae* reveals pathways affected by arsenic toxicity, Genomics 94 (2009) 294–307.

22

TOXICOKINETICS AND PHARMACOKINETIC MODELING OF ARSENIC

ELAINA M. KENYON[1] AND HARVEY J. CLEWELL, III[2]

[1] *US EPA, NHEERL, Research Triangle Park, NC, USA*

[2] *Hamner Institutes for Health Sciences, Research Triangle Park, NC, USA*

22.1 INTRODUCTION

Understanding the uptake and distribution of a toxicant is essential to understanding its toxicology. Toxicokinetics describe the rate of uptake of a chemical, how it is distributed within the body, and the rate of excretion or metabolic inactivation/activation to less/more toxic forms. The variation of these processes across species and between individuals is an important determinant of the toxicological outcome of exposures. Much research has been devoted to elucidating the toxicokinetics of arsenic in animal models and humans in order to gain a deeper understanding of its toxicology and to provide a quantitative estimate of dose to the target tissues with the goal of providing a sound biological basis for dose–response analysis. This chapter provides an overview of arsenic toxicokinetics and physiologically based pharmacokinetic (PBPK) modeling with particular emphasis on key factors needed for development of a model useful for dose–response analysis, applications of arsenic models, as well as research needs.

Arsenic: Exposure Sources, Health Risks, and Mechanisms of Toxicity, First Edition.
Edited by J. Christopher States.
© 2016 John Wiley & Sons, Inc. Published 2016 by John Wiley & Sons, Inc.

22.2 OVERVIEW OF ARSENIC TOXICOKINETICS

The metabolism and disposition of inorganic arsenic (iAs) is largely dependent on its valence state and both pentavalent (iAs^{V}) and trivalent (iAs^{III}) inorganic arsenic compounds are the arsenicals to which humans are most likely to be exposed in the environment [1]. Inorganic arsenic is readily absorbed from the gastrointestinal tract in humans based on the rapid excretion of inorganic arsenic and its methylated metabolites in urine in controlled human exposure studies [2, 21]. Studies in mice suggest that iAs^{III} is more rapidly absorbed and distributed compared to iAs^{V} following a relatively low single dose (0.4–0.5 mg (As)/kg), whereas the reverse appears to be true following a higher single dose (4–5 mg (As)/kg) [12, 36].

Inorganic arsenic is rapidly and widely distributed in tissues following acute oral exposure in laboratory animals [13]. Limited data available in humans suggest that chronic low-level exposure is also characterized by relatively widespread accumulation in tissues with most arsenic present being in the inorganic form, less in the form of dimethyl arsenic (DMA) and monomethyl arsenic (MMA), these latter being generally undetectable except in the liver and kidney [39]. Studies in rodents indicate that the distribution of arsenic and its methylated metabolites in tissues is both organ-specific and dose-dependent, and not necessarily reflective of overall flux through the metabolic pathways as measured by cumulative urinary elimination of metabolites. For example, mice subchronically exposed to iAs^{V} in drinking water accumulate arsenic in kidney, urinary bladder, and lung to a much greater extent compared to blood, liver, and skin. Although kidney, bladder, and lung tissue preferentially accumulate arsenic, the distribution of metabolites differs markedly in urinary bladder and lung compared to kidney. MMA is the predominant metabolite in kidney, whereas DMA is the predominant metabolite in lung and bladder, particularly at higher exposure levels [15]. Increasingly in recent years, specific binding proteins and transporters have been identified, which may account for organ-specific patterns of accumulation [34].

In most mammalian species, including humans and rodents, iAs is rapidly transformed to methylated metabolites, which are excreted principally in the urine [1]. Historically, arsenic methylation has been characterized as a series of sequential reduction and oxidative methylation steps to produce mono-, di-, and tri-methylated metabolites [34]. A single enzyme, arsenic(+3)methyltransferase (AS3MT), that is capable of catalyzing all the oxidation and reduction steps leading to the production of these metabolites has been isolated and extensively characterized, including the identification of genetic polymorphisms [35]. Two alternative metabolic schemes have been proposed in recent years: one that differs principally in the reversible formation of arseno-glutathione complexes that are intermediate metabolic products [10] and another that proposes a reductive methylation pathway [29]. Based on the overall capacity, the liver is the major site of arsenic methylation [11, 23]. However, various other organs, including kidney, lung, and testes in mice have the capability to methylate arsenic [11] and AS3MT is also expressed in all regions of mouse brain [31]. Interestingly, both Healy et al. [11] and Sanchez-Pena et al. [31] report that exposure of mice to arsenite did not induce its metabolism. In the rat, mRNA expression levels were high for cyt19 (former name for AS3MT) in heart, liver, and testis, and low in spleen, lung, and skeletal muscle, while levels of the protein were high in both heart and liver [19]. Interestingly, Lin et al. [23] also reported detectable levels of cyt19 mRNA in both lung and urinary bladder, which are target organs for arsenic-induced cancer in humans. The chemistry and metabolism of arsenic is discussed in detail in Chapter 4.

22.3 ROLE OF PBPK MODELING

Studies of chemical kinetics in both pharmacology and toxicology arise from the need to relate internal concentrations of active compounds at their target sites to the doses of the compound administered to an animal or human subject. The reason for this interest is a fundamental tenet in pharmacology and toxicology—that both beneficial and adverse responses to compounds are related to the free concentrations of active compounds reaching target tissues, rather than the amount of compound at the site of absorption. The relationship between tissue dose and administered dose can be complex, especially at high doses, with repeated daily dosing, or when metabolism or toxicity at routes of entry alter pharmacokinetic processes for various routes of exposure. Pharmacokinetic models of all kinds are primarily tools to assess internal dosimetry in target tissues as a result of the absorption, distribution, metabolism, and excretion of the compound. PBPK models are particularly useful because they can include specific compartments for tissues involved in exposure, toxicity, biotransformation, and clearance processes. Tissues in a PBPK model are connected by blood flows, and compartments and blood flows are described using physiologically meaningful parameters. Because PBPK models use realistic parameters for tissue volumes and kinetic processes, they can be used for extrapolation across doses, exposure routes, and species. The mechanistic basis of PBPK models allows for applications such as understanding species differences in target tissue chemical exposure, determining whether results from different experimental designs are consistent, and exploring possible mechanisms responsible for unexpected or unusual data. These attributes have led to widespread development of PBPK models in recent years [3].

The trade-off for the greater predictive capability of physiologically based models is the requirement for an increased number of parameters and equations compared to the more empirical models. However, values for many parameters, particularly physiological parameters, are readily available in the literature. Furthermore, many *in vitro* techniques have been developed for rapid determination of compound-specific parameters, such as those describing tissue partitioning and metabolism. An important advantage of PBPK models is that they provide a biologically meaningful quantitative framework wherein *in vitro* data can be more effectively utilized to predict *in vivo* toxicokinetics. There is even a prospect that predictive PBPK models can be developed based almost entirely on data obtained from *in vitro* studies and quantitative structure–activity relationship (QSAR) modeling, eliminating the need for the use of animals in toxicokinetic analyses. This effort to eliminate the need for animal-based kinetic studies through the use of PBPK models, QSAR modeling, and *in vitro* data collection is generally referred to as "*in vitro* to *in vivo* extrapolation" or IVIVE [42].

PBPK models are often developed in support of chemical risk assessments [27]. A properly validated PBPK model can be used to perform the high-to-low dose, dose route, and interspecies extrapolations that are necessary for estimating human risk from toxicology studies in laboratory animals [17]. PBPK models are also used for examining the effects of changing physiology on target tissue dosimetry, as in the case of early life exposure where rapid organ growth and development of important metabolic enzymes can affect chemical kinetics [4].

22.4 PHARMACOKINETIC MODELING FOR ARSENIC—NEEDS AND ISSUES

A PBPK model for arsenic has many potential applications, which may be divided into two broad categories: (i) mechanistic pharmacokinetic models for use in generating and testing alternative hypothesis to assist in the design of experiments, and (ii) risk assessment applications, including the assessment of internal dose–response relationships evaluating the impact of genetic polymorphisms in AS3MT on target tissue dosimetry, and exposure reconstruction. Model structure and detail are conditioned upon both the intended use of the model and the availability (and acceptable uncertainty) of information available to parameterize it. As with any PBPK model, arsenic PBPK model development requires decisions about inclusion of (i) routes of exposure and absorption kinetics, (ii) organs/tissues, (iii) mechanisms of tissue transport (e.g., blood flow limited vs. membrane limited), (iv) routes of excretion, (v) metabolic pathways and their kinetic behavior, and (vi) whether the distribution/disposition of metabolites needs to be described (submodels). These general considerations as they apply specifically to arsenic are discussed in this section.

With respect to arsenic, ingestion is the major route of exposure associated with human health effects worldwide—mainly via contaminated drinking water, although ingestion in foods and from contaminated soil can also be important in specific situations. In the case of soils and foods, the bioavailability of arsenic is also a significant consideration. In general, studies indicate that oral bioavailability of arsenic in soil or dust is considerably lower compared to the pure soluble salts typically used in toxicity studies [1]. Oral absorption kinetics has generally been described as a simple first-order process in most arsenic models, reflecting the rapid appearance of arsenic in the blood following oral administration. Concern with inhalation exposure for arsenic has historically been limited to occupational situations (e.g., smelters) and use of high arsenic coal. Inhaled arsenic would be in particulate form or adsorbed onto particulates thus requiring consideration of issues specific to lung anatomy and particle dosimetry such as mean particle diameter and the distribution of particle diameters [1].

Organs and tissues included in an arsenic PBPK model will generally need to be those that have a significant role in metabolism, excretion, and storage, or are target tissues. In the case of arsenic, this will include the liver for metabolism and potentially a number of target organs such as the lung, kidney, and bladder. Description of membrane transport into these tissues may be specified as either perfusion-limited or diffusion-limited, depending upon the balance of processes governing diffusion across the cell membrane. When the cell membrane permeability coefficient (PA—a chemical-specific parameter) is much greater than the blood flow rate to a specific tissue, then diffusion into the tissue is limited by the rate at which blood containing the chemical is delivered to the tissue; that is, blood flow or perfusion-limited. This description requires fewer chemical-specific parameters, that is, only a tissue solubility parameter or partition coefficient (PC). On the other hand, if uptake into tissues is controlled or rate-limited by cell membrane permeability and total membrane area, then transport is diffusion-limited; this requires a somewhat more complex mathematical description and two chemical-specific parameters, PA and PC. In practice, a simpler perfusion-limited description is usually used initially and more complex uptake kinetic descriptions added if a simpler formulation does not adequately describe available data.

Urine is the primary excretory route for all arsenicals in mammalian species and is routinely collected during human biomonitoring studies. Inclusion of urinary excretion in arsenic PBPK models is thus necessary to both analyze pharmacokinetic data and particularly if the model is to be used for exposure reconstruction applications (reverse dosimetry). Biliary excretion of certain arsenical complexes has been demonstrated in rats [1]. However, it appears that at least for the major metabolites of arsenic in humans and mice, that the urine accounts for quantitatively 90–95% of metabolites excreted following animal dosing or controlled human exposures.

Arsenic methylation was once believed to simply be a means to facilitate excretion and transform arsenicals to less toxic metabolites based on the relatively low acute toxicity of pentavalent methylated arsenicals compared to iAs^{III} and iAs^{V}. It is now widely understood based on many lines of evidence that the trivalent methylated arsenicals are uniquely toxic compared to their pentavalent counterparts and mediate arsenic toxicity in ways that may be unique for specific tissues and forms of toxicity (see Chapters 4 and 13). The requirement of methylation for arsenical toxicity, as well as evidence that tissue accumulation can be both organ-specific and dose-dependent, indicates the necessity of modeling arsenic biotransformation pathways, as well as the importance of including submodels linked by metabolism that describe the distribution of arsenic metabolites in tissues.

Incorporation of arsenic methylation pathways and linked submodels for various arsenicals increases the utility of an arsenic PBPK model for both risk assessment and mechanistic hypothesis testing. However, there are also many challenges associated with this undertaking due to the added complexity in terms of numbers of chemical-specific parameters required, availability of data appropriate to estimate these parameters, and the difficulty posed by limitations in analytical chemistry methodology in terms of reliably distinguishing trivalent versus pentavalent methylated metabolites in tissues (as opposed to urine). *In vitro* studies using hepatocytes or subcellular systems (e.g., cytosol) are a very valuable tool to obtain estimates of metabolic rate parameters provided that appropriate companion data are available to enable them to be reliably scaled to the *in vivo* situation [17, 42]. This can provide independent estimates for model parameters and limit the problems associated with parameter estimation using data that are not sufficiently sensitive to the parameter being estimated.

22.5 TOXICOKINETIC MODELS OF ARSENIC METHYLATION

Recognition of the central importance of arsenic methylation and its complexity has led a number of investigators to develop models that specifically describe arsenic methylation using kinetic data generated from hepatic subcellular fractions and hepatocytes [6, 14, 32, 33]. Kenyon et al. [14] developed a kinetic model describing arsenite methylation as a sequential process that included first-order reversible protein binding of iAs^{III}, MMA (MMA^{III} +MMA^{V}), and DMA (DMA^{III} + DMA^{V}) based on data describing these processes in rat liver cytosol. This model also incorporated uncompetitive inhibition of the second methylation step by iAs^{III}; this was the only inhibition mechanism that was consistent with both the time-course data available and empirical data indicating an increasing time lag to detection of DMA with increasing initial arsenite concentration. The purpose of this modeling effort was to facilitate more directed experimental design and planning for purposes of parameter estimation. Easterling et al. [6] developed a model describing the sequential methylation of iAs^{III} to MMA and DMA in rat hepatocytes. This model also incorporated reversible first-order protein-binding of iAs^{III} as well as transport of iAs^{III} and MMA into and out of the hepatocyte and DMA out of the hepatocyte. The model also incorporated inhibition of the second methylation step by iAs^{III}; however, the inhibition mechanisms could not be clearly distinguished based on available data, although uncompetitive inhibition did explain the observed data somewhat better. The purpose of this model development effort was also to aid in experimental planning and design. It was also notable that this early mechanistic kinetic model identified the importance of transport parameters. One limitation of both the Kenyon et al. [14] and Easterling et al. [6] models is that they could not incorporate the distinction between trivalent and pentavalent methylated arsenicals due to limitations in analytical chemistry methodology and available data.

Stamatelos et al. [32] developed a toxicokinetic model describing iAs^{III} transport across the cell membrane and arsenic metabolism in human hepatocytes based on an alternative metabolic pathway. This pathway differs from the traditional sequential reduction–oxidative methylation metabolic scheme in that it incorporates reversible formation of iAs^{III}–glutathione complexes as necessary intermediates in arsenic methylation [10]. The parameterized schematic for this model explicitly distinguishes all reduction and oxidative methylation steps as well as the reversible complex formation between glutathione and trivalent arsenicals. In addition, the uptake of iAs^{III} into hepatocytes via cellular membrane pores was described as an ion channel conductance-based phenomenon mediated by electrochemical potentials; the biological basis for this description was that iAs^{III} uptake via aquaglyceroporin transporters is governed by electrochemical potentials [40]. The authors acknowledge a number of limitations inherent in a quantitative model formulation of this complexity, most notably parameter identifiability. The difficulty is that available experimental data are not generally sufficient to uniquely estimate all or most parameters for specific processes, but rather describe aggregations of steps, which does not allow the relative contribution of specific pathways to be evaluated [32]. This in turn limits the ability to insert these cellular-level models into whole organism toxicokinetic models.

These same investigators integrated the toxicokinetic model referenced earlier with a toxicodynamic model describing steps in the production of oxidative DNA damage and subsequent DNA repair [33]. While largely theoretical in nature, this model emphasizes the importance of feedback loops when describing this type of response process. Cellular-level models of this type are also a necessary initial step in the process of developing truly mechanistic, as opposed to simply empirical, biologically based dose–response models that will ultimately be useful for risk analysis applications [16, 18].

22.6 INORGANIC ARSENIC PBPK MODELS

A number of arsenic PBPK models for both humans and experimental animals have been published since the mid-1990s [7–9, 25, 26, 43–45]. These models vary in their structure and assumptions as summarized in Table 22.1, as well as their general utility. The Yu models are essentially theoretical in nature because the publications do not provide information on whether model predictions were evaluated against data not used to develop the models. In other words, complete information on model evaluation was not provided.

The original Mann et al. [25] hamster/rabbit model was developed with the goal of scaling it to humans [26]. The strengths of these models are that the basis for model structure and assumptions are clearly identified and the origin of the parameters well documented. In addition, separate data were used for model calibration (parameter estimation) and model evaluation and these data are clearly identified. The Gentry et al. [9] mouse model is essentially an extension of the Mann et al. model and shares many of its strengths. The model was developed for the purpose of evaluating whether pharmacokinetic differences among mouse strains might explain differences in response in mouse cancer bioassays; the authors concluded that pharmacokinetic factors did not account for differences in response [9].

The El Masri and Kenyon [7] model has a number of strengths that were derived from both the general approach used in model development and the availability of an additional decade of mechanistic research focused heavily on arsenic methylation. The model development process emphasized the use of human *in vitro* and *in vivo* data for derivation of chemical-specific parameters whenever possible and avoided problems with parameter identifiability by careful selection of model calibration data that was most sensitive to changes in the parameter being estimated. This model also

TABLE 22.1 Summary of Published Arsenic PBPK Models

Model	Description
Mann et al. [25]	
Species	Rabbits, hamsters
Routes/absorption	Inhalation, oral, intratracheal (IT), intraperitoneal (IP); first-order absorption for lung (multiple regions) and gastrointestinal (GI) tract
Compartments	Blood (plasma, RBC), liver, skin, lung, kidney, GI, residual
Submodels	Four—As^{V}, As^{III}, MMA, and DMA linked by reduction/methylation
Tissue uptake	Unique diffusion-limited mechanism using inputs of pore size, total pore area, and capillary thickness, As^{V}, MMA, and DMA assumed diffusion across membrane through capillary pores since these arsenicals are ionized at physiological pH, and As^{III} assumed to occur across entire capillary area due to being nonionized at physiological pH
Excretion	Urine (based on glomerular filtration rate (GFR) from plasma), feces, keratin (binding rate for As^{III})
Methylation	Sequential ($As^{III} \rightarrow$ MMA $\rightarrow$ DMA) in liver, Michaelis–Menten
Notes	Reduction/oxidation of $As^{V} \leftrightarrow As^{III}$ occurs in the plasma; reduction of As^{V} to As^{III} also occurs in kidney
Mann et al. [26]	
Species	Human
Routes/absorption	Inhalation, oral, IT, IP; first-order absorption for lung (multiple regions) and GI tract
Compartments	Blood (plasma , RBC), liver, skin, lung, kidney, GI, residual
Submodels	Four—As^{V}, As^{III}, MMA, and DMA linked by reduction/methylation
Tissue uptake	Same as Mann et al. [25], values for rabbit/hamster assumed for human
Excretion	Urine (based on GFR from plasma), feces, keratin (binding rate for As^{III})
Methylation	Sequential ($As^{III} \rightarrow$ MMA $\rightarrow$ DMA) in liver, Michaelis–Menten
Notes	Reduction/oxidation of $As^{V} \leftrightarrow As^{III}$ occurs in the plasma; reduction of As^{V} to As^{III} also occurs in kidney
Yu [43]	
Species	Mouse, rat
Routes/absorption	Oral (acute exposure), first-order uptake
Compartments	Lung, fat, vessel rich group (VRG), kidney, muscle, skin, liver, intestine
Submodels	None
Tissue uptake	Perfusion limited
Excretion	Urine, feces, bile; first order
Methylation	Sequential ($As^{III} \rightarrow$ MMA $\rightarrow$ DMA) in liver, Michaelis–Menten
Notes	Treats As^{III} and As^{V} as a single entity; MMA and DMA are formed in the liver, but do not circulate
Yu [44, 45]	
Species	Human
Routes/absorption	Oral, first-order uptake
Compartments	Lung, fat, kidney-VRG, muscle, skin, liver, intestine

TABLE 22.1 (*Continued*)

Model	Description
Submodels	Four—As^V, As^{III}, MMA, and DMA linked by reduction/methylation
Tissue uptake	Perfusion limited
Excretion	Urine, feces, bile; first order
Methylation	$As^{III} \rightarrow$ MMA, $As^{III} \rightarrow$ DMA, MMA $\rightarrow$ DMA; liver and kidney-VRG, Michaelis–Menten
Notes	Reduction of As^V to As^{III} assumed to occur in all perfused tissue groups via the first-order process; the Yu [45] model is an expanded description of the Yu [44] model with some updated parameters
Gentry et al. [9]	
Species	Mouse
Routes/absorption	Oral, inhalation; first-order absorption for lung (multiple regions) and GI tract
Compartments	Blood (plasma, RBC), liver, skin, lung, kidney, GI, residual
Submodels	Four—As^V, As^{III}, MMA, and DMA linked by reduction/methylation
Tissue uptake	Diffusion limited as in Mann et al. [25]
Excretion	Urine (based on GFR from plasma), feces, bile (first order), keratin (binding rate for As^{III})
Methylation	Sequential ($As^{III} \rightarrow$ MMA $\rightarrow$ DMA) in liver, Michaelis–Menten, includes colocality for metabolism of MMA to DMA and uncompetitive inhibition of formation of DMA from MMA
Notes	Reduction/oxidation of $As^V \leftrightarrow As^{III}$ occurs in the plasma; reduction of $As^V \rightarrow As^{III}$ also occurs in kidney and urine; this model was based on Mann et al. [25]
El Masri and Kenyon [7]	
Species	Human
Routes/absorption	Oral; first-order absorption
Compartments	Lung, liver, GI tract, kidney, muscle, brain, skin, heart, blood
Submodels	Four—As^V, As^{III}, MMA, and DMA linked by reduction/methylation
Tissue uptake	Perfusion limited
Excretion	Urine (first order)
Methylation	Sequential ($As^{III} \rightarrow$ MMA $\rightarrow$ DMA, $As^{III} \rightarrow$ DMA) in liver and kidney, Michaelis–Menten, noncompetitive inhibition of formation of DMA from MMA by As^{III}
Notes	Includes reversible reduction/oxidation of $As^V \leftrightarrow As^{III}$ in lung, liver, and kidney; reduction of As^V to As^{III} occurs predominantly in GI tract lumen

incorporated an updated and expanded description of arsenic methylation and the ability to predict formation and urinary excretion of methylated trivalent arsenicals. This model has been translated for use as part of a PBPK model "tool kit" intended for use in evaluating potential exposures from contaminated environmental media and to facilitate interpretation of site-specific biomonitoring data [30].

22.7 EXAMPLE APPLICATIONS OF INORGANIC ARSENIC MODELS

There exist in the literature several examples of specific applications of both PBPK and compartmental arsenic models [20, 22, 24, 38]. Ling and Liao [24] linked the human PBPK model developed by Yu [44, 45] to an empirical three-parameter Hill equation-based pharmacodynamic model to evaluate the risks of consumption of farmed tilapia for the development of lung and bladder cancer. They concluded that consumption of farmed tilapia posed no significant risk for bladder and lung cancer when consumed in amounts averaging 5–17 meals per month or 2–6 g/day. Liao et al. [22] used the same PBPK model linked with a Weibull dose–response function to assess the risk of arsenic-induced skin lesions in children consuming arsenic in drinking water. They used the linked models to develop recommended safe drinking water levels for arsenic based on specific endpoints (hyperpigmentation and keratosis) and response levels (e.g., $ED_{0.1}$).

In another type of model application, Xue et al. [38] used probabilistic exposure modeling in combination with a PBPK model to assess the contribution of food relative to drinking water to total arsenic exposure including evaluation of model predictions for biomarker (total arsenic in urine) data using NHANES data. Using this analytical approach, the authors were able to identify the major dietary contributors to total arsenic exposure and concluded that, for the general US population, diet is a greater contributor to total and inorganic arsenic exposure relative to drinking water.

Lawley et al. [20] developed a compartmental model to analyze and interpret data on the effects of folate supplementation on arsenic methylation. This type of analysis is an example of a model to evaluate the potential effects of a public health intervention (i.e., folate supplementation) on an arsenic-exposed population. Compartmental models such as this are simpler than PBPK models in that there are fewer compartments and parameters. However, the compartments have no physiological meaning and thus do not provide a basis for extrapolation outside the conditions of the data used to develop (or calibrate) the compartmental model [3]. Nonetheless, if sufficient human population-relevant data are available to develop a compartmental model, they can provide a means to evaluate more readily and interpret additional similar data sets and can be a useful refinement compared to the use of only external exposure estimates (e.g., water or air concentration) in dose–response analysis.

None of the example model applications for inorganic arsenic discussed earlier utilize a PBPK model for more refined internal dose–response analysis for specific endpoints or for high- to low-dose extrapolation. While this type of model application is clearly valuable and needed for stronger biologically based dose–response analysis, limitations in available data and more detailed understanding of mode of action for arsenic are clearly needed to accelerate this type of application.

22.8 CURRENT CHALLENGES AND FUTURE DIRECTIONS

Metals generally, and arsenic in particular, provide unique challenges in PBPK model development and health risk assessment. This is because metal speciation—the occurrence of an element in separate and identifiable forms—profoundly influences both toxicokinetic and toxicodynamic behavior [41]. Arsenic, as a metalloid with toxicologically active trivalent methylated metabolites, poses an additional challenge because it has historically been difficult to obtain dose–response and time-course data for formation of these metabolites because of limitations in extraction methodology to maintain these metabolites reliably in their trivalent form. Thus, the specific data needed to develop more detailed and toxicologically relevant descriptions for arsenic methylation have frequently been unavailable. Continuing advances in extraction techniques and analytical methodology will be necessary to bridge this gap.

O'Flaherty summarized the issues that are distinctive to metals in PBPK model development compared to volatile organic chemicals. Among these are specific binding to macromolecules that can profoundly alter distribution and sequestration, unique cellular uptake mechanisms, as well as metal–metal interactions at many levels, including toxic metal and essential metal competition for binding sites on transport proteins and at other metabolic and mechanistic control points [28]. A number of authors have acknowledged that their models as implemented describe the general kinetic behavior of arsenic, but may not reflect in detail the pharmacokinetic mechanisms involved [7, 32]. For example, while uptake and elimination of the arsenic compounds by cells are associated with specific transporters [40], the descriptions used in PBPK models have generally used equations that are more consistent with thermodynamic partitioning or diffusion. Recent mechanistic studies have identified transporters such as GLUT2 and MRP2 that can affect cellular uptake and efflux of iAs^{III} and its methylated metabolites in a dose-dependent manner, thus effectively altering the capacity of human hepatocytes to metabolize iAs^{III} [5]. Quantitative mechanistic description of these processes is a prerequisite to predicting the target tissue dosimetry of active trivalent methylated arsenicals.

The capabilities of the currently available PBPK models are restricted by the nature and extent of the experimental data on which they rely. In particular, most of the data available for model development and evaluation do not distinguish the trivalent and pentavalent arsenical. Of particular concern is the lack of data on MMA^{III}, which binds much more avidly than the other forms of arsenic and may be prone to accumulation in cells. Accumulation of MMA^{III} could be an important factor in the toxicity of arsenic. While the acute toxicity and lethality of iAs occurs at similar doses across species, humans appear to be significantly more sensitive than experimental animals to the chronic noncancer toxicity of iAs. Toxicity in chronic animal studies has only been observed for iAs doses above 1 mg/kg/day, while multiple effects of chronic exposure in humans, particularly vascular effects, have frequently been associated with drinking water exposures of less than 0.1 mg/kg/day [1]. The possibility that these differences in toxicity result from differences in target tissue exposure to MMA^{III} is suggested by the fact that humans excrete more MMA than other animal species, as well as by data from a number of epidemiological studies showing an association between a higher ratio of urinary MMA to DMA with a greater risk of arsenic toxicity and tumorigenicity [37]. Unfortunately, there are very few data on the individual arsenic species concentrations in tissues associated with arsenic exposures in either experimental animals or humans. If tissue concentrations of the different arsenicals were more frequently examined in humans and experimental animals, correlations between toxicity and particular species of arsenic could be more easily developed and hypotheses for causation of toxic effects could be more readily tested.

Another limitation in the data available for developing more biologically based PBPK models is information on the metabolism of arsenic in the target tissues for toxicity and cancer. Due to the avidity with which it binds to proteins, exposure to MMA^{III} in the target cells is more likely to result from *in situ* reduction of MMA^{V} than from systemic circulation of MMA^{III}. Studies are needed to determine the nature and extent of arsenic metabolism (both methylation and reduction) in the various target tissues for the toxicity and carcinogenicity of arsenic, as a function of circulating concentrations of MMA^{V}, in order to estimate local concentrations of MMA^{III} in the target tissues that could then be linked with tissue responses in the dose–response analysis. Of particular concern would be tissues that are able to reduce MMA^{V} to MMA^{III}, but lack the capacity for methylation of MMA^{III} to DMA^{V}, resulting in an increased potential for MMA^{III} interactions with key cellular proteins.

Collection of additional data in exposed human populations is needed to estimate the target tissue exposures to the various species of arsenic (particularly arsenite, MMA^{III}, and DMA^{III}) associated with increased tumor incidence. Given the large populations currently exposed to high concentrations of iAs in drinking water, it should be possible to obtain valuable information on arsenic tissue dosimetry. Potential sources of such data include skin and liver biopsies, exfoliated bladder cells, and tissues removed during surgery. It would also be of interest to obtain additional data on tissue concentrations of the various arsenic species in experimental animals, both following chronic exposure and in naive animals, in order to correlate tissue concentrations with specific toxic effects of various forms of arsenic.

REFERENCES

[1] ATSDR, in: U.S. DHHS (Ed.), Toxicological Profile for Arsenic, Agency for Toxic Substances and Disease Registry (ATSDR), Atlanta, GA, 2007.

[2] J.P. Buchet, R. Lauwerys, H. Roels, Comparison of the urinary excretion of arsenic metabolites after a single oral dose of sodium arsenite, monomethylarsonate and dimethylarsinate in man, Int. Arch. Occup. Environ. Health 48 (1981) 71–79.

[3] H.J. Clewell, M.B. Reddy, T. Lave, M.E. Andersen, Physiologically based pharmacokinetic modeling, in: C., Shane Gad (Ed.), Preclinical Development Handbook: ADME and Biopharmaceutical Properties, John Wiley & Sons, Inc., New York, 2008.

[4] R.A. Clewell, E.A. Merrill, J.M. Gearhart, P.J. Robinson, T.R. Sterner, D.R. Mattie, H.J.I. Clewell, Perchlorate and radioiodide kinetics across life-stages in the human: using PBPK models to predict dosimetry and thyroid inhibition and sensitive subpopulations based on developmental stage J. Toxicol. Environ. Health A 70 (2007) 408–428.

[5] Z. Drobna, F. Walton, D.S. Paul, X. Weibing, D.J. Thomas, M. Styblo, Metabolism of arsenic in human liver: the role of membrane transporters, Arch. Toxicol. 84 (2010) 3–16.

[6] M.R. Easterling, M. Styblo, M.V. Evans, E.M. Kenyon, Pharmacokinetic modeling of arsenite uptake and metabolism in hepatocytes—mechanistic insights and implications for further experiments, J. Pharmacokinet. Pharmacodyn. 29 (2002) 207–234.

[7] H.A. El-Masri, E.M. Kenyon, Development of a human physiologically based pharmacokinetic (PBPK) model for inorganic arsenic and its mono- and di-methylated metabolites, J. Pharmacokinet. Pharmacodyn. 35 (2008) 31–68.

[8] M.V. Evans, S.M. Dowd, E.M. Kenyon, M.F. Hughes, H.A. El-Masri, A physiologically based pharmacokinetic model for intravenous and ingested dimethylarsinic acid in mice, Toxicol. Sci. 104 (2008) 250–260.

[9] P.R. Gentry, T.R. Covington, S. Mann, A.M. Shipp, J.W. Yager, H.J. Clewell, Physiologically based pharmacokinetic modeling of arsenic in the mouse, J. Toxicol. Environ. Health A 67 (2004) 43–71.

[10] T. Hayakawa, Y. Kobayashi, X. Cui, S. Hirano, A new metabolic pathway of arsenite: arsenite-glutathione complexes are substrates for human arsenic methyltransferase Cyt 19, Arch. Toxicol. 79 (2005) 183–191.

[11] S.M. Healy, E.A. Casarez, F. Ayala-Fierro, H. Aposhianm, Enzymatic methylation of arsenic compounds. V. Arsenite methyltransferase activity in tissues of mice, Toxicol. Appl. Pharmacol. 148 (1998) 65–70.

[12] M.F. Hughes, E.M. Kenyon, B.C. Edwards, C.T. Mitchell, D.J. Thomas, Strain-dependent disposition of inorganic arsenic in the mouse, Toxicology 137 (1999) 95–108.

[13] IPCS, Environmental Health Criteria for Arsenic and Arsenic Compounds, EHC No. 224 World Health Organization, International Programme on Chemical Safety (IPCS), Geneva, 2001.

[14] E.M. Kenyon, M. Fea, M. Styblo, M.V. Evans, Application of modelling techniques to the planning of in vitro arsenic kinetic studies, Altern Lab Anim 29 (2001) 15–33.

[15] E.M. Kenyon, M.F. Hughes, B.M. Adair, J.H. Highfill, E.A. Crecelius, H.J. Clewell, J.W. Yager, Tissue distribution and urinary excretion of inorganic arsenic and its methylated metabolites in C57BL/6 mice following subchronic exposure to arsenate (AsV) in drinking water, Toxicol. Appl. Pharmacol. 232 (2008) 448–455.

[16] E.M. Kenyon, W.T. Klimecki, H. El-Masri, R.B. Conolly, H.J. Clewell, B.D. Beck, How can biologically-based modeling of arsenic kinetics and dynamics inform the risk assessment process?—A workshop review, Toxicol. Appl. Pharmacol. 232 (2008) 359–368.

[17] E.M. Kenyon, Interspecies extrapolation, in: B. Reisfield, A. Mayeno (Eds.), Methods in Molecular Biology—Computational Toxicology, Humana Press, New York, 2012, Vol. 1, pp. 501–521.

[18] K.T. Kitchin, R. Conolly, Arsenic-induced carcinogenesis-oxidative stress as a possible mode of action and future research needs for more biologically based risk assessment, Chem. Res. Toxicol. 23 (2010) 327–335.

[19] Y. Kobayashi, T. Hayakawa, S. Hirano, Expression and activity of arsenic methyltransferase Cyt19 in rat tissues, Environ. Toxicol. Pharmacol. 23 (2007) 115–120.

[20] S.D. Lawley, M. Cinderella, M.N. Hall, M.V. Gamble, H.F. Nijhout, M.C. Reed, Mathematical model insights into arsenic detoxification, Theor. Biol. Med. Model. 8 (2011) 31.

[21] E. Lee, A physiologically based pharmacokinetic model for the ingestion of arsenic in humans, University of California, Irvine, CA, 1999, Ph.D. thesis.

[22] C.-M. Liao, T.-L. Lin, S.-C. Chen, A Weibull-PBPK model for assessing risk of arsenic-induced skin lesions in children, Sci. Total Environ. 392 (2008) 203–217.

[23] S. Lin, Q. Shi, F.B. Nix, M. Styblo, M.A. Beck, K.M. Herbin-Davis, L.L. Hall, J.B. Simeonsson, D.J. Thomas, A novel S-adenosyl-L-methionine:arsenic(III) methyltransferase from rat liver cytosol, J. Biol. Chem. 277 (2002) 10795–10803.

[24] M.-P. Ling, C.-M. Liao, A human PBPK/PD model to assess arsenic exposure risk through farmed tilapia consumption, Bull. Environ. Contam. Toxicol. 83 (2009) 108–114.

[25] S. Mann, P.O. Droz, M. Vahter, A physiologically based pharmacokinetic model for arsenic exposure .1. Development in hamsters and rabbits, Toxicol. Appl. Pharmacol. 137 (1996) 8–22.

[26] S. Mann, P.O. Droz, M. Vahter, A physiologically based pharmacokinetic model for arsenic exposure .2. Validation and application in humans, Toxicol. Appl. Pharmacol. 140 (1996) 471–486.

[27] E.D. McLanahan, H. El Masri, L.M. Sweeney, L. Kopylev, H.J. Clewell, J. Wambaugh, P.M. Schlosser, Physiologically based pharmacokinetic model use in risk assessment—why being published is not enough, Toxicol. Sci. 126 (2012) 5–15.

[28] E.J. O'Flaherty, Physiologically based models of metal kinetics, Crit. Rev. Toxicol. 28 (1998) 271–317.

[29] K. Rehman, H. Naranmandura, Arsenic metabolism and thioarsenicals, Metallomics 4 (2012) 881–892.

[30] P. Ruiz, B.A. Fowler, J.D. Osterloh, J. Fisher, M. Mumtaz, Physiologically based pharmacokinetic (PBPK) tool kit for environmental pollutants—metals, SAR QSAR Environ. Res. 21 (2010) 603–618.

[31] L.C. Sánchez-Peña, P. Petrosyan, M. Morales, N.B. González, G. Gutiérrez-Ospina, L.M. Del Razo, M.E. Gonsebatt, Arsenic species, AS3MT amount, and AS3MT gene expression in different brain regions of mouse exposed to arsenite, Environ. Res. 110 (2010) 428–434.

[32] S.K. Stamatelos, C.J. Brinkerhoff, S.S. Isukapalli, P.G. Georgopoulos, Mathematical model of uptake and metabolism of arsenic(III) in human hepatocytes—incorporation of cellular antioxidant response and threshold-dependent behavior, BMC Syst. Biol. 5 (2011) 16.

[33] S.K. Stamatelos, I.P. Androulakis, K. Ah-Ng Tony, P.G. Georgopoulos, A semi-mechanistic integrated toxicokinetic-toxicodynamic (TK/TD) model for arsenic(III) in hepatocytes, J. Theor. Biol. 317 (2013) 244–256.

[34] D.J. Thomas, Molecular processes in cellular arsenic metabolism, Toxicol. Appl. Pharmacol. 222 (2007) 365–373.

[35] D.J. Thomas, J. Li, S.B. Waters, W. Xing, B.M. Adair, Z. Drobna, V. Devesa, M. Styblo, Arsenic (+3 oxidation state) methyltransferase and the methylation of arsenicals, Exp. Biol. Med. 232 (2007) 3–13.

[36] M. Vahter, H. Norin, Metabolism of ^{74}As-labeled trivalent and pentavalent inorganic arsenic in mice, Environ. Res. 21 (1980) 446–457.

[37] M. Vahter, Mechanisms of arsenic biotransformation, Toxicology 181 (2002) 211–217.

[38] J. Xue, V. Zartarian, S.-W. Wang, S.V. Liu, P. Georgopoulos, Probabilistic modeling of dietary arsenic exposure and dose and evaluation with 2003–2004 NHANES data, Environ. Health Perspect. 118 (2010) 345–350.

[39] H. Yamauchi, Y. Yamamura, Concentration and chemical species of arsenic in human tissue, Bull. Environ. Contam. Toxicol. 32 (1983) 416–421.

[40] H.C. Yang, H.L. Fu, Y.F. Lin, B.P. Rosen, Pathways of arsenic uptake and efflux, Curr. Top. Membr. 69 (2012) 325–358.

[41] R.A. Yokel, S.M. Lasley, D.C. Dorman, The speciation of metals in mammals influences their toxicokinetics and toxicodynamics and therefore human health risk assessment, J. Toxicol. Environ. Health B Crit. Rev. 9 (2006) 63–85.

[42] M. Yoon, J.L. Campbell, M.E. Andersen, H.J. Clewell, Quantitative in vitro to in vivo extrapolation of cell-based toxicity assay results, Crit. Rev. Toxicol. 42 (2012) 633–652.

[43] D.H. Yu, Uncertainties in a pharmacokinetic modeling for inorganic arsenic, J. Environ. Sci. Health A 33 (1998) 1369–1390.

[44] D.H. Yu, A pharmacokinetic modeling of inorganic arsenic: a short-term oral exposure model for humans, Chemosphere 39 (1999) 2737–2747.

[45] D.H. Yu, A physiologically based pharmacokinetic model of inorganic arsenic, Regul. Toxicol. Pharmacol. 29 (1999) 128–141.

23

CONSIDERATIONS FOR A BIOLOGICALLY BASED RISK ASSESSMENT FOR ARSENIC

HARVEY J. CLEWELL, III[1], P. ROBINAN GENTRY[2] AND JANICE W. YAGER[3]

[1] *Hamner Institutes for Health Sciences, Research Triangle Park, NC, USA*
[2] *Environ International Corporation, Monroe, LA, USA*
[3] *Department of Internal Medicine, University of New Mexico, Albuquerque, NM, USA*

23.1 INTRODUCTION

Arsenic presents unique challenges for conducting a human health risk assessment. While chronic human exposures in drinking water have repeatedly been associated with increases in skin, lung, and bladder cancer, inorganic arsenic has typically not produced tumors in standard animal bioassays conducted at much higher concentrations, except when administered during early development or in conjunction with another carcinogen. Although the human studies provide unequivocal evidence of both cancer and noncancer effects from exposure to arsenic in drinking water at concentrations above 100 ppb, there is still significant uncertainty regarding the potential for effects at lower concentrations. This chapter describes the variety of factors that must be considered in order to derive a biologically sound risk assessment for arsenic, including both cancer and noncancer endpoints.

Arsenic: Exposure Sources, Health Risks, and Mechanisms of Toxicity, First Edition.
Edited by J. Christopher States.
© 2016 John Wiley & Sons, Inc. Published 2016 by John Wiley & Sons, Inc.

23.2 BACKGROUND: RISK ASSESSMENT ISSUES FOR ARSENIC

The most striking feature of inorganic arsenic (iAs) toxicity and carcinogenicity is the large number of studies unambiguously demonstrating effects in humans. The availability of human data reflects the widespread distribution of iAs-rich rock formations around the globe, together with the frequent occurrence of human activities that result in exposures. Contamination of well water by natural iAs has been reported on several continents, including Asia, North America, and South America. In addition, some ores for important metals, including copper, zinc, and lead, also contain iAs, resulting in the release of iAs-containing dusts and vapors during processing. Moreover, iAs's properties have made it desirable for uses in medicines and pesticides throughout history. In the late 1960s and 1970s, epidemiological studies demonstrated that chronic oral iAs exposures were associated with an increased risk of developing skin lesions and skin cancer, while chronic inhalation exposures were associated with an increased risk of developing lung cancer. More recently, human exposure to iAs in drinking water has also been associated with increased risks of internal cancers of the lung and bladder [2, 43].

Despite the strong evidence that exposure to high concentrations of iAs in air or in drinking water is associated with an increased risk of cancer and noncancer health effects in human populations, there is uncertainty regarding the nature of the dose–response relationship for these effects in the studied populations, especially in the low-dose region. There is even more uncertainty regarding the extrapolation of the dose–response relationship observed in highly exposed populations to populations in other parts of the world, such as the United States, where exposures are much lower and genetic background and lifestyle factors (e.g., diet and smoking) may be very different. In the case of drinking water exposures, iAs concentrations clearly associated with skin cancer in Taiwan (above 100 ppb) are only about an order of magnitude higher than the current maximum contaminant level (MCL) for iAs in drinking water in the United States, 0.01 mg/L (10 ppb). This apparently small margin of safety makes it imperative that there is a solid basis of experimental data and scientific understanding prior to policy decisions regarding the extent of hazards associated with continued use of contaminated water supplies.

Daily intakes of total arsenic in food are in the order of 0.05 mg/day, of which a maximum of 20–25% is iAs [9]. For comparison, at a standard drinking water ingestion rate of 2 L/day, the ingestion of iAs at the MCL of 0.01 mg/L would represent a total exposure of 0.02 mg/day. Thus certain foods (e.g., rice) may represent a significant source of exposure. A number of years ago, USEPA [99] estimated that consumption of drinking water containing iAs at 10 ppb entails a lifetime risk of skin cancer on the order of 0.5 per thousand, based on dose–response data for skin cancer in a population chronically exposed to high concentrations (>100 ppb) of iAs in drinking water in Taiwan [94, 95]. Given the increasing evidence that iAs exposure is associated with numerous cancer and noncancer health effects in humans [2, 43], the USEPA has for many years been weighing the costs relative to health benefits to the public by reducing the MCL for iAs from the current MCL of 10 ppb. Much of the controversy surrounding USEPA's reassessment of the MCL has focused on concerns regarding the appropriateness of the USEPA methods for estimating cancer risk at the relatively low iAs exposures associated with the consumption of drinking water at the MCL and below [71].

USEPA estimates of carcinogenic potency for iAs have traditionally been based on the default assumption of low-dose linearity, that is, that any dose, no matter how small, poses some level of risk [98]. However, recent investigations in a number of areas have fueled speculation that the dose–response for iAs carcinogenicity is nonlinear and that, mechanistically, a "threshold" for iAs carcinogenicity may indeed exist [1, 32, 81]. If this is the case, then characterization of the dose–response relationship for tumors and extrapolation of that relationship to other populations may require a significant departure from traditional risk assessment practices. To be useful, a risk assessment for iAs should include explicit consideration of the underlying biological processes, and integrate and quantify the dose of the active arsenic species to the target tissue with an evaluation of the dose–response for those cellular interactions that are thought to enhance tumor formation.

Thus far, the available pharmacokinetic and mechanistic information for arsenic has had little impact on quantitative cancer risk assessments for iAs, which uniformly followed default guidelines rather than relying on chemical-specific mechanistic data [98]. However, the regulatory climate has recently changed, particularly with the development of new USEPA cancer guidelines [102]. The new guidelines represent a major departure from the previous USEPA [98] approach for cancer risk assessment. Important features include the possibility of providing both linear and nonlinear (threshold) approaches for low-dose extrapolation, with the mode of action of the chemical determining both the conditions under which the chemical should be considered a cancer hazard for humans and the appropriate low-dose extrapolation approach. For chemicals such as iAs that have been hypothesized to cause cancer by indirect genomic mechanisms, the use of a nonlinear, margin-of-exposure (MOE) approach can lead to acceptable environmental levels that are very different from those that would result from a traditional linear default approach. However, departure from the linear default requires experimental information regarding the shape of the dose–response curve for the carcinogenicity of iAs at exposure levels below those associated with observable tumor incidences.

To the extent that the interactions between iAs and tissues leading to noncancer effects are related to, or are precursors of, the interactions leading to carcinogenicity, it might be possible to utilize information on the dose–response for noncancer effects as a surrogate for the cancer dose–response. Species differences in the dose–response for noncancer effects may also help to explain the marked species differences in carcinogenic potency. Further appreciation of the dose–response for the noncancer effects of iAs is also necessary in order to appropriately evaluate the potential impact of a revised iAs cancer assessment on drinking water standards (e.g., MCL). If a nonlinear or MOE approach were selected for cancer risk assessment for iAs, an exposure limit or standard based a noncancer health effect could result in a more conservative value.

23.3 METABOLISM AND DISPOSITION OF INORGANIC ARSENIC

A detailed discussion of the pharmacokinetics and metabolism of iAs is provided in the previous chapters (Chapters 4 and 22), so only a few aspects relevant to risk assessment will be discussed here. The predominant metabolism of iAs in mammals is methylation [104]. Methylation occurs for iAs in its trivalent state [90, 106], but pentavalent arsenate is readily reduced to trivalent arsenite *in vivo*. Methylation occurs primarily in the liver, via a saturable enzymatic process, initially forming monomethylarsonic acid (MMA^{V}) and then dimethylarsinic acid (DMA^{V}). The methylated arsenic metabolites are less reactive with tissue components than the inorganic forms and are readily excreted in the urine [104]. However, the trivalent methylated metabolites (MMA^{III} and DMA^{III}) are more reactive and play a significant role in the toxicity of iAs [32, 87].

Humans are thought to have a higher urinary fraction of MMA than other animal species [104]. It has also been demonstrated that with increasing iAs intake in humans, there is a saturation of the second methylation step (from MMA to DMA) in short-term exposures [14, 15], with evidence for the saturation of DMA production occurring following ingestion of approximately 0.2–0.5 mg iAs. Increased organoarsenical ratios (MMA:DMA) have been reported in humans chronically exposed to iAs concentrations exceeding 0.4 mg/L (400 ppb) in drinking water [17, 36, 112], and it has been suggested that changes in the MMA:DMA ratio may in some way be correlated with the observation of carcinogenesis. Clearly, a biologically motivated risk analysis must take into account the metabolism and disposition of iAs and the toxicity of its trivalent metabolites.

23.4 EVIDENCE OF ARSENIC CARCINOGENICITY

Oral exposures of experimental animals to iAs alone have generally failed to elicit a cancer response unless the exposure includes the early developmental period [92]. In contrast, there is abundant epidemiological evidence of arsenic carcinogenicity to human populations, including cases where the exposure did not include early life [96]. Reviews of the epidemiological evidence for the toxicity and carcinogenicity of arsenic are provided elsewhere in this text. Here discussion focuses on the small number of epidemiological studies that provide insights into the possible dose–response for arsenic carcinogenicity. The most widely studied iAs-exposed population is located along the southwest coast of Taiwan and includes 42 villages in six townships [94, 95]. Because of the high salinity of shallow wells in this area, the studied population used deep artesian wells for drinking water [18]; alternative water sources did not become generally available until the 1970s [49]. The artesian well water contained iAs concentrations ranging from 0.01 to 1.82 mg/L [94]. An analysis of the dose–response for lung and bladder cancer in this population [68] is the basis for the current arsenic MCL.

Studies of larger populations have allowed further evaluation of the evidence for the carcinogenicity of arsenic as a function of exposure concentration. Guo et al. [46]

performed an epidemiological drinking water study that included 243 townships throughout Taiwan with a combined population of 11.4 million and evaluated arsenic ingestion in drinking water from over 80,000 wells obtained from a nationwide census survey. Positive associations were noted in transitional cell carcinoma (TCC), kidney, ureter, and all urethral cancers combined at arsenic levels greater than 640 ppb in both men and women. However, no association was seen between arsenic drinking water concentrations less than 330 ppb and any of the cancer endpoints.

In another study, the occurrence of urinary tract cancers and TCC in northeastern Taiwan was investigated in a cohort of 8102 residents [24]. Arsenic concentrations in drinking water ranged from less than 0.15 to 3590 ppb. At arsenic concentrations of greater than 100 ppb, the relative risk for both urinary tract cancer and TCC was statistically significantly increased, but no statistically significant increase in the incidence of urinary tract cancer or TCC was observed following exposures of less than 100 ppb. There was a statistically significant higher risk of urinary tract cancers and TCC in cigarette smokers, but there was no significant synergistic association between arsenic exposure and cigarette smoking. In a 12-year follow-up study conducted in the northeastern Taiwan population [23], relative risks for all urinary cancers and urothelial cancers did not reach statistical significance until arsenic water concentration reached greater than 100 μg/L. Similarly, a number of other epidemiological studies conducted in Europe and the United States found no evidence of increased cancer incidence with chronic exposure to low concentrations (<100 ppb) of inorganic arsenic in drinking water [5, 6, 47, 54, 58, 59, 67, 84].

Although medicinal exposures to arsenicals are not necessarily identical to drinking water exposures, studies of patient populations provide additional data on cancer following oral intake of iAs; in particular, cancers of the skin and internal organs have been reported [7]. A significant excess in the incidence of bladder cancer mortality and a weak dose–response trend for respiratory cancer have been reported among patients administered more than 0.5 g of potassium arsenite, in the form of Fowler's solution, for periods ranging from 2 weeks to 12 years [34, 35]. It was also noted in these studies that among a subgroup of patients examined for dermatological signs of arsenicism, all cancer deaths occurred among those showing evidence of skin disease [34, 35].

Inhalation of iAs dusts and vapors in occupational settings represents another major route of exposure. Studies of copper smelter worker populations (Tacoma, Washington; Anaconda, Montana; Ronnskar, Sweden; and Magma, Utah), who were exposed to iAs (arsenic trioxide plus other compounds, e.g., copper and sulfur dioxide) via inhalation, reported associations between occupational iAs exposure and increased lung cancer mortality rates [22]. Similar studies involving workers at pesticide manufacturing or packaging facilities have also shown excess lung cancer mortality among exposed workers [62, 72].

Analyses of the data regarding respiratory cancer mortality of 2802 men who worked at a copper smelter in Tacoma, Washington, for 1 year or more during the period from 1940 to 1964, demonstrated a relationship between the incidence of respiratory, kidney, and bone cancers and iAs exposure [39, 40]. iAs exposures were estimated based on urinary arsenic levels, rather than airborne iAs levels [40]. A further study of 527 men retired from this copper smelter found that the mortality rate among arsenic trioxide-exposed workers was 12.2% higher than that of a comparably unexposed population [74]. A reanalysis of the Tacoma cohort data found nonlinearity in the dose–response for lung cancer related to year of entry into the study [107].

A study of cause-specific mortality of workers at a Swedish smelter found a positive dose–response relationship between cumulative iAs exposure and lung cancer mortality [52, 53]. Interestingly, lung cancer mortality was related to estimated average iAs exposure, but not to duration.

A relationship between respiratory cancer and iAs exposure was also observed among 1800 men employed at a copper smelter in Anaconda, Montana [113]. Exposure levels were categorized as low ($<0.1\,mg/m^3$), medium ($0.1–0.499\,mg/m^3$), high ($0.5–4.999\,mg/m^3$), and very high ($>5.0\,mg/m^3$). Potentially confounding factors that were considered included smoking habits and exposures to sulfur dioxide and asbestos. A clear dose–response relationship between iAs exposure and respiratory cancer mortality was established, with men exposed to very high iAs concentrations having a sevenfold excess above the expected respiratory cancer mortality. Ceiling iAs exposure was found to be more important than time-weighted average exposure. A study of mortality patterns in a US copper smelter showed overall deficits in deaths from all cancers as well as most causes of death categories examined [65].

23.4.1 Cancer Risk Estimates

Taken together, the many epidemiological studies conducted on iAs-exposed populations clearly demonstrate an increased risk for cancer of the skin, lung, bladder, and perhaps some other internal organs associated with exposure to iAs. However, the quantitative relationship between iAs exposure and cancer incidence remains a subject of significant debate. The USEPA [97] used data from the epidemiological studies of the Anaconda (Montana) and ASARCO (Washington) smelters to estimate inhalation risks based upon lung cancer mortality. Based on these studies of smelter workers exposed to iAs in dust at concentrations in the order of $1\,mg/m^3$, the USEPA [97] derived a cancer risk of 4.3 per thousand for lifetime inhalation of iAs at $0.001\,mg/m^3$.

In two separate published analyses, the multistage model of carcinogenesis was used to analyze inhalation data, in an attempt to determine whether iAs affected the earlier or later stages in the carcinogenic process. In both cases, the effects of iAs were found most likely to be at late stages of the cancer process [10, 11, 66].

The oral cancer slope factor for iAs calculated by the USEPA [99], 1.5(mg/kg/day)-1, is based upon the incidence of skin cancer reported in the early Taiwanese drinking water studies [94, 95]. The associated lifetime risk predicted for ingestion of drinking water containing 0.05 mg/L iAs (the MCL at that time) was 2.5 per thousand. At the time it was developed, it was felt that there was not sufficient dose–response data to develop a risk estimate based upon the incidence of internal tumors. Upper bound estimates of the overall cancer risk associated with lifetime consumption of drinking water containing 0.05 mg/L iAs range as high as 13 per thousand [79].

Evaluation of the dose–response for cancers in the Taiwanese drinking water exposures is complicated by limitations in the exposure assessments performed in those studies [13]. Nevertheless, a careful reevaluation of the dose–response for tumors in the Taiwanese population [12] concluded that there was no evidence of excess risk of cancer mortality for exposure to drinking water iAs concentrations below 0.1 mg/L. More recent studies of bladder and kidney cancer incidence in the Taiwanese population, in which a more detailed exposure assessment was performed, also found evidence of a highly nonlinear dose–response [45], as well as a significant increase in these cancers at exposure levels greater than 100 ppb [23, 24], fueling speculation that a "threshold" for the carcinogenicity of iAs may exist [1, 81].

In 2001, the MCL for arsenic in drinking water was revised from 0.05 to 0.01 mg/L [101]. This value was derived based on the estimated dose–response based on risk distributions for bladder and lung cancer reported by Morales et al. [68] in a population in Taiwan chronically exposed to concentrations of arsenic in drinking water as high as 1.75 mg/L. The dose–response calculations were performed under the standard regulatory default assumption of low-dose linearity.

23.5 NONCANCER EFFECTS OF INORGANIC ARSENIC

In general, trivalent iAs compounds produce effects due to the inhibition of critical sulfhydryl-containing enzymes [83]. This inhibition is typically reversible; the toxicity of trivalent iAs can be reduced by the addition of a free thiol such as glutathione, and glutathione depletion increases toxicity. Binding of iAs to dithiols is more difficult to reverse. For example, the critical acute effects of iAs result from the inhibition of the pyruvate oxidase system, which produces acetyl coenzyme A. The inhibition results from the binding of trivalent iAs to the dithiol, lipoic acid, a necessary cofactor for pyruvate oxidase. Because of the avidity of the binding of iAs to this dithiol, treatment with another dithiol, British anti-Lewisite, is required to reverse the toxicity. In contrast to the binding-dependent activity of trivalent iAs, pentavalent iAs compounds appear to produce toxicity primarily through the similarity of arsenate to phosphate. This similarity results in the formation of unstable arsenate esters in place of phosphate in the important energy source, adenosine triphosphate, leading to an uncoupling of oxidative phosphorylation known as arsenolysis [88]. As discussed in the section on metabolism and disposition, pentavalent iAs is readily reduced to trivalent *in vivo*, complicating interpretation of experimental data. Although tissue speciation data from autopsies suggest that the pentavalent state predominates *in vivo*, acute iAs toxicity generally appears to result from the more potent effects of the trivalent form.

While the acute toxicity and lethality of iAs occurs at similar doses across species, humans appear to be significantly more sensitive than experimental animals to the chronic noncancer toxicity of iAs. Toxicity in chronic animal studies has only been observed for iAs doses above 1 mg/kg/day, while multiple effects of chronic exposure in humans, particularly vascular effects, have frequently been associated with drinking water exposures of less than 0.1 mg/kg/day [2]. Moreover, based on an *in vitro* clonal cytotoxicity assay [77], normal human diploid fibroblasts appear to be much more sensitive to the cytotoxicity of sodium arsenite (ID_{50} = 0.6 μM) than Chinese hamster V79 cells (ID_{50} = 12.5 μM) or ovary cells (ID_{50} = 25 μM). These investigators suggest that human cells are less resistant to the chronic toxicity of iAs due to the lack of an inducible, efflux-mediated tolerance to arsenite [77] as observed in hamster cells [109, 111]. The inducible arsenite tolerance observed in hamster cells involves *de novo* mRNA and protein synthesis and appears to be distinct from the "heat shock" response observed for iAs and other stressors [110]. Differences in cellular arsenite tolerance could possibly explain the greater susceptibility of humans to the noncancer effects of chronic iAs exposure and, perhaps, to the carcinogenic effects as well. However, the relevance of these *in vitro* observations to chronic exposure *in vivo* remains to be determined.

Among the Taiwanese population chronically exposed to iAs via drinking water, an association between iAs exposure levels and increased incidence of peripheral vascular and cardiovascular diseases has been reported [94, 116]. Nine people per 1000 were found to have Blackfoot disease, a peripheral vascular disorder resulting in gangrene of the extremities. Interestingly, highly exposed populations in some geographic locations (e.g., Bangladesh) have not reported an elevated incidence of Blackfoot disease. In persons with Blackfoot disease, the walls of the arteries thicken, harden, and loose elasticity [95]. The incidence of Blackfoot disease increases with age and iAs dose. A clear dose–response was observed using standard mortality ratios from Blackfoot disease in iAs-exposed populations compared to populations in areas using shallow wells [20]. Increases in hyperpigmentation and keratosis (the formation of horny skin growths) were also observed among this population [95]. More recently, alterations in skin pigmentation were found in 21.6% of the individuals examined in a town with an iAs drinking water concentration of 0.41 mg/L, as compared to 2.2% of the individuals examined from a town with an iAs drinking water concentration of 0.007 mg/L [17]. Of the highly exposed individuals found to have pigment alterations, 17.6% had hypopigmentation, 12.2% had hyperpigmentation, 11.2% had palmoplantar keratosis, and 5.1% had papular keratosis (Bowen's disease). Alterations in peripheral vascular circulation were also reported in the highly exposed population.

Chen et al. [21] conducted a study of residents of three villages on the southwestern coast of Taiwan. The residents were evaluated to determine possible associations between long-term iAs exposure through drinking water and hypertension. Factors considered in the study included age, sex, diabetes mellitus, proteinuria, body mass index, and serum triglyceride level. Increasing levels of iAs in the drinking water were associated with a greater number of cases of hypertension. In another study, a significant dose–response was observed between iAs concentration in well water and the prevalence of cerebrovascular disease [26]. In this study, the multivariate adjusted odds ratios for cerebral infarction were 1.0, 3.4, 4.5, and 6.9 for the consumption of well water with iAs concentrations of 0, 0.1 to 50, 50–300, and >300 ppb, respectively. Other studies have demonstrated associations between chronic iAs exposure and diabetes mellitus [57], ischemic heart disease [19], and peripheral vascular disease [93]. A recent review of the noncancer effects of arsenic concluded that particular attention should be given to the potential for developmental effects [69].

A prospective study of respiratory symptoms associated with chronic arsenic exposure in Bangladesh demonstrated a significant positive association between respiratory symptoms and exposure to arsenic in drinking water [73]. In particular, a significant increase was observed for exposure to drinking water at 7–40 ppb as compared to those exposed to less than 7 ppb. There was also evidence of a positive interaction with smoking.

Noncancer effects have also been reported for chronic inhalation of iAs. Among workers at a Swedish smelter, mortality from cardiovascular disease was increased twofold in a dose-related manner following occupational iAs exposure [4]. However, Jarup et al. [53] found no relationship between iAs exposure and either ischemic heart disease or cerebrovascular disease in this same cohort. A more recent study of workers at a copper smelter [65] found limited evidence of increasing mortality risk from cerebrovascular disease with increasing duration and cumulative arsenic exposure, but the authors could not rule out possible confounding by co-exposures or other factors correlated with arsenic exposure.

Several of the epidemiological studies have evaluated relationships between other iAs-induced diseases and cancers [25, 95, 116]. Observable lesions have been a particular focus because they might be useful biomarkers of adverse health effects. Cancer incidence has been reported to be significantly higher among those with Blackfoot disease after adjustment for iAs exposure [25]. Arsenic-induced skin lesions also appear to be a predictor of higher cancer risk [35, 96]. However, data available to date indicate that it is unlikely that any of the noncancer effects of iAs, such as skin keratoses, are precursors to neoplastic lesions.

23.5.1 Noncancer Reference Dose

The USEPA's oral Reference Dose (RfD) for iAs is 0.0003 mg/kg/day, which equates to about 0.01 mg/L assuming a 70-kg person drinks 2 L of water per day [100]. The RfD is based on skin changes and possible vascular complications observed in the Taiwanese population exposed to a mean concentration of iAs in drinking water of 0.17 mg/L [94, 95]; these effects were not observed in the control population exposed to a mean concentration of 0.009 mg/L. Important assumptions used in deriving the RfD included estimates that the Taiwanese population drank 4.5 L of water per day, weighed 55 kg, and ingested 0.002 mg arsenic per day in food. These assumptions were used with the mean drinking water concentrations to derive a no-observed-adverse-effect level (NOAEL) of 0.0008 mg/kg/day and a lowest-observed-adverse-effect level (LOAEL) of 0.014 mg/kg/day. In deriving the RfD, an uncertainty factor of 3 was applied to the NOAEL to reflect concerns about human variability and inadequate data on the possibility of reproductive toxicity being a more sensitive endpoint. No inhalation Reference Concentration (RfC) has been determined for iAs.

23.6 EVIDENCE FOR THE MODE OF ACTION FOR INORGANIC ARSENIC

While there is currently no completely satisfactory description of the mode of action for the carcinogenic effects of arsenic, it appears to involve alterations of cell cycle control, replication, and differentiation [50, 78, 80]. The biological effects of arsenic may result, at least in part, from the avidity with which the trivalent arsenic species, iAs^{III}, and the trivalent methylated metabolites, MMA^{III} and DMA^{III}, bind to vicinal dithiols in cellular proteins [56, 121]. iAs is clastogenic, producing chromosomal aberrations, but does not produce point mutations at single gene loci, suggesting the possibility of a nonlinear dose–response [28]. Although chronic exposure to iAs alone has infrequently been associated with tumors in animal studies [70, 92, 108], iAs has repeatedly been shown to act as a comutagen *in vitro* [115, 118] and as a cocarcinogen *in vivo* (Burns et al. [16]; Tokar et al. [91]). Moreover, iAs inhibits a number of DNA repair processes [37, 51, 60, 61, 86, 119, 121], providing a possible explanation for its observed comutagenicity and cocarcinogenicity.

Paradoxically, arsenic trioxide has been shown to be an effective antineoplastic agent for acute promyelocytic leukemia and multiple myeloma [8, 120], yet iAs chemotherapy entails procarcinogenic side effects [82]. Antitumorigenic activity of iAs has also been reported in animal studies [3, 42].

Concentration–response alterations of selected genes and proteins in short-term *in vitro* studies [80, 86] provide evidence of reversals in the change of direction of expression between low and high concentrations. Cellular responses at submicromolar concentrations are adaptive (protective), while those at higher concentrations are toxic. An integration of available gene expression data from both *in vitro* and *in vivo* studies [42] provides clear evidence of a concentration- and time-dependent change in the expression of various genes or proteins following exposures to iAs compounds. In primary cells incubated for 24 hours with arsenite, upregulation of genes associated with oxidative stress (i.e., superoxide dismutase 1 and nicotinamide adenine dinucleotide phosphate (NADPH) oxidase quinine oxidoreductase) was observed at the lowest concentrations tested (~0.01 μM). In addition, transcripts representing genes that are modulated in response to increased level of reactive oxygen species (i.e., thioredoxin reductase and glutaredoxin) were upregulated, with downregulation in gene expression of selected DNA repair genes (mutS homolog 5 protein MSH5 and DNA damage-binding protein DDB2) noted at low concentrations. With increasing concentrations of arsenite, a transition in the expression of genes related to DNA repair, cell cycle control, and apoptosis was noted in primary cells with downregulation reversing to upregulation at concentrations between 0.1 and 1.0 μM, suggesting a transition or threshold from expression changes in pathways associated with adaptive responses to those potentially relevant to the mode of action for iAs carcinogenicity [42].

In an *in vivo* drinking water study conducted in mice [31], significant changes in the expression of genes similar to those observed in *in vitro* studies [42] were noted in mouse bladder cells following 1 or 12 weeks of exposure. The changes in gene expression were bimodal in nature, with substantial changes in expression following exposure to the lowest concentration (0.5 mg As/L) and the two highest concentrations (10 and 50 mg As/L), but few significant changes observed following exposure to 2 mg As/L. Based on the arsenite measured in the urine of the treated mice (~0.05–16 μg/L), the bladder cells in this *in vivo* study would be expected to be exposed to similar concentrations *in vivo* to those evaluated *in vitro* (0.01–10 μM) as reviewed by Gentry et al. [42]. However, cells *in vivo* would also be expected to be exposed to a mixture of arsenic and methylated metabolites, making direct comparisons difficult. This *in vivo* mouse study provides additional evidence suggesting that there is a transition in gene expression consistent with *in vitro* data [42], with a transition/threshold concentration at approximately 0.1 μM.

A recent *in vitro* study with primary human uroepithelial bladder cells from 15 normal individuals provides information on the potential impact of gene expression changes in the human bladder [117]. Analyses of the gene expression changes were conducted following incubation for 24 hours with a mixture of arsenite and its metabolites, representative of the mixture expected to be present in the human bladder based on evaluation of human urine samples [48, 105]. In this study, significant changes in gene expression for the most common genes affected across individuals were observed at concentrations in the range of approximately 0.1–1.0 μM for both trivalent and pentavalent arsenic mixtures. Evidence of upregulation (adaptation) in the vicinity of 0.1 μM total trivalent arsenic (As^{III} + MMA^{III} + DMA^{III}) dose followed by downregulation (toxicity) at doses greater than 0.1 μM was observed for genes related to DNA damage sensing, thioredoxin regulation, and immune response. These results are consistent with the changes noted in the review of studies of primary cells *in vitro*, as well as in mouse bladder cells *in vivo*, providing further evidence of a threshold or transition concentration critical to the potential carcinogenicity of arsenic compounds. The most common pathways affected in individuals [117], in agreement with the integrated *in vitro* data [42] and *in vivo* mouse data [31], were genes related to oxidative stress response (i.e., heme oxygenase-1 (HMOX1), thioredoxin reductase, thioredoxin, and metallothionine regulation), protein folding (FKBP5), DNA damage sensing (DDB2), cell adhesion, growth regulation (LGALS8), and immune response (THBD). Benchmark dose analyses on the gene expression results in primary human bladder cells indicated benchmark dose lower confidence limits (BMDLs) were in the approximate range of 0.09–0.58 μM for total arsenic in trivalent arsenical mixtures, and 0.35–1.7 μM for total arsenic in pentavalent mixtures. Benchmark doses (BMDs) and BMDLs only varied by an approximate factor of three across individuals.

23.7 RISK ASSESSMENT CONSIDERATIONS

The effects of iAs can to a large extent be understood as the superposition of highly specific direct interactions of arsenite (or MMA^{III}) with critical proteins containing vicinal di-thiols [56], overlaid on a background of chemical stress, including proteotoxicity and depletion of nonprotein sulfhydryls, that induce an array of genomic responses common to many other types of cellular stressors (Fig. 23.1).

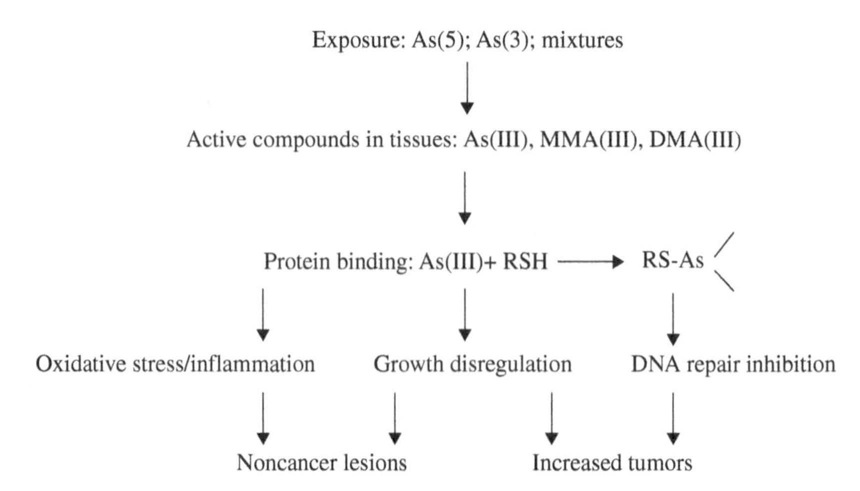

FIGURE 23.1 Components in the mode of action for the cancer and noncancer effects of inorganic arsenic.

This mode of action is equally applicable to both cancer and noncancer effects of iAs. At low concentrations, the cell appears to enter an essentially adaptive state characterized by inflammatory/survival signaling coupled with an apparent overresponse to oxidative stress [31, 80]. At higher concentrations, induction of genes associated with apoptosis (*in vitro*) reflects a cellular response to toxicity, while increased expression of cell cycle checkpoint control genes suggests cellular recognition of unrepaired DNA damage. These cellular stresses are further exacerbated at higher concentrations by the increasing expression of genes associated with proliferative signaling, exerting pressure for cell division, together with inhibition of DNA repair.

The cellular dose–response for iAs is remarkably consistent based on a variety of evidence under different exposure conditions. The carcinogenic effects of arsenic have been observed in populations exposed to very high drinking water concentrations (in the order of 600 μg/L, resulting in a daily dose in the order of 1–2 mg). Interestingly, the dose–response for noncancer vascular proliferative effects of arsenic (e.g., blackfoot disease) was similar to that of cancer effects [30]. The observation of a similar dose–response for the cancer and noncancer effects of arsenic is consistent with the genomic evidence indicating induction of vascular endothelial growth factor (VEGF) in the same concentration range (1–10 μM) as the inhibition of DNA ligase I. It has been suggested that stimulation of VEGF-mediated angiogenesis may represent an axis for the progression of arsenic-induced vascular disease [82], while inhibition of DNA repair, and specifically ligase activity, has repeatedly been identified as a possible factor in arsenic tumorigenicity [51, 60, 61, 81, 115, 118].

It has been suggested that the dose–response for the carcinogenicity of arsenic is highly nonlinear, with an effective threshold, due to the apparent mode of action involving inhibition of DNA repair [81]. Based on the analyses of epidemiological and *in vitro* data, this threshold has been suggested to be approximately 0.1 μM from *in vitro* assays (e.g., [117]), or in the order of 10–100 ppb in drinking water from epidemiological studies, roughly an order of magnitude below the concentrations at which tumors have been observed [24, 81, 117]. On the other hand, the seemingly counterintuitive effectiveness of arsenic as an antineoplastic agent for some cancers, when administered at the even higher dose levels used in chemotherapy (10 mg/day), is consistent with the dominance of the apoptotic gene response at concentrations in the order of 100 μM.

23.7.1 Risk Assessment Extrapolations

While the *in vitro* data in both humans and animals are consistent with a dose-dependent transition concentration for arsenic carcinogenicity in the order of 0.1 μM, several challenges remain in incorporating these results into the risk assessment paradigm. The integration of the *in vivo* and *in vitro* information is critical, but also requires the application of assumptions or models to interpret the concentrations related to the *in vitro* results to determine comparable *in vivo* exposures. Pharmacokinetic models, such as those developed by El-Masri and Kenyon [38] and Mann et al. [63, 64], could be used to extrapolate from the concentrations used in the *in vivo* mouse studies and the *in vitro* human studies to the equivalent human drinking water concentrations. The models can also assist in the estimation of exposure concentrations based on urine concentrations of arsenic metabolites reported in human populations [41, 55, 76, 85]. It is critical that individual exposures in epidemiological studies be characterized accurately. Crump [33] has demonstrated that exposure misclassification in epidemiology studies transforms an exposure response in the direction toward supralinear (i.e., tends to linearize a threshold).

In addition to *in vitro* to *in vivo* extrapolation, consideration of duration effects is needed for the interpretation of the effects observed following acute *in vitro* exposures in human uroepithelial cells [117] in the context of a risk assessment focused on effects resulting from chronic exposure. Both *in vivo* [31] and *in vitro* studies [42] suggest changes in dose–response between acute and chronic exposure. Therefore, extending *in vitro* studies to evaluate these duration-related changes is critical. Nevertheless, the available evidence for iAs supports a common mode of action for cancer and noncancer effects (Fig. 23.1), in which the specific and nonspecific binding of the trivalent arsenicals to key cellular proteins induces oxidative stress and inflammatory responses, growth dysregulation, and inhibition of DNA repair. These disruptions of normal cellular function, if maintained chronically, can lead to both neoplastic and nonneoplastic proliferative lesions, particularly under conditions where there is coexposure to environmental mutagens.

23.7.2 Population Variability

In the determination of an acceptable exposure, dose–response assessment provides an estimate of the approximate mean value for consideration in a population. What is critical to consider in conjunction with that value is the potential impact of human individual variability. Previous studies of the variability of human pharmacokinetics

(PK) using physiologically based pharmacokinetic (PBPK) modeling have concluded that PK variability is typically, in the order of a factor of 3, between average and sensitive individuals [29]. Similarly, the available information from the *in vitro* data with primary human uroepithelial cells [117] suggests that pharmacodynamics (PD) variability is in the order of a factor of 3. Based on the BMD analysis, BMDLs in that study for gene expression changes observed in cells from different individuals ranged from approximately 0.1 to 0.6 μM total arsenic for trivalent arsenical mixtures and 0.35 to 1.7 μM total arsenic for pentavalent arsenical mixtures. Taking half of these ranges as an estimate of the difference between average and sensitive individuals results in a PD factor of about 3. Together, these PK and PD results provide evidence of an overall (PK plus PD) individual variability of roughly 10-fold, consistent with default expectations.

Considering the potential impact of interindividual variability on the population dose–response curve, the nonlinearity for the population would perhaps extend over roughly an order of magnitude. For example, in Figure 23.2, a linear risk estimate for arsenic carcinogenicity, such as that used in the current USEPA [103] dose–response assessment, is depicted by the dashed straight line. The heavy solid curve provides an example of a more plausible nonlinear dose–response for arsenic in an average individual, based on the integration of the epidemiological and toxicological data. The strongly nonlinear nature of this curve reflects the transition that might be expected to occur in a particular individual's cells, from concentrations of arsenic below the transition or threshold concentration to those at which key events in the development of cancer become evident. Factors that can impact both the pharmacokinetics and pharmacodynamics of arsenic in an individual impact the potential shift

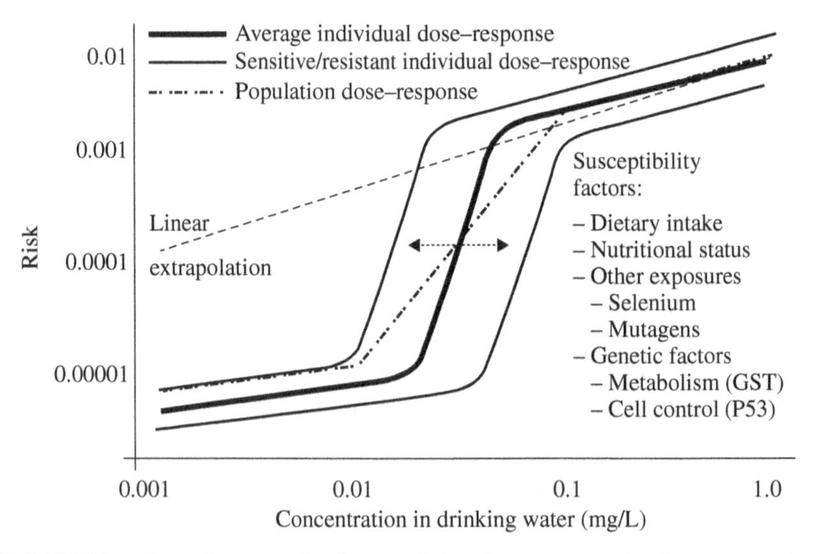

FIGURE 23.2 Illustrative example of a population dose–response for the cancer and noncancer effects of inorganic arsenic (Adapted from Clewell [27]).

of the dose–response curve in either direction from the population mean. The dose–response for the population would therefore be a curve with a shallower slope, transitioning smoothly between the curves for the more and less sensitive individuals. This "true" population risk is illustrated by the heavy dashed curve in Figure 23.2. Thus, the slope of the dose–response associated with the nonlinearity would depend on both the magnitude of the overall risk reduction in an individual due to the nonlinearity in the mode of action, and the breadth of the population transition resulting from interindividual variability in sensitivity. The net result, as illustrated by comparing the curved and straight dashed lines, would be that lower arsenic exposures would result in a much greater reduction in the population risk than would otherwise be predicted by linear estimates, even when taking into consideration the potential interindividual variability in response.

23.7.3 Potential Risk Assessment Approach

In the absence of a complete understanding of mechanism(s) underlying the carcinogenicity and toxicity of iAs, it would be of value to be able to use an MOE approach that maintains a biological basis as an alternative to the default linear option. It has been suggested that the dose–response component of cancer risk assessment could be based on quantitation of molecular endpoints, or "bioindicators" of response, selected on the basis of their association with obligatory precursor events for tumorigenesis [75]. Recent studies have demonstrated that the threshold for cellular genomic responses can be used to predict both cancer and noncancer endpoints [89]. A comparison of transcriptional bench mark dose values for the most sensitive pathway with BMD values for the noncancer and cancer apical endpoints of a number of chemicals showed a high degree of correlation. The average ratio of apical-to-transcriptional BMD values was less than twofold, suggesting that transcriptional perturbation provides a useful surrogate for apical responses. Based on the existing data on the cellular effects of arsenic, it appears that it may be possible to develop an approach for conducting a MOE risk assessment for iAs using *in vitro* human dose–response information on the expression of genes (or groups of genes) associated with the key elements of the interaction of arsenic with target cells. Potential contributory factors in arsenic carcinogenesis include DNA repair inhibition under conditions of oxidative stress, inflammation, and proliferative signaling, leading to a situation in which the cell is no longer able to maintain the integrity of its DNA prior to division. Importantly, the genomic dose–response data on iAs is generally consistent with data on other chemical stressors, in that there appears to be a transition from an adaptive state at low concentrations to an impaired state at higher concentrations [44, 81, 114]. In the proposed risk assessment approach, the concentration at which this transition occurs in the human target cells (e.g., bladder epithelium) is approximately 0.1 μM and may be used as the point of departure for the MOE determination. The MOE would then be selected to consider both variability in the human population (e.g., polymorphic variation) and uncertainty in the mode of action.

REFERENCES

[1] C.O. Abernathy, W.R. Chappell, M.E. Meek, H. Gibb, H.R. Guo, Is ingested inorganic arsenic a "threshold" carcinogen? Fundam. Appl. Toxicol. 29 (1996) 168–175.

[2] Agency for Toxic Substances and Disease Registry. Toxicological Profile for Arsenic. http://www.atsdr.cdc.gov/toxprofiles/tp2.pdf. 2007. Atlanta, U.S. Department of Health and Human Services. Accessed April 10, 2015.

[3] G.J. Ahlborn, G.M. Nelson, R.D. Grindstaff, M.P. Waalkes, B.A. Diwan, J.W. Allen, K.T. Kitchin, R.J. Preston, A. Hernandez-Zavala, B. Adair, D.J. Thomas, D.A. Delker, Impact of life stage and duration of exposure on arsenic-induced proliferative lesions and neoplasia in C3H mice. Toxicology 262 (2009) 106–113.

[4] O. Axelson, E. Dahlgren, C.D. Jansson, S.O. Rehnlund, Arsenic exposure and mortality: a case-referent study from a Swedish copper smelter. Br. J. Ind. Med. 35 (1978) 8–15.

[5] R. Baastrup, M. Sorensen, T. Balstrom, K. Frederiksen, C.L. Larsen, A. Tjonneland, K. Overvad, O. Raaschou-Nielsen, Arsenic in drinking-water and risk for cancer in Denmark. Environ. Health Perspect. 116 (2008) 231–237.

[6] M.N. Bates, A.H. Smith, K.P. Cantor, Case-control study of bladder cancer and arsenic in drinking water. Am. J. Epidemiol. 141 (1995) 523–530.

[7] M.N. Bates, A.H. Smith, C. Hopenhayn-Rich, Arsenic ingestion and internal cancers: a review. Am. J. Epidemiol. 135 (1992) 462–476.

[8] J.R. Berenson, H.S. Yeh, Arsenic compounds in the treatment of multiple myeloma: a new role for a historical remedy. Clin. Lymphoma Myeloma 7 (2006) 192–198.

[9] D.R. Borum, C.O. Abernathy Human oral exposure to inorganic arsenic. in: W.R. Chappell, C.O. Abernathy, and C.R. Cothern (Eds.), Arsenic Exposure and Health, Science and Technology Letters, Northwood, UK, 1994, pp. 21–29.

[10] C.C. Brown, K.C. Chu, A new method for the analysis of cohort studies: implications of the multistage theory of carcinogenesis applied to occupational arsenic exposure. Environ. Health Perspect. 50 (1983) 293–308.

[11] C.C. Brown, K.C. Chu, Implications of the multistage theory of carcinogenesis applied to occupational arsenic exposure. J. Natl. Cancer Inst. 70 (1983) 455–463.

[12] K.G. Brown, C.J. Chen, Significance of exposure assessment to analysis of cancer risk from inorganic arsenic in drinking water in Taiwan. Risk Anal. 15 (1995) 475–484.

[13] K.G. Brown, H.R. Guo, T.L. Kuo, H.L. Greene, Skin cancer and inorganic arsenic: uncertainty-status of risk. Risk Anal. 17 (1997) 37–42.

[14] J.P. Buchet, R. Lauwerys Inorganic arsenic metabolism in humans. in: W.R. Chappell, C.O. Abernathy, and C.R. Cothern (Eds.), Arsenic Exposure and Health, Science and Technology Letters, Northwood, UK, 1994, pp. 181–190.

[15] J.P. Buchet, R. Lauwerys, P. Mahieu, A. Geubel, Inorganic arsenic metabolism in man. Arch. Toxicol. Suppl. 5 (1982) 326–327.

[16] F.J. Burns, A.N. Uddin, F. Wu, A. Nadas, T.G. Rossman, Arsenic-induced enhancement of ultraviolet radiation carcinogenesis in mouse skin: a dose-response study. Environ. Health Perspect. 112(5) (2004) 599–603.

[17] M.E. Cebrian, A. Albores, G. Garcia-Vargas, L.M. Del Razo Chronic arsenic poisoning in humans: the case of Mexico. in: J. Nriagu (Ed.), Arsenic in the Environment. Part II: Human Health and Ecosystem Effects, John Wiley & Sons, Inc, New York, 1994, pp. 93–107.

[18] C.J. Chen, C.W. Chen, M.M. Wu, T.L. Kuo, Cancer potential in liver, lung, bladder and kidney due to ingested inorganic arsenic in drinking water. Br. J. Cancer 66 (1992) 888–892.

[19] C.J. Chen, H.Y. Chiou, M.H. Chiang, L.J. Lin, T.Y. Tai, Dose-response relationship between ischemic heart disease mortality and long-term arsenic exposure. Arterioscler. Thromb. Vasc. Biol. 16 (1996) 504–510.

[20] C.J. Chen, Y.C. Chuang, T.M. Lin, H.Y. Wu, Malignant neoplasms among residents of a blackfoot disease-endemic area in Taiwan: high-arsenic artesian well water and cancers. Cancer Res. 45 (1985) 5895–5899.

[21] C.J. Chen, Y.M. Hsueh, M.S. Lai, M.P. Shyu, S.Y. Chen, M.M. Wu, T.L. Kuo, T.Y. Tai, Increased prevalence of hypertension and long-term arsenic exposure. Hypertension 25 (1995) 53–60.

[22] C.J. Chen, L.J. Lin Human carcinogenicity and atherogenicity induced by chronic exposure to inorganic arsenic. in: J. Nriagu (Ed.), Arsenic in the Environment. Part II: Human Health and Ecosystem Effects, John Wiley & Sons, Inc, New York, 1994, pp. 109–131.

[23] S.H. Chen, Y.W. Wang, J.L. Hsu, H.Y. Chang, C.Y. Wang, P.T. Shen, C.W. Chiang, J.J. Chuang, H.W. Tsai, P.W. Gu, F.C. Chang, H.S. Liu, N.H. Chow, Nucleophosmin in the pathogenesis of arsenic-related bladder carcinogenesis revealed by quantitative proteomics. Toxicol. Appl. Pharmacol. 242 (2010) 126–135.

[24] H.Y. Chiou, S.T. Chiou, Y.H. Hsu, Y.L. Chou, C.H. Tseng, M.L. Wei, C.J. Chen, Incidence of transitional cell carcinoma and arsenic in drinking water: a follow-up study of 8,102 residents in an arseniasis-endemic area in northeastern Taiwan. Am. J. Epidemiol. 153 (2001) 411–418.

[25] H.Y. Chiou, Y.M. Hsueh, K.F. Liaw, S.F. Horng, M.H. Chiang, Y.S. Pu, J.S. Lin, C.H. Huang, C.J. Chen, Incidence of internal cancers and ingested inorganic arsenic: a seven-year follow-up study in Taiwan. Cancer Res. 55 (1995) 1296–1300.

[26] H.Y. Chiou, W.I. Huang, C.L. Su, S.F. Chang, Y.H. Hsu, C.J. Chen, Dose-response relationship between prevalence of cerebrovascular disease and ingested inorganic arsenic. Stroke 28 (1997) 1717–1723.

[27] H.J. Clewell, Research requirements for a biologically realistic cancer risk assessment for arsenic. in: W.R. Chappell, C.O. Abernathy, and R.L. Calderon (Eds.), Arsenic Exposure and Health Effects IV, Elsevier Science, Oxford, UK, 2001.

[28] H.J. Clewell, P.R. Gentry, H.A. Barton, A.M. Shipp, J.W. Yager, M.E. Andersen, Requirements for a biologically realistic cancer risk assessment for inorganic arsenic. Int. J. Toxicol. 18 (1999) 131–147.

[29] H.J. Clewell, III, M.E. Andersen, Use of physiologically based pharmacokinetic modeling to investigate individual versus population risk. Toxicology 111 (1996) 315–329.

[30] H.J. Clewell, K.S. Crump, Quantitative estimates of risk for noncancer endpoints. Risk Anal. 25 (2005) 285–289.

[31] H.J. Clewell, R.S. Thomas, E.M. Kenyon, M.F. Hughes, B.M. Adair, P.R. Gentry, J.W. Yager, Concentration- and time-dependent genomic changes in the mouse urinary bladder following exposure to arsenate in drinking water for up to 12 weeks. Toxicol. Sci. 123 (2011) 421–432.

[32] S.M. Cohen, L.L. Arnold, B.D. Beck, A.S. Lewis, M. Eldan, Evaluation of the carcinogenicity of inorganic arsenic. Crit. Rev. Toxicol. 43 (2013) 711–752.

[33] K.S. Crump, The effect of random error in exposure measurement upon the shape of the exposure response. Dose Response 3 (2005) 456–464.

[34] J. Cuzick, S. Evans, M. Gillman, D.A. Price Evans, Medicinal arsenic and internal malignancies. Br. J. Cancer 45 (1982) 904–911.

[35] J. Cuzick, P. Sasieni, S. Evans, Ingested arsenic, keratoses, and bladder cancer. Am. J. Epidemiol. 136 (1992) 417–421.

[36] L.M. Del Razo, G.G. Garcia-Vargas, H. Vargas, A. Albores, M.E. Gonsebatt, R. Montero, P. Ostrosky-Wegman, M. Kelsh, M.E. Cebrian, Altered profile of urinary arsenic metabolites in adults with chronic arsenicism. A pilot study. Arch. Toxicol. 71 (1997) 211–217.

[37] W. Ding, W. Liu, K.L. Cooper, X.J. Qin, P.L. de Souza Bergo, L.G. Hudson, K.J. Liu, Inhibition of poly(ADP-ribose) polymerase-1 by arsenite interferes with repair of oxidative DNA damage. J. Biol. Chem. 284 (2009) 6809–6817.

[38] H.A. El-Masri, E.M. Kenyon, Development of a human physiologically based pharmacokinetic (PBPK) model for inorganic arsenic and its mono- and di-methylated metabolites. J. Pharmacokinet. Pharmacodyn. 35 (2008) 31–68.

[39] P.E. Enterline, R. Day, G.M. Marsh, Cancers related to exposure to arsenic at a copper smelter. Occup. Environ. Med. 52 (1995) 28–32.

[40] P.E. Enterline, V.L. Henderson, G.M. Marsh, Exposure to arsenic and respiratory cancer. A reanalysis. Am. J. Epidemiol. 125 (1987) 929–938.

[41] C. Fillol, F. Dor, L. Labat, P. Boltz, B.J. Le, K. Mantey, C. Mannschott, E. Puskarczyk, F. Viller, I. Momas, N. Seta, Urinary arsenic concentrations and speciation in residents living in an area with naturally contaminated soils. Sci. Total Environ. 408 (2010) 1190–1194.

[42] P.R. Gentry, T.B. McDonald, D.E. Sullivan, A.M. Shipp, J.W. Yager, H.J. Clewell, III, Analysis of genomic dose-response information on arsenic to inform key events in a mode of action for carcinogenicity. Environ. Mol. Mutagen. 51 (2010) 1–14.

[43] H. Gibb, C. Haver, D. Gaylor, S. Ramasamy, J.S. Lee, D. Lobdell, T. Wade, C. Chen, P. White, R. Sams, Utility of recent studies to assess the National Research Council 2001 estimates of cancer risk from ingested arsenic. Environ. Health Perspect. 119 (2011) 284–290.

[44] M.I. Gilmour, M.S. Jaakkola, S.J. London, A.E. Nel, C.A. Rogers, How exposure to environmental tobacco smoke, outdoor air pollutants, and increased pollen burdens influences the incidence of asthma. Environ. Health Perspect. 114 (2006) 627–633.

[45] H.R. Guo, H.S. Chiang, H. Hu, S.R. Lipsitz, R.R. Monson Arsenic in drinking water and urinary cancers: a preliminary report. in: W.R. Chappell, C.O. Abernathy, and C.R. Cothern (Eds.), Arsenic Exposure and Health, Science and Technology Letters, Northwood, UK, 1994, pp. 119–124.

[46] H.R. Guo, H.S. Chiang, H. Hu, S.R. Lipsitz, R.R. Monson, Arsenic in drinking water and incidence of urinary cancers. Epidemiology 8 (1997) 545–550.

[47] Y.Y. Han, J.L. Weissfeld, D.L. Davis, E.O. Talbott, Arsenic levels in ground water and cancer incidence in Idaho: an ecologic study. Int. Arch. Occup. Environ. Health 82 (2009) 843–849.

[48] C. Hopenhayn-Rich, A.H. Smith, H.M. Goeden, Human studies do not support the methylation threshold hypothesis for the toxicity of inorganic arsenic. Environ. Res. 60 (1993) 161–177.

[49] Y.M. Hsueh, G.S. Cheng, M.M. Wu, H.S. Yu, T.L. Kuo, C.J. Chen, Multiple risk factors associated with arsenic-induced skin cancer: effects of chronic liver disease and malnutritional status. Br. J. Cancer 71 (1995) 109–114.

[50] Y. Hu, X. Jin, E.T. Snow, Effect of arsenic on transcription factor AP-1 and NF-kappaB DNA binding activity and related gene expression. Toxicol. Lett. 133 (2002) 33–45.

[51] Y. Hu, L. Su, E.T. Snow, Arsenic toxicity is enzyme specific and its affects on ligation are not caused by the direct inhibition of DNA repair enzymes. Mutat. Res. 408 (1998) 203–218.

[52] L. Jarup, G. Pershagen, Arsenic exposure, smoking, and lung cancer in smelter workers: a case-control study. Am. J. Epidemiol. 134 (1991) 545–551.

[53] L. Jarup, G. Pershagen, S. Wall, Cumulative arsenic exposure and lung cancer in smelter workers: a dose-response study. Am. J. Ind. Med. 15 (1989) 31–41.

[54] M.R. Karagas, T.D. Tosteson, J.S. Morris, E. Demidenko, L.A. Mott, J. Heaney, A. Schned, Incidence of transitional cell carcinoma of the bladder and arsenic exposure in New Hampshire. Cancer Causes Control 15 (2004) 465–472.

[55] M.L. Kile, E. Hoffman, Y.M. Hsueh, S. Afroz, Q. Quamruzzaman, M. Rahman, G. Mahiuddin, L. Ryan, D.C. Christiani, Variability in biomarkers of arsenic exposure and metabolism in adults over time. Environ. Health Perspect. 117 (2009) 455–460.

[56] K.T. Kitchin, K. Wallace, Arsenite binding to synthetic peptides based on the Zn finger region and the estrogen binding region of the human estrogen receptor-alpha. Toxicol. Appl. Pharmacol. 206 (2005) 66–72.

[57] M.S. Lai, Y.M. Hsueh, C.J. Chen, M.P. Shyu, S.Y. Chen, T.L. Kuo, M.M. Wu, T.Y. Tai, Ingested inorganic arsenic and prevalence of diabetes mellitus. Am. J. Epidemiol. 139 (1994) 484–492.

[58] S.H. Lamm, A. Engel, M.B. Kruse, M. Feinleib, D.M. Byrd, S. Lai, R. Wilson, Arsenic in drinking water and bladder cancer mortality in the United States: an analysis based on 133 U.S. counties and 30 years of observation. J. Occup. Environ. Med. 46 (2004) 298–306.

[59] D.R. Lewis, J.W. Southwick, R. Ouellet-Hellstrom, J. Rench, R.L. Calderon, Drinking water arsenic in Utah: a cohort mortality study. Environ. Health Perspect. 107 (1999) 359–365.

[60] J.H. Li, T.G. Rossman, Inhibition of DNA ligase activity by arsenite: a possible mechanism of its comutagenesis. Mol. Toxicol. 2 (1989) 1–9.

[61] S. Lynn, H.T. Lai, J.R. Gurr, K.Y. Jan, Arsenite retards DNA break rejoining by inhibiting DNA ligation. Mutagenesis 12 (1997) 353–358.

[62] K. Mabuchi, A.M. Lilienfeld, L.M. Snell, Lung cancer among pesticide workers exposed to inorganic arsenicals. Arch. Environ. Health 34 (1979) 312–320.

[63] S. Mann, P.O. Droz, M. Vahter, A physiologically based pharmacokinetic model for arsenic exposure. I. Development in hamsters and rabbits. Toxicol. Appl. Pharmacol. 137 (1996) 8–22.

[64] S. Mann, P.O. Droz, M. Vahter, A physiologically based pharmacokinetic model for arsenic exposure. II. Validation and application in humans. Toxicol. Appl. Pharmacol. 140 (1996) 471–486.

[65] G.M. Marsh, N.A. Esmen, J.M. Buchanich, A.O. Youk, Mortality patterns among workers exposed to arsenic, cadmium, and other substances in a copper smelter. Am. J. Ind. Med. 52 (2009) 633–644.

[66] S. Mazumdar, C.K. Redmond, P.E. Enterline, G.M. Marsh, J.P. Costantino, S.Y. Zhou, R.N. Patwardhan, Multistage modeling of lung cancer mortality among arsenic-exposed copper-smelter workers. Risk Anal. 9 (1989) 551–563.

[67] J.R. Meliker, M.J. Slotnick, G.A. AvRuskin, D. Schottenfeld, G.M. Jacquez, M.L. Wilson, P. Goovaerts, A. Franzblau, J.O. Nriagu, Lifetime exposure to arsenic in drinking water and bladder cancer: a population-based case-control study in Michigan, USA. Cancer Causes Control 21 (2010) 745–757.

[68] K.H. Morales, L. Ryan, T.L. Kuo, M.M. Wu, C.J. Chen, Risk of internal cancers from arsenic in drinking water. Environ. Health Perspect. 108 (2000) 655–661.

[69] M.F. Naujokas, B. Anderson, H. Ahsan, H.V. Aposhian, J.H. Graziano, C. Thompson, W.A. Suk, The broad scope of health effects from chronic arsenic exposure: update on a worldwide public health problem. Environ. Health Perspect. 121 (2013) 295–302.

[70] J.C. Ng, A.A. Seawright, L. Wi, C.M. Garnett, B. Cirswell, M.R. Moore Tumours in mice induced by exposure to sodium arsenate in drinking water. in: W.R. Chappell, C.O. Abernathy, and C.R. Cothern (Eds.), Arsenic Exposure and Health, Science and Technology Letters, Northwood, UK, 1994, pp. 217–223.

[71] D.W. North, H.J. Gibb, C.O. Abernathy Arsenic: past, present, and future considerations. in: C.O. Abernathy, R.L. Calderon, and W.R. Chappell (Eds.), Arsenic Exposure and Health Effects, Chapman & Hall, New York, 1997, pp. 406–423.

[72] M.G. Ott, B.B. Holder, H.L. Gordon, Respiratory cancer and occupational exposure to arsenicals. Arch. Environ. Health 29 (1974) 250–255.

[73] F. Parvez, Y. Chen, P.W. Brandt-Rauf, V. Slavkovich, T. Islam, A. Ahmed, M. Argos, R. Hassan, M. Yunus, S.E. Haque, O. Balac, J.H. Graziano, H. Ahsan, A prospective study of respiratory symptoms associated with chronic arsenic exposure in Bangladesh: findings from the Health Effects of Arsenic Longitudinal Study (HEALS). Thorax 65 (2010) 528–533.

[74] S.S. Pinto, P.E. Enterline, V. Henderson, M.O. Varner, Mortality experience in relation to a measured arsenic trioxide exposure. Environ. Health Perspect. 19 (1977) 127–130.

[75] R.J. Preston, DNA reactivity as a mode of action and its relevance to cancer risk assessment. Toxicol. Pathol. 41 (2013) 322–325.

[76] Z. Rivera-Nunez, J.R. Meliker, J.D. Meeker, M.J. Slotnick, J.O. Nriagu, Urinary arsenic species, toenail arsenic, and arsenic intake estimates in a Michigan population with low levels of arsenic in drinking water. J. Expo. Sci. Environ. Epidemiol. 22 (2012) 182–190.

[77] T.G. Rossman, E.I. Goncharova, T. Rajah, Z. Wang, Human cells lack the inducible tolerance to arsenite seen in hamster cells. Mutat. Res. 386 (1997) 307–314.

[78] K. Salnikow, M.D. Cohen, Backing into cancer: effects of arsenic on cell differentiation. Toxicol. Sci. 65 (2002) 161–163.

[79] A.H. Smith, C. Hopenhayn-Rich, M.N. Bates, H.M. Goeden, I. Hertz-Picciotto, H.M. Duggan, R. Wood, M.J. Kosnett, M.T. Smith, Cancer risks from arsenic in drinking water. Environ. Health Perspect. 97 (1992) 259–267.

[80] E.T. Snow, M. Shuliga, S. Chouchane, Y. Hu. Sub-toxic arsenite induces a multi-component protective response against oxidative stress in human cells. Proceedings of the Fourth International Conference on Arsenic Exposure and Health Effects IV, June 18–22, 2000. 265–275. 2001. San Diego, CA.

Ref Type: Conference Proceeding

[81] E.T. Snow, P. Sykora, T.R. Durham, C.B. Klein, Arsenic, mode of action at biologically plausible low doses: what are the implications for low dose cancer risk? Toxicol. Appl. Pharmacol. 207 (2005) 557–564.

[82] N.V. Soucy, M.A. Ihnat, C.D. Kamat, L. Hess, M.J. Post, L.R. Klei, C. Clark, A. Barchowsky, Arsenic stimulates angiogenesis and tumorigenesis in vivo. Toxicol. Sci. 76 (2003) 271–279.

[83] K.S. Squibb, B.A. Fowler The toxicity of arsenic and its compounds. in: B.A. Fowler (Ed.), Biological and Environmental Effects of Arsenic, Elsevier, Amsterdam, 1983, pp. 233–269.

[84] C. Steinmaus, Y. Yuan, M.N. Bates, A.H. Smith, Case-control study of bladder cancer and drinking water arsenic in the western United States. Am. J. Epidemiol. 158 (2003) 1193–1201.

[85] G. Sun, Y. Xu, X. Li, Y. Jin, B. Li, X. Sun, Urinary arsenic metabolites in children and adults exposed to arsenic in drinking water in Inner Mongolia, China. Environ. Health Perspect. 115 (2007) 648–652.

[86] P. Sykora, E.T. Snow, Modulation of DNA polymerase beta-dependent base excision repair in cultured human cells after low dose exposure to arsenite. Toxicol. Appl. Pharmacol. 228 (2008) 385–394.

[87] D.J. Thomas, Molecular processes in cellular arsenic metabolism. Toxicol. Appl. Pharmacol. 222 (2007) 365–373.

[88] D.J. Thomas, Arsenolysis and thiol-dependent arsenate reduction. Toxicol. Sci. 117 (2010) 249–252.

[89] R.S. Thomas, S.C. Wesselkamper, N.C. Wang, Q.J. Zhao, D.D. Petersen, J.C. Lambert, I. Cote, L. Yang, E. Healy, M.B. Black, H.J. Clewell, III, B.C. Allen, M.E. Andersen, Temporal concordance between apical and transcriptional points of departure for chemical risk assessment. Toxicol. Sci. 134 (2013) 180–194.

[90] D.J. Thompson, A chemical hypothesis for arsenic methylation in mammals. Chem. Biol. Interact. 88 (1993) 89–114.

[91] E.J. Tokar, L. Benbrahim-Tallaa, J.M. Ward, R. Lunn, R.L. Sams 2nd, M.P. Waalkes, Cancer in experimental animals exposed to arsenic and arsenic compounds. Crit. Rev. Toxicol. 40(10) (2010) 912–927.

[92] E.J. Tokar, B.A. Diwan, J.M. Ward, D.A. Delker, M.P. Waalkes, Carcinogenic effects of "whole-life" exposure to inorganic arsenic in CD1 mice. Toxicol. Sci. 119 (2011) 73–83.

[93] C.H. Tseng, C.K. Chong, C.J. Chen, T.Y. Tai, Dose-response relationship between peripheral vascular disease and ingested inorganic arsenic among residents in blackfoot disease endemic villages in Taiwan. Atherosclerosis 120 (1996) 125–133.

[94] W.P. Tseng, Effects and dose: response relationships of skin cancer and blackfoot disease with arsenic. Environ. Health Perspect. 19 (1977) 109–119.

[95] W.P. Tseng, H.M. Chu, S.W. How, J.M. Fong, C.S. Lin, S. Yeh, Prevalence of skin cancer in an endemic area of chronic arsenicism in Taiwan. J. Natl. Cancer Inst. 40 (1968) 453–463.

[96] T. Tsuda, A. Babazono, E. Yamamoto, N. Kurumatani, Y. Mino, T. Ogawa, Y. Kishi, H. Aoyama, Ingested arsenic and internal cancer: a historical cohort study followed for 33 years. Am. J. Epidemiol. 141 (1995) 198–209.

[97] U.S. Environmental Protection Agency (USEPA). Health Assessment Document for Inorganic Arsenic. EPA-600/8-83-021F. 1984. Washington, DC, USEPA.

Ref Type: Report

[98] U.S. Environmental Protection Agency (USEPA). Risk Assessment Guidelines of 1986. EPA/600/8-87/045. 1987. Washington, DC, Office of Research and Development.

[99] U.S. Environmental Protection Agency (USEPA). Special Report on Ingested Inorganic Arsenic. Skin Cancer; Nutritional Essentiality. EPA 625/3-87-013. 1988. Washington, DC, Risk Assessment Forum.

[100] U.S. Environmental Protection Agency (USEPA). Integrated Risk Information System (IRIS), Arsenic. CASRN 7440-38-2. 1998. Washington, DC, U.S. Environmental Protection Agency (USEPA).

[101] U.S. Environmental Protection Agency (USEPA). National primary drinking water regulations: arsenic and clarifications to compliance and new source contaminants monitoring. Final Rule. 40 CFR Parts 141 and 142. Federal Register 66 (2001) 6976–7066.

[102] U.S. Environmental Protection Agency (USEPA). Guidelines for Carcinogen Risk Assessment. EPA/630/P-03/001B. 2005. Washington, DC, Risk Assessment Forum.

[103] U.S. Environmental Protection Agency (USEPA). Toxicological Review of Inorganic Arsenic. 2010. Washington, DC, U.S. Environmental Protection Agency (USEPA).

[104] M. Vahter Species differences in the metabolism of arsenic. in: W.R. Chappell, C.O. Abernathy, and C.R. Cothern (Eds.), Arsenic Exposure and Health, Science and Technology Letters, Northwood, UK, 1994, pp. 171–180.

[105] M. Vahter, Methylation of inorganic arsenic in different mammalian species and population groups. Sci. Prog. 82 (Pt 1) (1999) 69–88.

[106] M. Vahter, J. Envall, In vivo reduction of arsenate in mice and rabbits. Environ. Res. 32 (1983) 14–24.

[107] J. Viren, A. Silvers, Nonlinearity in the lung cancer dose-response for airborne arsenic: apparent confounding by year of hire in evaluating lung cancer risks from arsenic exposure in Tacoma smelter workers. Regul. Toxicol. Pharmacol. 30 (1999) 117–129.

[108] M.P. Waalkes, J. Liu, J.M. Ward, B.A. Diwan, Animal models for arsenic carcinogenesis: inorganic arsenic is a transplacental carcinogen in mice. Toxicol. Appl. Pharmacol. 198 (2004) 377–384.

[109] Z. Wang, S. Dey, B.P. Rosen, T.G. Rossman, Efflux-mediated resistance to arsenicals in arsenic-resistant and -hypersensitive Chinese hamster cells. Toxicol. Appl. Pharmacol. 137 (1996) 112–119.

[110] Z. Wang, G. Hou, T.G. Rossman, Induction of arsenite tolerance and thermotolerance by arsenite occur by different mechanisms. Environ. Health Perspect. 102 Suppl 3 (1994) 97–100.

[111] Z. Wang, T.G. Rossman, Stable and inducible arsenite resistance in Chinese hamster cells. Toxicol. Appl. Pharmacol. 118 (1993) 80–86.

[112] M.L. Warner, L.E. Moore, M.T. Smith, D.A. Kalman, E. Fanning, A.H. Smith, Increased micronuclei in exfoliated bladder cells of individuals who chronically ingest arsenic-contaminated water in Nevada. Cancer Epidemiol. Biomarkers Prev. 3 (1994) 583–590.

[113] K. Welch, I. Higgins, M. Oh, C. Burchfiel, Arsenic exposure, smoking, and respiratory cancer in copper smelter workers. Arch. Environ. Health 37 (1982) 325–335.

[114] J.D. West, L.J. Marnett, Alterations in gene expression induced by the lipid peroxidation product, 4-hydroxy-2-nonenal. Chem. Res. Toxicol. 18 (2005) 1642–1653.

[115] J.K. Wiencke, J.W. Yager, Specificity of arsenite in potentiating cytogenetic damage induced by the DNA crosslinking agent diepoxybutane. Environ. Mol. Mutagen. 19 (1992) 195–200.

[116] M.M. Wu, T.L. Kuo, Y.H. Hwang, C.J. Chen, Dose-response relation between arsenic concentration in well water and mortality from cancers and vascular diseases. Am. J. Epidemiol. 130 (1989) 1123–1132.

[117] J.W. Yager, P.R. Gentry, R.S. Thomas, L. Pluta, A. Efremenko, M. Black, L.L. Arnold, J.M. McKim, P. Wilga, G. Gill, K.Y. Choe, H.J. Clewell, Evaluation of gene expression changes in human primary uroepithelial cells following 24-hr exposures to inorganic arsenic and its methylated metabolites. Environ. Mol. Mutagen. 54 (2013) 82–98.

[118] J.W. Yager, J.K. Wiencke, Enhancement of chromosomal damage by arsenic: implications for mechanism. Environ. Health Perspect. 101 Suppl 3 (1993) 79–82.

[119] J.W. Yager, J.K. Wiencke, Inhibition of poly(ADP-ribose) polymerase by arsenite. Mutat. Res. 386 (1997) 345–351.

[120] P.Z. Zheng, K.K. Wang, Q.Y. Zhang, Q.H. Huang, Y.Z. Du, Q.H. Zhang, D.K. Xiao, S.H. Shen, S. Imbeaud, E. Eveno, C.J. Zhao, Y.L. Chen, H.Y. Fan, S. Waxman, C. Auffray, G. Jin, S.J. Chen, Z. Chen, J. Zhang, Systems analysis of transcriptome and proteome in retinoic acid/arsenic trioxide-induced cell differentiation/apoptosis of promyelocytic leukemia. Proc. Natl. Acad. Sci. U. S. A. 102 (2005) 7653–7658.

[121] X. Zhou, X. Sun, K.L. Cooper, F. Wang, K.J. Liu, L.G. Hudson, Arsenite interacts selectively with zinc finger proteins containing C3H1 or C4 motifs. J. Biol. Chem. 286 (2011) 22855–22863.

24

TRANSLATING EXPERIMENTAL DATA TO HUMAN POPULATIONS

J. CHRISTOPHER STATES

Department of Pharmacology and Toxicology, University of Louisville, Louisville, KY, USA

24.1 INTRODUCTION

As detailed in previous chapters, human arsenic exposure is quite varied throughout the world as are the disease manifestations of these exposures. The mechanisms of toxicity leading to the different disease outcomes may be common or may be specific to each disease state. The challenge for experimentalists is to define the human exposure and disease outcome they are modeling with their experimental systems be they animal models or cell cultures. As discussed in earlier chapters of this volume, a plethora of models are in use including rodents (most often mice), nonmammalian systems including fish and invertebrates (worms, flies), yeast, and mammalian and human cells in culture. This chapter will address issues confounding experimental models and will discuss approaches to reconciling these issues.

24.2 EXPOSURE AND DOSE

Determining dose is important in any toxicological study. Most human studies report exposures, for example, arsenic concentrations in drinking water. These exposures have been used as surrogates for dose. Similar problems exist with data from animal studies. Doses can be estimated knowing the average amount of contaminated water consumed per day and individual body weight. However, consumption data in human

Arsenic: Exposure Sources, Health Risks, and Mechanisms of Toxicity, First Edition.
Edited by J. Christopher States.
© 2016 John Wiley & Sons, Inc. Published 2016 by John Wiley & Sons, Inc.

studies usually are available only as estimates based on study subjects' recollections of consumption. Furthermore, the consumption is reported as a mean across the population under study. Thus, the data are confounded by variability in individual consumption, in specific water supplies (often individuals throughout a population do not share a single water source, and individuals may use multiple water sources), and in accuracy of individual recollections. Determining actual consumption in animal studies is possible, but often cost prohibitive at the individual animal level. Mice are normally housed three to five per cage and consumption per animal is derived from a total for the cage. Housing mice individually for prolonged studies is often beyond the fiscal means available to most investigators. Short-term individual housing in metabolic cages may be performed if additional data are collected. Hence, the best that can be accomplished under most circumstances is an estimate of dose based on average exposure and average consumption.

Biomarkers of exposure are also used as surrogates for internal dose. Arsenic concentration in blood, urine, hair, or toe nails is commonly used. Each of these biological samples has its own advantages and disadvantages. Blood arsenic level can be used to estimate an acute dose. However, because arsenic is cleared from the blood quite rapidly, blood arsenic level is not a good measure of chronic exposure. Urine arsenic is a better indicator of chronic exposure and offers the possibility of speciation. The speciation often has an impact on the interpretation and can provide information on metabolism. Metabolite ratios are different in humans and mice reflecting differences in metabolism of arsenic by the two species. Humans generally have higher levels of MMA in their urine [41], whereas most rodents metabolize it further to DMA. Urine arsenic also can be confounded by other organic species such as arsenobetaine as a consequence of fish consumption. Thus, speciation of urine arsenic is important to rule out this potential confounding factor. Hair and toe nail samples may be confounded by airborne or dust contamination. Accessible biomarkers such as arsenic levels in urine versus hair/toenail can exhibit variation in concentration in spite of similar exposure.

Determining appropriate exposures for nonmammalian models including invertebrates can be even more problematic. The zebrafish has become popular as a model system. Genes and gene function are remarkably conserved between zebrafish and humans. Notably in terms of arsenic metabolism, zebrafish express aquaglyceroporins aqp3, aqp3l, aqp9a, aqp9b, and aqp10, which are closely homologous to human AQP3, AQP9, and AQP10 and facilitate the uptake of arsenite (iAs3) and monomethylarsonous acid (MMA3) [14]. Zebrafish also express an arsenic 3 methyltransferase (AS3MT), which is highly homologous to human AS3MT in all tissues and this enzyme is capable of methylating iAs3 to MMA3, monomethylarsinic acid (MMA5), and dimethylarsinic acid (DMA5) [15]. Relatively high sodium arsenite exposures ranging from 10 to 400 mg/L disturbed vascular development in zebrafish embryos [25]. These concentrations were orders of magnitude greater than those needed to disturb epithelial to mesenchymal transition in murine coronary progenitor cells [1]. More studies are needed to determine whether these differences in required exposure levels are due to tissue/cell type differences in sensitivity, or species differences in uptake and disposition.

24.3 FOOD AS A SOURCE OF EXPOSURE

In general, the major source of arsenic exposure is contaminated drinking water. However, arsenic in food can be a substantial fraction of the total dose in areas where arsenic concentrations in water are low, but verges on inconsequential in areas where the arsenic concentration in water is very high. Thus, the contribution of food to the total arsenic body burden is inversely related to the arsenic concentration in the drinking water. Similar considerations apply to animal studies. The fractional contribution by food will be less if the exposure via water is high than if the water arsenic concentration is low. Little is known regarding the bioavailability of arsenic in organic compounds of food stuffs. Some foods, notably seafood and algae, can have high levels of these types of compounds, the major species being arsenobetaine in seafood and arsenosugars in algae [22].

The influence of arsenic in the food on gene expression in mice given water with low arsenic (100 ppb) has been reported [20]. In this case, a cereal-based nonpurified diet (LRD-5001) containing 390 ppb total arsenic induced differential gene expression profiles in liver and lung as compared with mice fed a purified diet (AIN-76A) containing 20 ppb total arsenic. Cereal grains are well known to bioaccumulate heavy metals including arsenic and cadmium. However, phytochemicals contained in the nonpurified diet may confound the results by altering arsenic uptake or metabolism. Thus, although the arsenic levels in the two diets were determined, it is not clear whether the effects on gene expression were due solely to the differences in arsenic. The effect of high dietary arsenic would have been more clear if this study had been performed using purified diet that had been supplemented with arsenic of known speciation. Studies comparing effects on gene expression in mice given water with high (≥1 ppm) arsenic levels have not been published. Clearly, well-controlled studies in which only the arsenic in the diet is altered need to be done to clarify how much influence arsenic in food affects the toxicology. The impact on suitability of a given animal model is that the speciation of arsenic in the diet as well as the absolute amount are important factors that must be considered in light of the drinking water exposure being used in the model. Hence, it is important for investigators to be aware of the arsenic in the animal chow, and to account for it appropriately.

The diet can not only contribute to the body burden of arsenic but can also influence response to arsenic. Some food stuffs are known to contain arsenic in higher levels. Seafood often has high levels of total arsenic, but it is usually in a form that is not readily bioavailable. In contrast, rice can accumulate arsenic from ground and water contamination, and this arsenic is readily bioavailable, contributing to the arsenic burden. The contribution of arsenic in food stuffs to total arsenic burden has received some attention and is dependent on a number of variables. These variables include the locale in which the rice has been grown as well as the rice cultivar. Some regions have naturally high levels of arsenic in the water used for irrigation. This problem is frequent in regions like Bangladesh and West Bengal where ground water from contaminated wells is used for irrigation. Other regions have soil contaminated from previous use of arsenicals as pesticides. The use of these pesticides was common in the United States to protect cotton and tobacco crops from insects. Thus, the soil became contaminated with arsenic. When these regions are converted to rice paddies, such as that occurred in the south-central region of the United States, the rice accumulates the arsenic leached from the contaminated soil. Research has shown that different cultivars accumulate arsenic to different levels. The consequence of cultivar dependence is that the US long-grain rice may have high levels of arsenic depending on the conditions of growth [46] whereas Basmati rice usually has low levels because it does not bioaccumulate arsenic [45]. There are also differences in the relative amounts of arsenic species in different rice cultivars. Arsenic species detected in extracts of rice grains included iAs3, iAs5, and DMA5 [45]. The US long grain has DMA5 as the predominant species whereas European, Bangladeshi, and Indian rice have predominantly iAs species [45]. Further complicating the situation is that arsenic is not distributed evenly throughout the grain. Several studies have reported that highest concentrations are in the bran, with little in polished rice (see [5] for a review). Hence, a dietary analysis for arsenic content including speciation would be helpful information for human studies that could guide diet preparation for experimental studies.

Micronutrients have been implicated in the response to arsenic. In particular, low dietary selenium levels have been correlated with sensitivity to arsenic. Selenium is an essential cofactor for enzymes in the thioredoxin systems. Thus, it is essential for proper functioning of cellular redox systems and maintenance of cellular redox state. On the other hand, excess selenium is toxic. Hence, selenium levels may play an important role and are a consideration in choosing a diet to be fed to experimental animals. However, we need to know more about the specific role of selenium in response to arsenic and how dietary supplementation may affect the response.

24.4 FOLATE

Folate is a water-soluble B vitamin essential for one-carbon metabolism. Folate plays essential roles in the conversion of homocysteine to methionine in the synthesis of *S*-adenosylmethionine and in the conversion of deoxyuridylate to thymidylate necessary for DNA synthesis (http://ods.od.nih.gov/factsheets/Folate-HealthProfessional/). *S*-adenosylmethionine plays an essential role in methylation reactions including the methylation of inorganic arsenic species and in DNA methylation. It has been proposed that arsenic metabolism may compete for and deplete *S*-adenosylmethionine levels, thus depriving DNA methyltransferase of the needed cofactor. Disruptions to DNA methylation have epigenetic consequences that may contribute to carcinogenesis. Hence, folate supplementation has been suggested as a potential protection against arsenic-induced carcinogenesis. Epidemiological studies of folate influence on cancer risk suggest a possible link between excess folate and colon cancer risk [17], although a later meta-analysis contradicts this finding [9]. However, the data are not definitive and more studies are needed to understand fully the role of dietary folate in carcinogenesis. Furthermore, several studies report that the effect of arsenic on DNA methylation is mediated via decreased DNA methyltransferase expression and activity and not by depletion of

S-adenosylmethionine [11, 27, 30]. Thus, the mechanism of DNA hypomethylation caused by chronic arsenic exposure remains controversial.

24.5 *IN VITRO* TO *IN VIVO* TRANSLATION

Mechanistic studies are much easier to perform with cultured cells. Thus, a great many studies have used cultured cells exposed to arsenicals. Choosing a concentration that has physiological relevance can be troublesome. One might consider matching blood arsenic concentrations. However, there are not many studies reporting blood arsenic concentrations for environmental studies. A study showing inhibition of proliferation of peripheral blood mononuclear cells from an arsenic exposed Mexican population reported blood arsenic levels less than 1 μM [42]. Thus, *in vitro* studies using arsenic concentration above 1 μM may lack physiological relevance. On the other hand, peak blood levels in acute promyelogenous leukemia patients may reach approximately 5 μM [26]. Thus, the specific purpose of the study may allow for low micromolar concentrations and be physiologically relevant. For studies of bladder carcinogenesis, even higher arsenic concentrations may be reasonable because urine arsenic concentrations can be even higher. However, speciation becomes an issue when modeling urinary exposures.

24.6 SCALING

Scaling is an important concept in converting dose between species. The reader is referred to an excellent in-depth presentation of scaling for a more complete discussion of the parameters involved in scaling [4]. For the sake of the present discussion, the two-most commonly accepted means to convert dose across species or among patients are normalization to either body weight (mg/kg) or body surface area (mg/m^2). Directly converting dosages by these measures ignores potential differences in metabolism and clearance between the two species [41]. Differences in metabolism of a toxicant can have dramatic effects on the relative toxicity in different species. The examples to be discussed will show how differences in arsenic metabolism and tissue retention in mice and humans make dose conversion difficult. These are important concepts that are ignored by many investigators.

The most common approach to "dose" is by varying exposure, for example, parts per billion arsenic in drinking water, and to attempt to equate exposure in the experimental animal to human exposures. This confusion of exposure with dose can lead to serious misinterpretation of the relevance of the study. The reasons behind the problem lie in understanding the dramatic differences in water consumption per kilogram body weight between small experimental animals and humans, and the differences in metabolism. In addition to the fundamentally dramatic differences between species, there is also a large variation in water consumption among strains of mice [2]. As an example, let us consider an exposure of 100 ppb arsenic (as sodium arsenite) in drinking water, a not uncommon exposure of concern in many parts of the world. The dose is the product of the exposure (e.g., 100 μg/L) times the consumption (e.g., mL water/day) divided by the body weight. Thus, a 20-g mouse consuming 7 mL daily of water containing 100 μg arsenic/L would receive a dose of 35 μg/kg/day. The equivalent daily dose for a 70-kg human, converted on a body weight basis, would be 2.45 mg. If the average water consumption is 2 L/day, then the arsenic concentration would need to be 1.23 mg/L or 1.23 ppm. This level of exposure in humans has been reported to result in acute toxicity [8]. If we calculate the dose conversion on the basis of body surface area, then the dose per mouse is 93.3 μg/m^2/day. Converting that to a human with body surface area equal to 1.85 m^2, the daily dose would be 172.7 μg, obtainable by consuming 2 L of water containing arsenic at 87 ppb, a far different outcome than the calculation by body weight. This exposure scenario seems reasonable until we take into consideration the differences in arsenic metabolism and clearance. This difference is best exemplified by analyzing an exposure in mice of 49 ppm used in transplacental carcinogenesis [44] and atherogenesis [36] studies. Liver arsenic in the pregnant dams was reported by the latter group in a later study [38]. They found that the tissue arsenic levels (1.2 μg/g) corresponded to levels reported in livers of humans (mean: 1.46 ± 0.42 mg/kg; maximum: 6 mg/kg) exposed to arsenic in drinking water in West Bengal [13]. The arsenic levels in the groundwater used by 90% of these patients for drinking ranged from 0.05 to 3.2 mg/L. The population also included 20 patients consuming water with industrial arsenic contamination with levels of 5.05–14.2 mg/L. Unfortunately, no details on the correlation of arsenic concentrations in drinking water and liver arsenic levels were provided. However, even if we assume that the maximum liver levels corresponded to the maximum industrial exposure, the liver level is five times higher in the human population than in the mice that received a minimum 3.5-fold higher exposure! Thus, humans exhibit a minimum 15-fold increase in tissue retention relative to exposure compared to mice indicating that neither dose conversion method is accurate. If this relationship holds at the lower exposures, then the 100-ppb exposure used in the example mentioned earlier would correlate with a 6- to 7-ppb exposure in humans. Clearly, more data on human and mouse tissue retention are needed.

24.7 STRAIN DIFFERENCES IN SENSITIVITY

In a model of acute arsenic intoxication, strain differences were seen in sensitivity to arsenic-induced renal injury in C57BL/6 and BALB/c mice [18]. BALB/c mice were more sensitive showing severe hemorrhages, acute tubular necrosis, neutrophil infiltration, cast formation, and disappearance of periodic acid-Schiff (PAS)-positive brush borders within 10 h after subcutaneous injection with sodium arsenite at 13.5 mg/kg. Within 24 h after injection, half the BALB/c mice were dead and all the C57BL/6 were still alive. Investigation of four heavy metal-inducible genes (multidrug transporter protein 1 (Mrp-1, Abcc1), multidrug resistance protein (Mdr-1), metallothionein (Mt)-1, and arsenite inducible, cysteine- and histidine-rich RNA-associated protein (Airap)) showed that mRNAs for Mt-1, Mdr-1, and Airap were all induced by the arsenite injection equally in BALB/c and C57BL/6 kidneys, and that Mrp-1 was highly induced in C57BL/6 but not in BALB/c kidneys. This differential Mrp-1 induction was confirmed at the protein level by immunohistochemistry of renal tubular epithelium. The tissue arsenic levels decreased much more rapidly in C57BL/6 mice than in BALB/c mice, most likely a consequence of the differential expression of MRP-1. Thus, in an acute exposure model, there is a clear strain difference in iAs3 clearance.

In another acute toxicity study comparing C57BL/6J and SWV/Fnn mice, a strain difference was observed in teratogenicity of sodium arsenite (10mg/kg) injected intraperitoneally [24]. In this case, the C57BL/6J mice were more sensitive than the SWV/Fnn mice exhibiting neural tube defects and embryo lethality in a much higher fraction of the embryos. The spectrum of malformations induced was dependent on gestational time of exposure and the neural tube defects and other malformations were increased when the splotch allele (mutation of Pax-3) was introduced in the C57BL/6 mice. The specific mechanism of the strain difference was not determined. Nonetheless, this study provided clear evidence of strain-specific differences in sensitivity to acute arsenic exposure.

Strain differences in disposition of inorganic arsenic have been studied in mice [16]. Strain differences in disposition were observed but these differences were dependent on dose, arsenic species, and route of exposure. When radiolabeled arsenite or arsenate was administered in water by oral gavage, differences between strains in disposition of arsenate was observed at both doses tested but greater differences were seen at the higher dose and shorter times. C57BL6N mice exhibited slower clearance than either B6C3F1 or C3H mice. However, when the arsenic was administered by intraperitoneal injection, the strain differences in arsenate clearance were not seen. These data suggest that there may be strain differences in absorption. In addition, to the differences observed in clearance, differences in metabolites were also seen suggesting strain-dependent differences in methylation. Thus, in addition to species-dependent differences in arsenic disposition between humans and mice, there are considerations to be made concerning route of exposure and strain when designing experiments in mice.

In addition to differences in absorption, mice show strain-specific differences in body weight, water, and food consumption [2]. In a study of 28 different mouse strains, adult body weight varied over a twofold range from approximately 15g to greater than 30g. Food consumption varied not only in absolute amounts but also when normalized to body weight. Once again, these differences varied over a twofold range. Water consumption showed a similar variability both in absolute amounts and in amounts normalized to body weight. There was a positive correlation between body weight and intake of both food and water and an inverse correlation of body weight and intake normalized to body weight. Water and food intake showed good correlation independent of normalization to body weight. Thus, although exposures may be the same in studies with different strains of mice, the actual dosages are likely different when water and food are provided *ad libitum*. This confounder makes it important for measures of arsenic dosage to be made so that comparisons across studies are more meaningful.

24.8 IMPORTANCE OF DEVELOPMENTAL EXPOSURE ON DISEASE OUTCOME

Although inorganic arsenic has been a known human carcinogen for decades, induction of cancer by exposure of adult animals to inorganic arsenic has not been demonstrated. It was not until the Waalkes group showed that *in utero* exposure of C3H mice to inorganic arsenic could induce cancer in the exposed pups [44] was there an animal model for inorganic arsenic-induced cancer (see Chapter 19). This group has gone on to show that "whole-life" exposure, starting by exposing the dams before conception, results in increased carcinogenesis at lower exposures in CD1 mice [40]. A more recent publication shows an increase in lung tumorigenesis in CD1 mice at even lower exposures (<1 ppm) [43]. This latter paper also reports an inverse exposure response in contrast to previous data with exposures of 1 ppm and greater. This result suggests that, at least for lung cancer, the arsenic dose–response may not be monotonic. All together, these data indicate that although adult exposure alone is not sufficient to induce cancer in mice, when combined with the *in utero* exposure as a triggering event, adult exposure greatly enhances carcinogenesis.

Long-term study of the Chilean population exposed *in utero* and/or early childhood showed increased incidence of lung disease including cancer and bronchiectasis [35]. A follow-up study showed dose–response in increased susceptibility to both lung and bladder cancer in this population [39] and concluded that early life arsenic exposure played an important role in the susceptibility to later life cancer. This group also has documented higher incidence of childhood liver cancer in arsenic-exposed children [23]. The arsenic exposure in this population was very high, raising some concern that cancer effects were only of concern with exposures well in excess of 100 μg/L. However, this group also has shown that exposures below 100 μg/L also increase the risk of lung cancer in young adults [12]. These data reinforce the hypothesis that humans are extraordinarily sensitive to early life arsenic exposure.

Early life arsenic exposure also has been associated with an increased risk of cardiovascular disease. Evidence of the impact of early life arsenic exposure on coronary artery atherosclerosis is in the reports of myocardial infarction due to coronary artery blockage in infants in the Chilean population exposed to high levels of arsenic [31, 32]. The specific role of early life arsenic exposure in the development of cardiovascular disease has not been studied in humans. However, animal studies have since shown increases in atherosclerosis in adult ApoE-knockout mice after an *in utero* arsenic exposure [36]. This group reported that the *in utero* arsenic exposure induced gene expression changes suggesting a proinflammatory state [38]. Investigation of gene expression changes in cord blood mononuclear cells of newborns of mothers exposed to arsenic in Thailand showed similar inflammatory gene expression changes [10]. The commonality of findings in these two reports suggests that *in utero* arsenic exposure predisposes the affected individual to inflammation that may be the initiating factor for disease. The long latency of gene expression changes observed in the mouse system also suggests that the *in utero* exposure has induced epigenetic changes that influence developmental gene expression program. Thus, it appears that developmental exposure is an important factor in the likelihood of adult disease manifestation, an earlier age of onset, and the severity of the disease. More research is clearly needed to learn the impact of early life arsenic exposure on cardiovascular disease, which is the major cause of mortality both in the population at large as well as in the arsenic-exposed population.

In contrast to the necessity for *in utero* exposure for arsenic-induced carcinogenesis, adult arsenic exposure is sufficient to increase atherogenesis in the ApoE-knockout mouse model [21, 34, 37] and in the ApoE/LDLR double knockout [3]. The study by Srivastava et al. [37] shows positive exposure response with exposures of 1 ppm and greater. Curiously, the study of Lemaire et al. [21] showed an inverse dose–response with exposures of 1 ppm and 200 ppb. These results suggest a nonmonotonic exposure response for arsenic-induced atherogenesis, similar to the observations discussed earlier for arsenic-induced lung carcinogenesis in CD1 mice. Clearly, more research is needed to determine whether indeed the responses are nonmonotonic and whether the apparent nonmonotonic responses are specific to the particular models. Nonetheless, there are similarities between arsenic-induced carcinogenesis and atherogenesis in that *in utero* exposure alone makes the animals more susceptible to the disease. The difference appears to be in that adult exposure alone is sufficient to accelerate atherogenesis in a susceptible animal but not carcinogenesis. It is also clear that *in utero*/early life exposure plays a role in carcinogenesis in both mice and humans.

Early life exposure also appears to be important in nonmalignant lung disease. In addition to the elevated incidence of bronchiectasis noted earlier, lung function shows an inverse relationship with arsenic exposure. A large study conducted in Bangladesh showed that for every standard deviation increase in baseline water-arsenic exposure, a lower level of FEV1 (−46.5 mL; $P<0.0005$) and FVC (−53.1 mL; $P<0.01$) was observed in regression models adjusted for age, sex, body mass index, smoking, socioeconomic status, betel nut use, and arsenical skin lesions status [28]. Susceptibility to influenza virus infection was also increased in mice exposed to arsenic *in utero* and during the first week of life [29]. These authors reported that viral clearance was reduced in arsenic-exposed mice, although there was an increase in inflammatory markers. This increase in inflammatory markers is similar to that reported in studies of the gene expression modulation in livers of mice exposed to arsenic *in utero* [38]. A recent review of the impact of *in utero* and early life arsenic exposure on disease found that animal studies supported the human studies and together provide strong evidence that *in utero* exposure plays a role in carcinogenesis, atherogenesis, and respiratory diseases [7]. Clearly, there is a need for more research to define the mechanisms of these effects in order to develop approaches to remediation. In the meantime, greater effort to protect pregnant women and children from arsenic exposure is needed.

24.9 ROLE OF ARSENIC METABOLISM IN MEDIATING DISEASE PROCESSES

The affinity of trivalent arsenicals for vicinal thiol groups is well established. In proteins, vicinal thiols occur when the sulfhydryl groups of cysteines are arranged in proximity by the three-dimensional structure of the folded protein. These vicinal cysteine thiols are present in AS3MT (the major enzyme responsible for arsenic methylation), in the ArsR repressor that mediates arsenite signaling in *Escherichia coli*, and in metallothioneins that bind heavy metals as well as arsenite [33]. A large class of proteins with vicinal thiols are zinc finger proteins. The zinc ions are coordinated in a tetrad of electron donors that can include the sulfhydryls of cysteines, the nitrogen in the imidazole group of histidines, or the carboxy group of aspartates. Thus, the classes of zinc finger proteins are designated as C4, C3H1, C2H2, and C2H1D1. The differential affinities of trivalent arsenicals for these zinc fingers have been discussed in Chapter 13 and in a recent article [49]. It is clear that iAs3 binds both C4 and C3H1 fingers but not C2H2 fingers (and likely not C2H1D1) because arsenite has coordinating valences. MMA3 will bind C2H2 fingers (and likely C2H1D1) in addition to C3H1 and C4 fingers. Thus, the potential disruption caused by MMA3 includes an extra class of zinc finger proteins and is therefore much greater than the potential disruption by iAs3. Zinc fingers are essential components of a very large number of regulatory proteins including transcription factors, ubiquitin ligases, and DNA methyl transferases, among others. Disrupting function of these regulatory factors is likely a major mode of action of trivalent arsenicals resulting in alterations in gene and protein expression, as well as epigenetic modifications.

Differential binding of the zinc finger classes by iAs3 and MMA3 indicates that iAs3 methylation capacity will alter toxic response. Various cell types have different capacity to metabolize iAs to MMA3 [19]. Thus, the precise mechanisms of arsenic toxicity may be cell type specific. Use of AS3MT-knockout mice may help identify whether iAs3 or MMA3 effects are responsible for a specific toxicity. For instance, urothelial cell cytotoxicity and hyperplasia are enhanced in AS3MT-knockout mice compared to wild-type mice given iAs3 in drinking water [47, 48]. However, caution must be exercised in interpreting these results because As3mt-knockout mice retain higher levels of total arsenicals and particularly iAs3 in their tissues [6]. Thus, although iAs3 exhibits preferential binding to C4 and C3H1 zinc fingers, at higher concentrations, it may disrupt C2H2 zinc fingers as well. Systematic investigation of the concentration dependence of zinc finger inactivation by iAs3 and MMA3 is needed to address these questions.

24.10 SUMMARY AND CONCLUSIONS

Differences between experimental animal models and humans in metabolism and disposition of arsenic, as well as difficulties in extrapolating from results *in vitro* to *in vivo* effects in humans, make translating experimental results to human application difficult. Another complication arises when differences between adult and developmental exposures are considered. Both epidemiological studies and animal experiments point to significant impact of both *in utero* and early life exposures on disease development in later life. Data collected from most studies, both experimental and epidemiological, often include measurements of arsenic exposure but not dose. Thus, means to convert exposures in animal studies (mostly in mice) to specific human exposures is needed. Dose conversion between humans and mice is complicated by dramatic differences in arsenic metabolism and disposition. Thus, biomarkers of equivalence are needed for translation of experimental findings to human pathology. These biomarkers would allow phenotypic anchoring and could include tissue dosimetry, or changes in gene expression, epigenetic marks, or proteome profiles.

REFERENCES

[1] P. Allison, T. Huang, D. Broka, P. Parker, J.V. Barnett, T.D. Camenisch, Disruption of canonical TGFbeta-signaling in murine coronary progenitor cells by low level arsenic. Toxicol. Appl. Pharmacol. 272 (2013) 147–153.

[2] A.A. Bachmanov, D.R. Reed, G.K. Beauchamp, M.G. Tordoff, Food intake, water intake, and drinking spout side preference of 28 mouse strains. Behav. Genet. 32 (2002) 435–443.

[3] M. Bunderson, D.M. Brooks, D.L. Walker, M.E. Rosenfeld, J.D. Coffin, H.D. Beall, Arsenic exposure exacerbates atherosclerotic plaque formation and increases nitrotyrosine and leukotriene biosynthesis. Toxicol. Appl. Pharmacol. 201 (2004) 32–39.

[4] E.J. Calabrese, Principles of Animal Extrapolation, Lewis Publishers, Chelsea, 1991.

[5] A.M. Carey, E. Lombi, E. Donner, M.D. de Jonge, T. Punshon, B.P. Jackson, M.L. Guerinot, A.H. Price, A.A. Meharg, A review of recent developments in the speciation and location of arsenic and selenium in rice grain. Anal. Bioanal. Chem. 402 (2012) 3275–3286.

[6] B. Chen, L.L. Arnold, S.M. Cohen, D.J. Thomas, X.C. Le, Mouse arsenic (+3 oxidation state) methyltransferase genotype affects metabolism and tissue dosimetry of arsenicals after arsenite administration in drinking water. Toxicol. Sci. 124 (2011) 320–326.

[7] S.F. Farzan, M.R. Karagas, Y. Chen, In utero and early life arsenic exposure in relation to long-term health and disease. Toxicol. Appl. Pharmacol. 272 (2013) 384–390.

[8] E.J. Feinglass, Arsenic intoxication from well water in the United States. N. Engl. J. Med. 288 (1973) 828–830.

[9] J. Fife, S. Raniga, P.N. Hider, F.A. Frizelle, Folic acid supplementation and colorectal cancer risk: a meta-analysis. Colorectal Dis. 13 (2011) 132–137.

[10] R.C. Fry, P. Navasumrit, C. Valiathan, J.P. Svensson, B.J. Hogan, M. Luo, S. Bhattacharya, K. Kandjanapa, S. Soontararuks, S. Nookabkaew, C. Mahidol, M. Ruchirawat, L.D. Samson, Activation of inflammation/NF-kappaB signaling in infants born to arsenic-exposed mothers. PLoS Genet. 3 (2007) e207.

[11] H.Y. Fu, J.Z. Shen, Y. Wu, S.F. Shen, H.R. Zhou, L.P. Fan, Arsenic trioxide inhibits DNA methyltransferase and restores expression of methylation-silenced CDKN2B/CDKN2A genes in human hematologic malignant cells. Oncol. Rep. 24 (2010) 335–343.

[12] C. Steinmaus, C. Ferreccio, Y. Yuan, J. Acevedo, F. González, L. Perez, S. Cortés, J.R. Balmes, J. Liaw, A.H. Smith, Elevated lung cancer in younger adults and low concentrations of arsenic in water. Am J Epidemiol. 180 (2014) 1082–1087.

[13] D.N. Guha Mazumder, Arsenic and liver disease. J. Indian Med. Assoc. 99 (2001) 311, 314–311, 320.

[14] M. Hamdi, M.A. Sanchez, L.C. Beene, Q. Liu, S.M. Landfear, B.P. Rosen, Z. Liu, Arsenic transport by zebrafish aquaglyceroporins. BMC Mol. Biol. 10 (2009) 104.

[15] M. Hamdi, M. Yoshinaga, C. Packianathan, J. Qin, J. Hallauer, J.R. McDermott, H.C. Yang, K.J. Tsai, Z. Liu, Identification of an *S*-adenosylmethionine (SAM) dependent arsenic methyltransferase in Danio rerio. Toxicol. Appl. Pharmacol. 262 (2012) 185–193.

[16] M.F. Hughes, E.M. Kenyon, B.C. Edwards, C.T. Mitchell, D.J. Thomas, Strain-dependent disposition of inorganic arsenic in the mouse. Toxicology 137 (1999) 95–108.

[17] Y.I. Kim, Folate: a magic bullet or a double edged sword for colorectal cancer prevention? Gut 55 (2006) 1387–1389.

[18] A. Kimura, Y. Ishida, T. Wada, H. Yokoyama, N. Mukaida, T. Kondo, MRP-1 expression levels determine strain-specific susceptibility to sodium arsenic-induced renal injury between C57BL/6 and BALB/c mice. Toxicol. Appl. Pharmacol. 203 (2005) 53–61.

[19] C. Kojima, D.C. Ramirez, E.J. Tokar, S. Himeno, Z. Drobna, M. Styblo, R.P. Mason, M.P. Waalkes, Requirement of arsenic biomethylation for oxidative DNA damage. J. Natl. Cancer Inst. 101 (2009) 1670–1681.

[20] C.D. Kozul, A.P. Nomikos, T.H. Hampton, L.A. Warnke, J.A. Gosse, J.C. Davey, J.E. Thorpe, B.P. Jackson, M.A. Ihnat, J.W. Hamilton, Laboratory diet profoundly alters gene expression and confounds genomic analysis in mouse liver and lung. Chem. Biol. Interact. 173 (2008) 129–140.

[21] M. Lemaire, C.A. Lemarie, M.F. Molina, E.L. Schiffrin, S. Lehoux, K.K. Mann, Exposure to moderate arsenic concentrations increases atherosclerosis in ApoE-/- mouse model. Toxicol. Sci. 122 (2011) 211–221.

[22] W. Li, C. Wei, C. Zhang, M. van Hulle, R. Cornelis, X. Zhang, A survey of arsenic species in Chinese seafood. Food Chem. Toxicol. 41 (2003) 1103–1110.

[23] J. Liaw, G. Marshall, Y. Yuan, C. Ferreccio, C. Steinmaus, A.H. Smith, Increased childhood liver cancer mortality and arsenic in drinking water in northern Chile. Cancer Epidemiol. Biomarkers Prev. 17 (2008) 1982–1987.

[24] A.F. Machado, D.N. Hovland, Jr., S. Pilafas, M.D. Collins, Teratogenic response to arsenite during neurulation: relative sensitivities of C57BL/6J and SWV/Fnn mice and impact of the splotch allele. Toxicol. Sci. 51 (1999) 98–107.

[25] C.W. McCollum, C. Hans, S. Shah, F.A. Merchant, J.A. Gustafsson, M. Bondesson, Embryonic exposure to sodium arsenite perturbs vascular development in zebrafish. Aquat. Toxicol. 152 (2014) 152–163.

[26] J. Ni, G. Chen, Z. Shen, X. Li, H. Liu, Y. Huang, Z. Fang, S. Chen, Z. Wang, Z. Chen, Pharmacokinetics of intravenous arsenic trioxide in the treatment of acute promyelocytic leukemia. Chin Med J (Engl) 111 (1998) 1107–1110.

[27] K. Nohara, T. Baba, H. Murai, Y. Kobayashi, T. Suzuki, Y. Tateishi, M. Matsumoto, N. Nishimura, T. Sano, Global DNA methylation in the mouse liver is affected by methyl deficiency and arsenic in a sex-dependent manner. Arch. Toxicol. 85 (2011) 653–661.

[28] F. Parvez, Y. Chen, M. Yunus, C. Olopade, S. Segers, V. Slavkovich, M. Argos, R. Hasan, A. Ahmed, T. Islam, M.M. Akter, J.H. Graziano, H. Ahsan, Arsenic exposure and impaired lung function. Findings from a large population-based prospective cohort study. Am. J. Respir. Crit. Care Med. 188 (2013) 813–819.

[29] K.A. Ramsey, R.E. Foong, P.D. Sly, A.N. Larcombe, G.R. Zosky, Early life arsenic exposure and acute and long-term responses to influenza A infection in mice. Environ. Health Perspect. 121 (2013) 1187–1193.

[30] J.F. Reichard, M. Schnekenburger, A. Puga, Long term low-dose arsenic exposure induces loss of DNA methylation. Biochem. Biophys. Res. Commun. 352 (2007) 188–192.

[31] H.G. Rosenberg, Systemic arterial disease with myocardial infarction. Report on two infants. Circulation 47 (1973) 270–275.

[32] H.G. Rosenberg, Systemic arterial disease and chronic arsenicism in infants. Arch. Pathol. 97 (1974) 360–365.

[33] S. Shen, X.F. Li, W.R. Cullen, M. Weinfeld, X.C. Le, Arsenic binding to proteins. Chem. Rev. 113 (2013) 7769–7792.

[34] P.P. Simeonova, T. Hulderman, D. Harki, M.I. Luster, Arsenic exposure accelerates atherogenesis in apolipoprotein E(-/-) mice. Environ. Health Perspect. 111 (2003) 1744–1748.

[35] A.H. Smith, G. Marshall, Y. Yuan, C. Ferreccio, J. Liaw, O. von Ehrenstein, C. Steinmaus, M.N. Bates, S. Selvin, Increased mortality from lung cancer and bronchiectasis in young adults after exposure to arsenic in utero and in early childhood. Environ. Health Perspect. 114 (2006) 1293–1296.

[36] S. Srivastava, S.E. D'Souza, U. Sen, J.C. States, In utero arsenic exposure induces early onset of atherosclerosis in ApoE-/- mice. Reprod. Toxicol. 23 (2007) 449–456.

[37] S. Srivastava, E.N. Vladykovskaya, P. Haberzettl, S.D. Sithu, S.E. D'Souza, J.C. States, Arsenic exacerbates atherosclerotic lesion formation and inflammation in ApoE-/- mice. Toxicol. Appl. Pharmacol. 241 (2009) 90–100.

[38] J.C. States, A.V. Singh, T.B. Knudsen, E.C. Rouchka, N.O. Ngalame, G.E. Arteel, Y. Piao, M.S. Ko, Prenatal arsenic exposure alters gene expression in the adult liver to a proinflammatory state contributing to accelerated atherosclerosis. PLoS One 7 (2012) e38713.

[39] C. Steinmaus, C. Ferreccio, J. Acevedo, Y. Yuan, J. Liaw, V. Duran, S. Cuevas, J. Garcia, R. Meza, R. Valdes, G. Valdes, H. Benitez, V. VanderLinde, V. Villagra, K.P. Cantor, L.E. Moore, S.G. Perez, S. Steinmaus, A.H. Smith, Increased lung and bladder cancer incidence in adults after in utero and early-life arsenic exposure. Cancer Epidemiol. Biomarkers Prev. 23 (2014) 1529–1538.

[40] E.J. Tokar, B.A. Diwan, J.M. Ward, D.A. Delker, M.P. Waalkes, Carcinogenic effects of "whole-life" exposure to inorganic arsenic in CD1 mice. Toxicol. Sci. 119 (2011) 73–83.

[41] M. Vahter, Methylation of inorganic arsenic in different mammalian species and population groups. Sci. Prog. 82 (Pt 1) (1999) 69–88.

[42] L. Vega, P. Ostrosky-Wegman, T.I. Fortoul, C. Diaz, V. Madrid, R. Saavedra, Sodium arsenite reduces proliferation of human activated T-cells by inhibition of the secretion of interleukin-2. Immunopharmacol. Immunotoxicol. 21 (1999) 203–220.

[43] M.P. Waalkes, W. Qu, E.J. Tokar, G.E. Kissling, D. Dixon, Lung tumors in mice induced by "whole-life" inorganic arsenic exposure at human-relevant doses. Arch. Toxicol. 88 (2014) 1619–1629.

[44] M.P. Waalkes, J.M. Ward, J. Liu, B.A. Diwan, Transplacental carcinogenicity of inorganic arsenic in the drinking water: induction of hepatic, ovarian, pulmonary, and adrenal tumors in mice. Toxicol. Appl. Pharmacol. 186 (2003) 7–17.

[45] P.N. Williams, A.H. Price, A. Raab, S.A. Hossain, J. Feldmann, A.A. Meharg, Variation in arsenic speciation and concentration in paddy rice related to dietary exposure. Environ. Sci. Technol. 39 (2005) 5531–5540.

[46] P.N. Williams, A. Raab, J. Feldmann, A.A. Meharg, Market basket survey shows elevated levels of As in South Central U.S. processed rice compared to California: consequences for human dietary exposure. Environ. Sci. Technol. 41 (2007) 2178–2183.

[47] M. Yokohira, L.L. Arnold, K.L. Pennington, S. Suzuki, S. Kakiuchi-Kiyota, K. Herbin-Davis, D.J. Thomas, S.M. Cohen, Severe systemic toxicity and urinary bladder cytotoxicity and regenerative hyperplasia induced by arsenite in arsenic (+3 oxidation state) methyltransferase knockout mice. Toxicol. Appl. Pharmacol. 246 (2010) 1–7.

[48] M. Yokohira, L.L. Arnold, K.L. Pennington, S. Suzuki, S. Kakiuchi-Kiyota, K. Herbin-Davis, D.J. Thomas, S.M. Cohen, Effect of sodium arsenite dose administered in the drinking water on the urinary bladder epithelium of female arsenic (+3 oxidation state) methyltransferase knockout mice. Toxicol. Sci. 121 (2011) 257–266.

[49] X. Zhou, X. Sun, C. Mobarak, A.J. Gandolfi, S.W. Burchiel, L.G. Hudson, K.J. Liu, Differential binding of monomethylarsonous acid compared to arsenite and arsenic trioxide with zinc finger peptides and proteins. Chem. Res. Toxicol. 27 (2014) 690–698.

INDEX

Arsenic: Exposure Sources, Health Risks, and Mechanisms of Toxicity, First Edition.
Edited by J. Christopher States.
© 2016 John Wiley & Sons, Inc. Published 2016 by John Wiley & Sons, Inc.